Hydroinformatics in Hydrology, Hydrogeology and Water Resources

Edited by:

IAN D. CLUCKIE
YANGBO CHEN
VLADAN BABOVIC
LENNY KONIKOW
ARTHUR MYNETT
SIEGFRIED DEMUTH
DRAGAN A. SAVIC

Proceedings of Symposium JS.4 at the Joint Convention of the International Association of Hydrological Sciences (IAHS) and the International Association of Hydrogeologists (IAH) held in Hyderabad, India, 6–12 September 2009.

Publication of this volume was sponsored by:

The Association of Hydrologists of India

and

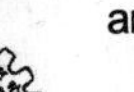

The National Geophysical Research Institute, Hyderabad, India

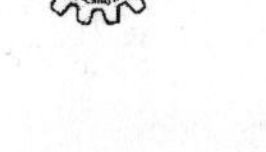

IAHS Publication 331
in the IAHS Series of Proceedings and Reports

Published by the International Association of Hydrological Sciences 2009

IAHS Publication 331

ISBN 978-1-907161-02-5

British Library Cataloguing-in-Publication Data.
A catalogue record for this book is available from the British Library.

The papers included in this volume have been reviewed and some were extensively revised by the Editors, in collaboration with the authors, prior to publication.

IAHS is indebted to the employers of the Editors for the invaluable support and services provided that enabled them to carry out their task effectively and efficiently.

Publications in the series of Proceedings and Reports are available from:
IAHS Press, Centre for Ecology and Hydrology, Wallingford, Oxfordshire OX10 8BB, UK
tel.: +44 1491 692442; fax: +44 1491 692448; e-mail: jilly@iahs.demon.co.uk

Printed in India by Vamsi Art Printers Pvt. Ltd., Hyderabad.

Preface

Hydroinformatics is a reflection of the intense development that has occurred in the application of information technology in the areas of Hydrology, Hydraulics and Water Resources. Despite having started within the computational hydraulics community and been an object of seminal support by the International Association of Hydraulics Research (IAHR), in the last two decades it has grown to embrace the whole of the water community. Many other areas of science have seen a similar progression ranging from health informatics to robotic systems. This widening interest in the water community was reflected by the foundation of the Joint Committee on Hydroinformatics at the Cardiff International Conference on Hydroinformatics held in the UK in July 2002. This was the fifth conference in a continuing series and recognised the support of the IWA and IAHS, who joined with IAHR to form the Joint Committee on Hydroinformatics.

Hydroinformatics has seen significant growth as an area of interest in the IAHS community and this has been reflected in the attendance by many hydrologists at Hydroinformatics events over recent years. The joint symposium between IAHS and the IAH recognised the increasing strength of interest in this area and sought to focus on applications of Hydroinformatics in Hydrology, Hydrogeology and Water Resources. It was the first time that IAHS had directly sponsored this area at either a scientific assembly, or indeed at any IUGG Congress, and clearly signals its future importance to the Association.

This publication comprises a collection of peer-reviewed papers presented at the joint symposium JS.4: *Hydroinformatics in Hydrology, Hydrogeology and Water Resources* that was held during the 8th IAHS Scientific Assembly and 37th IAH Congress, in Hyderabad, India, 6–12 September 2009. The symposium was jointly sponsored by the following IAHS commissions and working groups: HYINF, ICSW, ICWRS and ICRS, and also IAH. The volume contains 60 papers from more than 20 countries, reflecting the international dimension of the symposium.

The editors would like to thank all symposium participants for their scientific contributions. The invaluable support of Marghi Peacock from the UK - Flood Risk Management Research Consortium (FRMRC) was critical and the editors and participants owe her a great debt of gratitude. We also express special thanks to Cate Gardner, Penny Perrins and Frances Watkins from IAHS Press for their professional advice and help with the processing of the manuscripts and the production of this seminal IAHS publication.

Editor-in-chief:

Ian Cluckie
School of Engineering, Swansea University, UK

Co-editors:

Yangbo Chen
Centre of Water Resources and Environment, Sun Yat-Sen University, Guangzhou, China

Vladan Babovic
Department of Civil Engineering, National University of Singapore, Singapore

Lenny Konikow
US Geological Survey, Reston, Virginia, USA

Arthur Mynett
UNESCO-IHE, Delft University of Technology and Deltares, Delft, The Netherlands

Siegfried Demuth
UNESCO Water Sciences Division, Paris, France

Dragan Savic
School of Engineering, Computer Science and Mathematics, University of Exeter, UK

Contents

2 Hydrological Applications of Hydroinformatics

3 Hydrogeological Applications and Modelling Large Systems

1 Whole System Modelling and Uncertainty

Keynote Paper

Whole System Modelling and Hydroinformatics

IAN CLUCKIE[1] & YANGBO CHEN[2]

1 *School of Engineering, Swansea University, Swansea, UK*
i.d.cluckie@swansea.ac.uk[1]

2 *Centre of Water Resources and Environment, Sun Yat-Sen University, China*

Abstract The increasing trend towards the coupling of complex model systems in a cascade is a particular feature of future "Whole System Modelling" approaches. Major advances in numerical weather prediction (NWP) have made it possible to provide rainfall forecasts, along with many other meteorological data fields at high spatial and temporal resolutions. The incorporation of high-resolution mesoscale model output directly into real-time flood forecasting systems provides extended lead-times and also allows the development of a "Whole Systems Approach" to the treatment of uncertainty. The uncertainties inherent in the NWP can be propagated into hydrological and hydraulic domains, and may be magnified by the scaling process. As ensemble weather forecasts become operationally available, it is of particular interest to note the potential and implications of ensemble inputs to real-time hydroinformatic modelling systems in terms of uncertainty propagation. This paper discusses the use of ensemble forecasts from a short-range high-resolution mesoscale weather model (MM5). The results derive from a series of applications using a fully distributed rainfall–runoff model (GBDM) and include a special study of extreme flooding in the Thames Estuary due to a North Sea storm surge. The concluding comments indicate the importance attached to emerging techniques for uncertainty handling in complex model cascades and acknowledge the ongoing HEPEX (Hydrological Ensemble Prediction Experiment) initiative.

Key words uncertainty; hydroinformatics; whole system modelling; numerical weather prediction

INTRODUCTION

Numerical weather prediction (NWP) has advanced considerably in recent years, and it is now possible to generate very high-resolution rainfall forecasts at catchment scale and therefore, more real-time flood forecasting systems are tending to utilise quantitative precipitation prediction (QPF) from mesoscale models in order to extend the forecast lead-time. This is particularly true in the flash flooding area where model performance is highly dependent on the rapid availability of knowledge on the rainfall distribution in advance (Ferraris *et al.*, 2002). Many efforts have been made to utilise the QPF in the context of real-time flood forecasting in which one or more state-of-the-art QPFs are integrated into the whole system (e.g. De Roo *et al.*, 2003; Bartholmes & Todini, 2005; Kobold & Sušelj, 2005; Xuan *et al.*, 2005; Verbunt *et al.*, 2006; Xuan & Cluckie, 2006). A particular feature of mesoscale models is that the effects of QPF uncertainty on the whole system can be easily appreciated either intuitively or via case studies (Xuan *et al.*, 2005). Recent research on integrating the QPF directly into the real-time flood forecasting domain has revealed that direct coupling of the QPF with the hydrological model can result in large bias and uncertainty that can result in not only severe underestimation (Bartholmes & Todini, 2005; Kobold & Sušelj, 2005) but also in over-prediction, especially in mountainous areas (Verbunt *et al.*, 2006). It is logical to divide such a complex coupled system into at least two components, the weather modelling system (NWP) and the hydroinformatic system, through which the uncertainties from the weather domain are propagated into the final system output space.

FORECASTING WEATHER AT HIGH SPACE AND TIME SCALES

Weather models that are routinely operated in national weather centres have such a coarse spatial resolution that hydroinformatic systems have difficulty in applying the results from them as an effective input. A downscaling procedure is needed to bridge the scale gap between the large-scale weather forecast domains and catchment-sized flood-forecasting domains. A dynamical-downscaling approach is often applied to resolve the dynamics over smaller grids. The forecasts/

analyses from global weather models can be used to settle the initial and lateral boundary conditions (IC/LBC) of a mesoscale model that is then able to benefit from resolvable terrain features and physics at high resolutions. This type of mesoscale model is often referred to as a local area model (LAM).

Lorenz (1963, 1993) introduced the concept that the time evolution of a nonlinear, deterministic dynamical system, to which the atmosphere (essentially a boundary layer process) and the equations that describe air motion belong, are very sensitive to the initial conditions of the system. The uncertainties inherited in the larger scale model that provides the parent domain with the IC and LBC, can also be propagated to the nested mesoscale models. Given the large number of degrees of freedom of the atmospheric state-space, it is impractical to directly generate solutions of the probabilistic equations from the initial states (Kalnay, 2002). The ensemble forecast method runs the model separately over a probabilistically generated ensemble of initial states, each of which represents a plausible and equally likely state of the atmosphere, and projects them into future state-space. As such the future state-space is represented by the statistics of the derived ensemble. The hydrological model, like weather models, is subject to the same factors regarding the uncertainty sources in weather modelling systems. For a stand-alone hydrological model, although uncertainties in data inputs, e.g. measurements or observations, have been regarded as one of the main sources of uncertainty, the model structure and parameterisations have drawn more attention from hydrologists (Wagener *et al.*, 2001). Simply speaking, the uncertainty related to hydrological modelling is categorised as: the inability to observe the state of the system and that due to inefficient modelling of reality. As for a coupled system it is important to recognise that the weather input uncertainty may outweigh the impact of the cascade model structure and can be magnified by propagation through a particular coupling process.

Considerable research effort has been made on both NWP-based QPF and the application of NWPs in flood forecasting (e.g. Smith & Austin, 2000). This paper focuses on the uncertainties in ensemble-based flood forecasting when driven by high-resolution ensemble rainfall forecasts from a NWP and attempts to understand the implications of the spatial/temporal variability of these forecasts in the flood forecasting environment. To achieve this, a simple distributed hydrological model was employed to investigate the distribution effect of rainfall forecasts and a high-resolution rainfall ensemble prediction system was established to provide rainfall forecasts at the catchment scale.

THE BRUE CATCHMENT

The Brue catchment, located in the southwest of England, UK, is an ideal experimental site for research on weather radar, quantitative precipitation forecasting and rainfall–runoff modelling, as it has been facilitated with a dense raingauge network, as well as coverage by three weather radars. Numerous studies (Bell & Moore, 2000; Cluckie *et al.*, 2000; Moore *et al.*, 2000; Pedder *et al.*, 2000; Chen, 2004) have been conducted regarding the catchment, notably during the period of the UK NERC Hydrological Radar EXperiment (HYREX).

Figure 1 shows the location of the Brue catchment and the gauging stations. The River Brue rises in clay uplands to the east of the catchment. The major land-use is pasture on clay soil and there are some patches of woodland in the upper eastern catchment, which forms part of the unique landscape of the Somerset Levels and Moors. The catchment has a drainage area of 135 km^2, with an average annual rainfall of 867 mm and an average mean river flow of 1.92 m^3/s, for the period from 1961 to 1990. In addition to the weather radars, a dense raingauge network comprised of 49 tipping-bucket raingauges, with a recording resolution of 10 s (Moore *et al.*, 2000) had been established during the HYREX programme. The network provided at least one raingauge in each of the 2-km grid squares that lie within the catchment. Data sets from HYREX were available through the British Atmospheric Data Centre (BADC) for the period from 1993 to 2000. A simplified grid-based distributed rainfall–runoff model (GBDM), based on the Kinematic Wave Approximation, was chosen for this study. This model has been successfully applied in different catchments of Taiwan (e.g. Yu & Jeng, 1997; Yu *et al.*, 2001) following its original development at the University of Birmingham in the 1980s (Yu & Cluckie, 1988). In this study, the model was calibrated and verified using 17 historical storm events using the shuffled complex evolution (SCE) method (Duan *et al.*, 1992, 1993, 1994), which is a general-purpose global optimisation strategy designed to solve the various response surface problems in calibrating a nonlinear simulation model.

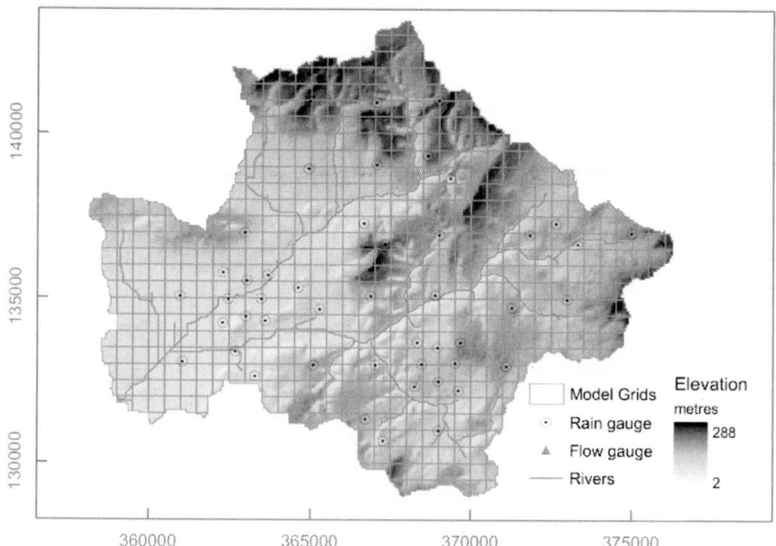

Fig. 1 The Brue catchment and mesh configured for the GBDM model.

THE SHORT-RANGE ENSEMBLE RAINFALL PREDICITION SYSTEM

A short-range ensemble QPF system was used to produce ensemble rainfall forecasts that were utilised by the hydrological model to generate the river flow forecasts. The ensemble QPF system was initially implemented in the EU FP5 FLOODRELIEF project and has since been used in several case studies (Xuan *et al.*, 2005). The system consisted of: the PSU/NCAR mesoscale model (MM5) (Dudhia *et al.*, 2003), the global forecast data sets from the European Centre for Medium-range Weather Forecasts (ECMWF, Persson, 2003), the model physics selector and a post-processing system. Using the ensemble approach, the system is capable of representing uncertainties due to perturbations in initial and boundary conditions and the efficiency of the model structure. In this study, the system was configured to produce ensembles over a short time range, i.e. 24 h. The model structure was not perturbed in terms of changing physics parameterisations, as previous research revealed that uncertainties in rainfall forecasts due to parameterisations is not always significant when compared with the inaccuracy of initial and/or boundary conditions (Xuan *et al.*, 2005). The ensembles comprised only the direct downscaling of perturbed background fields from ECMWF, i.e. the 50 members of the ECMWF ensemble prediction system, and one operational forecast member that represented the best estimate of the atmospheric state.

In order to reach the spatial scale complementary to the rainfall–runoff modelling system, four spatial nested domains were used with the inner-most domain having a resolution of about 2 km and covering a region around 100 km × 100 km (Fig. 2). The domains were deliberately set so that the target catchment was well centred inside all the nests in order to largely reduce the effects of spatial distortion due to the different map projections used in weather models and rainfall–runoff models.

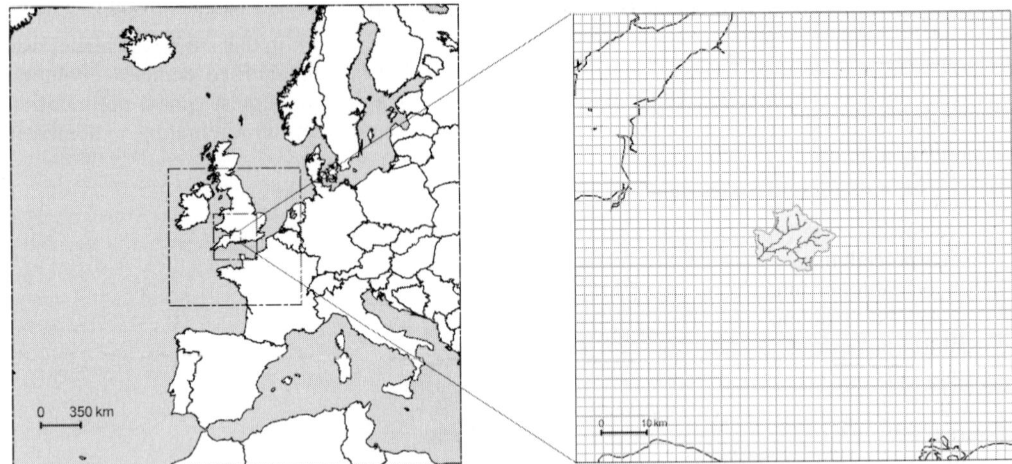

Fig. 2 The domain configuration of the MM5 (left with each domain indicated by dash-dot lines) and the detailed view of domain 4 (the Brue Catchment shown with shade near the centre).

Table 1 The verification of GBDM with four historic storm events.

Event no.	Date	Coefficient of model efficiency (CE)
14	23/10/1999 09:00 – 28/10/2000 08:00	0.88
15	17/12/1999 04:00 – 21/12/1999 03:00	0.69
16	31/31/2000 23:00 – 05/02/2000 14:00	0.92
17	02/04/2000 15:00 – 07/04/2000 14:00	0.87

THE SIMULATION EXERCISE

The distributed hydrological model, GBDM, was calibrated with 13 historical storm events and verified using four events (see Table 1) that occurred over the Brue catchment from 1993 to 2000, covering different seasons.

The parameters of GBDM were categorised into two groups, i.e. those physically-based parameters which can be generated directly from topographic, soil, and vegetation maps, and those that needed to be calibrated from historical rainfall and flow data. Three calibration parameters were calibrated by applying both an optimisation technique and an objective function. In this study, the SCE method was adopted for model calibration. The SCE method is a general-purpose global optimisation strategy designed to solve the various response surface problems encountered in calibrating a nonlinear simulation model. Details can be found in Duan *et al.* (1992, 1993, 1994) and Chi (2006). GBDM uses the raingauge data for its calibration and verification. Although the model is grid based and has a grid size of 500 m in this case, the rainfall value for each grid was actually obtained using the Thiessen polygon method. When the rainfall forecasts are used to drive the hydrological model, the rainfall values from the weather model were first subject to a projection transformation (from Lambert Conformal projection to the UK National Grid), and then linear interpolation was adopted to transfer the rainfall from 2-km weather model grids to the hydrological grids which had a 500-m grid size.

It has been long recognised that there are always considerable uncertainties regarding the NWP forecast of rainfall locations and timings. In order to account for the fact that the rainfall distribution was not correctly positioned, a "best match" approach was introduced to determine the location of the forecast that best resembled the rainfall pattern in the catchment. A similar method (Ebert & McBride, 1993) has been used to verify the weather forecast of precipitation and other spatially-distributed variables where it was found that the model performance was not sufficiently evaluated by the grid-by-grid criteria, e.g. critical success index (CSI), false alarm ratio (FAR), etc. This approach involved three steps: (1) Extract a grid mask of the catchment that represents the shape of the catchment and the relative position of each grid within the catchment. (2) Determine the distribution of rainfall accumulation over the catchment grid for a given period of time. A 24 h rainfall accumulation from raingauges was adopted as a reference distribution. (3) Within the weather model domain, move the catchment grid mask in both x (east–west) and y (south–north) directions step-by-step. For each step, calculate the correlation coefficient of the rainfall forecast extracted by the mask and the reference distribution obtained in the previous step. Thus, a series of correlation coefficient values were obtained in respect of different offsets of x and y. After this search procedure, a "best-matched" area was identified and the data values within this area were then available for future processing.

Another problem related to the application of the QPF in a hydrological model, where the final forecast can be severely underestimated if applied with a hydrological model calibrated using point rainfall from raingauges. As the grid value for the weather model represents a box-average of the grid where the sub-grid dynamics play an important role in contributing a great deal of spatial variability, even within a small grid (2 km in this case), there is a high chance of a gauge-calibrated hydrological model producing a simulation much lower in magnitude than if the rainfall forecast was applied directly. Again, a 24 h rainfall accumulation was used to adjust the rainfall forecast bias that was performed through the following procedure: (1) Determine the total catchment rainfall over the specified 24 h period that will be used as a reference value. (2) Determine the ratio of the amount obtained in step (1) to the total catchment rainfall value from the weather model in the same period. (3) Adjust the hourly rainfall value from the weather model by the factor obtained in step (2). It should be noted this simple bias correction is a sort of "posterior" method; it is used here for evaluation only and is not suitable for forecasting.

TEST EVENTS FOR THE ENSEMBLE FORECASTS

Storm events no. 15 and no. 16 were selected for uncertainty analysis. While event 15 contains several sub-events, event 16 seems to have one single continuous event.

The original verification results from these events are shown in Fig. 3. It can be seen that the model performed very well for event 16, but under-predicted the main flow peak of event 15. One reason for this is that the GBDM has difficulty identifying the transition state of the catchment in the interval between two consecutive heavy rainfall inputs.

In order to evaluate the uncertainty of the rainfall forecasts regarding the location, the results of the "best-match" in terms of maximum correlation coefficient are plotted in Fig. 4. Under the current configuration, the catchment mask can move within the MM5 innermost domain with a maximum distance in x and y directions of about 80 km. The origins in Fig. 4 are the initial settings, i.e. the original position of catchment. The highest value, R_{max}, and the lowest one, R_{min}, from all the "best-matched" ensemble members are also depicted in the figure for both events. Figures 5 and 6 give the 24 h ensemble rainfall forecasts over the catchment in forms of box plots. Note that in terms of areal rainfall, the median values of the ensembles still resemble each other after undergoing the location correction as mentioned above.

The ensemble rainfall forecasts are then utilised in a time window of 24 h for both events. For event 15, the ensemble rainfall forecast was initialised on 18 December 1999 12:00 UTC and the hourly rainfall forecasts were produced up to 19 December 1999 12:00 UTC, corresponding to the time window from the 34th step to the 57th step. The original rainfall values within this window have then been substituted by the ensemble rainfall forecasts. The same procedure was also applied to event 16, but with a time start of 01 February 2000 12:00 UTC and the corresponding window from step 14 to step 37. The ensemble forecasts are displayed in Figs 7 and 8 where the shaded areas represent the spread of all ensemble members between the quantiles $q_{0.1}$ and $q_{0.9}$. Also shown are the control forecasts, based on the "best estimate" rainfall forecast results. Note that the rainfall values here are referred to the average of all ensemble members.

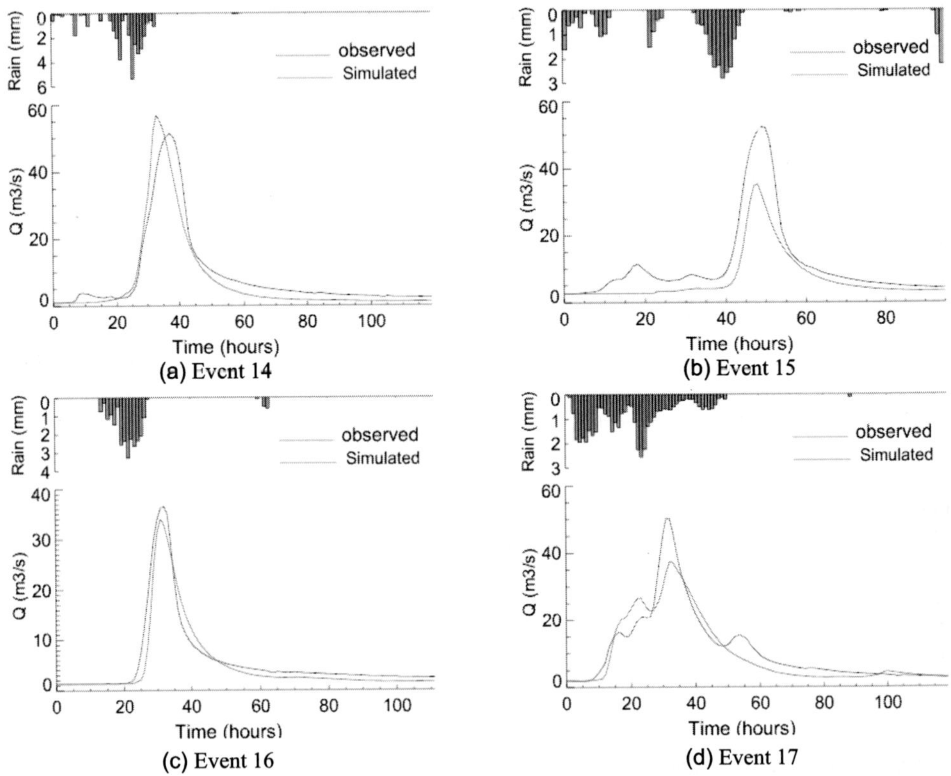

Fig. 3 The verifications of the discharge forecasts for the four events with the observed hourly areal rainfall shown on the top.

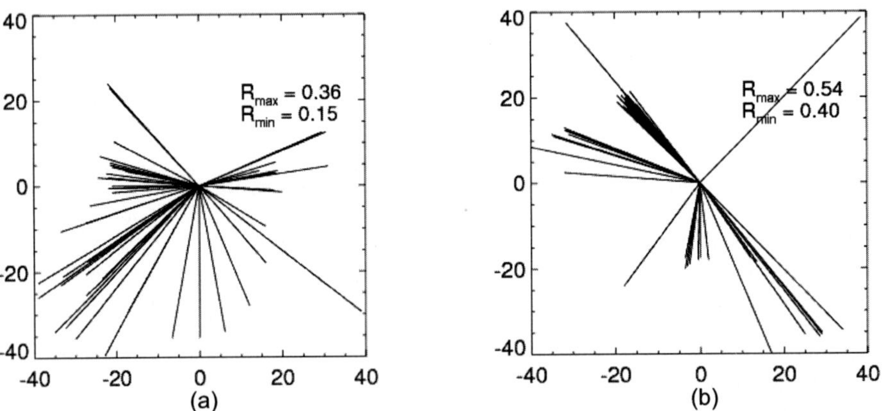

Fig. 4 The shift of mask frame to get the highest correlation coefficient between the rainfall ensemble members and the observed catchment rainfall distribution for events.15 (a) and 16 (b).

As shown in Figures 7 and 8, for both events the direct utilisation of the rainfall ensemble by the hydrological model results in strong bias – a severe underestimation of river flow. Although the river flow ensemble does express a considerable spread, it fails to encompass either the observed value, or the original forecast with raingauge inputs. As shown in Figs 5 and 6, the

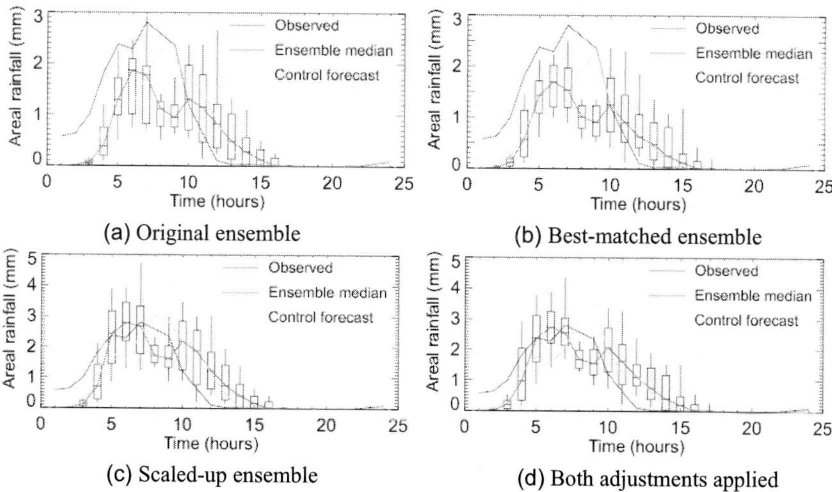

Fig. 5 The box plots of the areal rainfall ensemble forecast of event 15 subject to different adjustments. The lower and upper bounds of the boxes correspond to quantiles $q_{0.1}$ and $q_{0.9}$.

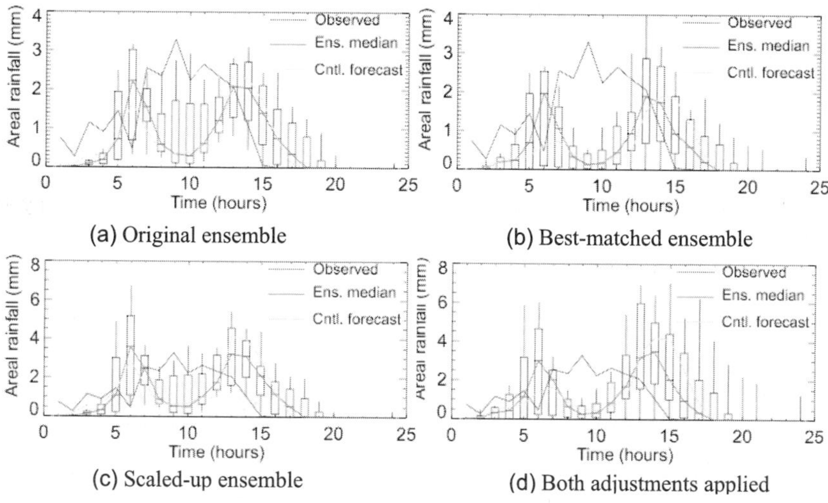

Fig. 6 The same as Fig. 5, but for event 16.

difference of hourly catchment average rainfall between rainfall forecasts and the gauge network can be easily appreciated and again it represents the weather model's limitation in predicting spatial variability. The temporal distribution of catchment average rainfall, on the other hand, resembles that from raingauges for event 15, and therefore the hydrographs generated by the ensemble mean are similar to the original, although they underestimate the latter.

The introduction of the "best-match" rainfall field pattern provides a new prospective on the uncertainty of the QPF regarding location. Although this method is defined by the maximum correlation between the rainfall forecast and the reference one, it also reveals the spatial relationship of all ensemble members. Members that agree with each other must have locations close to each other in the "best-match" map, as in Fig, 4; therefore the dispersion of those locations to some extent can be seen as a indicator of the agreement among ensemble members and as such, imply the spread of the trajectories of members in the model's phase space.

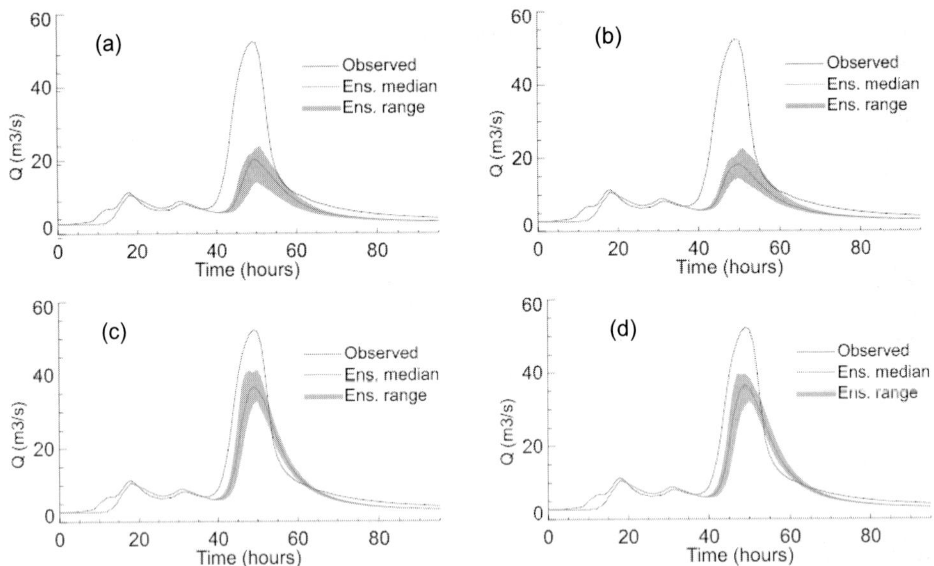

Fig. 7 The results of the flow ensemble forecast for event 15 which include: (a) the initial ensemble without any correction of rainfall; (b) the ensemble with "best-matched" rainfall; (c) the ensemble with the amount scaled; and (d) the scaled-up version after matching. The shaded area is the ensemble spreads in between quantiles $q_{0.1}$ and $q_{0.9}$ of the members.

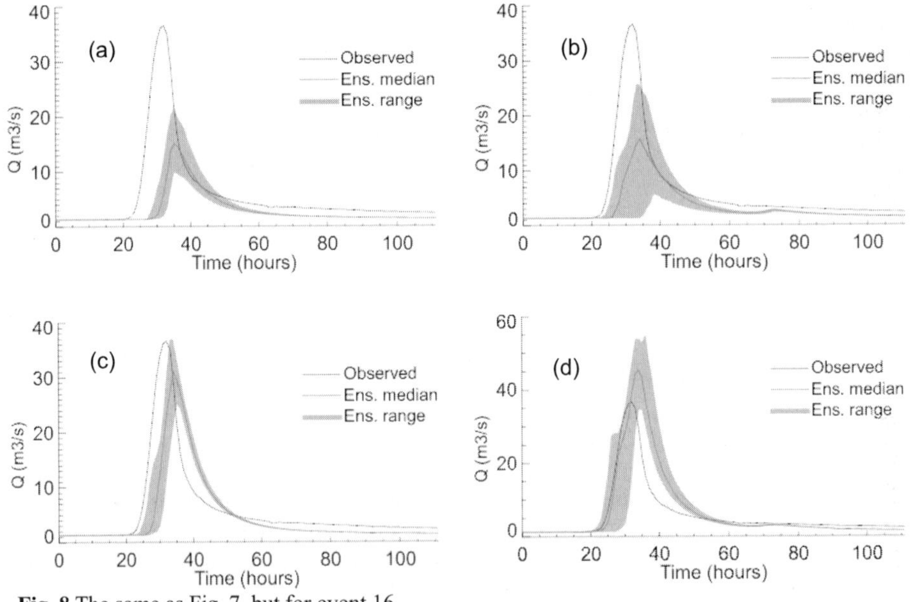

Fig. 8 The same as Fig. 7, but for event 16.

It is worth noting that due to the high nonlinearity of the precipitation process, the correlation coefficient in a linear sense can only reveal a small fraction of the behaviour of ensemble forecasts. Interestingly, the rainfall ensemble provides a contrasting dispersion result for event 15 and event 16. There are several cluster-like areas for event 16 where several members are supposed to produce similar results, while a quite dispersive set was obtained for event 15 where no clear pattern could be found.

Two distinct results were produced after applying location correction and bias adjustment to both events. For event 16, the corrected version has clearly improved the quality of flow forecasts – the flow ensemble has successfully covered the observation, as shown in Fig. 8(d). As to event 15, the correction actually has not made things better as can be seen from Fig. 7(d) where the median forecast has under-predicted the river flow, even with the average catchment rainfall being corrected to the level of the raingauge. One important effect regarding the implication of the rainfall pattern can be found if the flow ensembles are compared with the results of the catchment average rainfall ensembles that are shown in Figs 5 and 6. The temporal evolution of rainfall forecast in terms of the spatial average follows closer to the gauge value for event 15 than event 16. This is of particular interest because it reveals that the ensemble members have probably not captured or approached the rainfall pattern that really happened, even if the catchment average rainfall looks similar to the real one. The maximum correlation coefficient for this event, as seen in Fig. 4, is much lower than for event 16, which agrees with this speculation.

TIDAL STORM SURGE SIMULATION USING 24 ENSEMBLE MEMBERS

A second case study is now briefly introduced that focuses on a tidal surge application in the Thames Estuary as a key element in the process of determining coastal flooding risk. The forecast water level for the Thames is given by predicted values at Sheerness by the STFS (Storm Tide Forecasting Service). The water levels at Sheerness (the seaward boundary) were derived from the sum of the astronomic time-series for the site together with the forecast surge residual. The forecast surge residual for Sheerness was provided by the North Sea Model – which is a cut-down version of the POL (Proudman Oceanographic Laboratory) CS3 model. The model was run twice daily by the UK Meteorological Office to produce 36-hourly surge residual forecasts. In March 2006 a unique co-location workshop was held that focused on the forecasting of a simulated 1000-year storm and its impact on the Thames Estuary and London. The surge forecasts 3-days ahead, 1-day ahead and on that day itself (29 September 2015), together with wind speed and wind direction at that time, were input as the downstream boundary condition in MIKE 11 while monthly average inflow at Teddington Weir was set as the upstream boundary condition. Figure 9 shows the 24 3-day ahead surge forecast ensemble forecast water levels at the Thames Barrier, given the Forecast Ensemble at Sheerness from the STFS.

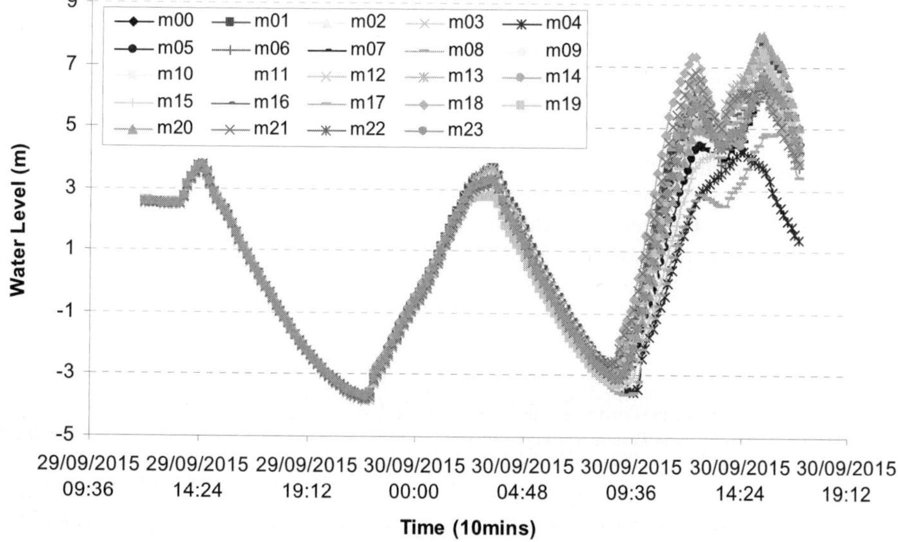

Fig. 9 Ensemble forecast water level at the Thames barrier – MIKE 11.

In general, the MIKE 11 surge tide model presents a means of exploring a number of potential scenarios through using alternative input tide level data as boundary conditions. The most important use of the model is to propagate the uncertainty and simulate the relative water levels at various points along the Thames and to provide the crucial real-time data to support the flood management system. The benefit of having access to the high-resolution spatial and temporal data from the mesoscale model was that the CS3 model could be refined, the wind speed data could be used to generate a wind set-up correction and the uncertainty of all can be estimated from the ensemble data.

CONCLUDING COMMENTS

The development of a Whole System Modelling approach to modelling complex hydrological hydroinformatic systems is now becoming a reality, and the ability to attack the problem of propagating uncertainty through complex model cascades a necessity. The study draws attention to the meteorological model and hydrological model boundary through which the transport of uncertainties can be investigated with the help of the mesoscale ensemble short-range rainfall forecasts that can expose the uncertain nature of weather systems. Also the distributed hydrological modelling system is capable of reflecting the effects of the spatio/temporal variability of the rainfall fields. Ensemble forecasts can reveal the uncertain nature of the modelling system. The ensemble rainfall inputs, however, need to be well adjusted for bias in rainfall amount and properly shifted to match the rainfall pattern. Considerable uncertainty can be propagated from weather domains where a more chaotic weather system is concerned. In this case, even the short-range rainfall ensemble can end up with a spread of forecasts that may become unacceptable after the model coupling process. The proposed "best matching" procedure reflects the uncertain nature of the ensemble itself, and the inability of the current weather models to resolve sub-grid scale precipitation features. The bias, specifically the common underestimates of rainfall at fine scale, can result in unrealistic low riverflow forecasts; this has to be properly dealt with separately, not simply attributed to the system's uncertainties. Despite these difficulties, the development of coupled models that allow the propagation of uncertainty are an increasingly important development in flow forecasting systems. In terms of the current capabilities of NWPs and ensemble procedures, data fields such as windspeed and direction, pressure and temperature can be predicted with some skill. This is good from the point of view of coastal and estuarial flood forecasting systems. The difficult physics is that associated with precipitation processes where the nonlinearities and spatio-temporal problems are extreme. However, even here the ability to incorporate dynamic downscaling and the generation of unbiased ensembles promises great advances in the near future.

Acknowledgements The authors would like to thank British Atmospheric Data Centre (BADC), ECMWF, NCAR for provision of data sets of HYREX, the meteorological background fields and the MM5 modelling system, respectively. This study was funded by the EU FP5 FLOODRELIEF project. Additional support was provided from EPSRC FRMRC (grant ref: GR/s76304/01). Particular thanks are also due to Dr Yunqin Xuan and Dr Yuchi Wang.

REFERENCES

Bell, V. A. & Moore, R. J. (2000) Short period forecasting of catchment-scale precipitation. Part II: A water-balance storm model for short-term rainfall and flood forecasting. *Hydrol. Earth Syst. Sci.* **4**, 635–651.

Chen, Y. *et al.* (eds) (2004) *GIS and Remote Sensing in Hydrology, Water Resources and Environment* (ICGRSHWRE, Three Gorges Dam, China, September 2003). IAHS Publ. 289. IAHS Press, Wallingford, UK. ISSN 0144-7815.

Cluckie, I. D., Griffith, R. J., Lane, A. & Tilford, K. A. (2000) Radar hydrometeorology using a vertically pointing weather radar. *Hydrol. Earth Syst. Sci.* **4**, 565–580.

Cluckie, I. D., Rico-Ramirez, M. A., Xuan, Y. & Szalinska, W. (2006) An experiment of rainfall prediction over the Odra catchment by combining weather radar and a numerical weather model. In: *7th International Conference on Hydroinformatics* (Nice, France).

Bartholmes, J. & Todini, E. (2005) Coupling meteorological and hydrological models for flood forecasting. *Hydrol. Earth Syst. Sci.* **9**, 333–346.

de Roo, A. P. J., Gouweleeuw, B., Thielen, J. *et al.* (2003) Development of a European flood forecasting system. *Int. J. River Basin Manage.* **1**, 49–59.

Duan, Q., Sorooshian S. & Gupta, V. K. (1992) Efficient global optimization for conceptual rainfall–runoff models. *Water Resour. Res.* **28**, 1015–1031.

Duan, Q., Gupta, V. K. & Sorooshian, S. (1993) A shuffled complex evolution approach for effective and efficient global minimization. *J. Optimization Theory Appl.* **76**(3), 501–521.

Duan, Q., Sorooshian S. & Gupta, V. K. (1994) Optimal use of the SCE-UA global optimization method for calibrating watershed models. *J. Hydrol.* **158**, 265–284.

Dudhia, J., Gill, D., Manning, K. *et al.* (2003) PSU/NCAR Mesoscale Modelling System Tutorial Class Notes and User's Guide: MM5 Modelling System Version 3. Mesoscale and Micro-scale Meteorology Division, National Centre for Atmospheric Research.

Ebert, E. E. & McBride, J. L. (2000) Verification of precipitation in weather systems: determination of systematic errors. *J. Hydrol.* **239**, 179–202.

Ferraris, L., Rudari, R. & Siccardi, F. (2002) The uncertainty in the prediction of flash floods in the northern Mediterranean environment. *J. Hydrometeorol.* **3**, 714–727.

Kalnay, E. (2002) *Atmospheric Modeling, Data Assimilation and Predictability*. Cambridge University Press, Cambridge, UK.

Kobold, M. & Sušelj, K. (2005) Precipitation forecasts and their uncertainty as input into hydrological models. *Hydrol. Earth Syst. Sci.* **9**, 323–332.

Lorenz, E. N. (1963) Deterministic non-periodic flow. *J. Atmos. Sci.* **20**, 130–141.

Lorenz, E. N. (1993) *The Essence of Chaos*. University of Washington Press, Seattle, USA.

Moore, R. J., Jones, D. A., Cox, D. R. & Isham, V. S. (2000) Design of the HYREX raingauge network. *Hydrol. Earth Sys. Sci.* **4**, 523–530.

Pedder, M. A., Haile, M. & Thorpe, A. J. (2000) Short period forecasting of catchment-scale precipitation. Part I: the role of Numerical Weather Prediction. *Hydrol. Earth Syst. Sci.* **4**, 627–633.

Persson, A. (2003) User Guide to ECMWF forecast products. *Meteorol. Bull.* M3.2, ECMWF.

Singh, V. P. (1978) Mathematical modelling of watershed runoff. In: *International Conference on Water Resources Eng.*, 10–13 (Bangkok, Thailand).

Smith, K. T. & Austin, G. (2000) Nowcasting precipitation: a proposal for a way forward. *J. Hydrol.* **239**, 34–45.

Verbunt, M., Zappa, M., Gurtz, J. & Kaufmann, P. (2006) Verification of a coupled hydrometeorological modelling approach for alpine tributaries in the Rhine basin. *J. Hydrol.* **324**, 224–238.

Wagener, T., Boyle, D. P., Lees, M. J., Wheater, H. S., Gupta, H. V. & Sorooshian, S. (2001) A framework for development and application of hydrological models. *Hydrol. Earth Syst. Sci.* **5**, 13–26.

Xuan, Y. (2007) Uncertainty propagation in complex coupled flood risk models using numerical weather prediction and weather radars. PhD Thesis, University of Bristol, Bristol, UK.

Xuan, Y. & Cluckie, I. D. (2008) Uncertainty propagation in ensemble rainfall prediction systems used for operational real-time flood forecasting. In: *Practical Hydroinformatics: Computational Intelligence and Technological Developments in Water Applications* (ed. by R. J. Abrahart, L. M. See & D. P. Solomatine), Chapter 31, 437–448. Springer-Verlag, Water Sci., Tech., no. 68, ISBN: 978-3-540-79880-4.

Xuan, Y., Cluckie, I. D. & Han, D. (2005) Uncertainties in application of NWP-Based QPF in real-time flood forecasting. In: *Proc. EU MITCH/FLOODRELIEF Symposium, Innovation, Advances and Implementation of Flood Forecasting Technology* (Tromso, Norway). ISBN. 1898485148.

Yu, P. S. & Jeng, Y. C. (1997) A study on grid based distributed rainfall runoff model. *J. Water Resour. Manage.* **11**, 83–99.

Yu, P. S., Yang, T. C. & Chen, S. J. (2001) Comparison of uncertainty analysis methods for a distributed rainfall–runoff model. *J. Hydrol.* **244**, 43–59.

Ensemble of radar and MM5 precipitation forecast models with M5 model trees

G. A. CORZO[1,2], Y. XUAN[1], C. A. MARTINEZ[1] & DIMITRI SOLOMATINE[1,2]

1 *UNESCO-IHE Institute for Water Education, PO Box 3015, 2601DA Delft, The Netherlands*
g.corzo@unesco-ihe.org

2 *Delft University of Technology, Faculty of Civil Engineering and Geosciences, PO Box 5048, 2600 GA Delft, The Netherlands*

Abstract The use of weather radars for precipitation forecasting in fast response river basins is very important for early warning systems. Some of the most used techniques, such as radar echo tracking, which is based on patterns and correlations, do not contemplate the decay/growth mechanism of rainfall systems. Numerical weather prediction models do consider much more information than radars, but their accuracy is less than the radar forecast; at least for a lead time of up to several hours. In this paper we explore the integration of the two forecasting approaches. The MM5 numerical weather prediction model is set up for a region in southeast England. The Nimrod radar data, for the same region, from the British Atmospheric Database Centre (BADC) is adapted and used as reference. The M5 prime regression model tree is used for the integration of both models. The results show that the proposed integration reduced the error of tracking significantly. The results presented here have been evaluated on an hourly time step, but further research will be done in order to extend and generalize the methodology.

Key words weather radar echo tracking; MM5 model; model trees; model integration; numerical weather prediction; nowcasting

INTRODUCTION

Weather radars are used to obtain rainfall information with high spatial and temporal resolution, which nowadays is integrated into hydrological flow models for forecasting. One of the main advantages of weather radars is that they are able to scan large areas and to obtain the rainfall distribution information in a very short time. Previous studies have been carried out using different methods on precipitation forecasting and radar rainfall data. Pessoa *et al.* (1993), for instance, used the weather radar for flood forecasting in combination with a distributed rainfall–runoff model. His goal was to explore the sensitivity analysis of different flood situations to radar-derived rainfall forecast. Sun *et al.* (2000) predicted flooding using radar and raingauge data showing that the estimated rainfall improves considerably because it combines both the raingauge and radar data. Li *et al.* (2004) coupled weather radar rainfall data with a distributed hydrological model for real time forecasting. The results simulated by the weather radar data were better than those simulated by raingauge data, but in terms of errors the result based on the weather radar are almost the same as the result from the raingauge data. Cluckie *et al.* (2006) combined rainfall prediction from radar and a high resolution mesoscale model to explore the advantages from both sides, obtaining high accuracy forecast field locations from radar and NWP models.

Since it is known that the radar forecast is primarily elaborated on the basis of a pattern displacement, and not on any kind of mass or energy formulation, the decay of the mass is not considered. However, numerical weather prediction (NWP) models do consider this information for forecasting, but their accuracy is limited and depends on a warming period of the numerical scheme and the availability of boundary conditions data. Although in operation the NWP model is based on the physical representation of the weather, and therefore on an extended lead time, it is more accurate than radar tracking (Xuan, 2007). Xuan *et al.* (2009) analysed the uncertainty issues associated with the application of NWP in hydrological forecasting by utilising the ensemble methods. In this paper, we present an integrated attempt to use model trees to combine the forecast results of a MM5 model and a radar tracking technique. The results show that the combined approach of both models using a model tree is more accurate than the single radar echo-tracking model for all lead-times tested.

The radar and numerical weather forecasting models are explained initially followed by a description of model trees in general, and the M5 prime algorithm is described next. Performance criteria for the analysis are explained before the results and conclusions are presented.

RADAR FORECASTING

The information used in this study comes from the UK Met Office Nimrod system (Golding, 1998). This system is oriented for nowcasting system purposes that analyse weather variables, including precipitation amount, in a very short-range prediction. The data is already processed, and the data available is formatted on an hourly basis. This information includes remote-sensed imagery data.

The processing of the Nimrod radar data (Golding, 1998) used in this study includes image quality control, echo removal and vertical profile correction. The imagery is calibrated against the radar composite and combined with lightning data. The Nimrod technique has been tried for extrapolating the precipitation distribution by selecting the optimum vector for each precipitation object, from either cross-correlation or the NWP model winds. This study uses 5-km grid resolution data using the radar data located in the southeast of England (100–250 km N, 500–650 km E in the UK-based NGR coordinates) and a temporal resolution of 15 minutes. The area is gridded into 31 by 31 cells of 5 km each. Radar information from 27 May 2007 is used for training and verification. The information is integrated with a moving average for mapping this information with the available weather prediction model.

RADAR ECHO TRACKING

The radar echo tracking is used as a model reference in this study; this methodology is based on the persistence of the rainfall pattern (Lagrangian persistence). It permits us to predict the new development, displacement and dissipation of the rainfall in existing cells by means of a cross-correlation between consecutive radar scans. Its maximum correlation shows the displacement of the rainfall event. The radar tracking echo is successful when it has been predicted one or two steps ahead, but the error increases with longer lead times. Previous studies using this method have shown that an improved QPF with a three hour lead time can be reached (Cluckie *et al.*, 2006).

The MATLAB function called xcorr2 is applied to return the cross-correlation of matrices A and B. Matrix A has dimensions ($m1$, $n1$) and matrix B has dimensions ($m2$, $n2$). Equation (1) shows the two-dimensional discrete cross-correlation:

$$C(i,j) = \sum_{m=0}^{(m1-1)} \sum_{n=0}^{(n1-1} A(m,n) * conj(B(m+i,n+j)) \tag{1}$$

where $0 \le i < m1 + m2 - 1$ and $0 \le j < n1 + n2 - 1$.

MM5 NUMERICAL WEATHER PREDICTION MODELS

The Fifth-Generation NCAR/Penn State Mesoscale Model (Dudhia, 2003), mostly known as MM5, is one of the mesoscale numerical weather models extensively used in research and operational areas. In this study, the MM5 model is configured to generate rainfall prediction for the study area with sufficiently high resolution both spatially and temporally. The most used configurations were adopted to customise the MM5 model with 23 vertical layers and the finest horizontal spatial resolution is 2 km that can match the Nimrod data sets.

The MM5 model was initialised with the initial condition (IC) and lateral boundary condition (LBC) using the deterministic forecast data set from the European Centre for Medium Weather Forecast (ECMWF). Hourly rainfall accumulation was also produced in order to compare the hourly available radar and tracking data. An extra interpolation procedure was adopted to unify the different map projections used by the MM5 and the Nimrod data sets.

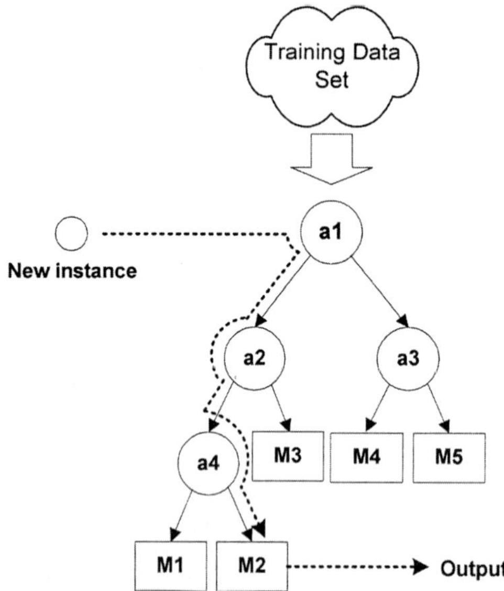

Fig. 1 Training M5 model trees and their operation. a1…a3 are data partitioning rules; M1...M5 are multiple linear regression models (source: Solomatine & Dulal, 2003).

MODEL TREES

Model trees are data-driven and can be classified as modular models (Haykin, 1999). To train them, a set of data may be split into subsets corresponding to a particular sub-process to be modelled, and then each module (as seen in Fig. 1) is trained on these non-intersecting subsets. After the model is trained, each new input vector is first classified to one of the regions for which the modules were trained, and then only one module is run to produce the prediction. Classes of such methods employing consecutive progressive splits are typically referred to as "trees"; examples are: decision trees, regression trees (Breiman, 1984; Friedman, 1991; Solomatine & Xue, 2004), M5 model trees (Quinlan, 1992) (Fig. 1).

Since for each data instance (input vector) only one local model is used for prediction, there is a problem of compatibility at the boundary between the regions for which the modules are responsible: for the two neighbouring input vectors the predicted outputs could be very different. A solution could be in updating the local models to make them compatible at the boundaries, as is done in the M5 model tree algorithm through smoothing. Wang & Witten (1997) presented the M5 algorithm based on the original M5 algorithm but able to deal with enumerated attributes, to treat missing values and using a different splitting termination condition. Several advantages of using the model tree are that it is a non black-box model, understandable, easy to use and to learn, fast in training, robust when dealing with missing data, able to handle a large number of features, and able to tackle tasks with very high dimensionality.

PERFORMANCE ANALYSIS

There are a wide variety of statistical forecasting measures for the performance analysis; here we consider the root mean squared error (RMSE) and a detailed analysis of the variability of the results in each cell of the forecast grid.

In this study the radar information is used as the reference value of the measured precipitation. The RMSE is used to compare the differences between the radar measured and the rainfall forecast obtained from the tracking forecast and the integrated model.

METHODOLOGY

The MM5 model is initialised at 00 UTC for 24-hour forecast of the selected event, 27 May 2007. The objective of the forecast is to obtain 4-hours forecast, each hour, in the selected date. For this, rainfall prediction for the MM5 is obtained from several model parameters. The boundary and initial conditions of the MM5 are obtained from the operational forecasts of European Centre for Medium Weather Forecast (ECMWF). This information is fed once and is run over the entire prediction interval. On the other hand, the domains of the radar information and MM5 model are geo-referenced and fit into the same scale. Part of the geo-referenced process included the interpolation of the MM5 into the radar domain using the nearest-neighbour method; as MM5 does not use the local projection of the radar. The radar data obtained from the Nimrod system is obtained from British Atmospheric Data Centre (BADC) in an hourly accumulated rainfall format.

The rainfall prediction in the case of the weather radar is obtained by implementing an extrapolation of hourly-accumulated radar images (as mentioned before). The maximum cross-correlation of two radar scans at time t and $t - 1$ indicates the displacement of the rainfall event.

The data set is split into a training (9130 instances) and verification (9130 instances) set to be able to have a better assessment of the performance results.

After the split is done, the M5 model tree is constructed with rainfall prediction from radar tracking and the MM5 model output. For the model integration the radar and MM5 models were indexed in such a way that the results of each forecasted image can be graphically evaluated independently. Figure 2 shows the indexed mapping used.

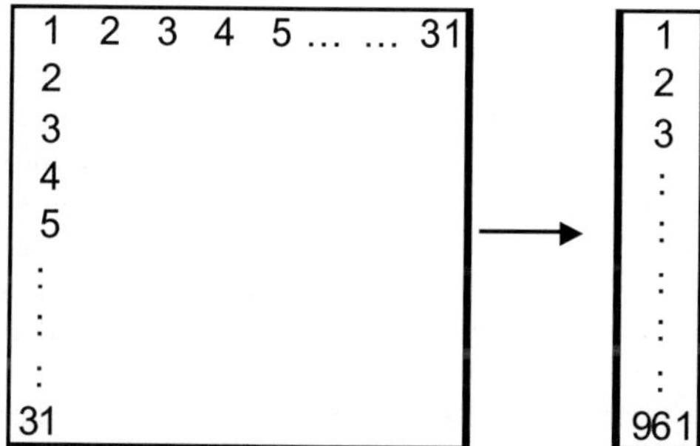

Fig. 2 Radar and MM5 image split.

RESULTS

The RMSE of the verification results is used to asses the performance of the integration. The M5 tree model integration is better able to reduce the RMSE values for all the forecast steps (Fig. 3). The results show values of around 28% reduction in all the forecasts made.

A more detailed analysis is made to analyse the cell results for the period considered. Each image is analysed independently and the cell's results for each hour forecast are plotted. Since the results have been indexed and added together, only a fragment of the forecast is shown in Fig. 4. Here we can see the first 100 cells results (e.g. first image forecast), for a 3-hours forecast. The figure shows the M5P model tree, radar measured and tracking radar forecast. Although the prediction results with the model tree underestimate the rainfall, it can be seen that the integrated model captures the time events.

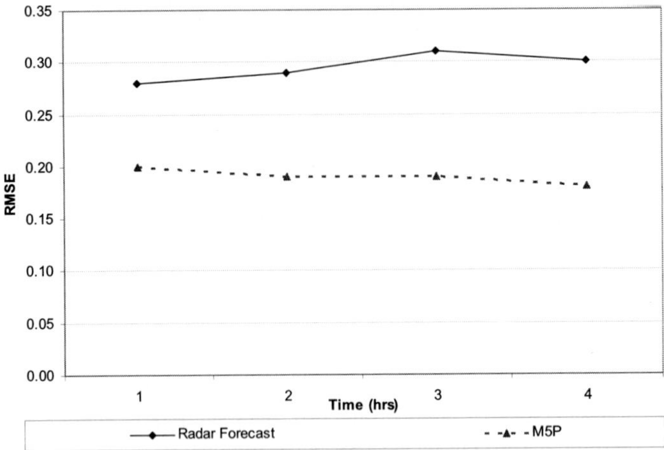

Fig. 3 RMSE Performance comparison between radar forecast and M5 Model Tree.

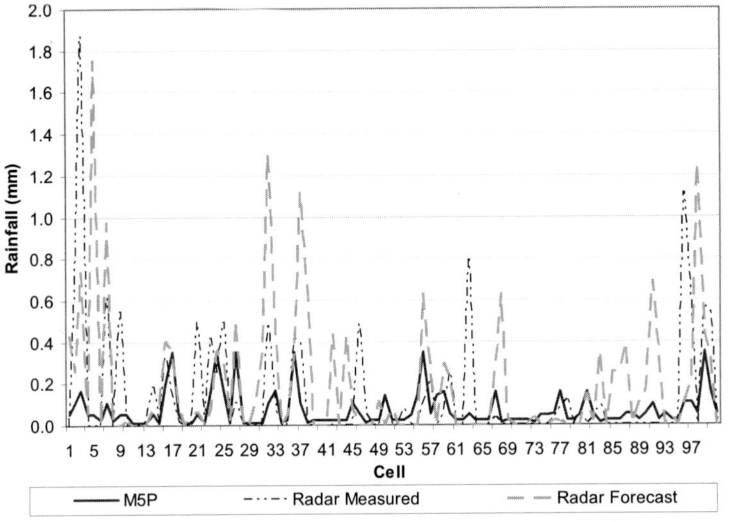

Fig. 4 Rainfall forecast made for 3 hours (Indexed image in 961 cells).

CONCLUSIONS

The use of the information from a numerical weather prediction models in combination with a radar tracking forecast model is explored in this study. Although a simple set-up with an MM5 model alone does not have a very good performance, it seems that information contained in its results is valuable for complementing a radar technique. It would appear that the information from the numerical weather model helped in reducing the magnitude of the individual error in each cell of the forecast image. It would appear that the reason is not only the fact that the MM5 model did not generate high cell precipitation values, but also its ability to represent well the phenomena in a low scale of values. The results showed similar results for all the forecast horizons tested.

The studies point to extending the research not only on the generalization of the process with more data and a deeper analysis, but also on improving the MM5 model and the integration process with a more detailed modular scheme based on regimes.

Acknowledgements The authors would like to thank Dr Miguel A. Rico-Ramirez at the University of Bristol for the valuable discussion and data support. The authors are also grateful for the data sets provided by ECMWF that were used to initialise the MM5 mesoscale model.

REFERENCES

Breiman, L. (1984) *Classification and Regression Trees*. Chapman & Hall/CRC, USA.

Cluckie, I., Rico-Ramirez, M., Xuan, Y. & Szalinska, W. (2006) An experiment of rainfall prediction over the Odra catchment by combining weather radar and a numerical weather model. 7th International Conference on Hydroinformatics (Nice France 2006).

Dudhia, J., Gill, D., Manning, K. & Bruyere, C. (2003) PSNU/NCAR Mesoscale Modelling System Tutorial Class Notes and User's Guide: MM5 Modelling System Version 3. Mesoscale and Micro scale Meteorology Division, National Centre for Atmospheric Research, USA.

Friedman, J. (1991) Multivariate adaptive regression splines. *Ann. Statistics* **19**(1), 1–141.

Golding, B. (1998) Nimrod: A system for generating automated very short range forecasts. *Met. Applications* **5**(01), 1–16.

Haykin, S. (1999) *Neural Networks. A Comprehensive Foundation*. Prentice Hall, New York, USA.

Li, Z., Ge, W., Liu, J. & Zhao, K. (2004) Coupling between weather radar rainfall data and a distributed hydrological model for real-time flood forecasting. *Hydrol. Sci. J.* **49** (945–958).

Pessoa, M., Bras, R. & Williams, E. (1993) Use of weather radar for flood forecasting in the Sieve river basin: a sensitivity Analysis. *J. Appl. Met.* **32**(3), 462–475.

Quinlan, J. R. (1992) Learning with continuous classes. In: *The 5th Australian Joint Conference on AI*, 343–348. World Scientific, Singapore.

Solomatine, D. P. & Dulal, K. N. (2003) Model trees as an alternative to neural network in rainfall–runoff modelling. *Hydrol. Sci. J.* **48**(3), 399–411.

Solomatine, D. P. & Xue, Y. (2004) M5 model trees compared to neural networks: application to flood forecasting in the upper reach of the Huai River in China. *J. Hydrologic Engng* **9**(6), 491–501.

Sun, X., Mein, R., Keenan, T. & Elliott, J. (2000) Flood estimation using radar and raingauge data. *J. Hydrol.* **239**(1-4), 4–18.

Wang, Y. & Witten, I. (1997) Inducing model trees for continuous classes. Poster Paper at the 9th European Conference on Machine Learning (ECML 97), 128–137.

Xuan, Y. (2007) Uncertainty propagation in complex coupled flood risk models using numerical weather prediction and weather radars. PhD Thesis, University of Bristol, Bristol, UK.

Xuan, Y., Cluckie, I. D. & Wang, Y. (2009) Uncertainty analysis of hydrological ensemble forecasts in a distributed model utilising short-range rainfall precipitation. *Hydrol. Earth System Sci.* **13**, 293–303.

A hydroinformatic approach to development of design temporal patterns of rainfall

CARLOS VARGA[1,2], JAMES E. BALL[1] & MARK BABISTER[2]

1 *School of Civil and Environmental Engineering, University of Technology Sydney, PO Box 123, Broadway, New South Wales 2085, Australia*

2 *WMAwater Pty Ltd, Sydney, Level 2, Clarence St, Sydney, New South Wales 2000, Australia*
varga@wmawater.com.au

Abstract Estimation of the design rainfall for design flood estimation remains a problem for many engineering hydrologists despite many research studies into appropriate methodologies. An important aspect of flood flow estimation through catchment simulation is the design rainfall. Presented herein is a new approach for estimation of the temporal pattern of rainfall during a hypothetical design storm. The basis of the approach is a conditional random walk in non-dimensional space to create a finite number of storm patterns based on the probability that the storm event is convective or frontal and the probability that the storm centre of mass is located at the beginning, middle, or end of the storm event. It is shown that the resultant storm patterns more closely reflect historical patterns than alternative methods for estimating the design temporal pattern of rainfall.

Key words temporal patterns; rainfall; design; flood

INTRODUCTION

Estimation of the design rainfall for design flood estimation remains a problem for many engineering hydrologists. In many situations, advice is required regarding flood magnitudes for design of culverts and bridges for roads and railways, design of urban drainage systems, design of flood mitigation levees and other flood mitigation structures, design of dam spillways, and many other situations.

Where data are available, this problem of flood flow estimation can be undertaken using at-site flood frequency techniques such as those presented by Kuczera (1999), Jin & Stedinger (1989), Martins & Stedinger (2000) and Wang (2001). However, for the far more common case of an absence of suitable data, catchment simulation techniques are commonly used to provide predictions of the desired flood quantile.

An important aspect when flood quantiles are estimated using catchment simulation is the design rainfall used to predict the catchment response. There are three components to the design rainfall:

(a) Average intensity of rainfall (or its counterpart which is the rainfall depth) over the storm burst duration;
(b) Spatial distribution of the rainfall; and
(c) Temporal distribution of the rainfall.

The concern here is the third aspect of design rainfall, namely that of estimating the temporal distribution of rainfall during the design storm burst. While the assumption of a constant rainfall intensity over the storm burst can be adopted, Ball (1994) showed that this assumption resulted in a lower bound to potential flood quantiles estimated from the peak flow of the resultant predicted hydrograph.

For this reason among others, the temporal pattern of the storm burst to be used in design flood estimation has been the focus of many studies; examples are Keifer & Chu (1957), Huff (1967), and Pilgrim & Cordery (1975). The approach adopted for development of the design rainfall hyetograph in these studies varied between the studies. Keifer & Chu (1957) aimed at developing a hyetograph that ensured that the design rainfall intensity for varied catchment response times was experienced by each component of the system. The study by Huff (1967) considered the probabilities associated with cumulative rainfall totals with the implicit assumption that rainfall early in the storm event, or burst, was the worst situation for design. Of these studies, only that by Pilgrim & Cordery (1975) explicitly stated that the aim of the hyetograph developed was to ensure that the annual exceedence probability (AEP) of the rainfall depth or intensity was transformed into the AEP of the resultant peak flow. To achieve this stated aim, Pilgrim & Cordery (1975) proposed that a storm pattern with average variability would not impact on the AEP transformation and, hence, developed a technique for development of design rainfall hyetographs which incorporated average variability in patterns. These techniques were used subsequently in the development of the recommended temporal patterns for design flood estimation in Australia (Pilgrim, 1987).

However, while the temporal patterns presented in *Australian Rainfall and Runoff* (Pilgrim, 1987) have been adopted for design flood estimation, their application in practical situations has resulted in the need to consider alternatives. Of particular concern are:

- the dominance of particular durations which bear no relationship to catchment response times;
- internal bursts within the storm which have a higher AEP than the design AEP for the full storm burst; and
- the increasing need to consider flood volume – particularly where the volume of flood storage being considered is a significant fraction of the flood hydrograph.

To mitigate some of these issues, Phillips (1994) and Rigby *et al.* (2003) proposed the use of embedded design storms instead of using only the peak storm burst. This approach is based on embedding the design burst inside a historical storm burst. When using this approach, the influence of the historical portion of the storm event on the resultant flood hydrograph is assumed not to be significant and, therefore, not to influence the translation of the rainfall probability into the flood flow probability.

Presented herein is a new approach for estimation of the temporal pattern of rainfall during a hypothetical design storm with the aim of overcoming the problems noted above. However, rather than developing a single temporal pattern of rainfall, a suite of potential patterns result which enables the designer to assess the sensitivity of the peak flow estimate to the temporal pattern of rainfall.

RAINFALL DATA

The present study is limited to two sites (Sydney and Richmond) in eastern New South Wales, Australia (Fig. 1). Rainfall characteristics for these two sites from the Australian Bureau of Meteorology (2008) are used: Sydney has an average annual rainfall of 1215.2 mm with the highest rainfall occurring between February and June; and Richmond has an average annual rainfall of 810.4 mm, with the highest rainfalls occurring between December and April.

The rainfall data for the two sites comes from the digitized pluviograph records for Sydney Observatory Hill and Richmond RAAF weather stations. These stations were chosen due to the length of record; the record for Sydney Observatory Hill begins in October 1913 while that for Richmond RAAF begins in May 1953. While data were available for Sydney Observatory Hill to December 2005, that for Richmond RAAF ended in December 1993. Hence 92 years of record were available at Sydney Observatory Hill and 40 years of record at Richmond RAAF. For both sites, the original records were digitised from pluviographs in 6-minute increments and hence the data were available only in 6-minute time periods.

One of the issues affecting the resultant database of storm events when the information is available only at discrete time intervals is the occurrence of an intense burst within the storm of interest covering two intervals; this problem is illustrated in Fig. 2. The burst of rainfall occurring in the time period prior to and post 12:12 h is disaggregated into two parts, with one part being in the period 12:06 h to 12:12 h and the second part in the period 12:12 h to 12:18 h. This problem of restricted time periods is documented in the literature (e.g. see Dwyer & Reed, 1995) when the time period of interest is 1 day. While this issue is important for implementation of the proposed methodology for practical applications, the focus herein was on "proof of concept" and hence no modification to the database was undertaken.

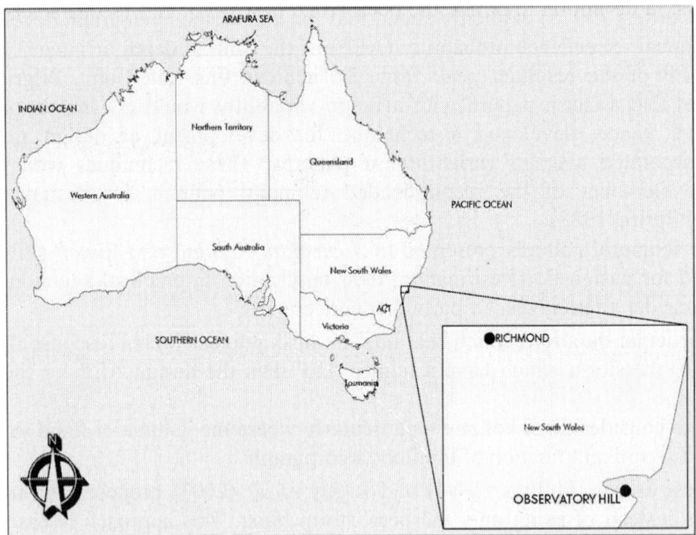

Fig. 1 Location of case study sites.

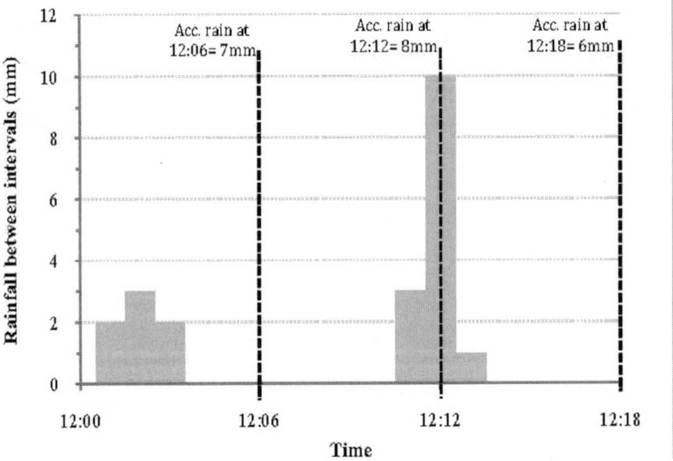

Fig. 2 Problem of restricted time periods.

Several other issues regarding data quality were also found during consideration of the data. Firstly, large periods of rainfall (up to 9 hours in some cases) with constant intensities of rainfall were found in the database; these periods suggest disaggregation using linear techniques had occurred. As storms do not have long consecutive periods of uniform rainfall intensity, a filter was applied with the criteria of selecting storms where four or more consecutive values of the same rain depth occurred; these storms were then deleted from the subsequent analysis.

STORM DEFINITION

A clear and agreed definition of a storm event and its start and end points of a storm event is not available in the literature, with many researchers using alternative definitions to suit the purpose of their investigations. Hence, a definition to define the occurrence of a storm and its characteristics was developed.

For the purpose of this study, a storm was defined as a rainfall event having a total depth greater than 5 mm with more than 1 mm occurring within a 30-minute period. Furthermore, a minimum period of 1 hour was assumed between different events; bursts of rainfall occurring either side of a period greater than 1 hour were assumed to be separate storm events.

Inter-event times (the time between different storm events) of two hours have been used in several studies (e.g. see Eagleson, 1978; Heneker *et al.*, 2001). However, when this parameter was used it was found that many storms had significant periods of low rainfall with the suspicion that multiple storm generation mechanisms were present. Furthermore, storms lasting more than 24 hours have a high probability of originating from more than one storm generation process (Pilgrim *et al.*, 1969) with, typically, short duration storms embedded within a longer duration storm. As the focus of the study reported herein was based on single storm events, it was desired that these multiple storms within a storm be discriminated. This discrimination was achieved with the shorter inter-event time of 1 hour and, hence, this value was adopted.

For each of the storms identified in the continuous rainfall records, the highest rainfall intensities were calculated for periods of 6, 12, 18, 24, 30 minutes and 1, 1.5, 2, 2.5, 3, 3.5, 4, 4.5, 5.5, and 6 hours. The storms having the 50 highest intensities for each period were then extracted for further analysis. For the studied period, 128 storms were extracted from the Sydney Observatory Hill record and 156 storms were extracted from the Richmond RAAF record.

STORM CATEGORISATION

As part of the analysis of storm patterns, the selected storms were subdivided into those with a frontal origin and those with a convective origin. Storms originating from frontal events tended to have lower intensities but occurred over longer durations, while storms classified as convective had higher intensities but were of shorter durations. In order to achieve this classification, the following storm characteristics were extracted from each storm:

- date of occurrence;
- total rainfall (mm);
- duration (hours);
- average intensity (mm/h) and the most intense rain increment (mm per 6 minute period);
- proportion of total rainfall depth in the most intense period;
- the period in the storm when the temporal semi-variogram obtained the value of 1;
- median rain increment;
- root mean squared error (RMSE) of the storm rainfall; and
- beta parameter. as defined by Llasat (2001).

Semi-variograms have been used previously by Umakhantan & Ball (2005) to assess rainfall homogeneity in space and time within the Sydney area. Following that approach, a standardised semi-variogram in the temporal dimension can be defined by:

$$\gamma^*(k) = \frac{1}{2(T-k)\sigma_i^2} \sum_{t=1}^{T-k} [P_i(t) - P_i(t+k)]^2 \qquad (1)$$

where T represents the total number of time intervals within a storm, k is the lag time, σ_i^2 is the variance of the storm rainfall depth increments, and $P_i(t)$ is the rainfall increment at time t. When the semi-variogram is plotted against different lag times, a rapid rise of the function to the value of 1 will indicate a non-homogeneous storm in time, while a slower rise will indicate a more uniform storm.

Convective storms tend to have higher temporal variation and hence the correlation structure of the storm will be lower, as indicated by the semi-variogram obtaining a value of 1 with fewer lags than would be the case for frontal storms. Figures 3 and 4 show the hyetographs and semi-variograms for two typical storm events at Richmond. The first storm corresponds to a frontal

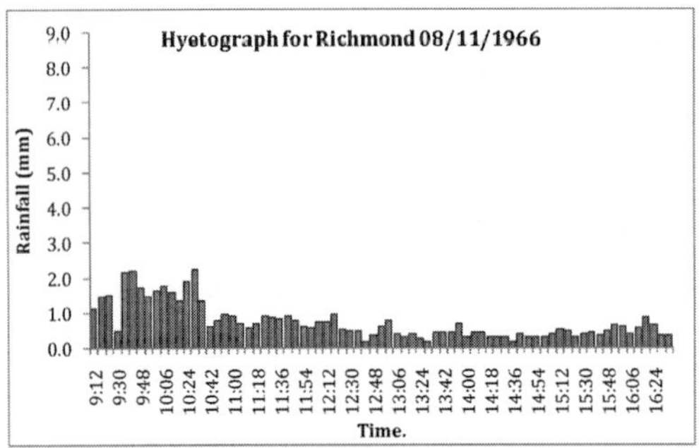

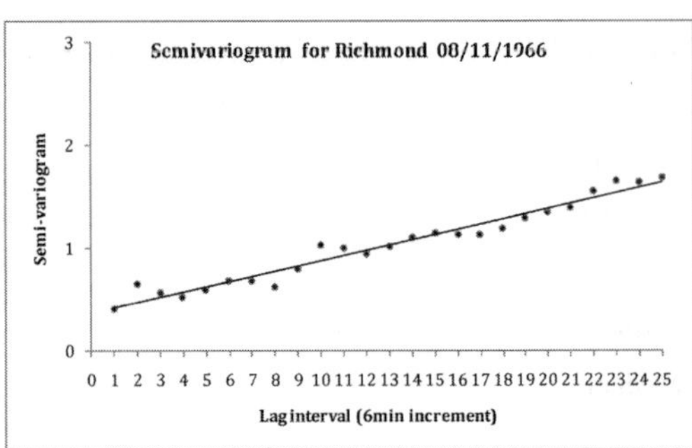

Fig. 3 Storm at Richmond on 8 November 1966.

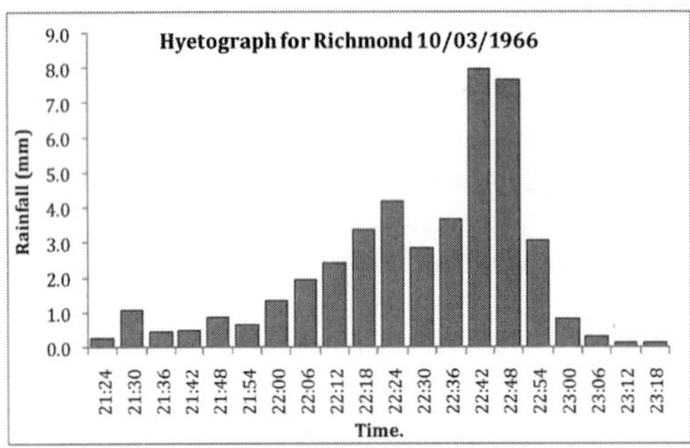

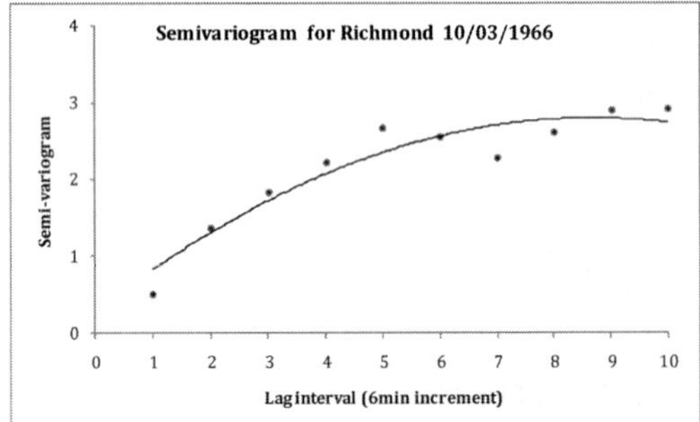

Fig. 4 Storm at Richmond on 10 March 1966.

event which occurred on 8 November 1966, while the second event represents a convective event which occurred on 10 March 1966. For the first event, the semi-variogram reaches the value of 1 after 13 time lags, or 78 minutes, while for the second event, the semi-variogram function reaches a value of 1 approximately at the first lag time. These results illustrate the longer persistence and greater uniformity for frontal storm events and the greater heterogeneity that occurs during convective events.

Llasat (2001) proposed the use of a Beta parameter to characterise convective storm events. In essence, this parameter relates the rainfall depth in periods during the storm where the depth is greater than a predefined threshold to the total rainfall over the storm duration. It is defined by:

$$\beta = \sum_{i=1}^{N} I(t_i)\,\theta(I-L) \Big/ \sum_{i=1}^{N} I(t_i) \tag{2}$$

where L is the threshold depth fixed at 3.5 mm per 6-minute period (the time increment of 6 minutes corresponds to the pluviograph record), I is the rain increment for the time t_i, and $\theta(I-L)$ is the Heaviside function defined as:

$\theta(I-L) = 1$ if $I \geq L$; and

$\theta(I-L) = 0$ if $I < L$.

Any storm that had a β value greater than 0.3 was considered to be of convective origin.

A third methodology used to categorise the storms was based on the rainfall variability in the temporal dimension. To define this variability, the RMS variation from the average intensity was calculated for each storm. Consistent with the previous approaches, a higher rainfall variability was expected to occur in convective events when compared to frontal events. For frontal storm events, the average RMS was found to be 0.6 mm at Richmond and 1.0 mm at Sydney Observatory Hill while, for convective storm events, the average RMSE was found to be 2.5 mm at Richmond and 2.9 mm at Sydney Observatory Hill. The effect of storm duration on the

Table 1 Effect of storm duration on storm variability. D = Duration.

	Sydney Observatory Hill									
Duration	D ≤ 1h		1h < D ≤ 2h		2h < D ≤ 3h		3h < D ≤ 6h		6h < D ≤ 24h	
No. of storms	14		15		21		34		44	
Prob.	11%		12%		16%		27%		34%	
RMS	mm	%	mm	%	mm	%	mm	%	mm	%
Median	3.48	10.65	2.74	7.68	2.27	4.17	1.36	1.99	0.81	0.88
Average	3.76	10.74	3.08	7.17	2.36	4.38	1.59	2.11	0.89	0.93
	Richmond RAAF									
	D ≤ 1h		1h < D ≤ 2h		2h < D ≤ 3h		3h < D ≤ 6h		6h < D ≤ 24h	
No. of storms	42		21		14		37		42	
Prob.	27%		13%		9%		24%		27%	
RMS	mm	%	mm	%	mm	%	mm	%	mm	%
Median	2.26	12.18	1.62	5.18	0.84	3.26	0.48	1.52	0.33	0.70
Average	2.36	13.09	1.66	6.52	1.04	3.35	0.60	1.62	0.47	0.80

variability of the storms is shown by the RMS in Table 1 where it can be seen that, as the duration of the storm increases, the variability within the storms decreases.

Ball (1994) showed that the pattern of rainfall can have a significant influence on the magnitude and location of the peak and on the shape of the hydrograph. Therefore, the storm events were further categorised into front-loaded, middle-loaded and back-loaded categories according to the location of centre of rainfall mass:

front-loaded – centre of mass located within first 33% of storm duration;
middle-loaded – centre of mass located within middle third of storm duration; and
back-loaded – centre of mass located within last 33% of storm duration.

Shown in Table 2 are the number of storms in each category and the resultant probabilities of occurrence for each category. The frontal storms have a higher probability of occurring at both locations with the probability at Richmond being slightly higher than Sydney. At both Sydney Observatory Hill and Richmond RAAF sites, there is a preference for the storm centre to occur in the middle third of the duration with approximately 2/3 of the selected storms being middle loaded storms.

Table 2 Number of storms and probabilities of categories.

Location	Sydney Observatory Hill		Richmond RAAF	
	No. of storms	Prob.	No. of storms	Prob.
Convective	61	47.7%	63	40.4%
Frontal	67	52.3%	93	59.6%
Convective front loaded	18	29.5%	25	39.7%
Convective mid loaded	34	55.7%	30	47.6%
Convective back loaded	9	14.8%	8	12.7%
Frontal front loaded	3	4.5%	4	4.3%
Frontal mid loaded	53	79.1%	79	84.9%
Frontal back loaded	11	16.4%	10	10.8%

NON-DIMENSIONAL RAINFALL MASS CURVES

As part of the analysis undertaken, all storms were non-dimensionalised in terms of storm duration and storm depth; for this purpose, τ is the non-dimensional time defined by $\tau = t / t_d$ and δ is the non-dimensional rainfall depth defined by $\delta = d(t) / d(t_d)$ where $d(t)$ is the rainfall depth at time t and t_d is the storm duration. These non-dimensional mass curves preserved the following characteristics:

the mass curves started at $\tau = 0, \delta = 0$ and ended at $\tau = 1, \delta = 1$; and

the slope between any two points will be either 0 or some positive value.

Shown in Fig. 5 are the resultant non-dimensional mass curves for the frontal and convective events at both Sydney Observatory Hill and Richmond RAAF. Also shown are the average patterns for all six subcategories at the two locations. For purposes of generating alternative temporal patterns for storms, these non-dimensional mass curves were divided into 10 equally-spaced temporal intervals.

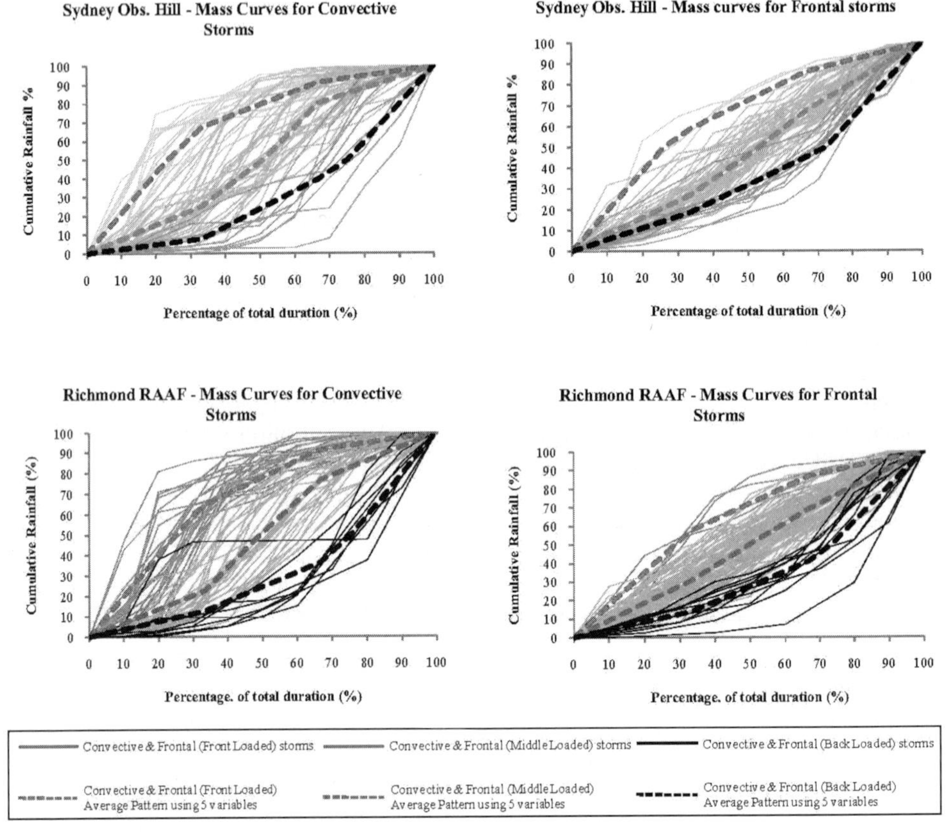

Fig. 5 Non-dimensional storm patterns.

GENERATION OF DESIGN STORMS

The procedure for the creation of random design storms was based on a uniform random selection of a conditional walk jump at each time decile. The number of possible combinations when creating random storms in this manner is the number of storms in each sub-category raised to the power of 9. This method does not need to resample the jump at each time because it will never exceed the upper limit ($1 - d_i$). However, when sampling random storms to allocate them in the front-loaded, middle-loaded and back-loaded "bins", a reduced number of them (less than 3%) did not reproduce the correct location of the centre of mass. Therefore, these storms were eliminated from the sample.

Conditional random walk theory has been used to develop non-dimensionalised mass curves to enable the creation of storm patterns. The movement of any randomly generated storm within the dimensionless mass curve, from one point to another, is described by a discrete stochastic process. The limits for this movement are bounded from a positive jump (rainfall occurrence at time step) or no jump at all (no rainfall). The basis of this comes from the self-similarity concept studied by Woolhiser & Osborn (1985) and Koutsoyiannis (1993). Storms should preserve their internal structure conditions regardless of rainfall amount and duration.

For every storm, according to their sub-category type, a conditional walk jump was obtained at each decile by:

$$C_{jump} = \frac{(d_{i+1} - d_i)}{1 - d_i} \tag{3}$$

where C_{jump} is the jump from the dimensionless mass curve at a time step i, d is the value of the rain in the dimensionless mass curve. This formula represents the ratio of jump taking into account the remaining depth in the dimensionless mass curve. Periods of no rain are mixed with the values of the jumps at any time step, when they are present in the historical storms. Therefore, for any random storm created, the random dimensionless mass curve will be constructed as follows:

$$Rd_{i+1} = C_{jump} \times (1 - Rd_i) + Rd_i \tag{4}$$

where Rd_i is the random dimensionless mass curve increment at time I.

RESULTS

Using the techniques described above, a number of storms in each of the six categories were generated for the Sydney and Richmond locations. At each location, 210 frontal and 210 convective storms were generated. Within these categories, the probabilities of front-loaded, middle-loaded and back-loaded events were used to ensure the generated storms had similar characteristics to the historical storms.

Shown in Fig. 6 are the generated non-dimensional storm patterns for Sydney Observatory Hill with bounds developed from the maximum and minimum values of the historical storm patterns. Reproduction of the historical patterns is evident on inspection of this figure. In general, all of the generated storms fall within the boundaries of the historical storms; the exception is the front-loaded frontal storm category. This exception is considered to be the result of only a small number of historical storms for generating the random storm patterns under this category.

In order to evaluate the generated non-dimensional storm patterns, the non-dimensional storm patterns were converted into storm events using durations of 1, 3 and 6 hours with rainfall depths given by Average Recurrence Intervals (ARIs) of 5 and 10 years. These dimensional storms were analysed for internal rainfall intensities using fixed increments of the storm duration, namely 0.1, 0.2 and 0.3 of the storm duration.

While the storms had similar characteristics to the historical storm events, it was found that the internal rainfall intensities were greater than the ARI determined using the total storm duration. For example, the convective storms generated for durations of 1 and 3 hours for both ARIs contained internal bursts where the rainfall intensities were higher than the rainfall intensity for an ARI of 50 years obtained from the IFD diagram; this result occurred at both sites. This result, however, is consistent with many historical storm events as shown when the internal burst intensities are plotted on an IFD diagram; an example of this shown in Fig. 7 where the internal burst intensities for a number of storms at Sydney Observatory Hill are plotted on an IFD diagram. As shown in this figure, it is common for the intensity during an internal storm burst to be greater

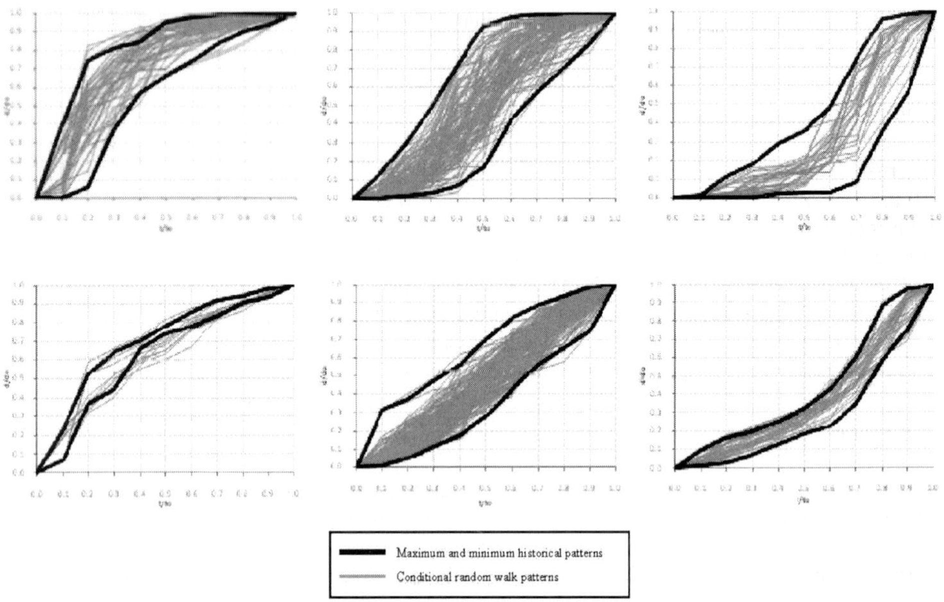

Fig. 6 Generated storm patterns at Sydney Observatory Hill.

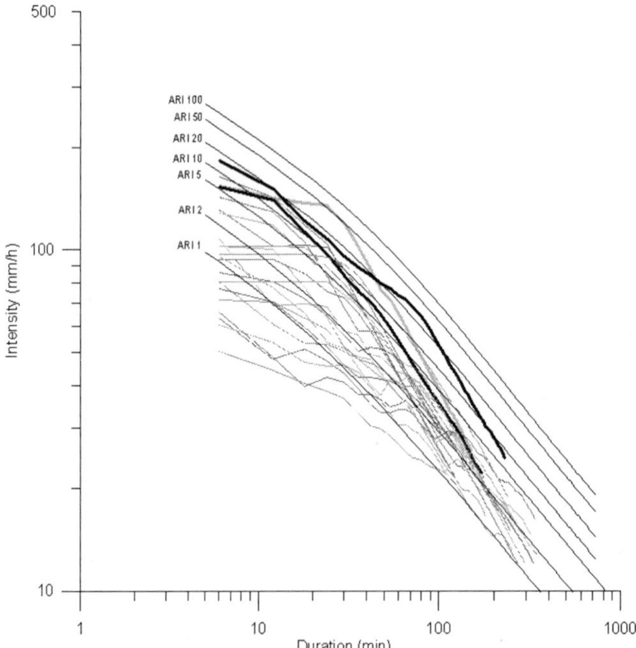

Fig. 7 Typical storm burst intensities.

than the average intensity over the storm duration as occurred in the generated storm events. Hence, it is considered that the appearance of bursts with an intensity equivalent to a rarer event (i.e. larger ARI) should not result in these events being rejected as a feasible realisation of a design storm event.

CONCLUSIONS

Average patterns do not properly represent storm pattern variability within the region. An additional alternative was presented by the use of a conditional random walk as a method to generate storms, based on the self-similarity concept that preserves the characteristics of historical storms.

Some issues regarding IFD characteristics of the conditional walk patterns arose in this study. Internal burst frequencies within ARR temporal patterns do not exceed the frequency of occurrence of the total storm. However, in the design of constrained random walk patterns, for both locations, it is possible to observe bursts with a rarer ARI than the design ARI of the storm (higher chance to occur in Sydney). Nonetheless, these situations often occur in real storms.

A change in the conditions of the storm selection filter may allow retrieval of a larger number of storms to be analysed. One of the most important factors that affected the selection of storms was the presence of consecutive rain intervals with the same value. As stated before, this is unlikely to exist in real life. Another possibility to improve the number of storms is going back to the raw data on those days when large storms occurred and consecutive values were found, and to extract the values again. By using a regional approach, the number of storms extracted should increase. This may lead to a higher variability in the random patterns produced in order to obtain a better representation of the temporal patterns in each zone, thus using an ARI threshold as a filter for selecting storms.

Another approach for conditional random walk might involve allocation of non-dimensional storms to different "bins" according to duration to resolve time scale problems. This will improve the selection of storm patterns. Intense bursts that occurred in short duration storms will not be dimensionalised into longer duration convective storms. In this case, a relatively long and good quality series record is needed.

REFERENCES

Australian Bureau of Meteorology, (2008) *Climate Statistics for Australian Locations*, Retrieved 1 August 2008, from www.bom.gov.au/climate/averages/.

Ball, J. E. (1994) The influence of temporal patterns on catchment response. *J. Hydrol.* **158**(3/4), 285–303.

Dwyer, I. J. & Reed, D. W. (1995) Allowance for discretization in hydrological and environmental risk estimation (ADHERE). Institute of Hydrology, Wallingford, UK.

Eagleson, P. (1978) Climate, soil and vegetation 2. The distribution of annual precipitation derived from observed storm sequences. *Water Resour. Res.* **14**(5), 713–721.

Heneker, T. M. (2002) An improved engineering design flood estimation technique: removing the need to estimate initial loss. PhD Dissertation, Department of Civil and Environmental Engineering, University of Adelaide, Australia.

Huff, F. A. (1967) Time distribution of rainfall in heavy storms. *Water Resour. Res.* **3**(4), 1007–1019.

Jin, M. & Stedinger, J. R. (1989) Flood Frequency Analysis with Regional and Historical Information. *Water Resour. Res.* **25**(5), 925–936.

Keifer, C. J. & Chu, H. H. (1957) Synthetic storm pattern for drainage design. *ASCE, J. Hydraulics Div.* **83**(HY4), 1–25.

Koutsoyiannis, D. (1993) A stochastic disaggregation method for design storm and flood synthesis. *J. Hydrol.* **156**, 193–225.

Kuczera, G. (1999) Comprehensive at-site flood frequency analysis using Monte Carlo Bayesian inference. *Water Resour. Res.* **35**(5), 1551–1558.

Llasat, M. C. (2001) An objective classification of rainfall events on the basis of their convective features: Application to rainfall intensity in the northeast of Spain. *Int. J. Climatology* **21**, 1385–1400.

Martins, E. S. & Stedinger, J. R. (2000) Generalized Maximum Likelihood GEV quantile estimators for hydrologic data. *Water Resour. Res.* **36**(3), 737–744.

Phillips, B. C. (1994) Embedded design storms – An improved procedure for design flood level estimation. *Proc. 1994 Hydrology and Water Resources Symposium* (Adelaide, Australia), 235–240.

Pilgrim, D. H. (ed.) (1987) *Australian Rainfall & Runoff – A Guide to Flood Estimation.* Institution of Engineers, Australia, Barton, ACT, Australia.

Pilgrim, D. H. & Cordery, I. (1975) Rainfall temporal patterns for design flood estimation. *ASCE, J. Hydraulics Div.* **100**(HY1), 81–95.

Pilgrim, D. H., Cordery, I. & French, B. E. (1969) Temporal patterns of design rainfall for Sydney. *IE Aust. Civil Engineering Transactions* **11**(1), 9–14.

Rigby, E. H., Boyd, M. J., Ruso, S. & Van Drie, R. (2003) Storms, storms bursts and flood estimation - a need for review of the AR&R procedures. *Proc. 28th International Hydrology and Water Resources Symposium* (Wollongong, NSW), 17–24.

Umakhantan, K. & Ball, J. E. (2005) Rainfall models for catchment simulation. *Australian J. Water Resources* **9**(1), 55–67.

Wang, Q. J. (2001) A Bayesian joint probability approach for flood record augmentation. *Water Resour. Res.* **37**(6), 1707–1712.

Woolhiser, D. A. & Osborn, H. B. (1985) A stochastic model of dimensionless thunderstorm rainfall. *Water Resour. Res.* **21**(4), 511–522.

Uncertainty propagation in hydrological forecasting using ensemble rainfall forecasts

SARA LIGUORI[1], MIGUEL RICO-RAMIREZ[1] & IAN CLUCKIE[2]

1 *University of Bristol, Department of Civil Engineering, Bristol BS8 1TR, UK*
s.liguori@bristol.ac.uk

2 *Swansea University, Swansea SA2 8PP, UK*

Abstract Rainfall forecasts provided by Numerical Weather Prediction (NWP) models are affected by different sources of uncertainty, as the models aim to simulate a chaotic non-linear system that is highly sensitive to small changes in the initial conditions. Therefore, there is a need to move towards an operational probabilistic approach and this is often based upon the concept of ensemble forecasts. This paper attempts to assess the feasibility of the use of high-resolution precipitation ensemble forecasts as a direct input to hydrological models used in operational real-time flood forecasting and warning systems. The mesoscale model MM5 was utilised to run 48-hour lead-time forecasts. The MM5 model, like all mesoscale models, requires conditioning by initial and boundary conditions taken from a global model that generally operates at coarse spatial and temporal resolutions. On this occasion these were obtained from ECMWF (European Centre for Medium-Range Forecast), which also provides the perturbed ensemble forecasts. The MM5 model was set up using a four-nested domain configuration with the smallest domain having a 2-km grid resolution. The domains dynamically downscale and were centred on the southeast of England, in the Upper Medway Catchment. An established rainfall–runoff model, the Probability Distributed Model (PDM), has initially been implemented: this receives ensembles of forecasted rainfall generated by the MM5 model in order to predict the runoff generation at the catchment outlet 48 hours ahead. This paper presents an initial assessment of the probabilistic forecasts produced using a coupled-modelling approach and contributes to the development of Hydroinformatics in the context of Operational Hydrology.

Key words probabilistic forecasts; ensembles; high-resolution NWP; coupled models

INTRODUCTION

Numerical Weather Prediction (NWP) models aim to simulate the atmosphere, which is a chaotic non-linear system highly sensitive to small changes in the initial conditions. Rainfall forecasts by NWP models are therefore affected by the uncertainty in the estimation of the initial state of the atmosphere, as well as by the limited understanding of dynamic and physical processes of the atmosphere. Thus, instead of aiming at getting the best estimate of the future state of the atmosphere through a single deterministic forecast, it is now becoming an operational probabilistic forecasting approach.

Probabilistic forecasts aim to forecast the forecasts skill (Person & Grazzini, 2007). They are based upon the notion that deterministic forecasts are imperfect and the need to assess the degree of uncertainty in the forecast. The value of probabilistic forecasts compared to deterministic forecasts derives from the fact that they allow us to estimate the probability of occurrence of the forecasted events, therefore also that of extreme and rare events (Buizza, 2008).

Probabilistic forecasts make use of ensemble systems, which are sets of forecasts valid at the same time. They are not just able to produce the most likely outcome (as a single deterministic forecast), but also able to give information about the probability of occurrence of any event that is particularly relevant for rare and extreme events.

In order to assess the value of the use of probabilistic predictions in real time flood forecasting systems, hydrological ensemble forecasts can be generated from ensemble rainfall forecasts by routing the output of a NWP model through a hydrological model, in a coupling weather/rainfall–runoff modelling approach. This process transfers the quantitative information about the uncertainty of the forecast to the hydrological forecast (Zappa *et al.*, 2008) and requires high resolution rainfall forecasts at the catchment scale.

NWP models tend to analyse different meteorological features with respect to the simulated scales: on the synoptic scale, models analyse meteorological features at scales exceeding 2000 km, while on the regional scale models deal with features ranging from near the synoptic scale (mesoscale-alpha) to individual cloud cells with dimensions of 1–20 km (mesoscale-gamma) (COMET®).

The horizontal resolution of a numerical weather model is related, for grid-point models, to the spacing between the grid points. The higher the resolution, the smaller the features the model can simulate. Mesoscale NWP models tend to analyse features at small scales, therefore they need a high resolution definition which, in a coupling modelling perspective, is functional to producing the input to a hydrological model.

When analysing mesoscale features, non-hydrostatic processes cannot be neglected as at these scales the vertical velocities can even exceed horizontal velocities. Therefore mesoscale models need to include the parameterization of those sub-grid processes which have non-hydrostatic effects (COMET®).

Previous uncertainty analyses using ensemble forecasts in coupled models have highlighted the need to address the inability of the model to resolve the sub-grid scale precipitation features (Xuan *et al.*, 2009). Thus, another source of uncertainty in predicting rainfall is related to the parameterization of the convective cells. Therefore, ensemble rainfall forecasts from NWP can be generated in different ways, and in particular by perturbing: (a) the initial conditions around the true initial state of the atmosphere, (b) the convective cells' parameterization, which is relevant in terms of rainfall generation.

This paper attempts to assess the feasibility of the use of high-resolution precipitation ensemble forecasts as a direct input to hydrological models used in operational real-time flood forecasting and warning systems. The study employs a coupling system having two components: a high resolution mesoscale model giving short-range ensemble forecasts and a lumped rainfall–runoff model receiving in input the ensemble rainfall forecasts from the weather model.

Ensemble forecasts

The estimation of uncertainty is a crucial component in a flood forecasting system. The sources of uncertainty are in the data observations, model structure and model parameters (see e.g. Pappenberger *et al.*, 2006). Ensemble forecasts were introduced in order to quantify the possible uncertainties in the forecasts (Cheung, 2001). The operational analysis is the best estimate of the initial state of the atmosphere based on available observations. Unlike climate researchers, who perturb the model parameters within certain ranges of uncertainty and integrate the model to discard unrealistic climates (Allen, 1999), weather forecasters produce the ensemble forecasts by adding small perturbations or errors to the operational analysis. The ensemble QPFs include the control forecast, which is the forecast obtained with the operational analysis without perturbations, and a set of ensemble members obtained with the perturbed operational analysis. Every ensemble member should represent an equally probable state of the atmosphere and therefore the initial perturbations are crucial in ensemble generation. There are several methods for calculating the optimal perturbations. The European Centre for Medium-range Weather Forecast (ECMWF) Ensemble Prediction System (EPS) calculates the perturbations using singular or optimal vectors, which are the fastest-growing perturbations over the first 48 hour of the forecasts (Palmer *et al.*, 1997). These perturbations are added to the operational analysis. The ensemble forecasting at the National Centers for Environmental Prediction (NCEP) uses a different method known as the breeding method to calculate the fastest-growing perturbations (Toth & Kalnay, 1997). This process consists in adding a small perturbation to the operational analysis at day t_0. Then the model is integrated with perturbed and unperturbed initial conditions during a short period ($t_1 - t_0$) and the forecasts are subtracted and the difference is scaled down as the initial perturbation. This perturbation is added to the analysis corresponding to day t_1 and the process is repeated. In this way, this method breeds the perturbations that grow faster and after a few cycles the decaying components will become negligible (Toth & Kalnay, 1997). Other methods to calculate the perturbations, such as the Monte Carlo and Kalman filtering methods, are described by Cheung (2001). The ensemble mean forecast should provide a smaller error than the control forecast, given that each ensemble member represents the real state of the atmosphere (Cheung, 2001). The number of ensemble members must represent accurately the probability distribution of the state of the atmosphere in order to improve the skill of the control forecast by ensemble averaging. Toth & Kalnay, (1997) found that minimal improvement is obtained beyond 20 members, even though the error continues to improve up to 40 members.

The 5th Generation NCAR/Penn State Mesoscale model MM5

The mesoscale model utilized for this research work is the nonhydrostatic version of the PSU/NCAR MM5 (Dudhia, 1993; Grell *et al.*, 1994; Dudhia *et al.*, 2005). The model requires initial and lateral boundary conditions to solve the mathematical equations expressing the physical laws of motion and conservation of energy, as well as the parameterization of sub-grid non resolvable processes, such as radiation, moisture fluxes, turbulence, convection, condensation, evaporation and surface heat. The use of convective parameterization schemes accounts for the lack of understanding of these non resolvable processes and reduces atmospheric instability in the model (COMET®).

Initial and lateral boundary condition need to be provided by a coarse-resolution global model. The meteorological input data used to initialize the model are horizontally interpolated from a latitude–longitude grid to mesoscale, rectangular gridded domains and vertically interpolated from pressure levels to terrain following σ-coordinates required by the numerical integration.

The probability distributed rainfall–runoff model

The rainfall–runoff model employed for this study is the probability distributed lumped rainfall–runoff model, PDM. The model receives in input rainfall and potential evaporation data and derives the flow at the catchment outlet (CEH Wallingford, 2005; Moore, 2007). The runoff generation process is assumed to be a saturation excess runoff process, controlled by the absorption capacity of the soil, including soil, surface and canopy.

The PDM assumes a distribution of store depths across the catchment with depths ranging from a c_{min} (minimum depth) to a c_{max} (maximum depth): the most commonly used distribution is the Pareto distribution.

The overflow from the soil stores enters the surface storage component, while the subsurface storage receives input from the groundwater recharge coming from the soil-moisture store as a nonlinear function of the effective store contents with a time constant k_g (CEH Wallingford, 2005).

The surface storage is conceptualized as a cascade of two linear reservoirs in series with time constants k_1 and k_2, while the groundwater storage is assumed to be a cubic store with time constant k_b.

The output from surface and subsurface storages yields the surface runoff and the baseflow components of the total flow (Moore, 2007).

THE CATCHMENT AND MODELS SET UP

Hydrological data

The NWP model was focused on the Upper Medway Catchment (Fig. 1) for the hydrological purpose of this study. The catchment is located in the southeast of England, in the county of Kent and covers an area of 256 km^2. The catchment is mainly covered in grass, according to the classification carried out by the National Resources Institute in Cranfield, UK. The catchment response is controlled by the land cover and other factors, such as catchment wetness, soil moisture, infiltration and evapotranspiration (Reichel *et al.*, 2009).

Nine real-time TBRs (tipping-bucket raingauges), operated by the Environment Agency (Fig. 1) are located within the catchment and provide rainfall data at 15-min time intervals. The average rainfall over the catchment was calculated with the Thiessen polygon method.

Discharge observations at 15 min time resolution were obtained from the flow gauge located at the catchment outlet at Chafford. The daily potential evapotranspiration data were provided by the UK Met Office.

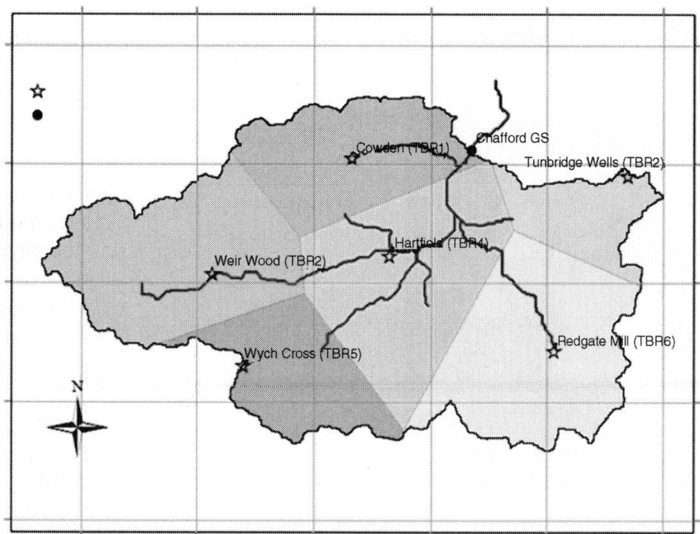

Fig. 1 Upper Medway Catchment and gauges network (from Reichel *et al.*, 2009, copyright 2009, *Water Management*).

Radar data set for forecasts verification

The meteorological forecasts were verified against 15-min radar observations at 5-km resolution. The data set was provided by the UK Meteorological Office through the British Atmospheric Data Centre (BADC).

Radar rainfall estimates are subject to two types of errors: random and systematic. The random error usually tends to average out at large spatial and temporal scales, whereas systematic errors remain. For hydrological applications, the systematic error needs to be reduced (Reichel *et al.*, 2009). Correction methodologies were applied by the UK Meteorological Office radar processing system to account for the different errors in estimation of precipitation from radar (Harrison *et al.*, 2007). The sources of error in radar estimations of rainfall are radar calibration, signal attenuation and errors in conversion of radar reflectivity Z into rainfall intensity R (Z–R relationship), variations in drop size distribution as well as ground clutter, anomalous propagation, variation of vertical reflectivity of precipitation (VPR), range effects, vertical air motions, beam overshooting shallow precipitation and sampling effects (Rico-Ramirez *et al.*, 2007).

MM5 model set up

The mesoscale model MM5 was set up with a four-nested domain configuration and centred on the Upper Medway Catchment, as shown in Fig. 2. The four domains have increasing resolutions with respect to the nesting ratio required by the model (3:1), the smallest domain having a 2-km resolution.

The data used as initial and lateral boundary conditions were provided by the global model developed at the European Centre for Medium Range Weather Forecast (ECMWF). The meteorological fields were selected from the Ensemble Prediction System (EPS), which is part of the Operational Archive and comprise the control forecast and the perturbed forecasts (plus the equivalent deterministic products).

The ECMWF ensemble system applies a perturbation technique which is based on a singular vector decomposition method (Buizza & Palmer, 1995): the perturbations are added to the analyses to represent the uncertainty of the initial conditions, creating a range of slightly different initial conditions. The control forecast is the forecast run from unperturbed analyses, whereas the ensemble forecasts are run from the alternative analyses. The ensemble forecasts are equally likely

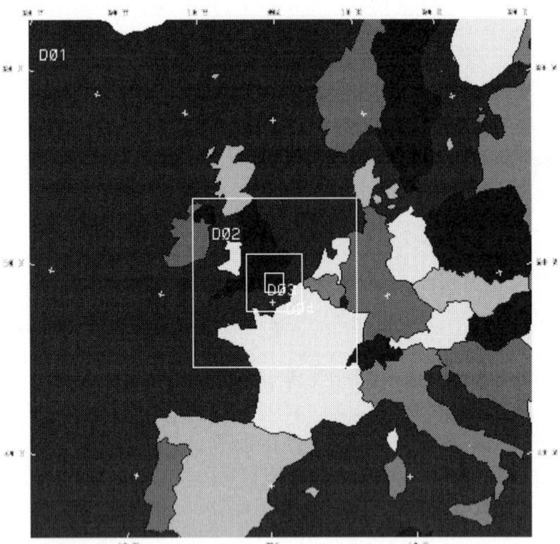

Fig. 2 MM5 domains configuration.

among themselves, but on average less skilful than the control forecast that runs from the unperturbed analysis, being the magnitude of the perturbations comparable with the analysis error (Person & Grazzini, 2007). The data were retrieved from the EPS database at 1° × 1° lat./long. grid resolution at 6-h intervals.

The physics parameterizations in MM5 include the use of the MRF Planetary Boundary Layer parameterization (Hong & Pan, 1996), a mixed phase explicit moisture scheme (Reisner et *al.*, 1998), a Rapid Radiative Transfer Model long wave scheme for atmospheric radiation and a Five-Layer Soil model for surface temperature prediction (Dudhia, 1996).

On the occasion of this study different Cumulus Parameterization Schemes (CPS), representing different closure assumptions and scale representativeness and widely used in numerical weather prediction models, have been investigated. The CPSs here investigated can all be integrated in MM5 and will be referred to as Kain-Fritsch (Wang & Seaman, 1996), Anthes-Kuo (Anthes, 1977), Grell (Grell *et al.*, 1994), Bett-Miller (Betts, 1986) and Kain-Fritsch 2 (Kain, 2002).

The parameterization schemes were implemented in domain 1 and domain 2 but not in domain 3 and domain 4, as, with respect to the grid resolutions, the scale of resolvable features on domain 3 and domain 4 becomes closer to the scale of the parameterized process.

PDM set-up

The PDM model was calibrated and validated for the Upper Medway Catchment. The model was calibrated for the period 1 February–31 December 2006 and validated for the period 1 February 2006–14 December 2007, based on the hydrological data sets previously described.

The model was calibrated (at 15-min time steps) using a mixture of manual and automatic calibration, fitting the baseflow response before the fast response. The objective function employed in this study for calibration purposes is the RMSE, which is an error measure of the modelled flow compared to the observed flow and has the advantage of giving reasonable weight to errors in high flows compared to errors in low flows:

$$RMSE = \sqrt{\frac{1}{N}\sum_{i=1}^{N}(Q_i - q_i)^2}$$

where Q_i is the modelled flow and q_i the observed flow and N the number of data points. Table 1 summarizes the optimized PDM parameters.

Table 1 PDM model parameters.

F	c_{min} (mm)	c_{max} (mm)	b	be	k_1 (h)	k_2 (h)	k_b (h mm m^{-1})	k_g (h mm bg^{-1})	b_g
0.9	0	120	0.5	2.5	4.0022	11	193.59	1.1	1.9

f: rainfall factor; c_{min}, c_{max}: minimum and store capacity; *b*: exponent of Pareto distribution; *be*: exponent in actual evaporation function; k_1, k_2: time constants of the two linear reservoirs; k_b: baseflow time constant; k_g: groundwater recharge time constant; b_g: exponent of recharge function.

Simulations

Two sets of rainfall ensemble simulations with MM5 have been run:

(a) 48-h lead time ensemble forecasts based on physics perturbations;

(b) 48-h lead time ensemble forecasts based on perturbed initial and boundary conditions.

Set (a) has been obtained by employing five different CPSs in the weather model component MM5: Kain-Fritsch (KF), Anthes-Kuo (AK), Grell (GR), Bett-Miller (BM) and Kain-Fritsch 2 (KF2). As initial and lateral boundary conditions, the control forecasts (unperturbed initial condition) from the ECMWF EPS global model were employed.

Set (b) has been obtained by assimilating the 50 members from the ECMWF EPS as initial conditions to MM5 and using one of the CPSs simulated in set (a).

Several convective and frontal events occurring during the year 2007 were simulated and the significant results from some of them will be discussed and compared as follows:

From sets (a) and (b) two sets of hydrological ensemble simulations with the PDM have been run:

(c) 48-h lead time ensemble forecasts based on the set (a) of rainfall ensemble as input to the PDM;

(d) 48-h lead time ensemble forecasts based on the set (b) of rainfall ensemble as input to the PDM.

Verification measures

The normalised bias (*NB*) and the normalised error (*NE*) were employed as a measure of performance to verify the rainfall forecasts with respect to reference values. The normalised bias is:

$$NB = \frac{\sum_{i=1}^{N} F_i - O_i}{\sum_{i=1}^{N} O_i}$$

The normalised bias (*NB*) gives a measure of overestimation ($NB > 0$) or underestimation ($NB < 0$) of the forecasted values (F_i) against the observations (O_i). N is the number of observed and forecast values. The normalised error is given by:

$$NE = \frac{\sum_{i=1}^{N} |F_i - O_i|}{\sum_{i=1}^{N} O_i}$$

The normalised error (*NE*) is a measure of the average magnitude of the errors normalised with respect to the observations (O_i).

RESULTS

The rainfall ensemble forecasts have been verified against the radar observations. Both forecasted and observed values are averaged on the catchment domain (N is also the number of grid points covering the catchment domain at 5-km resolution). The verification has been made on the accumulated rainfall from T = 0 to T = 48 h.

Three events representative of different meteorological conditions have been evaluated. They occurred on the 20 July 2007, 22 January 2007 and 29 June 2007. The 20 July 2007 storm was a heavy frontal event but characterized by embedded convective elements. The monthly rain in different counties across the UK was exceptionally high and major flooding occurred in Tewkesbury. The other two simulated events were characterized by predominant frontal elements and lower rainfall intensity. Figures 3 to 5 show NB and NE for the rainfall ensemble forecasts obtained from different CPSs.

For the 22 January 2007 event (Fig. 3) the different CPSs did not produce good results during the first 8–10 h producing rainfall underestimation, as indicated by the negative NB values. In terms of bias, GR seems to produce the smallest underestimation for short lead-times (15–20 h) while BM is the best among the schemes from T = 20 h to the end of the forecast.

In terms of normalized error, GR behaves better than the other schemes, except that from T = 20 to T = 25 h, where BM produces a smaller error. KF produces the worst results throughout the entire duration of the forecast and highly underestimates the rainfall amount.

For the 29 June 2007 event (Fig. 4) the results show that the CPSs performance skill improves from 5 h ahead. Some of the schemes produce overestimation and some underestimation. BM produce better results from 5 to 10 h, based on the NB and NE values but overestimate the rain from 12 h ahead, producing an overall bigger error for longer lead times. Better results for long lead times are given by GR, while AK underestimates the rainfall more than the other schemes.

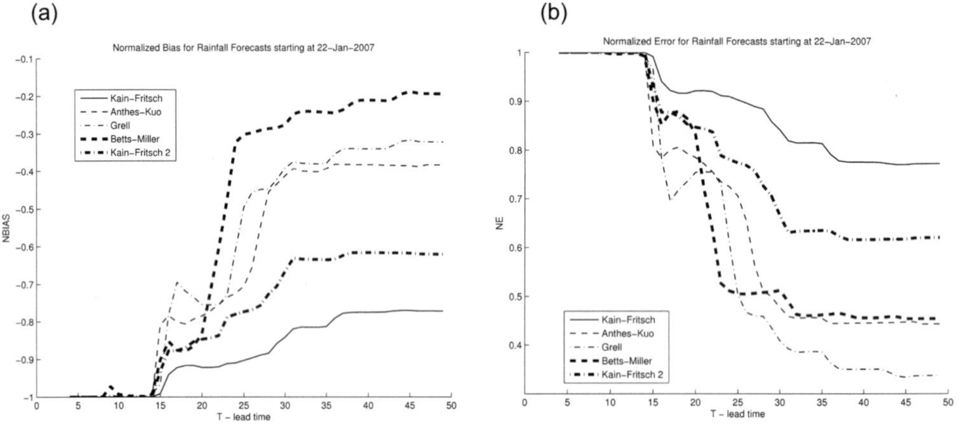

Fig. 3 (a) Normalised bias, and (b) normalised error, for the CPS rainfall ensemble forecast starting at 22 January 2007-00:00.

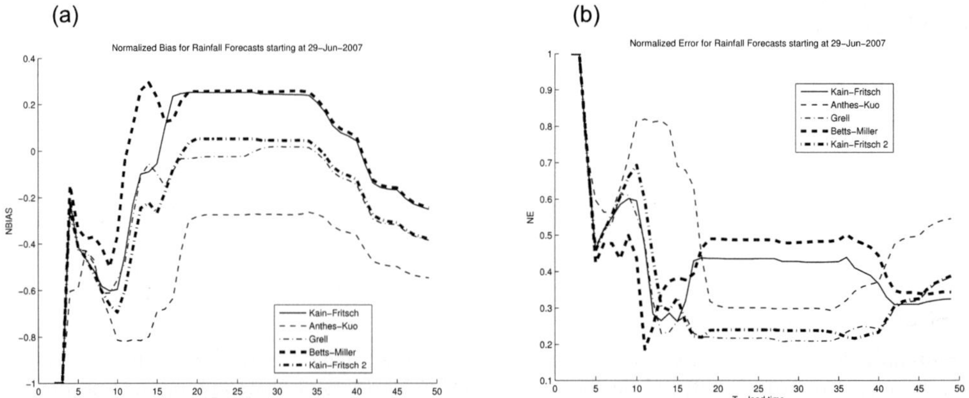

Fig. 4 (a) Normalised bias, and (b) normalised error, for the CPS rainfall ensemble forecast starting at 29 June 2007-00:00.

The 20 July 2007 event results (Fig. 5) show that KF predictions are closer to the observations from 10 h ahead, but are affected by the worst error between T = 5 h and T = 10 h. The five schemes are very close before T = 10 h. Afterwards BM seems to produce overestimation and the other schemes underestimation.

The hydrological ensemble forecasts derived from the CPS ensemble forecasts show consistent results (Figs 6 and 7). For 22 January (Fig. 6(a)) the KF generated hydrograph is far from the one generated from raingauge data, and BM is the closest among the other schemes. For 20 July, the best results are given by BM (Fig. 7).

The hydrological ensemble derived from the 50 rainfall ensemble forecasts using GR, simulating equally likely rainfall pattern for the 48 h of the forecasts, present different characteristics for the different events.

The 29 June ensemble captures the event and the ensemble stream is spread around the observation hydrograph (Fig. 8).

Figure 9(a) shows the 20 July ensemble results from the MM5 model settings previously described, while Fig. 9(b) shows the ensemble derived by slightly changing some of the MM5 settings. The maximum domain resolution of MM5 was increased to 1 km (as well as the other domains dimensions and resolution as a consequence) and a different CPS was used for the simulations (KF).

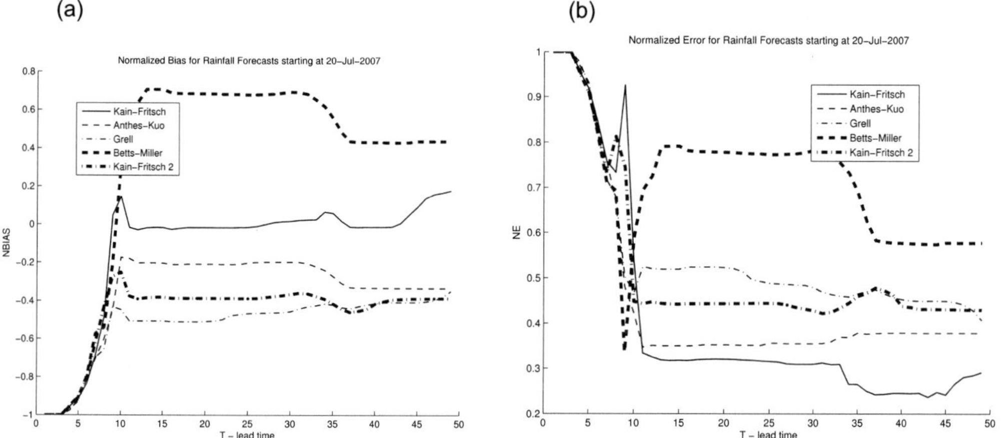

Fig. 5 (a) Normalised bias, and (b) normalised error, for the CPS rainfall ensemble forecast starting at 20 July 2007-00:00.

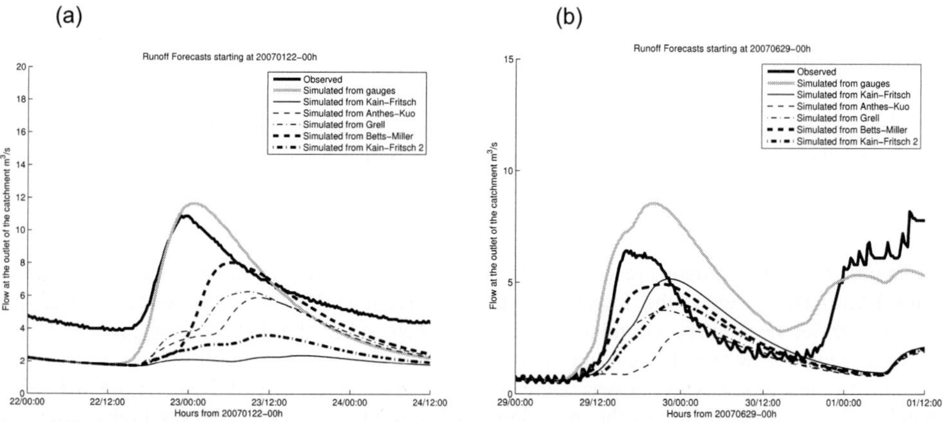

Fig. 6 PDM CPS ensemble forecast starting at (a) 22 January 2007-00:00, and (b) 29 June 2007-00:00.

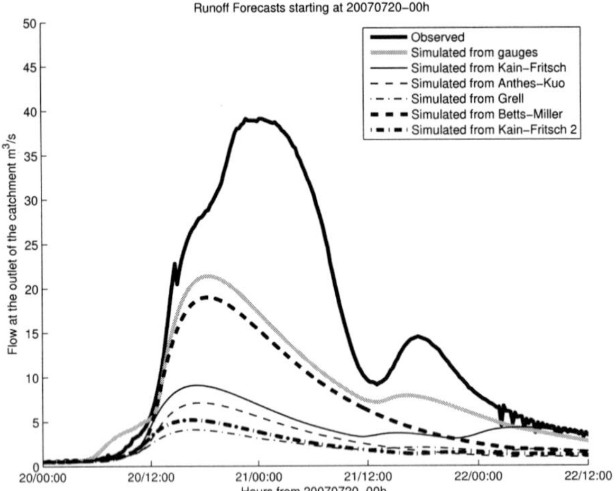

Fig. 7 PDM CPS ensemble forecast starting at 20 July 2007-00:00.

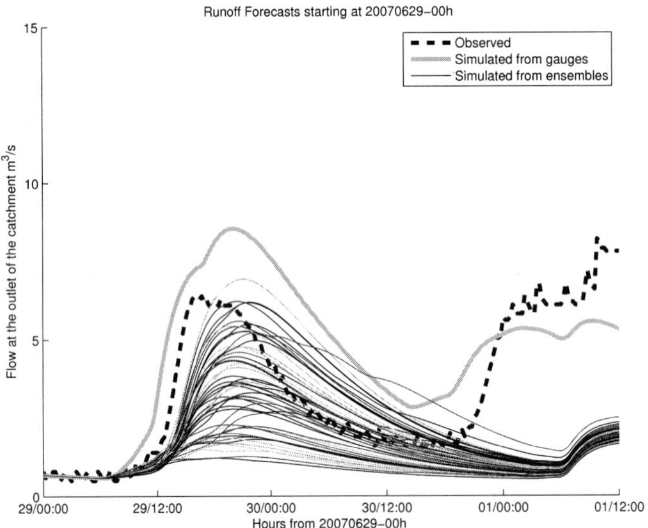

Fig. 8 PDM EPS ensemble forecast starting at 29 June 2007-00:00.

For the 2-km simulation (Fig. 9(a)) the ensemble does not capture the event and all the ensemble members lie below the observations and the raingauge-driven runoff simulation, while the 1-km ensemble (Fig. 9(b)) captures the event showing the spread of the ensemble around the reference simulation. The results highlight the dependence of the hydrological ensemble on the parameters of the rainfall prediction from NWP.

Changing the MM5 domains resolutions implies a change in the catchment location within the domains boundaries, even if the location of the catchment within the outer domain does not change. The differences in the boundary conditions developed in the downscaling process affect the rainfall prediction within the domain box, and these differences are amplified by the increase in the domains resolution when downscaling from the outer domain to the inner domains. This ultimately affects the estimation of the average rainfall input to the PDM from the high resolution domain. The results also show that by using different convective schemes the amount of explicit

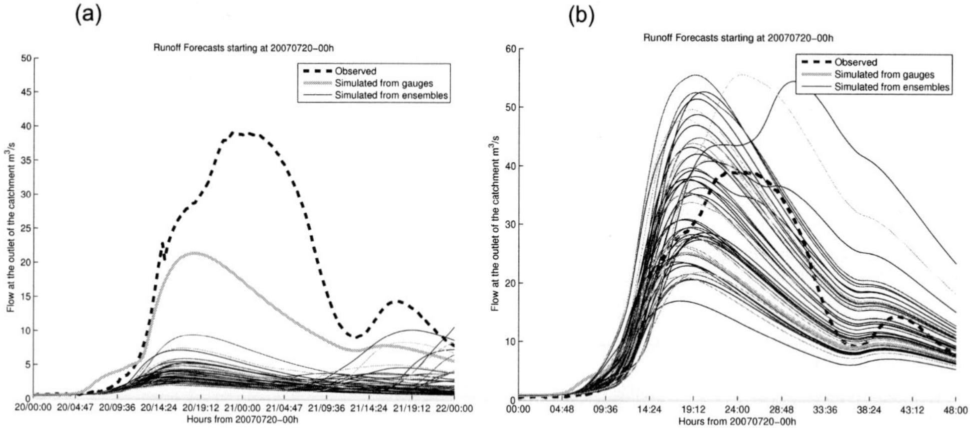

Fig. 9 PDM EPS ensemble forecast starting at 20 July 2007-00:00 at: (a) 2-km and (b) 1-km grid resolutions.

precipitation forecasted on the high resolution domain can considerably change and produce very different results in terms of runoff.

CONCLUDING COMMENTS

The test events ensemble simulations show that:

- analyses do not show a general trend in the CPS's behaviour and further investigation is required;
- the results on the catchment scale are affected in terms of explicit precipitation from the type of parameterization applied in the outer domains; and
- the variation of the NWP settings in terms of domains location, resolution and parameterization considerably affect the results of the hydrological simulations, showing a great dependence of the model behaviour on the rainfall input estimation uncertainties.

Acknowledgements The authors would like to thank ECMWF, BADC, UK Met Office and UK Environment Agency for providing some of the data sets used in this study.

REFERENCES

Allen, M. (1999) *Do-It Yourself Climate Prediction*. Macmillan Ltd, UK.

Anthes, R. A. (1977) A cumulus parameterization scheme utilizing a one-dimensional cloud model. *Monthly Weather Rev.* **105**(3), 270–286.

Betts, A. K. (1986): A new convective adjustment scheme. Part I: Observational and theoretical basis. *Quart. J. Roy. Met. Soc.* **112**, 677–692.

Betts, A. K. & Miller, M. J. (1986) A new convective adjustment scheme. Part II: Single column tests using GATE wave, BOMEX, ATEX and Arctic air-mass data sets. *Quart. J. Roy. Met. Soc.*, **112**, 693–709.

Buizza, R. (2008) The value of probabilistic prediction. *Atmos. Sci. Lett.* **9**, 36–42.

Buizza, R, & Palmer, T. N. (1995). The singular-vector structure of the atmospheric global circulation. *J. Atmos. Sci.* **52**(9), 1434–1456.

CEH Wallingford (2005) PDM Rainfall–Runoff Model. Version 2.2. Centre for Ecology & Hydrology, Wallingford. (Includes Guide, Practical User Guides, User Manual and Training Exercises).

Cheung, K. K. W. (2001) A review of ensemble forecasting techniques with a focus on tropical cyclone forecasting. *Meteorological Applications* **8**, 315–332.

COMET® Mesoscale Meteorology: A Primer (A Module Collection). University Corporation for Atmospheric Research (UCAR). Website at http://meted.ucar.edu/.

Dudhia, J. (1993) A Nonhydrostatic version of the Penn State-Ncar Mesoscale Model: validat6ion tests and simulation of an Atlantic cyclone and cold front. *Monthly Weather Rev.* **121**(5), 1493–1513.

Dudhia, J. (1996) A multi-layer soil temperature model for MM5. *Preprints, The Sixth PSU/NCAR Mesoscale Model Users' Workshop*, July 1996, Boulder, Colorado, 49–50. http://www.mmm.ucar.edu/mm5/mm5v2/whatisnewinv2.html.

Dudhia, J., *et al.* (2005) PSU/NCAR Mesoscale Modelling System. Tutorial Class Notes and User's Guide: MM5 Modelling System Version 3. Mesoscale and Microscale Meteorology Division National Center for Atmospheric Research. http://www.mmm.ucar.edu/mm5/documents/tutorial-v3-notes.html.

Grell, G. A., Dudhia J. & Stauffer, D. R. (1994) A description of the Fifth-generation Penn State/NCAR mesoscale model (MM5). *NCAR Technical Note*, NCAR/TN 398+STR, http://www.mmm.ucar.edu/mm5/documents/mm5-desc-doc.html.

Harrison, D. L., Driscoll, S. J. & Kitchen, M. (2000) Improving precipitation estimates from weather radar using quality control and correction techniques. *Meteorological Applications* **7**, 135–144.

Hong, S. Y. & Pan, H.-L. (1996) Nonlocal boundary layer vertical diffusion in a medium-range forecast model. *Monthly Weather Rev.* **124**(10), 2322–2339.

Kain, J. S. (2002) The Kain–Fritsch convective parameterization: an update. *J. Appl. Met.* **43**(1), 170–181.

Moore R. J. (2007) The PDM rainfall–runoff model. *Hydrol. Earth Syst. Sci.* **11**(1), 483–499.

Palmer, T. N., Barkmeijer, J., Buizza, R. & Petroliagis, T. (1997) The ECMWF Ensemble Prediction System, *Meteorological Applications* **4**, 301-304.

Pappenberger F. *et al.* (2006) Implementation Plan for Library of Tools for Uncertainty Evaluation. *FRMRC Research Report.* http://www.floodrisk.org.uk.

Person A. & Grazzini F. (2007) User guide to ECMWF forecast products. *ECMWF Meteorological Bulletin M3.2.* http://www.ecmwf.int/products/forecasts/guide/user_guide.pdf.

Reichel, F., Verworn, H.-R., Kramer, S., Cluckie, I. & Rico-Ramirez, M. A. (2009) Radar-based flood forecasting for river catchments. *Water Manage.* **162**, 159–168.

Reisner, J., Rasmussen, R. J. & Bruintjes, R. T. (1998) Explicit forecasting of supercooled liquid water in winter storms using the MM5 mesoscale model. *Quart. J. Roy. Met. Soc.* **124**, 1071–1107.

Rico-Ramirez, M. A., Cluckie, I. D., Shepherd, G.& Pallot, A. (2007) A high-resolution radar experiment on the island of Jersey. *Meteorological Applications* **14**, 117–129.

Toth, Z. & Kalnay, E. (1997) Ensemble forecasting at NCEP and the breeding method. *Monthly Weather Rev.* **125**, 3297–3319.

Xuan, Y., Cluckie, I. D. & Wang, Y. (2009) Uncertainty analysis of hydrological ensemble forecasts in a distributed model utilising short-range rainfall prediction. *Hydrol. Earth Syst. Sci.* **13**, 193–303.

Wang, W. & Seaman, N. (1996) A comparison study of convective parameterization schemes in a mesoscale model. *Monthly Weather Rev.* **125**, 252–278.

Zappa, M., Rotach, M. W., Arpagaus, M., Dorninger, M., Hegg, C., Montani, A., Ranzi ,R., Ament, F., Germann, U., Grossi, G., Jaun, S., Rossa, A., Vogt, S., Walser, A., Wehrhan, J. & Wunram, C. (2008) MAP D-PHASE: Real-time demonstration of hydrological ensemble prediction systems. *Atmos. Sci. Lett.* **9**, 80–87.

Rainfall–runoff modelling using a wavelet-based hybrid SVM scheme

RENJI REMESAN, MICHAELA BRAY, MUHAMMAD ALI SHAMIM & DAWEI HAN

Water and Environmental Management Research Centre, Department of Civil Engineering, University of Bristol, Lunsford House, Cantocks Close, Clifton, Bristol BS81UP, UK

renji.remesan@bristol.ac.uk

Abstract Efficient flood forecasting based on rainfall–runoff modelling is an important non-structural approach for flood mitigation. The support vector machine, a novel artificial intelligence-based method developed from statistical learning theory, is adopted herein in conjunction with wavelets to establish a real-time flood forecasting model. We compared them with another hybrid model called the neuro-wavelet model (NW). The methods were tested using the data from a small watershed (the Brue catchment in southwest England, UK), for which 7 years of records were available. The results reveal that the wavelet-based hybrid models can provide accurate runoff estimates for flood forecasting in the Brue catchment. In this study the training data length and input data structure were determined using another novel technique, the gamma test.

Key words hybrid models; gamma test; neural networks; flood; model selection; support vector machine

INTRODUCTION

Artificial intelligence (AI) based techniques offer many popular data-driven models that have been used extensively in the past couple of decades in streamflow forecasting and rainfall–runoff modelling. A comprehensive review by ASCE Task Committee on Application of Artificial Neural Networks in Hydrology (ASCE, 2000) shows the acceptance of the artificial neural network (ANN) approach among hydrologists. The major criticism against AI techniques in hydrology is their limited ability to account for any physics of the hydrological processes in a catchment. That concern was partially alleviated by Jain *et al.* (2004), whose study proved that the distributed structure of ANNs is able to capture certain physical properties. Support vector machines (SVMs) based modelling is an equally popular soft computing technique to ANNs in hydrology, which was developed at the AT&T Bell Laboratory (USA) by Vapnik and co-workers in the early 1990s (Boser *et al.*, 1992). SVMs were initially used for classification only; SVMs for regression were first introduced in Vapnik (1995). Their first applications were reported in the late 1990s (Drucker *et al.*, 1997). A detailed literature review of SVM-based hydrological modelling and forecasting can be found in Yu *et al.* (2006). Wavelet analysis is a well defined concept witnessing increasing applications to the quantitative explanation of time series and it has recently been identified as a useful tool for analysing both rainfall and runoff time series (Lane, 2007). By combining these AI techniques with one another, or with other dynamic predictive models, the individual strengths of each approach can be exploited in a synergistic manner for the construction of powerful intelligent systems (Nayak *et al.*, 2004).

Despite successful applications of SVM and ANNs in hydrology, there are still many issues unsolved, particularly in the selection of training data length and data structure. A recently introduced technique, called the gamma test (GT) by Agalbjörn *et al.* (1997), has the potential to successfully overcome these issues, as has been demonstrated in water level and flow modelling in the River Thames (Durrant, 2001), and in daily solar radiation prediction (Remesan *et al.*, 2008). GT would help us to identify the best embedded structure and reasonable data length for training of any smooth model before modelling. Thus we can minimise the guesswork associated with nonlinear modelling techniques.

In this paper we present two alternative data-driven hybrid models for rainfall–runoff modelling based on a wavelet decomposition methodology in conjunction with ANNs and SVMs, namely the neuro-wavelet (NW) model and wavelet-support vector machines (W-SVMs). The

SVM has a network architecture similar to multilayer perceptrons in artificial neural networks (ANNs), so in this study the modelling ability of the hybrid SVM was compared with that of the hybrid ANN. The results obtained using the proposed NWs and W-SVMs are compared with another popular and basic AI model, the neural network auto-regressive with exogenous input (NNARX), and basic benchmark models such as a naïve model (in which the predicted runoff value is equal to the latest measured value) and a trend model (in which the predicted runoff value is based on a linear extrapolation of the two previous runoff values). The gamma test is used in identifying the required input data set and data length for rainfall–runoff modelling along with the hybrid neural method.

DATA AND METHODS

Brue catchment

The data for this study were obtained from the NERC (Natural Environment Research Council) funded HYREX project (Hydrological Radar Experiment) which ran from May 1993 to April 1997 (its data collection was extended to 2000). The Brue catchment is located in Somerset (51.0750N and 2.580W) with a drainage area of 135 km^2 (Fig. 1). It is a predominantly rural catchment of modest relief with spring-fed headwaters rising in the Mendip Hills and Salisbury Plain. The rain-gauge network consists of 49 Cassella 0.2-mm tipping-bucket type raingauges. An automatic weather station and automatic soil water station located in the catchment recorded global solar radiation, net radiation and other physical parameters, such as wind speed, wet and dry bulb temperatures and atmospheric pressure at hourly intervals. Eight years (1993–2000) of daily rainfall–runoff data were collected from the Brue catchment and used in this study.

Gamma test (GT)

This concept was developed as the gamma test by Agalbjörn *et al.* (1997). A formal proof for the gamma test can be found in Evans (2002) and Evans & Jones (2002). This novel technique enables us to quickly evaluate and estimate the best mean-squared error that can be achieved by a smooth model on unseen data for a given selection of inputs, prior to model construction. The basic idea is quite distinct from earlier attempts at nonlinear analysis. Suppose we have a set of data observations of the form:

$$\{(x_i, y_i), 1 \le i \le M\} \tag{1}$$

where the inputs $x \in \Re^m$ are vectors confined to some closed bounded set $C \in \Re^m$ and, without loss of generality, the corresponding outputs $y \in \Re$ are scalars. Where we presume that the vectors $\boldsymbol{x}$ contain predicatively useful factors influencing the output y. The only assumption made is that the underlying relationship of the system under investigation is of the following form:

$$y = f(x_1 \ldots . x_m) + r \tag{2}$$

where f is a smooth function and r is a random variable that represents noise. Without loss of generality it can be assumed that the mean of the distribution of r is zero (since any constant bias can be subsumed into the unknown function f and that the variance of the noise var(r) is bounded. The domain of possible models is now restricted to the class of smooth functions which have bounded first partial derivatives. The gamma statistic (Γ) is the estimate of that part of the variance of the output that cannot be accounted for by a smooth data model.

The gamma test is based on $N[i, k]$, which are the kth ($1 \le k \le p$) nearest neighbours $x_{N[1,k]}(1 \le k \le p)$ for each vector $x_i(1 \le k \le p)$. Specifically, the gamma test is derived from the delta function of input vectors:

$$\delta_M(k) = \frac{1}{M}\sum_{i=1}^{M}\left|x_{N(1,k)} - x_i\right|^2 \qquad (1 \le k \le p) \tag{3}$$

where $|...|$ denotes Euclidean distance, and corresponding gamma function of output values:

$$\gamma_M(k) = \frac{1}{2M}\sum_{i=1}^{M}\left|y_{N(1,k)} - y_i\right|^2 \qquad (1 \le k \le p) \tag{4}$$

where $y_{N[1,k]}$ is the corresponding y value for the kth nearest neighbour of x_i in equation (3). In order to compute Γ, a least squares fit regression line is constructed for the p points ($\delta_M(k)$, $\gamma_M(k)$).

Neural network autoregressive with exogenous inputs (NNARX) model

In this study, a neural network autoregressive with exogenous inputs (NNARX) model is built on the basis of a linear ARX model. The discrete linear ARX model can be expressed as a simple linear difference equation. A simple linear ARX model for rainfall–runoff modelling with one exogenous input (rainfall) can be expressed as follows:

$$Q(t) = -\sum_{n=1}^{n_a} a_n Q(t-n) + \sum_{m=1}^{n_b} b_m P(t-m) + e(t) \tag{5}$$

where Q is the runoff, t represents time step, a_n, and b_m denote the model parameters, n_a and n_b are the orders associated with the output (Q) and the input (P), and $e(t)$ is the mapping error. The choice of the appropriate ARX model structure (the appropriate n_a and n_b) is very important and can be guided with Akaike's Information Criterion (AIC); the best input combination selection is selected by the gamma test. In this study we established a three-layer feed-forward neural network (one input layer, one hidden layer and one output layer). This topology has proved its ability in modelling many real-world functional problems. The Levenberg-Marquardt training algorithm was used to adjust the weights of the feed-forward neural network.

Support vector machines (SVM)

Just like ANNs, SVMs can be represented as two-layer networks (where the weights are nonlinear in the first layer and linear in the second layer).

Mathematically, a basic function for the statistical learning process is:

$$y = f(x) = \sum_{i=1}^{M} \alpha_i \varphi_i(x) = w\varphi(x) \tag{6}$$

where the output is a linearly weighted sum of M. The nonlinear transformation is carried out by $\varphi(x)$.

The decision function of SVM is represented as:

$$y = f(x) = \left\{\sum_{i=1}^{N} \alpha_i K(x_i, x)\right\} - b \tag{7}$$

where K is the kernel function, α_i and b are parameters, N is the number of training data, $\boldsymbol{x}_i$ are the vectors used in training process and $\boldsymbol{x}$ is the independent vector. The parameters α_i and b are derived by maximising their objective function.

The role of the kernel function simplifies the learning process by changing the representation of the data in the input space to a linear representation in a higher-dimensional space called a feature space. A suitable choice of kernels allows the data to become separable in the feature space despite being non-separable in the original input space. Four standard kernels are usually used in classification problems and also used in regression cases: linear, polynomial, radial basis and sigmoid:

Linear: $u' \times v$

Polynomial: $(\gamma \times u'v + coef)^{\text{degree}}$

Radial basis: $e^{-\gamma|u-v|^2}$

Sigmoid: $\tanh(\gamma \times u'v + coef)$

A number of support vector machine software is now available. The software used in this project was LIBSVM developed by Chih-Chung Chang and Chih-Jen (Bray & Han, 2004), and supported by the National Science Council of Taiwan. The basic algorithm is a simplification of both SMO by Platt and SVMLight by Joachims. The source code is written in C++. The choice of this software was made on its ease of use and dependability. It has been tried and tested in several research institutions worldwide including the Computer and Information Sciences Department, University of Florida, USA and the Institute for Computer Science, University of Freiburg, Germany.

The LIBSVM is capable of C-SVM classification, one-class classification, ν-SV classification, ν-SV regression and ε-SV regression. The software firsts trains the support vector machine with a list of input vectors describing the training data. It outputs a model file, which contains a list of support vectors and hence the description of the hypothesis for the particular regression problem. The model file is then used alongside a list of input vectors for making predictions. Figure 1 illustrates the flowchart describing the processes carried out in this study.

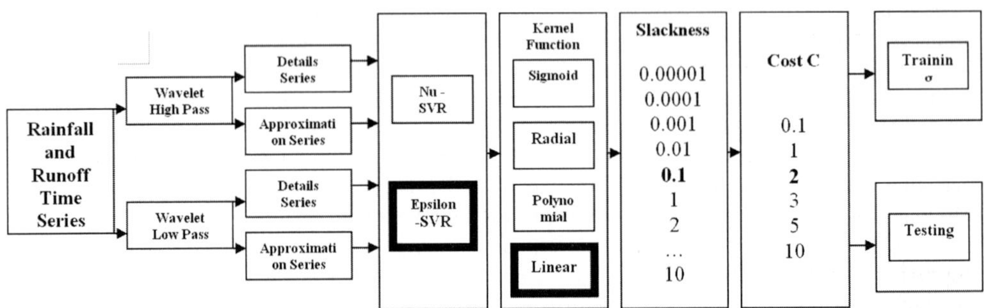

Fig. 1 Flowchart of the proposed W-SVM modelling scheme used in this study.

In Stage 1, data of antecedent rainfall and runoff of the Brue catchment were decomposed to three series of detailed coefficients (*D*) and three series of approximation (*A*) sub-time series. There are four possible kernels. To narrow down the kernel functions, the parameters of each kernel were kept at their default values and consequently a fair comparison of each SVM could be made. In the next stage, we tried SVM modelling with different kernel functions and different SVR types (ν-SV regression and ε-SV regression). The results from support vector machines with ε-SV regression and linear kernel performed considerably better than the remaining models, so linear kernel function was used for this study.

The deviation between the target value and the function describing the hypothesis found by the support vector machine is controlled by the ε parameter. In this study, the value of ε is set as 0.1, as suggested by Bray & Han (2004).

The cost of error assigns a penalty for the number of vectors falling between the two hyperplanes. If the data are of good quality, the distance between the two hyperplanes is narrowed down. If the data are noisy, it is preferable to have a smaller value of *C* which will not penalise the vectors. In this study, the cost value was chosen to be 2, as recommended by Bray & Han (2004). It produced the least error and was most economical on computing processing time in this study.

Neuro-wavelet (NW) model

The wavelet was introduced by Morlet & Grossman (1984) as a time-scale analysis tool with non-stationary signals for analysing irregular data with discontinuities and nonlinearity. The wavelet transform could be used as a tool that splits up data or functions into different frequency components, and then one could study each component with a certain resolution. Flexibility is the key advantage of wavelets over Fourier analysis; this helps us to study the signal of interest at various resolutions. Fourier transform techniques are widely used for linear decomposition of static time series. The problem with the Fourier approach is that information in a given time series is represented as a function of frequency with no specific time information. However, a wavelet analysis works with translation and dilation of a single local function ψ, generally referred to as the mother wavelet. In other words, the advantage of wavelet transform is that the wavelet coefficient reflects in a simple and precise manner the properties of the underlying function. Wavelet transform theory and its application to multi-resolution signal decomposition has been thoroughly developed and well documented over the past decades (Daubechies, 1988, 1992). Daubechies (1988) introduced the concept of orthogonal wavelet; generally referred to as Daubechies wavelets. Indeed, multi-resolution analysis and discrete wavelet transform have a strong potential for exploring various dynamic features of any time dependent data. Mathematical details of wavelet decomposition may be found in Daubechies (1988, 1992).

In this paper, we use the above multi-resolution data in the prediction model. Given a signal $f(t)$ of length N, the DWT consists of log2N stages, at most. The first step produces, starting from $f(t)$, two sets of coefficients: approximation coefficients $A1$, and detail coefficients $D1$. These vectors are obtained by convolving $f(t)$ with the low-pass filter $h1$ for approximation, and with the high-pass filter $g1$ for the detail. The next step splits the approximation coefficients $A1$ into two parts using the same scheme ($h2$ and $g2$), replacing $f(x)$ by $A1$, and producing $A2$ and $D2$, and so on. In other words, we wish to build a multi-resolution representation based on the differences of information available at two successive resolutions $2J$ and $2J$+1.

In this study, a multi-layer feed-forward network type of artificial neural network (ANN) and discrete wavelet transform (DWT) model were combined to obtain a neuro-wavelet (NW) model. The discrete wavelet transfer model functions through two set of filters: high-pass and low-pass filters, which decompose the signal into two set of series, namely detailed coefficients (D) and approximation sub-time series (A). In the proposed NW model, these decomposed sub-series obtained from DWT on the original data are directly used as inputs to the ANN model. The NW structure employed in the present study is shown in Fig. 2. Antecedent rainfall and runoff data of the Brue catchment were decomposed to three series of detailed coefficients (D) and three series of approximation (A) sub-time series. The present value of runoff was estimated using the three resolution levels of antecedent runoff and rainfall information (i.e. runoff and rainfall time series of 2-day mode (D_q^j1, D_p^i1), 4-day mode (D_q^j2, D_p^i2), 8-day mode (D_q^j3, D_p^i3) and approximate mode A_q^j1, A_p^i1, where q denotes runoff, p denotes rainfall and i and j denote the number of antecedent data sets of rainfall and runoff, respectively.

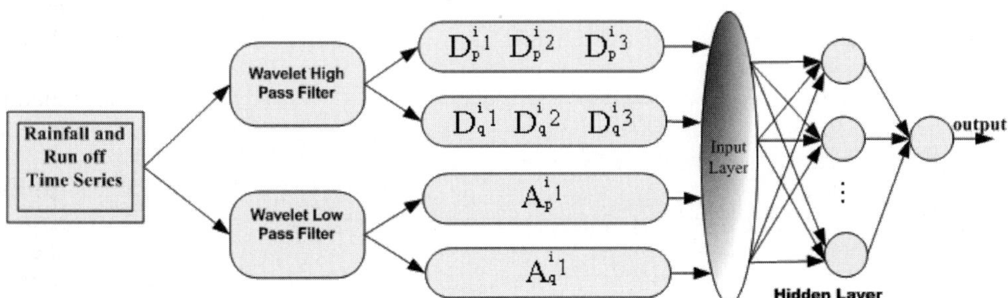

Fig. 2 The proposed hybrid NW modelling scheme.

Statistical parameters

The performances of the developed models (NNARX, NNARX and NW) and the basic benchmark models in this study were compared using various standard statistical performance evaluation criteria. The statistical measures considered were: correlation coefficient (CORR), mean absolute percentage error (MAPE), efficiency (E), root mean squared error (RMSE), mean bias error (MBE) and variance of the distribution of differences with MBE (S_d^2).

RESULTS AND DISCUSSION

Gamma test for data length selection and input identification

The gamma test analysis can provide vital information which would help in the modelling. There are $2^n - 1$ possibly meaningful combinations of inputs; from which, the best one can be determined by observing gamma statistic values. Different combinations of inputs evaluated in this study are shown in Table 1. The least gamma value is observed when we used the input data combination equivalent to [4, 4], i.e. four antecedent runoffs and four antecedent rainfalls as input. But *M* test results showed high complexity and low predictability for [4, 4], because of relatively high values of gradient (*A*) and *V* ratio. However, the next best value of gamma statistic (Γ), *A* and *V* ratio were observed when we used the three-steps antecedent runoff values ($Q(t-1)$, $Q(t-2)$, $Q(t-3)$), one-step antecedent rainfall ($P(t-1)$) and current rainfall information ($P(t)$). The quantity of available input data to predict the desirable output was analysed, using the *M* test. The gamma statistic (Γ) and standard error (SE) variation with unique data points obtained from *M* test analysis is shown in Fig. 3. The test produced an asymptotic convergence of the gamma statistic to a value of 0.0192 at around 1056 data points (i.e. $M = 1056$). The SE corresponding to $M = 1056$ is relatively small at ~0.0073.

Table 1 The gamma test results on the rainfall–runoff data.

Parameters	Different input combinations*						
	[1,1]	[2,2]	[3,3]	[2,1]	[3,2]	[2,3]	[4,4]
Gamma (Γ)	0.08218	0.07604	0.08512	0.07883	0.07601	0.08922	0.07582
Gradient (*A*)	0.09989	0.05999	–0.01286	0.04883	0.04139	0.04937	0.05095
Standard error	0.00615	0.01038	0.00787	0.01042	0.01009	0.00971	0.00726
V ratio	0.32875	0.30302	0.34048	0.31533	0.30575	0.35691	1.36331
Near neighbours	10	10	10	10	10	10	10
M	2236	2236	2236	2236	2236	2236	2236
Mask	10100000	11101000	11111010	10101000	11101010	11111000	11111111

* [*a*,*b*], where *a* denotes the number of antecedent runoff data sets used for the *M* test and *b* denotes the corresponding number of antecedent rainfall data and present rainfall information.

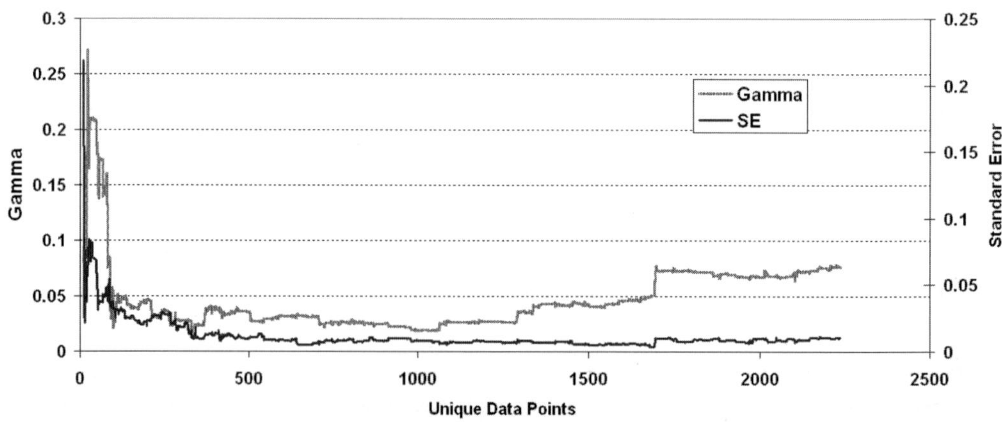

Fig. 3 Variation of gamma statistic and SE with unique data points.

The embedding 11101010 model (five input set and one output set of I/O pairs) was identified as the best structure for the following reasons: its low noise level (Γ value), the rapid fall off of the *M* test SE graph, relatively low *V* ratio value (indicates the existence of a reasonably accurate smooth model), the regression line fitted with slope $A = 0.04139$ (low enough as a simple non-linear model with less complexity) and good fit with SE of 0.01009. These values give a clear indication that an adequate nonlinear predictive model can be constructed using around 1056 data points, using the remainder of the 2240 data points as the validation data set.

Nonlinear modelling with wavelet support vector machines (W-SVMs)

The DWT and SVM techniques are combined to make a W-SVM model for rainfall–runoff modelling. The GT is used to identify input structure (three previous runoff values, one previous and present rainfall value) used for modelling. DWT decomposed the input data sets into three wavelet decomposition levels (2–4–8). The three detailed coefficient series and first approximate series of original data of runoff and rainfall used in this study are presented in Fig. 4.

In this study, the W-SVM was compared with the "multi-layer feed-forward network" type of artificial neural network using the Levenberg-Marquardt training algorithm. The ANN used in this study consists of an input layer with five inputs equivalent to the input structure of an ARX (3,2) model. The inputs are: $Q(t-1)$, $Q(t-2)$, $Q(t-3)$, $P(t-1)$ and $P(t)$, which were identified with the gamma test. In order to compare the ability of these models with that of the basic benchmark models, a naïve model and trend model were constructed. The performances of these models in terms of seven model efficiency evaluation criteria, namely CORR, slope, RMSE, MAPE, MBE, efficiency and variance of the distribution of differences about MBE (S_d^2) are presented in Table 2. Table 2 shows that the performance analysis of both the NW model and the W-SVM models perform remarkably well in both validation and training of data.

The W-SVM model showed an efficiency of 90.0% (an increase of 6.76 % from the NNARX model) for the training data, and a validation efficiency of 77.0% (an increase of 7.39% compared to NNARX). The correlation coefficient between the computed and observed was found to be 0.90 during training and 0.75 during validation. The observed and estimated runoff values of the W-SVM model for the training data are given in Fig. 5(a) in the form of a time series plot, and corresponding values for the validation data are given Fig. 5(b) in the form of a scatter plot. The RMSE for the NNARX model is lower (0.558 m^3/s; 27.9%) compared to the W-SVM model (0.37 m^3/s; 18.8%) during training; the corresponding values are very high for the basic bench-mark models (naïve model and trend model) with values of 2.03 m^3/s (101.6%) and 3.33 m^3/s (155.6%), respectively.

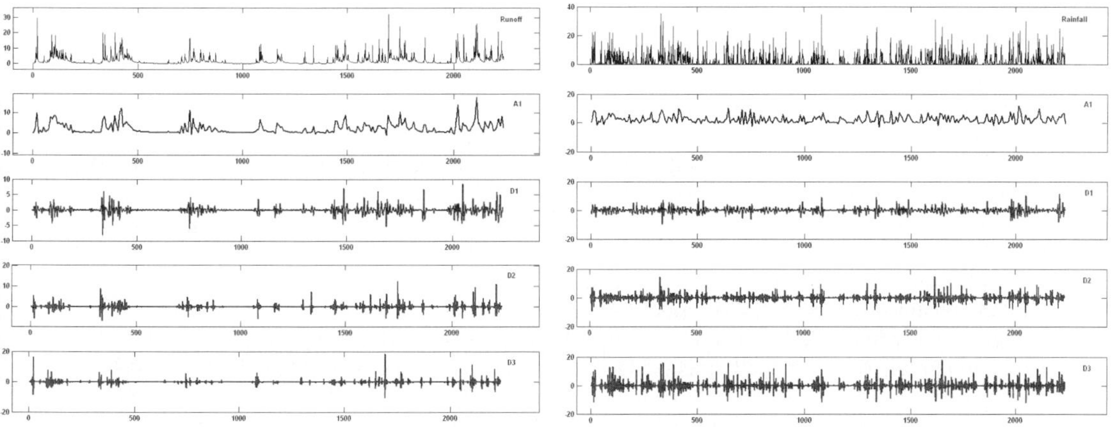

Fig. 4 Decomposed wavelet sub-time series components and approximation series of original runoff (left) and rainfall (right) data of the Brue catchment.

Table 2 Comparison of some basic performance indices of different models employed in the study for rainfall–runoff modelling.

Models used	Training data (1056 data points): RMSE (m^3/s, %)	CORR	Slope	MBE (m^3/s)	MAPE (m^3/s)	E	S_d^2 (m^3/s)	Validation data: RMSE (m^3/s, %)	CORR	Slope	MBE (m^3/s)	MAPE (m^3/s)	E	S_d^2 (m^3/s)
Naïve	2.03 (101.6)	0.51	0.81	0.078	0.432	0.471	4.11	2.70 (112.5)	0.39	0.78	–0.20	0.299	0.313	7.30
Trend	3.33 (155.6)	0.39	0.92	0.22	0.379	–0.265	9.83	4.24 (176.8)	0.30	0.90	0.15	0.494	–0.693	17.96
NNARX	0.558 (27.9)	0.83	0.90	–0.007	0.335	0.843	0.468	0.877 (36.4)	0.68	0.80	–0.085	0.297	0.717	1.152
W-SVM	0.370 (18.8)	0.90	0.90	–0.051	0.097	0.90	0.137	0.670 (27.2)	0.75	0.83	–0.112	0.167	0.770	0.497
NW	0.275 (13.6)	0.96	0.97	0.00025	0.188	0.962	0.075	0.699 (28.6)	0.81	0.92	–0.0068	0.205	0.805	0.496

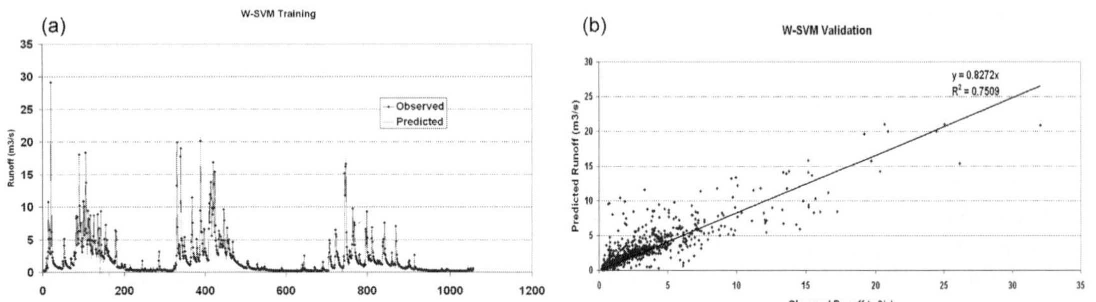

Fig. 5 Observed and predicted values of daily runoff using W-SVM model: (a) training data, and (b) validation data.

From the MBE value, one can deduce that the NNARX model showed underestimation for both the training data and validation data. The variance of the distribution differences of S_d^2 about MBE values are shown in Table 2 with the distinct variation of prediction by both NNARX and W-SVM models. We can see higher values of S_d^2 observed for the NNARX model (0.468 and 1.152 m^3/s) compared to those of the W-SVM model (0.137 and 0.447 m^3/s) for both training and validation phases.

Modelling with the NW model

The DWT and ANN techniques are combined to make a NW model for rainfall–runoff modelling. The neuro-wavelet model performance indices are summarised in the Table 2. The performance of the NW model in predicting runoff values is shown to be superior to the conventional NNARX and hybrid wavelet-based SVM model. The observed and estimated runoff values of the NW model for the training data are given in Fig. 6(a) in the form of a time series plot and corresponding values for the validation data are given in Fig. 6(b) in the form of a scatter plot. The runoff prediction is underestimated by all the models including NW model for both training and validation phases, as indicated by the MBE values in Table 2. The worst value in the validation phase is obtained by the N-SVM model (–0.112 m^3/s), followed by the NNARX (–0.085 m^3/s); though both are much higher than that of the NW model. The performance efficiency of the NW model in rainfall–runoff modelling is 4.54% higher than that of the N-SVM model for validation, and the corresponding value for training data is 6.88% higher. Compared with the NNARX model, the efficiency values of the NW model are 14.1% and 12.27% higher for training and validation data, respectively. MAPE values are found to be 0.188 and 0.205 m^3/s during the training phase

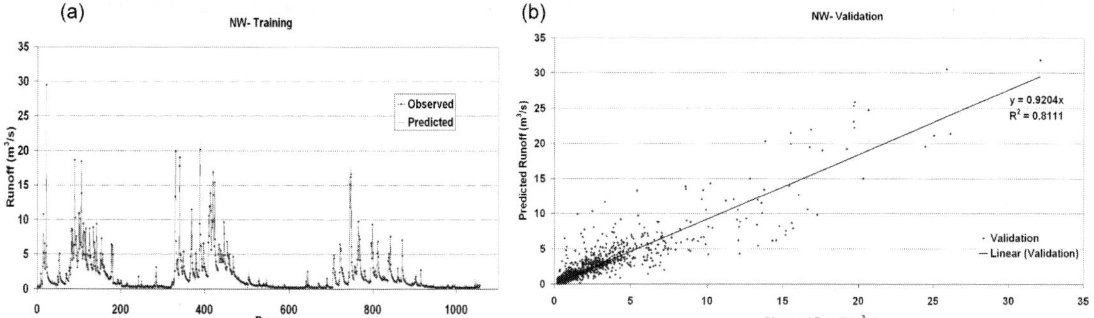

Fig. 6 Observed and predicted values of daily runoff at Brue catchment using the NW model: (a) training data; and (b) validation data.

and validation phase, respectively. The corresponding values for NNARX are found to be higher for both validation and training phases, whereas for the W-SVM model, the MAPE values observed were lower than that of the NW model. In terms of MBE and S_d^2, the performance of the NW model outperforms all other tested models in both training and validation phases.

CONCLUSIONS

Effective runoff prediction is one of the significant aspects of successful water resources management and flood forecasting. We have shown applications of wavelet decomposition in conjunction with support vector machines (SVMs) and artificial neural networks (ANNs) in predicting daily streamflows in the Brue catchment, UK. We describe a new approach of rainfall–runoff modelling from antecedent rainfall and runoff data sets with the gamma test in combination with nonlinear modelling techniques, such as NW and W-SVMs. The study successfully demonstrated the informative capability of the gamma test in identifying relevant antecedent input variables for runoff prediction prior to modelling. The information on the quantity of data required to construct a reliable model was determined using the M test, which has identified $M = 1056$ as the "sufficient data" scenario. The input combination identified by the GT includes three-step antecedent runoff values ($Q(t-1)$, $Q(t-2)$, $Q(t-3)$), one-step antecedent rainfall ($P(t-1)$) and the current rainfall information ($P(t)$), which is equivalent to a input combination of ARX(3,2) model, as the best input combination. Seven statistical indices, namely correlation coefficient (CORR), mean absolute percentage error (MAPE), efficiency (E), root mean squared error (RMSE), mean bias error (MBE) and variance of the distribution of differences with MBE (S_d^2), were considered for the performance evaluation of models under rainfall–runoff modelling. The study indicates that both the NW model and the W-SVM model have high standard of statistical properties in comparison to the traditional NNARX model. It is also observed that the hybrid neural network model outperformed the support vector machine-based model in one-step runoff prediction in the Brue catchment. Two benchmark models, namely the trend model and the naïve model, were evaluated to see how much our results differ from them.

REFERENCES

Agalbjörn, S., Koncar, N. & Jones, A. J. (1997) A note on the gamma test. *Neural Comput. & Applic.* **5**(3), 131–133.

ASCE Task Committee on Application of Artificial Neural Networks in Hydrology (2000) Artificial neural networks in hydrology. II: hydrologic applications. *J. Hydrol. Engng* **5**(2), 124–137.

Boser, B. E., Guyon, I. & Vapnik, V. (1992) A training algorithm for optimal margin classifiers. In: *Proceedings Fifth ACM Workshop on Computational Learning Theory* (Pittsburgh, Pennsylvania), 144–152. ACM Press, USA.

Bray, M. & Han, D. (2004) Identification of support vector machines for runoff modelling. *J. Hydroinf.* **6**(4), 265–280.

Daubechies, I. (1988) Orthonormal bases of compactly supported wavelets. *Commun. Pure Appl. Math.* **41**, 909–996.

Daubechies, I. (1992) *Ten Lectures on Wavelets*, Ch. 3–5. SIAM, Philadelphia, USA.

Durrant, P. J. (2001) winGamma: A non-linear data analysis and modelling tool with applications to flood prediction. PhD Thesis, Department of Computer Science, Cardiff University, Cardiff, UK.

Drucker, H., Burges, C. J. C., Kaufman, L., Smola, A. & Vapnik, V. (1997) Support vector regression machines. In: *Advances in Neural Information Processing Systems* (ed. by M. Mozer, M. Jordan & T. Petsche), vol. 9, 155–161. MIT Press, Cambridge, Massachusetts, USA.

Evans, D. (2002) Data derived estimates of noise using near neighbour asymptotics. PhD Thesis, Department of Computer Science, Cardiff University, Cardiff, UK.

Evans, D. & Jones, A. J. (2002) A proof of the gamma test. *Proc. Roy. Soc. A* **458**(2027), 2759–2799.

Jain, A., Sudheer, K. P. & Srinivasulu, S. (2004) Identification of physical processes inherent in artificial neural network rainfall runoff models. *Hydrol. Processes* **18**, 571–581. doi:10.1002/hyp.5502.

Lane, S. N. (2007) Assessment of rainfall–runoff models based upon wavelet analysis. *Hydrol. Processes* **21**, 586–607. doi::0.1002/hyp.6249.

Morlet, J. M. & Grossman, A. (1984) Decomposition of Hardy functions into square integrable wavelets of constant shape.. *SIAM J. Math. Anal.* **15**(4), 723–736.

Nayak, P. C., Sudheer, K. P., Rangan, D. M. & Ramasastri, K. S. (2004) A neuro-fuzzy computing technique for modelling hydrological time series. *J. Hydrol.* **291**, 52–66. doi:10.1016/j.jhydrol.2003.12.010.

Remesan, R., Shamim, M. A. & Han, D. (2008) Model data selection using gamma test for daily solar radiation estimation. *Hydrol. Processes* **22**, 4301–4309.

Remesan, R., Shamim, M. A. & Han, D. (2009) Runoff prediction using an integrated hybrid modelling scheme. *J. Hydrol.* **372**(1-4), 48–60.

Vapnik, V. N. (1995) *The Nature of Statistical Learning Theory*, Springer-Verlag, New York, USA.

Yu, P.-S., Chen, S.-T. & Chang, I.-F. (2006) Support vector regression for real-time flood stage forecasting. *J. Hydrol.* **328**(3/4), 704–716.

Appropriate data normalization range for daily river flow forecasting using an artificial neural network

ADARSH SINGH, RABINDRA K. PANDA & NIRANJAN PRAMANIK
Agricultural and Food Engineering Department, Indian Institute of Technology, Kharagpur, West Bengal 721 302, India
rkpanda@iitkgp.ac.in

Abstract Advance time-step river flow forecasting is of paramount importance in controlling flood damage. For the last few years, artificial neural networks (ANN) have been used extensively in river flow forecasting and have proven to be an efficient technique. Various statistical techniques have been used to pre-process the input data sets before using them in ANN model training to reduce the forecasting error. Normalizing the input–output data sets between certain ranges is one of the methods of pre-processing the raw data; this improves input–output generalization to yield accurate forecasts. In the present study, 1-day, 2-day and 3-day advance flow forecasts were carried out at a site on the Brahmani River, on the east coast of India, using the ANN approach, utilizing the data sets normalized within four different ranges (–1.0 to +1.0; 0.0–1; 0.1–0.9 and 0.2–0.8). The results obtained using the normalized data of three different ranges were compared in terms of root mean square error (RMSE) and modelling efficiency (*E*) to determine the best range of data normalization. The study revealed that normalizing the data sets within any of the three ranges (0.0–1; 0.1–0.9 and 0.2–0.8), yielded significantly improved results compared to the range –1 to +1. In the case of the 1-day ahead forecast, all three data ranges produced comparable and satisfactory results, whereas for 2-day and 3-day ahead forecasts, the ANN model trained using the normalized data of the range 0.2–0.8 yielded the best result.

Key words river flow forecasting; data normalization; artificial neural network

INTRODUCTION

The ability to predict river flows quickly and accurately is of crucial importance in flood forecasting operations. Hydrodynamic models provide a sound physical basis for this purpose and are capable of simulating a wide range of flow situations. However, these models require accurate river geometric data, which may not be available in many locations. It is also not possible to integrate observed data directly at desired locations to improve the model result. Moreover, physically-based models require high computation time, thereby making them incapable of producing real-time forecasting, which is required for flood warning to save lives and property. The artificial neural network (ANN) provides a quick and flexible approach for data integration and model development for real-time forecasting. The ANN approach has been used by many researchers in the field of water resources engineering (Dibike & Solomatine, 2001; Dolling & Veras, 2002; Zhang *et al.*, 2002; Sudheer *et al.*, 2003; Shrestha *et al.*, 2005). The underlying concepts of the ANN and its application to hydrological problems are presented by, for example, the ASCE Task Committee (2000a,b), Imrie *et al.* (2000) and Dawson *et al.* (2002). Several methodologies have been proposed to improve the forecasting capability outside the range of training data sets (Kisi, 2004). In ANN application, the input data are generally normalized in the range 0–1.0. As revealed from the literature, hardly any attempt was made in the past to normalize the input data sets in the range –1 to 1 for daily flow forecasting using ANN. However, a few studies suggested that the normalization range could be between $-\infty$ and $+\infty$, if a sigmoidal activation function is used (Burian *et al.*, 2001). Therefore, in the present study, an attempt was made to normalize the data sets in the range –1 to 1, which is within the range of $-\infty$ to $+\infty$, for ANN modelling (Shanker *et al.*, 1996). To accommodate the data beyond the training range, alternative normalization ranges such as: 0–1, 0.1–0.9 and 0.2–0.8 have been suggested. Thus, the present study aimed at assessing the appropriate data normalization range for obtaining improved flood forecasts at a site on the Brahmani River in eastern India.

Artificial neural network

An artificial neural network (ANN) is a computing system made up of an interconnected set of simple information processing units, which are analogous to biological neurons. The neurons collect inputs from both a single and multiple sources and produce output in accordance with a certain transfer function. Some inputs to the neuron may have more relevance than others, and this is modelled by weighting. Generally, the main principle of neural computing is the decomposition of the input–output relationship into a series of linearly separable steps using hidden layers (Haykin, 1994).

There are four distinct steps in developing an ANN-based solution to a certain problem, including flood forecasting. The first step is data transformation, scaling or normalization. Large variations in the input data can slow down or even prevent the training of the network. To overcome this potential problem, the data are usually scaled using statistical, min-max, sigmoidal or principal component transformations (Priddy & Keller, 2005). It is also important that absolute input values are scaled to avoid asymptotic issues (Haykin, 1994). The second step is the network architecture definition, where the number of hidden layers, the number of neurons in each layer, and the connectivity between the neurons are set. There are three basic layers or levels of data processing units namely, the input layer, the hidden layer and the output layer. Each of these layers consists of processing neurons. The number of input neurons, output neurons and the neurons in the hidden layer depends upon the problem being studied. If the number of neurons in the hidden layer is small, the network may not have sufficient degrees of freedom to learn the process correctly. If the number is too high, the training will take a long time and the network may sometimes over-fit the data (Karunanithi *et al.*, 1994). Obtaining the optimum number of hidden neurons in the ANN architecture is a trial-and-error process, which depends on the quality of data and the type of problem. In the third step, a learning algorithm is used to train the network to respond correctly to a given set of inputs. The training goal is to find parameters that result in the best performance of the network with unfamiliar data. An ANN is better trained when more input data are available. During the training process, the connection weights are updated. At the beginning of training, the initial value of weights can be assigned randomly or based on experience. In this process, the learning algorithm systematically changes the weights to correctly perform a desired input–output relationship. The process of training is said to be finished when the difference between the ANN output and the actual output meets the preset error goal. Finally, the validation step in which the performance of the trained ANN model is tested using some selected validation data sets. Validation can be employed to monitor the network error during training to determine the optimal number of training iterations. It can also be used to find out the optimal number of hidden neurons.

STUDY AREA

The present study was carried out at Jenapur gauging site on the Brahmani River located in eastern India (Fig. 1). The Brahmani basin lies between latitude 20°28′–23°35′N and longitude 83°52′–87°30′E. The delta region of the basin starts at the Pankapa 1 gauging site and ends at the Bay of Bengal where it drains into the sea. The deltaic region, which is densely populated and covered with major crops of the area, is highly flood prone. The average annual rainfall of the basin is about 1305 mm. During the rainy season (July–September), the runoff from the basin accumulates and causes floods. The maximum flood in the river has been recorded as 24 246 m^3/s. The highest gauge level at the site during that event was recorded as 24.78 m, against the danger level of 23.00 m. The location map of the Brahmani River basin showing Jenapur gauging site is shown in Fig. 1.

METHODOLOGY

Data preparation

A remarkable property of ANNs is their ability to handle nonlinear, noisy, and nonstationary data. However, with suitable data preparation beforehand, it is possible to improve the performance further (Maier & Dandy, 2000; Bray & Han, 2004). Data preparation involves a number of

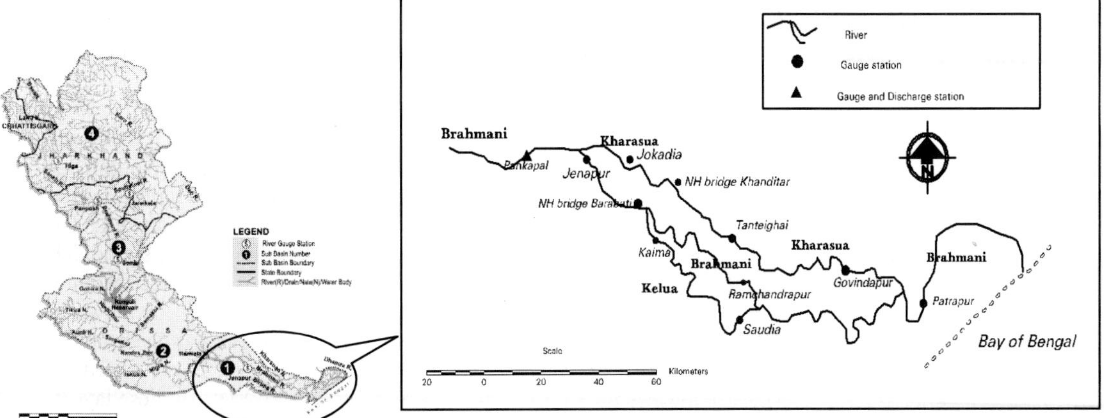

Fig. 1 Location map of the Brahmani Delta showing rivers and gauging sites.

Table 1 Basic statistics of the training and testing data sets.

Basic statistics	Training (1999–2003)	Testing (2004–2005)
Sample size	1826	731
Minimum flow (m^3/s)	37.0	56.535
Maximum flow (m^3/s)	10077.0	10314.0
Flow range (m^3/s)	10040.0	10257.0
Mean flow (m^3/s)	568.63	463.37
Std. deviation (m^3/s)	957.45	748.48
Std. error (m^3/s)	25.049	27.683
Skewness	4.5754	6.658
Kurtosis	28.416	67.981

processes such as data collection, data division and data pre-processing. Daily flow data at Jenapur gauging site were collected for a period of seven years (1999–2005) from the Central Water Commission (CWC), Bhubaneswar, India. Many researchers have discussed the data division in the process of application of ANN (ASCE, 2000a). The major portion of the data sets, that is 75% of the total data length, is used for training and remaining 25% are used for testing. In this study daily flow data for the period 1999–2003 were used for ANN training and the data of the years 2004 and 2005 were used for ANN model testing. The basic statistics of the training and testing data sets are presented in Table 1.

In order to ensure that all variables receive equal attention during the training process, normalization was carried out to improve the weight optimization (Gunn, 1998). The data sets were normalized between four different ranges: –1.0 to +1.0; 0.0– 1; 0.1–0.9 and 0.2–0.8. The data were normalized using the following equation:

$$x_n = (b-a)\left[\frac{x_0 - x_{\min}}{x_{\max} - x_{\min}}\right] + a \tag{1}$$

where x_n is a normalized variable; a is the minimum range of the variable; b is the maximum range of the variable; $x_{\max}$ is the maximum flow value in the time series and $x_{\min}$ is the minimum flow value in the time series.

Input selection

Partial autocorrelation analysis of the time series was employed to study the effect of preceding flows. The partial autocorrelation statistics and the corresponding 95% confidence bands from zero

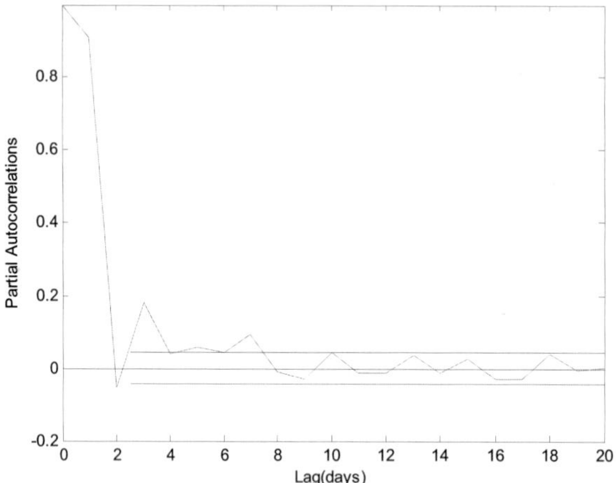

Fig. 2 Partial autocorrelation plot of the daily flow series.

Table 2 ANN models for 1-day, 2-day and 3-day ahead flood forecasting.

Models	Output	Inputs
Model 1	$Q_{(t+1)}$	$f(Q_t, Q_{(t-1)}, Q_{(t-2)}, Q_{(t-3)}, H_t)$
Model 2	$Q_{(t+2)}$	$f(Q_t, Q_{(t-1)}, Q_{(t-2)}, Q_{(t-3)}, H_t)$
Model 3	$Q_{(t+3)}$	$F(Q_t, Q_{(t-1)}, Q_{(t-2)}, Q_{(t-3)}, H_t)$

time lags to 20-day time lag were estimated, and are presented in Fig. 2. It was observed that there is significant correlation up to three time lags in the time series of the flow data, which indicates the antecedent flow values up to three-day lag could be used as input in the models for advance time-step forecasting. The rapid decaying pattern of partial autocorrelation function confirms the dominance of autoregressive process, relative to the moving average process. Therefore, in the present study three antecedent flow values were selected for developing ANN models.

Three models were proposed for 1-day, 2-day and 3-day ahead flood forecasting, as presented in Table 2. Time series of water level corresponding to time t (H_t) was also incorporated in all three models. In the case of all three models, antecedent flow values up to 3-day time lags i.e. Q_t, $Q_{(t-1)}$, $Q_{(t-2)}$ and $Q_{(t-3)}$, along with H_t, were included in the input data matrix. The Levenberg-Marquart algorithm was used to train the ANN models and the Matlab code was developed to carry out the computation.

Performance evaluation of ANN models

The statistics of the error comparison were used in the study. The overall performances of the trained networks were judged with respect to the following criteria: modelling efficiency (E), correlation coefficient (R), index of agreement (IOA) and root mean square error (RMSE). These coefficients are independent of the scale of data used and are useful in assessing the goodness of fit of the model (Dawson & Wilby, 1999; Dawson *et al.*, 2002). The difference between the observed and the ANN computed peak flow (DPF) was used to evaluate the prediction capability of the trained networks. The expressions of the goodness-of-fit indices (GFIs) used in the study are presented in Table 3.

RESULTS AND DISCUSSION

The ANN architecture with one hidden layer was found adequate to generalize the input–output mapping. Three hidden neurons in the hidden layer were decided by trial and error in the case of

Table 3 Mathematical expressions of goodness-of-fit indices (GFI).

GFI	Abbreviation	Formula
Index of agreement	IOA	$1-\dfrac{\sum_{i=1}^{n}(O_i-P_i)^2}{\sum_{i=1}^{n}\left[\lvert P_i-\overline{O}_i\rvert+\lvert O_i-\overline{O}_i\rvert\right]^2}$
Correlation coefficient	R	$\dfrac{\sum_{i=0}^{n}(O_i-\overline{O}_i)(P_i-\overline{P}_i)}{\sqrt{\sum_{i=1}^{n}(O_i-\overline{O}_i)^2\sum_{i=1}^{n}(P_i-\overline{P}_i)^2}}$
Modelling efficiency	E	$1-\dfrac{\sum_{i=1}^{n}(O_i-\overline{O}_i)^2}{\sum_{i=1}^{n}(O_i-\overline{O}_i)^2}$
Root mean square error	RMSE	$\sqrt{\dfrac{1}{n}\sum_{i=1}^{n}(O_i-P_i)^2}$

O_i: observed flow; P_i: predicted flow; $\overline{O}$: average of the observed flow; n: number of data sets.

Table 4 GFIs obtained for 1-day ahead flood forecasting using all four normalization ranges.

Scaling range	ANN structure	GFI: RMSE (m^3/s)	R	E	IOA	DPF (m^3/s)
−1 to 1	5-3-1	214.34	0.981	0.9218	0.9757	3652
0 to 1	5-3-1	104.69	0.994	0.9813	0.9949	1756
0.1 to 0.9	5-2-1	98.33	0.994	0.9835	0.9955	1599
0.2 to 0.8	5-2-1	95.76	0.994	0.9844	0.9958	1547

1-day and 2-day flood forecasting. The performance of Model 1 for all normalization ranges obtained during model testing is summarized in Table 4. It was observed that the overall results of the model obtained using the normalization ranges 0–1, 0.1–0.9 and 0.2–0.8 were found to be better than that obtained using the range −1 to +1. Simpler architectures with only two hidden neurons, in the case of normalization ranges 0.1–0.9 and 0.2–0.8, yielded better results. The computed RSME between the ANN-predicted and observed flow values were quite low for the range of 0.2–0.8. The difference between the observed peak flow and ANN-computed peak flow was also reduced when the normalization range decreases. The observed and ANN-computed flood hydrographs obtained from model testing using all normalization ranges for 1-day ahead flood forecast during the flooding periods are presented in Fig. 3(a).

The results of Model 2 obtained using the data of all four normalization ranges showed the same trend as those obtained from Model 1. The results obtained using data within the normalized range 0.2–0.8 were found to be superior to those in the other normalizations ranges. Table 5 presents the GFI statistics of Model 2. Although there is an overall increase in the predictability of the model using the normalized data sets, the peak flow still needs to be improved. The time series comparison in the case of 2-day ahead forecasting is presented in Fig. 3(b). The model performance using the data within the ranges 0–1 and 0.2–0.8 was found to be similar and comparable.

The results of Model 3, used for 3-day ahead forecasting, are presented in Table 6. The RMSE was found to be lower when the normalization range is decreased, but peak flow is estimated poorly. The model with the data within the range 0–1 yielded the peak flow with less deviation as

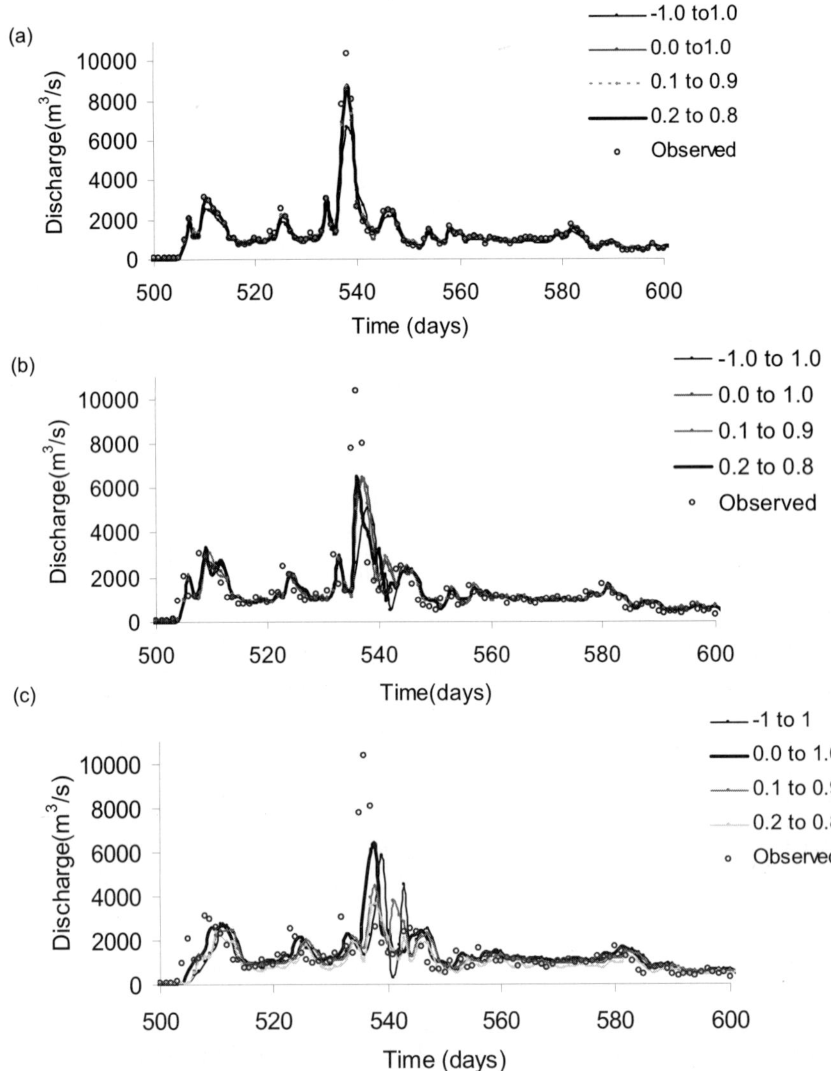

Fig. 3 Temporal variation of observed and ANN-computed discharge for: (a) 1-day ahead, (b) 2-day ahead, and (c) 3-day ahead forecasting at Jenapur site for different normalization ranges.

Table 5 GFIs obtained for 2-day ahead flood forecasting using all four normalization ranges.

Scaling range	ANN structure	GFI: RMSE (m^3/s)	R	E	IOA	DPF (m^3/s)
–1 to 1	5-3-1	490.95	0.768	0.5896	0.8505	5229
0 to 1	5-3-1	421.55	0.836	0.6978	0.9009	3854
0.1 to 0.9	5-2-1	433.40	0.830	0.6802	0.8907	3924
0.2 to 0.8	5-2-1	406.57	0.849	0.7189	0.9081	3845

compared to other ranges. The time series comparison for Model 3, for 3-day ahead forecasting, is presented in Fig. 3(c).

Table 6 GFIs obtained for 3-day ahead flood forecasting using all four normalization ranges.

Scaling range	ANN structure	GFI: RMSE (m^3/s)	*R*	*E*	IOA	DPF (m^3/s)
–1 to 1	5-3-1	590.21	0.644	0.4071	0.761	4515
0 to 1	5-3-1	449.26	0.675	0.4368	0.797	3970
0.1 to 0.9	5-2-1	442.32	0.697	0.4839	0.795	5861
0.2 to 0.8	5-2-1	432.92	0.71	0.4999	0.8139	6670

CONCLUSIONS

Normalizing the data sets between the ranges of 0.0 to 1; 0.1 to 0.9 and 0.2 to 0.8, rather than the range of –1.0 to +1.0, improves the predictability considerably. In case of 1-day ahead forecast, all three data normalization ranges 0.0–1, 0.1–0.9 and 0.2–0.8 produced comparable and satisfactory results, as revealed from their low values of RMSE and DPF. It was concluded that normalizing the data within a narrow range greatly reduces the prediction error. However, the normalization has very little impact on reducing peak flow prediction error beyond 1-day ahead forecasting. This deviation could be reduced by selecting the inputs containing a higher number of peak flow values in the training data sets.

REFERENCES

ASCE Task Committee (2000a) Artificial neural networks in hydrology. I: preliminary concepts. *J. Hydrol. Engng ASCE* **5**, 115–123.

ASCE Task Committee (2000b) Artificial neural networks in hydrology. II: hydrologic applications. *J. Hydrol. Engng ASCE* **5**, 124–135.

Burian, S. J., Durrans, S. R., Nix, S. J. & Pitt, R. E. (2001) Training artificial neural networks to perform rainfall disaggregation. *J. Hydrol. Engng ASCE* **6**, 43–51.

Bray, M. & Han, D. (2004) Identification of support vector machines for runoff modelling. *J. Hydroinf.* **6**, 265–280.

Dawson, C. W. & Wilby, R. B. (1999) A comparisons of artificial networks flow forecasting. *Hydrol. Earth System Sci.* **3**, 529–540.

Dawson, C. W., Harpharan, C., Wilby, R. B. & Chen, Y. (2002) Evaluation of artificial neural network techniques for flow forecasting in river Yangtze, China. *Hydrol. Earth System Sci.* **6**, 619–626.

Dibike, Y. B. & Solomatine, D. P. (2001) Hydrological modeling using artificial neural network *.Phys. Chem. Earth* **B** *Hydrology Ocean Atmos.* **26**(1), 1–7.

Dolling, O. R. & Veras, E. A. (2002) Artificial neural networks for streamflow prediction. *J. Hydraul. Res.* **40**, 547–554.

Gunn, R. R. (1998) Support vector machines for classification and regression. Tech. report, University of Southampton, UK.

Haykin, S. (1994) *Neural Networks.* Macmillan, Englewood Cliffs, New Jersey, USA.

Imrie, C. E., Durucan, S. & Korre, A. (2000) River flow prediction using artificial neural networks: generalization beyond the calibration range. *J. Hydrol.* **233**, 138–153.

Karunanithi, N., Grenney, W. J., Whitley, D. & Bovee, K. (1994) Neural networks for river flow prediction. *J. Comput. Civil Engng ASCE* **8**(2), 201–220.

Kisi, O. (2004) River flow modeling using artificial neural networks. *J. Hydrol. Engng ASCE* **9**(1), 60–63.

Maier, H. R. & Dandy, G. C. (2000) Neural networks for the prediction and forecasting of water resources variables: a review of modelling issues and applications. *Environ. Modeling & Software* **15**, 101–124.

Priddy, L. K. & Keller, E. P. (2005) *Artificial Neural Networks, An Introduction.* SPIE Press, Bellingham, Washington, USA.

Shanker, M., Hu, M. Y. & Hung, M. S. (1996) Effect of data normalization on neural network training. *Omega, Int. J. Manage Sci.* **24**, 385–397.

Shrestha, R. R., Theobald, S. & Nestmann F. (2005) Simulation of flood flow in a river system using artificial neural networks. *Hydrol. Earth System Sci.* **9**(4), 313–321.

Sudheer, K. P., Gosain, A. K. & Ramasastri, K. S. (2003) Estimating actual evapotranspiration from limited climatic data using neural network computing technique. *J. Irrig. Drain. Engng* **129**(3), 214–218.

Zhang, Y., Pulliainen, J., Koponen, S.& Hallikainen, M. (2002) Application of an empirical neural network to surface water quality estimation in the Gulf of Finland using combined optical data and microwave data. *Remote Sens. Environ.* **81**, 327–336.

Hydrological impact assessment of climate change on water resources of the Niger River basin using a water balance model and ANNs

J. N. OKPARA[1] & M. PERUMAL[2]

1 *Department of Applied Met. Services, Nigerian Met. Agency, Abuja, Nigeria*
juddyokpara@gmail.com

2 *Department of Hydrology, Indian Institute of Technology, Roorkee, India*

Abstract The study explored and assessed the potential impacts of climate change on water resources of the Nigerian sector of the Niger River basin of sub-Saharan West Africa, with simulation models, the Thornthwaite water balance accounting scheme and artificial neural networks (ANNs). The water balance model uses mean monthly climate variables of precipitation, temperature, net radiation, as well as soil water holding capacity and potential evapotranspiration, for estimating water surpluses of five Niger sub-basins, linking the water surpluses of these sub-basins with the corresponding observed monthly average discharges using the ANNs model. Ninety-six months of data have been used for calibrating the ANNs and 24 months data were used for verification. Areally averaged temperature and precipitation changes from formulated climate change synthetic scenarios were imposed on each sub-basin for assessing climate change impacts on runoff (water surplus). ArcView GIS was vital for the hydrological modelling, especially the building of the grid containing the water holding capacity parameter of the soil, creating climatic surfaces and merging various spatial themes (data layers) that were useful in interpretation, analysis (soil water budgeting) and change detection of spatial structures and objects. Results show discernible evidence and the sensitivity of the Niger River basin to the changing climate; annual water surplus is highest over the south of Niger (1241.2 mm), followed by Lower Benue (973.6 mm), Lower Niger (729.4 mm), Upper Benue (495.3 mm) sub-basins and the least value is observed over Upper Niger sub-basin (360.7 mm). It is further observed that the water surplus is much more sensitive to the accuracy of potential evaporation estimate (that depends on temperature) in the humid climate than the arid climate. In addition, when temperature increases by 2°C, the mean monthly runoff on the average is expected to change by –10 to –50%, –5 to –40% and 15 to 60%, respectively, for precipitation changes of –20%, 0% and 20%. By implication then, water, hydro-power, health and food security of the country may be seriously threatened by climate change, unless rainfalls turnout to be on the increase and adequate management strategies are put in place. Necessary adaptation measures and nonconventional sources of water that can be exploited in the future are also highlighted.

Key words ANN; Niger River; water balance; climate change

INTRODUCTION

It has been projected that, by mid-century, annual average river runoff and water availability will decrease by 10–30% over some dry regions at mid latitudes and dry tropics, while increasing by 10–40% at high latitudes and in some wet tropics areas (Milly *et al.*, 2005), and more pronounced changes are likely by the end of this century. Hence, many of the presently water stressed semi-arid and arid areas of sub-Saharan West Africa, Brazil, etc. are likely to suffer from decreased water resources availability due to climate change, as both river flows and groundwater recharge decline (Döll & Flörke, 2005). Kundzewicz *et al.* (2008) and Rosenzweig *et al.* (2007) also document that changes in the distribution of river flows and groundwater recharge over space and time are factored by changes in temperature, evaporation and, crucially, precipitation, while Kundzewicz *et al.* (2007) assessed the potential climate change on water quality. Also, the IPCC opined that yields from rain-fed agriculture in Africa could be reduced by as much as 50% by 2020. Water shortages and decrease in land suitable for agriculture would cause other social and political disruptions, including forced migration and conflict. It is widely recognized also, that the economic development of much of Africa and other third world countries is strongly dependent on the characteristics and proper management of available water resources, which is often subject to large spatio-temporal variability (IPCC 1996, 2001, 2007). Over the Lake Chad basin, Okpara (2007) has observed a decreasing trend of rainfall and a southward shift in the isohyets and opined that climatic change impact, apart from some other human activities (irrigation), was the key factor responsible for the shrinkage of the lake size from 25 000 km^2 in the 1960s to the present barely 2000 km^2.

The Niger basin is being studied because it is not providing enough food for its population, and water resources development projects and study is conceptualized as a means of improving the availability of water for food production. Hence, evaluating such a programme will require the knowledge or assessment of the amount and spatial pattern of available water resources and the impact of development projects and postulated climate change upon it.

Water resource assessment and management requires climate-driven hydrological models. In most climate impact assessment studies, conceptual rainfall–runoff (CRR) models have been widely used (Wilby, 2005; Murphy & Charlton, 2006; Charlton *et al.*, 2006). Arora (2001) observed that general circulation models (GCMs) currently in use (Canadian General Circulation Model (CGCM), UK Hadley GCM, and ECHAM) are generally considered to be the most comprehensive models for investigating the physical and dynamic processes of the Earth surface–atmosphere system and they provide plausible patterns of global climate change. However, it is not yet possible to make reliable predictions of regional hydrological changes directly from climate models, due to the coarse resolution of GCMs and the simplification of the hydrological cycle in climate models. The work of Leavesley (1994), as cited by Jackson (2007), has also shown that the use of conceptual models, based on effective parameters, to investigate future scenarios is questionable, because they are based on historic data and therefore only represent the relationship between the input stimuli and output response for the period in which they were developed. However, Jiang *et al.* (2007) still hold the view that conceptual soil water balance models are preferable in assessing the impact of climate change on regional water resources. For example, they asserted that, for assessing water resources management on a regional scale, monthly rainfall–runoff (water balance) models were useful for identifying hydrological consequences of changes in temperature, precipitation and other climate variables. The calibration of these models is much easier than for models with shorter time steps, due to easy access to the required input data. Yates & Strzepek (1998) developed and applied a monthly water balance model of the Nile River basin (WBNILE) for assessing potential climatic change impacts on Nile runoff. The WBNILE model consists of 12 Nile sub-catchments, including characterization of the Lakes Region of equatorial Africa and the Sudd swamp.

A number of researchers have also investigated the potentials of neural networks in modelling watershed runoff based on rainfall inputs. In a preliminary study, Halff *et al.* (1993) designed three-layer feed-forward artificial neural networks (ANNs) using the observed rainfall hyetographs as inputs and hydrographs recorded by the US Geological Survey (USGS) at Bellvue, Washington, as outputs.

Dieulin *et al.* (2006) show the contribution of GIS to hydrological modelling, especially towards the building of grids containing the water holding capacity (WHC) parameter of soils. Also, Maidment & Reed (1996) and Maidment *et al.* (1996, 1997) demonstrated how simple potential evaporation and soil-water budget calculations can be made within ArcView/ArcInfo GIS using monthly data sets for a region in West Africa (Morocco) and East Africa (Uganda) under the FAO/UNESCO Water Balance for Africa project. The monthly soil water surplus is computed with ArcView GIS using the Thornthwaite soil water balance model. Against this background, the study attempts to estimate the water surplus or runoff that indicates the available water resources of the basin and examines how the postulated climate change will impact on the surplus, using simulation models, the Thornthwaite water balance accounting scheme and artificial neural networks (ANNs).

THE STUDY AREA

The study area is the Nigerian sector of the Niger River basin. Nigeria is located in the tropical zone of sub-Saharan West Africa between latitudes 4°N and 14°N and longitudes 2°2′E and 14°30′E, and has a total area of 923 770 km^2. The Niger basin area is about 2.3 million km^2; the river rises in Guinea and flows for a total length of about 4200 km through Mali, Niger and Nigeria before reaching the Atlantic Ocean (Fig. 1(a)). The basin covers five out of the eight hydrological areas of the country otherwise known as sub-basins (Fig. 1(b)). Within Nigeria, the

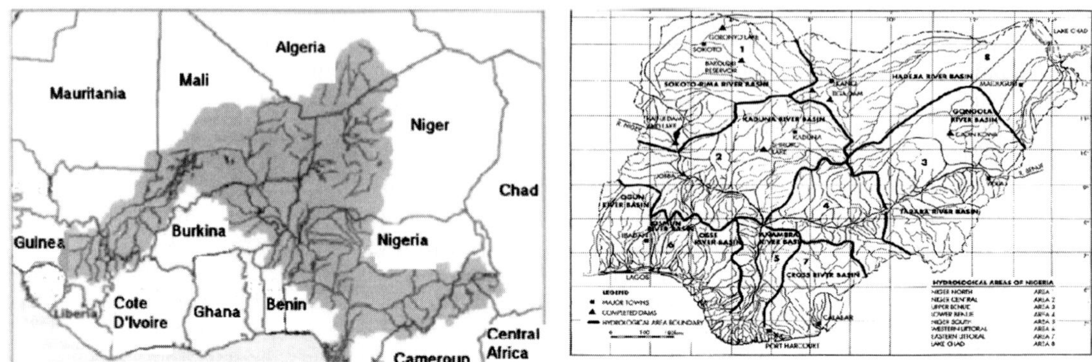

Fig. 1 (a) Catchment area of Niger River. (b) Hydrological zones: Upper Basin Niger[1], Lower Niger[2], Upper Benue[3], Lower Benue[4], NigerSouth5, Western Littoral[6], Eastern Littoral[7], Lake Chad[8]

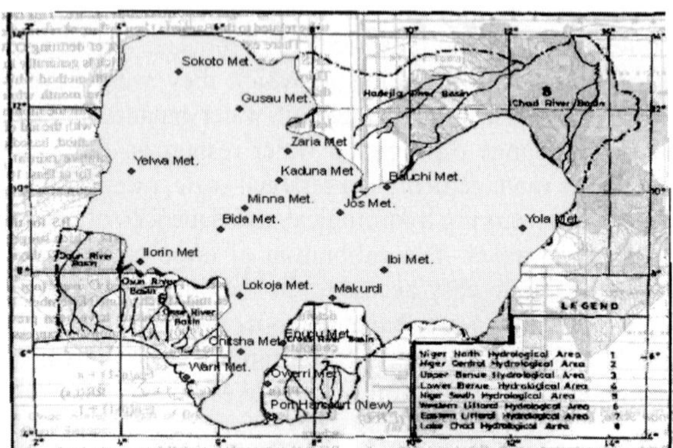

Fig. 2 Basin area coverage and networks of meteorological stations.

Niger River basin covers two thirds (approx. 595 489 km^2) of the total land mass of the country (Fig. 2). The land cover ranges from thick mangrove forests and dense rain forests in the south to a near-desert condition in the northeastern corner of the country. The River Niger is the third longest river in Africa and the fifth largest in terms of discharge.

The climate of the Niger basin is tropical, characterized by high temperatures and humidity, as well as marked wet and dry seasons, though there are variations between south and north. River Niger flows through all the major climatic zones of West Africa, from the wet equatorial belt along the coast to the semi-arid region in the interior. The rainfall pattern of the zones located north of Latitude 8°N exhibits a unimodal rainfall pattern that stretches from March to November with very small amounts of rainfall towards the onset and cessation periods, while the zones located south of latitude 8°N exhibit a bimodal pattern of rainfall that stretches from January to December, but with significant amount of rainfalls towards the onset and cessation periods (Fig. 3). The total rainfall decreases rapidly from the coast northwards. The south has an annual rainfall ranging between 1500 and 4000 mm and the extreme north between 500 and 1000 mm. By contrast, actual evapo-transpiration increases inland so that water deficits are often experienced for varying lengths of time in all but a narrow coastal belt of West Africa. The rainfall regime indeed exercises a direct influence on river regimes in West Africa.

The hydrology of Nigeria is dominated by two great river systems, the Niger-Benue and the Chad systems. The Niger-Benue and Chad river systems are separated by a primary watershed extending northeast and northwest from the Bauchi Plateau which is the main source of their principal tributaries. The River Niger has been particularly prone to the recurrent episodes of

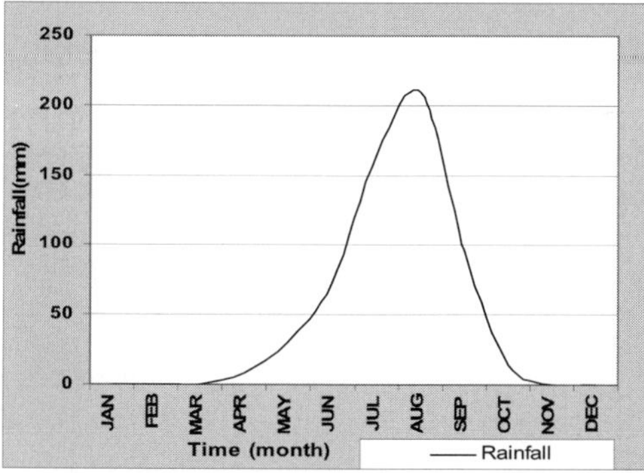

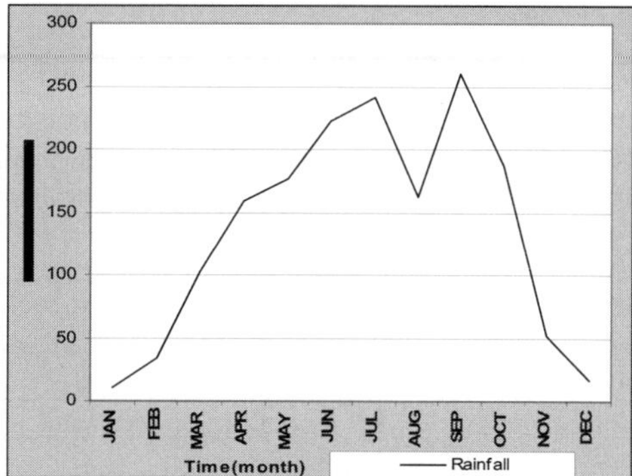

Fig. 3 Monthly rainfall distribution pattern over Nigeria (1940–2001): (a) north of Lat. 08°N; (b) south of Lat. 08°N.

drought and dry spells that have long afflicted the Sudano-Sahelian zone of West Africa. In the recent past too, flood episodes have become a recurrent issue particularly over the lower Niger side. There are two distinct flood phenomena that occur annually in the basin. The first is the "black flood" which originates from high rainfall area in the headwaters. This flood arrives at Kainji (Nigeria) in November and lasts until March at Jebba (Nigeria) after attaining a peak rate of about 2000 m^3/s in February (Oyebande *et al.*, 1980). The second flood, which is only more prominent downstream of Sabongari, soon after the river enters Nigeria, is the "white flood", usually heavily laden with silt and other suspended particles. The flood derives its flow from the local tributaries and reaches Kainji in August in the pre-Kainji Dam on the Niger, attaining peak flows of 4000–6000 m^3/s in September–October in Jebba.

STUDY APPROACH

In this study, the Thornthwaite simple bucket water balance model and artificial neural networks (ANNs) model (Neuralpower software) were used, in combination with hypothetical climate change scenarios and observed historical time series, for the simulation of the hydrological conditions of the Niger basin and assessment of the direct impacts of climate change on water resources and the water resources system by examining the potential changes projected for the river systems; i.e. the current status of the Niger River basin, as well as the projected future state, is considered. The soil water balance model is developed to work with ArcView Avenue programs, and is used to estimate the water surplus or runoff of the basin in the same manner as Maidment *et al.* (1996, 1997) used the model under the FAO/UNESCO project on assessment of water resources of Africa, except in the calibration approach where ANNs are used. The model uses a simple accounting scheme to predict soil-water storage, evaporation and water surplus. The surplus precipitation, which does not evaporate or remain in the soil storage, goes to both surface and sub-surface runoff.

Water balance model description

The water balance model was developed by Thornthwaite (1948) and later revised by Thornthwaite & Mather (1955).The method is essentially a book-keeping procedure, which estimates the balance between the inflow and outflow of water. As shown in Fig. 5, the main inputs into the model are precipitation and potential evaporation, while the main outputs are actual evaporation and water surplus or runoff. The model estimates the potential evaporation using the Priestley-Taylor technique described below. Estimation of evaporation is based upon knowledge of the potential evapotranspiration, available water-holding capacity of the soil, and a moisture extraction function. The general structure of this model is often represented as below to include the monthly time scale:

$$S(t+1) = S(t) + P(t) - E(t) - Q(t) \tag{1}$$

where $S(t)$ represents the amount of soil moisture stored at the beginning of time interval t; $S(t+1)$ represents the storage at the end of that interval; flow across the control surface during the interval consists of precipitation $P(t)$, actual evapotranspiration $E(t)$ and soil water surplus $Q(t)$.

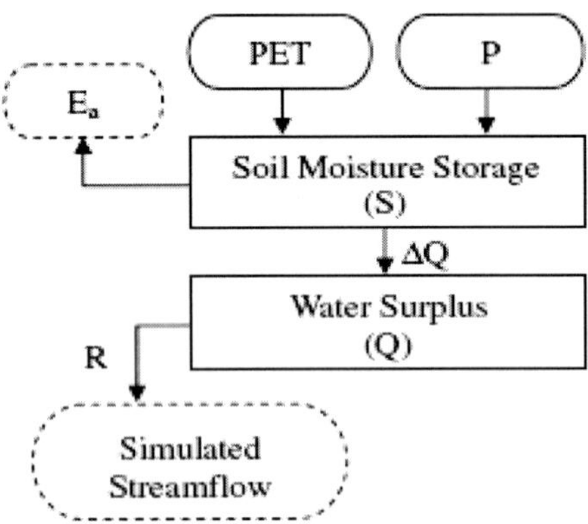

Fig. 4 Schematic representation of the Thornthwaite model.

Sources of data

Gridded digital data sets of mean monthly temperature and precipitation, radiation, runoff and political boundaries were required and obtained from ftp.crwr.utexas.edu in the directory /pub/crwr/gishydro/Africa (The Centre for Research in Water Resources (CRWR), University of Texas at Austin, USA, has an anonymous ftp site where anyone can log in and download data). The precipitation, temperature and radiation data are interpolated to a 0.5 degree grid spanning the Earth (720 cells east–west and 360 cells north–south, 259 200 cells in all), covering both the land and the oceans. On each grid cell is presented the mean monthly and mean annual values of each variable in ArcView GIS shapefiles (a special data format that was created for ArcView and which is not readily handled by Arc/Info). These global data sets are from the "Global Air Temperature and precipitation Data Archive" compiled by D. Legates and C. Willmott (Legates & Willmott, 1990; Maidment *et al*., 1997).

Soil water budget methodology

Where detailed data about soil layers, depth to groundwater and vegetation are not available, hydrologists have often resorted to simple bucket models and budgeting schemes to model near-surface hydrology. The conservation of mass equation for soil water can be written as follows (Thornthwaite, 1948):

$$Q = P - E - \Delta w/\Delta t \tag{2}$$

where Q is the surplus, P is precipitation, E is evaporation, w is soil moisture, and t is time.

An Avenue script (Reed, 1996), based on WATBUG (Willmott, 1977), is used to calculate the soil water budget. The actual calculations are done on a daily basis in the program. Values of the monthly precipitation and potential evaporation are divided by the number of days in a month to give average daily values. The runoff or water surplus in each day is found as the product of precipitation and the ratio of the actual soil moisture level to the soil moisture capacity. The evaporation on each day is the product of the potential evaporation and the ratio of actual soil moisture level to the soil moisture capacity. A trial calculation is done first to find the end of day storage and if this exceeds the soil moisture capacity, the excess soil moisture is added to the water surplus and storage is reset to the soil moisture capacity before the next day's computations are done. The climate data used are mean monthly values for the period of record. The computations are done for the 12-month representative period several times in sequence, until the computed soil moisture levels in each month stabilize to constant values. These values represent expected monthly soil moisture levels. Surpluses are excess rainfalls that are available for streamflow generation. The generated surpluses can be transferred into subwatersheds of the Niger basin and routed through its river system using the Muskingum-Cunge method or response function approach, or the two-step flow routing approach (Maidment *et al.*, 1997). In this study, however, an artificial neural network (ANN) has been employed in place of the water surplus routing process. ANN, being an empirical and black-box model, has an in-built capability that takes into consideration the catchment characteristics during simulations. Further details on this are given below under calibration procedure adopted.

Methods for model calibration and validation

Since some of the parameters of the model cannot be directly determined from field measurement or estimated from catchment characteristics, they have been estimated by model calibration. There are many objective functions (OF) that could be used for model calibration, but in this study, the model parameters are optimized by minimizing the values of the objective function given by the equation:

$$\mathrm{OF} = \sum_{t=1}^{n} (qobs_t - qsim_t)^2 \tag{3}$$

where $qobs_t$ and $qsim_t$ are the observed streamflow and simulated streamflow, respectively of month t, and n is the total number of months of simulation.

The Simplex method is used in non-linear programming form in this study, for the model parameter optimization. The Simplex method is a local, direct search algorithm that has been commonly used for conceptual hydrological models of this type (Johnson & Pilgrim, 1976; Gan & Biftu, 1996; Hendrickson *et al.*, 1988; Jiang *et al.*, 2007).

It is well-known that the first requirement of a model is that it should have the ability to reproduce the mean value of observed streamflow. However, it is also a known fact that the mean value does not fully indicate how well the individual simulated values match observed values. In order to overcome this limitation, the root mean square error (RMSE) and/or coefficient of efficiency or Nash-Sutcliffe efficiency E (Nash & Sutcliffe, 1970) is considered. RMSE is simply the average of the squared errors for all simulation results and provides an objective measure of the difference between observed and simulated values:

$$\mathrm{RMSE} = 1/n\sqrt{\sum_{t=1}^{1} (qobs_t - qsim_t)^2} \tag{4}$$

$$E = \sum_{t=1}^{n} (qobs_t - \overline{qobs})^2 - \sum_{t=1}^{n} (qobs_t - qsim_t)^2 \bigg/ \sum_{t=1}^{n} (qobs_t - \overline{qobs})^2 \tag{5}$$

where $qobs$ is the mean of observed discharge, E is a dimensionless coefficient for measuring the degree of association between the observed and simulated values, i.e. efficiency. The value of E is always less than unity; a value equal to unity would represent a perfect agreement between observed and simulated streamflows.

Notably, there are a number of ways of modelling or linking the relationship between the causative factor, rainfall, to runoff observed at a particular site of a basin by using equations which describe the major physical processes involved in the transformation or by representing these component processes in a conceptual manner, or alternatively using the neural networks approach. Considering the impossibility of representing the component processes of transformation process using a physics-based approach, one may opt for either the conceptual representation or the neural network modelling approach. While calibrating the conceptual model or the neural network model using the past recorded input with the corresponding outflows observed at a specific location of a river, it is implicit that when the calibrated model is applied for future predictions, the input that would be used in the model is in the same range of the input used for the calibration.

However, it is important to state that when the range of future input that would be applied in the calibrated model is far away from that of the input used for calibration, then the calibrated models of both types would not serve the intended purpose. In this case, the conceptual model parameters require modification and in the case of a neural network model, the entire network structure of the model needs to be changed.

For this reason, a simpler approach of model-to-model calibration technique, i.e. using one model to calibrate another model, has been adopted. In the same manner, Reeds & Maidment, (1996) and Maidment *et al.* (1997) linked an estimated water surplus or runoff of a water balance model to a conceptual watershed model calibrated using observed average monthly discharges. The estimated water surpluses from the same type of water balance model have been linked to the observed average monthly discharges with an artificial neural networks (ANNs) calibrated using the observed average monthly river discharges of the sub-basins. To arrive at this point, observed mean monthly values of river discharges of the 96 month period, 1991–1998, together with other mean monthly values of climatic variables of the Niger basin have been used as an input to calibrate the ANN model parameters. Thereafter, the verification of ANN simulation output values obtained using the same input data (i.e. rainfall, actual evaporation and change in soil storage, 1991–2000) of the water balance model were in turn used for the optimization of the water balance model parameters. The verification output in this case, is taken to be as good as observed runoff from the Niger basin.

The choice of ANNs and justification of this approach is based on the following reasons:

- ANNs being empirical models, are black-box and/or deterministic models, which do not consider randomness.
- Since a deterministic model does not consider randomness, a given input always produces the same output.
- Most hydrological variables have little or no randomness, especially when dealing with large river basins like the Niger basin, due to basin storage effects. A large basin area entails having a large storage which minimizes any randomness.
- ANNs by their deterministic nature mimic the behaviour of the basin by linking the causative factor, i.e. rainfall, excess rainfall, etc., with the effect factor, i.e. the runoff at the basin outlet. Hence, it does not require going through the rigorous and long process of using two different models, i.e. a watershed model for the transformation of the water surplus or runoff to local runoff and surface flow routing model for the routing of the runoff through the river systems.

For the training of the networks, the learning algorithm used is the standard back propagation, also called incremental back propagation (IBP). The network connection type is multi-layer normal feed-forward, while the total number of layers is 15. The node number of input layer is 14 (consisting of first hidden layer = 7, and second hidden layer = 7) and the output layer is 1. The transfer function used is sigmoid and transfer function slope is 1; while RMSE is the objective function. To further verify the ANNs performance and efficiency, Nash-Sutcliffe coefficient (E) was computed as well. Figure 5 shows the neural networks structure and weight distribution used in the training of the networks during the calibration simulation. The simulation is terminated when a reasonable coefficient of correlation (R), say 0.96 and above, and a reasonably low value of RMSE are achieved.

Climate change scenarios

In this study, synthetic or hypothetical climate change scenarios are adopted, and applied in the same manner as Jiang *et al.* (2007). The hypothetical scenarios are shown in Table 1. Synthetic scenarios describe techniques where particular climatic (or related) elements are changed by a realistic but arbitrary amount, often according to a qualitative interpretation of climate model simulations for a region. Most studies have adopted synthetic scenarios of constant changes

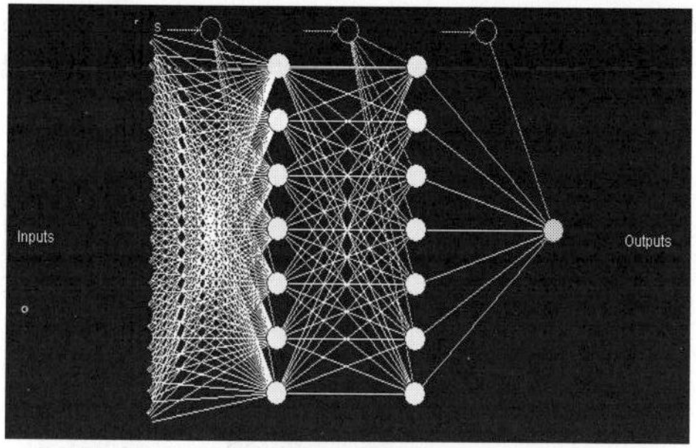

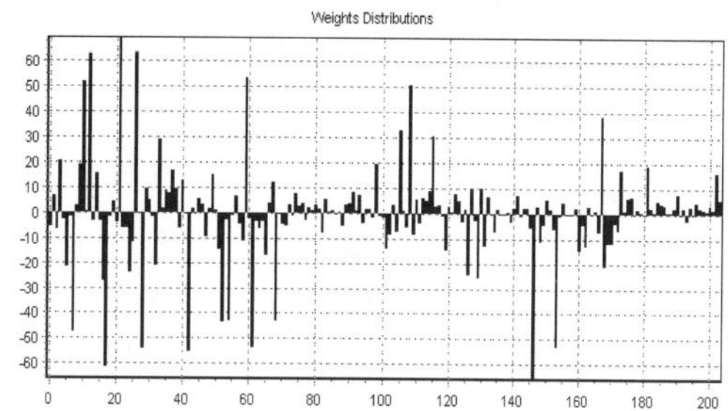

Fig. 5 (a) Typical architecture of the neural networks. (b) Typical weight distribution of neural networks structure.

Table 1 Hypothetical climate change scenarios (source: Jiang *et al.*, 2007).

Scenario no.	1	2	3	4	5	6	7	8	9	10	11	12	13	14	15
ΔT (°C)	1	1	1	1	1	2	2	2	2	2	4	4	4	4	4
ΔPrec (%)	–20	–10	0	10	20	–20	–10	0	10	20	–20	–10	0	10	20

throughout the year (e.g. Rosenzweig *et al.*, 1996), but some have introduced seasonal and spatial variations in the changes (e.g. Rosenthal *et al.*, 1995), and others have examined arbitrary changes in inter-annual, within-month and diurnal variability as well as changes in the mean (e.g. Mearns *et al.*, 1992, 1996; Semenov & Porter, 1995).

The expression for calculating the percentage change or differences in the runoff is:

Change in Runoff = Predicted Runoff – Baseline Runoff

Percentage (%) Change in Runoff = (Change in Runoff / Baseline Runoff) × 100

The baseline Runoff was estimated using the WMO standard reference years or baseline (1961–1990) period temperature, rainfall, evaporation, and change in soil storage values as input. The temperature parameter was used to estimate evaporation, which now served as the required input along with rainfall and change in soil storage.

RESULTS AND DISCUSSION

The results of the water balance simulation for the various sub-basins in Fig. 6 show a direct link between rainfall and water surplus generated over the Niger basin. The peak of the rainfall coincides with the same month as the peak of water surplus and *vice versa.* Since water surplus indicates the available water resources of the region, this reveals how sensitive the water resources are to the climate of the region. Over the Upper Niger sub-basin, a peak rainfall of 265 mm in the month of August, produced a runoff or water surplus of 160.5 mm; while over Lower Niger and Niger South sub-basins, peak rainfall of 294 mm and 347 mm yielded water surplus of 185 mm and 238 mm, respectively. Also, over the Lower and Upper Benue, peak rainfalls of 318 and 233 mm yielded runoff of 209.9 and 130 mm, respectively. Furthermore, it is observed that the predicted water surplus volumes are much more sensitive to the accuracy of potential evaporation estimate in the humid climate (Niger South) than the arid climate (Upper Niger). The surplus is more when the potential evapotranspiration is less and *vice versa*; unlike over the arid region. Conversely, the actual evapotranspiration is seen to have a direct relationship with the soil moisture storage. Further results show that annual water surplus is highest over Niger South (1241.2 mm), followed by the Lower Benue sub-basin (973.6 mm), Lower Niger sub-basin (729.4 mm), Upper Benue sub-basin (495.3 mm), and the smallest value is observed over the Upper Niger sub-basin (360.7 mm).

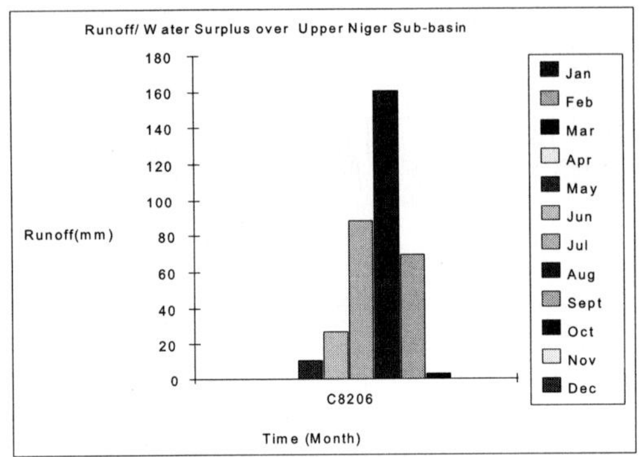

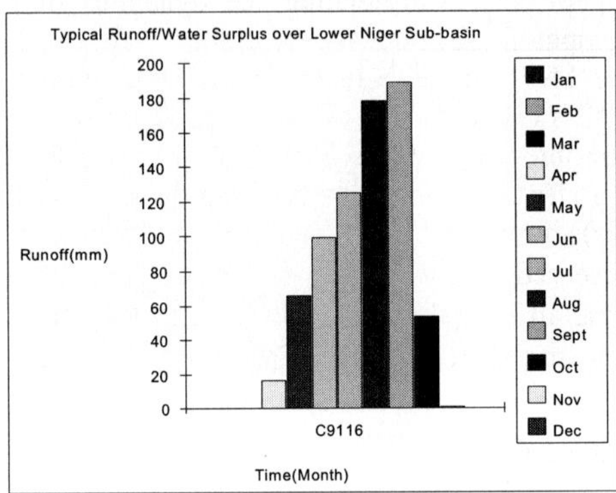

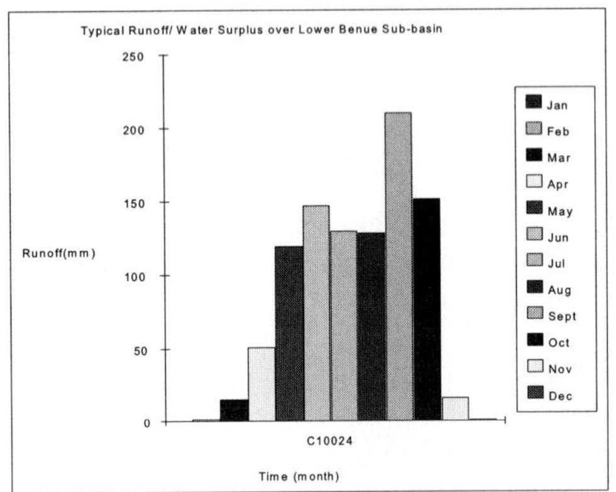

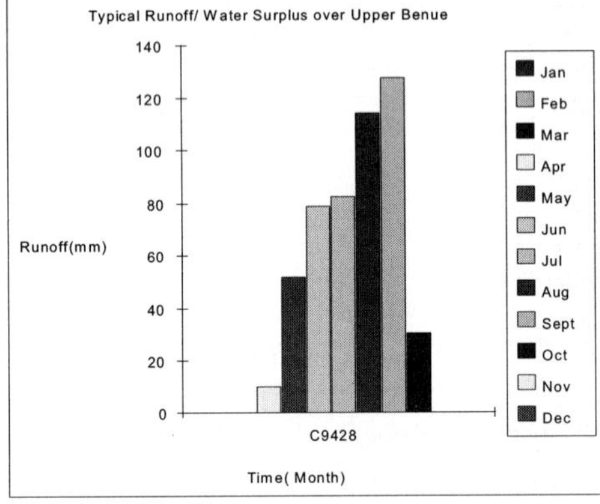

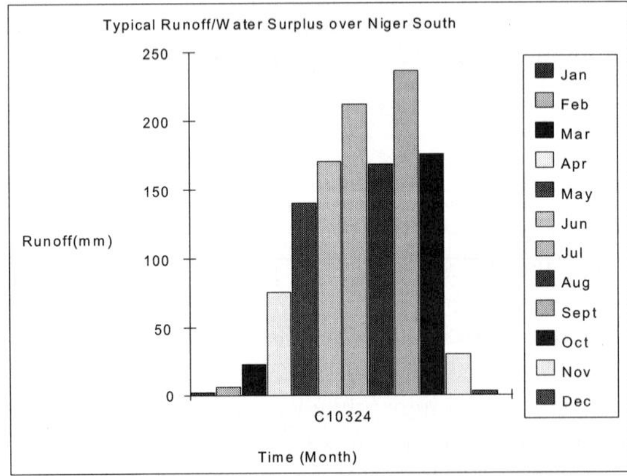

Fig. 6 Typical simulated runoff over the Niger River sub-basins.

Both statistical and visual comparisons of observed and simulated values were conducted, and the performance of the models evaluated. Table 2 gives a summary of the statistics of simulations for both calibration and validation or verification periods. The calibration and optimization results show that the models performance has been quite good, as adjudged by the values of RMSE and *E* in Table 2 for both calibration and validation periods. The highest RMSE values and lowest *E* values were observed over the Niger South and Lower Niger sub-basins, respectively. These results seemed to demonstrate the basic capability of the model to reproduce long-term mean monthly runoff in the Niger River basin. Also, in order to further demonstrate the models' capabilities in simulating the dynamics of monthly runoff series, the monthly runoff values computed using the water balance model is analysed and correlated with the runoff estimated by the ANNs model using a linear regression equation $Y = mX + c$. In the equation, Y represents the ANNs calculated runoff and X is the observed runoff and m and c are constants representing slope and intercept, respectively. The scatter plot and the results of regression analysis are shown in Fig. 7(a), while the visual comparison of the calibration simulation is shown in Fig. 7(b). Figure 7(a) shows that: (1) as far as R^2 values are concerned, the model's predictions correlated well with observed values resulting in R^2 values ≥ 0.95 in all the sub-basins; and (2) the values of c are generally less than 10% of the mean monthly values, and the slopes (m) are all within ±10% of the 1:1 line (i.e. angle of 45°). It suffices to say, therefore, that the models have shown good

Table 2 Results of the model in reproducing time series runoff in Niger Basin (1991–1998).

Basin	Equations	Coeff. Corr. (R)	RMSE	NS-efficiency (E)
Niger	$X = 0.85P - 0.70E - 0.49D - 0.1060$	0.9679	11.663	0.9453
Linear regression fitting				
Niger	$Y = 0.8725X + 15.617$	0.9632	15.6578	0.9991
Sub-basin				
Upper Niger	Model verification (1999–2000)		12.6909	0.9362
Lower Niger			8.4665	0.9578
Lower Benue			3.8248	0.9982
Upper Benue			7.3819	0.9689
Niger South			4.1315	0.9987

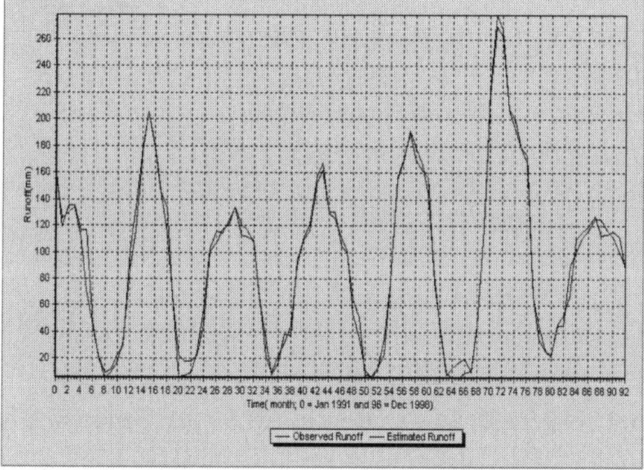

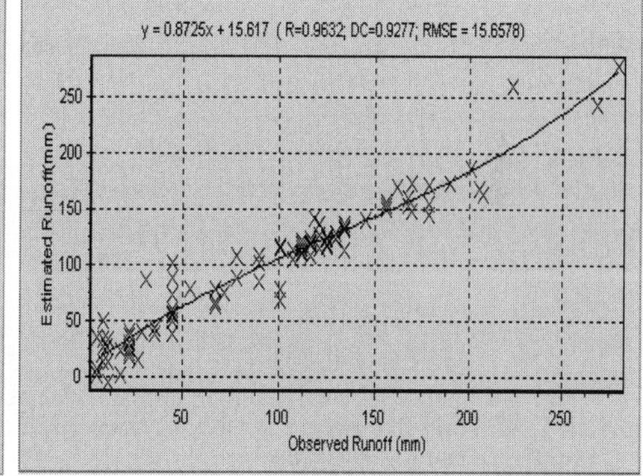

Fig. 7 (a) ANNs model calibration simulation output. (b) Scatter plot of observed and estimated runoff over Upper Niger.

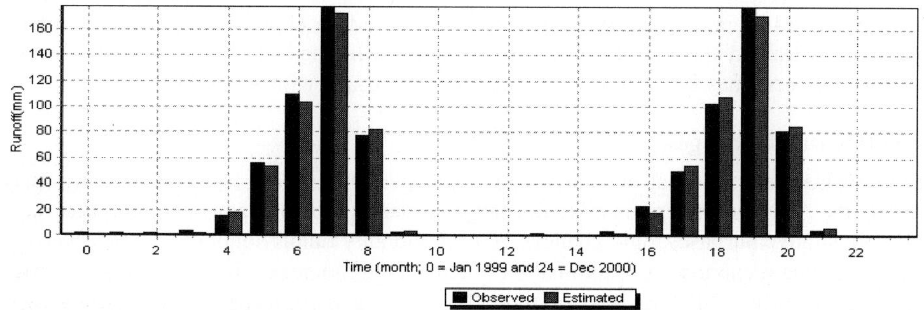

Fig. 8 Model verification result over the Upper Niger.

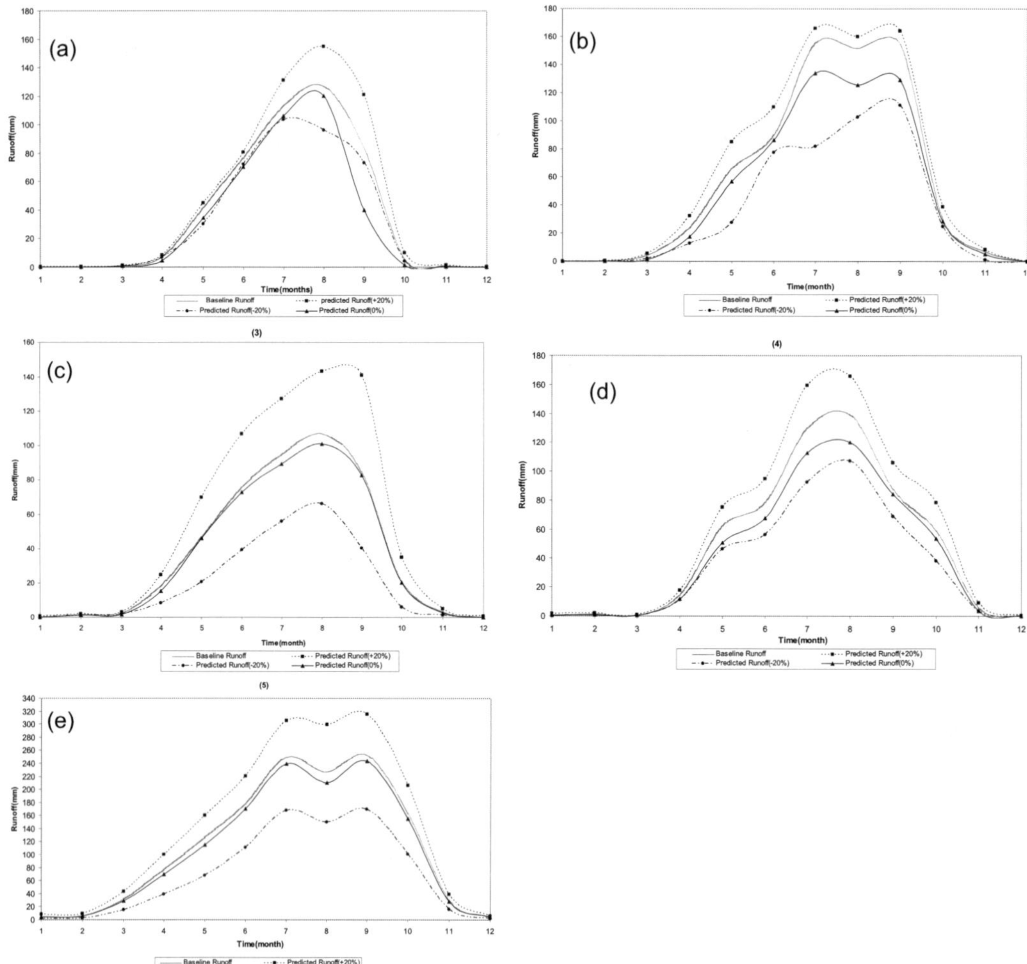

Fig. 9 Changes in runoff under climate change scenarios of temperature = +2°C and precipitation = $\pm$ 20% and 0% over: (a) Upper Niger, (b) Lower Niger, (c) Lower Benue, (d) Upper Benue, and (e) Niger South, sub-basins.

capabilities to reproduce historical monthly runoff series with an acceptable accuracy proved by verification results (Fig. 8). However, the assumptions made and the absence of intermediate component processes of rainfall–runoff transformation should be kept in mind; hence estimated observed values obtained through the ANN model have been used in place of actual observed runoff. Therefore, there may be need to further verify these results with actual field data.

Mean monthly runoff changes

In order to evaluate the seasonal and inter-annual changes, differences in mean monthly runoff simulated by the ANNs with various climate change scenarios are compared with baseline runoff values. For illustrative purposes, only changes in mean monthly runoff simulated by the ANNs for three climate change scenarios (i.e. combination of temperature increases by 2°C ($\Delta T = +2°C$) and precipitation changes by ±20% and 0%) are plotted in Figs 9 and 10, respectively. Figure 9 shows various changes in runoff expected over the Niger basin in the face of the postulated climate

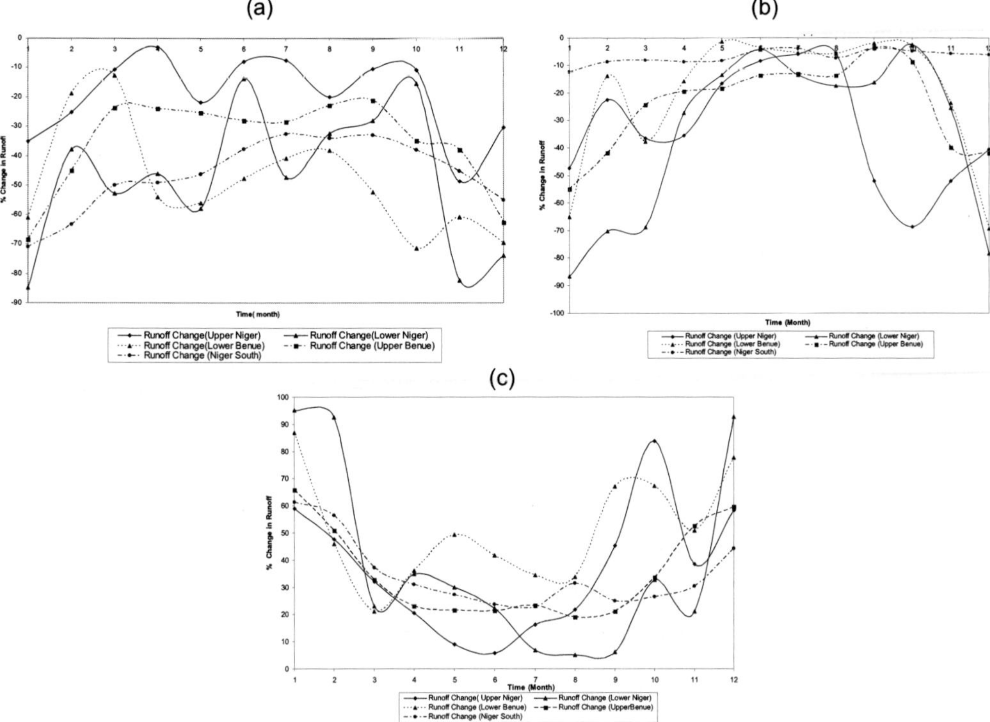

Fig. 10 Comparison of mean monthly changes in runoff simulated by ANNs for three climate change scenarios: (a) ΔT = +2°C and ΔPrec = –20%, (b) ΔT = +2°C and ΔPrec = 0%, and (c) ΔT = +2°C and ΔPrec = 20%.

Table 3 Summary of expected future changes in runoff.

Sub-basins	ΔT = +2°C ΔPrec = –20%	ΔT = +2°C ΔPrec = 0%	ΔT = +2°C and ΔPrec = 20%
Upper Niger	–5 to –50%	–3 to –45%	5 to 65%
Lower Niger	–10 to –55%	–5 to –50%	5 to 60%
Lower Benue	–10 to –50%	–2 to –50%	2 to 55%
Upper Benue	–20 to –55%	–4 to –55%	20 to 65%
Niger South	–30 to –50%	–4 to –13%	20 to 60%

change; while Fig. 10 shows: (a) larger differences in the percent changes in runoff during the dry season (winter period), this may be attributed to smaller absolute values of runoff during the dry season; and (b) the model responded very well in predicting changes in monthly runoff. Average mean monthly runoff changes are shown in Table 3. These results are also in agreement with the findings of Jiang *et al.* (2007) over Dongjiang basin, China.

From Figs 9 and 10, and Table 3, it is obvious that flood magnitude may increase most over the Lower Benue, followed by the Upper Niger and Niger South sub-basins in future, if the precipitation increases by 20% as projected, especially between the months of July and September. Taking account of the projected population growth of Nigeria, with the population of the region projected to double by 2020, existing primary sources of water supply are likely to become further stretched over the coming years. Also, the current prevailing critical drought conditions in the Sahel and Sudan savanna regions of Nigeria, may be exacerbated by the projected climate change, if the precipitation decreases by (–20%) in the future. It is further observed that the sensitivity of the sub-basins to climate change varies; with Upper Niger sub-basin being highly sensitive followed by Lower Benue and Upper Benue sub-basins.

ADAPTATION ACTIONS

In order to minimize the negative impact of climate change on the ecosystems, water resources and socio-economic domains, a number of adaptation measures are open to Nigeria. These adaptation measures are mainly centred on new water resources and conservation efforts. The main adaptation measures and nonconventional sources of water that can be exploited in the future are summarized in Table 4.

Table 4 Technical adaptation measures and nonconventional water resources.

Adaptation measure	Potential benefits	Best uses
Conservation	Curbs water demand increase	Domestic, industrial, agricultural demand reduction
Weather and climate forecast needs	Helps to curb or reduce flood and drought disasters	Efficient and effective dams operation in water releases and agriculture
Use of surplus rainy season runoff	Collectable runoff can constitute up to 10% rainfall (Bou-Zeid & El-Fadel, 2002)	Irrigation, aquifer recharge
Wastewater reclamation	All collected wastewater can be re-used	Irrigation, sanitation use
Seawater/brackish water desalinization	Provides unlimited water supply	Domestic, industrial
Rainfall enhancement by clouds seeding with silver iodide crystals	Can increase precipitation up to 15% in arid regions (Bou-Zeid & El-Fadel, 2002)	Irrigation, aquifer recharge
Ecological civilization	Helps in being conscious of the available Earth resources and avoiding ecological disasters and unnecessary wastages or over-exploitation	Pollution control and resources management

CONCLUSION

It is observed that when temperature increases by 2°C, the mean monthly runoff on the average changes by –10 to –50%, –5 to –40% and 15 to 60% respectively for precipitation changes of –20%, 0% and 20%. If the prevailing climatic conditions of increasing temperature trend and decreasing rainfall trends continue unabatedly, postulated climate change may exacerbate its impacts on the water resources of the region, resulting in water stress condition.

REFERENCES

Arora, V. K. (2001) Streamflow simulations for continental-scale river basins in a global atmospheric general circulation model. *Adv. Water Resour.* **24**, 775–791.

Charlton, R., Fearly, R., Moore, S., Sweeney, J. & Murphy, C. (2006) Assessing the Impact of climate change on water supply and flood hazard in Ireland using statistical downscaling and hydrological modeling techniques. *Climatic Change J.* **74**(4), 475–491.

Dieulin, C., Boyer, J. F., Ardoin-Bardin, S. & Dezetter, A. (2006) The contribution of GIS to hydrological modelling. In: *Climate Variability and Change – Hydrological Impacts* (ed. by S. Demuth *et al.*) (Proc Fifth FRIEND World Conference, Havana, Cuba, November 2006), 68–74. IAHS Publ. 308. IAHS Press, Wallingford, UK.

Döll, P. & Flörke, M. (2005) Global scale estimation of diffuse groundwater recharge. Frankfurt Hydrology paper 03, Institute of Physical Geography, Frankfurt University, Germany. http://www.geo.uni-frankfurt.de/jpg/ag/dl/f_publikationen/2oo5/FHP_03 Doell_Floerke_2005.pdf [last accessed 8 January 2008].

Gan, T. Y., & Biftu, G. B. (1996) Automatic calibration of conceptual rainfall–runoff models: optimization algorithms, catchment conditions, and model structure. *Water Resour. Res.* **32**(12), 3513–3524.

Halff, A. H., Halff, H. M. & Azmoodeh, M. (1993) Predicting runoff from rainfall using neural networks. In: *Proc. Engng Hydrol. ASCE* 768–765.

Hendrickson, J., Sorooshian, S. & Brazil, L. E. (1988) Comparison of Newton-type and direct search algorithms for calibration of conceptual rainfall-runoff models. *Water Resour. Res.* **24**(5), 691–700.

Intergovernmental Panel on Climate Change, Working Group I (IPCC-WGI) (1996a) *Climate Change 1995: Impacts, Adaptations and Mitigation of Climate Change, Scientific and Technical Analysis*. Cambridge University Press, Cambridge, UK.

Intergovernmental Panel on Climate Change, Working Group I (IPCC-WGI) (1996b) *The Science of Climate Change*. Cambridge University Press, Cambridge, UK

IPCC (2001a) *Climate Change 2001: Impacts Adaptations and Vulnerability*, Contribution of Working Group II to the Third Assessment Report, Cambridge University Press, Cambridge, UK.

IPCC (2007) Summary for policy-makers. In: *Climate Change 2007: The Physical Science Basis*. Contributions of Working Group I to the Fourth Assessment Report of the Intergovernmental Panel on Climate Change.

Jackson, C. (2007) *Effects of Climate Change on Groundwater Resources – Previous and future Research*. British Geological Survey, Natural Environment Research Council, UK.

Jiang, T., Chen, Y. D., Xu, C.-Y., Chen, X. H., Chen, X. & Singh, V. P (2007) Comparison of hydrological impacts of climate change simulated by six hydrological models in the Dongjiang Basin, South China. *J. Hydrol.* **336**, 316–333.

Johnston, P.R., & Pilgrim, D.H., (1976): Parameter optimization for watershed models. *Water Resour. Res.* **12**, 477–486.

Kundzewicz, Z. W., Mata, L. J., Arnell, N., Döll, P., Kabat, P., Jiménez, B., Miller, K., Oki, T., Şen, Z. & Shiklomanov, I. (2007) Freshwater resources and their management. In: *Climate Change 2007: Impacts, Adaptation and Vulnerability*, Contribution of Working Group II to the Fourth Assessmnent Report of IPCC (ed. by M. L. Parry, O. F. Canziani, J. P. Palutikof, P. J. Van der Linden & C. E. Hanson), 173–210. Cambridge University Press, Cambridge, UK. http://www.ipcc.ch/pdf/assessment-report/ar4/wg2/ar4-wg2-chapter3.pdf [last accessed 8 January 2008].

Kundzewicz, Z. W., Mata, L. J., Arnell, N. W., Döll, P., Jimenez, B., Miller, K., Oki, T., Sen, Z. & Shiklomanov, I. (2008) The implications of projected climate change for freshwater resources and their management. *Hydrol. Sci. J.* **53**(1).

Leavesley, G. H. (1994) Modeling the effects of climate change on water resources: A review. *Climate Change* **28**, 159–177.

Legates, D. R. & Willmott, C. J. (1990) Mean seasonal and spatial variability in gauge-corrected global precipitation. *Int. J. Climatol.* **10**, 111–127.

Maidment, D. R. & Reed, S. (1996) Introduction to ArcView GIS in West Africa. FAO/UNESCO Water Balance of Africa

Maidment, D. R., Reed, S., Olivera, F. & Martinez, K. (1996) Soil water balance in West Africa. FAO/UNESCO Water Balance of Africa.

Maidment D. R., Reed, S. M., Akmansoy, S., Mckinney, D. C., Olivera, F. & Ye, Z. (1997) Water balance of the Niger River Basin in West Africa.

Maidment D. R., Reed, S. & Olivera, F. (1997) Soil water balance in East Africa. FAO/UNESCO Water Balance of Africa.

Mearns, L. O., Rosenzweig, C. & Goldberg, R. (1992) The effect of changes in interannual climatic variability on Ceres-wheat yields: sensitivity and 2 CO studies. *J. Agric. For. Met.* **62**, 159–189.

Mearns, L. O., Rosenzweig, C. & Goldberg, R. (1996) The effect of changes in daily and interannual climatic variability on Ceres-wheat: a sensitivity study. *Climatic Change* **32**, 257–292.

Milly, P. C. D., Dunne, K. A. & Vecchia, A. V. (2005) Global pattern of trends in streamflow and water availability in a changing climate. *Nature* **438**, 347–350.

Murphy, C. & Charlton, R. (2006) Climate change impact on catchment hydrology and water resources for selected catchments in Ireland. National Hydrology Seminar, 2006.

Nash, J. E. & Sutcliffe, J. V. (1970) River flow forecasting through conceptual models: Part 1, A discussion of principles. *J. Hydrol.* **10**(3), 282–290.

Okpara, J. N. (2007) Monitoring the impacts of climate change/variability and anthropogenic activities on Lake Chad, Northeastern Nigeria, using remote sensing satellite. Proceedings of IUGG2007 International Conference, Perugia, Italy.

Oyebande, L., Sagua, V. O. & Ekpenyong, J. L. (1980) Effects of Kainji Dam on the hydrological regime water balance and water quality of the Niger River. In: *The Influence of Man on the Hydrological Regime with Special Reference to Representative and Experimental Basins*, 215–220. IAHS Publ. 130. IAHS Press, Wallingford, UK. http://www.iahs.info/redbooks/130.htm.

Reed, S. (1996) Soil-Water Budget part I: Methodology and results. http://www.ce.utexas.edu/ prof/maidment/GISHydro/seann /explsoil/method.htm.

Reed, S. M. & Maidment, D. R. (1996) Water balance model for the Souss Basin. Water Balance of Africa. FAO / UNESCO.

Rosenthal, D. H., Gruenspecht, H. K. & Moran, E. A. (1995) Effects of global warming on energy use for space heating and cooling in the United States. *Energy J.* **16**(2), 77–96.

Rosenzweig, C., Phillips, J., Goldberg, R., Carroll, J. & Hodges, T. (1996) Potential impacts of climate change on citrus and potato production in the US. *Agric. Systems* **52**, 455–479, doi:10.1016/0308-521X (95)00059-E.

Rosenzweig, C., Casassa, G., Imeson, A., Karoly, D. J., Liu Chunzhen, Menzel, A., Rawlins, S., Root T. L., Seguin, B. & Tryjanowski, P. (2007) Assessment of observed changes and responses in natural and managed systems. In: *Climate Change 2007: Impacts, Adaptation and Vulnerability*, Contribution of Working Group II to the Fourth Assessmnent Report of IPCC (ed. by M. L. Parry, O. F. Canziani, J. P. Palutikof, P. J. Van der Linden & C. E. Hanson), 79–131. Cambridge University Press, Cambridge, UK.

Semenov, M. A. & Porter, J. R. (1995) Climatic variability and the modelling of crop yields. *J. Agric. For. Met.* **73**, 265–283.

Thornthwaite, C. W. (1948) An approach towards a rational classification of climate. *Geophys. Rev.* **38**, 55–94.

Thornthwaite, C. W. & Mather, J. R. (1955) The water balance. *Publications in Climatology* **8**(l). C. W. Thornthwaite & Associates, Centerton, New Jersey, USA.

Wilby, R. L. (2005) Uncertainty in water resource model parameters used for climate change impact assessment. *Hydrol. Processes* **19**(16), 3201–3219.

Willmott, C. J. (1977) WATBUG: A FORTRAN IV algorithm for calculating the climatic water budget. *Publications in Climatology* **30**(2). C. W. Thornthwaite & Associates, Centerton, New Jersey, USA.

Yates, D. N. & Strzepek, K. M. (1998) Modeling the Nile under climatic change. *J. Hydrol. Engng ASCE* **3**(2), 98–103.

Integration of hydrological modelling with artificial intelligence tools for an agricultural watershed in India

T. V. RESHMIDEVI[1], T. I. ELDHO[2] & R. JANA[3]

1 *Centre for Inter-disciplinary Studies in Environment and Development, ISEC campus, Nagarbhavi, Bangalore 560072, India*
reshmidevi@isec.ac.in

2 *Department of Civil Engineering, Indian Institute of Technology Bombay, Powai, Mumbai 400076, India.*

3 *Environmental Assessment Division, Bhabha Atomic Research Center, Mumbai 400085, India.*

Abstract In this paper, a fuzzy membership-based approach is presented to estimate the surface water potential of an area for supplementary irrigation in terms of both soil moisture stress and feasibility for supplementary irrigation from the nearby surface water bodies. The model is applied to the Gandheshwari sub-watershed in West Bengal, India. With the help of a geographic information system (GIS), a raster-based analysis is performed. The area is divided into cells of 20 m × 20 m spatial resolution. Remote sensing satellite imageries are used to identify the agricultural areas as well as to locate the spatially distributed water bodies in the area. The Natural Resources Service (previously known as Soil Conservation Service) Curve Number-based method is modified and used for the simulation of soil moisture conditions in the agricultural fields. Further, a fuzzy membership-based approach is used to aggregate various attributes related to the surface water potential.

Key words artificial intelligence; fuzzy logic; geographic information system; soil moisture; supplementary irrigation

INTRODUCTION

Continuously depleting water resources and dry spells occurring within the crop growth period are making sustainable management of water resources a challenging task for the decision makers. Water resource development and management includes several hydrological, environmental and social concerns, and hence is identified as a multi-criteria evaluation (MCE) problem. It often needs extensive analysis of the data supported by complex modelling techniques to understand the implications of any proposed management action. Incorporation of heuristic and indigenous information improves the local acceptability of the scientific recommendations. Due to their capability in representing information by using linguistic terms, artificial intelligence (AI) tools such as fuzzy logic are now widely used to develop decision support tools in land and water resource management (Wang *et al.*, 1990; Ahamed *et al.*, 2000; Bojörquez-Tapia *et al.*, 2001; Sicat *et al.*, 2005; Gemitzi *et al.*, 2006; Shaban *et al.*, 2006; Nasiri *et al.*, 2007; Shreshta *et al.*, 2007). This paper presents the application of a fuzzy membership-based approach for identifying the suitability of the agricultural fields for supplementary irrigation by considering the surface water potential alone.

Due to the erratic distribution of rainfall, soil moisture stress and the resulting crop loss are critical problems in rainfed agricultural fields, especially in the semi-arid and sub-humid regions. With almost 60% of the agricultural areas under rainfed cultivation, agricultural drought is a serious problem in India, resulting in crop loss of millions of rupees every year. Dry spells occurring during the critical crop growth period often make supplementary irrigation unavoidable, even during the monsoon period. Runoff harvesting and supplementary irrigation from harvested water have been found effective in reducing the soil moisture stress during the critical crop growth period over many parts of the world (Fox & Rockstorm, 2003; Panigrahi *et al.*, 2005; Oweis & Hachum, 2006; Kar *et al*, 2006; Ashraf *et al.*, 2007). The success of the supplementary irrigation depends on the water availability at the time when irrigation is needed, its availability near the demand point as well as the fertility of the land. Other than complex mathematical models representing the hydrology of the watershed, heuristic information is usually essential to arrive at decisions related to supplementary irrigation assessment and management. Such information is generally expressed in qualitative terms, rather than precise numeric values. Due to the uncertainty in the suitability of attribute values as well as in the suitability criterion itself, suitability evaluation becomes difficult to achieve using traditional modelling approaches.

Using the concept of partial membership to represent the uncertainty, fuzzy membership-based approaches have been widely applied for developing decision support tools related to land suitability evaluation. Wang *et al.* (1990) proposed a fuzzy membership-based approach to evaluate the land suitability for different crops. In this study, the sharp boundary between the suitable and non-suitable classes was replaced by different suitability classes. Input attributes and the suitability criteria were classified into these classes by using the membership values. The concept of Euclidean distance in the vector feature space was used to aggregate the attribute suitability criteria and to find an overall suitability index. The fuzzy set-based approach was further modified by using a weighted nonlinear aggregation method (Yager's aggregation method), where weights are assigned to the attributes to represent their relative importance to decide the overall suitability (Yager & Filev, 1994). Weighted Euclidian distance (Ahamed *et al.*, 2000), weighted linear aggregation (Bojórquez-Tapia *et al.*, 2001; Prakash, 2003; Braimoh *et al.*, 2004) and the use of fuzzy logical operators (Sicat *et al.*, 2005) are some other methods used for estimating an overall suitability index by using many attributes. In these attribute aggregation methods, the relative importance of the attributes is assumed to be the same throughout the study area. However, in the real world, the relative importance of the attributes varies from one location to another depending on the location characteristics as well as the value of these attributes itself (Reshmidevi *et al.*, 2009).

In this paper, an attribute aggregation method using fuzzy logical operators is presented to evaluate the surface water potential. The fuzzy membership-based model is integrated with a catchment hydrological model for the estimation of the soil moisture stress in the areas and hence the need for supplementary irrigation. This paper briefly describes the concept of fuzzy logic. Further, the proposed fuzzy membership-based system is explained in detail with a case study application in a sub-humid watershed in India, where rainfed paddy is the main crop.

CONCEPT OF FUZZY SETS

The concept of fuzzy set was first proposed by Zadeh (1965). In a fuzzy set, as distinct from the classical set (crisp), the sharp boundary between two classes is replaced by a gradual variation from one class to the other. The degree to which a point belongs to a fuzzy set is represented by its membership value – complete membership is represented by 1, non-membership is represented by 0, and partial membership is represented by any value ranging from 0 to 1. Fuzzy membership functions are used to map the elements of a universe into the range [0,1]. This process of mapping attributes of different scale into the range [0,1] is called fuzzification, which is expressed as:

$$\mu_{\tilde{X}}(r) \in [0,1] \tag{1}$$

where, $\mu_{\tilde{X}}(r)$ is the membership of the attribute r in the fuzzy set X. The concept of partial membership in fuzzy logic helps in modelling involving uncertainties.

METHODOLOGY

In this study, as a preliminary step, agricultural land areas and the water bodies in the catchment are identified by using the land-use/land-cover map of the area. A raster-based analysis is performed, wherein the entire study area is divided into cells of 20 m × 20 m dimension. Since the objective is to identify the feasibility of supplementary irrigation, the analysis is carried out only for the cells identified as agricultural field. In order to identify the surface water potential for supplementary irrigation, soil moisture stress during the crop growth period, distance of the field from the nearest surface water body (WB_{prox}) and the relative elevation of the field with respect to the water body (*elevation*) are considered.

The present study considers the surface water potential in terms of the relative location of the surface water bodies with respect to the demand point, presuming sufficient water is available to meet the demand. Estimation of the quantity of surface water available and the capability to meet the irrigation demand are beyond the scope of this study.

Hydrological modelling to estimate soil moisture stress

Attribute *DD* is used to represent the soil moisture stress in the analysis. Soil moisture stress periods are identified by comparing the soil moisture in the crop root zone and the allowable soil moisture deficit. Soil moisture content below which the crops suffer water stress affecting the yield is termed as the critical soil moisture. Simulation of spatio-temporal variation in soil moisture helps to identify the areas and periods of significant soil moisture stress in the crop lands. Soil moisture deviation of up to 20% of the saturation soil moisture content is assumed tolerable in the paddy fields in dry areas (Jensen *et al.*, 1993; Panigrahi *et al.*, 2001).

The soil moisture balance equation in the crop root zone, considering rainfall, surface runoff, infiltration, deep percolation and crop evapotranspiration is used for the simulation of soil moisture variation. Field bunds are commonly used in the rainfed paddy fields in the area to store the surface runoff so as to increase the soil moisture availability for the crops. In order to incorporate the effect of field bunds and the resulting increase in soil moisture, the widely used Natural Resources Conservation Service (previously known as Soil Conservation Service) Curve Number (SCS-CN) method is modified to develop the SCS-CN-SMS model (Reshmidevi *et al.*, 2008). Using this model, the daily variation in the soil moisture is simulated for each cell in the study area. The average number of days in which the soil moisture falls below the critical limit during the crop growth period is estimated for each cell and is presented using the attribute *DD* in this study.

Framing the fuzzy membership-based model for estimating surface water potential

For the proposed raster-based analysis, attribute values are estimated for each cell and presented in map format with uniform resolution. Map layers corresponding to all the three proposed attributes are aggregated during the analysis and the resulting surface water potential is estimated for each cell. Development of the proposed fuzzy membership-based model includes two steps: fuzzification of the attributes and aggregation of the attribute membership values to find an overall suitability index. Figure 1 shows the schematic representation of the methodology.

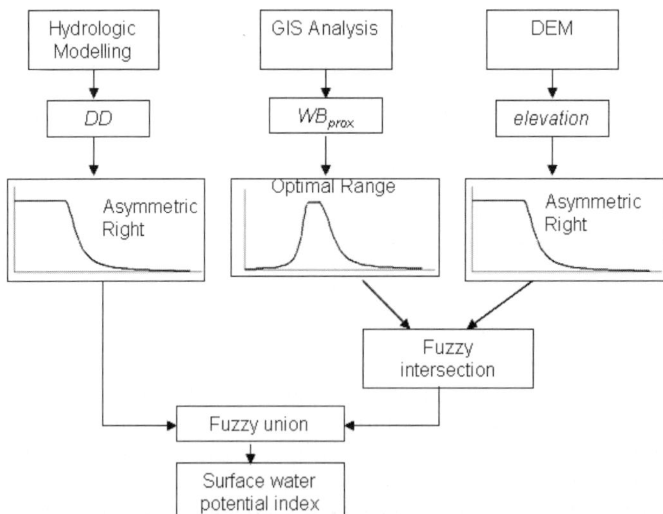

Fig. 1 Schematic representation of the model for the surface water potential evaluation.

Step 1: Fuzzification of the attributes

Fuzzification is the process in which the attributes measured at different scales are mapped into the range [0,1] by using membership functions. The attributes used in the current study are measured continuously and hence semantic import (SI) functions (Baja *et al.*, 2002) of various shapes are used depending upon the association of each attribute towards the suitability criteria (Fig. 1). Optimal range (OR), asymmetric left (AL) and asymmetric right (AR) are the three types of SI functions used for mapping the attributes. The AL type SI function is used when the suitability of the attribute increases with the increase in the attribute value. If the suitability decreases with increase in the attribute value, AR type SI functions are used for the attribute mapping. In order to show the middle suitability range of attributes, OR type functions are used. Membership functions generally used for each of the attributes are shown in Fig. 1.

Based on the suitability criterion, a gradual variation from non-suitability (zero membership, $\mu = 0$) to suitability ($\mu = 1$) is represented by using four functional parameters $r_{(0.5,L)}$, $r_{(0.5,R)}$, $t_{(L)}$ and $t_{(R)}$, where $r_{(0.5,L)}$, $r_{(0.5,R)}$ are the attribute values corresponding to $\mu = 0.5$ at the left and right sides, respectively and, $t_{(L)}$, $t_{(R)}$ are the ranges of attribute values for which the membership changes from 0.5 to 1 on the left and right sides, respectively (Braimoh *et al.*, 2004).

Step 2: Aggregation of the attributes membership values

The attribute *DD* is used in the present study to indicate the soil moisture stress in the agricultural fields in terms of average number of soil moisture stress days. Attributes WB_{prox} and *elevation* together decide the feasibility for supplementary irrigation in terms of accessibility of the nearest surface water body. If the average number of dry days in any field is found to be less than the critical limit, *DD* is assumed to be good. Supplementary irrigation is not required in that field and therefore the suitability of the attributes WB_{prox} and *elevation* are not significant. However, if *DD* is not suitable, indicating that the number of dry days is more than the critical limit that the crop can sustain, supplementary irrigation is essential. From the heuristic information, it is found that if the area is near to any surface water body and is at a suitable elevation with respect to the water body, there is scope for improvement of the area through supplementary irrigation. Hence, the suitability of the attributes WB_{prox} and *elevation* becomes significant. Fuzzy logical operators are used to reproduce this relative importance of the attributes as:

$$\mu_{\underset{\sim}{X}}(s) = \mu_{\underset{\sim}{X}}(DD) \cup \left[\mu_{\underset{\sim}{X}}(WB_{prox}) \cap \mu_{\underset{\sim}{X}}(elevation) \right] \tag{2}$$

Fuzzy intersection operation is used to aggregate the parameters WB_{prox} and *elevation*, and the output is aggregated with *DD* using the fuzzy union operator. The aggregated output is the fuzzy output with values in the range [0,1] showing the surface water potential of the area. Suitability index 1 indicates good surface water potential, and 0 shows poor potential. For each cell of 20 m × 20 m dimension, the surface water potential index is estimated.

MODEL APPLICATION IN THE CASE STUDY AREA

The model was tested in the Gandeshwari sub-watershed (up to the confluence point of Dhajajor nullah) in West Bengal, India (Fig. 2).

It is an agricultural watershed with rainfed paddy as the major crop. The area falls under the sub-humid agro-climatic region (latitude: 21°13′–23°32′N and longitude: 86°53′–87°8′E). The area of the watershed (up to the confluence point of Dhajajor nullah) is 105.6 km^2. Annual average rainfall of the area is about 1400 mm, which occurs mainly during the monsoon period from June to October. Due to the erratic distribution of the monsoon rainfall, the rainfed agricultural fields in the area face severe water stress even during the monsoon crop period (kharif season). The area is characterized by a large number of small water bodies distributed over the entire watershed.

The model was applied in the Gandheshwari sub-watershed to study the surface water potential with regard to supplementary irrigation from the local water bodies. The entire study area

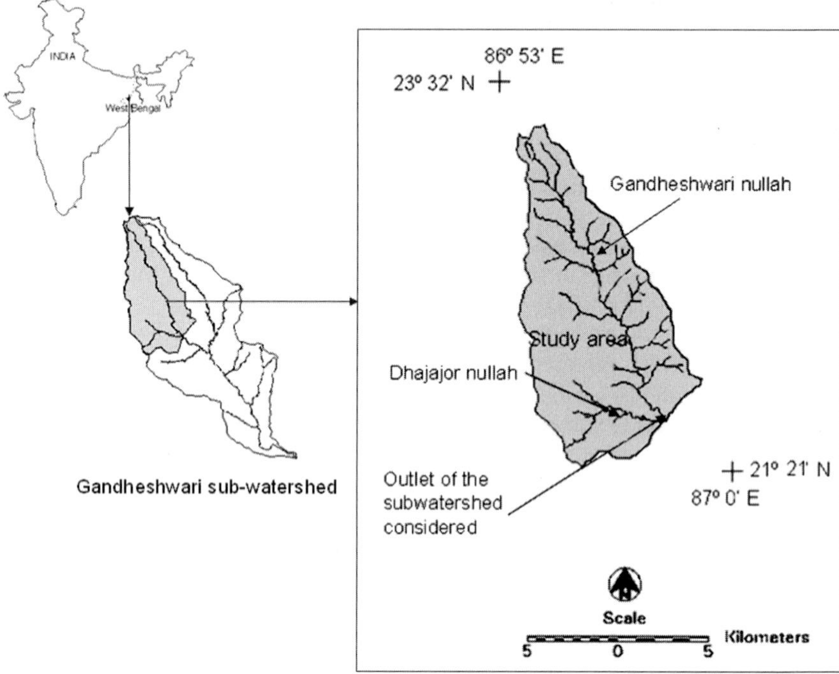

Fig. 2 Location map of the Gandheshwari sub-watershed.

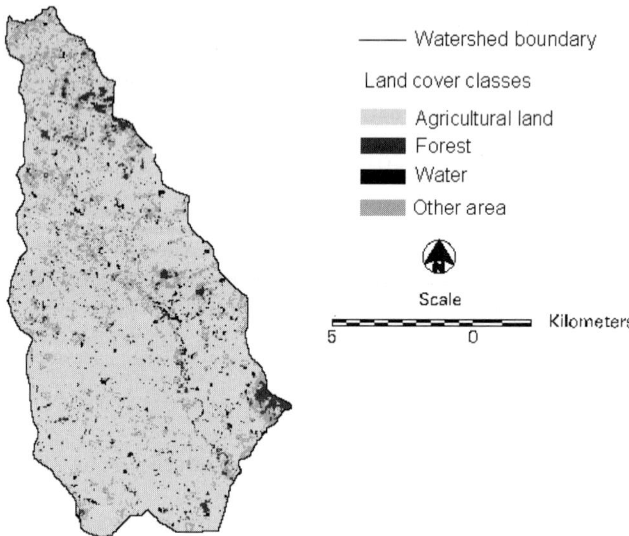

Fig. 3 Land cover map of the study area.

was divided into cells of 20 m × 20 m spatial resolution and geo-referenced raster attribute layers were prepared by using ERDAS IP 9.1. Agricultural areas and the water bodies in the area were identified with the help of the remote sensing satellite imagery (IRS LISS III imageries, dated April 1998) through visual interpretation (Fig. 3).

Since the objective was to assess the surface water potential with regard to supplementary irrigation, the analysis was carried out only for cells that were identified as agricultural fields. Attribute WB_{prox} was obtained using a region growing method in GIS, in which the distance of each cell to the nearest surface water body was estimated. The elevation of each cell in the watershed was obtained from the digital elevation model (DEM). The SCS-CN-SMS model was used to estimate *DD* for each cell in the watershed, which was identified as an agricultural land.

Once the attribute data layers were prepared, these attribute values were fuzzified and mapped into the range [0,1] using SI functions. The type of SI membership functions used for mapping each attribute and the corresponding values of the function parameters are given in Table 1. The area has a large number of scattered surface water bodies. Significant elevation difference was not observed between the fields and the nearby surface water bodies. Therefore, when the scope for irrigation was studied, attribute elevation was assumed as good and a corresponding membership value of 0.9 was assigned to *elevation* throughout the watershed.

Table 1 Membership function characteristics.

Attribute	Type of SI function and function parameters
DD	Asymmetric right: (c2,t2) = (8,3)
WB_{prox}	Optimal range: (c1,t1,c2,t2) = (20,10,300,100)
elevation	Assumed 0.9 throughout

RESULTS AND DISCUSSION

Raster data layers of the attributes *DD*, WB_{prox} and *elevation* from the study area were used in the model to assess the surface water potential. The final surface water potential index obtained for each cell is presented in the map format in Fig. 4. In order to discuss the spatial distribution of the index over the watershed, three sample locations (G1, G2 and G3) are selected as shown in Fig. 4. Membership values of the three attributes and the resulting surface water potentials at these locations are given in Table 2.

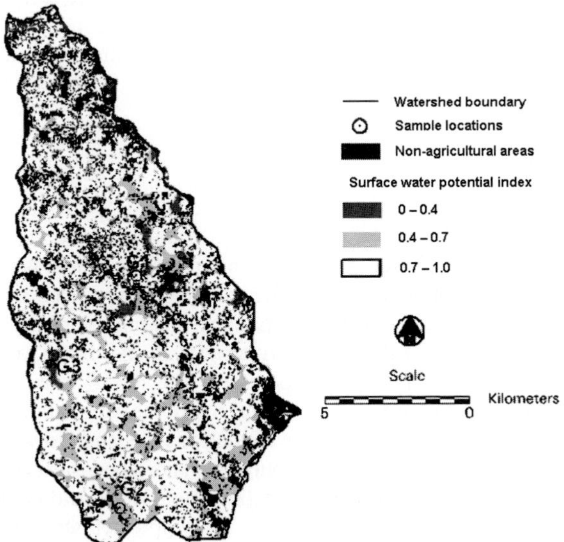

Fig. 4 Surface water potential index distribution in the study area.

Table 2 Attribute membership values and the surface water potential indices at the sample locations.

Location	Attribute membership values:			Surface water potential index
	DD	WB_{prox}	*elevation*	
G1	0.02	1.0	0.9	0.9
G2	0.45	0.07	0.9	0.45
G3	0.02	0.12	0.9	0.12

The poor membership values of the attribute *DD* at locations G1 and G3 indicate high soil moisture stress. Supplementary irrigation is therefore inevitable to ensure healthy crops. Therefore, the surface water potential of these two locations depends on the feasibility for supplementary irrigation in terms of access to the surface water body. With the attribute *elevation* assumed suitable throughout the watershed, feasibility for supplementary irrigation depends on the attribute WB_{prox}. Location G1, being close to the surface water body, as indicated by the membership value of the attribute WB_{prox}, was therefore found to have higher surface water potential. Overall surface water potential index of G1 was found to be 0.9. However, WB_{prox} being unsuitable at location G3, it was found to have poor surface water potential, as indicated by the overall surface water potential index of 0.12.

From the attribute *DD*, soil moisture stress during the crop growth period at location G2 was found to be moderate, as indicated by the membership value 0.45. Therefore, supplementary irrigation requirement and hence the attributes WB_{prox} and *elevation* are less important at location G2. Consequently, location G2 was found to have moderate surface water potential even though WB_{prox} was found unsuitable.

From the analysis, 3% of the study area was found to have poor surface water potential (surface water potential index less than 0.4), which was either due to poor soil moisture condition or due to poor access to surface water bodies for supplementary irrigation. Another 20% of the area was found to have a surface water potential index between 0.4 and 0.7. Areas having surface water potential index higher that 0.7 were assumed to have good surface water potential. Approximately 77% of the area was identified with good surface water potential either in terms of soil moisture availability or in terms of access to surface water bodies.

From the study it was found that, even though the erratic monsoon rainfall and poor water holding capacity of the soil creates severe soil moisture stress in the area, the small field ponds distributed over the entire watershed were found to be effective in reducing the soil moisture stress through supplementary irrigation. However, further studies are required to quantitatively compare the irrigation demand with the water availability in these water bodies.

SUMMARY

This paper presents an integrated approach to estimate the surface water potential of an agricultural watershed, wherein the hydrological model based on the physics of the catchment hydrology is integrated with a fuzzy membership-based model. The hydrological model is used to estimate the soil moisture stress in the catchment, where as the fuzzy membership-based model is used to aggregate the indigenous knowledge about the water potential by making use of the model results. In order to consider the spatial variation of the suitability criteria in the model, attribute membership values were aggregated by using fuzzy logical operators. The model thus helps to identify the agricultural areas badly affected by soil moisture stress and the areas that can be improved by providing supplementary irrigation. The model was applied to the Gandheshwari sub-watershed in India. From the analysis, 77% of the area was found to have good surface water potential; 3% of the area was found to have poor surface water potential due to the combined effect of poor soil moisture and poor access to surface water bodies for supplementary irrigation. The method thus shows the integration of hydrological models with artificial intelligence tools in developing decision support tools for land and water management.

Acknowledgements This study was partially sponsored by the Department of Science and Technology, Government of India, through the research project carried out at CSRE, Indian Institute of Technology Bombay, and is gratefully acknowledged.

REFERENCES

Ahamed, T. R. N., Rao, K. G. & Murthy, J. S. R. (2000) GIS-based fuzzy membership model for crop-land suitability analysis. *Agric. Systems.* **63**, 75–95.

Ashraf, M., Kahlown, M. A. & Ashfaq, A. (2007) Impact of small dams on agriculture and groundwater development: a case study from Pakistan. *Agric. Water Manage.* **92**, 90–98.

Baja, S., Chapman, D. M. & Dragovich, D. (2002) A conceptual model for defining and assessing land management units using a fuzzy modeling approach in GIS environment. *Environ. Manage.* **29**(5), 647–661.

Bojórquez-Tapia, L., Díaz-Mondragón, S. & Ezcurra, E. (2001) GIS-based approach for participatory decision making and land suitability assessment. *Geograph. Information Sci.* **15**(2), 121–153.

Braimoh, A. K., Vlek, P. L. G. & Stein, A. (2004) Land evaluation for maize based on fuzzy set theory and interpolation. *Environ. Manage.* **33**(2), 226–238.

Fox, P. & Rockstorm, J. (2003) Supplemental irrigation for dry-spell mitigation of rainfed agriculture in the Sahel. *Agric. Water Manage.* **61**, 29–50.

Gemitzi, A., Petalas, C., Tsihrintzis, V. A. & Pisinaras, V. (2006) Assessment of groundwater vulnerability to pollution: a combination of GIS, fuzzy logic and decision making techniques. *Environ. Geol.* **49**, 653–673.

Jensen, J. R., Mannan, S. M. A. & Uddin, S. M. N. (1993) Irrigation requirement of transplanted monsoon rice in Bangladesh. *Agric. Water Manage.* **23**, 199–212.

Kar, G., Verma, H. N. & Singh, R. (2006) Effects of winter crop and supplemental irrigation on crop yield, water use efficiency and profitability in rain-fed rice based cropping system of Eastern India. *Agric. Water Manage.* **79**, 280–292.

Nasiri, F., Maqsood, I., Huang, G. & Fuller, N. (2007) Water quality index: a fuzzy river-pollution decision support expert system. *J. Water Resour. Plan. Manage. ASCE* **133**(2), 95–105.

Oweis, T. & Hachum, A. (2006) Water harvesting and supplemental irrigation for improved water productivity of dry farming systems in West Asia and North Africa. *Agric. Water Manage.* **80**, 57–73.

Panigrahi, B., Panda, S. N. & Mull, R. (2001) Simulation of water harvesting potential in rainfed rice lands using water balance model. *Agric. Systems* **69**, 165–182.

Panigrahi, B., Panda, S. N. & Agrawal, A. (2005) Water balance simulation and economic analysis for optimal size of on-farm reservoir. *Water Resour. Manage.* **19**, 233–250.

Prakash, T. N. (2003) Land suitability analysis for agricultural crops: A fuzzy multicriteria decision making approach. MSc Thesis, International Institute for Geoinformation Science and Earth Observation, The Netherlands.

Reshmidevi, T. V., Jana, R. & Eldho, T. I. (2008) Geo-spatial estimation of soil Moisture in rain-fed paddy fields using SCS-CN-based model. *Agric. Water Manage.* **95**(4), 447–457.

Reshmidevi, T. V., Eldho, T. I. & Jana, R. (2009) A GIS-integrated fuzzy rule-based inference system for land suitability evaluation in agricultural watersheds. *Agric. Systems* (in press).

Shaban, A., Khawlie, M. & Abdallah, C. (2006) Use of remote sensing and GIS to determine recharge potential zones: the case of Occidental Lebanon. *Hydrogeol. J.* **14**, 433–443.

Shrestha, R. R., Bardossy, A. & Rode, M. (2007) A hybrid deterministic–fuzzy rule based model for catchment scale nitrate dynamics. *J. Hydrol.* **342**, 143–156.

Sicat, R. S., Carranza, E. J. M. & Nidumolu, U. B. (2005) Fuzzy modeling of farmers' knowledge for land suitability classification. *Agric. Systems* **83**, 49–75.

Wang, F., Hall, G. B. & Subaryono (1990) Fuzzy information representation and processing in conventional GIS software: database design and application. *Int. J. Geograph. Information Sci.* **4**(3), 261–283.

Yager, R. R. & Filev, D. P. (1994) *Essentials of Fuzzy Modeling and Control.* John Wiley & Sons, Inc., New York, USA.

Zadeh, L. A. (1965) Fuzzy sets. *Information and Control* **8**, 338–353.

Comparison of Genetic Algorithm and Ant Colony optimization methods for optimization of short-term drought mitigation strategies

S. MOHAMMAD MORTAZAVI N., GEORGE KUCZERA & LIJIE CUI
School of Engineering, University of Newcastle, New South Wales 2308, Australia
seyedmohammad.mortazavinaeini@studentmail.newcastle.edu.au

Abstract Drought represents a major threat to the security of water supplies. There is a variety of short- and long-term options to mitigate drought impacts. In this paper, short-term strategies are explored with the goal of finding the best strategies that minimize cost. The performance of two heuristic optimization methods, Genetic Algorithm and Ant Colony, is investigated. The results show that the Ant Colony method is sensitive to the choice of method representing the decision space. It was found that Ant Colony methods were more robust and efficient than Genetic Algorithms, particularly when the number of function evaluations is limited.

Key words drought management; optimization; genetic algorithms; ant colony optimization

INTRODUCTION

As Mays (2007) observes, the world's population is expanding rapidly, yet there is no more fresh-water on Earth than there was 2000 years ago, when the population was less than three percent of the current six billion. Water availability is one of the most important issues facing society and will continue to be so well into the future. Therefore, it is important in urban water management to supply water in a sustainable manner in terms of quality and quantity. The challenge is particularly severe during periods of drought when sustainability of supply is threatened on a short-term basis and, if not managed properly, could affect the long-term viability of the urban area.

A variety of options is available to mitigate drought impacts. These options can be very diverse and have different effectiveness in terms of economic and social outcomes. However, only a limited number of options are practicable and potentially effective at an acceptable cost (Garrote *et al.*, 2007). The challenge is to identify such options. Drought impact mitigation is a complex management problem, usually involving conflicting management objectives, alternative actions and different players (Traore & Fontane, 2007). The management of complex water supply systems requires advanced tools to reduce the uncertainties in hydrological and management issues, and to improve the quality of decision making for different climate conditions (Merabtene *et al.*, 2002).

Generally actions which are useful to mitigate drought impacts can be categorized into two types, long-term and short-term. The actions taken before drought are long-term strategies. On the other hand, the actions taken after initiation of drought are referred to as short-term strategies. The most common short-term strategy in Australia is to impose restrictions on consumption.

The main aim of this paper is to identify short-term strategies which minimize cost. This is tackled by evaluating two promising optimization methods, Genetic Algorithm and Ant Colony. A case study involving the Canberra headworks system, in Australia, is used to demonstrate the capabilities of the optimization methods.

OPTIMIZATION APPROACHES

This section focuses on two optimization approaches, the well-established Genetic Algorithm (GA) and the more recent Ant Colony Optimization (ACO). Both use probabilistic methods to conduct the search and thus cannot guarantee finding a global optimum. Importantly they can interface with any simulation model and therefore do not suffer from the often severe limitations of classical mathematical programming approaches.

In this particular problem, function evaluations are very expensive, especially if long stochastic flow sequences are used. For instance, Cui & Kuczera (2005) report a typical simulation for the Sydney headworks system taking approx 5 CPU minutes. To manage this computational problem, it is important to minimize the size of the search space by carefully selecting decision variables and also to employ an efficient and robust search method. The latter motivates the focus on the efficiency and robustness of the GA and ACO methods.

Genetic algorithm

The Genetic Algorithm (GA) optimization method is based on Darwin's theory of evolution. It is a population based algorithm with every generation consisting of chromosomes which, in turn, consist of genes. There are three main operations in GAs, selection, crossover and mutation. Selection involves identifying fit parents for the crossover and mutation steps. Crossover involves a pair of parent chromosomes exchanging a portion of their bit sequence at randomly set crossover points. Mutation consists of randomly flipping each gene. Crossover and mutation operations help the GA to jump out from local optima and search the decision space more vigorously and efficiently (Cui & Kuczera, 2003).

Ant colony optimization

Ant colony optimization (ACO) is a more recent development than GA. ACO was inspired by the fact that some kinds of ants are blind yet they can find the minimum path between their nest and food. This is because of chemical substances called pheromones which ants deposit when they travel on a route. Based on this behaviour of real ants, Dorigo (1992) developed an ant colony optimization (ACO) method called Ant System (AS) to solve the travelling salesman problem (TSP) (Dorigo & Stutzle, 2004).

A variety of ACO methods have been introduced in the literature. Different methods for updating pheromones and different techniques to discipline ants to find the optimum have motivated a variety of ACO methods such as ACS, Max-Min AS, and AS-ranked.

Dorigo *et al.* (1996) used the first ant colony optimization method, called Ant System (AS), to solve the travelling salesman problem (TSP) and job-shop scheduling problem (JSP). The results showed that ACO dominated all other classical or heuristic methods and found the minimum route obtained by other methods. Dorigo & Gambardella (1997) introduced a new method based on AS called Ant Colony System (ACS). Stutzle & Hoos (1997) applied Max-Min AS to solve the travelling salesman problem. All of these methods have been successfully applied to a number of benchmark combinatorial optimization problems (Dorigo & Di Caro, 1999). ACO has been used in water engineering applications. For example, Zecchin *et al.* (2005) applied ACO to optimize water distribution systems. They undertook a parametric study for ACO parameters and compared the performance of GA and ACO for two case studies. In both case studies, ACO had a far shorter search-time than GA. However, in one case study ACO found high quality solution while in the other the solution was poor.

CANBERRA HEADWORKS SYSTEM CASE STUDY

To investigate the potential of GA and ACO to handle drought management plan optimization problems, the Canberra headworks water supply system was chosen as the case study. The Canberra system, which serves Australia's capital city, was chosen because it had enough complexity to warrant optimization yet could be simulated on a single CPU in a reasonable time. In Fig. 1 a schematic of the Canberra system is shown. The system consists of four reservoirs referred to as Corin, Bendora, Cotter and Googong, which are in service, and one proposed reservoir called Tennent. Also, there are several pump stations to convey water from rivers and reservoirs to Canberra.

WathNet was used to model the operation of the Canberra system. WathNet is an example of a generalized simulation model using network linear programming (NLP) to simulate the operation of a wide range of water supply headworks configurations (Cui, 2003).

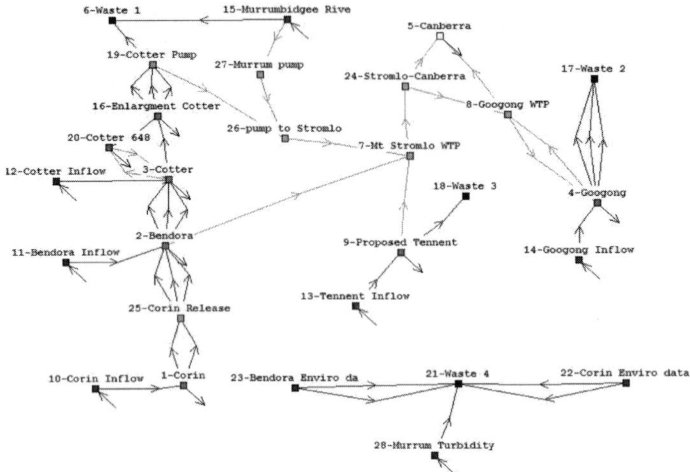

Fig. 1 Schematic of Canberra water supply headworks system.

Decision variables

In most drought management plans, imposing restrictions is a key element. It is usual to impose restrictions using a trigger level based on reservoir storage. Because the trigger levels must form a decreasing sequence, some care is required in specifying the decision variables. In Fig. 2 the schematic shows the sequence of trigger levels. For example, if the first trigger level is 0.6, the second trigger level must be less than 0.6. This makes the problem a constrained one.

Corin and Googong reservoirs are the two main sources of water supply for Canberra. Optimizing the balance between these reservoirs can help to better mitigate drought. To achieve this balance two decision variables, base gain (BG) and incremental gain (IG) are introduced in WathNet. To explain how they work, suppose there are N carryover arcs leaving a reservoir with capacity C. The penalty assigned to each arc is:

$$\cos t(i) = -BG - IG(i-1), \qquad i = 1,\ldots,N \tag{1}$$

The capacity of each arc is C/N. Because the penalty is negative the NLP tries to fill the arc to capacity. By varying BG and IG for different reservoirs, a wide range of reservoir storage target profiles can be simulated. The reservoir with higher gains (more negative carryover costs) will tend to carryover water for the next season in preference to a reservoir with lower gains.

In this study, there are six decision variables, four restriction storage trigger levels, Googong's base gain and incremental gain.

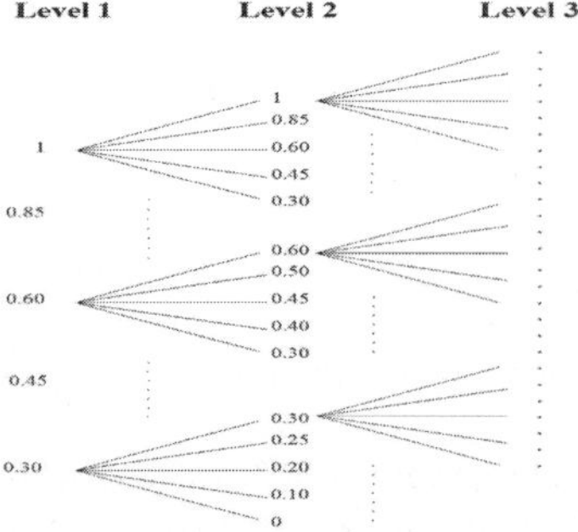

Fig. 2 The sequence of storage trigger levels for imposing restrictions on water usage.

Objective function

In the work reported here, single objective optimization was considered. The objective is to minimize total expected cost:

$$O.F. = C_O + P_R + P_S \quad (2)$$

where C_O is the expected operational cost, P_R is the penalty associated with imposing restrictions and P_S is the penalty for unplanned shortfall. Equation (3) is used to calculate restriction penalty (Cui, 2003):

$$P_R = \frac{\varepsilon}{\varepsilon+1} p_1 D \left[1-(1-R)^{\frac{\varepsilon}{\varepsilon+1}} \right] \qquad \text{for } \varepsilon \neq -1 \quad (3)$$

where p_1 is the current price of water, D is the unrestricted demand for water, ε is the price elasticity and R is the demand reduction, assumed to be 0.95, 0.8, 0.7 and 0.65 for the four restriction levels, respectively. In this study the following values were considered: $\varepsilon = -0.25$ and $p_1 = \$600/\text{ML}$.

The penalty for unplanned shortfalls in supply is applied to discourage the socially and politically unacceptable scenario in which the system runs out of water. Lack of water for sanitation would impact heavily on commercial activity. Moreover, many industries are dependent on water and their productivity would be reduced if supply was cut. In this study, it is assumed that the consequence of unplanned water shortage is loss of gross income which is taken as the penalty for unplanned water shortages. Noting that Canberra's gross income for year 2006–07 was $21 billion and its water consumption was 56 148 ML, one can infer a penalty of $37 400/ML assuming loss of gross income is proportional to unplanned shortfall.

GA implementation

The GA formulation used in this case study was described by Cui & Kuczera (2003). A binary GA with tournament selection, one-point crossover and mutation in conjunction with elitism was applied to the Canberra case study.

There are two approaches for handling constraints in GAs. One is to impose a penalty on infeasible solutions and using this to augment the objective function. The other is to use a mapping to transfer the constrained search space into an unconstrained space consisting of a unit hypercube. Cui (2003) developed a mapping method to generate trigger levels using the following equation:

$$Level(j) = Level(j+1) + [100 - level(j+1)] \times Decision(j), j = N,...,1 \quad (4)$$

where *Level*() denotes as storage trigger level, *Decision*() is the gene produced by GA and N is maximum number of trigger levels.

ACO implementation

The ACO was implemented in the case study by first representing the search space as a graph. In the literature, two methods have been used to represent the search space. Abbaspour *et al.* (2001) and Kumar & Reddy (2006) used a graph similar to Fig. 3(a) and Maier *et al.* (2003) presented a graph similar to Fig. 3(b). In both methods, the decision variable range is split into a specific number of segments. In the first graph each variable value is represented by a node but in the second by a route. Ants travel between nodes corresponding to different variables to define a route. In the first approach, called the nodal method, the route is denoted as (k,i,j), which means an ant travelled from node i of variable k to node j of variable k+1 while in the second one, referred to as the link method, the route is denoted as (k,i) which means an ant travelled along route i from node k which corresponds to variable k. To handle constraints in ACO, a taboo list is defined to prevent ants travelling on infeasible routes.

The next step is to define a transition rule which defines how ants should find their route. The transition rule for the nodal method is:

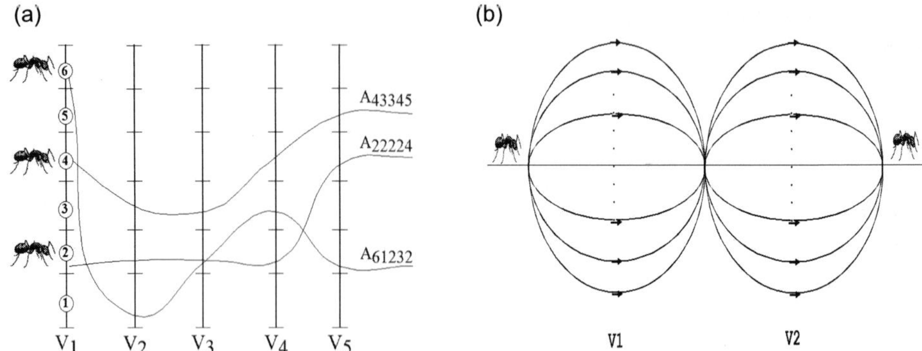

Fig. 3 (a) Schematic of ant routes travelling between five discrete variables (nodal method) and (b) between two discrete variables (link method).

$$P_{k\,ij} = \frac{[\tau_{k\,ij}]^{\alpha}[\eta_{kij}]^{\beta}}{\sum_{i=1}^{N}[\tau_{kij}]^{\alpha}[\eta_{k\,ij}]^{\beta}} \tag{5}$$

where P_{kij} is the probability the ant at node i will travel to node j; and τ_{kij} is the pheromone trail strength and η_{kij} is heuristic information. The pheromone trail strength encodes a long-term memory about the entire ant search process, and is updated by the ants themselves. In contrast, the heuristic information represents *a priori* information about the problem or run-time information provided by a source different from the ants. The parameters α and β are introduced to control the relative importance of the pheromone and heuristic information respectively.

A review of the literature shows there is no general rule for defining heuristic information. In this particular problem, the heuristic information for BG and IG was defined to force the ants to keep the same proportion of storage in Googong and other reservoirs. For instance, Googong's BG is varied between 8000 and 12 000 while BG for other reservoirs is 10 000. To maintain balance between Googong and the other reservoirs it is desirable to assign the same BG for all of them. In Fig. 4(a) a graph is presented to show how heuristic information is defined for BG. The maximum heuristic information is assigned to BG equal 10000 and falls off as the bounds are approached

It is important to keep reservoirs as full as possible to mitigate drought most effectively since, when reservoirs are full, it is possible to supply water for a longer period or possibly avoid imposing severe restrictions. For this reason, the heuristic information for storage trigger levels was defined so as to keep reservoirs as full as possible. The maximum heuristic information is assigned to the maximum storage trigger level and the minimum heuristic information to the minimum storage trigger level.

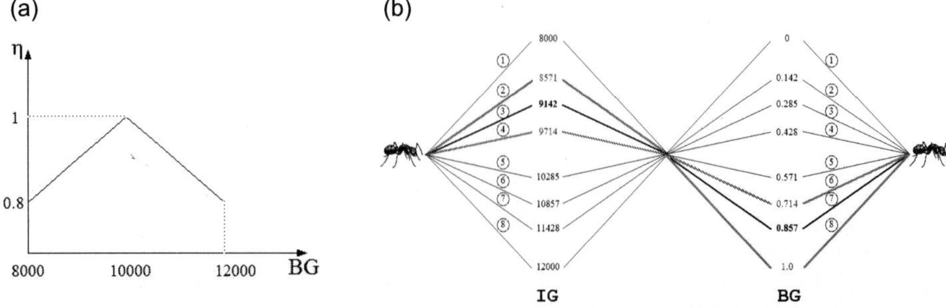

Fig. 4 (a) Relation between heuristic information and different BG values. (b) Schematic of local search.

In this study the Max-Min AS was used. Max-Min AS differs from other ACO methods in the way it updates the pheromone. In this method, the pheromone is updated just for the best iteration route or best-so-far route based on the following equations:

$$\tau_{kij}(t+1) = (1-\rho)\tau_{kij} + \Delta\tau_{kij}^{best} \tag{6}$$

where τ_{kij} is the amount of pheromone on the route between node i and j, ρ is a coefficient representing the evaporation of pheromone on the trail between iteration t and $t + 1$, and $\Delta\tau_{kij}^{best}$ is defined as:

$$\Delta\tau_{kij}^{best} = \frac{c}{f_{best}} \tag{7}$$

where c is a positive constant and f_{best} is the optimum objective function value at iteration t or the best objective function found so far.

In the Max-Min AS method the strategy is to deposit pheromone on just the best route. Unfortunately, this strategy may result in all ants following the same route because of the excessive growth of pheromone on a good, although suboptimal, tour. To counteract this effect, lower and upper limits $\tau_{\min}$ and $\tau_{\max}$ on the possible pheromone values on any arc are imposed to avoid search stagnation:

$$\tau_{\min} \le \tau \le \tau_{\max} \tag{8}$$

$$\tau_{\max} = \frac{1}{\rho * f_{best}} \tag{9}$$

$$\tau_{\min} = \frac{\tau_{\max}}{z} \tag{10}$$

where z is a parameter which is assigned 150 in this study.

In this study, the objective function varies between 2×10^6 and 300×10^6. In this case, if the actual objective function was used to calculate $\tau_{\max}$ or $\Delta\tau$ these parameters could be very small. For this reason the objective function is divided by 10^7 before calculating those values.

A local search can be used to improve the efficiency of Max-Min AS. When ants find the best-so-far solution, routes adjacent to that solution would be checked to find any possible improvement. In Fig. 4(b) a local search schematic is illustrated. For instance, if the third and seventh routes are assumed as the best-so-far routes, the adjacent routes, namely the second and fourth routes for IG and sixth and eighth routes for BG, are considered as a local search routes.

RESULTS AND DISCUSSION

Input data

The simulation model was run for 100 years using data generated from a stochastic model fitted to historic data. To explore the behaviour of a heavily stressed system the demand was taken as that corresponding to 175% of the current population.

Results for the variants of ACO

Because ACO and GA are probabilistic search methods, an optimization run with a different initial random number seed may produce a different optimum. This raises the question of robustness of the search. A robust method is one that shows little variation in the optimum between runs. This issue was addressed by running all methods 10 times and reporting the distribution of optimized costs.

The Min-Max AS with local search was used. Moreover, an elitist strategy in which one ant is forced to pass the best-so-far route was employed. The parameters shown in Table 1 were applied to both nodal and link implementations.

Table 1 Parameters for nodal and link ACO methods.

Ants	Iteration	Segment	α	β	ρ	z	C
30	500	63	1	2	0.1	150	0.5

In Fig. 5 objective function plots for the nodal and link ACO methods are presented for a single run. The link method displays superior performance with the nodal method unable to improve the objective function after about 50 iterations.

To understand this, consider Fig. 6 which presents box-plots for the objective function at each iteration. The tops and bottoms of each "box" are the 25th and 75th percentiles of the objective function values sampled by the ants, respectively. The line in the middle of each box is the data median. Outliers are displayed with a + sign. The aim of these figures is to show how ants explore the search space. In the link method the size of boxes was large over the first 50 iterations, after which the boxes shrunk but the ants continued to explore the search space. In the nodal method, however, the boxes had shrunk almost immediately with the consequence that the ants performed less vigorous exploration of the search space resulting in premature convergence.

In Fig. 7 box-plots are presented for the link method when there is no limitation on maximum pheromone. The ants converged very quickly in this case but at the expense of a greater chance of converging to a local optimum.

In Fig. 8 box-plots of the optimal cost for 10 runs for the nodal and link methods are presented. Each of the 10 runs started with a different random number seed. Figure 11 confirms the superior performance of link method. Of particular importance is that the median link cost is

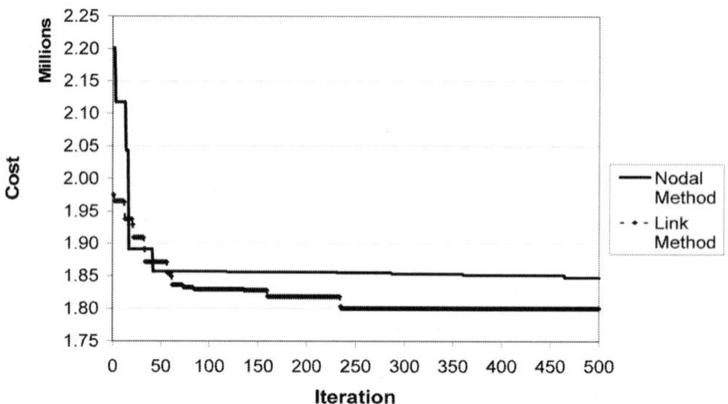

Fig. 5 Plot of objective function *versus* iterations for a single run.

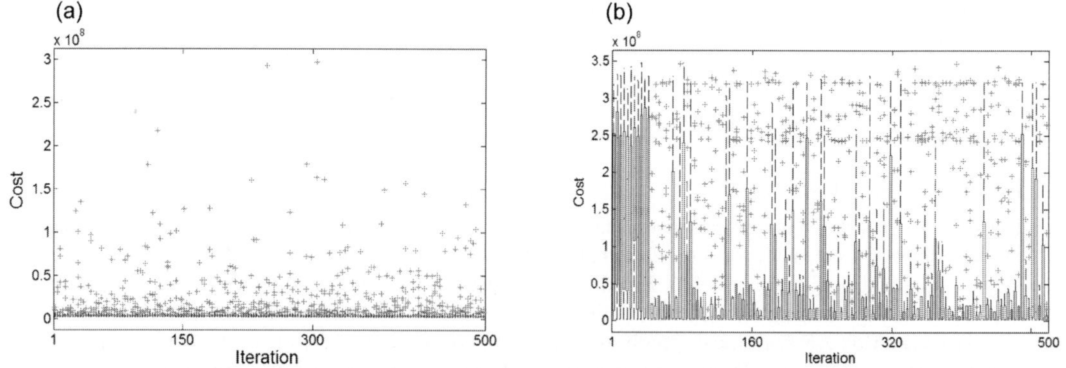

Fig. 6 Box-plot of objective functions for: (a) nodal method for a single run, and (b) link method for a single run.

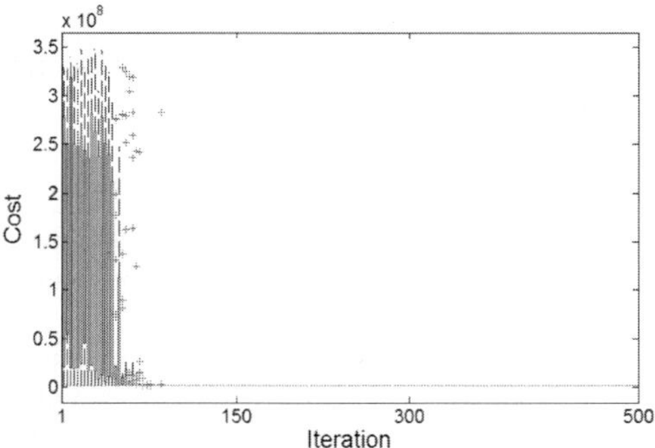

Fig. 7 Box-plot for link method results without any limitation for τ_{max} and τ_{min}.

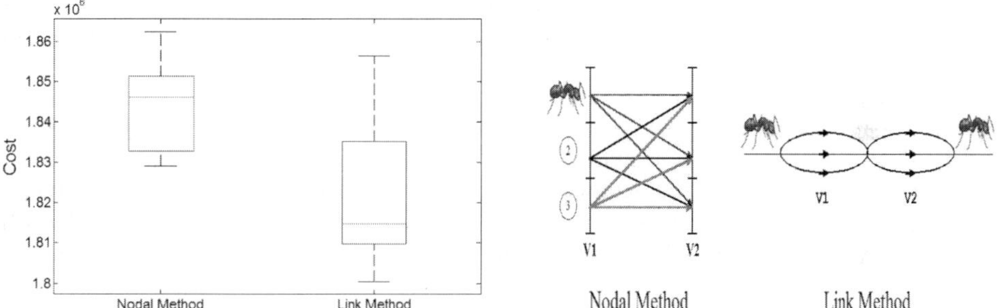

Fig. 8 (a) Box-plot of optimized costs based on 10 runs. (b) Different possible routes for nodal and link methods.

very close to the lowest cost. This behaviour is attributed to the fact that the number of possible routes in the link method is less than in the nodal method and so ants can find the best route more efficiently. To illustrate this, Fig. 8(b) shows the possible routes for the nodal and link methods when there are three segments. It shows there are nine possible routes for the nodal method and six possible routes for the link method. This difference becomes more significant when the number of segments is increased. Based on the superior performance of the link method, it is used in the reminder of the comparison.

To investigate the role of heuristic information, different values of β were tested with the results shown in Fig. 9. This graph shows that use of heuristic information generally improves the ACO results. There is less variation for β equal to 1 while the best performance occurred for β equal to 5. Moreover, for β equal to 5, the median of the costs is closest to the lowest cost.

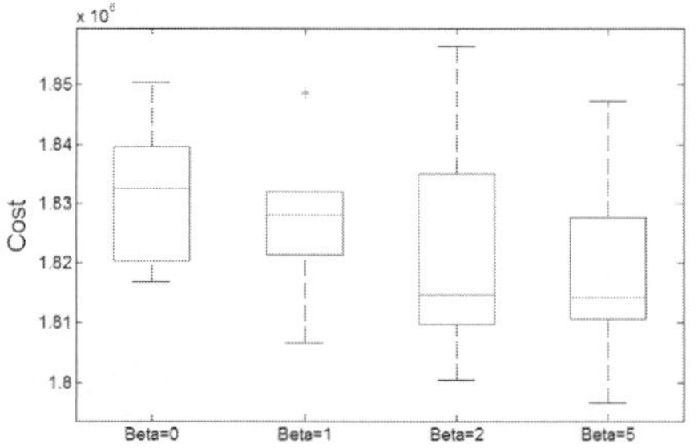

Fig. 9 Assessment the role of heuristic information.

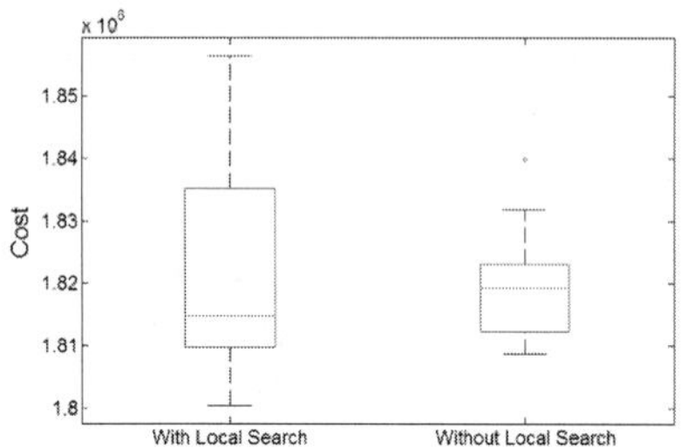

Fig. 10 Role of local search to improve link ACO method performance.

In Fig. 10 the ACO results with and without local search are presented. It is shown that applying local search can improve ACO results. The minimum objective function achieved with local search is less than the minimum without local search. Also, with local search, the median optimized cost is closer to the best cost. However, applying local search produces greater variation between runs.

It can therefore be summarized that ACO, like other heuristic optimization methods, is sensitive to the selection of tuning parameters. To tune these parameters, sensitivity analysis must be done. Figure 11 presents the results of a sensitivity analysis. In the last graph in Fig. 11, the product of the number of ants and iterations is fixed to give a total of 15 000 evaluations. It is clear that in this problem ACO performs better with fewer ants and more iterations. The best parameter values identified by the sensitivity analysis are summarized in Table 2.

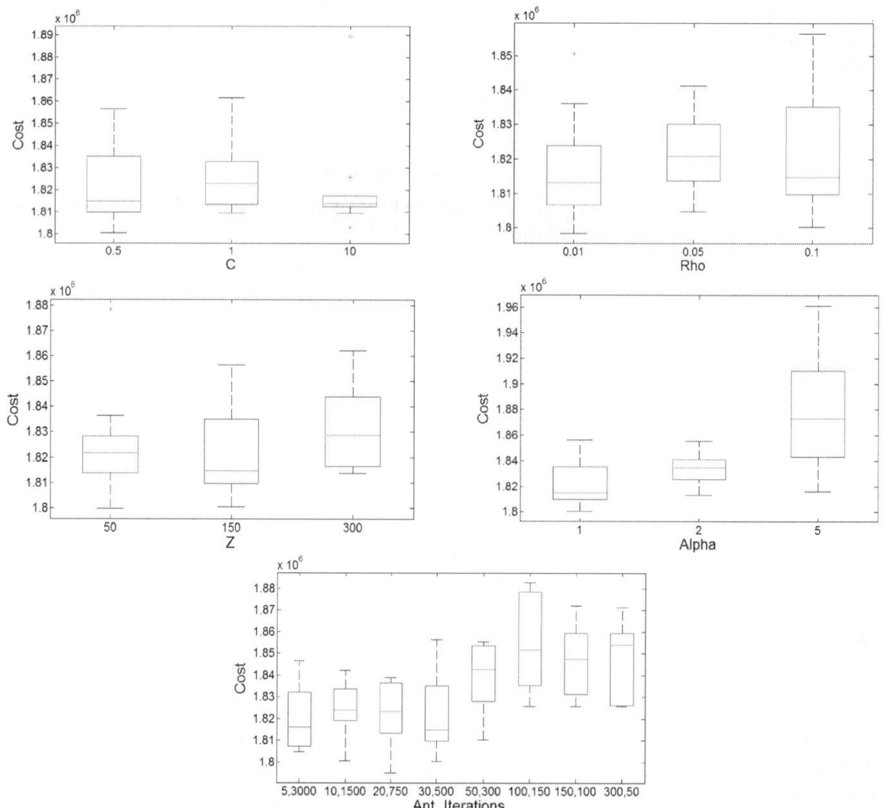

Fig. 11 Sensitivity analysis results.

Table 2 Best parameters based on sensitivity analysis.

Ants	Iteration	α	β	ρ	z	C
20	750	1	2	0.01	50	0.5

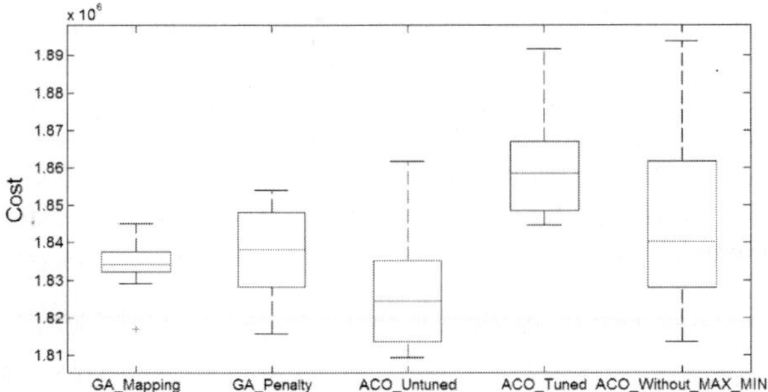

Fig. 12 Box-plots of optimized cost of ACO (tuned and untuned), two GA methods and ACO method without any limitation for τ_{max} and τ_{min} for 10 runs and 5000 evaluations.

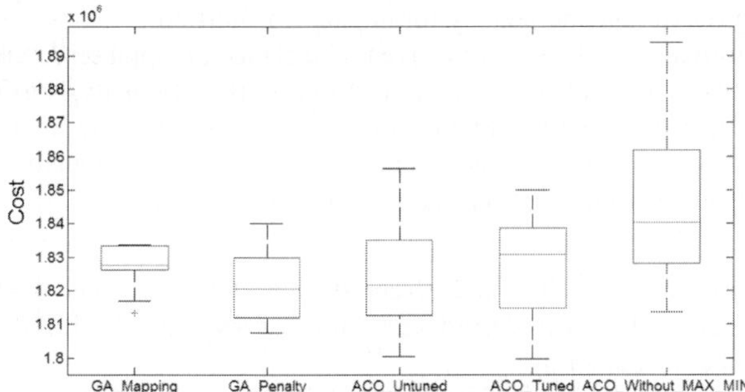

Fig. 13 Box-plots of optimized cost of ACO (tuned and untuned), two GA methods and ACO method without any limitation for τ_{max} and τ_{min} for 10 runs and 10 000 evaluations.

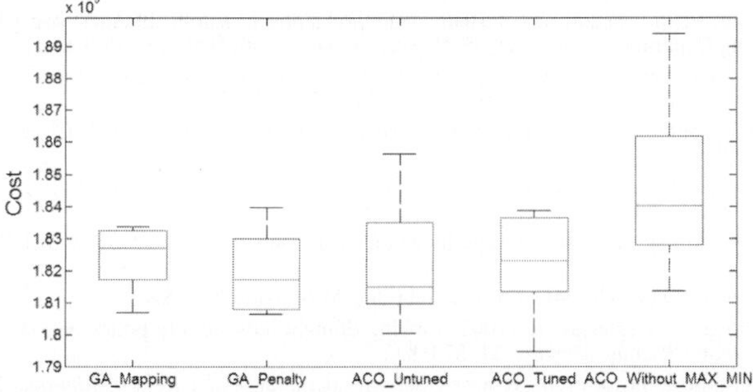

Fig. 14 Box-plots of optimized cost of ACO (tuned and untuned), two GA methods and ACO method without any limitation for τ_{max} and τ_{min} for 10 runs and 15 000 evaluations.

Comparison of ACO and GA methods

This section compares the performance of both ACO and GA methods. In this study, a 6-bit binary GA was implemented with one point crossover. The GA parameters suggested by Cui & Kuczera (2003) were used as default values for this case study: crossover rate = 0.9; mutation rate = 1/(population size); inversion rate = 0.7; and population size = 200. In Figs 12–14, five methods are compared for 10 runs and evaluations fixed at 5000, 10 000 and 15 000, respectively. The two

GA methods differed in the way the constraints on restriction triggers were handled. For the three ACO variants, two used Max-Min AS with untuned parameters (Table 1) and tuned parameters (Table 2), while the third had no limit on it.

For the 5000 evaluation case, the untuned ACO was unambiguously the best method with the best median and minimum costs. For the other two cases, the untuned ACO still performed strongly and would probably be judged the best of the five. It is clear that not imposing limits on τ_{max} and τ_{min} resulted in the greatest variability suggesting premature convergence. However, the behaviour in Fig. 7 suggests convergence, though suboptimal, may occur rapidly. For computationally expensive simulations, it may be worth trading off loss of robustness for greater efficiency. It is interesting to note that the untuned ACO performed better than the tuned ACO. This raises questions about the worth of sensitivity analysis. Its main weakness is that it does not explicitly allow for interactions between parameters.

CONCLUSION

The main aim of this study was to investigate the performance of two heuristic methods, GA and ACO, to optimize the short-term drought mitigation strategies. The Canberra headworks system was chosen as a case study. The objective function was the total expected cost consisting of operating cost, penalty for imposing restriction and penalty for unplanned shortfall.

The results showed that the performance of ACO is sensitive to the choice of graph representing the decision tree. However, the results cast doubt on the value of using sensitivity analysis to tune ACO parameters. Overall, ACO performed more robustly than GA, especially for the case involving the fewest number of evaluations. Because drought management simulation is computationally expensive, this is a significant finding and warrants further assessment of the potential of ACO.

Acknowledgements This work was supported by funds from the eWater CRC. The authors gratefully acknowledge the assistance of Richard Barratt and Tim Purves from ACTEW who provided data and advice on the Canberra case study.

REFERENCES

Abbaspour, K. C., Schulin, R. & van Genuchten, M. T. (2001) Estimating unsaturated soil hydraulic parameters using ant colony optimization. *Adv. Water Resour.* **24**(8), 827–841.

Cui, L.-J. (2003) Optimization of urban water supply headworks system using probabilistic search methods and parallel computing. PhD Thesis, Civil Engineering Department, University of Newcastle, New South Wales, Australia.

Cui, L.-J. & Kuczera, G. (2005) Optimizing water supply headworks operating rules under stochastic inputs: Assessment of genetic algorithm performance. *Water Resour. Res.* **41**. W05016, doi:10.1029/2004WR003517.

Dorigo, M. & Di Caro, G. (1999) Ant colony optimization: a new meta-heuristic. Proceedings of the 1999 Congress on Evolutionary Computation, 1999. CEC 99.

Dorigo, M. & Gambardella, L. M. (1997) Ant colony system: a cooperative learning approach to the traveling salesman problem. *Evolutionary Computation, IEEE* **1**(1), 53–66.

Dorigo, M., Maniezzo, V. & Colorni, A. (1996) Ant system: optimization by a colony of cooperating agents. *Systems, Man & Cybernetics, Part B, IEEE* **26**(1), 29–41.

Dorigo, M. & Stutzle, T. (2004) *Ant Colony Optimization.* The MIT Press, Cambridge, Massachusetts, USA.

Garrote, L., Martin-Carrasco, F., Flores-Montoya, F. & Iglesias, A. (2007) Linking drought indicators to policy actions in the Tagus Basin drought management plan. *Water Resour. Manage.* **21**, 873–882.

Kumar, D. & Reddy, M. (2006) Ant colony optimization for multi-purpose reservoir operation. *Water Resour. Manage.* **20**(6), 879–898.

Maier, H. R., Simpson, A. R., Zecchin, A. C., Foong, W. K., Phang, K. Y., Seah, H. Y. & Tan, C. L. (2003) Ant colony optimization for design of water distribution systems. *J. Water Resour. Plan. Manage.* **129**(3), 200–209.

Mays, L. W. (2007) *Water Resources Sustainability*, McGraw-Hill, USA.

Merabtene, T., Kawamura, A., Jinno, K. & Olsson, J. (2002) Risk assessment for optimal drought management of an integrated water resources system using a genetic algorithm. *Hydrol. Processes* **16**(11), 2189–2208.

Stutzle, T. & Hoos, H. (1997) MAX-MIN Ant System and local search for the traveling salesman problem. IEEE International Conference on Evolutionary Computation, 1997.

Traore, Z. N. T. & Fontane, D. G. (2007) Managing drought impacts: case study of Mali, Africa. *J. Water Resources Plan. Manage.* **133**(4), 300–308.

Zecchin, A. C., Simpson, A. R., Maier, H. R. & Nixon, J. B. (2005) Parametric study for an ant algorithm applied to water distribution system optimization. *Evolutionary Computation, IEEE* **9**(2), 175–191.

Neuro-fuzzy inference system for operation of a multi-purpose reservoir

PRAKASH C. SWAIN[1] & UMAMAHESH V. NANDURI[2]

1 *Department of Civil Engineering, University College of Engineering, Burla University, Burla, Orissa, India*

2 *Department of Civil Engineering, National Institute of Technology, Warangal 506004, AP, India,* mahesh@nitw.ac.in

Abstract In recent years soft computing techniques have been used increasingly by water resources engineers to model complex water resources systems. These techniques have the ability to mimic the human way of reasoning and decision making, and thus supplement the conventional modelling techniques. The present paper presents a neuro-fuzzy inference system for management of the Hirakud Reservoir on the River Mahanadi, in India, with the objectives of efficient flood control, irrigation and power generation. The objectives are considered as vaguely defined and hence are treated as fuzzy. The neuro-fuzzy inference system is used to capture the historical operation policy. The model developed is used to simulate the operation of reservoir and the performance of the reservoir is evaluated with reference to the identified fuzzy objectives. The performance of the model is found to be satisfactory and can be used as a rule curve for operating the reservoir.

Key words reservoir operation; soft computing techniques; neuro-fuzzy inference system

INTRODUCTION

Multi-purpose reservoir operation involves various interactions and trade-offs between purposes, which are sometimes complementary, but often competitive or conflicting. Reservoir operation may be based on the conflicting objectives of maximizing the amount of water available for conservation purposes and maximizing the amount of empty space for storing future flood waters to reduce the downstream damage. Studies of long-term storage reallocations and designing seasonal rule curves are two important types of reservoir system modelling and analysis applications (Wurbs, 1996). With the growing pressure on available supplies of water, the problem of conservation and allocation has assumed importance in recent decades.

Water resources projects are managed in an environment characterized by a varying degree of risk, often by relying on the manager's judgment, experience, vision and intuition. Traditional techniques, such as optimisation techniques, have serious limitations because they may not be able to incorporate the manager's experience and judgment, which are usually qualitative in nature. Hence there is a need for development of new techniques relying on advanced mathematical tools to model the expert opinion and experience, and automate the management of complex water resources projects.

In recent years, there has been increasing interest in neuro-fuzzy modelling techniques. Neuro-fuzzy techniques or soft computing techniques, which mimic the human way of reasoning and decision making, and can supplement conventional mathematical techniques for dealing with complex problems. Fuzzy logic is based on the mathematics of fuzzy set theory where the classical notion of binary set membership has been modified to include partial membership ranging between 0 and 1. The capability of fuzzy sets to express gradual transitions from membership provides a meaningful and powerful representation of measurement uncertainties and vogue concepts expressed in natural language. Thus fuzzy logic and fuzzy reasoning model the human way of thinking and reasoning (Jang *et al.*, 1997).

An Artificial Neural Network (ANN) is a massively parallel distributed system that has certain performance characteristics resembling the biological neural network of the human brain. The ANNs can be used as computational tools to solve large complex problems such as pattern recognition, nonlinear modelling, classification, association and control (Haykin, 1994). Neuro-fuzzy models integrate neural networks and fuzzy logic by incorporating low-level learning and computational power of neural networks into fuzzy systems.

Gupta & Rao (1994) described the basic notions of biological and computational neuronal morphologies. They also described the principles and architectures of fuzzy neural networks. They developed a fuzzy neural architecture based upon the notion of T-norm and T-co norm connectives. Buckley & Hayashi (1995) explained the process of representing fuzzy expert systems and fuzzy controllers as neural nets and as fuzzy neural nets. They arrived at the conclusion that fuzzy neural nets produce a more compact representation of fuzzy systems. Jang *et al.* (1997) discussed the theory and applications of fuzzy inference systems and neuro-fuzzy modelling. The models developed were compared with conventional models. Purvis *et al.* (1997) described a neuro-fuzzy architecture for environmental modelling. A case study of golf courses in the South Island of New Zealand was presented for illustration. See & Openshaw (1999) applied soft computing technology for flood forecasting. They used a self-organizing map (SOM) neural net to preclassify the water level data. The fuzzy logic model was used to integrate the individual MLPs based on current river levels. They optimized the fuzzy logic model with a genetic algorithm. Jain *et al.* (1999) applied artificial neural networks for inflow prediction and reservoir operation. They selected the Upper Indravati project in Orissa, India, as the focus area. The ANN performed better than the ARIMA model for flow prediction. A dynamic programming model was used to develop the reservoir operating policy and the optimal releases obtained were related with storage, inflow and demand through linear and nonlinear regression and the ANN. The ANN was found to perform better.

The objective of the present study is to develop a neuro-fuzzy inference system for operation of a multi-purpose reservoir, the Hirakud Reservoir on Mahanadi River in Orissa State, India. The fuzzy inference system model attempts to capture the expertise gained from the past experience of operating the reservoir. The operation of the reservoir is simulated and its performance is evaluated.

ADAPTIVE NEURO FUZZY INFERENCE SYSTEMS (ANFIS)

The Adaptive Neuro Fuzzy Inference System (ANFIS) incorporates the concepts of neural network learning in fuzzy inference systems and has the ability to model any nonlinear function (Jang *et al.*, 1997). ANFIS models are being employed in a wide variety of applications of modelling, decision making, signal processing and control. The basic features of ANFIS are described in the following sections.

Fuzzy inference systems

The fuzzy inference system is a popular computing framework based on the concepts of fuzzy logic. The basic structure of a fuzzy inference system consists conceptually of three components namely a rule base, a database, and a reasoning mechanism. The rule base contains a selection of fuzzy if–then rules, while the database defines the membership functions used in the fuzzy rules. The reasoning mechanism performs the inference procedure upon the rules and given facts to derive a reasonable output or conclusion. The fuzzy inference system used in this study is the Sugeno fuzzy model. A typical fuzzy rule in a two-inputs Sugeno fuzzy model has the form:

$$\text{if } x \text{ is } A \text{ and } y \text{ is } B \text{ then } z = f(x,y) \tag{1}$$

where A and B are fuzzy sets over inputs x and y, respectively, while $z = f(x,y)$ is a crisp function which is the output of the system. If $f(x,y)$ is a first order polynomial of x and y, then the fuzzy inference system is called a first-order Sugeno fuzzy model. In the present study the architecture of the ANFIS model used represents a first-order Sugeno fuzzy model.

Architecture of ANFIS Model

ANFIS, proposed by Jang (1993), is based on the first-order Sugeno fuzzy model. The neural network paradigm used is a multi-layer feed-forward back propagation network. For simplicity, the fuzzy inference system under consideration is assumed to have two inputs, x and y, and one output z. For a first-order Sugeno fuzzy model, a typical rule set with two fuzzy if–then rules can be expressed as:

Rule 1: If x is A_1, and y is B_1, then $f_1 = p_1x + q_1y + r_1$
Rule 2: If x is A_2 and y is B_2 then $f_2 = p_2x + 1_2y + r_2$

Figure 1(a) illustrates the fuzzy model and the corresponding equivalent ANFIS architecture is shown in Fig. 1(b). In the ANFIS, nodes in the same layer have similar functions, as described below. The output of node i in Layer l is denoted as O_{li}.

– Layer 1: Every node in this layer is an adaptive node with a node output defined as:

$$O_{1,i} = \mu_A(x) \text{ for } i = 1,2 \quad (2)$$

$$O_{1,i} = \mu_{Bi-2}(y) \text{ for } i = 3,4 \quad (3)$$

where x (or y) is the input to the node; and A_i (or B_{i-2}) is fuzzy set associated with this node.

– Layer 2: Every node in this layer is a fixed node labelled Π, which multiplies the incoming signals and outputs the product. For instance:

$$O_{2,i} = w_i = \mu_{Ai}(x) \times \mu_{Bi}(y),\ I = 1,2 \quad (4)$$

Each node output represents the firing strength of a rule.

– Layer 3: Every node in this layer is a fixed node labelled **N**. The i-th node calculates the ratio of ith rule's firing strength to the sum of all rules' firing strengths:

$$O_{3,i} = \mathbf{W}_i = \frac{w_i}{w_1 + w_2},\ i=1,2 \quad (5)$$

– Layer 4: Every node in this layer is an adaptive node with a node function:

$$O_{4,i} = \mathbf{W}_i f_i = \mathbf{W}_i (p_i x + q_i y + r_i) \quad (6)$$

where $\mathbf{W}_i$ is output of Layer 3 and $\{p_i, q_i, r_i\}$ is the parameter set.

– Layer 5: The single node in this layer is fixed node labelled ∑, which computes the overall output as the summation of the incoming signals:

$$O_{5.1} = \text{overall output} = \sum_i \mathrm{W}if_i = \frac{\sum_i wif_i}{\sum w_i} \quad (7)$$

Thus, an ANFIS network is functionally equivalent to a Sugeno fuzzy model. This network can easily be extended to a Sugeno fuzzy model with multiple inputs and rules.

The output f of an ANFIS network shown in Fig. 1 can be written as:

$$f = \frac{w_1}{w_1 + w_2} f_1 + \frac{w_2}{w_1 + w_2} f_2 = \mathbf{W}_1 f_1 + \mathbf{W}_2 f_2 = (\mathbf{W}_1 x)p_1 +$$

$$(\mathbf{W}_1 y)q_1 + (\mathbf{W}_1)\, r_1 + (\mathbf{W}_2 x)p_2 + (\mathbf{W}_2 y)q_2 + (\mathbf{W}_2)r_2 \quad (8)$$

where p_1, q_1, r_1, p_1, q_2 and r_2 are the parameters of the model. From equation (8), it is observed that the output is linear in the parameters p_1, q_1, r_1, p_2, q_2 and r_2, which are known as consequent parameters.

The nodes in Layer 1 are adaptive nodes with a node function given by equation (2). The output, $O_{1,i}$ of node i in this layer is the membership grade of a fuzzy set A (=A_1, A_2, B_1 or B_2) and it specifies the degree to which the given input x (or y) satisfies the quantifier A. The membership function for A can be any appropriate parameterized membership function. If a generalized bell function is used, the membership function is given by:

$$\mu_{Ai}{=}(x) = \frac{1}{1 + \left|\dfrac{x - c_i}{a_i}\right|^{2b_i}} \quad (9)$$

where $\{a_i, b_i, c_i\}$ is the parameter set. These parameters are referred to as premise parameters. The output of the network f is obviously nonlinear in premise parameters. Thus the set of total

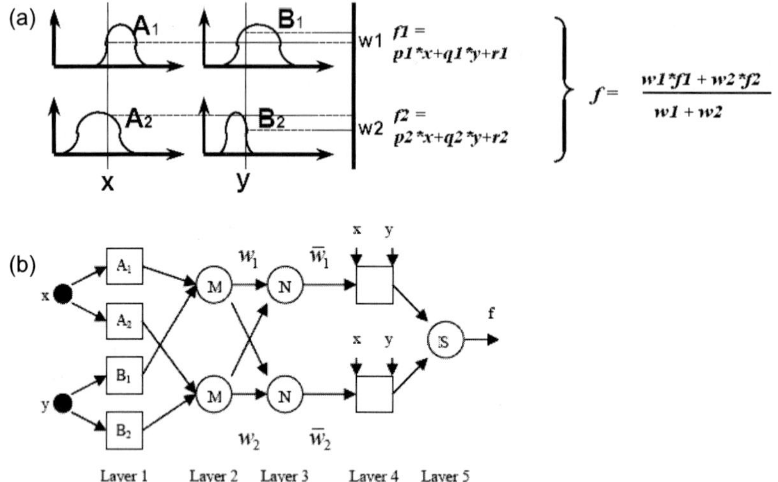

Fig. 1 (a) Fuzzy inference system with two inputs and two fuzzy rules. (b) Typical architecture of a two-input ANFIS Model.

parameters S can be partitioned into two subsets: a set of premise (nonlinear) parameters S_1 and a set of consequent (linear) parameters S_2.

Learning algorithm

Jang (1993) proposed a hybrid learning algorithm for training ANFIS. The learning takes place in two stages. In the forward pass of the hybrid learning algorithm, functional signals go forward until Layer 4, and the consequent parameters are identified by the least squares estimate. In the backward pass, the error rates propagate backward and the premise parameters are updated by the gradient descent, similar to a back-propagation algorithm. The hybrid approach is much faster in training the network than the strict gradient descent or back propagation algorithm.

Jang *et al.* (1997) have shown that ANFIS has unlimited approximation power for matching any nonlinear function arbitrarily well, provided the number of rules is not restricted. Thus ANFIS can be considered as a universal approximator. This property of ANFIS is very useful in modelling highly nonlinear systems. ANFIS can also be used in control systems. Jang (1993) and Jang *et al.* (1997) presented four examples to demonstrate the applicability of ANFIS. In the first two examples, ANFIS is used to model two nonlinear functions and the results are compared with those achieved by neural networks and fuzzy modelling. In the third example, ANFIS is used for online identification of a nonlinear component in a discrete control system. In the last example, a chaotic time series is predicted using ANFIS, and its superiority over standard statistical and neural network methods is demonstrated. All these examples demonstrate the capabilities of ANFIS.

STUDY AREA

The ANFIS model discussed in the previous section is used to develop the operating rules for the Hirakud Reservoir on the River Mahanadi in the State of Orissa, India. The Hirakud Dam is a multi-purpose project built across the River Mahanadi at 21°32′N latitude, 83°52′E longitude. The project provides 155 635 ha of kharif (wet season) and 108 385 ha of rabi (winter season) irrigation in the districts of Sambalpur, Baragarh, Subarnapur and Bolangir of Orissa State. Installed capacity for generation of hydropower by this project is 307.5 MW through its two power houses at Burla (at right bank toe of the dam) and at Chipilima (22.5 km downstream of the dam). Besides this, the project provides flood protection to 9500 km^2 of the Mahanadi Delta in Cuttack, Puri, Khurda and Kendrapara districts.

Hirakud Reservoir is conceived as a multi-purpose reservoir with irrigation, flood control and hydropower generation as the main objectives. The reservoir gets most of its inflows during the monsoon season, which usually commences in the third week of June and continues up to the end of October. The flows into the reservoir are significantly high during August and September, and critically effect the operation of the reservoir.

MODEL DEVELOPMENT

In the present study an ANFIS model is developed for the operating policy of Hirakud Reservoir on the River Mahanadi. The 10-day flows into the reservoir, reservoir levels and releases made for irrigation and hydropower are available for 19 years (from 1981 to 1999). The database is divided into two periods: monsoon and non-monsoon. The monsoon period (rainy season) is from 21 June to 31 October (13 10-day time periods, i.e. June III to October III). The non-monsoon is considered as a single time period (from November I to June II). The reservoir operation during the monsoon is complicated due to conflicting objectives like flood control, irrigation, power generation and conservation at the end of the period. The non-monsoon operation is relatively simple. Hence only reservoir operation for the monsoon period is considered in this study.

The model developed for operation of the reservoir consists of three modules. The first module, Neuro-Fuzzy Release Model (NFRM) is an ANFIS model with two inputs, namely, reservoir inflow and reservoir level during the time period, and one output, the total release made from the reservoir during the period. The second module, Neuro-Fuzzy Irrigation Release Model (NFIRM), takes reservoir inflow, level and total release as inputs and gives the release made for irrigation as output. In the third module, the Neuro-Fuzzy Power Generation Model (NFPGM) inflow, initial storage and the total release made from the reservoir are taken as inputs, while the power generation during the time period is the output of the model. These three ANFIS models are trained using the available past data. Out of the total 247 data sets available (10 daily data for monsoon period for 19 years), 226 data sets are used for training the ANFIS models, while the remaining 21 data sets are used for testing the models.

The inputs of all the ANFIS models are classified into three fuzzy sets. The hybrid learning algorithm proposed by Jang (1993) is used to train the ANFIS models with the training data sets. The trained ANFIS networks are tested using the testing data sets and the performance of the models is evaluated.

RESULTS AND DISCUSSIONS

In the model NFRM, the inflows and the initial storage levels are considered as input variables, and total release made during the time period is the output. The model has 35 nodes, 45 parameters (27 linear and 18 nonlinear), with 9 fuzzy rules. This network gives a root mean squared error of 0.0283 in training, and 0.0264 in testing. The correlation coefficient between modelled and observed output is 0.9736 in training and 0.9582 in testing phases. The release plan with the knowledge of inflow and initial storage level of the reservoir, is shown in Fig. 2 in the form of a decision surface. The operation rules for total release are presented in Table 1. In the model NFIRM, inflow, initial storage level and total release are considered as input variables, and irrigation releases made during the time period are considered as the output. The model has 78 nodes, 135 parameters (108 linear and 27 nonlinear), with 27 fuzzy rules. This network gives a root mean squared error of 0.0634 in training, and 0.0550 in testing. Similarly the correlation coefficients are 0.9522 and 0.8907 in training and testing, respectively. In the model NFPRM, the inflows, the initial storage levels, and total release are considered as input variables, and power generation made during the time period is taken as the output. The third network NFPRM has 78 nodes, 135 parameters (108 linear and 27 nonlinear), with 27 fuzzy rules. This network gives a root mean squared error of 0.1117 in training and 0.0859 in testing. The correlation coefficients are 0.9655 and 0.9853 for training and testing, respectively.

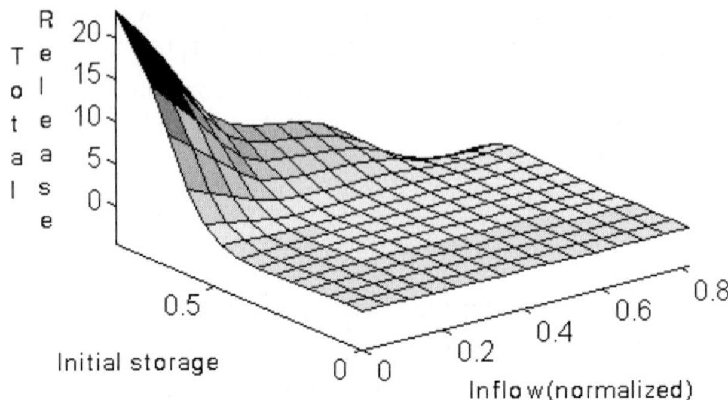

Fig. 2 Operation rule generated by NFRM model.

Table 1 Reservoir operation rules for total release.

Rule No.	Initial storage	Inflow	Total Release $Z^* = f(\times 1, \times 2) = (p^* \times 1 + q^* \times 2 + r)$
1	Low	Low	$Z = (0.4848^* \times 1 + 0.2403^* \times 2 + 0.05939)$
2	Low	Medium	$Z = (6.784^* \times 1 - 4.462^* \times 2 + 1.017)$
3	Low	High	$Z = (-49.71^* \times 1 + 45.58^* \times 2 - 12.27)$
4	Medium	Low	$Z = (-0.1885^* \times 1 + 0.913^* \times 2 + 0.1085)$
5	Medium	Medium	$Z = (0.7683^* \times 1 + 0.5439^* \times 2 - 0.1376)$
6	Medium	High	$Z = (-1.344^* \times 1 + 13.4^* \times 2 - 5.084)$
7	High	Low	$Z = (-0.02633^* \times 1 + 0.3606^* \times 2 - 0.09069)$
8	High	Medium	$Z = (-1.46^* \times 1 + 0.5615^* \times 2 + 0.4764)$
9	High	High	$Z = (6.545^* \times 1 - 30.71^* \times 2 + 14.75)$

* ×1 is initial storage and ×2 is inflow (both in million cubic metres).

The reservoir is simulated over 19 years using the historical data in time steps of 10 days in the monsoon period. Knowing the storage level at the beginning of the period and inflow during the period, the NFRM model is used to determine the total release to be made from the reservoir. Subsequently the models NFIRM and NFPGM are used to determine irrigation release and power generated during that time period. This process is repeated over all time periods in each year. After simulating over the last time period of the monsoon season, the storage level at the end of the monsoon period is determined. Using the results of the simulation, the performance of the reservoir is evaluated with reference to the four objectives of the reservoir. The four objectives of the reservoir are as follows:

- To keep the reservoir "as empty as possible" to absorb anticipated floods.
- To keep the reservoir "as full as possible" by the end of monsoon.
- To "satisfy" the irrigation demand at different time periods.
- To generate "sufficient amounts" of hydropower in each time period.

These four objectives are treated as fuzzy, and fuzzy membership functions are developed to evaluate the level of achievement of these objectives. A sigmoid membership function is used which is expressed mathematically as:

$$\mu(x) = \frac{1}{1 + \exp[-a(x - c)]}$$

where $\mu(x)$ is the value of membership for a given x, a and c are the parameters, where c is the centroid of the curve. The maximum and minimum values of the objective for each time period from historical data are noted. Then c and a are found by trial-and-error so as to yield the fittest membership function. For the flood control objective, the lowest level of the reservoir in a given time period, corresponds to "Highest satisfaction" with membership value of 1.0 and the reservoir full condition corresponds to "Lowest satisfaction" with membership value of 0.0. The objective for filling the reservoir at the end of the monsoon period considers the reservoir empty condition as the "Lowest satisfaction" with membership value of 0.0, and the reservoir full condition as the "Highest satisfaction" with membership value of 1.0. For irrigation objective, fulfilment of the total irrigation demand with "Highest satisfaction level" corresponds to the membership value of 1.0. Supply of <50% of the irrigation demand is considered as the "Lowest satisfaction level", which corresponds to a membership value 0.0. Likewise the power generation objective considers the generation of full potential power during that period as the "Highest satisfaction level" with a membership value 1.0. No power generation corresponds to "Lowest satisfaction level" with the membership value of 0.0. The level of achievement of these objectives is evaluated by simulating the reservoir with the proposed operating policy.

The first objective, the flood control, requires that the reservoir should be kept empty to absorb any anticipated flood. The results of simulation indicate that the performance of the reservoir with reference to this objective is highly satisfactory, with an average level of satisfaction of 88.62% and a standard deviation of 0.2380. The degree of achievement of this objective is above 90% in all time periods, except during the 6th year and 15th year of simulation, the reason being, the high inflows in these years.

The second objective is to keep the reservoir as full as possible by the end of monsoon. This objective is achieved with a consistently high degree of satisfaction. The average level of satisfaction is 94%, with a very low level of standard deviation of 0.038. The lowest level of satisfaction is 85.5% observed in the 9th, 12th and 14th years. Thus it can be concluded that the present operating policy is able to achieve this objective with a very high degree of satisfaction.

The third objective is to satisfy the irrigation demand at different time periods. Based on the results obtained from simulation, it is observed that the level of achievement of this objective is not as high as that of the other two objectives. The average level of achievement for this objective is only 79.91%, with a relatively high standard deviation of 0.2703. The lowest level of satisfaction achieved is 7%, observed in the 5th, 6th, 8th, 9th, and 10th time periods.

The fourth objective is to generate a sufficient amount of hydropower in each time period. Based on the results obtained from simulation, it is observed that this objective is achieved with a medium level of satisfaction of 77.32%, with a relatively high standard deviation of 0.2893. The level of achievement of this objective is lowest among all the objectives considered. The lowest level of satisfaction achieved is 7.1% in the 3rd, 9th and 15th years. The level of satisfaction of this objective is observed to be very low in these three years. The reason for the very low level of satisfaction during these three years is attributed to medium inflows. The overall efficiency is low compared to the first three objectives. Based on these results it can be observed that there is scope for improving the operation of the reservoir with reference to irrigation and hydropower.

REFERENCES

Buckley, J. J. & Hayashi. Y. (1994) Fuzzy neural networks: a survey. *Fuzzy Sets & Systems* **66**, 1–3.

Gupta, M. M. & Rao, D. H. (1994) On the principles of fuzzy neural networks. *Fuzzy Sets & Systems* **61**, 1–18.

Haykin, S. (1994) *Neural Networks*. Macmillan College Publishing Co., New York, USA.

Jain, S. K., Das, A. & Srivastava, D. K. (1999) Application of ANN for reservoir inflow prediction and operation. *J. Water Resour. Plan. Manage.* **125**(5), 263–271.

Jang, J.-S. R. (1993) ANFIS: Adaptive-network-based fuzzy inference system. *IEEE Trans. Syst., Man., Cybern.* **23**(3), 665–684.

Jang, J. S. R., Sun, C.-T. & Mizutani, E. (1997) *Neuro-Fuzzy and Soft Computing*. Prentice Hall, Upper Saddle River, New Jersey, USA.

Purvis, M., Kasabov, N., Benwell, G., Zhou, Q. & Zhang. F. (1997) Neuro-fuzzy methods for environmental modelling. In: *Environmental Software Systems 2*, IFIP TC 5 WG5.11 International Symposium on Environmental Software Systems (ISESS'97), British Columbia, Canada.

See, L. & Openshaw, S. (1999) Applying soft computing approaches to river level forecasting. *Hydrol. Sci. J.* **44**(5), 763–778.

Wurbs, R. A. (1996) *Modelling and Analysis of Reservoir System Operations*. Prentice Hall PTR, New Jersey, USA.

Daily discharge forecasting using WANNs coupled with nonlinear bias correction techniques

MUKESH K. TIWARI & C. CHATTERJEE

Agricultural and Food Engineering Department, Indian Institute of Technology, Kharagpur, India

mukesh@agfe.iitkgp.ernet.in

Abstract A wide variety of rainfall–runoff models have been developed and applied for discharge forecasting. Hydrological prediction usually relies on incomplete and uncertain process descriptions that are deduced from sparse and noisy data sets. While conventional (deterministic) ANNs have been traditionally used in these studies, the use of wavelet-based ANNs (WANNs) is a more recent development. Discharge data are decomposed into wavelet components via discrete wavelet transform, and the new wavelet series, consisting of the sum of selected wavelet components, is used as input for the ANN model. Seven years data of daily discharge from seven upstream gauging stations are used for daily discharge forecasting for 1–5 day lead time forecasts at Naraj site in Mahanadi River basin, India. Performances of the neuro-wavelet hybrid forecasting model provided satisfactory results for up to 2-day lead time forecasts. Performance of the wavelet-ANN model is found to be better compared to traditional ANN for all the lead times. The resulting outputs are corrected for bias using nonlinear correction techniques. Nonlinear bias corrections are found to be very suitable tools for daily discharge forecasting using ANN, but the wavelet-ANN performance does not improve using the nonlinear bias correction technique showing that the wavelet technique itself reduces the systematic shifting between the observed and predicted discharge.

Key words neural networks; wavelet; forecasting, discharge

INTRODUCTION

Daily discharge forecasting is desirable for water resources planning. A wide variety of rainfall–runoff models have been developed and applied for discharge forecasting. Most of the rainfall–runoff models have been developed based either on a mechanistic approach or on a systems theoretic approach. Hydrological prediction usually relies on incomplete and uncertain process descriptions that have been deduced from sparse and noisy data sets. The spatially distributed modelling approach has been criticized for resulting in models that are overly complex, leading to problems of over parameterization and equifinality (Beven, 2006), which may manifest itself in large prediction uncertainty (Uhlenbrook *et al.*, 1999), while in the system theoretic approach the concern is with the system operation, not the nature of the system itself or the physical laws governing its operation.

Use of artificial neural networks (ANNs) based on the systems theory approach has gained momentum in last few decades for river flow forecasting. The ability of ANNs in mapping complex nonlinear rainfall–runoff relationships has increased the number of applications in rainfall–runoff modelling and river discharge forecasting (Jain & Srinivasulu, 2004). Substantial literature on ANN have been reported in ASCE (2000a,b). Input selection is one of the most important phases in the ANN process. Wavelet transforms, which provide information in both the time and frequency domains of the signal, give considerable information about the physical structure of the data, and wavelet analysis provides a time–frequency representation of a signal at many different periods in the time domain (Daubechies, 1990).

Recently, the wavelet-ANN model has been successfully employed in some hydrology and water resource studies. For example, Wang & Ding (2003) proposed a wavelet network model with a combination of the wavelet transform and the ANN; they decomposed the original time series into periodic components by wavelet transform. Later, sub-time series were used as the inputs for ANNs, and the resulting model was applied to forecast the original time series. This approach was used for monthly groundwater level and daily discharge forecasting. Kim & Valdes (2003) presented a hybrid neural network model combined with dyadic wavelet transforms to improve the forecasting of regional droughts. The researchers applied this model to the monthly and annual inflow and rainfall data and showed that the model significantly improved the neural network's forecasting performance. Anctil & Tape (2004) used a wavelet–neural network model successfully for one-day ahead rainfall–runoff forecasting. Partal & Cigizoglu (2008) predicted the suspended sediment load in rivers by a combined wavelet–ANN method. Measured data are decomposed into wavelet components via discrete wavelet transform, and the new wavelet series, consisting of the sum of selected wavelet components, is used as input for the ANN model. The wavelet–ANN model provided a good fit to observed data for the testing period. In this study, the wavelet transforms and the ANN are employed collectively in order to estimate the daily discharge forecasts for 1–5 day lead times.

THEORY

Artificial neural networks

Artificial neural networks are information processing systems composed of simple processing elements (nodes) linked by weighted synaptic connections (Muller & Reinhardt, 1991). Neural network models are developed by training the network to represent the relationships and processes that are inherent within the data. Being essentially nonlinear regression models, they perform an input–output mapping using a set of interconnected simple processing nodes or neurons. They reconstruct the complex nonlinear input/output relations by combining multiple simple functions, by analogy with the functioning of the human brain. This approach is fast and robust in noisy environments, flexible in the range of problems it can solve, and highly adaptive to newer environments.

The wavelet analysis

Wavelet analysis emerged as a filtering and data compression method in the 1980s. Wavelet analysis transforms a time series simultaneously in the depth or time domain and scale or frequency domain by using various shapes and sizes of short filtering functions called wavelets. Wavelet transforms (WTs) allow for the automatic localization of periodic-signals, gradual shifts and abrupt interruptions, trends and onsets of trends in time series. WT uses narrow band analysis windows at high frequencies, and wide analysis windows at low frequencies, in contrast to the Sliding-Window Fourier transform that uses shifting analysis windows of constant width. Thus, trend analysis using the proposed CWT method involves separating the trend and stochastic component using wavelet coefficients.

Assuming the signal $f(t) \in L^2(R)$ (set of signals of finite energy) the continuous wavelet transform is:

$$W_f(a,b) \leq f, \psi_{a,b} \geq |a|^{-\frac{1}{2}} \int_R f(t)\psi\left(\frac{t-b}{a}\right)dt \quad (1)$$

where $W_f(a,b)$ is the wavelet coefficient and a and b are real numbers. Thus, the wavelet transform is a function of two variables, a and b. The parameter "a" can be interpreted as a dilation ($a > 1$) or contraction ($a < 1$) factor of the wavelet function $f(t)$ corresponding to different scales of observation. The parameter "b" can be interpreted as a temporal translation or shift of the function $f(t)$, which allows the study of the signal $f(t)$ locally around the time b.

Nonlinear bias correction

Linear bias correction, as applied by Hansen & Ines (2005) and Jana *et al.* (2007), provides a proportional shifting effect to the ANN-predicted curve and brings the mean of the ANN-predicted values closer to that of the target values. Matching of cumulative distribution functions (CDFs) is a technique that has been used to correct for nonlinear bias. The technique is based on the idea of obtaining the predicted parameter values corresponding to the probability of occurrence of values on the CDF of the target parameter. Statistical tests are conducted to test for the type of distribution (e.g. normal, lognormal, gamma) of the parameters. CDFs are then obtained for the target and predicted values for each parameter based on the type of distribution they follow. For a particular predicted discharge value, there exists a particular probability of occurrence. Similarly, for a particular probability of occurrence, there exists a corresponding target discharge value. CDF matching is achieved by forcing the predicted discharge value with a particular occurrence probability toward the corresponding target discharge value.

Nash-Sutcliffe efficiency (E), root mean square errors (RMSE), mean absolute errors (MAE), and percentage peak deviation (P_{dv}) performance measures are used to evaluate the accuracy of the bootstrap based ANN.

Table 1 Statistics of the data set of monsoon period for daily river flow forecasting (1 July to 10 October).

Year	Statistics	Khairmal	Hirakud Release	Salebhata	Kesinga	Kantamal	Tikarpara	Naraj
2000	Mean (m^3/s)	1461.5	615.4	22.4	300.6	473.9	1621.7	1873.9
	Standard deviation (m^3/s)	941.9	641.6	21.4	322.5	569.1	959.1	892.2
	Maximum (m^3/s)	5720.0	4601.3	114.0	2452.0	3661.0	4774.8	5050.0
	Minimum (m^3/s)	305.8	132.9	0.7	40.0	63.8	320.0	344.1
2001	Mean (m^3/s)	6933.1	3657.1	319.0	1240.7	1871.8	5929.8	9417.1
	Standard deviation (m^3/s)	6543.9	4224.8	519.5	1862.9	2269.7	5566.4	8756.6
	Maximum (m^3/s)	30468.9	21567.3	3225.0	12822.0	12770.0	26700.0	36714.7
	Minimum (m^3/s)	781.5	350.8	6.4	90.2	301.5	818.6	1567.8
2002	Mean (m^3/s)	1730.1	963.5	100.4	186.5	346.4	1780.2	2615.8
	Standard deviation (m^3/s)	2454.5	1674.2	212.3	290.5	518.0	2112.4	3104.4
	Maximum (m^3/s)	14724.8	10329.4	1545.4	1990.9	2900.0	12305.6	16630.3
	Minimum (m^3/s)	305.8	43.2	1.1	10.0	20.2	436.8	215.1
2003	Mean (m^3/s)	7095.3	3910.2	517.1	1057.9	1327.8	5957.4	9309.0
	Standard deviation (m^3/s)	7632.2	4729.3	1012.8	1209.2	1817.5	6001.0	9191.1
	Maximum (m^3/s)	34150.1	24267.2	7916.0	8908.4	12915.1	25062.0	35253.2
	Minimum (m^3/s)	611.6	373.0	0.0	100.0	98.3	522.2	1028.6
2004	Mean (m^3/s)	3680.9	2036.6	155.2	618.5	808.9	3415.8	4398.8
	Standard deviation (m^3/s)	4117.1	2844.0	209.5	589.4	656.3	3447.7	4871.6
	Maximum (m^3/s)	20331.5	14952.0	1004.0	3240.6	4014.0	17744.4	22465.4
	Minimum (m^3/s)	781.5	308.3	2.1	104.5	143.6	820.7	720.7
2005	Mean (m^3/s)	4399.6	2717.9	108.9	520.9	764.6	3918.1	6355.6
	Standard deviation (m^3/s)	4988.1	3261.9	245.7	1031.9	1528.0	3810.0	6510.8
	Maximum (m^3/s)	26249.7	13196.6	1924.0	8120.8	11030.1	19000.0	24399.1
	Minimum (m^3/s)	747.6	393.9	1.0	100.0	113.2	491.3	356.7
2006	Mean (m^3/s)	6195.6	2422.3	404.2	1340.5	2074.8	5807.9	9653.2
	Standard deviation (m^3/s)	5782.3	2895.9	742.9	2365.2	3315.6	5265.9	8659.3
	Maximum (m^3/s)	27099.2	11870.3	4681.4	21192.0	17500.0	29000.0	33979.2
	Minimum (m^3/s)	509.7	91.8	0.7	90.0	134.0	424.0	421.4

STUDY AREA AND DATA USED

Mahanadi River basin, which is the fourth largest river basin in India, is selected for this study. The Mahanadi River flows to the Bay of Bengal in east–central India, draining an area of 141 589 km^2, and has a length of 851 km. It lies between east longitudes 80°30′ to 86°50′ and north latitudes 19°21′ to 23°35′. Numerous dams, irrigation projects, and barrages are present in the Mahanadi River basin, the most prominent of which is Hirakud Dam. The middle reaches of the lower Mahanadi River basin, located in Orissa between 82°E 19°N and 86°E 22°N, and encompassing a geographical area of 47 558.6 km^2, forms the study area (Fig. 1). The main river reach extends from Hirakud Dam to Naraj, having a total length of 358.4 km. The main soil types found in the study area are red and yellow soils. The normal annual rainfall is 1458 mm and temperature in this region varies from 14°C to 40°C. The average monthly pan evaporation of the area varies from 2.4 mm to 14.6 mm. Most of the rainfall and river flow occur during the monsoon season, between June and September. In the delta region of the Mahanadi River basin, water management and estimation of discharge is critical during monsoon seasons. Naraj, which is situated at the mouth of the delta, is selected for daily discharge forecasting.

The data used for the study consists of daily discharge data of seven upstream gauging stations during the period 2000–2006. Some of the statistical properties of the discharge data are presented in Table 1. The location of the gauging stations is shown in Fig. 1.

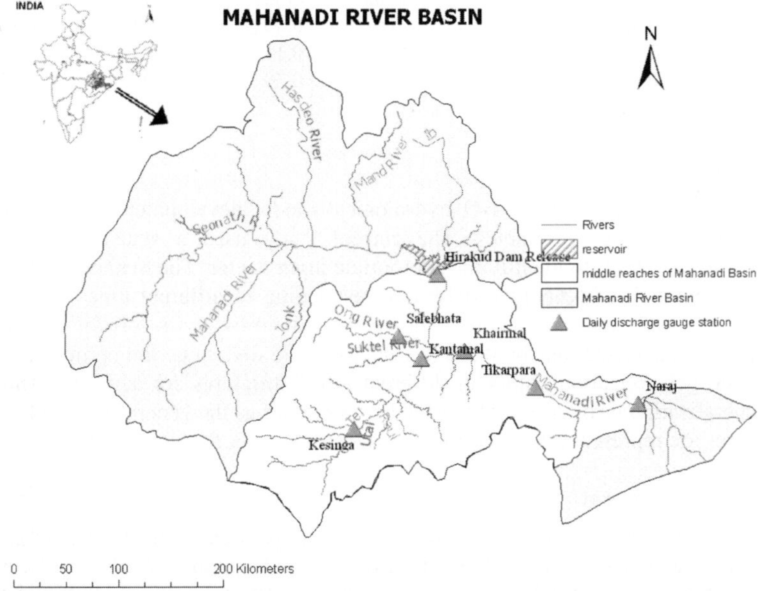

Fig. 1 Index map of the middle reaches of Mahanadi River basin showing location of different gauging stations.

METHODOLOGY

Wavelet decomposition

First, the original time series are decomposed into periodic components by DWT. Each of the components is a detailed time series representing the original data. The correlation coefficients between the wavelet components (inputs) and observed discharge at Naraj is calculated and the appropriate components that have significant correlation with Naraj are determined, and a new series for each variable is constituted by summing the determined components. The new series is

used as inputs of the wavelet-ANN model. The new decomposed components are the time series and the present variations on the different periods. Each of the wavelet components plays a distinct role in the original time series and has different effects on the observed discharge data. The wavelet decomposition is carried out for three levels of wavelet decomposition. Three levels of decomposition and approximation for Khairmal discharge data are shown in Fig 2. The effective wavelet components are determined using the correlation coefficients between the wavelet components and the observed discharge at Naraj.

Table 2 shows that the significant wavelet coefficients are A3 and D1 for Khairmal, Hirakud release data and Kantamal; A3, D2 and D3 for Tikarpara and Naraj; and only approximation A3 for Salebhata and Kesinga constituted the new wavelet discharge series. The significant wavelet components of a particular gauging station are added and employed to constitute the inputs of the wavelet-ANN for daily discharge forecasting.

Table 2 The correlation coefficients between the discrete wavelet components and the observed discharge data at Naraj.

Discrete wavelet components	Gauging stations: Khairmal	Hirakud Release	Salebhata	Kesinga	Kantamal	Tikarpara	Naraj
A3	0.91	0.87	0.75	0.60	0.66	0.92	0.93
D1	–0.04	–0.04	–0.01	–0.04	–0.02	0.01	0.14
D2	0.06	0.03	0.00	–0.01	0.00	0.13	0.19
D3	0.21	0.13	0.07	0.09	0.11	0.23	0.27
Original	0.88	0.81	0.52	0.39	0.50	0.93	1.00

ANN model development

One of the most important steps in the ANN hydrological model development process is the determination of significant input variables. The current study used a statistical approach suggested by Sudheer *et al.* (2002) to identify the appropriate input vector. The method is based on the heuristic that the potential influencing variables corresponding to different time lags can be identified through statistical analysis of the data series that uses cross-correlations, auto-correlations, and partial autocorrelations between the variables. To determine the optimal number of hidden neurons, the generalization ability of different ANN structures are compared and the model with least generalization error is selected as the best structure as the generalization errors of various ANN structures are representative of the ANN performance (Jia & Culver, 2006).

Wavelet-ANN model development

The wavelet-ANN and ANN models are developed using the most significant inputs, which are linearly scaled to the range (0, 1) for ANN modelling (e.g. Campolo *et al.*, 1999). Log-transformation is used for conventional ANN model development. The computational efficiency of the training process is an important consideration for ANN modelling. A computationally efficient second-order training method, the Levenberg–Marquardt method (Hagan & Menhaj, 1994), is used to minimize the mean squared error between the forecast and the observed gauges. Daily discharge data of seven years for the monsoon period from 2000 to 2006 (714) are used for this study. Five years of data sets from 2000 to 2004 (510) are used for training and the remaining two years of training data sets, 2005–2006 (204) are used for testing the developed wavelet-ANN models. There are several methods to avoid over-fitting in ANNs, which are summarized by Giustolisi & Laucelli (2005). In our study, we employed the "early stopping" technique, in which the training set is split into two subsets: the estimation and the validation. In our study, 2/3 of the first 510 values in the training set (340 values) are considered for estimation, and the last 170 values are used for validation. The mean square error (MSE) is computed at each training step by means of

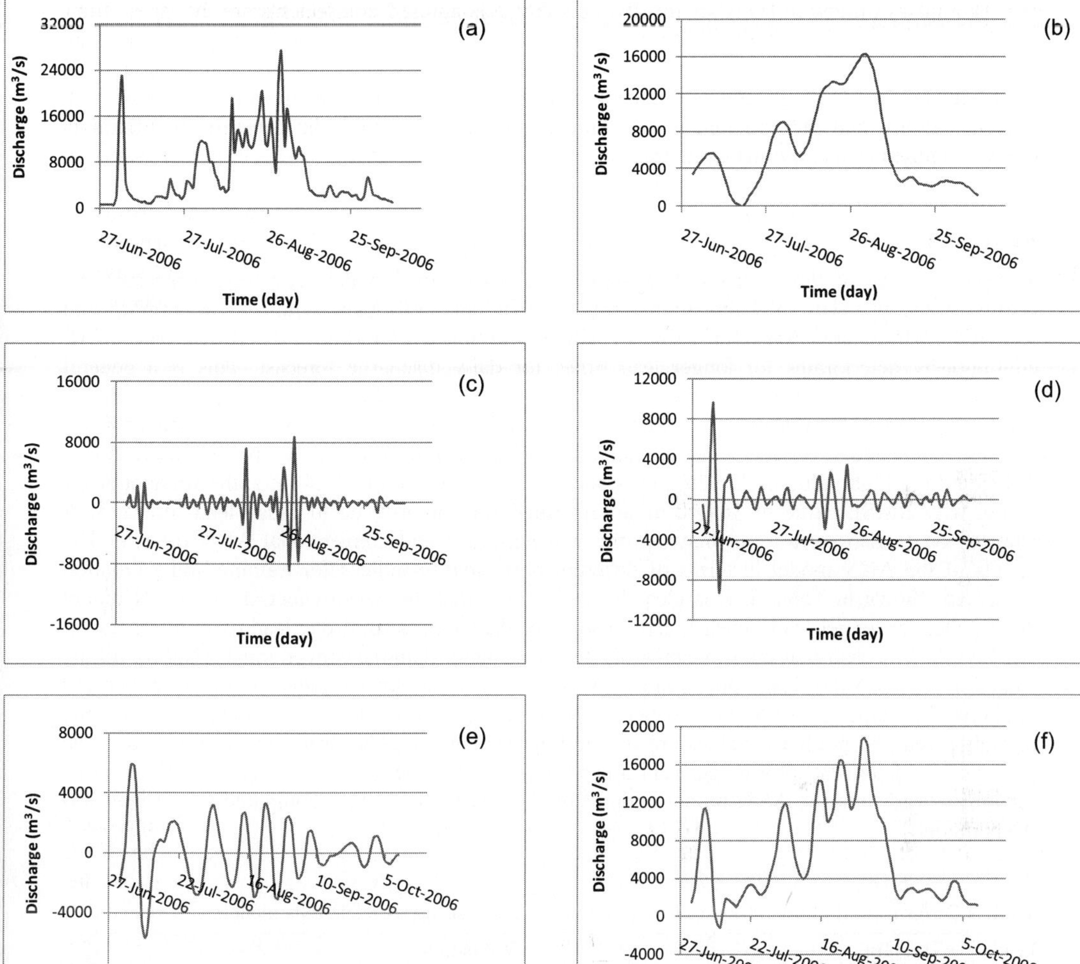

Fig. 2 The discrete wavelet components of the discharge series of Khairmal for the year 2006: (a) original; (b) A1; (c) D1; (d) D2; (e) D3; (f) A3 + D3.

Table 3 Most significant input variables selected using cross-correlation statistics (CCF, ACF, PACF).

Stations	Lags (days)
Naraj	1–2
Tikarpara	1–2
Khairmal	1–3
Kantamal	2–3
Hirakud Release	1–3
Salebhata	1–3
Kesinga	2–3

the validation subset, while the search direction is computed by means of the estimation subset. The training is stopped as soon as the validation error rate starts to go up.

The cross-correlation statistics for daily discharge is applied to select significant inputs from the discharge data of seven discharge gauging stations for daily river discharge forecasting. The total input vectors identified are presented in Table 3. The optimum numbers of hidden neurons are calculated using the generalization error. The ANN and WANN structures are tested for 1–6 hidden neurons, and the structure with 4 hidden neurons, for which the generalization error is lower, is chosen as the optimal structure.

RESULTS

The performance of the wavelet-ANN model is checked for 1–5 day lead time forecasts. The results of the wavelet-ANN model in terms of different performance indices for training and validation period are shown in Table 4. The performance in terms of E, RMSE and MAE continuously deteriorates for longer lead times for daily discharge forecast. This is a general tendency as for longer lead time the last river flow values contain less information than for the shorter lead times for mapping the flow values at longer time horizon. The RMSE statistic is a measure of residual variance and is indicative of the model's ability to predict high flows. Considering a high value of 33 979.2 m^3/s and a very low value of 421.4 m^3/s, the wavelet-ANN model with RMSE value of 3697.2 m^3/s performed satisfactorily up to 2-day lead forecasts. A comparison is also carried out between the wavelet-ANN and conventional ANN models. The results of the ANN model in terms of different performance indices for training and validation period are shown in Table 5. It is clear from the table that the performance of the ANN model deteriorates for longer lead times, except in some of the cases as for 4-day lead time in terms of E and RMSE. It is clear that the wavelet-ANN performs better than the conventional ANN for all the lead time up to 5-day lead time. Figures 3 and 4 show the scatter plots of the observed and predicted discharge for ANN and wavelet-ANN models, respectively. It is observed from the table that observed and predicted values are in good agreement for 1-day lead time forecasts and the performance deteriorates for longer lead times for both the models. It is clear from the figures that for longer lead time the ANN model underestimates the higher values compared to the wavelet-ANN model.

Table 4 Performance indices for 1- to 5-day lead time forecasts using wavelet-ANN model.

Lead time (day)	Training				Validation			
	E	RMSE (m^3/s)	MAE (m^3/s)	Peak deviation (%)	E	RMSE (m^3/s)	MAE (m^3/s)	Peak deviation (%)
1	98.4	767.0	521.0	8.3	90.9	2359.5	1572.5	21.7
2	97.8	911.6	602.9	4.0	77.6	3697.2	2423.6	20.5
3	90.2	1942.8	1060.3	4.5	67.7	4438.6	2753.9	–1.6
4	93.6	1578.0	1163.5	13.4	57.4	5048.9	3583.4	5.2
5	91.7	1791.6	1072.6	15.6	40.4	5982.3	3500.2	–37.1

Table 5 Performance indices for 1- to 5-day lead time forecasts using conventional ANN model.

Lead time (day)	Training				Validation			
	E	RMSE (m^3/s)	MAE (m^3/s)	Peak deviation (%)	E	RMSE (m^3/s)	MAE (m^3/s)	Peak deviation (%)
1	99.1	567.3	293.1	–4.0	89.9	2478.7	1468.6	–1.0
2	90.7	1881.2	809.9	10.4	67.7	4444.4	2338.2	–16.8
3	80.5	2739.4	1222.9	4.1	51.5	5441.5	3237.1	1.2
4	73.8	3186.6	1508.1	23.6	46.4	5661.0	3617.5	3.2
5	69.3	3439.0	1612.4	26.2	25.1	6709.8	4658.5	26.6

The CDF mapping method, which provides corrections to the first four moments of the distribution, is used for nonlinear bias correction. The different lead time forecasts are hypothesized to be normally distributed.

The ANN and wavelet-ANN predicted discharge values for 2005 and 2006 are split into two halves: one for the model calibration and the other for the validation of the bias correction scheme. The cumulative probabilities for each point value are computed using the mean and standard deviation of the calibration data set. In order to effectively scale the model predicted values to the measured data set, we now find θ_i^{pred} such that:

$$CDF(\theta_i^{pred}) = CDF(\theta_i^{target}) \quad (2)$$

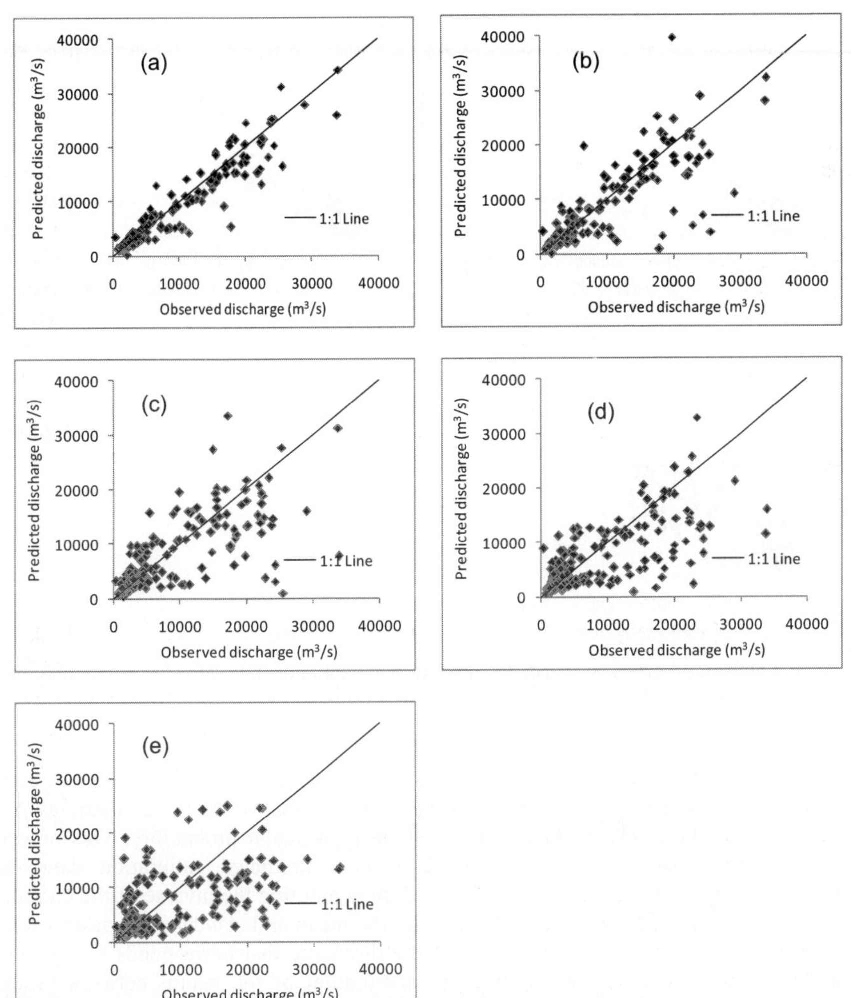

Fig. 3 Scatter plots of observed and predicted discharge of validation datasets of ANN models for 1 to 5 lead time L (day): (a) L = 1 day; (b) L = 2 day; (c) L = 3 day; (d) L = 4 day; (e) L = 5 day.

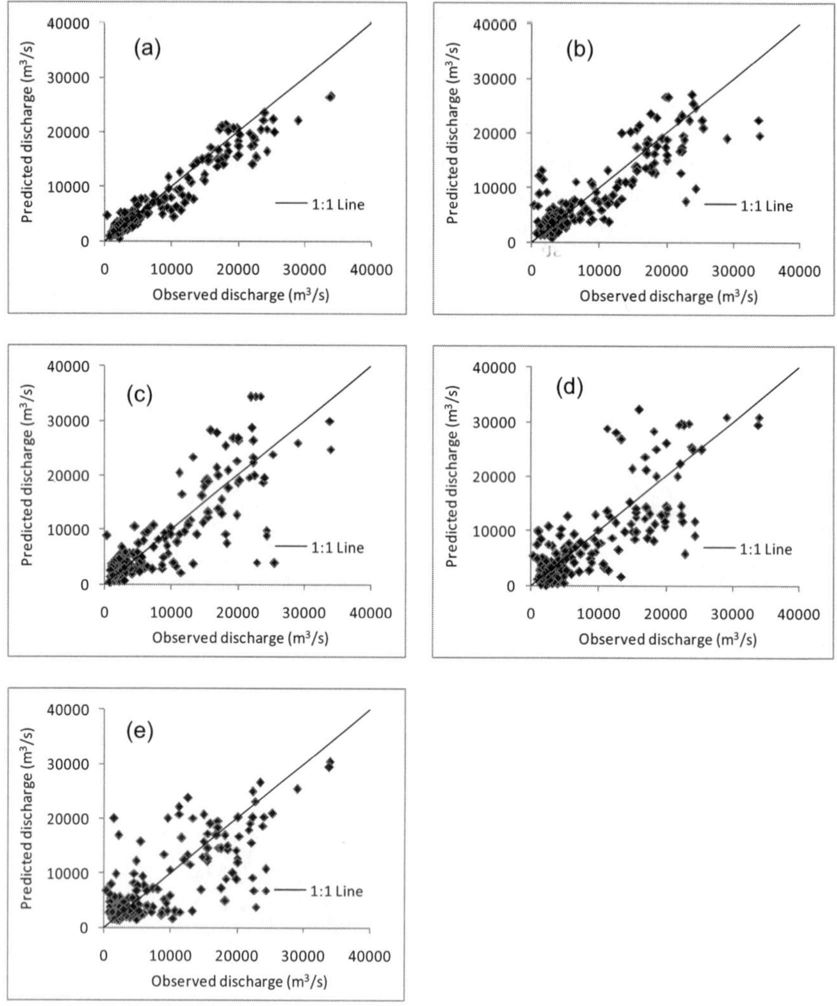

Fig. 4 Scatter plots of observed and predicted discharge of validation data sets of wavelet-ANN models for 1- to 5-lead time L (day): (a) L = 1 day; (b) L = 2 day; (c) L = 3 day; (d) L = 4 day; (e) L = 5 day.

This is achieved by computing the inverse of the cumulative probability values for the calibration data set, but with the mean and standard deviation values of the target distribution. The inverse is the value of the discharge that corresponds to a particular probability. This procedure effectively scales the distribution of the neural network predicted calibration data set to approximate that of the target values. This is achieved by computing the inverse of the cumulative probability values for the calibration data set, but with the mean and standard deviation values of the target distribution. The inverse is the value of the discharge that corresponds to a particular probability. This procedure effectively scales the distribution of the neural network predicted calibration data set to approximate that of the target values. The validated (bias corrected) discharge values are plotted in Fig. 5 for 5-day lead time forecasts of ANN and wavelet-ANN models. It can be seen that observed and bias-corrected prediction are in better agreement than the predicted discharge data sets for ANN model forecasts, but no improvement is seen with wavelet-ANN model forecasts. This shows that the wavelet technique itself has the capability to reduce the systematic shifting between observed and predicted discharge values.

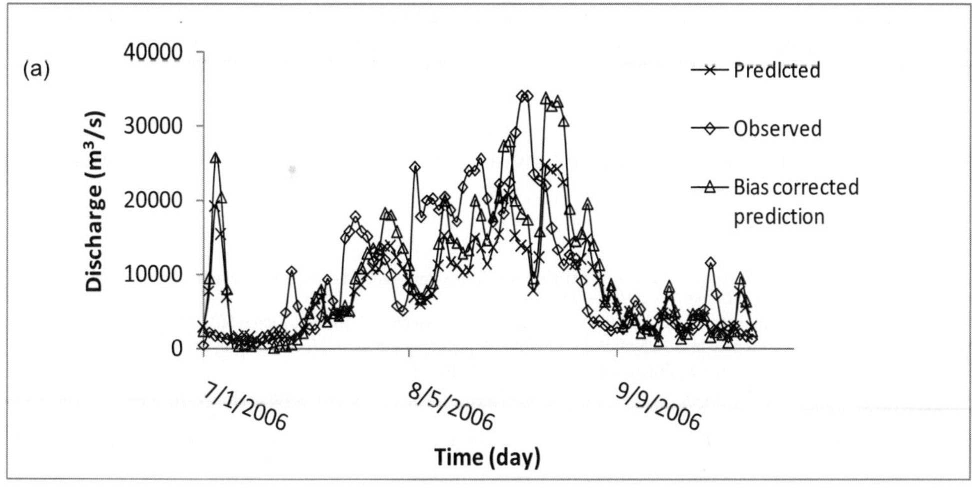

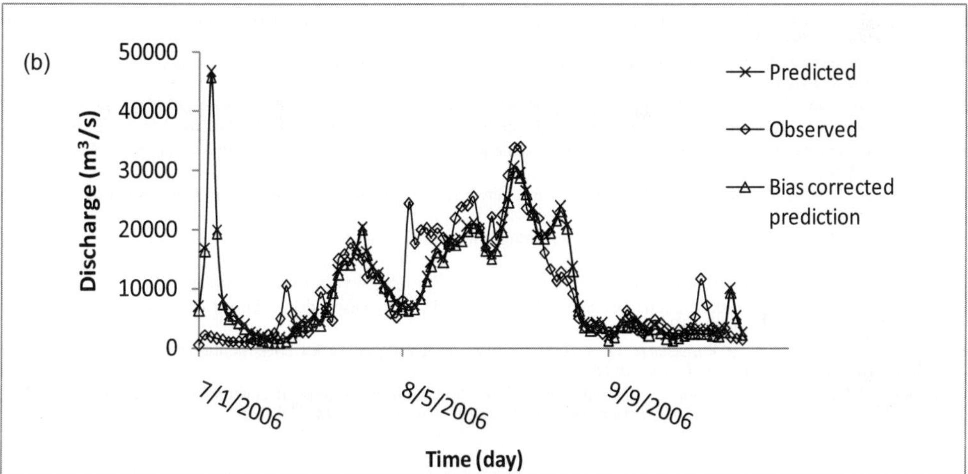

Fig. 5 Observed, predicted and bias corrected predictions by: (a) ANN and (b) wavelet-ANN models.

SUMMARY AND CONCLUSIONS

The study is the first application of the wavelet-ANN model to forecast daily discharge with nonlinear bias correction. The decomposed time series of the observed data present different resolution intervals and detailed information about variations in the inter-annual scale for the different periods. Each of the wavelet components makes a distinct contribution to the original time series. The contributions of the appropriate wavelet components are tested against the ANN model's ability, and a new series is composed by adding the dominant components for each time series: A3 and D1 for Khairmal, Hirakud release data and Kantamal, A3, D2 and D3 for Tikarpara and Naraj are added to each and only approximation A3 for Salebhata and Kesinga constituted the new wavelet discharge series. These series are employed to constitute the inputs to the wavelet–ANN applications in the discharge forecasting simulations. The inclusion of the dominant wavelet components in the ANN input layer enabled consideration of the effective periodic components in the discharge forecasting simulations. This brings a new perspective to the criticism of ANNs: that they do not consider the underlying physical processes in a watershed. The wavelet-ANN supports the recent studies about the physics involved in the ANNs. The wavelet-ANN model is found to be superior to the conventional ANN in terms of the selected performance criteria for all lead forecast times. Nonlinear bias correction is found to be a very suitable tool for daily discharge forecasting using ANNs, but the wavelet-ANN performance did not improve using the nonlinear bias correction technique and showed that the wavelet technique itself reduces the systematic shifting between the observed and predicted discharge. This study shows that the wavelet-ANN method is quite an appropriate tool, together with the conventional ANN method, for modelling discharge.

REFERENCES

Anctil, F. & Tape, D. G. (2004) An exploration of artificial neural network rainfall–runoff forecasting combined with wavelet decomposition. *J. Environ. Engng Sci.* **3**, 121–128.

ASCE (2000a) Artificial neural networks in hydrology. I. Preliminary Concepts. *J. Hydrol. Engng* **5**(2), 115–123.

ASCE (2000b) Artificial neural networks in hydrology. II. Hydrologic applications. *J. Hydrol. Engng* **5**(2), 124–137.

Beven, K. (2006) A manifesto for the equifinality thesis. *J. Hydrol.* **320**, 18–36.

Campolo, M., Andreussi, P. & Soldati, A. (1999) River flood forecasting with a neural network model. *Water Resour. Res.* **35**(4), 1191–1197.

Daubechies, I. (1990) The wavelet transform, time-frequency localization and signal analysis. *IEEE Trans. Information Theory* **36**(5), 6–7.

Giustolisi, O. & Laucelli, D. (2005) Improving generalization of artificial neural networks in rainfall–runoff modeling. *Hydrol. Sci. J.* **50**(3), 439–457.

Hagan, M. T. & Menhaj, M. B. (1994) Training feed forward techniques with the Marquardt algorithm. *IEEE Trans.Neural Networks* **5**(6), 989–993.

Hansen, J. W., & Ines, A. V. M. (2005) Stochastic disaggregation of monthly rainfall data for crop simulation studies. *Agric. For. Meteorol.* **131**, 233–246.

Jain, A. & Srinivasulu, S. (2004) Development of effective and efficient rainfall–runoff models using integration of deterministic, real-coded genetic algorithms and artificial neural network techniques. *Water Resour. Res.* **40**, W04302.

Jana, R. B., Mohanty, B. P. & E. P. Springer (2007) Multiscale pedotransfer functions for soil water retention. *Vadose Zone J.* **6**(4), 868–878.

Jia, Y & Culver, T. B. (2006) Bootstrapped artificial neural networks for synthetic flow generation with a small data sample. *J. Hydrol.* **331**, 580–590.

Kim, T. W. & Valdes, J. B. (2003) Nonlinear model for drought forecasting based on a conjunction of wavelet transforms and neural networks. *J. Hydrol. Engng* **6**, 319–328.

Muller, B. & Reinhardt, J. (1991) *Neural Networks – An Introduction.* Springer-Verlag, Berlin, Germany.

Partal, T. & Cigizoglu, H. K. (2008) Estimation and forecasting of the daily suspended sediment data using wavelet-neural networks. *J. Hydrol.* **358**(3–4), 317–331.

Sudheer, K. P., Gosain, A. K. & Ramasastri, K. S. (2002) A data-driven algorithm for constructing artificial neural network rainfall–runoff models. *Hydrol. Processes* **16**, 1325–1330.

Uhlenbrook, S., Seibert, J. Leibundgut, C. & Rodhe, A. (1999) Prediction uncertainty of conceptual rainfall–runoff models caused by problems to identify model parameters and structure. *Hydrol. Sci. J.* **44**, 779–798.

Wang, D. & Ding, J. (2003) Wavelet network model and its application to the prediction of hydrology. *Nat. Sci.* **1**, 67–71.

WANFIS model for monthly runoff forecasting

HUIFANG GUO[1], ZENGCHUAN DONG[1] & XIXIA MA[2]

1 *State Key Laboratory of Hydrology-Water Resources and Hydraulic Engineering, Hohai University, Nanjing, China*
guohuifang514@126.com

2 *College of Environment & Hydraulic Engineering, Zhengzhou University, Zhengzhou, China*

Abstract Wavelet analysis has good time-frequency analysis ability. Artificial Fuzzy Neural Network (ANFIS) is the combination of Artificial Neural Network and fuzzy theory. It can fit complex input and output, linear and nonlinear mapping relations. So it is especially applicable to complex nonlinear hydrological systems. This paper combines their abilities, and establishes the WANFIS model. It uses this model to forecast monthly runoff. The result shows that this model can achieve good results in simulating and predicting monthly runoff.

Key words wavelet analysis; ANFIS; WANFIS; runoff forecasting

INTRODUCTION

As the runoff formation process is not only affected by the certainty factors, but also by the uncertainty factors, so the changes of runoff are very complicated (Xiaohui Jiang, 2004). Its future is very difficult to describe accurately (Shouyu Chen, 1997). Long-term runoff forecast research is still at the developing stage and, compared with short-term hydrological forecasting, it seriously lags behind actual needs (Qitai Cao, 2006). Recently there have been many medium and long-term runoff forecast methods, but if runoff time series have large internal annual changes, they are difficult to forecast, and analysis using conventional methods and forecasting accuracy is poor.

Wavelet analysis is a good multi-resolution frequency method; by multi-scale sequence analysis it can effectively recognize the main frequency components and local information. ANFIS has strong nonlinear approximation functions and self-learning, adaptive characteristics. So they can be combined, and give full play to the advantage of both (Wensheng Wang, 2005).

Wavelet analysis

Wavelet analysis has a fixed window, but the shape can vary (variable bandwidth and when wide) changing in the time-frequency analysis method. It has an adaptive time-frequency window: high frequency, frequency-domain window increased, the time window reduced, the time window expands, and the frequency domain window reduced. Wavelet analysis is the key to satisfying certain conditions, the introduction of the basic wavelet function Ψ(t) to replace the Fourier transform of basis functions $e^{-i\omega t}$. The next is stretching and translation functions:

$$\Psi_{a,b}(t) = |a|^{-1/2}\,\Psi\left(\frac{t-b}{a}\right) \quad a,b \in R, a \neq 0$$

in which: $\Psi_{a,b}$ is called analysis wavelet or continuous wavelet; *a* is measurements (telescopic) factor, in a sense, is corresponding to frequency ω; *b* is the time (translation) factor, it reacts time's translation.

ANFIS

ANFIS (Artificial Fuzzy Neural Network) (Jang, 1993; Zhixing Zhang, 2000) is the combination of artificial neural network and fuzzy theory. It uses ANN to construct the fuzzy system. According to the input and output sample, it can automatically design and adjust the design parameters of the fuzzy system, and then it can achieve fuzzy system's self-learning and adaptive function. It can fit to complex input and outputs, linear and nonlinear mapping relations. So it is especially applicable to complex nonlinear hydrological systems (Xixia Ma, 2008). The network structure of ANFIS is shown as Fig. 1.

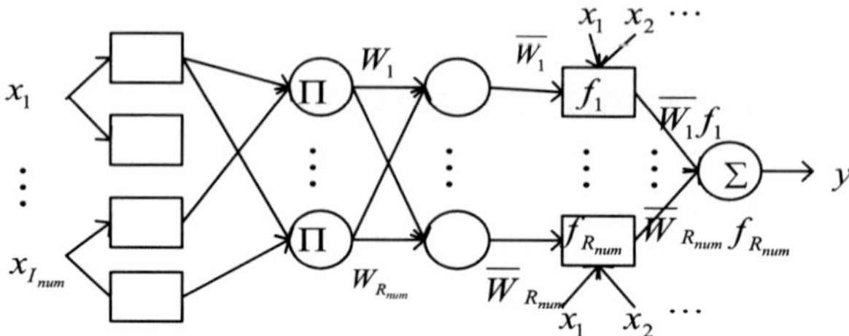

Fig. 1 ANFIS's network structure.

In Fig. 1, the connecting line between the nodes only shows the flow of signal, it does not associate with weight. Square nodes represent nodes, which have adjustable parameters; circular nodes represent nodes which do not have adjustable parameters. ANFIS's structure can be divided into five levels:

- Level 1: Membership$Z_{i,j} = \mu_i(x_j,\theta_i)$, I = 1,2,L, M_{num}, j = 1,2,L,I_{num}, $M_{num}$is the number of membership functions, I_{num} is the number of input variable, μ( ) is generalized membership function, the commonly used membership functions are triangle membership function, trapezoidal membership function, Gaussian membership function and bell-shaped membership function. The form of triangle and trapezoidal membership functions is simple, and their computing efficient is high. However, because their membership function is constructed by linear line, the corner points of some specified parameters are not smooth enough. Gaussian and bell-shaped membership functions have smooth and simple representations, so they are the most common forms used to define fuzzy sets. Equation 1 is the bell-shaped membership function:

$$\mu_i(x,\theta_i)=1/\left[1+\left[\frac{x-c_i}{a_i}\right]^{2b_i}\right]$$

x and $\theta_i = [a_i\ b_i\ c_i]$ are ith membership function's input and original reasoning parameter set, respectively.

- Level 2: kth incentive intensity, in which R_{num} is the number of fuzzy rules.

$$W_k = \prod^{I_{num}} Z_{ij}\ ,\ i \in [1,2,\cdots,M_{num}]\ ,\ i = 1,2,\cdots,R_{num}$$

- Level 3: normalized incentive intensity is the ratio of the rule's incentive intensity and the sum of all of the rule's incentive intensity.

$$\overline{W_k} = \frac{W_k}{\sum_{j=1}^{R_{num}} W_j}$$

, k = 1,2, …, R_{num}. Its vector form is: $\overline{W} = \left[\overline{W}_1, \overline{W}_2, \cdots, \overline{W}_{R_{num}}\right]^T$

- Level 4: fuzzy rule's conclusion, which is accurate output.

$$f_i = p_{i1}x_1 + p_{i2}x_2 + \cdots + p_{ij}x_i + r_i \qquad i = 1,2,\cdots,R_{num}\ ,\ j = I_{num}$$

The parameter set, which is composed by all of $\{p_{i,j}, r_i\}$, is called a consequent parameter set.

- Level 5: after weight-averaging, the overall output of the net can get: $y = \sum_{i=1}^{R_{num}} \overline{W}^T F$ in which F = [f_1,f_2,L,f_{Rnum}].

This paper uses a hybrid learning algorithm to optimally select the parameters of ANFIS; this gets the smallest sum of square error between the finally output result and the goal. The core idea of this algorithm is: in the forward calculation, it keeps the value of all of the original reasoning parameter unchanged, and improves the value of consequent parameter by recursive least squares; then, it keeps the value of all of the improved consequent parameter unchanged, and improves the value of original reasoning parameter by error back-propagation.

WANFIS

First this paper determines the most decomposition level, and then uses wavelet analysis to decompose the original sequence. This paper then uses continuous wavelet transform to analyse original sequence and the sequence of the results of the decomposition of the original sequence, then identifies their primary and secondary cycle. The sequences which contribute relatively more to the original sequence are then selected. T moment of these sequences as the input, t + T moment's original sequence as the output; this can make a networking structure. Based on experience and debugging, it identifies the membership function, the training error, training frequency, then trains this network, in the end uses the trained network to simulate and predict.

Calculation example

This paper uses the Zhaopingtai Reservoir as the research object, which is in upper part of the Shahe River of the Huaihe area. Its runoff changes in a very wide range. This paper uses the data of monthly runoff from 1952 to 1994 to stimulate, and the data from 1995 to 1996 as the test sequence.

This paper uses db4 as the function of wavelet decomposition, the largest decomposition layer is 7, and chooses a Mexican hat function as the function of continuous wavelet transform (CWT). The result of the 7-level decomposition of the original signal is delayed as shown in Fig. 2. The result of continuous wavelet transform of original signal is delayed as Fig. 3.

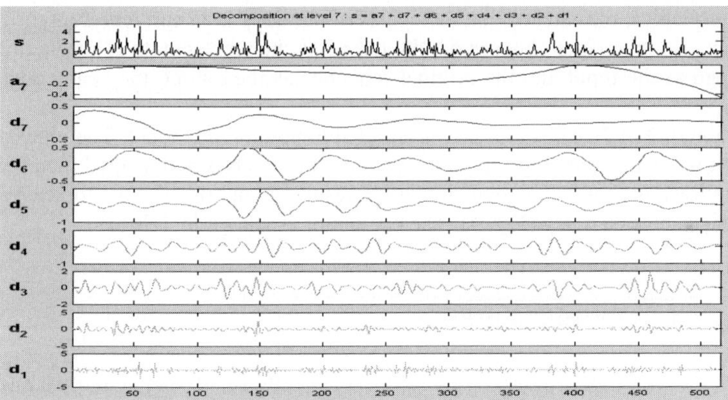

Fig. 2 Result of 7-level decomposition.

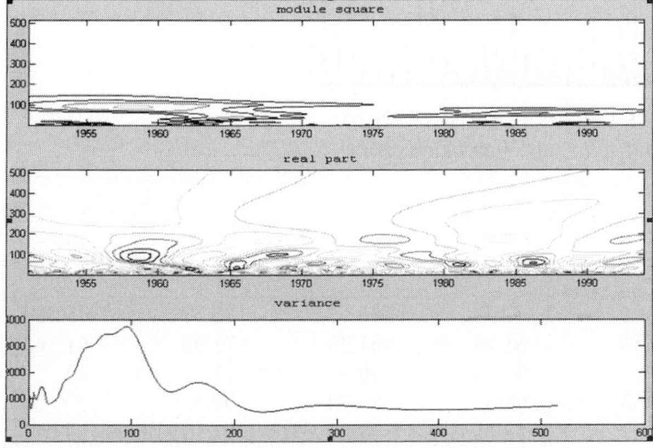

Fig. 3 Module square, real and variance of CWT.

After the 7 level decomposition and continuous wavelet transform, the results of every level cycles are displayed as Table 1. In this table, we still cannot clearly identify representatives of the original sequence, so we reconstruct the sequences. The cycles of the reconstruction sequences are shown in Table 2.

Table 1 Cycle of every level.

Sequence	Main cycle	Secondary cycle	a7	d7	d6
Cycle (month)	13, 95, 165	5, 90	239	114,207	61
Sequence	d5	d4	d3	d2	d1
Cycle (month)	35	17	11	5	2

Table 2 Cycle of reconstruction sequences.

Sequence	Main cycle	Secondary cycle	a7	a6	d6	a5
Cycle (month)	13, 95, 165	5, 90	239	113, 174, 300	61	94, 165, 290
Sequence	d5	a4	d4	d3	d2	d1
Cycle (month)	35	95,166, 290	17	11	5	2

From the above two tables, we can see that the cycles of d2, d3, d4, a5 can represent the cycle of the original sequence. So this paper chooses d2, d3, d4, a5 as the representatives of the original sequence, and uses them as the input and the original sequence as the output, they can composite the network structure.

The ANFIS function of MATLAB's tool box is adopted to predict the model. The input of the model is 4, every input has 3 membership functions, the type of which is bell-shaped membership function. After 50 iterations, we can get a good simulation result. Comparison of the forecast result and the original sequence shows a delay (Fig. 4). The evaluation is shown in Table 3.

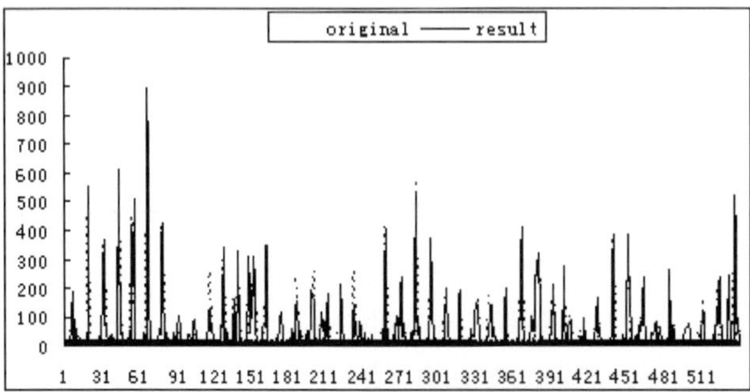

Fig. 4 Comparison result of original signal and simulation result.

Table 3 Evaluation of the model

Month	1	2	3	4	5	6
DS (%)	100.00	95.56	97.78	86.67	86.67	73.33
DVS (%)	81.82	79.55	68.18	61.36	75.00	61.36
Month	7	8	9	10	11	12
DS (%)	93.33	93.33	93.33	80.00	95.56	100.00
DVS (%)	72.73	77.27	88.64	77.27	72.73	84.09

CONCLUSION

This article combines wavelet analysis and ANFIS. It uses multi-resolution analysis functions of wavelet decomposition to decompose runoff series into high-frequency and low-frequency sequences. Next it uses wavelet analysis' continuous wavelet transform function to define the original sequence's large contribution components; then it uses these components as the input of ANFIS, and uses the original sequence as the output. It can then establish the WANFIS model, and finally it trains the model and applicants in the trained network into simulating and predicting the original signal. The result shows that this model can achieve good results in monthly runoff forecasting. The research area has 540 data and is very variable. If a simple method is used to simulate and predict it, a satisfactory result cannot be obtained. From the results in this paper, we can see that the WANFIS model can achieve a good result for us. So in the future, this model may be extensively used in monthly runoff forecasting.

REFERENCES

Jang, J. S. R. (1993) ANFIS: Adaptive-network-based fuzzy inference system. *IEEE Trans. System, Man and Cybernetics* **23**(3), 665–685.

Qitai Cao, Wensheng Wang & Chengyou Tang (2006) A new wavelet network prediction model portfolio. *People of the Yangtze River* **11**, 65–67.

Shouyu Chen (1997) Comprehensive analysis of long-term hydrological forecasting model and method. *Water J.* **8**, 15–21.

Xiaohui Jiang & Changming Liu (2004) Radial basis function networks based on wavelet analysis for the annual flow forecast. *J. Appl. Sci.* **9**, 411–414.

Xixia Ma, Haoze Mu & Huifang Guo (2008) Reservoir monthly runoff forecast model based on wavelet – ANFIS Analysis. *Water Resources and Power* **2**, 26–30.

Wensheng Wang, Jing Ding & Yueqing Li (2005) *Hydrological Wavelet Analysis.* Chemical Industry Publishing House, Beijing, China.

Zhixing Zhang, Chunzai Sun, Sunguyinger, *et al.* (2000) neuron fuzzy and soft computing. Xi'an Communication University Press, Xi'an, China.

Effectiveness of complex physics and DTM based distributed models for flood risk management of the River Tone, UK

JONG-SOOK PARK[1], QIWEI REN[2], YANGBO CHEN[2], IAN DAVID CLUCKIE[1], MICHAEL BUTTS[3] & DOUGLAS GRAHAM[3]

1 *School of Engineering, Swansea University, Singleton Park, Swansea, Wales SA1 8PP, UK*
j.park@swansea.ac.uk

2 *Department of Water Resources & Environment, Sun Yat-Sen University, Guangzhou, 51027, China*

3 *DHI, Agern Alle 5, DK-2970 Horsholm, Denmark*

Abstract This study investigates the effectiveness of an existing complex physics-based fully distributed model (MIKE SHE) and a DTM derived distributed model used to provide a better understanding of flood risk, in particular the hydrological impact of land-use change in the catchment of the River Tone in the South West region of the United Kingdom. Sustainable catchment management must consider the impact of remarkable floods in the context of recent and forecast climate change. These distributed models are implemented in order to reflect the physical and hydrological features of the catchment and to aid the evaluation of catchment flood response subject to a variety of land-use management practices. The findings can contribute to a better understanding of integrated water and land-use management when considering increasingly more extreme events that can cause damage to property and adversely affect livelihoods, creating local tensions and demands for more immediate and effective action in flood risk management. The River Tone is an integral part of the world renowned Somerset Levels that make up one of the UK's largest wetland habitats.

Key words model effectiveness; fully distributed model; MIKE SHE; flood risk management (FRM); whole catchment modelling (WSM); land use change; measurement uncertainty

INTRODUCTION

There are growing concerns that land-use changes, particularly changing agricultural practices, may have increased the risk of flooding in the UK (Environment Agency, 2000; English Nature *et al.*, 2003; O'Connell *et al.*, 2004; Wheater & Peach, 2004). The rainfall of September–December 2000 was the highest of any 4-month rainfall sequence in England and Wales since 1766 (Marsh, 2001). Even prior to this event, the Somerset Levels had experienced significant floods, such as occurred in spring 1999 (Environment Agency, 2000).

In this context it is necessary to know what can be learned about the flood impacts of changes in rural land use and management that have taken place in the past. In addition, to manage a river system as a unit, specifically for an agriculturally-dominated area, it is necessary to consider more flexibly the impact of land-use change and the introduction of active water management practice in order to mitigate the effects of flood risk under hypothetical conditions that represent the critical parameters in terms of holistic river basin management (Park & Cluckie, 2006).

Two different distributed models were developed for this study in order to examine existing rainfall–runoff records and isolate and quantify flood impacts with a view to reducing the uncertainties caused by poor input data and particular model structures. Also the hydrological response to land-use change that may be brought about by human activities in the form of agricultural practices within the study area, the River Tone, located on the South West Peninsula of the United Kingdom, is evaluated.

COMPLEX AND SIMPLE DISTRIBUTED MODEL STRUCTURES

The complex physics-based modelling method was chosen because this study requires a model that can reflect various physical features and their temporal and spatial interaction at catchment scale, particularly for medium and large areas with physical and biological heterogeneity. Also it is able to provide a better understanding of process in order to manage water resources, flood risk and the water environment (Graham & Butts, 2005).

The two fully distributed models developed in the course of this study are referred to for convenience as the "complex distributed model" and the "simple distributed model" considering model complexity, computed parameters, required input data and model run time as illustrated in Fig. 1. The complex model was built using MIKE SHE, and the simple distributed model was developed as a generic DTM based model for this particular study as an extension of mathematical structures originally developed for modelling catchments in the Pearl River Basin in southern China.

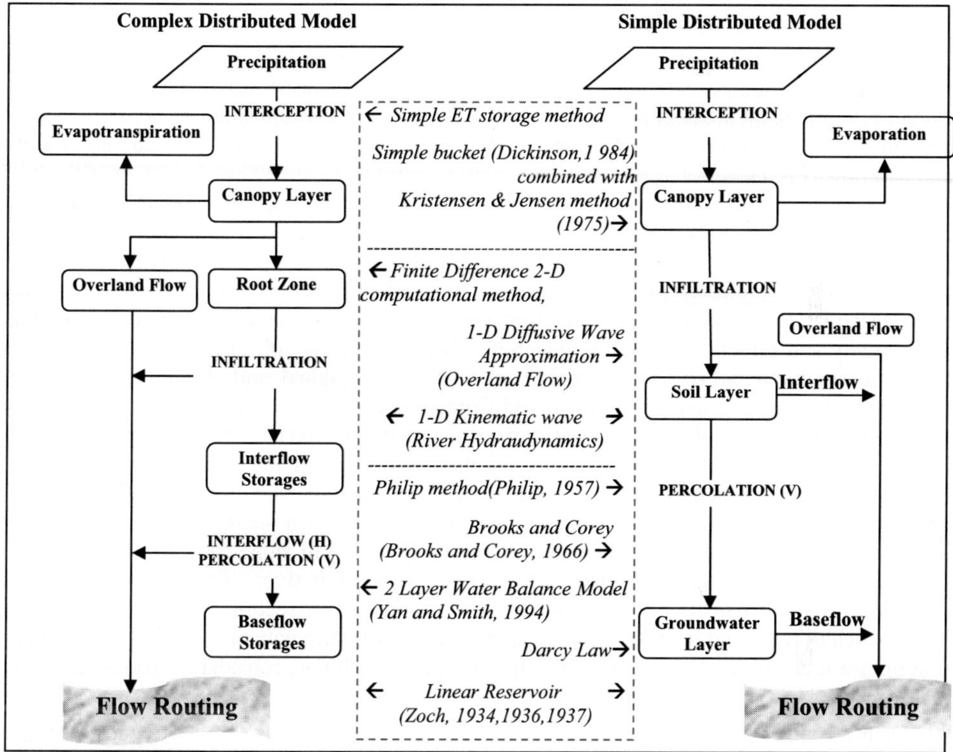

Fig. 1 Comparison of complex and simple physics-based distributed model structure with applied formula in procedures

* '→' represents formula/method applied for the model; '← →'used for both models

Complex distributed model scheme

The MIKE SHE model has continuously developed since it was originally introduced in the 1980s as "SHE" developed by three European organizations – the UK Institute of Hydrology, the Danish Hydraulics Institute and the French consulting company SOGREAH (Abbott *et al.*, 1986). The SHE model is a physics-based fully distributed modelling system used for models of all or any part of the land phase of the hydrological cycle in any geographical area. MIKE SHE is a development of the original SHE system and is therefore also a process-based model that is able to describe the processes in the land phase of the hydrological cycle and includes both pre- and post-processing of the modules of overland flow, unsaturated and saturated flow, evapotranspiration and their various interactions (Graham & Butts, 2006).

The complex distributed model utilizes the finite difference 2-D computational method for computing overland sheet flow, and MIKE 11 is applied to implementing dynamic river hydraulics. The river dynamic MIKE 11 model was coupled with MIKE SHE and computes the unsteady river flow using a full hydrodynamic solution. Also, the two-layer water balance model

(Yan & Smith, 1994) is used for computing actual evapotranspiration and the amount of water that recharges the saturated zone. This method is useful for areas with a shallow ground water table, such as wetlands areas, where the actual evapotranspiration rate is close to the reference rate, although this is not practically representing the flow dynamics in the unsaturated zone. The Linear reservoir model was used to compute the water movement in the saturated zone through shallower and deeper groundwater due to the scarcity of existing surveyed or measured data (Fig. 1).

Simple distributed model scheme

The simple distributed model has been applied to act as a comparison model, and is composed of three layers: the Canopy, Soil and Groundwater layers as illustrated in Fig. 1. This model is also discretized into equal areas in the soil horizon. Each cell has a combination of rainfall, vegetation type and soil properties in vertical layers. Three types of flow are computed – overland, interflow and baseflow (Fig.1).

The canopy layer initially intercepts the rainfall and a simple bucket approach is then employed which computes the bucket capacity depending upon the Leaf Area Index (LAI). This is then used to calculate the canopy layer interception (Dickinson, 1984). Then the potential infiltration rate is calculated by Philip's equation to generate surface runoff via excess water on the surface (Philip, 1957). Meanwhile, later flow and percolation are computed considering hydraulic conductivity and effective saturation, which are calculated using soil physical features: saturated hydraulic conductivity, soil water content at saturation, residual soil water content and pore disconnectedness index, using the Brooks and Corey relationship (1966).

Initial surface flow directions were derived from IfSAR (Interferometric Synthetic Aperture Radar) DTM (Digital Terrain Model) data, and then used to create the routing order. A post-order traverse algorithm, which is the classic tree traverse algorithm, was then applied to obtain the sequential order for cells from upstream to downstream; this is then incorporated as a routing order algorithm. Runoff routing was considered in two categories: hill slope routing and river routing. Hill slope flow was governed by the 1-D kinematic wave. For river routing, a one-dimensional diffusive wave approximation was employed. The channel cross-section is generally assumed to be an isosceles trapezoid in shape, and the applied diffusive wave equations for hill slope and river routing are then solved by the Newton-Raphson method without considering back-water effects (Chow *et al.*, 1988).

STUDY SITE AND PARAMETERISATIONS

Study site: River Tone

The River Tone (414 km^2) originates at Beverton Pond in the northwest of the catchment, and is about 33 km long and falls 370 m to the confluence with the River Parrett (Fig. 2). The Tone is inherently a flashy river that at times contributes a significant sediment load to the lower reaches of the Somerset Levels.

Most of the land in the upper reaches of the catchment is permanent pasture with woodland on the steeper valley sides. As the valley widens, land use becomes more intensive, with improved and re-seeded grassland, maize cultivation and potatoes. Sheep and cattle grazing are common with increasing numbers of horses. In the lower reaches of the Tone, the flood plain is open moorland with improved permanent pasture, re-seeded grassland, withy beds and some maize cultivation. Land use in the tributaries tends to be a mixture of improved grassland and arable, and is predominantly rural (Environment Agency, 2004).

The annual rainfall of the Tone is about 1000 mm, and there are big differences between the east and west of the catchment. The western part usually receives more than 200 mm more than the east based on 40 years (1961–2000) of daily rainfall records. The western part of the catchment is close to the Bristol Channel and therefore is heavily influenced by the maritime environment (Park, 2006).

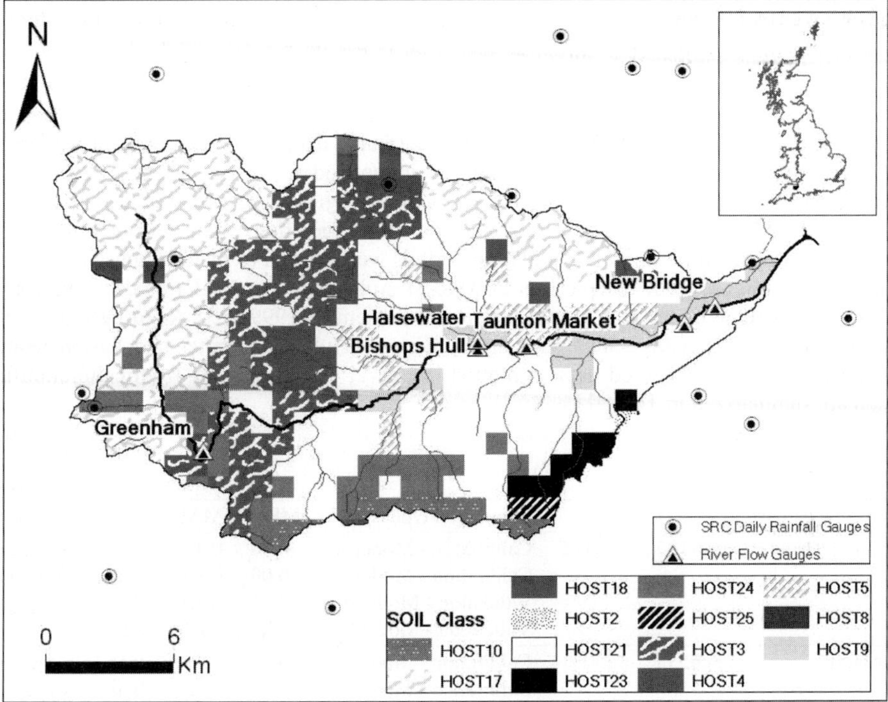

Fig. 2 The Tone catchment – daily rain and riverflow gauges.

Input parameters

Distributed models require an extensive parameterization exercise as part of the calibration process, and many parameters are also interpreted as averages over the fundamental discrete model units due to lack of measured data or inadequacy of existing information. Spatial data including catchment area and sub-divisions, topography, soil type and depth, vegetation cover, stream slopes and impermeable areas were divided into a number of computational cells (200 m × 200 m for the complex model, and 100 m × 100m for the simple model) for the numerical solution of the governing equations.

Commonly for complex and simple distributed models, a horizontal 5-m spatial resolution of IfSAR DTM data obtained from the Environment Agency (UK) is used for the topography with the slope distributions computed directly from the topographic data. Daily rainfall was used as the main input data initially, and daily mean flow data used as reference data to compare with the model output. The raingauges and flow gauges used are depicted in Fig. 2. Soil data was obtained from the National Soil Research Institute (NSRI) at 1-km resolution with attributes including field capacity, wilting point, soil water content at saturated conditions and soil structure classification data originally drawn up from fieldwork during the period 1978–1983 over England and Wales.

The HOST (Hydrology of Soil Types) is a classification established by hydrologists of the Macaulay Institute (a UK land-use research centre), CEH (Centre for Ecology and Hydrology) and the Cranfield University. The classification is based on the fact that the physical properties of soils have a major influence on catchment hydrology. Thirteen HOST classes are represented in the Tone catchment (Fig. 2). Cultivated crops/vegetation types were considered with their growing stages and potential evapotranspiration was calculated based on the UK Meteorological Office MORECS and MOSES systems (Hough *et al.*, 1997; Smith *et al.*, 2004). The surface roughness was computed using the Stickler roughness coefficient and the resistance values were obtained from Chow's work by reference (Chow, 1959).

MODEL EVALUATIONS

Most parameters were estimated as surveyed or measured data was inadequate; thus an interactive process of model parameterization and calibration has been evolved. Quantitative criteria – Mean error (ME), Mean absolute error (MAE) and Nash-Sutcliffe coefficient – were used as desirable targets prior to the calibration because of the lack of measured data, uncertainties in measurements and estimated inputs and possible limitations inherent in the model.

In order to see the impact of land-use changes on hydrological processes it is inevitable to run models for a long time to cover the whole cycle of crop/vegetation. Also considering extreme floods in recent years, calibration and validation periods were chosen to include unusual floods. A calibration period was set from 1 September 2000 to 31 August 2001, considering data availability. Validation with extreme floods was carried out from 1 September 1999 to 31 August 2000.

Data from four flow gauge sites: Greenham, Bishops Hull, Halsewater and Knapp Bridge, were compared with the observed data as plotted in Fig. 2, and the results from calibration and validation are summarized in Table 1.

Table 1 Statistical model efficiencies.

Observed flow gauges	X	Y	Simulation type	ME	MAE	Nash-Sutcliffe
Greenham	308126	120422	Calibration - Model A	0.00	0.41	0.68
			Calibration - Model B	–0.09	0.35	0.66
			Validation - Model A	0.06	0.43	0.70
			Validation - Model B	–0.20	–0.46	0.61
Bishops Hull	320533	125031	Calibration - Model A	–0.27	1.31	0.58
			Calibration - Model B	–0.16	1.02	0.58
			Validation - Model A	0.24	1.59	0.52
			Validation - Model B	–0.84	1.56	0.39
Halsewater	320593	125302	Calibration - Model A	0.04	0.44	0.54
			Calibration - Model B	–0.09	0.35	0.58
			Validation - Model A	0.18	0.60	0.47
			Validation - Model B	–0.32	0.55	0.40
Knapp Bridge	330169	126044	Calibration - Model A	–0.92	1.94	0.54
			Calibration - Model B	0.38	1.57	0.62
			Validation - Model A	–0.45	2.42	0.57
			Validation - Model B	–0.40	1.20	0.50

* Calibration period: 01/09/2001 – 31/08/2002; Validation period: 01/09/1999 – 31/09/ 2000

* Model A – complex distributed model built using MIKE SHE; Model B – simpler distributed model applied for this study; X: x-coordinate; Y: y-coordinate; ME (mean error); MAE (mean absolute error); Nash-Sutcliffe (Nash Sutcliffe coefficient)

The calibration was carried out using a combined trial-and-error method and the overall results from the calibration showed the model efficiency was better at the upstream end of the catchment and was less effective downstream, varying from 68% (complex distributed model) to 66% (simple model) at Greenham, and 54% and 62% respectively at Knapp Bridge. This effect is also seen in the validation results. It was presumed that the daily rainfall did not well represent the rainfall distribution over the catchment, and particularly for heavy rainfall where the impact seemed to be accumulated towards the downstream catchment.

Figure 3 shows the results from the calibration, and both models had a good relationship with the observed flow data except for the January–February 2001 floods that were also affected by the rainfall bias issue.

The validation was done as a split-sample test and was conducted on an independent period of similar length (September 1999–August 2000) that included few severe floods, as illustrated in Fig. 4, and the performance was slightly less accurate than the results obtained through the calibration.

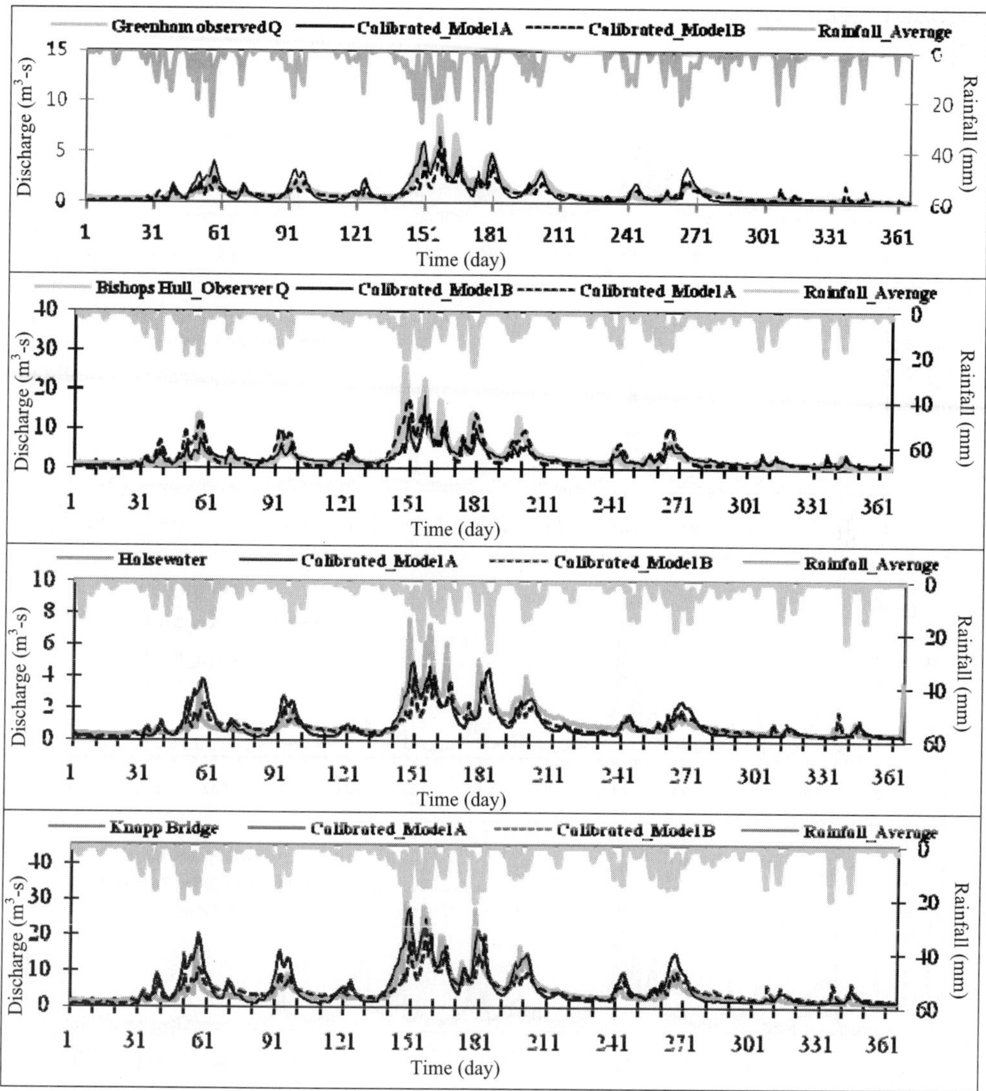

Fig. 3 Calibration results for the four observed sites from upstream to downstream.

During the validation period, there was an inconsistency between the measured rainfall and flow data. This inconsistency reflected the unusual peak flow for late September–early October. The results showed a good correlation with the observed flows but the more extreme period of December 1999–January 2000 was not matched by either of the models.

DISCUSSION AND CONCLUSIONS

This is an ongoing study as part of the UK Flood Risk Management Research Consortium (FRMRC) and aims to evaluate the impact of land-use changes on the hydrological cycle at the whole catchment scale using different physics-based models. A complex physics-based distributed model was built using MIKE SHE, and a simpler model structure was developed as a generic DTM based model. Considering the input data was daily rainfall with daily mean flow, both

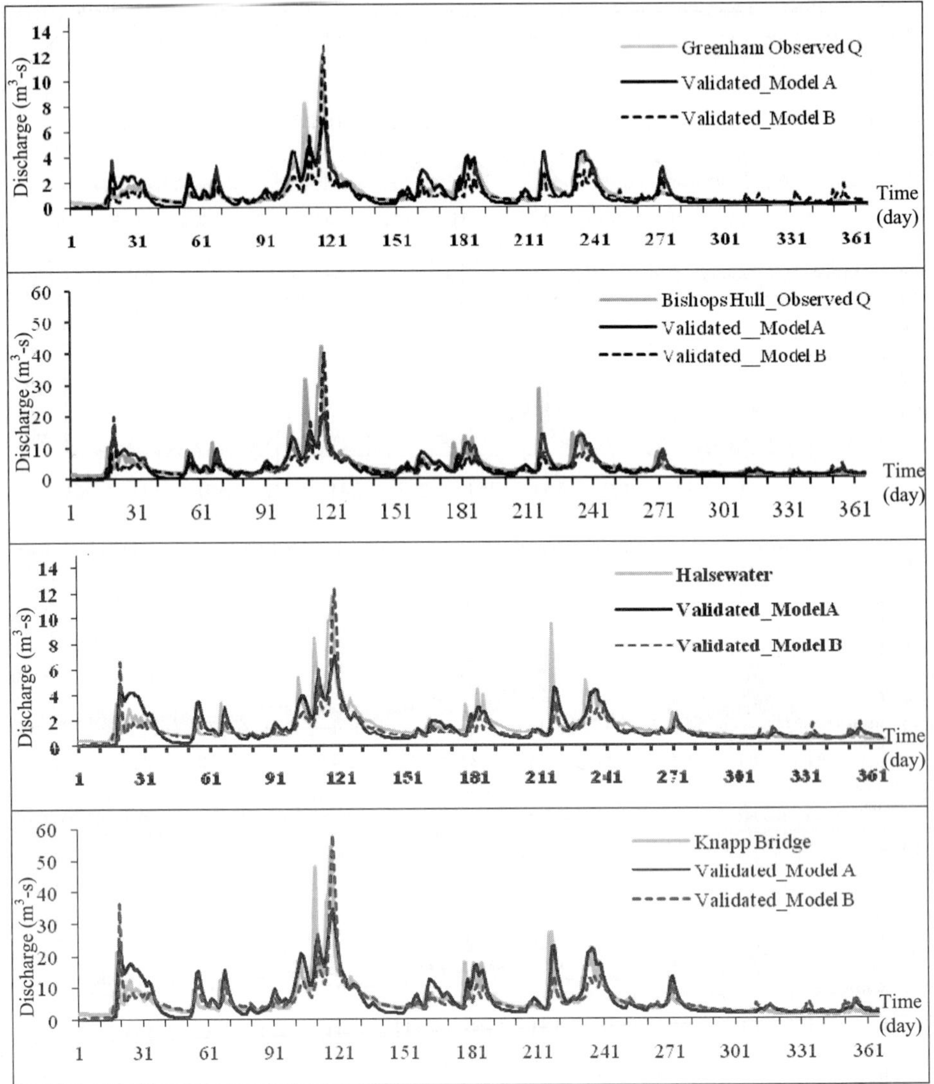

Fig. 4 Validation results of four observed sites from upstream to downstream.

models gave reasonable performances for both calibration and validation with similar values of the statistics ME and MAE. However, the comparison has been initially limited to the more extreme flows and the differences between measured and computed flows increase towards the downstream end of the catchment. This tends to confirm the suspicion that the daily rainfall does not effectively represent the spatio-temporal structure of the heavy rainfall systems.

Both complex and simple physics-based distributed models performed well in the model evaluations and are expected to provide detailed scientific understanding in order to reduce uncertainties caused from input data and model structures, even though it confirms that uncertainty and sensitivity analysis will be helpful to improve model efficiencies and may reveal the influence of model complexities.

However, in order to identify the potential impact of land-use change on the hydrological process at catchment scale, future studies will utilize high-resolution distributed hydrological models, weather radar data, and even ensemble based Numerical Weather Prediction Models (NWP) in order to reduce uncertainties in the utilization of the rainfall observations and address some of the space–time rainfall representivity issues. The ability to understand the impact of the fundamental model structure on land-use management issues is a crucial first step in the process of developing a capacity to understand climate impact given its intimate relationship with land use management processes. Indeed, the developing ability for utilizing complex model cascades in this context is slowly becoming a reality.

Acknowledgements This study has been funded through the Flood Risk Management Research Consortium sponsored by the Engineering and Physical Sciences Research Council (EPSRC) under grant EP/FO20511/1, with additional funding from the Environment Agency (EA)/ Department for Environment, Food and Rural Affairs (DEFRA) in the UK and the Northern Ireland Rivers Agency (DARDNI) and Office of Public Works (OPW), Dublin. Also the simple physics-based distributed model has been developed through research collaboration between Swansea University and Sun Yat-Sen University in Guangzhou, China, as part of a long standing collaboration in the context of FRMRC.

REFERENCES

Abbott, M. B., Bathurst, J. C., Cunge, J. A., O'Connell, P. E. & Rasmussen, J. (1986) An introduction to the European Hydrological System – Systeme Hydrologique Europeen, "SHE", 1: History and philosophy of a physically-based, distributed modeling system. *J. Hydrol.* **87**(1), 45–59.

Brooks, R. N. & Corey, A. T. (1966) Properties of porous media affecting fluid flow. *J. Irrig. Drain. Engng* **92**(IR2), 61–88.

Chow, V. T. (1959) *Open-channel Hydraulics*. McGraw-Hill, New York, USA.

Chow, V. T., Maidment, D. R. & Mays, L. W.(1988) *Applied Hydrology*. McGraw-Hill, New York, USA.

Dickinson, R. E.(1984) Modeling evapotranspiration for three-dimensional global climate models. In: *Climate Processes and Climate Sensitivity* (ed. by J. E. Hansen & T. Takahashi), 58–72. Geophysical Monograph 29, American Geophysical Union, Washington DC, USA.

English Nature, Environment Agency and DEFRA (2003) Wetlands, Land Use Change and Floods Management. A joint statement (unpublished).

Environment Agency (2002) The Parrett Catchment: Water management Strategy Action Plan. Environment Agency, Exeter, UK.

Environment Agency (2004) The Tone Catchment Abstraction Management Strategy. Environment Agency, UK.

Graham, D. N. & Butts, M. B. (2006) Flexible integrated watershed modeling with MIKE SHE. In: *Watershed Models* (ed. by V. P. Singh & D. K. Frevert), 245-272, CRC Press, Boca Raton, Florida, USA .

Hough, M., Palmer, S., Weir, A., Lee, M. & Barrie, I., (1997) The Meteorological Office Rainfall and Evaporation Calculation System: MORECS version 2.0. The Met. Office, UK,

Kristensen, K. J. & Jensen, S. E.(1975) A model for estimating actual evapotranspiration from potential evapotranspiration, *Nordic Hydrology* **6**, 170–188.

Marsh, T. J. (2001) The 2000/01 floods in the UK – a brief overview. *Weather* **56**, 343–345.

O'Connell, P. E., Beven, K. J., Carney, J. N., Clements, R. O., Ewen, J., Fowler, H. *et al.* (2004) Review of Impacts of Rural Land Use and Management on Flood Generation. DEFRA/Environment Agency Flood and Coastal Defence R&D Programme, R&D Technical Report FD2114.

Park, J.-S. (2006) Whole system modelling of the impact of land use management in the Parrett Catchment, PhD Thesis, University of Bristol, Bristol, UK.

Park, J.-S. & Cluckie, I. D. (2006) Whole Catchment Modelling Project – Technical Report, The Parrett Catchment Project Report (PCP/PCP/WEMRC – Technical Report).

Philip, J. R.(1957) The theory of infiltration: 4. sorptivity and algebraic infiltration equation. *Soil Science* **84**, 257–264.

Smith, R. N. B., Blyth, E. M., Finch, J. W., Goodchild, S., Hall, R. L. & Madry, S. (2004) Numerical Weather Prediction: Soil state and surface hydrology diagnosis based on MOSES in the Met Office Nimrod nowcasting system. The Met Office, UK Forecasting Research Technical Report No. 428.

Wheater, H. S. & Peach, D. (2004) Developing interdisciplinary science for integrated catchment management: The UK Lowland Catchment Research (LOCAR) Programme. *Water Resources Development* **20**(3), 369–385.

Yan, J. & Smith, K. R. (1994) Simulation of integrated surface water and ground water systems – model formulation. *Water Resources Bull.* **30**(5), 879–890.

Zoch, R. T. (1934) On the relation between rainfall and stream flow, *Monthly Weather Review* **62**(9), 315–322.

Zoch, R. T. (1936) On the relation between rainfall and stream flow-II, *Monthly Weather Review* **64**(4), 105–121.

Zoch, R. T. (1937) On the relation between rainfall and stream flow-III, *Monthly Weather Review* **65**(4), 135–147.

An integrated distributed watershed model with channel network for rainfall–runoff simulation

K. VENKATA REDDY[1], T. I. ELDHO[2], E. P. RAO[2] & A. T. KULKARNI[2]

1 *Department of Civil Engineering, National Institute of Technology, Warangal 506 004, Andhra Pradesh, India*
kvenkatareddy9@rediffmail.com

2 *Department of Civil Engineering, Indian Institute of Technology Bombay, Mumbai 400 076, India*

Abstract An integrated distributed watershed model with channel network for rainfall–runoff simulation is presented. Interception is estimated by an exponential model based on leaf area index (LAI). The Green-Ampt Mein Larson (GAML) model is used for the estimation of infiltration in the watershed. For runoff estimation, diffusion wave equations are solved using the finite element method (FEM), together with application of the channel network concept. Interflow is simulated using the FEM-based model. A geographic information system (GIS) was used to prepare the input data required for the distributed model. The developed channel network based distributed model was applied to Peacheater Creek watershed, USA. The model performed well on some validation events, but not on all of them. The problems may be due to topographic complexities, or to differences in the characteristics between the calibration storms and those in the validation data set. The developed model has to be further verified for watersheds at different geographic locations for better validation.

Key words channel network; diffusion wave model; event-based runoff; finite element method; GIS; interception; interflow; GAML infiltration model; watershed

INTRODUCTION

Over the years, the demand on water resources has increased due to population and industrial growth. This high demand and reduced availability of freshwater necessitate management of available water resources in a sustainable way. In a watershed perspective, spatial and temporal estimation of runoff is important for the sustainable management of water resources. The physical parameters that affect runoff, such as slope and Manning's roughness, vary spatially over the watershed. Spatial variation of these properties can be better incorporated into the model using a geographic information system (GIS).

Many researchers have used a simplified form of St Venant equations, such as kinematic and diffusion wave equations, and solved these using advanced numerical methods to simulate runoff in a watershed (Blandford & Ormsbee, 1993; Jain & Singh, 2005; Alhan & Medina, 2007). Few of them have also discussed the relative advantages of the diffusion wave model (Morris & Woolhiser, 1980; Hromadka II *et al.*, 1986).

Infiltration greatly influences the surface runoff in a watershed. The Green-Ampt model is one of the most widely-used infiltration models. Mein & Larson (1973) developed a two-stage infiltration model for constant rainfall intensity and homogeneous soil with uniform initial soil moisture content. Chu (1978) modified the traditional Green-Ampt model for unsteady rainfall events. Reddy *et al.* (2007) used the combined Green-Ampt Mein Larson (GAML) model to simulate the infiltration in a watershed. Interception can be incorporated through an interception model to simulate runoff. Jetten (2002) used the leaf area index (LAI) and cover fraction-based interception method in the LImburg Sediment Erosion Model (LISEM).

Based on the topography of the watershed, a dendritic channel network model may give more realistic runoff simulation. Akan & Yen (1981) developed a nonlinear diffusion wave model to simulate the unsteady flow in a dendritic channel network. Blandford & Ormsbee (1993) used a diffusion wave equation to develop a finite element model for a dendritic channel network. Garg & Sen (2001) presented a model which simulates the rainfall–runoff response of a catchment by considering flow through a network of channels fed with lateral overland inflows. Interflow is more likely to be the main form of drainage in steep humid catchments. Jayawardena & White (1977, 1979) developed a finite element-based distributed catchment model with an interflow component, and applied it to real catchments.

Integration of GIS techniques with modelling leads to improved accuracy, more flexibility, ease of data sharing and greater efficiency (Ogden *et al.*, 2001). Vieux (2001) discussed the use of the Finite Element Method (FEM) and GIS in watershed modelling. Reddy *et al.* (2007) developed a kinematic wave-based distributed watershed model using FEM, GIS and remotely sensed data.

In this paper, a diffusion wave-based distributed rainfall–runoff model with channel network using FEM and GIS is described for the simulation of event-wise surface runoff for a watershed. The model simulates runoff of a watershed by considering flow through a network of channels fed by lateral overland inflows. In the model, interception loss is estimated by an empirical model based on LAI, and infiltration loss by the Green-Ampt Mein Larson (GAML) model. The interflow is modelled using the FEM. The overland flow and channel network flow are estimated using diffusion wave-based FEM models and coupled to obtain the runoff at the outlet. The developed integrated model was calibrated and validated on some of the rainfall events of Peacheater Creek watershed, USA.

MATHEMATICAL FORMULATION

In this paper, all the hydrological processes are modelled in one-dimensional form. The hydrological processes considered for simulation are: interception, infiltration, overland flow, channel flow and interflow.

Interception model and GAML infiltration model

Interception is estimated based on the interception equations used in the LISEM model (Jetten, 2002). Infiltration has been simulated using the GAML model (Mein & Larson, 1973; Chu, 1978). The full formulation of interception model and infiltration model used in the present paper has been given in Reddy *et al.* (2007).

Surface runoff model

In a watershed, surface runoff can be divided into overland flow and channel flow. All channels in the watershed can be connected as a dendritic channel network which routes the flow to the outlet of the watershed. In the present model, the watershed has been divided into sub-watersheds, each of which contains one main channel. Overland and channel flow are formulated based on diffusion wave equations to route the runoff to outlet of the watershed.

The continuity and momentum equations for the diffusion wave for overland flow are given as follows (Singh, 1996):

$$\frac{\partial q}{\partial x} + \frac{\partial h}{\partial t} = r_e \tag{1}$$

$$\frac{\partial h}{\partial x} = S_o - S_f \tag{2}$$

where q is flow per unit width, h is depth of flow; r_e is excess rainfall intensity after interception and infiltrations loss, x is a variable representing space, t is a variable representing time, S_o is the slope of the overland flow plane, and S_f is the friction slope of the flow plane. The flow per unit width is given as $q = \alpha h^{\beta}$, where α and β can be derived by using Manning's equation and are given as $\alpha = (\sqrt{S_f})/n_o$ and $\beta = 5/3$, where n_o is the overland flow Manning's roughness coefficient.

The above governing equations are solved using initial and boundary conditions. The initial flow condition is no flow. The upstream boundary condition is assumed as zero inflow and it is given as $h = 0$ and $q = 0$ at all times t. The downstream boundary condition is of zero depth gradient (Morris, 1979). This condition can also be written with the end node M as $h_M = h_{M\text{-}1}$. Similarly, the continuity and momentum equations for channel flow can be expressed in terms of equations (1) and (2), where q is replaced by channel discharge (Q), h is replaced by area of flow in the channel, and r_e is replaced by q.

The equations of continuity and momentum are applied within the channels, and different relationships are used to link the flow variables at the channel junctions. The initial and boundary conditions in a general channel network are as follows:

Initial conditions In natural watersheds, constant initial depth of flow is assumed at all nodes of channel network. The routing in the channel network starts from most upstream channels and is carried towards downstream, channel by channel in sequence, satisfying the flow continuity requirement at junctions.

Upstream boundary conditions The upstream boundary conditions for starting channels which are not fed by any other channel in the network is of zero inflow condition. At the channel junction, the upstream boundary condition for the outflow channel is given as $\Sigma Q_k = Q_o$, where subscript k stands for any one of the inflowing channels and o represents the out-flowing channel. The depth of flow in the channel (H) can be obtained by assuming normal depth condition. For natural watersheds in hilly terrains having steep slopes, it is very difficult to implement the kinematic compatibility condition ($H_k = H_o$) given by Akan & Yen (1981). Therefore in the present channel network model, the kinematic compatibility condition is not considered.

Downstream boundary conditions For all channels in the network, the diffusion wave model downstream boundary condition of zero depth gradient is used.

Finite element formulation Applying the Galerkin finite element formulation (Reddy, 1993) to equation (1) gives:

$$[C]\{h\}^{t+\Delta t} = [C]\{h\}^{t} - \Delta t[B]\{(1-\omega)q^{t} + \omega q^{t+\Delta t}\} + \Delta t\{f\}\left((1-\omega)(r_e)^{t} + \omega(r_e)^{t+\Delta t}\right) \tag{3}$$

where $[C]$,$[B]$ and $\{f\}$ are element matrices; the superscripts t and $t+\Delta t$ indicate the variables at the previous time step and current time step, respectively; and ω is the factor that determines the type of finite difference scheme involved. In the present study, the Crank-Nicolson scheme with $\omega = 0.5$ is used. One-dimensional line elements are used for spatial discretization. Equation (3) is applied to all elements in the domain and assembled to form a system of equations. The system of equations is solved by Cholesky scheme after applying the boundary conditions for the unknown values of h. The solution of h requires iteration, due to the nonlinearity of equation (3). Iteration is continued until the convergence is reached at a specified tolerance value (ε). After convergence on h, the time step is incremented and the solution proceeds in the same manner by updating the time matrices and evaluating the new h values. During calculation of friction slope, the formulation given in explicit finite difference form for equation (2) by Jain & Singh (2005) has to be used. For channel flow, the finite element equation was formulated similar to equation (3).

Interflow model

The Interflow model was formulated based on the through flow equations given by Jayawardena & White (1977). The continuity equation for the interflow is as follows:

$$\frac{\partial q_i}{\partial x} + \eta \frac{\partial h_i}{\partial t} = I_i \tag{4}$$

where q_i is the interflow, h_i the saturated layer thickness in which interflow passes, η the porosity of the same interflow zone as in the infiltration model, and I_i the lateral recharge rate due to infiltration per unit area. The interflow q_i can be expressed by Darcy's law for flow in porous media, in a direction parallel to the overland flow plane slope (S_o) and is given as $q_i = KS_oh_i$, where K is the hydraulic conductivity of the interflow zone same as in infiltration model. Substituting q_i in equation (4) gives the interflow as:

$$\frac{\partial}{\partial x}(KS_o h_i) + \eta \frac{\partial h_i}{\partial t} - I_i = 0 \tag{5}$$

Finite element formulation By applying the Galerkin finite element formulation to equation (5), the final form of equation is as follows:

$$(\eta[C]^{(e)}+[B]^{(e)}\Delta t\omega KS_o\{h_i\}^{t+\Delta t}=(\eta[C]^{(e)}-(1-\omega)[B]^{(e)}\Delta tKS_o\{h_i\}^{t}+[f_i]^{(e)}\Delta t\{\omega(I_i)^{t+\Delta t}+(1-\omega)(I_i)^{t}\} \quad (6)$$

As in the case of overland flow, equation (6) is applied to all elements of overland flow plane and assembled to form a system of equations and solved.

MODEL DEVELOPMENT

Based on the model formulation, the overland and channel flow model and channel network model are developed. The computer program for the above models was developed using C programming language. These component models are coupled for the runoff simulation of the watershed.

In the present model, the whole watershed is divided into overland flow strips on either side of the channels. Each overland flow strip is discretized into overland flow elements. The channels are divided into channel flow elements which are equal to the number of strips. For a given rainfall event, the model calculates the interception, time step-wise and deducts from rainfall intensity to get effective rainfall for the rainfall duration. The infiltration module takes this effective rainfall and calculates the infiltration, time step-wise for the watershed. The infiltration is deducted from the effective rainfall to get excess rainfall and this becomes the input to the overland flow model. The overland flow is simulated node-wise for each strip. The simulation starts for overland flow strips from both sides of the channel, upstream to downstream. The output hydrograph of these overland flow strips becomes the lateral input to the channel flow model. Part of the infiltration is routed as interflow to the channel from the strips by the same procedure of overland flow as explained above.

In the channel network model, the uppermost channels in the watershed have an input hydrograph or zero inflow depending on discretization. The total discharge from the upstream channels becomes the discharge to the downstream channel.

MODEL APPLICATION

The developed integrated model was applied to Peacheater Creek watershed, USA. The data for Peacheater Creek watershed are available at the Distributed Model Intercomparison Project (DMIP) website (http://www.nws.noaa.gov/oh/hrl/dmip/index. html). Some data, such as watershed boundary, streamflow and gauging station information were also obtained through personal communication (Michael B. Smith and Seann M. Reed, DMIP, National Weather Service Office of Hydrologic Development, USA, 2005).

Study area and database preparation

The Peacheater Creek watershed is one of the experimental watersheds of the DMIP. It has an area of 65 km^2. It is a gauged sub-watershed of Baron Fork catchment, Oklahoma, USA. The gauging station for this watershed is located at Christie, Oklahoma. Annual rainfall and runoff values of the watershed are 1157 and 313 mm, respectively (Smith *et al.*, 2004). The watershed consists primarily of silty loam soils with forested (42%), grassy (57%) and urban (1%) areas (Vivoni *et al.*, 2005). Basin elevation varies from 248 m near the outlet to 432.5 m along an isolated ridge in the northeast. The ASCII files of the digital elevation model (DEM) (1 arc-second) and the soil and land-use map of the Baron Fork basin with 30-m grid resolution were converted into map format with appropriate projection information. The required DEM, soil and land-use maps for Peacheater Creek watershed were clipped, based on the boundary of the watershed. The percentage slope map was prepared from the DEM by using the slope option of Spatial Analyst in ArcMap.

The drainage map of the watershed was generated based on the DEM using Hydrology tools of ArcMap. Hourly rainfall data in binary format were obtained from the DMIP website (Stage 3 Next-Generation Weather Radar (NEXRAD) observations at 4 km × 4 km resolution). Based on the procedure explained at the DMIP website, the hourly rainfall maps were prepared and clipped for the watershed. The finite element grid map was prepared by dividing the whole watershed into

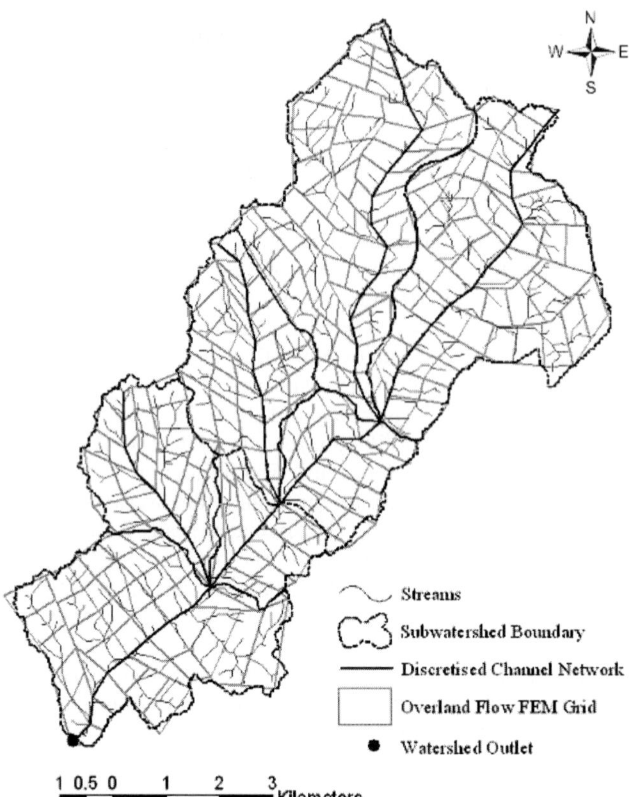

Fig. 1 Finite element grid map of Peacheater Creek watershed with channel network.

seven sub-watersheds, as shown in Fig. 1. Here, each sub-watershed is discretized into overland flow strips which are connected to one discretized channel. For this discretization, the overland flow element length varies from strip to strip. The discretized channels of all sub-watersheds are connected as a dendritic channel network. The weighted average rainfall for the watershed was considered in the present model.

RESULTS AND DISCUSSION

Since 99% of Peacheater Creek watershed consists of forest and grassy lands, the interception model was considered to simulate the rainfall events. The interflow model was used to simulate the runoff, to account for exfiltration in this watershed. The model was calibrated for four rainfall events and validated for two rainfall events in the watershed. Calibration was performed by altering the values of K_s, S_{av} and θ_i by trial and error, based on the best visual fit of the hydrographs. The parameters n_o and n_c were calibrated in the absence of data. Saturated moisture content (θ_s) of 0.41 was used. In view of the non-availability of data for the interception model, LAI of 1.5 and fraction of vegetation cover (c_p) of 0.5 were assumed. The calibrated parameters of the watershed are given Table 1. Validation of two rainfall events of Peacheater Creek watershed was carried out by taking the average parameters of four calibrated rainfall events. The runoff hydrographs for the calibration and validation events are shown in Figs 2 and 3, respectively. Simulation results are shown in Table 2. From the simulation results for the calibrated events, it is found that the volume of runoff was simulated within a variation of 34–64%. Peak runoff and time to peak runoff were simulated within a variation of 7–33% and 1–64%, respectively. From the

Table 1 Calibrated parameters for rainfall events of Peacheater Creek watershed.

Event	Saturated hydraulic conductivity, K_s (cm/h)	Initial water content, θ_i	Average suction head, S_{av} (cm)	Overland flow Manning's roughness coefficient, n_0	Channel flow Manning's roughness coefficient, n_c
6 February1999	0.7917	0.14	1.78	0.592	0.085
17 May 1999	0.6825	0.145	5	0.228	0.0575
28 June 2000	0.6875	0.15	3	0.52	0.279
23 February 2001	0.3125	0.22	6	0.2875	0.0575

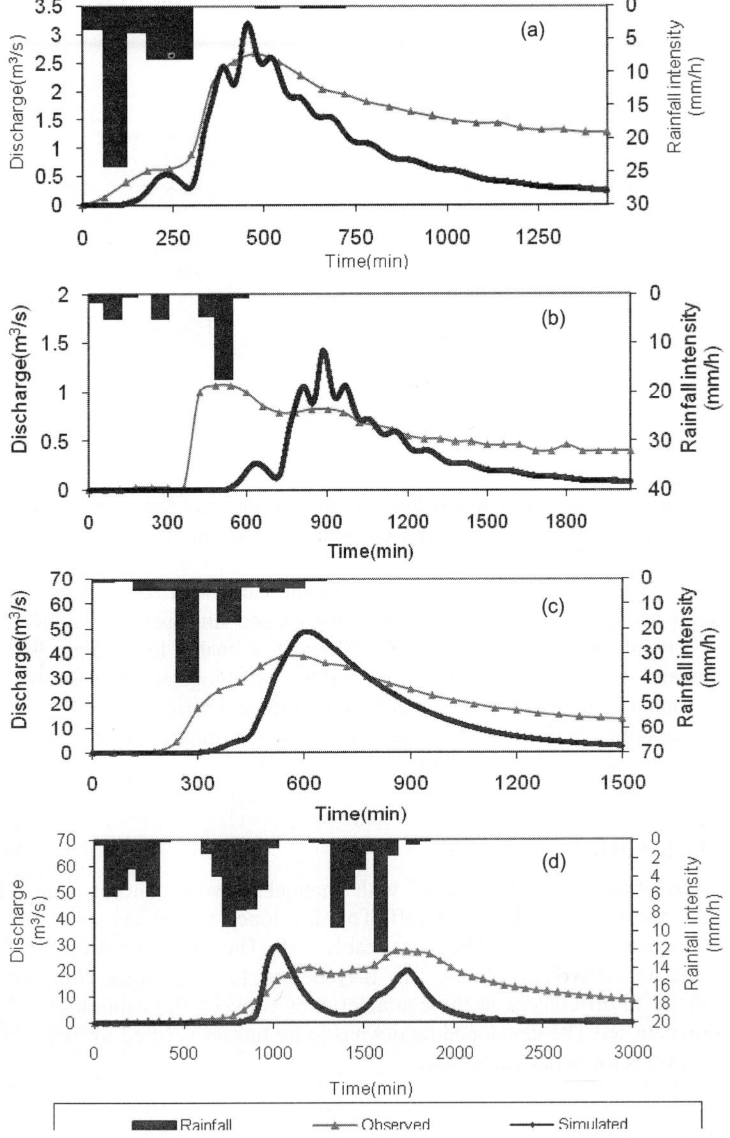

Fig. 2 Observed and simulated hydrographs for calibrated events of Peacheater Creek watershed: (a) 6 February 1999; (b) 17 May 1999; (c) 28 June 2000; (d) 23 February 2001.

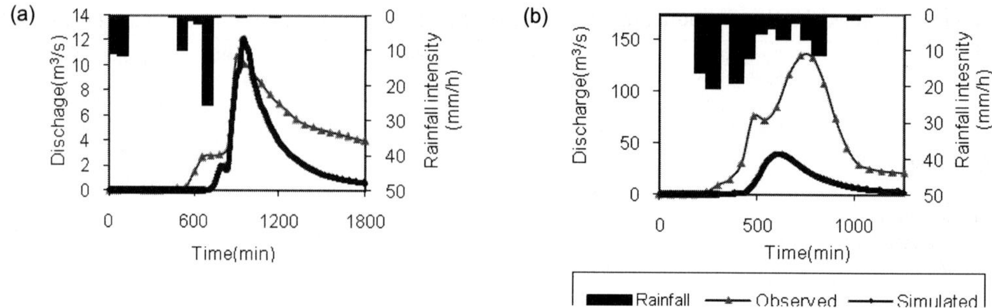

Fig. 3 Observed and simulated hydrographs for validated events of Peacheater Creek watershed: (a) 4 May 1999; and (b) 21 June 2000.

Table 2 Simulated results for Peacheater Creek watershed.

Date of rainfall event	Volume of runoff (mm)		Peak runoff (m^3/s)		Time to peak (min)	
	Observed	Simulated	Observed	Simulated	Observed	Simulated
Calibration event						
06 February 1999	2.025	1.219	2.668	3.185	480.0	457.5
17 May 1999	1.001	0.55	1.071	1.4216	540.0	885
28 June 2000	28.739	18.895	38.844	48.803	600.0	608.5
23 February 2001	35.033	12.725	27.431	29.458	1680.0	1018
Validation event						
4 May 1999	6.478	3.646	10.731	12.168	900.0	951
21 June 2000	57.673	12.509	135.159	39.978	720.0	608

simulation results of validation events, it is found that runoff volume was simulated within the range 43–78%. Peak runoff and time to peak runoff were simulated within 13–70% and 5–16%, respectively.

Topographic complexity in the Peacheater Creek controls the catchment response to rainfall via interactions between the active shallow aquifer, stream network and land surface (Vivoni *et al.*, 2005). Even though the present model has simulated interflow, it is unable to simulate perched return flow and groundwater exfiltration, which are important hydrological processes of this watershed. In addition, channel slope was assigned a constant value and different widths for sub-watersheds have also been assumed in the absence of data. These factors might have contributed to the discrepancy between observed and simulated flows.

SUMMARY AND CONCLUSIONS

A diffusion-wave-based distributed watershed model with channel network concept using FEM and GIS is presented here for the simulation of runoff. The developed model has simulated the hydrographs of runoff at the outlet of the watershed reasonably well. The model performed well on some validation events and not so well on others. This may be due to topographic complexities mentioned earlier, as well as to differences in the characteristics between the calibration storms and those in the validation data set. The developed model has to be further verified for watersheds at different geographic locations for better validation.

Acknowledgements The authors are thankful to the Indian Space Research Organisation (ISRO) for the financial assistance to carry out this work through Project no. 03IS007. We are also thankful Michael B. Smith and Seann M. Reed of the Distributed Model Intercomparison Project, National Weather Service, Office of Hydrologic Development, Silver Spring, USA for providing the database of Peacheater Creek watershed, Oklahoma, USA.

REFERENCES

Akan, A. O. & Yen, B. C. (1981) Diffusion-wave flood routing in channel networks. *J. Hydraul. Engng ASCE* **107**, 719–731.

Alhan, C. M. K. & Medina Jr, M. A. (2007) Kinematic and diffusion waves: Analytical and numerical solutions to overland and channel flow. *J. Hydraul. Engng ASCE* **133**(2), 217–228.

Blandford, G. E. & Ormsbee, E. L. (1993) A diffusion wave finite element model for channel networks. *J. Hydrol.* **142**(1/4), 99–120.

Chu, S. T. (1978) Infiltration during an unsteady rain. *Water Resour. Res.* **14**(3), 461–466.

Garg, N. K. & Sen, D. J. (2001) Integrated physically based rainfall–runoff model using FEM. *J. Hydrol. Engng ASCE* **6**, 179–188.

Hromadka II, T. V., Nestlinger, A. J. & DeVries, J. D. (1986) Comparisons of hydraulic routing methods for one-dimensional channel routing problems. In: *Hydraulic Engineering Software* (Proc. Second Int. Conf.) (ed. by M. Radojkovic, C. Maksmovic & C. A. Brebbia) (Hydrosoft 86, Southampton, UK, September 1986), 85–98. Springer-Verlag, Berlin, Germany.

Jain, M. K. & Singh, V. P. (2005) DEM-based modeling of surface runoff using diffusion wave equation. *J. Hydrol.* **302**(1/4), 107–106.

Jayawardena, A. W. & White, J. K. (1977) A finite element distributed catchment model. I. Analytical basis. *J. Hydrol.* **34**(3/4), 269–286.

Jayawardena, A. W. & White, J. K. (1979) A finite element distributed catchment model. II. Application to the real catchments. *J. Hydrol.* **42**(3/4), 231–249.

Jetten, V. (2002) LISEM *Limburg Soil Erosion Model. Windows version 2.x, User Manual 7-7*, Utrecht University, The Netherlands. (http://www.geog.uu.nl/lisem, last accessed 26 September 2005).

Mein, R. G. & Larson, C. L. (1973) Modeling infiltration during a steady rain. *Water Resour. Res.* **9**(2), 384–394.

Morris, E. M. (1979) The effect of the small slope approximation and lower boundary condition on solutions of Saint Venant equations. *J. Hydrol.* **40**(1/2), 31–47.

Morris, E. M. & Woolhiser, D. A. (1980) Unsteady one-dimensional flow over a plane: partial equilibrium and recession hydrographs. *Water Resour. Res.* **16**(2), 355–360.

Ogden, F. L., Garbrecht, J., DeBarry, P. A. & Johnson, L. E. (2001) GIS and distributed watershed models. II: Modules, interfaces and models. *J. Hydrol. Engng ASCE* **6**(6), 515–523.

Reddy, J. N. (1993) *An Introduction to the Finite Element Method.* McGraw-Hill, New York, USA.

Reddy, K. V., Eldho, T. I., Rao, E. P. & Hengade, N. (2007) A kinematic wave based distributed watershed model using FEM, GIS and remotely sensed data. *Hydrol. Processes* **21**, 2765–2777.

Singh, V. P. (1996) *Kinematic Wave Modeling in Water Resources.* Wiley-Interscience, New York, USA.

Smith, M. B., Seo, D., Koren, V. I., Reed, S. M., Zhang, Z., Duan, Q., Moreda, F. & Cong, S. (2004) The distributed model intercomparision project (DMIP): motivation and experiment design. *J. Hydrol.* **298**, 4–26.

Vieux, B. E. (2001) *Distributed Hydrologic Modeling Using GIS.* Kluwer, Dordrecht, The Netherlands.

Vivoni, E. R., Ivanov, V. Y., Bras, R. L. & Entekhabi, D. (2005) On the effects of triangulated terrain resolution on distributed hydrologic model response. *Hydrol. Processes* **19**, 2101–2122.

Analyse the sources of equifinality in hydrological model using GLUE methodology

LI LU[1,2], XIA JUN[1], XU CHONG-YU[3], CHU JIANJING[1,2] & WANG RUI[4]

1 *Key Lab. of Water Cycle & Related Land Surface Processes, Institute of Geographic Science and Natural Resources Research, Chinese Academy of Sciences, Anwai Datun Road A11, Beijing 100101, China*
marylilu@163.com

2 *Graduate University of Chinese Academy of Sciences, Beijing 100049, China*

3 *Department of Geosciences; University of Oslo, Oslo, Norway*

4 *Wuhan University, Wuhan 430072, China*

Abstract The equifinality problem has been universally found in hydrological models. The Generalized Likelihood Uncertainty Estimation methodology (GLUE) is widely used to quantify the parameter uncertainty in a variety of hydrological models. This study makes a comprehensive discussion about the sources of equifinality in a distributed conceptual hydrological model. The study is performed using the model DTVGM on Chao River basin in north China. It analyses the impacts of three aspects on the equifinality, including the number of parameters, the systematic errors of input data and the object function. The Monte Carlo scatter figures of parameters, the relative width of 95% confidence intervals and the percent of observations bracketed by the 95% confidence intervals are used as criteria to estimate uncertainties in parameters and the simulated stream flow. In addition, the study gives an example of reducing the equifinality in the model by using a baseflow separation method; this can be used in basins that only have discharge data. The results indicate that: (a) overparametrization, systematic errors of input data and too little constraint conditions are the sources of equifinality in model DTVGM; (b) systematic errors of input data are not only increasing the predictive uncertainty caused by equifinality, but also reduce accuracy of the model; (c) the uncertainty of parameters has been greatly reduced through the secondary model objective which uses baseflow as an additional condition to constrain the model parameters.

Key words GLUE; equifinality; hydrological model; DTVGM; multi-objective function

INTRODUCTION

The traditional calibration method for hydrological models aims to find an optimal set of parameter values, which makes a good fit to observed data. However, all model calibrations and subsequent predictions will be subject to uncertainty, which arises in that no rainfall–runoff model is a true reflection of the processes involved, and it is impossible to specify the initial and boundary conditions required by the model with complete accuracy. There are four important sources of uncertainties in hydrological modelling (Refsgaard & Storm, 1996; Engeland *et al.*, 2005): (a) uncertainties in input data; (b) uncertainties in output data used for calibration; (c) uncertainties in model parameters; (d) uncertainties in model structure. The conceptual model is not necessarily deterministic. The ultimate target of simulation is likely to find a stochastic–conceptual model, which contains the random variables that reflect the unavoidable uncertainty of data. However, almost all the conceptual models are conceptual–deterministic models (Kirkby, 1978). Hydrologists have also been exploring stochastic–conceptual models. However, it will be very complicated to study the modelling uncertainty, considering all of the four error sources. The problem can be simplified by considering these uncertainties separately based on certain assumptions. Beven & Binley (1992) and Beven & Freer (2001) proposed the Generalized Likelihood Uncertainty Estimation (GLUE) method in order to quantify the parameters uncertainty. Multiple sets of parameters can make an equally good or bad simulation, which is named equifinality in the literature. There are many different model structures and many different parameter sets in a model that may be acceptable in reproducing the observed data (Freer *et al.*, 1996; Zak & Beven, 1999).

In order to reduce the model uncertainty through reducing the equifinality, hydrologists carry out many studies on formulation of the objective function, parameters feasible scope, etc. Franks *et al.* (1997) used catchment saturated area extent, which is estimated by an approach based on the combination of the topographic index and the Saturation Potential Index from the ground truth data, as a secondary modelling objective for model evaluation. And through the specification of a secondary modelling objective, the parametric and predictive uncertainty has been greatly reduced, especially in constraining the catchment effective saturated transmissivity parameter. Mo & Beven (2004) used two data sets, which are the observations of CO_2 and heat fluxes, before and after an irrigation event in a wheat field to constrain the model. It showed that some parameters are strongly conditioned by the observed fluxes. Besides, the GLUE method was used to compare the uncertainty and parameter differences between the two-source canopy model and the three-source canopy model. Gallart *et al.* (2007) used water table records and the distribution of parameters obtained from point observations to reduce the uncertainty of predictions for both streamflow and groundwater contribution. Then, increasing the additional constraint information, which are independent from the original data, can reduce the equifinality of the model. Furthermore, Multi-objective and multi-criteria approach can be used in model calibration to reduce the equifinality (Franks *et al.*, 1998; Kuczera & Mroczkowski, 1998; Yapo *et al.*, 1998; Choi & Beven, 2007). There were also some studies on the equifinality of the Xin'anjiang model by GLUE approach in China (Zeng, 2005; Huang & Xie, 2007; Shu *et al.*, 2008). All these papers mostly focus on the parameter uncertainty, reducing the uncertainty of parameter and model, improving the accuracy of simulations, etc. There are only a few papers about the sources of equifinality. This paper seeks to provide a comprehensive discussion about the sources of equifinality in hydrological models and the impacts on parameters of hydrological models which are quantified using the GLUE methodology. A monthly distributed conceptual model called DTVGM was used.

METHODOLOGY

The DTVGM model

The monthly Distributed Time-Variant Gain Model (DTVGM), which was developed based on the Time-Variant Gain Model (TVGM) (Xia *et al.*, 1997), has been applied in the Chaobai River basin and the Heihe River basin (Xia *et al.*, 2003; Wang *et al.*, 2005). The primary equations of the model are presented in Table 1.

Table 1 Principal equations for the DTVGM model.

snowmelt	$S_M = MF \cdot (T_{av} - T_{mlt})$
actual evapotranspiration	$ET_a = \left[(1-KAW)\cdot f(P/ET_p) + KAW \cdot AW/WM\right]\cdot ET_p$
surface runoff	$\begin{cases} Rs = g1\cdot(AW/WM)^{g2}\cdot P & AW \ge WMi \\ Rs = g1(AWi/WM)^{g2}\cdot P & AW < WMi \end{cases}$
underground runoff	$Rg = Kr \cdot AW \cdot ThinkU$
total runoff	$R = Rg + Rs$

MF is snowmelt rate (mm°C^{-1}month^{-1}), T_{mlt} is the Snowmelt temperature (°C), T_{av} is mean monthly air temperature (°C), *ETp* is potential evapotranspiration (mm month^{-1}), *ThinkU* is soil depth.

The GLUE methodology

The parameter uncertainty was evaluated by using the GLUE (Generalized Likelihood Uncertainty Estimation) methodology (Beven & Binley, 1992). In this method a large number of model runs are made with many different randomly chosen parameter values selected from *a priori* probability distributions. The acceptability of each run is evaluated against observed values and, if the acceptability is below a certain subjective threshold, the run is considered to be "non-behavioural" and that parameter combination is removed from further analysis. In this method, likelihood values serve as relative weights of each parameter set or simulated value. It is noted that the likelihood function and the threshold are subjectively determined, and this was discussed by Freer *et al.*

(1996). In this study, the Nash-Sutcliffe efficiency was chosen as the likelihood function:

$$L(\underline{\theta}_i \mid \underline{Y}) = (1 - \sigma_i^2 / \sigma_{obs}^2)^N \qquad \sigma_i^2 < \sigma_{obs}^2 \tag{1}$$

where $L(\underline{\theta}_i \mid \underline{Y})$ is the likelihood measure, σ_i^2 is the variance of errors for given parameter set $\underline{\theta}_i$ and the observed discharge data set $\underline{Y}$, σ_{obs}^2 is the variance of the observed data set, and N is a parameter.

Criteria for the comparison

In this paper, two indices are used to compare the derived 95% confidence interval (95CI), which are the relative width of 95% confidence interval ($R - 95CI$) and the percent of observations bracketed by the 95CI ($P - 95CI$):

$$R - 95CI = \left(Q_{UB} - Q_{LB}\right) / Q_{Obs} \tag{2}$$

Q_{UB} is the upper boundary value of $95CI$, Q_{LB} is the lower boundary value of $95CI$, Q_{Obs} is the observed discharge.

$$P - 95CI = NQ_{Ob} / NQ_{ALL} \times 100\% \tag{3}$$

NQ_{Ob} is the number of observations bracketed by the 95CI, NQ_{ALL} is the total number of observations. The goodness of calibration uncertainty is judged on the basis of the closeness of the $R - 95CI$ to 0 and the $P - 95CI$ to 100%.

RESULTS AND DISCUSSION

Sensitivity analysis of model parameters

The study area was Chao River basin upstream of the Miyun reservoir with drainage area of 5300 km^2, accounting for 40% of the Miyun Reservoir catchment area, which is one of the most important surface water resources of Beijing water supply. The data from the 1973 to 1982 was used for simulation. The model was calibrated based on the observed discharges at the watershed outlet (Xiahui station). There is rarely snow in winter in Chao River basin because of the little precipitation in the winter season. So snowmelt is not considered in the simulation with monthly time step. The parameters ranges of the model are shown in Table 2. The prior distribution of parameters is supposed to be a uniform distribution. A total of 200 000 runs of DTVGM were made using Monte Carlo-based random sampling of the original parameter ranges. The threshold value of the Nash-Sutcliffe efficiency used to select the behavioural parameter set was chosen to be 0.6. The results in Figs 1 and 2 show the ranking of the parameters sensitivity is as follows: *g*1, *WM*, *g*2 > *Kaw*, *Kr* > *WMi*.

The impact of the numbers of parameters on equifinality

In this paper, the impact of the number of parameters on equifinality in DTVGM is studied by reducing the number of parameters gradually, starting from the least sensitive parameter based on

Table 2 The ranges of six parameters in DTVGM model.

Parameter	Physical meaning	Minimum	Maximum
*g*1	The coefficients of surface water	0	1
*g*2	The coefficients of surface water	0	∞
Kr	The outflow efficient of groundwater	0	1
WMi	Minimum soil-moisture storage	0	0.4
WM	Maximum soil-moisture storage	0.4	1
Kaw	Weight coefficient for evapotranspiration	0	1

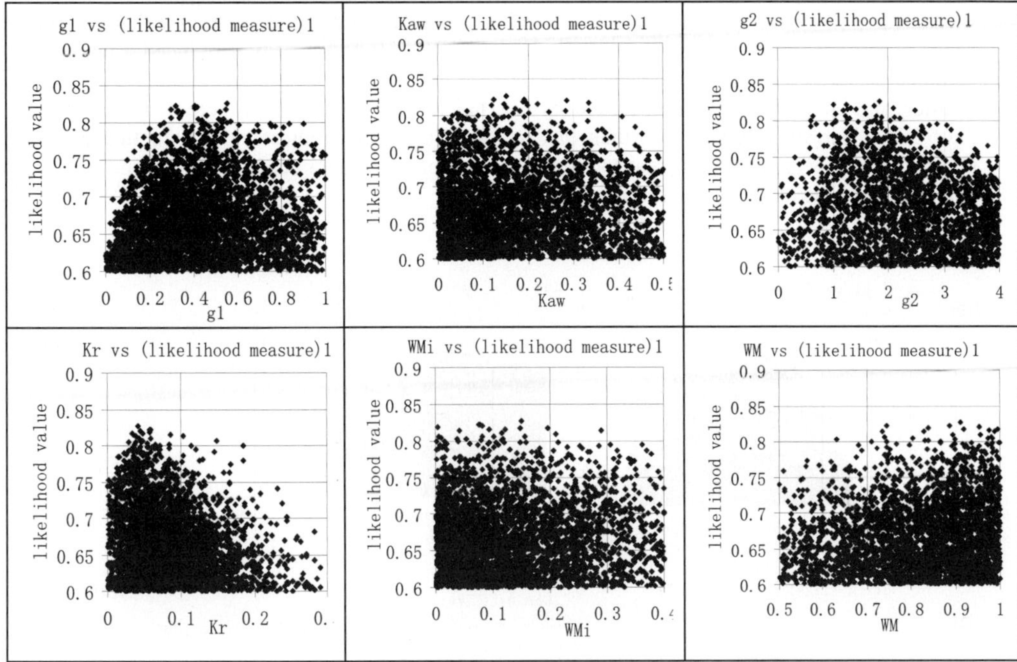

Fig. 1 Dotty plots for six DTVGM parameters against likelihood value.

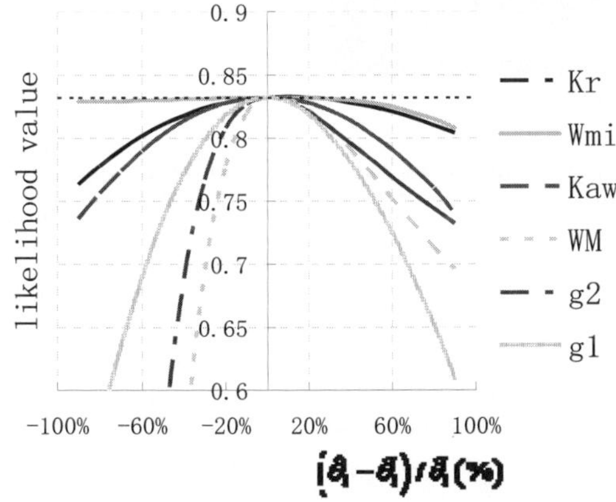

Fig. 2 Intersection of the hypersurface of the likelihood function. ($\hat{\theta}_i$ = optimised parameter value, $\tilde{\theta}_i$ = parameter values in the neighbourhood of $\hat{\theta}_i$).

the sensitivity analysis of parameters. The parameters *g*1 and *Kaw*, which are important and sensitive, are retained and chosen as example to demonstrate the effect of fixing other parameters on them. The parameters are fixed to the optimal values one by one in turn, by the sensitivity from weak to strong. The dotty plots of *g*1 and *Kaw* are shown in Fig. 3.

By comparison with Fig. 1, the dotty plots show nearly no change when the *WMi* parameter in Fig. 3(a) is fixed, which proves that *WMi* is a "non-sensitive" parameter. When considering hydrological model improvement, the non-sensitive parameters can be fixed by experience or experiment. So the hydrological model can be simplified and the time of model calibration shortened. With the increasing number of fixed parameters, the dotty plots become denser, and the peaks of response surface are sharper. So it demonstrates that equifinality is reduced. Figure 3 also demonstrates that as more parameters are fixed, the uncertainties of the other free parameters are smaller, and the probability of an efficiency coefficient bigger than 0.6 is increased. Furthermore, the 95*CI* is narrower in Fig. 4(b) than that in Fig. 4(a), which indicates the reduction of model uncertainty. This result reveals that too many parameters (overparametrization) are one reason for equifinality in the hydrological model.

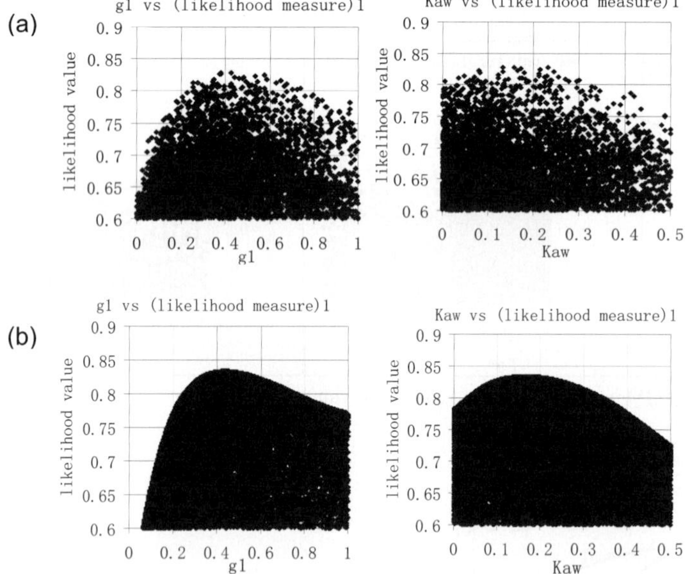

Fig. 3 Dotty plots of *g*1 and *Kaw* plotted against likelihood value, while (a) the parameter *WMi* of DTVGM is fixed to the optimal value, (b) the parameters *WMi*, *Kr*, *g*2 and *WM* of DTVGM are fixed to the optimal value.

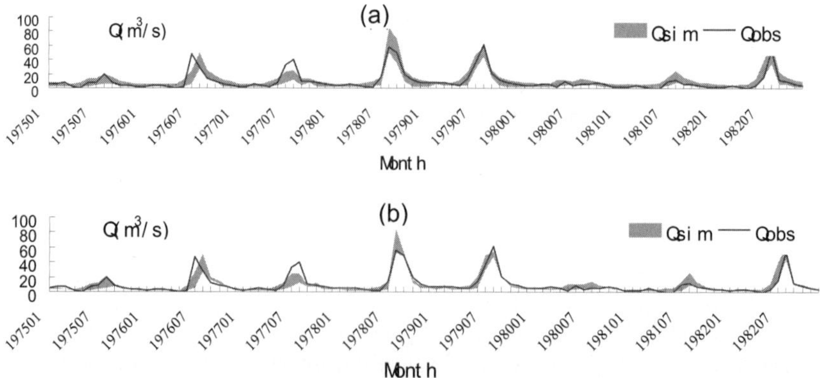

Fig. 4 The 95*CI* of monthly simulated discharge from 1975 to 1982 due to: (a) one parameter fixed just the same with Fig. 3(a), (b) four parameters fixed just the same with Fig. 3(b), for the Chao River basin. The Qsim is the 95*CI* of monthly simulated discharge and the Q_{obs} is the observed flow.

The impact of systematic errors of input data on equifinality

In the research, precipitation is used as an example of input data, and ±10% systematic errors are added to the precipitation. The same likelihood criteria and threshold value as used in Fig. 1 are used here. The simulated results are shown in Fig. 5 and Table 3 as follows:

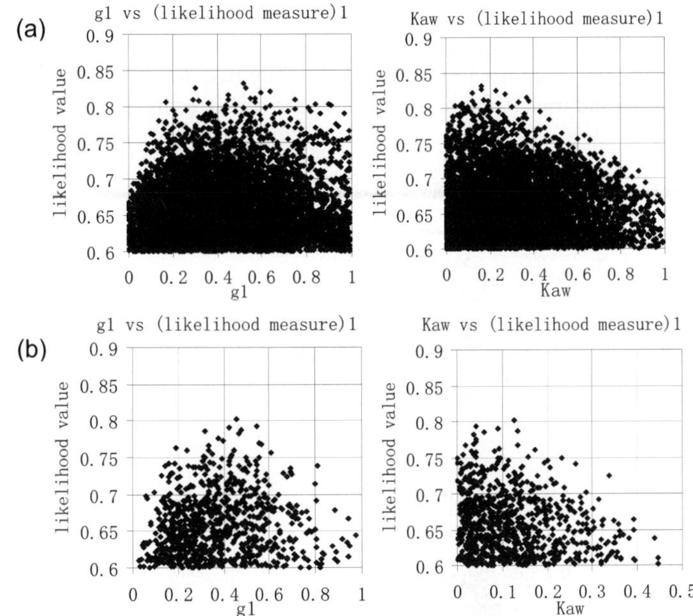

Fig. 5 Dotty plots of parameter g1 and Kaw for (a) precipitation reduced 10% systematically, (b) precipitation increased 10% systematically.

Table 3 The $R-95CI$ and the $P-95CI$ due to different input data of precipitation for Chao River basin.

Criterion	Original precipitation	Precipitation reduced by 10% systematically	Precipitation increased by 10% systematically
$R-95CI$	1.73	2.37	0.85
$P-95CI$ (%)	80.21	81.25	56.25

Figure 5 shows that the impacts of systematic error of input data on the two parameters are different. Parameter *Kaw* displays are more influenced. Why are they different? It is because the hydrological model calibration makes simulated runoff fit to observed runoff. As discussed before, the *Kaw* parameter controls the simulated evapotranspiration that is not included in the objective function, while the parameter *g*1 controls the simulated runoff which is constrained by the objective function. Accordingly, when the precipitation changed, the excess or missing water will be transferred to actual evapotranspiration, which results in the greater uncertainty of parameter *Kaw*. The difference in the impact of precipitation increase and reduction on the DTVGM model can also be seen from Table 3. The results indicate that the systematic error is likely to enhance the uncertainty of parameter values and reduces the accuracy of the simulation results. It should be noted that when the precipitation reduced by 10%, the response surface is more flat, and when the precipitation increased by 10% the overall simulation efficiency reduced. The results reveal that the systematic error of input data is another source of equifinality and uncertainty in DTVGM model.

The impact of insufficient constraint conditions on equifinality

The calibration of hydrological models is usually achieved with limited discharge data. In this study, the baseflow is used as additional data for model calibration in order to test the difference of multi- and single-objective functions. Many baseflow separation methods have been developed and applied to various catchments, such as the smoothed minimum method (Mazvimavi *et al.*, 2004), digital filter method (Lyne & Hollick, 1979; Weeks & Boughton, 1987), streamflow hydrograph separation (Ysep, 1996), etc., of which the digital filtering method is most widely used. In this research, three digital filtering methods developed by Lyne & Hollick (1979), Weeks & Boughton (1987), and Chapman (1991) are used to calculate the daily baseflow, which is then summed to get the monthly baseflow and used to calibrate the baseflow simulated by the model.

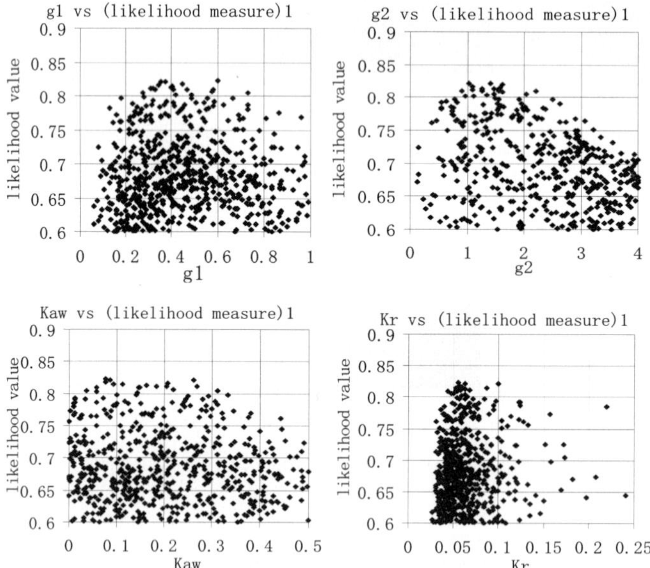

Fig. 6 Dotty plots of Nash and Sutcliffe coefficient against DTVGM parameter *g1*, *g2*, *Kaw* and *Kr* conditioning with GLUE based on 200 000 samples with threshold $R^2 \geq 0.06$ and $R^2_{base} \geq 0.5$ (R^2_{base} is the Nash and Sutcliffe coefficient of monthly baseflow).

Table 4 the $R-95CI$ of the monthly simulated discharge from 1975 to 1982 and the $P-95CI$ due to different constraint conditions of calibration for Chao River basin.

Criterion	Single-objective (total runoff)	Multi-objective (total runoff and baseflow)
$R-95CI$	1.73	1.06
$P-95CI$ (%)	80.21	78.08

Figure 6 shows that the number of behavioural simulations reduced obviously after additional evaluation of the simulations with respect to the baseflow. However, the optimal result was not affected. As can be seen, the dotty plots of parameter *Kr* changed significantly. That is because the parameter Kr controls the baseflow directly. Table 4 shows that $R-95CI$ reduced by 38.7%, from 1.73 to 1.06, while $P-95CI$ reduced by 2.1%. It illuminates that the uncertainty of the model is reduced with almost no impact on accuracy of the result after additional evaluation of the simulations with respect to the baseflow. With the increase in calibrated data and multi-objective functions, the uncertainty of relevant parameters and model uncertainty reduced significantly, which puts forward higher requirements on hydrological observation and data collection.

CONCLUSION

The study result indicates that too many parameters in the model (overparametrization), systematic errors of input data, and constraint conditions that are too little are the sources of equifinality in model DTVGM. By quantitative comparison of the sensitivity of parameters, the parameter of minimum soil moisture is found to be the most insensitive parameter in the model DTVGM, and has little effect on simulated results. This kind of insensitive parameters can be fixed against the physical characteristics of the basin to reduce the equifinality of the hydrological model. Furthermore, the systematic errors of input data not only increase the predictive uncertainty caused by equifinality, but also reduce the accuracy of the model. So it is important to have some idea of the accuracy in the input data. It's important to perform quality control on the input data before it is used to drive hydrological models. This paper also gives an example to reduce the equifinality in the model by using baseflow as an additional data for calibration. In this study, $R - 95CI$ and $P - 95CI$, as defined by equations (2) and (3), are used as two indices to compare the $95CI$ of the simulated result. The result illustrates that the uncertainty of parameters has been greatly reduced through the secondary model objective. This reduction in uncertainty of parameters translates into a marked reduction in the predictive uncertainty, most obviously in the constraint of the parameter *Kr* which is controlling baseflow simulation. So it is better to use as much information as possible to calibrate the model whenever the data are available.

It should be noted that although this study provides a comprehensive discussion about the sources of equifinality in the hydrological model and the impacts on parameters of the model are quantified using the GLUE methodology, it is only a preliminary study because only one model is used and the data are limited. More studies need to be done on other models and using more input and output data in order to generalize the findings obtained here.

Acknowledgements This study were supported by the Key Project of the Natural Science Foundation of China (No. 40730632), the Knowledge Innovation Key Project of the Chinese Academy of Sciences (Kzcx2-yw-126), the Special Fund of Ministry of Science and Technology, China (2006DFA21890) and the Key Project of International Cooperation in CAS (no. GJHZ06). We also thank Gangsheng Wang for providing the precipitation and discharge data used in the study.

REFERENCES

Beven, K. & Binley, A. (1992) Future of distributed models: Model calibration and uncertainty prediction. *Hydrol. Processes* **6**(3), 279–298.

Beven, K. & Freer, J. (2001) Equifinality, data assimilation, and uncertainty estimation in mechanistic modelling of complex environmental systems using the GLUE methodology. *J. Hydrol.* **249**(1–4), 11–29.

Chapman, T. G. (1991) Comment on "Evaluation of Automated Techniques for Base Flow and Recession Analyses" by R. J. Nathan & T. A. McMahon. *Water Resour. Res.* **27**(7), 1783–1784.

Choi, H. T. & Beven, K. (2007) Multi-period and multi-criteria model conditioning to reduce prediction uncertainty in an application of TOPMODEL within the GLUE framework. *J. Hydrol.* **332**(3–4), 316–336.

Engeland, K., Xu, C.-Y. & Gottschalk, L. (2005) Assessing uncertainties in a conceptual water balance model using Bayesian methodology. *Hydrol. Sci. J.* **50**(1), 45–63.

Franks, S. W., Beven, K. J., Chappell, N. A. & Gineste, P. (1997) The utility of multi-objective conditioning of a distributed hydrological model using uncertain estimates of saturated areas. In: *Proceedings of the International Congress on Modelling and Simulation*: *MODSIM 971*, 335–340.

Franks, S. W., Gineste, P., Beven, K. J. & Merot, P. (1998) on constraining the predictions of a distributed model: the incorporation of fuzzy estimates of saturated areas into the calibration process. *Water Resour. Res.* **34**, 787–797.

Freer, J., Beven, K. & Ambroise, B. (1996) Bayesian estimation of uncertainty in runoff prediction and the value of data: an application of the GLUE approach. *Water Resour. Res.* **32**(7), 2161–2174.

Gallart, F., Latron, J. Llorens, P. & Beven, K. J. (2007) Using internal catchment information to reduce the uncertainty of discharge and baseflow predictions. *Adv. Water Resour.* **30**(4), 808–823.

Huang, G. R. & Xie, H. H. (2007) Uncertainty analyses of watershed hydrological model based on GLUE method. *J. South China University of Technology (Natural Science Edition)* **35**(3), 137–142, 149 (in Chinese).

Kirkby, M. J. (1978) *Hillslope Hydrology*. Chichester, New York, USA.

Kuczera, G. & Mroczkowski, M. (1998) Assessment of hydrologic parameter uncertainty and the worth of multiresponse data. *Adv. Water Resour.* **34**(6), 1481–1489.

Lyne, V. & Hollick M. (1979) Stochastic time-variable rainfall–runoff modelling. In: *Institute of Engineers Australia National Conference*. Publ. 79/10, 89–93.

Mazvimavi, D., Meijerink, A. M. *et al.* (2004) Prediction of base flows from basin characteristics: a case study from Zimbabwe. *Hydrol. Sci. J.* **49**(4), 703–715.

Mo, X. & Beven, K. (2004) Multi-objective parameter conditioning of a three-source wheat canopy model. *Agricultural & Forest Meteorol.* **122**(1–2), 39–63.

Nash, J. E. & Sutcliffe, J. V. (1970) River flow forecasting through conceptual models. Part 1, A discussion of principles. *J. Hydrol.* **10**(3), 282–290.

Refsgaard, J. C. & Storm, B. (1996) Construction, calibration and validation of hydrological models. *Distributed Hydrological Modelling*, 41–54.

Ren, X. S., He, Shan, *et al.* (2007) Water resources assessment of Hai River Basin. Chinese Water Conservancy and Hydropower Press, Beijing, China (in Chinese).

Shu, Chang, Liu, S. X., Mo, X. G., Liang, Z. M. & Dai, Dong (2008) Uncertainty analysis of Xinanjiang model parameter. *Geographical Res.* **27**(2), 343–352 (in Chinese).

Wang, G. S. (2005) Theory and method of distributed time-variant gain model doctoral thesis. Grad. Sch. of the Chin. Acad. of Sci., Beijing, China (in Chinese).

Weeks, W. D. & Boughton, W. C. (1987) Tests of ARMA model forms for rainfall-runoff modelling. *J. Hydrol.* **91**(1/2).

Xia, J., O'Connor, K. M. *et al.* (1997) A non-linear perturbation model considering catchment wetness and its application in river flow forecasting. *J. Hydrol.* **200**(1–4), 164–178.

Xia, J., Wang, G. S., Lv, A. F. & Tan, G. (2003) A research on distributed time variant gain modelling. *Acta Geographica Sinica* **58**(05), 789–796 (in Chinese).

Yapo, P. O., Gupta, H. V. *et al.* (1998) Multi-objective global optimization for hydrologic models. *J. Hydrol.* **204**(1–4), 83–97.

Ysep, R. A. H. (1996) A computer program streamflow hydrograph separation and analysis. In: *USGS Water Resources Investigations Report*. USGS, Lemoyne, Pennsylvania, USA.

Zak, S. K. & Beven, K. J. (1999) Equifinality, sensitivity and predictive uncertainty in the estimation of critical loads. *Science Total Environ.* **236**(1–3), 191–214.

Zeng, Z. X. (2005) The study of equifinality of conceptual rainfall-runoff model parameter. Master Thesis, Wu Han University, Wu Han, China.

Comparaison de trois méthodes d'usage de la technique des voisins les plus proches en vue d'amélioration de la performance de l'algorithme SCE-UA appliqué pour le calage du modèle pluie–débit HBV

HAMMOUDA DAKHLAOUI[1], ZOUBEIDA BARGAOUI[1] & ANDRAS BARDOSSY[2]

1 *Ecole Nationale d'ingénieurs de Tunis, BP 37, 1002 Tunis le Belvédère, Tunisie*
hammouda.dakhlaoui@laposte.net; zoubeida.bargaoui@laposte.net

2 *Institute of Hydraulic Engineering, Universität Stuttgart, Pfaffenwaldring 61, D-70569 Stuttgart, Germany*
andras.bardossy@iws.uni-stuttgart.de

Résumé L'objectif de cette étude est d'améliorer l'efficacité de l'algorithme d'optimisation SCE-UA appliqué pour le calage automatique du modèle pluie–débit HBV. Ces améliorations sont basées sur l'usage de la méthode des voisins les plus proches KNN pour l'estimation de la fonction objective. Après un certain nombre d'itérations par SCE-UA, une base de données des jeux de paramètres explorés par l'algorithme d'optimisation et de leurs valeurs de fonction objective est constituée. En vue d'économiser le temps de calcul de la fonction objective par simulation, il est proposé de l'estimer par interpolation des voisins les plus proches. La question est ainsi de sélectionner les situations où la fonction objective est estimée directement en opérant le modèle HBV ou indirectement par KNN. Trois algorithmes sont proposés et testés. Le premier fait le choix de ces situations à travers un classement de la population explorée par ses valeurs de fonction objective, le deuxième fait une discrimination sur les bases des valeurs seuils de distance séparant les nouveaux points à leurs voisins les plus proches, et le dernier intègre la méthode KNN dans l'algorithme simplex évolutionnaire lui même. L'évaluation de la performance du modèle HBV est réalisée. Pour l'implémentation du KNN, différentes définitions de distance sont testées, notamment la distance euclidienne dans l'espace des paramètres normalisés et la distance euclidienne pondérée dans l'espace normalisé des paramètres. Il est proposé de relier ces poids à la sensibilité des paramètres à la fonction objective, en donnant plus d'importance aux paramètres les plus sensibles. Ces algorithmes hybrides sont testés sur des données synthétiques. L'analyse des résultats suggère une amélioration de la performance du SCE-UA de 30% au moins en terme de temps de calcul.

Mots clefs calage du modèle; SCE-UA; modèle HBV; KNN

Comparison of three methods using the *k*-Nearest Neighbours approach to improve the SCE-UA algorithm for calibration of the HBV rainfall–runoff model

Abstract The aim of this study is to improve the efficiency of the global optimisation method SCE-UA, applied for the calibration of the lumped HBV rainfall–runoff model. This improvement is performed using the *k*-Nearest Neighbours (KNN) approach in estimating the objective function. The motivation is that, when an objective function is evaluated by model simulation, the time of computation may be too long. Now, a database of the explored parameters and of their respective objective function values may be constituted as SCE-UA is running. In order to save computing time, it is therefore proposed to take advantage of this database by estimating the objective function using an interpolation by nearest neighbours in the parameter space. The problem is how to choose the situations where objective function will be estimated by the model simulation rather than by KNN. Three KNN implementation algorithms are thus proposed. In the first, choice is based on ranking of the explored population by their objective function values. In the second, discrimination is performed using threshold distances between the newly generated indvidual and its neighbours. In the last, the KNN method is fully integrated into the evolutionary simplex algorithm. Evaluation of the model performance was carried out. Several distance measures were adopted to perform the KNN technique, such as normalised Euclidian distance and normalised weighted Euclidian distance. Weights are evaluated taking account of the sensitivity of parameters to the objective function, i.e. giving more importance to the most sensitive parameters. These hybrid optimisation algorithms are tested for synthetic runoff. Analysis of the results suggests that an improvement on efficiency of more than about 30% is obtained.

Key words model calibration; SCE-UA; HBV model; KNN

1 INTRODUCTION

Les modèles pluie–débit sont essentiels dans la gestion des ressources en eaux, l'analyse du risque hydrologique, la prévision des crues, l'étude d'impact des changements climatique sur les cycles hydrologique, etc. Toutefois l'application de ces modèles dépend de la qualité de l'estimation de leurs paramètres. Bien que certains de ces paramètres puissent être estimés par des mesures *in situ*, ou à partir des caractéristiques du bassin versant (morphologie, pédologie, occupation du sol, etc.), la majorité des paramètres, et notamment dans le cas des modèles conceptuels, nécessite d'être estimée par calage. L'opération de calage consiste à jouer sur les paramètres du modèle pour obtenir la meilleure performance du modèle moyennant une confrontation des sorties simulées avec les observations.

Avec le développement des outils informatiques, le recours au calage automatique, basé sur des algorithmes d'optimisation est devenu de plus répandu. Le coût du calage en terme de temps de calcul reflété par exemple par le nombre de simulations nécessaires à la convergence de l'algorithme d'optimisation est un facteur important voire déterminant du choix et de la conception d'un algorithme d'optimisation. Durant les dernières décennies on a vu se développer des algorithmes d'optimisation de plus en plus efficaces, tels que: le Simplex recuit simulé (Bates, 1994), SCE-UA (Duan *et al.*, 1992, 1993), et les algorithmes génétiques (Goldberg, 1989). Ces algorithmes ont été appliqués aux problèmes de calage des modèles pluie–débit (Duan *et al.*, 1993; Franchini, 1996; Franchini & Galeati, 1997; Franchini *et al.*, 1998; Gupta *et al.*, 1999; Thyer *et al.*, 1999; Efstratiadis & Koutsoyiannis, 2002). L'adoption de fonctions d'approximation est une des voies de recherche visant la réduction du temps de calcul. Cette méthode consiste à estimer la fonction objective à partir de la base des données des jeux de paramètres explorés par l'algorithme d'optimisation au lieu de l'estimer par simulation directe à l'aide du modèle pluie–débit. Les méthodes d'approximation proposées dans la littérature sont, par exemple: l'interpolation polynomiale, l'interpolation géostatistique, la méthode des voisins les plus proches, les fonctions *radial basis*, etc. De telles techniques ont été, par exemple, introduites dans des algorithmes évolutionnaires pour l'optimisation des fonctions coûteuses en temps de calcul (Ong *et al.*, 2003, Ostfeld & Salomons, 2005, Zhou *et al.*, 2007). Toutefois, l'adoption des fonctions d'approximation peut altérer les performances et notamment l'effectivité de l'algorithme d'optimisation ou autrement son aptitude de trouver la solution optimale en orientant le mécanisme de recherche dans de mauvaises directions. La recherche directe a pour effet de donner une estimation de la valeur de la fonction objective mais d'augmenter les temps de calcul. L'introduction de fonctions d'approximation a pour effet de réduire le temps de calcul mais au détriment de la précision d'estimation de la fonction objective. L'idéal serait de réduire le nombre de simulations nécessaires à la convergence sans affecter l'aptitude de l'algorithme à reconnaître la solution optimale. Il s'agit de trouver un compromis entre, d'une part, le recours à l'estimation de la fonction objective par simulation via le modèle et, d'autre part, l'estimation de la fonction objective par des fonctions d'approximation. Plusieurs approches ont été proposées pour "optimiser" l'usage des fonctions d'approximation, et toujours elles ont été adaptées à l'algorithme d'optimisation employé, c'est-à-dire que leur mise en oeuvre est spécifique à l'algorithme d'optimisation (Giannakoglou, 2002; Ong *et al.*, 2003; Ostfeld & Salomons, 2005; Zhou *et al.*, 2007).

Dans ce travail, il est proposé d'adopter la méthode des voisins les plus proches (*k-Nearest Neighbours*, KNN) comme technique d'approximation de la fonction objective au sein de l'algorithme SCE-UA. On étudiera plus particulièrement son effet sur l'efficacité et l'effectivité de l'algorithme d'optimisation. Le problème d'optimisation est celui du calage du modèle HBV (Bergström, 1976). Pour cela on a choisi de travailler sur des données synthétiques provenant d'un *mimétisme* du bassin de Rottweil (affluent du Rhin, en Allemagne). Après une brève revue de littérature sur le modèle hydrologique adopté, et les méthodes SCE-UA et KNN (Section 2), nous présentons les approches proposées pour intégrer le KNN dans un SCE-UA attaché au modèle HBV (Section 3). Enfin, nous présentons et discutons les résultats et ceci après une brève présentation des données (Section 4).

2 REVUE DE OUTILS UTILISES

2.1 Le modele hydrologique HBV

C'est le modèle hydrologique HBV (Bergström, 1976) initialement proposé par l'Institut météorologique et hydrologique Suédois (SMHI). Depuis, ce modèle a connu des améliorations et

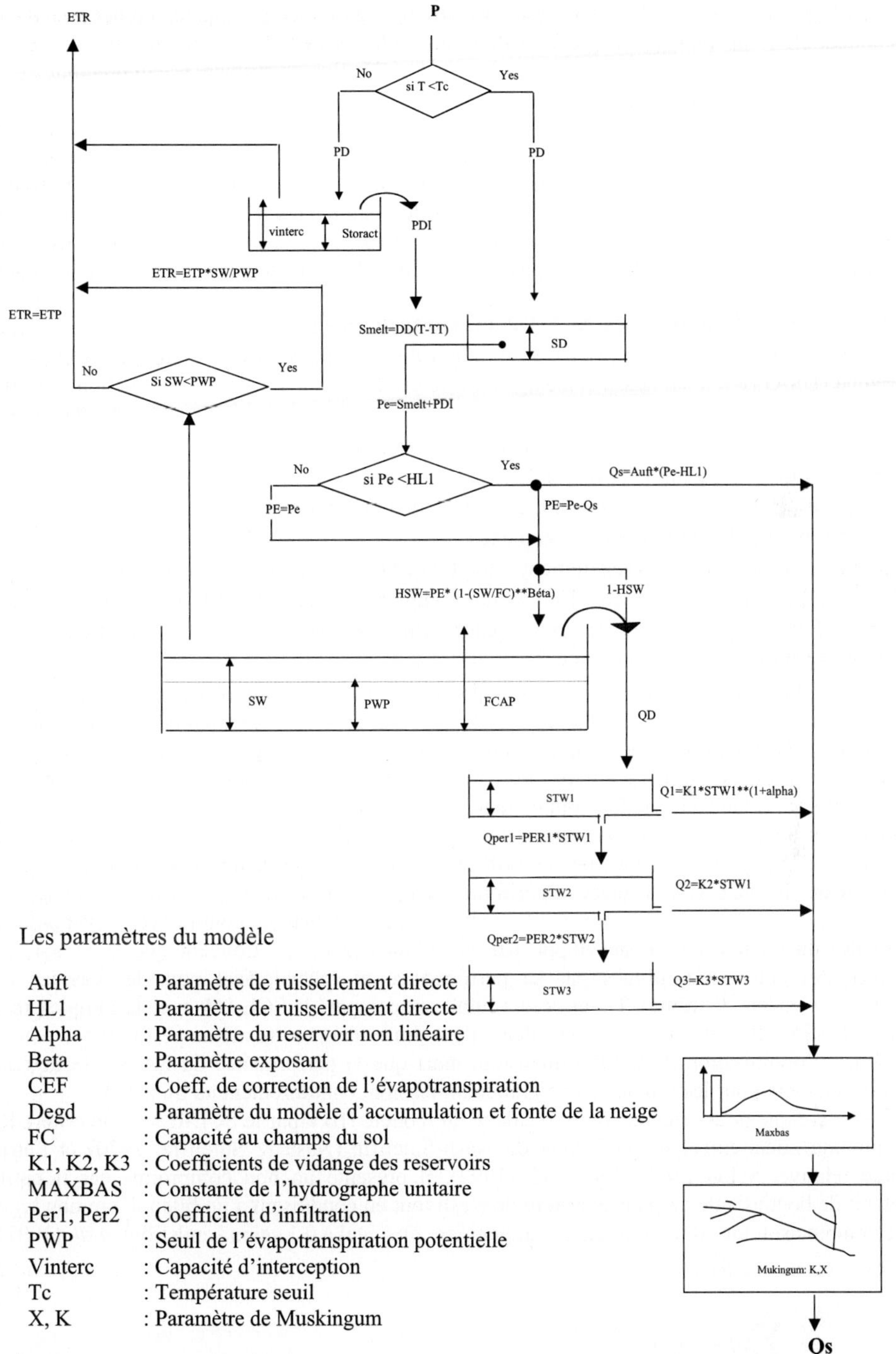

Fig. 1 Le modèle HBV.

modifications. Bardossy & Theisen (1999) proposent une version que nous utilisons dans le cadre de ce travail pour une modélisation globale. Le modèle fonctionne au pas de temps horaire. Le cycle de l'eau schématisé selon HBV est présenté en Fig. 1. Ses principales composantes sont: l'accumulation et la fonte de neige, l'infiltration, l'actualisation de l'humidité du sol, l'évapotranspiration, la percolation vers la nappe et le transfert des flux en surface. L'eau précipitée passe en premier lieu par une fonction d'interception dépendant de la couverture végétale, ensuite la fonte et l'accumulation de la neige sont calculées, ce qui permet de déterminer la pluie efficace (ou potentielle à l'écoulement). Après le passage par une fonction de répartition de l'écoulement direct sur les surfaces imperméables, l'eau s'infiltre dans le sol où le processus d'évapotranspiration est modélisé par une fonction non linéaire. Une partie de l'eau contenue dans le sol percole et passe par trois réservoirs de transfert disposés en série. Une fonction de transfert de surface basée sur la méthode de Muskingum modélise la propagation de l'écoulement dans le réseau hydrographique.

2.2 L'algorithme d'optimisation SCE-UA et critère de calage

Le *Shuffled Complex Evolution* (SCE) est un algorithme d'optimisation développé à l'Université d'Arizona (USA) (www.sahra.arizona.edu). Pour le cas de minimisation d'une fonction objective, les différentes étapes sont:

(a) la génération aléatoire d'un ensemble de points initiaux dans l'espace des paramètres (hypercube) et le calcul de la valeur de la fonction objective en ces points;
(b) la classification des points par valeur croissante de la fonction objective;
(c) la répartition des individus sur p complexes;
(d) l'évolution des complexes séparément selon l'algorithme Simplex évolutionnaire;
(e) le mélange et le réarrangement aléatoire des complexes en de nouveaux complexes; et
(f) la vérification des critères de convergence: s'il y a convergence alors l'optimisation est terminée sinon on revient à l'étape (b).

Selon la littérature, la méthode SCE-UA prend sa robustesse, son efficacité et son effectivité de quatre concepts sur lesquels elle est basée: (i) combinaison des méthodes déterministe et probabiliste, (ii) évolution systématique des points dans la direction d'amélioration de la fonction objective, (iii) évolution compétitive, et (iv) mélange aléatoire des complexes. Appliquant l'algorithme au calage du modèle pluie–débit, SFB, Kuczera (1997) a constaté qu'il présente "*a markedly superior convergence near the optimum*". Il conclut que cet algorithme a des performances supérieures à celles des algorithmes génétiques. De même, Franchini *et al.* (1998) ont comparé le SCE-UA à deux algorithmes d'optimisation (*Pattern Search* couplé avec *Sequential Quadratic Programming*, SQP, et un algorithme génétique couplé avec SQP) pour des cas réelles du modèle ADM, développé par Franchini (1996). Ils trouvent que le SCE-UA se distingue des autres algorithmes par sa performance et spécialement pour les cas les plus compliqués (modele distribué). Thyer *et al.* (1999) ont comparé le SCE-UA avec le Simplex recuit simulé (SA-SX). Ils ont trouvés que ces deux algorithmes donnent la même performance pour les problèmes d'optimisation de faible dimension, mais que la performance du SA-SX se détériore considérablement pour les problèmes de grande dimension, contrairement au SCE-UA.

Le critère que nous retenons pour le calage du modèle HBV par le SCE-UA est le critère RV, qui combine linéairement le coefficient du Nash-Sutcliffe (Nash & Sutcliffe, 1970) et l'erreur absolue relative. Selon Lindström *et al.* (1997), il présente un bon compromis car il permet d'obtenir de bonnes valeurs pour le critère de Nash tout en conduisant à un bilan d'eau non biaisé. Cette combinaison introduit un poids ω qui peut être pris égal à 0.1 selon Lindström *et al.* (1997):

$$\mathrm{RV} = \mathrm{Nash} - \omega\,|\mathrm{RD}| \tag{1}$$

$$\mathrm{Nash} = 1 - \frac{\sum_{i=1}^{n}(q_{ci} - q_{oi})^2}{\sum_{i=1}^{n}(q_{oi} - \bar{q}_o)^2} \tag{2}$$

RD est l'erreur absolue relative:

$$\text{RD}=\frac{1}{n}\sum_{i=1}^{n}\left(\frac{q_{ci}-q_{oi}}{q_{oi}}\right) \tag{3}$$

et $\bar{q}_o$ est la moyenne des débits observés:

$$\bar{q}_o=\frac{\sum_{i=1}^{n} q_{oi}}{n} \tag{4}$$

2.3 La méthode des voisins les plus proches

La méthode des voisins les plus proches, KNN, est une technique d'intelligence artificielle qui a été introduite par Fix & Hodges (1952) et Skellam (1952). Elle peut être utilisée comme un estimateur non paramétrique. Étant donnée une série $\{X_i, \text{F}(X_i)\}$ d'une fonction scalaire $\text{F}(X)$ définit par un vecteur d'attributs X. La méthode KNN permet d'estimer $\text{F}(X)$ pour un X différent des X_i présents dans la série et ce, moyennant une interpolation des valeurs de la fonction F à partir des voisins les plus proches. La détermination du voisinage fait intervenir une distance entre les attributs. La méthode KNN a trouvé récemment plusieurs applications intéressantes dans le domaine des sciences de l'eau: Brandsma & Können (2005) l'utilisent pour la simulation des variables climatiques; Khu & Madsen (2005) l'ont exploité pour la classification des solutions dans un algorithme de calage multi-objectif basé sur les algorithmes génétiques; Ostfeld & Salomons (2005) ont adopté la technique KNN pour améliorer la performance d'un algorithme génétique appliqué pour le calage d'un modèle environnemental. Quant à Bardossy & He (2006), ils l'ont utilisé pour la régionalisation des paramètres du modèle pluie–débit HBV.

3 METHODES D'INTEGRATION DU KNN DANS L'ALGORITHME DU SCE-UA ET LE CAS DU HBV

Après un certain nombre d'itérations par SCE-UA, une base de donnée des jeux de paramètres explorés ainsi que des valeurs de leur fonction objective est constituée. Cette base de données est exploitée pour approximer la fonction objective par la méthode KNN dans l'espace des paramètres. Soit FO_{KNN} la fonction objective estimée par KNN en un nouveau point généré et soit P_{new} ($P_{\text{new1}}, P_{\text{new2}}, \ldots, P_{\text{newn}}$) avec $P_{\text{new}i}$ ($i = 1, n$), les valeurs des paramètres au point P_{new}:

$$\text{FO}_{\text{KNN}} = \sum_{i=1}^{k} 1/d_i\,\text{FO}_i \tag{5}$$

où FO_i est la valeur de la fonction objective prise au voisin P_i; d_i est le distance entre le nouveau point généré et son voisin i; et k est le nombre de voisins les plus proches considérés lors de l'interpolation, qui est un paramètre à fixer.

Deux définitions de distance (d_i) sont proposées pour définir le voisinage. La première est la distance euclidienne dans l'espace des paramètres normalisés:

$$d_i=\frac{1}{n}\sqrt{\sum_{i=1}^{n}\left(\frac{P_{newi}-P_{\min}}{P_{i\max}-P_{i\min}}-\frac{P_i-P_{1\min}}{P_{i\max}-P_{i\min}}\right)^2} \tag{6}$$

où $P_{i\max}$ et $P_{i\min}$ sont les limites des paramètres P_i, qui définissent l'hypercube de l'espace de recherche.

Cette normalisation des paramètres permet de ne pas donner un poids plus important aux paramètres qui ont plus de marge de variation (comme FC qui varie entre 10 et 500) vis-à-vis des paramètres qui possèdent une faible marge de variation (comme les coefficients de vidange qui varient de 0 à 1).

La deuxième définition de distance proposée pour définir le voisinage est la distance euclidienne pondérée dans l'espace des paramètres normalisés. Les poids sont choisis conformément à la sensibilité régionale des paramètres à la fonction objective. Adopter des pondérations basées sur la sensibilité des paramètres donnerait en effct plus de chance aux paramètres les plus sensibles, de décider des points les plus proches. Plus précisément, après un certain nombre d'itérations, il est proposé de choisir les $2 \times N_{\text{wed}}$ points ayant les meilleures valeurs de la fonction objective où N_{wed} est un nombre à fixer. Le choix des meilleurs points pour estimer la sensibilité des paramètres se justifie par le fait que ces points représentent les zones d'attraction les plus importante et qui sont explorées par l'algorithme d'optimisation dans les itérations ultérieures. En effet, il semble plus avantageux d'étudier la sensibilité dans cette zone plutôt que dans celles où l'algorithme offrira moins de chance d'exploration. Les moyennes et les écarts-type des paramètres des deux ensembles constitués des N_{wed} premiers (meilleurs) points et des N_{wed} seconds points sont calculés. Si la moyenne et/ou l'écart type changent sensiblement entre ces deux ensembles, on admet que cela traduit une sensibilité du paramètre. La somme de l'écart relatif des moyennes et des écarts types des deux populations, notée w_{rsi} , est retenue comme une mesure de la sensibilité régionale qui est alors adoptée comme poids pour le calcul de la distance pondérée dp_i. Finalement il vient:

$$dp_i = \frac{1}{n}\sqrt{\sum_{i=1}^{n}\left(w_{rsi}\frac{P_{newi} - P_{\min}}{P_{i\max} - P_{i\min}} - \frac{P_i - P_{1\min}}{P_{i\max} - P_{i\min}})/\sum_{i=1}^{n} w_{rsi}\right)^2} \tag{7}$$

L'introduction des fonctions d'approximation peut perturber l'algorithme d'optimisation en orientant son mécanisme de recherche vers des mauvaises directions suite à une approximation trop forte. C'est pour cela qu'il est nécessaire de mettre en place un critère de décision concernant le moyen d'estimation de la fonction objective. Par simulation directe via le modèle, cela assure la valeur réelle de la fonction objective mais au détriment de l'efficacité. Par les fonctions d'approximation, on réduit le temps de calcul mais au détriment de la précision d'estimation de la fonction objective. Il s'agit autrement d'optimiser l'usage des fonctions d'approximation. Pour choisir les situations où on estime la fonction objective par KNN, trois façons de procéder sont proposées dans ce qui suit.

3.1 Algorithme 1

On commence par sélectionner une valeur critique de la fonction objective notée Fc. Quand la fonction objective du voisin le plus proche est inférieure à Fc, le point est considéré comme un "voisin intéressant", sinon il est considéré comme "inintéressant". Ceci permet de distinguer entre les régions où on doit raffiner la recherche et les régions qui sont moins intéressantes et à l'exploration desquelles il ne faut pas perdre de temps.

L'estimation de la fonction objective est obtenue par simulation si l'un des k voisins les plus proches est le meilleur point ou s'il n'est pas assez proche de ses voisins (Fig. 3). Pour évaluer si le voisin est *assez proche*, on a défini deux distances critiques CD1 et CD2. Quand la distance entre un nouveau point généré et son voisin le plus proche est plus petite que la distance critique, ce voisin est considéré assez proche. La distance critique CD1 est adoptée quand le voisin le plus proche est un point inintéressant, et CD2 pour le cas d'un voisin intéressant. Une grande valeur de CD1 se traduit par plus d'estimations de la fonction objective par KNN dans les régions les moins intéressantes de l'espace de recherche et induit donc un gain en temps de calcul. Au contraire, il y a intérêt à affecter à CD2 de faibles valeurs qui encouragent l'estimation de la fonction objective par simulation directe dans les zones les plus intéressantes.

A chaque fois que la fonction objective est estimée par simulation directe en un nouveau point, ce dernier est ajouté automatiquement à la base de données du KNN. La Fig. 2 résume les différents situations d'estimation de la fonction objective par KNN ou par simulation directe à travers le modèle HBV.

3.2 Algorithme 2

Une autre approche d'application du KNN pour le SCE-UA consiste en l'algorithme suivant:

– estimer moyennant la technique du voisin le plus proche la FO au niveau d'un nouveau point généré par le SCE-UA;

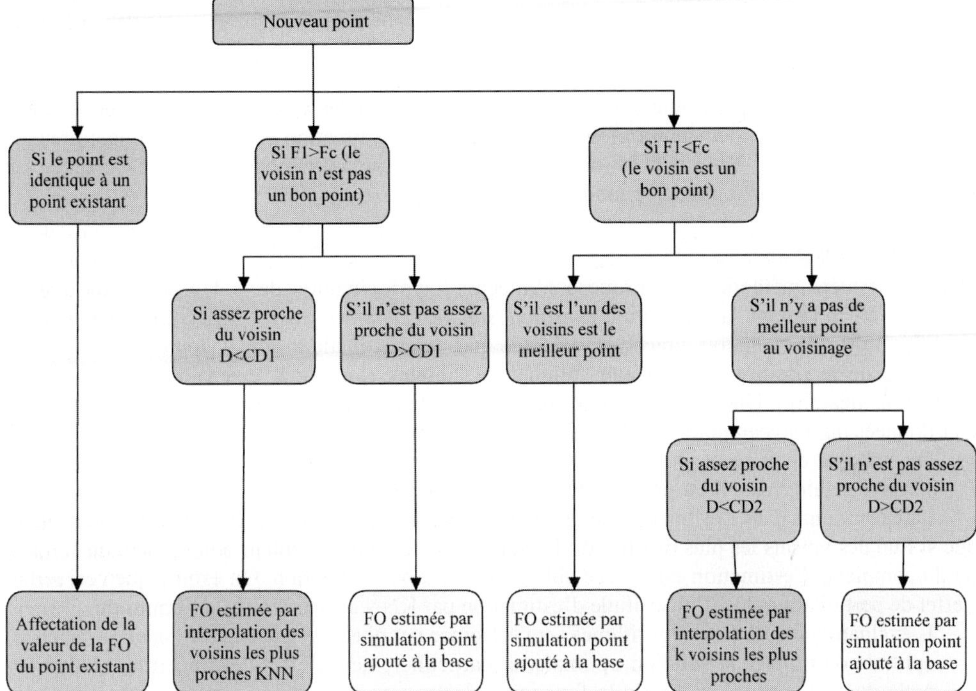

F1: fonction objective au niveau du voisin le plus proche; D1: distance entre le nouveau point et son voisin le plus proche; FC: Seuil de la fonction objective; CD1, CD2: seuils de distance

Fig. 2 Différents cas d'application de la méthode KNN pour l'estimation de la FO au niveau d'un nouveau point: cas de l'algorithme 1.

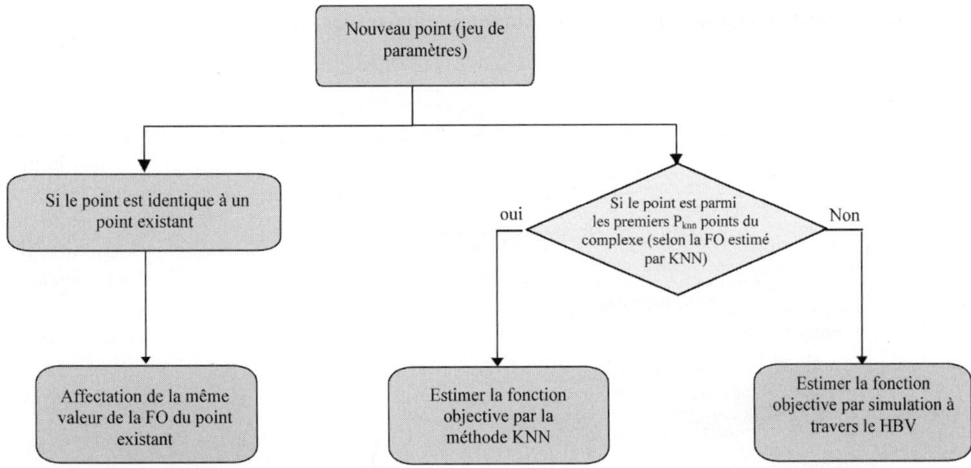

Fig. 3 Différents cas d'estimation de la fonction objective par KNN ou par simulation: cas de l'algorithme 2.

- classer le nouveau point dans la population du Complexe ($2n + 1$ points) selon les valeurs décroissantes de la FO; et
- s'il est parmi les P_{KNN} premiers points (des $2n + 1$ points du complexe), calculer la FO par simulation du modèle HBV, sinon garder l'estimation par KNN. Là P_{KNN} est un paramètre de KNN à déterminer.

Une contrainte supplémentaire est ajoutée. En effet si l'un des k voisins les plus proches est le meilleur ou second meilleur point global ou du simplex, l'estimation est réalisée obligatoirement par simulation. La Fig. 3 présente les différentes situations où la fonction objective est estimée par KNN ou par simulation à travers le modèle HBV.

3.3 Algorithme 3

Dans la dernière méthode il est proposé d'intégrer l'approximation de la FO par KNN dans le SCE-UA en modifiant la subroutine *downhill simplex évolutionnaire* (CCE). En effet le CCE modifié utilise la fonction objective calculée par approximation durant toute les étapes de recherche de la position du nouveau sommet du simplex (réflexion, extension, contraction ou génération aléatoire). Une fois que l'algorithme choisit le type de mouvement à faire et affecte les coordonnées du nouveau point, la FO est calculée par simulation par le modèle HBV.

On propose de faire usage du CCE modifié seulement dans une proportion des cas égale à P_{CCE}. Pour les autres cas, il est proposé de calculer par le CCE original.

Une contrainte dans l'estimation de la fonction objective est imposée. Elle consiste à considérer que si l'un des voisins les plus proches est le meilleur ou le second meilleur point global du simplex ou du complexe, l'estimation est opérée obligatoirement par simulation. On estime que ceci réduit l'effet de perturbation dû à l'incertitude d'estimation par KNN au voisinage de l'optimum.

L'évaluation de ces trois algorithmes sera effectuée sur plusieurs plans:

- aptitude de trouver la solution optimale, et particulièrement l'effet de l'incertitude additionnelle due à l'approximation de la fonction objective, ce qui revient à tester l'effectivité des algorithmes;
- nombre de simulations nécessaires à la convergence, qui caractérise l'efficacité des algorithmes; et
- nombre de paramètres internes de l'algorithme à fixer, et la difficulté de leur estimation.

4 RESULTATS ET INTERPRETATION

Afin de tester ces algorithmes d'optimisation, on a choisi de travailler sur des débits synthétiques. Ceci nous permet d'aborder un problème d'optimisation où les incertitudes sur les données de

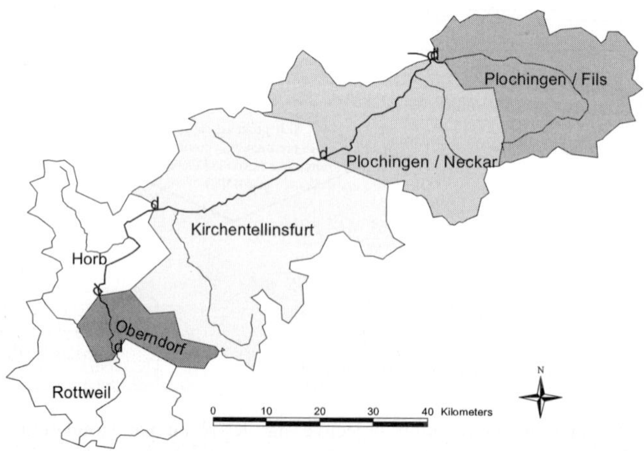

Fig. 4 Le bassin du Rottweil, Allemagne.

débits sont absentes et où la solution optimale est connue par avance. Les débits fictifs proviennent de la simulation du bassin de Rottweil qui est un affluent du Rhin en utilisant les données réelles de pluie et températures de l'année 1961. Le jeu de paramètres qui a permis cette simulation résulte d'un calage du modèle HBV en utilisant le SCE-UA dans sa version initiale. En Fig. 4 nous reportons la situation géographique du bassin de Rottweil. En Tableau 1 sont donnés les caractéristiques du bassin. En Fig. 5, sont présentés les données synthétiques relatives à l'année 1961 (pluie, débit et température) et qui sont adoptées pour la suite du travail.

L'optimisation portera sur les six paramètres du modèle HBV identifiés comme les plus sensibles: K_0, K_1, Per1, PWP, Beta et CEF. Les limites de l'espace de recherche sont choisies conformément à la littérature et sont présentées dans le Tableau 2.

Ce problème d'optimisation est adopté ainsi pour étudier les améliorations sur la performance obtenue par les trois algorithmes proposés. On teste aussi les deux définitions de distance et on étudie leur effet sur la performance des algorithmes. Pour chaque algorithme, on a procédé à 10

Tableau 1 Les caractéristiques du basin de Rottweil.

Superficie	456 km^2
Situation	Allemagne
Climat	Atlantique
Pluie annuelle moyenne	955 mm
Apport annuel moyen	163×10^6 m^3

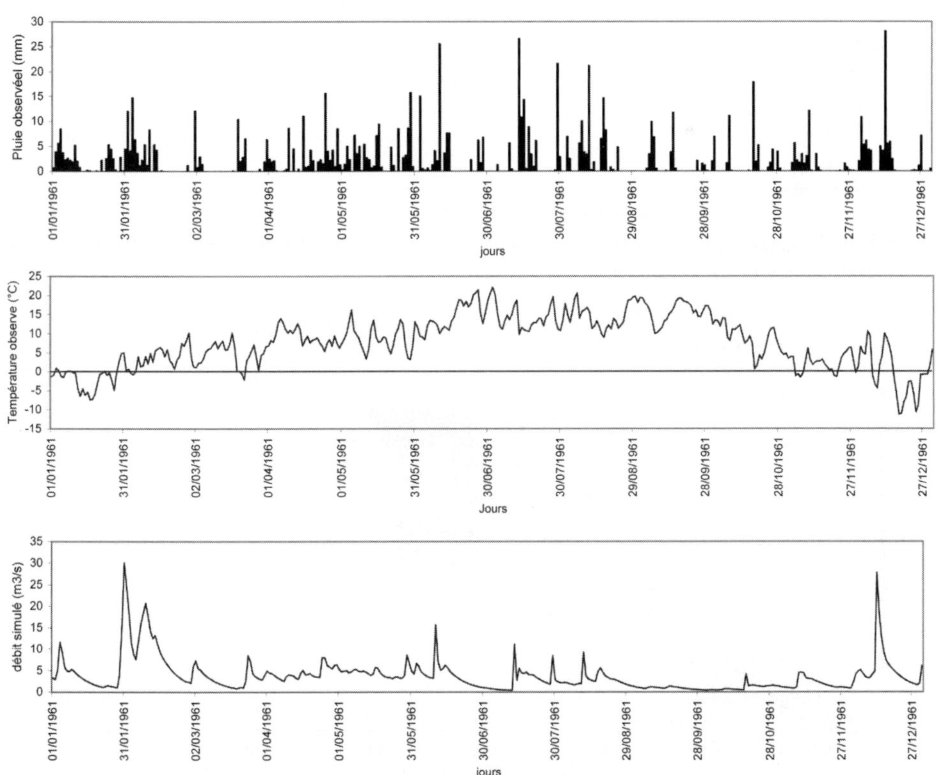

Fig. 5 Données du bassin de Rottweil pour l'année 1961: pluie observée, température observée, débits fictifs simulés par le modèle HBV avec le jeu de paramètres (Tableau 2).

Tableau 2 Les limites de l'espace de recherche avec SCE-UA des paramètres de HBV et la solution optimale pour le cas synthétique du Rottweil (1961).

Paramètre	Min	Max	Solution synthétique optimale
K_0	0	2	1
log(K1)	–5	0	–2.617
log(K2)	–5	0	–2.596
Log(PER1)	–5	–1	–1.507
TT (°C)	–1.5	2.5	0
DEGD (mm/°C/h)	1	10.0	2.6
MAXBAS (h)	1	6	1
PWP (mm)	10	500	16
FC (mm)	10	500	116
BETA	0	6	1.48
CEF	0	1	0.108

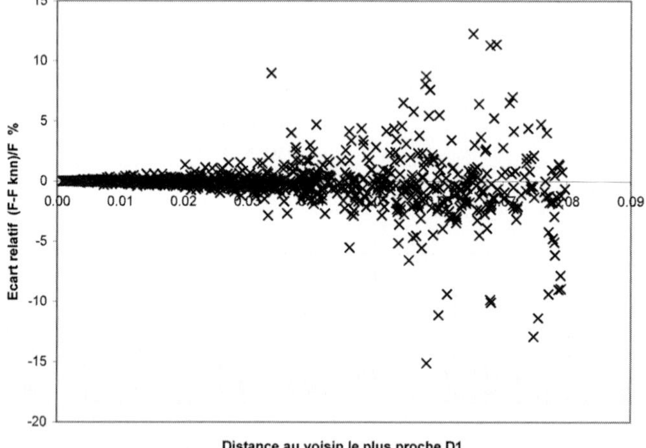

Fig. 6 Précision de l'estimation de la fonction objective par KNN (écart relatif entre la fonction objective estimée par KNN et par simulation) en fonction de la distance au voisin le plus proche D1.

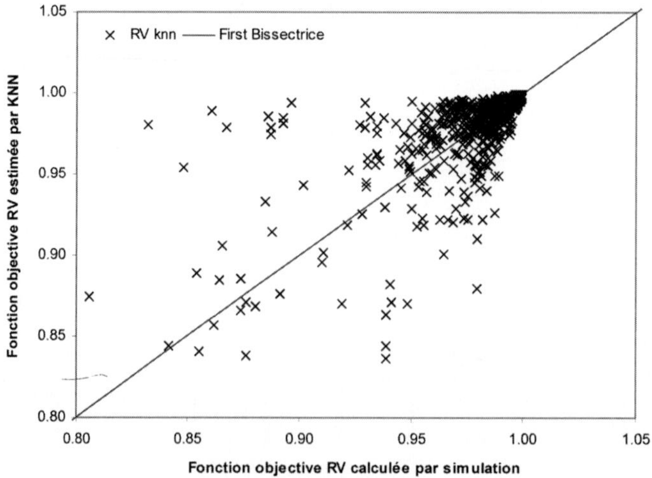

Fig. 7 Fonction objective estimée par KNN comparée à celle calculée par le modèle HBV. Les points correspondent aux jeux de paramètres explorés par l'algorithme 2 pour le cas 16 de la Fig. 8.

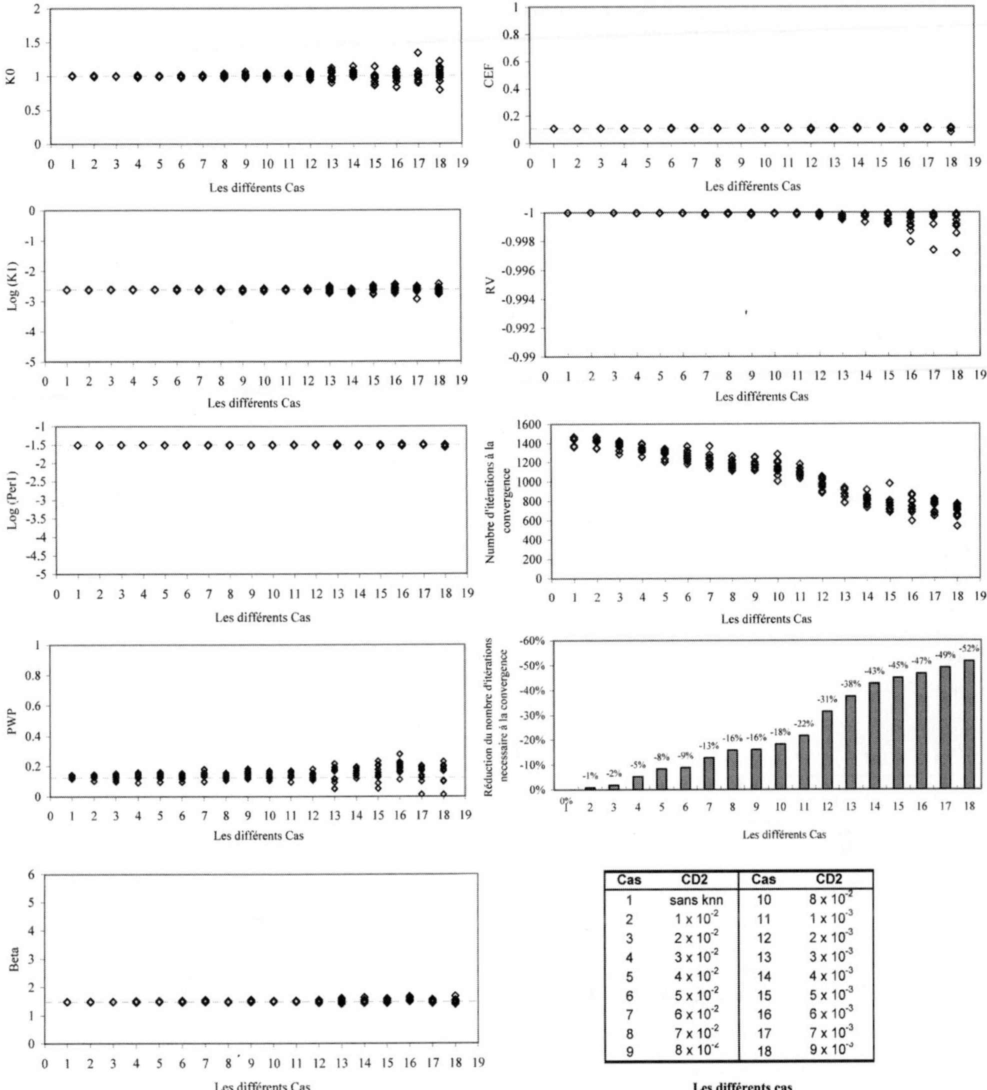

Cas	CD2	Cas	CD2
1	sans knn	10	8×10^{-2}
2	1×10^{-2}	11	1×10^{-3}
3	2×10^{-2}	12	2×10^{-3}
4	3×10^{-2}	13	3×10^{-3}
5	4×10^{-2}	14	4×10^{-3}
6	5×10^{-2}	15	5×10^{-3}
7	6×10^{-2}	16	6×10^{-3}
8	7×10^{-2}	17	7×10^{-3}
9	8×10^{-4}	18	9×10^{-3}

Fig. 8 Effet de la modification de CD2 sur le nombre d'itérations nécessaire à la convergence et sur la valeur de la fonction objective et paramètres obtenus à l'optimum, cas de l'algorithme 1. Chaque point représente la valeur de la fonction objective ou paramètre obtenu à la fin de l'essai de SCE-UA. La ligne discontinue représente la solution synthétique.

essais indépendants pour le calage des six paramètres les plus sensibles du modèle. Pour chaque essai on change les points de départs de l'algorithme d'optimisation (ou population initiale) et ceci par une nouvelle génération aléatoire. Le nombre de complexes est fixé à quatre et la population par complexe est choisie égale à 56 et ceci selon les recommandations de Duan *et al.* (1994). On présente en premier lieu les résultats obtenus en adoptant la distance euclidienne dans l'espace normalisé des paramètres et puis ceux obtenus avec la distance pondérée.

Concernant l'algorithme 1, des tests préliminaires (non montrés ici) suggèrent que des valeurs CD1 = 0.15 et FC = 0. La Fig. 6 montre que l'erreur d'estimation du KNN diminue considérablement

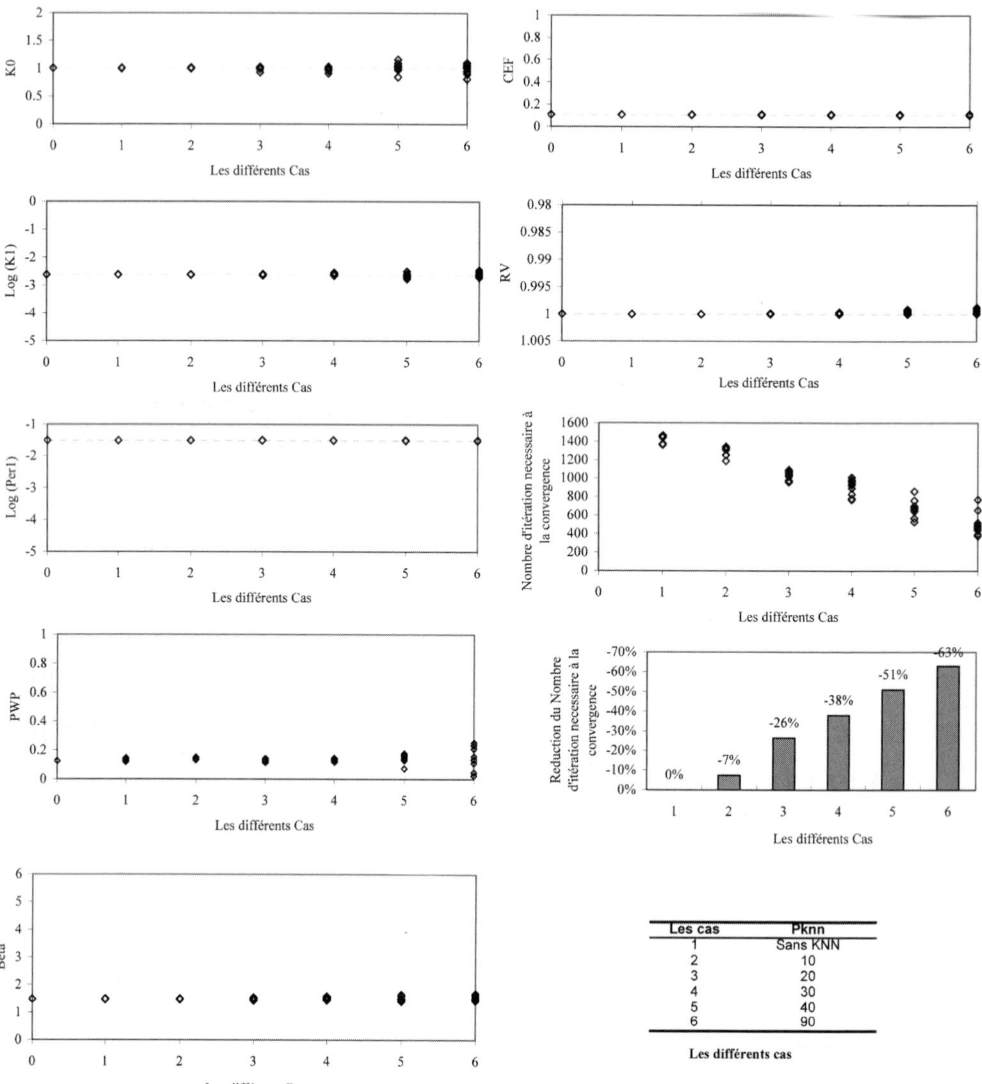

Les cas	Pknn
1	Sans KNN
2	10
3	20
4	30
5	40
6	90

Les différents cas

Fig. 9 Effet de la modification du paramètre P_{KNN} sur le nombre d'itérations nécessaire à la convergence et sur la valeur de la fonction objective et paramètres obtenus à l'optimum—la population du complexe est prise égale à 56 individus: cas de l'algorithme 2. Chaque point représente la valeur de la fonction objective ou paramètre obtenu à la fin de l'essai de SCE-UA. La ligne discontinue représente la solution synthétique.

pour les petites valeurs de RV. On peut déduire de cette figure que l'incertitude de l'estimation de la FO est réduite au voisinage de la solution optimale, cette incertitude augmente lorsqu'on est assez éloigné. Ceci minimise l'effet de cette incertitude sur la précision de la solution ou autrement l'effectivité de l'algorithme.

Comme le montre la Fig. 7, l'erreur d'estimation est proportionnelle à la distance qui sépare le nouveau point à son voisin le plus proche. C'est pour cette raison qu'on a choisi d'étudier l'effet du choix de la distance critique CD2 sur la performance de l'algorithme d'optimisation. La Fig. 8 montre l'effet de ce paramètre sur les résultats. La marge de variation de CD2 est 0.001–0.09. On

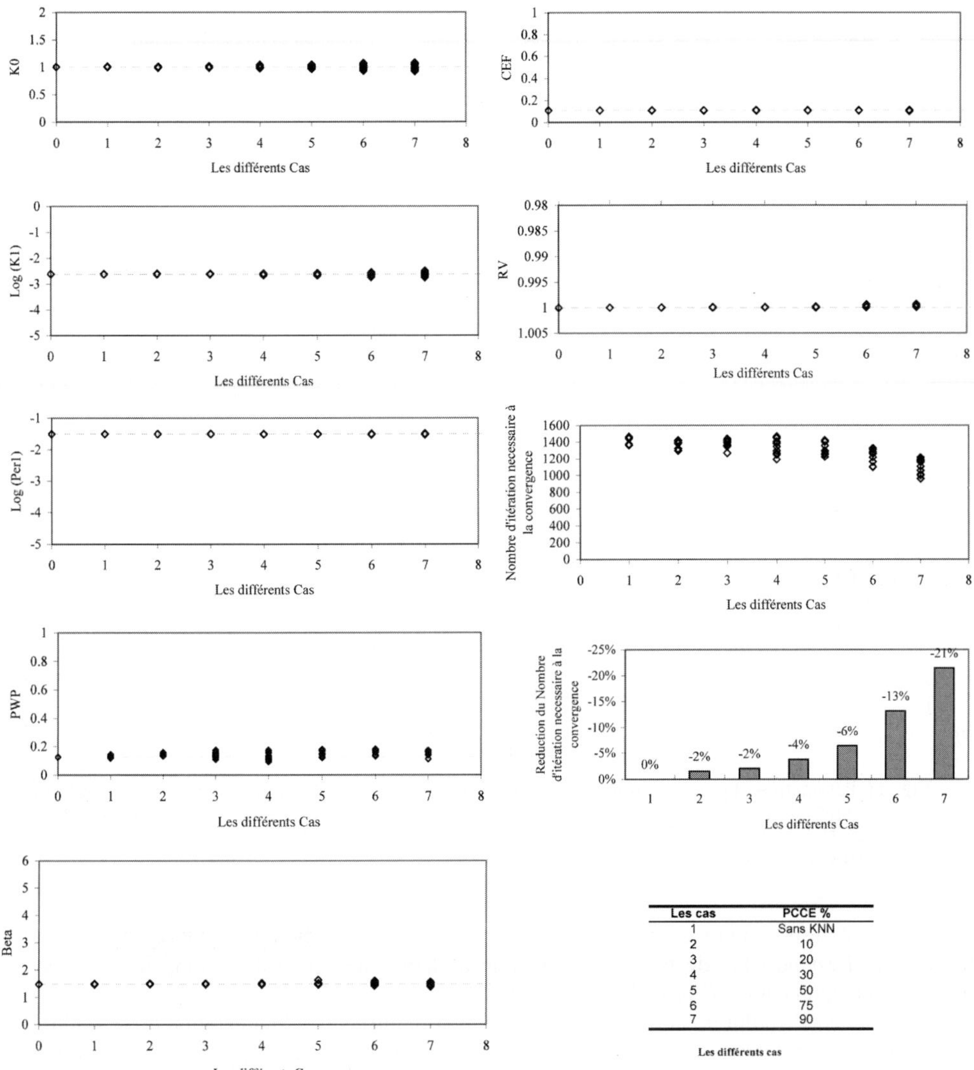

Les cas	PCCE %
1	Sans KNN
2	10
3	20
4	30
5	50
6	75
7	90

Les différents cas

Fig. 10 Effet de la modification du paramètre PCCE sur le nombre d'itérations nécessaire à la convergence et sur les valeurs des fonctions objectives et paramètres atteints à l'optimum: cas de l'algorithme 3.

remarque que lorsque la valeur de CD2 augmente le nombre d'itérations nécessaire à la convergence diminue. L'effectivité de l'algorithme peut être atteinte pour des valeurs grandes de CD2. Une valeur de CD2 égale à 0.02 semble donner une amélioration de l'efficacité de l'algorithme d'optimisation de l'ordre de 30% sans atteinte de son effectivité (Fig. 8).

Dans la Fig. 9 on présente les résultats de l'application de l'algorithme 2, en terme des valeurs des paramètres, de la fonction objective et du nombre d'itérations nécessaires à la convergence. On remarque que les grandes valeurs du paramètre P_{KNN} peuvent engendrer une perturbation de l'algorithme qui diminue son effectivité. Une valeur de 30 semble être raisonnable. En effet, en comparaison au SCE-UA initial, elle permet une réduction du nombre de simulations nécessaires de l'ordre de 30% sans effet sur l'effectivité de l'algorithme.

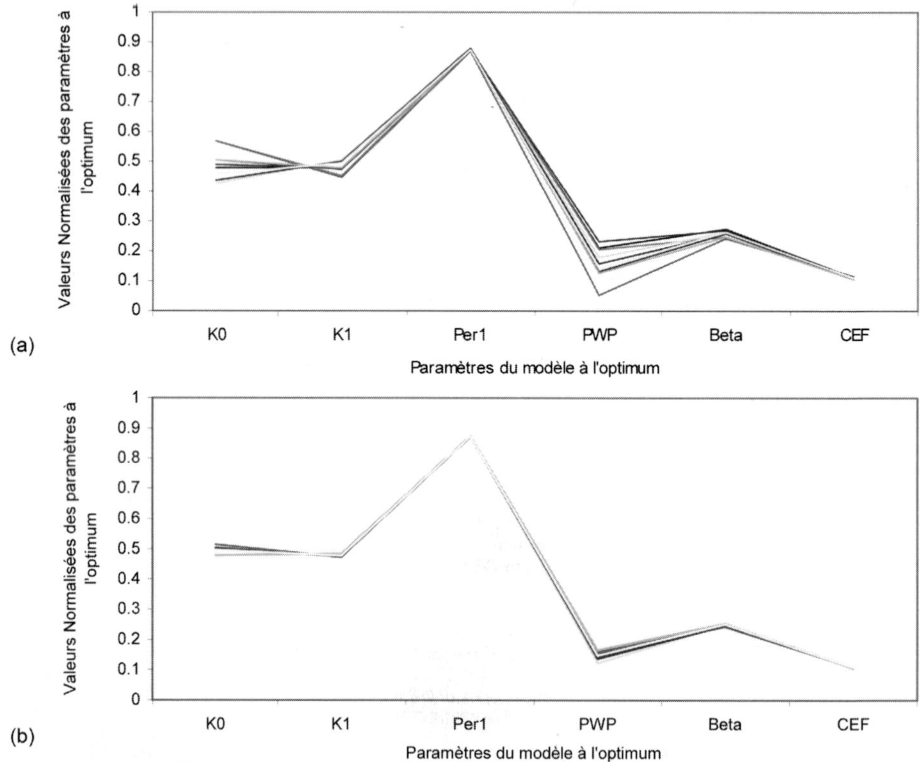

Fig. 11 Paramètres obtenus à l'optimum pour 10 tests différents de l'algorithme 1: (a) en utilisant la distance normalisée dans l'espace des paramètres; et (b) en utilisant la distance normalisée pondérée, avec CD1 = 0.15, CD2 = 0.05, N_{wed} = 50. Chaque ligne joint les paramètres optimaux obtenus à la fin d'un essai indépendant de l'algorithme d'optimisation.

L'algorithme 3 conduit à des résultats assez proche des précédents. En effet comme le montre la Fig. 10, il est possible de faire une économie de temps de calcul de l'ordre de 20% correspondant à un pourcentage P_{CCE} de l'ordre de 90%.

Finalement, l'emploi de la distance euclidienne normalisée avec pondération a été appliqué à dix essais de l'algorithme 2. La Fig. 11 présente les paramètres optimaux obtenus en utilisant les deux distances (pondérées et non pondérées). Les résultats suggèrent une différence dans l'effectivité de l'algorithme car il est clair que la distance euclidienne pondérée réduit considérablement l'incertitude sur les paramètres. Elle permet de faire usage de valeurs de CD2 plus grandes et donc une réduction plus importante en terme de temps de calcul qui peut atteindre 50%. Ainsi, la pondération des distances conduit à un algorithme deux fois plus performant que le SCE-UA initial.

5 CONCLUSIONS

On a pu montrer que l'usage de la technique des voisins les plus proches pour l'estimation de la fonction objective peut contribuer à réduire le temps de calcul de l'algorithme SCE-UA appliqué pour le calage du modèle pluie–débit HBV d'au moins 30%. Trois algorithmes basés sur des concepts différents ont été testés. Bien que ces algorithmes donnent des performances comparables, l'algorithme 2 et 3 se distinguent de l'algorithme 1 du fait qu'ils ne demandent de choisir qu'un seul paramètre interne (P_{KNN} ou P_{CCE}). L'algorithme 1 nécessite trois paramètres internes à fixer, et donc il est plus difficile à appliquer.

L'usage de la distance euclidienne pondérée pour le calcul des distances séparant le nouveau point à son voisinage semble réduire l'incertitude induite par l'estimation par KNN, et donne ainsi la possibilité de réduire le temps de calcul de moitié.

Des travaux futurs peuvent se rapporter à la détermination des paramètres internes de l'algorithme 1 (CD1, CD2 et Fc) et à leur liaison aux paramètres internes du SCE-UA ce qui réduit l'intervention de l'opérateur. L'usage d'autres distances semble aussi être un terrain fructueux. L'application de ces algorithmes pour des conditions climatiques différentes est aussi une voie à prospecter.

REFERENCES

Bates, B. C. (1994) Calibration of the SFB model using a simulated annealing approach. In: *Proc. Water Down Under 94 Conference*, Natl. Conf. Publ. 94/15, vol. 3, 1–6. Inst. of Eng., Canberra, A.C.T., Australia.

Begström, S. (1976) Development and application of a conceptual rainfall–runoff model for the Scandinivian catchment. SMHI RH07, Norrköping, Sweden.

Bardossy, A. & Theisen, H. W. (1999) Listing en Fortran du modèle HBV, version 0.3. IWS, Stuttgart, Germany.

Bardossy, A. & Yi He (2006) Application of a nearest method to a conceptual rainfall–runoff model. In: *Proc. Seventh Int. Conf. on HydroScience and Engineering* (Philadelphia, USA, 10–13 September 2006). Michael Piasecki and College of Engineering, Drexel University, 9/2006. ISBN: 0977447405. http://idea.library.drexel.edu/handle/1860/1408.

Brandsma, T. & Können, G. P. (2005) Application of nearest-neighbor resampling techniques for homogenizing temperature records on a daily to sub-daily level. *Int. J. Climatol.* 2006, **26**(1), 75-89,

Duan, Q., Sorooshian, S. & Gupta, V. (1992) Effective and efficient global optimization for conceptual rainfall–runoff models. *Water Resour. Res.* **28**(4), 1015–1031.

Duan, Q., Gupta, V. K. & Sorooshian, S. (1993) A shuffled complex evolution approach for effective and efficient global minimization. *J. Optimis. Theory Appl.* **76**(3), 501–521.

Duan, Q., Sorooshian, S. & Gupta, V. (1994) Optimal use of the SCE-UA global optimisation method for calibrating watershed models. *J. Hydrol.* **158**, 265–284.

Efstratiadis, A. & Koutsoyiannis, D. (2002) An evolutionnary annealing-simplex algorithm for global optimisation of water ressource systems. In: *Hydroinformatics 2002* (Proc. Fifth Int. Conf. on Hydroinformatics, Cardiff, UK), 1423–1428. IWA Publishing, UK.

Fix, E. & Hodges, J. L. (1952) Discriminatory analysis: non parametric discrimination. Small Sample Performance Project 21–49–004. Report no. 11, 28–32. USAF School of Aviation Medicine, Randolph Field, Texas, USA.

Franchini, M. (1996) Use of genetic algorithm combined with a local search method for the automatic calibration of conceptual rainfall–runoff models. *Hydrol. Sci. J.* **42**(3), 357–379.

Franchini, M. & Galeati, G. (1997) Comparing several genetic algorithm schemes for the calibration of conceptual rainfall–runoff models. *Hydrol. Sci. J.* **41**(1), 21–39.

Franchini, M. Galeati, G. & Berra, S. (1998) Global techniques for the calibration of conceptual rainfall–runoff models. *Hydrol. Sci. J.* **43**, 443–458.

Giannakoglou, K. C. (2002) Design of optimal aerodynamic shapes using stochastic optimisation methods and computational intelligence. *Progr. Aerospace Sci.* **38**(5), 43–76.

Goldberg, D. E. (1989) *Genetic Algorithms in Search, Optimization and Machine Learning*. Addison-Wesley, Reading, Massachusetts, USA.

Gupta, H. V., Sorooshian, S. & Yapo, P. O. (1999) Status of automatic calibration for hydrologic models: comparison with multilevel expert calibration. *J. Hydrol. Engng* **4**(2), 135–143.

Khu, T. S. & Madsen, H. (2005) Multiobjective calibration with Pareto preference ordering: an application to rainfall–runoff model calibration. *Water Resour. Res.* **41**(3), W03004, doi:10.1029/2004WR003041.

Kuczera, G. (1997) Efficient subspace probabilistic parameter optimization for catchment models. *Water Resour. Res.* **33**, 177–185.

Lindström, G., Johanson, B., Gardelin, M. P. M. & Bergström, S. (1997) Development and test of the distributed HBV-96 hydrological model. *J. Hy*drol. **201**(1-4), 272–288

Nash, J. E. & Sutcliffe, J. V. (1970) River flow forecasting through conceptual models. 1. A discussion of principles. *J. Hydrol.* **10**(3), 282–290.

Ong, Y. S., Nair, P. B. & Keane, A. J. (2003) Evolutionary optimization of computationally expensive problems via surrogate modelling. *Am. Inst. of Aeronautics and Astronautics J.* **41**(4), 687–696.

Ostfeld, A. & Salomons, S. (2005) A hybrid genetic-instance learning algorithm for CE*QAL-W2 calibration. *J. Hydrol.* **310**, 122–125.

Skellam, J. G. (1952) Studies in statistical ecology. I. Spatial patterns. *Biometrica* **39**, 346–362.

Thyer, M., Kuczera, G. & Bates, B. C. (1999) Probabilistic optimisation for conceptual rainfall–runoff models: a comparison of the shuffled complex evolution and simulated annealing algorithms. *Water Resour. Res.* **35**(3), 767–773.

Zhou, Z., Ong, Y. S., Nair, P. B., Keane, A. J. & Lum, K. Y. (2007) Combining global and local surrogate models to accelerate evolutionary optimization. *IEEE Trans. on Systems, Man & Cybernetics* (SMC), *Part C* **37**(1), 66–76. (doi:10.1109/TSMCC.2005.855506)

The robust detection of outlying rainfall observations

ZHAO CHAO[1,2,3], BAO WEI-MIN[2] & HONG HUA-SHENG[3]

1 *Water Resources and Environmental Institute, Xiamen University of Technology, Xiamen 361005, China*
zhaochao@xmut.edu.cn

2 *State Key Laboratory of Hydrology-Water Resources and Hydraulic Engineering, College of Hydrology and Water Resources , Hohai University, Nanjing 210098, China*

3 *State Key Laboratory of Marine Environmental Science, Environmental Science Research Center, Xiamen University, Xiamen 361005, China*

Abstract The detection of outliers is important for telemetric rainfall observations because outliers significantly distort the results of most hydrological models. In this paper a three-step robust statistical method combining robust statistical theory with distribution features of precipitation for the detection of outliers in a telemetry system is described. The proposed robust statistical method adopts the Tukey fence insensitive to outliers as identification bounds, and presents a three-step pattern to adapt the distribution of rainfall data. Moreover, the modified method based on dividing precipitation data into several groups further improves detection efficiency. Synthetic data are given to test the performance of the proposed method. The results illustrate that the proposed method produces reliable detection results. It is shown that the new method, based on distribution of the statistical features of rainfall, is suitable for hydrological needs.

Key words telemetry system; outlier; Tukey fence; distribution feature; three-stepwise robust detection

1 INTRODUCTION

Rainfall is the primary input to most hydrological models, and therefore the precision of rainfall data can have a key effect on the accuracy of hydrological forecasting and simulation.

Historically, manual observations at rainfall stations have been the main source of rainfall data. More recently, automatic acquisition via telemetric networks has played an increasingly important role in technologically developed countries, particularly with regard to provision of data in real time. It is obvious that the rainfall data obtained by telemetry systems is characterized as accurate and frequent. However, it is noted that in telemetric rainfall data there often are outliers resulting from instrument malfunctioning and false signal acquisition because of signal leak, collision and disturbance during signal transmission, in addition to unavoidable random errors normally distributed with zero mean and a small variance. The outliers have an unknown distribution with a much bigger variance, and appear to be inconsistent with the remainder of the data set (Barnet & Lewis, 1994; Han & Kamber, 2001) but are relatively large in magnitude.

In order to limit the influence of outliers on forecasting and simulation, it is essential to check and find out these outlying observations before they enter hydrological models.

The detection of outliers has been tackled by a variety of methods in hydrology. These can be classified into two kinds. One is called statistical test based on some hypotheses. For example, the method adopted by the US Water Resources Council (US-WRC) is the most commonly used hydrological technique for outlier detection in North America. This method is based on the principle of hypothesis testing with the underlying assumption of a log Pearson Type III probability distribution (Grubbs & Beck, 1952). The other is called comparison test based on functional interpolation and/or stochastic estimation. The approach used by the US Environmental Protection Agency (US-EPA) employing Mahalonobis distance for the detection of outliers belongs to this class. It should be admitted that the methods mentioned above are reasonably well used, and part of the outlying observations can be correctly located and removed. It is a pity that all the methods mentioned above are based on the classical least-squares estimation (Hu, 1987; Singh, 1991; Spencer & McCuen, 1996). And, it is well known that the least-squares estimation is not robust; even a single outlier can spoil the solution. Therefore, a new detection method insensitive to outliers is needed.

Usually, rainfall varies temporally and spatially in a watershed system. In order to correctly detect the outlying observations, the detection method has to depend on the statistical features of the variation. Considering synthetically those constraints, this paper makes an attempt to combine the robust statistical approach and distribution characteristics to increase the reliability for the detection of outliers. A three-step robust statistical detection method of outliers is proposed. The robust approach (Bounessah & Atkin, 2003; Zhou *et al.*, 2006, 2008; Daszykowski *et al.*, 2007) based on the inter-quartile range (IQR) is more tolerant of outliers. The three-step pattern agrees with the different distribution statistical features of hourly rainfall, hourly rainfall deviation from hourly areal mean rainfall (HRD), and hourly simulation error of hourly rainfall simulation (HSE).

This paper is organized as follows. First, a brief review of the statistical property of rainfall distribution is given in the next section, then the proposed method is presented in detail. A case study is conducted to demonstrate the proposed method and finally we provide some conclusions.

2 STATISTICAL FEATURES

For a watershed system, rainfall varies temporally and spatially. The statistical features of the variability can significantly influence the detection efficiency of abnormal data. It is the reason why the analysis of the statistical features is important.

A total of 133 000 pieces of hourly telemetric rainstorm data of 40 events during 1988–1997 from 43 raingauges in Qilijie basin of Fujian province of China, was selected to analyse the statistical features of distribution.

Hourly rainfall distribution is very complex, and depends upon air current, topography, elevation, geographic position, and many others. Precipitation data always locates in [0, maximum precipitation (P_{max})]. The frequency distribution of hourly precipitation is displayed in Fig. 1, where only the distribution for Wuyigong Station is depicted, since identical results were observed for other stations. The frequency distribution graph had a single peak, and 92.9% of precipitation data are located in [0, 4mm], which was the high frequency zone. The mean square deviation displaying dispersion degree of precipitation was 2.88.

HRD was obtained with the following equation:

$$hrd_i = P_i - P_{mean} \quad (1)$$

P_i is hourly rainfall of the ith raingauge; P_{mean} is hourly areal mean rainfall computed by weighted mean of rainfall of all raingauges except the ith one. The HRD frequency distribution (Fig. 2) displays a similar form to the precipitation distribution. 95.9% of HRD located in [–4 mm~4 mm], the high frequency zone. The mean square deviation of HRD was 2.24.

Rainfall simulation was a difficult task. This paper selected a simple quadric model by means of longitudes and latitudes of raingauges and P_{mean}, to simulate the hourly precipitation, HSE:

$$ee_i = P_i - P_{sim} \quad (2)$$

where, P_{sim} is the simulated hourly rainfall. The HSE frequency distribution is given in Fig. 3; 96.7% of HSE located in [–4 mm~4 mm]. The mean square deviation of HSE was 2.18.

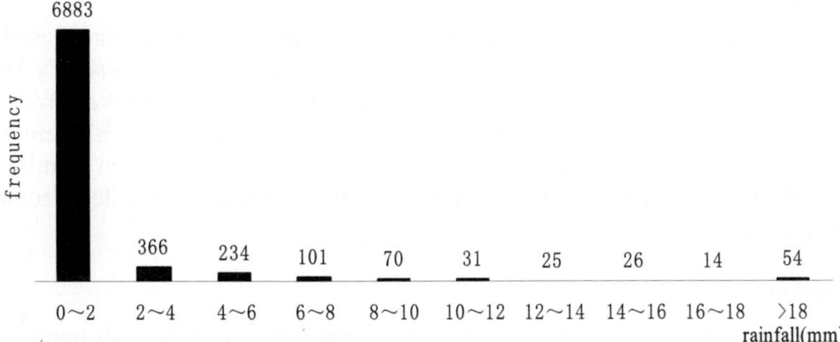

Fig. 1 Rainfall frequency distribution of Wuyigong Station.

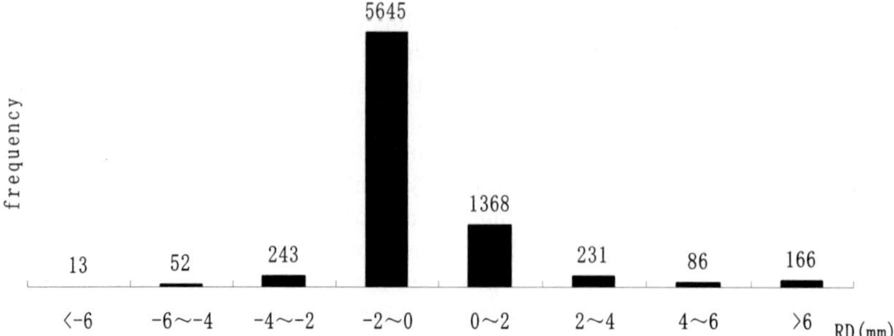

Fig. 2 RD frequency distribution of Wuyigong Station.

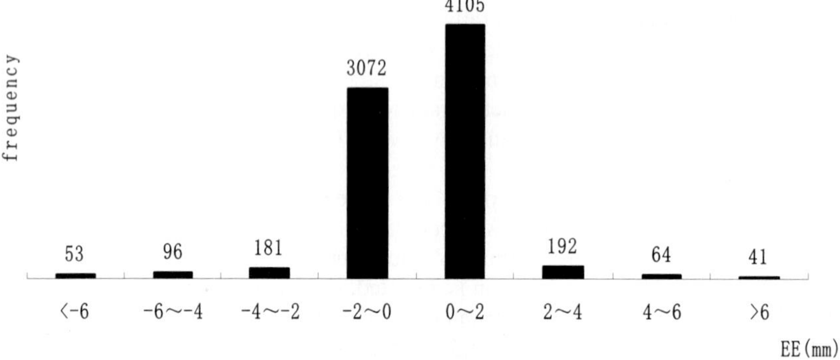

Fig. 3 EE frequency distribution of Wuyigong Station.

For all raingauges the distributions of precipitation, HRD and HSE had the following features. First, the distributions were limited in [0, maximum], and the distribution ranges were less and less from precipitation to absolute HRD, to absolute HSE. For example, for Wuyinggong Station, the distribution ranges of precipitation, absolute HRD and absolute HSE were [0, 36.9 mm], [0, 30.7 mm], [0, 28.6 mm], respectively. For 43 raingauges, the ranges of three distributions were [0, 65.2 mm], [0, 58.1 mm], [0, 50 mm], respectively. Secondly, for all raingauges, the mean square deviations of the three distributions were 2.38, 2.10 and 1.94, respectively, showing that the three distributions were closer and closer.

When we selected the detection method of abnormal data for telemetric rainfall, we had to take these features into account.

3 METHODS

As mentioned above, in hydrology the conventional detection methods for abnormal data based on least-squares (LS) procedure are not robust. It means that outliers may have an unusually large influence on the results of LS procedure. As a result, incorrect decisions may be derived from the statistical test based on the conventional methods. The advantage of a robust statistical method is that the negative effect of the outliers on the detection is greatly softened or even eliminated altogether when making the solution, though the detection efficiency is inferior to a least-squares detection when no outliers are present.

Robust statistical method

Let $X_1, X_2, \ldots, X_n$ be independent and identically distributed random variables with population mean $\overline{X}$, standard deviation σ. If abnormal data arises in samples, the $\overline{X}$ and σ are distorted. We need to search the new statistic. Let $X_{(1)} \le X_{(2)} \le \cdots \le X_{(n)}$ denote the ordered sequence. The *p*-quantile of samples has the form:

$$X_{(p)} = \begin{cases} X_{(np)} & np \in \text{int} \\ X_{[(np)+1]} & np \notin \text{int} \end{cases} \tag{3}$$

where, [ ] denotes the integer part of the number. The 1/4 quantile ($X_{(1/4)}$) is called the lower quartile, and 3/4 quantile ($X_{(3/4)}$) is called the upper quartile. As we know, the abnormal data is unusually large or small, and arranged at both ends of the ordered sequence. So the quantiles, for example $X_{(1/4)}$ and $X_{(3/4)}$, are less related to outliers than $\overline{X}$ and σ. Hence, we selected the robust statistical method based on quantiles as the detection tool.

The difference between $X_{(1/4)}$ and $X_{(3/4)}$ is named the inter-quartile range (IQR). The Tukey fence (Bounessah & Atkin, 2003; Zhou *et al.*, 2006; Daszykowski *et al.*, 2007) is the range from $[X_{(1/4)} - \beta * IQR,\ X_{(3/4)} + \beta * IQR]$. The values outside the Tukey fence are considered to be outliers. $X_{(1/4)}$ and $X_{(3/4)}$ are less related to outliers. Hence, the Tukey fence is not influenced by the outliers. β depends on the distribution; if the data comes from the normal distribution, $\beta = 1.5$, the interval contains 99.3% of the values. This is a robust statistical method of detecting outliers.

In order to improve the detection efficiency of the robust statistical method, this paper combined the aforementioned rainfall distribution statistical characteristics, and proposed a three-step robust statistical method to detect the abnormal data in telemetric rainfall sets. First, according to historic telemetric information of hourly precipitation, HRD, HSE, the upper quartiles, lower quartiles, IQRs and the Turkey fences of three distributions were computed, as criteria for detecting the abnormal data.

The three-stepwise robust statistical method is as follows:

Step 1 Judge whether the real-time hourly rainfall is within the Tukey fence of precipitation or not. If the data was not in, the rainfall observation was abnormal, or else entered into next step.

Step 2 Judge whether the real-time HRD is in its Tukey fence or not. If the data was not in, the rainfall observation was abnormal, or else entered into next step.

Step 3 Judge whether the real-time HSE is in its Tukey fence or not. If the data was not in, the rainfall observation was abnormal, or else ended the detection.

If the precipitation data was detected to be abnormal data, it would be replaced by the average value of other stations.

The three-stepwise method integrated the robust statistical theory and rainfall distribution features. It was a detection method adapted to hydrological needs.

However, in the process of computing the Turkey fences, we found that rainfall data of different magnitudes had large differences in the upper quartiles, lower quartiles and IQRs. So a modified detection method was proposed to further improve the detection efficiency.

Modified method

To consider the different magnitudes of rainfall sets, we divided the areal mean precipitation of historic data into several groups. For the every group, the upper quartiles, lower quartiles and IQRs of the three distributions were recalculated. The modified method was as follows:

Step 1 Compute the real-time hourly areal mean precipitation (P_{mean}), and select the corresponding group and Tukey fence of precipitation. Judge whether the real-time hourly rainfall is abnormal data nor not. If the observation was not in, the rainfall observation was abnormal, or else entered into next step.

Step 2 Recalculate the P_{mean}, and select the corresponding group and Tukey fence of HRD. Judge whether the data is abnormal data or not. If the observation was not in, the rainfall observation was abnormal, or else entered into next step.

Step 3 Recalculate the P_{mean}, and select the corresponding group and Tukey fence of HSE. Judge whether the data is abnormal data nor not. If the observation was not in, the rainfall observation was abnormal, or else ended the detection.

If the precipitation data was abnormal, it would be replaced by the average value of other stations.

4 EXPERIMENTAL RESULTS

Hourly precipitation data obtained by manual observation in 1998 was used in this study, in addition to the data described in Section 2. The data was from four events and 43 raingauges in 1998, totalling 23 000 pieces. The data in 1988 was obtained by manual observation, so we assumed it only contained random errors that could enter into hydrological models.

The performance of the proposed robust statistical method was demonstrated using synthetically generated data sets. The reason for using synthetic data was that the error structure was known and could be varied to test different hypotheses.

We added the following error distribution to the rainfall data sequence of single raingauge in 1988 in order to simulate the outlying distribution of telemetric system:

$$e_i = \begin{cases} re_{\max} & i = \text{int}(i/L)L \\ 0 & i \neq \text{int}(i/L)L \end{cases} \tag{4}$$

where, r is a random number; $e_{\max}$ is a constant that controls the maximum of e. L is the frequency of outliers. Adjusting $e_{\max}$, outliers of different sizes can be generated.

Performance of robust statistical method

The hourly telemetric rainstorm data of 40 events and 43 raingauges during 1988–1997 in Section 2 inevitably contained outliers. The Tukey fences of hourly precipitation, HRD and HSE distributions were calculated. In order to contain 99% of the values, β was 25 by means of a trial-and-error method.

We added the outliers using equation (4) to the precipitation data of one, two and five raingauges.

In evaluation of the robust statistical detection method, the following two performance measures (Narasimhan & Mah, 1987) were used in this study. One is detection efficiency (ee), the other is detection mistake (m) to detect true data as outliers.

$$ee = \frac{n_d}{n_o} \tag{5}$$

where n_d is the number of outliers correctly detected, n_o is the number of outliers.

$$m = \text{the number of outliers wrongly detected} \tag{6}$$

The detection efficiency of the proposed method is shown in Table 1. As a whole, the proposed method was suited to detect the abnormal data for telemetric rainfall sets, the mean detection efficiency was above 0.6. For the single raingauge, the detection efficiency improved as $e_{\max}$ increased: $ee1$ was 0.02 for 10 mm of $e_{\max}$ and 0.96 for 500 mm. Among the three steps, the detection efficiency of the first step was greatest, and that of the third step was least. For example, for one raingauge and $e_{\max}$ = 7 0 mm, the detection efficiency of the first, second and third steps were 0.5, 0.18, 0, respectively. For two and five raingauges, similar results could be obtained. As the number of contaminated stations increased, the efficiency of the first step reduced, the efficiency of the second and third step advanced. It meant that the performances of the three steps compensated one another. However, the changes of the detection efficiency were not large.

The detection mistakes are given in Table 2. It is obvious that the highest detection mistakes occurred in the second step, demonstrating that the Tukey fence of the second step was relatively little to true data. The changes of detection mistake were little as $e_{\max}$ increased. As the number of raingauges contaminated increased, the detection mistake of the first step reduced, and that of the second and third step increased in tendency.

Table 1 Detection efficiency of robust statistical method.

e_{max} (mm)	$e1$	e_11	e_21	e_31	$e2$	e_12	e_22	e_32	$e5$	e_15	e_25	e_35
10	0.02	0	0.02	0	0	0	0	0	0	0	0	0
20	0.06	0	0.06	0	0.13	0	0.13	0	0.04	0	0.04	0
30	0.36	0	0.36	0	0.43	0.01	0.41	0.01	0.32	0.02	0.3	0
40	0.68	0.32	0.36	0	0.53	0.14	0.39	0	0.53	0.17	0.34	0.02
50	0.66	0.32	0.34	0	0.69	0.4	0.29	0	0.6	0.3	0.29	0.01
60	0.76	0.54	0.22	0	0.7	0.48	0.21	0.01	0.71	0.5	0.21	0
70	0.68	0.5	0.18	0	0.63	0.46	0.17	0	0.76	0.54	0.22	0
80	0.72	0.58	0.14	0	0.81	0.67	0.14	0	0.8	0.64	0.16	0
90	0.78	0.6	0.18	0	0.85	0.64	0.21	0	0.79	0.65	0.14	0
100	0.88	0.82	0.06	0	0.8	0.59	0.21	0	0.81	0.65	0.15	0.01
200	0.92	0.82	0.1	0	0.95	0.87	0.08	0	0.91	0.87	0.04	0
500	0.96	0.96	0	0	0.98	0.95	0.03	0	0.98	0.95	0.03	0
Mean	0.62	0.46	0.17	0	0.63	0.43	0.19	0.00	0.60	0.44	0.16	0.00

$e1$, $e2$, $e5$: detection efficiency of data of 1, 2, 5 stations containing outliers using equation (4), respectively. Subscripts 1,2,3: the first, second, third step, respectively.

Table 2 Detection mistakes of robust statistical method.

e_{max} (mm)	$m1$	m_11	m_21	m_31	$m2$	m_12	m_22	m_32	$m5$	m_15	m_25	m_35
10	12	1	11	0	12	1	11	0	13	1	12	0
20	12	2	10	0	12	1	11	0	12	2	10	0
30	12	2	10	0	12	2	10	0	13	1	9	3
40	12	1	11	0	12	2	10	0	11	1	8	2
50	11	1	10	0	11	1	10	0	11	1	8	2
60	11	1	10	0	12	1	11	0	11	1	10	0
70	11	1	10	0	11	1	10	0	11	1	10	0
80	11	1	10	0	12	2	10	0	11	1	10	0
90	11	1	10	0	11	1	10	0	11	1	10	0
100	11	1	10	0	11	1	10	0	12	1	11	0
200	11	1	10	0	11	1	10	0	11	1	10	0
500	11	1	10	0	11	1	10	0	13	1	12	0

$m1$, $m2$, $m5$: detection mistake of data of 1, 2, 5 stations containing outliers using equation(4), respectively.

The performance of the three-stepwise robust statistical detection method was measured by $r_i j$ ($i = 1,2,3; j = 1,2,5$) (Bau *et al.*, 2003) computed by follows:

$$r_1 j = 1 - \sum|\hat{e}_1 j| \Big/ \sum|\hat{e}_0 j| \tag{7}$$

$$r_2 j = 1 - \sum|\hat{e}_2 j| \Big/ \sum|\hat{e}_1 j| \tag{8}$$

$$r_3 j = 1 - \sum|\hat{e}_3 j| \Big/ \sum|\hat{e}_2 j| \tag{9}$$

$$r_t j = 1 - \sum|\hat{e}_3 j| \Big/ \sum|\hat{e}_0 j| \tag{10}$$

where, e_0 is the outliers added to data using equation (4); e_1,e_2,e_3 are the errors of the first, second and third step detection, respectively; r_1,r_2,r_3 are the performance of the first, second and third step, respectively; and r_t is total performance of the method.

Table 3 Performances of robust statistical detection method.

e_{max} (mm)	$r1$	r_11	r_21	r_31	$r2$	r_12	r_22	r_32	$r5$	r_15	r_25	r_35
10	–0.151	–0.859	0	–1.141	–0.074	–0.451	0	–0.558	–0.027	–0.174	0	–0.206
20	–0.087	–0.333	0	–0.448	–0.031	0.015	0	–0.016	–0.015	–0.02	0.007	–0.029
30	–0.056	0.246	0	0.204	–0.022	0.447	0.019	0.445	0.004	0.431	–0.03	0.417
40	0.421	0.353	0	0.625	0.221	0.498	0	0.609	0.274	0.522	0.027	0.662
50	0.439	0.316	0	0.616	0.581	0.436	0	0.763	0.488	0.512	–0.005	0.749
60	0.703	0.099	0	0.733	0.693	0.336	0.028	0.802	0.716	0.49	0	0.855
70	0.711	0.052	0	0.726	0.713	0.207	0	0.772	0.751	0.521	0	0.88
80	0.783	–0.113	0	0.758	0.845	0.19	0	0.874	0.842	0.431	0	0.91
90	0.816	0.001	0	0.816	0.821	0.372	0	0.887	0.848	0.384	0	0.907
100	0.917	–0.583	0	0.869	0.818	0.389	0	0.889	0.869	0.431	0.040	0.929
200	0.95	–0.328	0	0.934	0.962	–0.069	0	0.959	0.971	0.031	0	0.972
500	0.991	–1.728	0	0.974	0.99	–0.645	0	0.984	0.991	–0.078	0	0.99
Mean	0.536	–0.240	0	0.472	0.543	0.144	0.004	0.618	0.559	0.290	0.003	0.670

It was noted that the performances resulted from the detection efficiency and mistake. The performances of increasing e_{max} from 10 mm to 500 mm are displayed in Table 3.

When data of only one raingauge contained outliers, the total performances increased as the e_{max} increased: r_t1 was –1.151 for 10 mm and 0.991 for 500 mm. For two and five raingauges, similar results were obtained. It meant the performances of the robust statistical method were obvious with an increase in e_{max}. Among the three steps, the performances of the first step were greatest, and the performances of the third step were least. For example, for one station of e_{max} = 70 mm, the performances of the first, second and third step were 0.711, 0.052 and 0, respectively. Moreover, as the number of raingauges contaminated increased, the performances of the second, third step were more obvious. These results agreed with those of detection efficiency and mistakes.

The negative values in Table 3 resulted from the detection mistakes. When e_{max} was small, and the detection efficiency was small, the response of detection mistake was obvious. When e_{max} increased, the detection mistake still occurred, but its response was concealed by the response of the detection efficiency.

Performances of the modified method

The historic hourly telemetric rainstorm data of 40 events and 43 rainfall stations during 1988–1997 was divided into five groups, i.e. [0–1 mm], (1 mm–2 mm], (2 mm–4 mm], (4 mm–6 mm], (>6 mm). The Tukey fences of every group were calculated.

The detection efficiency of the modified method is shown in Table 4. The mean detection efficiency was above 0.68. The detection efficiency increased as e_{max} increased, and it reduced as the number of raingauges contaminated increased. The most obvious efficiency occurred in the first step. Moreover, as the number of raingauges contaminated increased, the efficiency of the second and third step was larger and larger. Comparing the results in Table 1 and Table 4, the detection efficiency of the modified method was larger, particularly, when e_{max} was small.

The detection mistakes of the modified method are presented in Table 5. As the number of raingauges contaminated increased, the detection mistake of the first step reduced, and that of the second and third step increased. That agreed with detection efficiency. It meant that improvement of efficiency inevitably caused more mistakes for the same method. Comparing the results in Table 2 and Table 5, the detection mistakes reduced remarkably.

The performance features in Table 6 were similar with those of the detection efficiency. Comparing the results of Table 3 and Table 6, the performances of the modified method were better and the number of negative values was less. That was because that the detection efficiency was higher and mistakes were less.

Table 4 Detection efficiency of the modified method.

e_{max} (mm)	$E1$	E_11	E_21	E_31	$E2$	E_12	E_22	E_32	$E5$	E_15	E_25	E_35
10	0.100	0	0.100	0	0.060	0	0.060	0	0.070	0.004	0.068	0
20	0.460	0.360	0.100	0	0.520	0.340	0.180	0	0.410	0.192	0.184	0.036
30	0.580	0.560	0.020	0	0.650	0.500	0.150	0	0.460	0.260	0.160	0.04
40	0.740	0.720	0.020	0	0.670	0.570	0.100	0	0.660	0.360	0.264	0.036
50	0.760	0.740	0.020	0	0.730	0.650	0.080	0	0.680	0.396	0.228	0.052
60	0.860	0.860	0	0	0.780	0.680	0.100	0	0.720	0.464	0.228	0.028
70	0.780	0.780	0	0	0.780	0.660	0.120	0	0.780	0.488	0.268	0.024
80	0.840	0.800	0.040	0	0.840	0.750	0.090	0	0.830	0.516	0.276	0.036
90	0.820	0.800	0.020	0	0.890	0.800	0.080	0.010	0.830	0.512	0.284	0.036
100	0.900	0.900	0	0	0.830	0.760	0.070	0	0.820	0.504	0.272	0.048
200	0.880	0.880	0	0	0.950	0.850	0.100	0	0.930	0.708	0.196	0.028
500	1.000	1.000	0	0	0.990	0.920	0.070	0	0.970	0.904	0.068	0
Mean	0.727	0.700	0.027	0	0.724	0.623	0.100	0.001	0.680	0.442	0.208	0.030

$E1$,$E2$,$E3$: detection efficiency of data of 1, 2, 5 stations containing outliers using the modified method and equation(4), respectively.

Table 5 Detection mistakes of the modified method.

e_{max} (mm)	$M1$	M_11	M_21	M_31	$M2$	M_12	M_22	M_32	$M5$	M_15	M_25	M_35
10	3	2	1	0	3	2	1	0	3	1	2	0
20	3	2	1	0	3	1	2	0	1	0	1	0
30	3	2	1	0	3	1	1	1	2	0	1	1
40	3	1	2	0	3	1	2	0	2	0	2	0
50	3	0	3	0	2	0	2	0	2	0	2	0
60	3	2	1	0	3	0	3	0	2	0	2	0
70	3	2	1	0	2	1	1	0	3	0	2	1
80	3	0	3	0	3	0	3	0	2	0	2	0
90	3	1	2	0	3	0	3	0	3	0	2	1
100	3	0	3	0	3	0	3	0	2	0	1	1
200	3	0	3	0	3	0	3	0	3	0	2	1
500	3	0	3	0	3	0	3	0	3	0	2	1

$M1$,$M2$,$M3$: detection mistake of data of 1, 2, 5 stations containing outliers using equation(4), respectively.

Table 6 Performances of the modified method.

e_{max} (mm)	$R1$	R_11	R_21	R_31	$R2$	R_12	R_22	R_32	$R5$	R_15	R_25	R_35
10	−0.152	0.124	0	−0.010	−0.075	0.076	0	0.007	−0.009	0.097	0	0.089
20	0.458	0.131	0	0.529	0.493	0.277	0	0.633	0.323	0.307	0.087	0.572
30	0.709	0.007	0	0.711	0.651	0.246	−0.031	0.728	0.422	0.279	0.077	0.615
40	0.864	−0.090	0	0.852	0.760	0.166	0	0.799	0.553	0.486	0.107	0.795
50	0.830	−0.136	0	0.807	0.782	0.110	0	0.806	0.599	0.417	0.112	0.792
60	0.910	−0.065	0	0.904	0.853	0.166	0	0.878	0.677	0.450	0.065	0.834
70	0.904	−0.063	0	0.898	0.842	0.283	0	0.886	0.688	0.564	0.072	0.874
80	0.903	−0.131	0	0.890	0.872	0.166	0	0.893	0.703	0.681	0.159	0.920
90	0.921	−0.080	0	0.915	0.908	0.216	0.039	0.931	0.710	0.667	0.099	0.913
100	0.946	−0.280	0	0.931	0.925	0.147	0	0.936	0.737	0.654	0.138	0.921
200	0.974	−0.328	0	0.965	0.956	0.445	0	0.976	0.898	0.729	0.119	0.976
500	0.995	−0.696	0	0.992	0.986	0.384	0	0.992	0.982	0.554	−0.028	0.992
Mean	0.772	−0.134	0	0.782	0.746	0.223	0.001	0.789	0.607	0.490	0.084	0.774

CONCLUSION

The detection capability of the robust statistical algorithm and distribution features of the precipitation data were utilized in this paper to detect abnormal data in a telemetric system. The Tukey fences which were less sensitive to outlying observations were used. The case study on synthetic data has shown that the proposed method produces reliable detection results. Moreover, the modified method provided better results. That illustrates the importance of distribution features of the precipitation to the performance of the method.

As a part of future research, in addition to modifying the method agreement with the precipitation distribution, the detection risk of the method needs to be considered. An exhaustive research of detection risk should be carried out.

Acknowledgements The authors wish to acknowledge the financial assistance provided by Xiamen University of Technology (no. YKJ08015R). The authors also acknowledge the express their sincere thanks to the referees for the valuable suggestions which improved the final manuscript.

REFERENCES

Bao, W. M., Qu, S. M., Li, Q. S. & Huang, X. Q. (2003) Study of estimation methods of rainfall gauge errors in remote system. *J. Hydraulic Engng* **4**, 30–33 (in Chinese).

Barnett, V. & Lewis, T. (1994) *Outliers in Statistical Data*, third edn. John Wiley, Chichester, UK.

Bounessah, M. & Atkin, B. P. (2003) An application of exploratory data analysis (EDA) as a robust non-parametric technique for geochemical mapping in a semi-arid climate. *Applied Geochemistry* **18**, 1185–1195.

Daszykowski, M., Kaczmarek, K., Heyden, Y. V. & Walczak, B. (2007) Robust statistics in data analysis – a review of basic concepts. *Chemometrics and Intelligent Laboratory Systems* **85**, 203–219.

Grubbs, F. E. & Beck, G. (1972) Extension of sample sizes and percentage points for significance tests of outlying observations. *Technometrics* **4**(14), 847–853.

Han, J. W. & Kamber, M. (2001) *Data Mining: Concepts and Techniques.* Morgan Kaufmann Publishers, USA.

Hu, S. Y. (1987) Problems with outlier test methods in flood frequency analysis. *J. Hydrol.* **96**(1-4), 375–383.

Narasimhan, S. & Mah, R. (1987) Generalized likelihood ratio method for gross error identification. *AIChE J.* **33**, 1514–1521.

Singh, K. P. (1991) Flood frequency modeling and outliers. ASCE, USA.

Spencer, C. S. & McCuen, R. H. (1996) Detection of outliers in Pearson Type III data. *J. Hydrologic Engng* **1**(1), 2–10.

Zhou, Q., Li, S. N., Li, X. P., Wang, F. & Wang, Z. G. (2006) Detection of outliers and establishment of targets in external quality assessment programs. *Clinica Chimica Acta* **372**, 94–97.

Zhou, Q., Shen, Z. Y., Li, S. N., Li, X. P., Wang, W. & Wang, Z. G. (2008) Robust and traditional statistical methods in the establishment of immunoglobulin E target values in external quality assessment program. *Clinica Chimica Acta* **387**, 66–70.

Land use impacts on river hydrological regimes in Northern Asia

ALEXANDER ONUCHIN[1], TAMARA BURENINA[1], KULMURZA GAPAROV[2] & NATASHA ZIRYUKINA[1]

1 *V. N. Sukachev Institute of Forest Siberian Branch the Russian Academy of Sciences, 660036, Academgorodok, Krasnoyarsk, Russia*
institute@forest.academ.ru

2 *P. A. Gan Institute of Forest and Nut Plantation, Kyrgyzian Academy of Sciences, 720015, Bishkek, Kargachevaya rosha, 15, Kyrgyzstan*

Abstract River flow is vitally important to many human activities. River flow is influenced by climatic and land-cover changes. Land-use practices have a significant effect on water flow and quality. Land use can change surface runoff, which in turn can be used as an environmental indicator of a land use level of sustainability. Along with the regional climate, hydrological processes occurring in river basins in Siberia and mountainous Kyrgyzia are controlled by forest logging and afforestation. The method used to analyse annual river flow genesis to date allowed the onset of, and assessment of, the level of human activities in the watersheds. Moreover, river flow genesis can be used in land use decision-making. River flow reflects all watershed changes, which can have opposite effects, thus compensating for each other. This study confirmed that river flow changes in time, thus reflecting land cover changes in watersheds.

Key words hydrological regime; precipitation; river flow; river catchments; clear cuts; forest logging; afforestation

INTRODUCTION

The hydrological role of forests has been an international research challenge for many decades. The regional water resource response to changing watershed forest area and condition, as well as land uses, is of special interest. In this respect, investigating forest-harvesting impacts and post-logging forest recovery in the river basins of Low Angara region is critically important from a practical viewpoint, since these rivers contribute considerably to the Arctic Ocean freshwater budget. In central Asia, rivers develop in mountains and, as their water is used mostly for agriculture, a very important national economy sector in all central Asian countries, freshwater availability is an even greater concern here.

STUDY AREA

Five rivers (Taseyeva, Chadobets, Irkineyeva, Mura, and Karabula), with basins ranging from 4190 to 13 300 km^2 in area, were selected to study forest logging-caused river flow changes in the Low Angara region. The regional river flows develop in the southern taiga forest zone and are fed from different sources, mainly by melted snow water. Intensive forest logging began in the region in the late 1950s–early 1960s and has gradually extended northward since then.

Watershed forest area impacts on mountain river flows were studied in three watersheds ranging from 27 to 285 km^2 in size and found at elevations of 2100 to 2700 m a.s.l. located east of Issyk-kul Lake, central Asia. The watersheds are represented mainly by 25- to 30-grade slopes (even steeper in some places) covered by spruce (*Picea schrenkiana*) forest that has gradually increased in area since the 1950s as a result of sylvicultural treatments. Permanent stream channels fed by melted snow water and summer precipitation are found in each of the watersheds of interest. Although these watersheds contain neither glaciers, nor eternal snow, the stream channels might be contributed to by certain amounts of groundwater flowing down from the upper watersheds that receive water from melting glaciers. Our investigations were conducted at a research station (2036 m a.s.l.) located within a permanent study site of the P. A. Gan Institute of Forest and Nut Plantation established in the eastern part of the Terskey Ala-Too mountain range, adjoining the southern side of Issyk-kul Lake.

METHODS

In the Low Angara region flows of stream channels were studied based on reference data (Climate Guide, 1969; Surface Water Resources, 1973), while fire and logging dynamics was assessed using ground forest inventory information.

Flows of stream channels found in elementary watersheds near Issyk-Kul Lake were estimated by measuring water levels and discharge at stream channel outlets equipped with right-top-angle deltoid weirs. The measurements were done daily at 08:00 h and 17:00 h from 1980 to 2004. Multiple regression analysis (Lvovsky, 1988) was applied to process the data obtained and identify streamflow relationships with climate parameters and size of watershed forest area.

RESULTS

As river flow varies with climate and landscape and the latter can, in turn, be considerably altered by human activities, our goal was to assess human contribution to river flow variability. Clearly, this can be accomplished only by building appropriate models describing contributions of both climatic and landscape factors, and considering human-caused watershed changes. Attempts to develop such models have already been made for estimating river flow changes induced by agriculture and forest plantations (Onuchin *et al.* 2006, 2008) and the experience gained can apparently be used to assess post-logging and post-planting river flow changes, as these two activities result in watershed forest area reduction and increase, respectively.

Our analysis of the stream channel flow data, collected in the eastern Issyk-Kul region, was based upon developing the following equations describing the annual river flow relationship with hydro-climatic parameters and time since forest planting, for each watershed of interest.

For Adybaevo River:

$$Y = -49.3 + 20 \times \ln(X_l \times Y_p) \times \frac{T_8}{T_5 \times \ln(n)} + X_z \times (0.03 \times T_5 - 0.4) + 0.07 \times X_l \times \frac{(T_1 + 12)}{T_6} \quad (1)$$

$R^2 = 0.66$; $G = 8.9$; $F = 9.1$

For Ak-Tash River:

$$Y = -361 - 0.83 \times Y_p + 8.8 \times \ln(X_l \times Y_p) \times \frac{T_8}{T_5 \times \ln(n)} + 41 \times \ln(X_l \times Y_p) \quad (2)$$

$R^2 = 0.57$; $G = 15.2$; $F = 8.9$

For Bel-Kara River:

$$Y = 62.2 + 27.7 \times \ln(X_l \times Y_p) \times \frac{T_8}{T_5} - 517.3 \times \frac{T_5}{T_8} + 73.2 \times T_5 - 52 \times T_8 - 0.8 \times Y_p \quad (3)$$

$R^2 = 0.51$; $G = 21.5$; $F = 3.7$

where Y is annual river flow, mm; X_l is annual liquid precipitation (rain), mm; X_z is annual snow amount, mm; Y_p is previous year river flow, mm; n is time since forest planting, years; T_1, T_5, T_6, and T_8 are January, May, June, and August mean air temperatures, respectively, C^o; R^2 is multiple determination coefficient; G is standard error, mm; and F is Fisher criterion. All coefficients are significant at 95 confidence level.

As is clear from the above equations, river flow exhibited a complex relationship with climate and watershed forest area. River flow was found to increase with increasing liquid precipitation (rain) for all three stream channels. In two more-forested watersheds (Adybaevo and Ak-Tash), river flow appeared to decrease significantly over the time since forest plantations have been established there. While Bel-Kara-Su watershed forest area had increased from 5 to 21%, no significant river flow decrease was observed. A decrease in river flow with decreasing January and increasing June temperatures recorded for the most-forested watershed (Adybaevo) might be due to increasing soil freezing in January and evapotranspiration in summer.

The identified relationships are better understood from their graphs. River flow response to the changes of the factors under study was assessed, with other parameters being assumed constant.

Model (1) shows that river flow in Adybaevo watershed increases with increasing river flow in the previous year (Fig. 1(a)). The highest current-year increase rate occurs up to a previous-year surface runoff value of 30–40 mm, while a further increase in the previous year river flow, beyond this value, has a less pronounced effect on the current year river flow increase. This is probably attributable to soil water capacity that controls conservation of the water received by a watershed in the previous year, as well as the contribution of the water to the current year river flow.

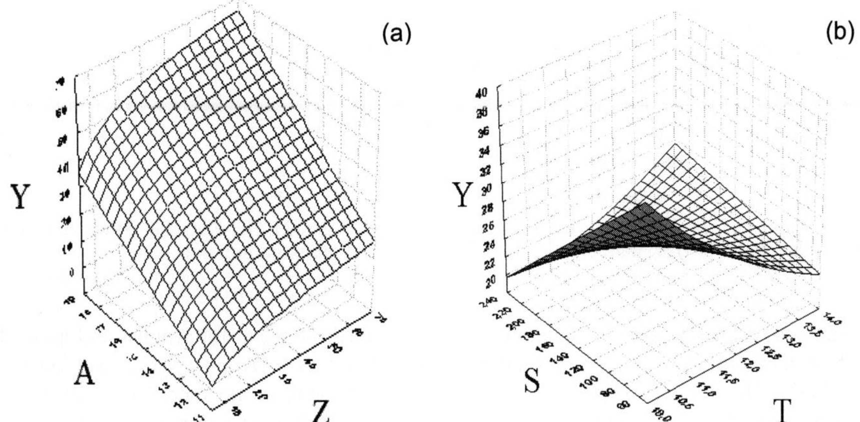

Fig. 1 A relationship of the annual river flow (Adybaevo River) with the previous year river flow, August air temperature (a), amount of snow and May air temperature (b). Y, annual river flow, mm; A, temperature in August, °C; Z, previous year river flow, mm; S, amount of snow precipitation; T, temperature in May, °C.

River flow was found to increase with increasing August air temperature. Glaciers and everlasting (non-melting) snow are known to feed Kirgizian mountain rivers (Bolshakov, 1962). The identified positive relationship between river flow and August air temperature was apparently due to addition of a part of groundwater flow of glacier melt water to the river flows. Khamaladze (1961) and Puzachenko (1987), in their study of seasonal river flow dynamics in the Caucasus (a place remarkable for its big glaciers) also attributed increasing river flow in June–September to high air temperatures. A number of Siberian rivers showed a similar trend of late-summer increases in their flows due to thawing of upper permafrost horizons (Onuchin, 2003).

Regarding river flow development, we found that river flow increased with increasing May air temperatures, provided that a snow layer was deep, and, conversely, it decreased at reduced snow cover depths (Fig. 1(b)). The complexity of this relationship is caused by characteristics of snow water redistribution between evaporation and river flow, as well as a negative correlation of snow amounts with May air temperatures. Deep snow requires much energy to melt and, thus, does not allow the air temperature to increase very much during this period of time.

The relationships determined for Adybaevo River appeared to be characteristic of the other two stream channels of interest. However, certain factors, such as January and June air temperatures and winter precipitation amounts, were not proved to be statistically significant. The narrow range of the forest area changes among the watersheds of interest (10%, 16%, and 29% for Ak-Tash, Bel-Kara-Su, and Adybaevo watersheds, respectively), the total watershed area differences, and other watershed-specific characteristics did not allow us to accurately quantify the influence of watershed forest area on river flow on a regional scale. However, our analysis of temporal river flow changes revealed that it tended to decrease, regardless of precipitation amount (Fig. 2).

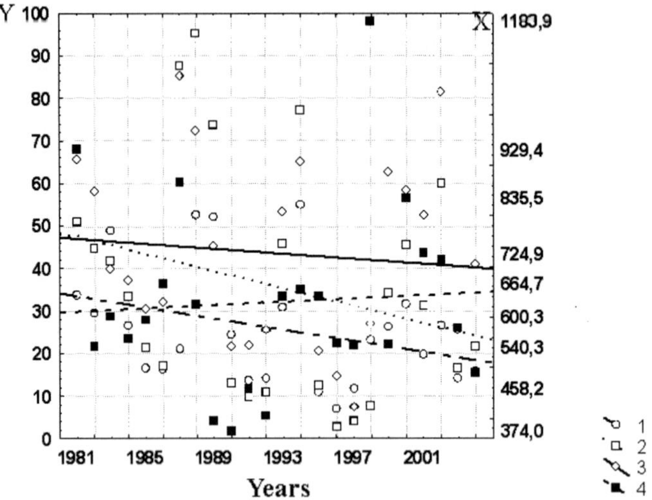

Fig. 2 Dynamics of annual river flow for Adybaevo (1); Ak-Tash (2); Bel-Kara-Su (3) watersheds; and the total precipitation (4). *Y*, annual river flow, mm; *X*, annual precipitation, mm.

Our analysis of air temperature dynamics showed a positive trend for August and a negative trend for June temperatures, and this should be accompanied by a river flow increase (equations (1), (2), and (3)). The annual river flow decreases observed for the watersheds under study might be due to landscape changes, primarily increasing forest area, rather than climatic factors.

This contradiction of a climate-based river flow development concept motivated us to integrate the available information and build a robust regional-scale model of river flow development. The model raw data included the results of the above mentioned long-term observation of annual channel flow dynamics, watershed forest area changes, as well as watershed total area data. To eliminate influences of numerous hydro-climatic parameters, moisture cycling that represents the annual river flow value averaged over all watersheds of interest was taken as a factor reflecting their integral influence. Our data analysis and integration yielded the following channel flow development model for the eastern Issyk-Kul region:

$$Y_0 = 29.2 + 2.8 \times V - 9.54 \times \ln(S) - 0.49 \times V \times \ln(L) + 4.4 \times \ln(L) \times \ln(S) \quad (4)$$

$R^2 = 0.89$; $G = 7.5$; $F = 148$;

where V is moisture cycling, mm; L is watershed forest area, %; and S is total watershed area, ha.

Equation (4) suggests complex river flow relationships with watershed total and forested areas, as well as precipitation amount. While surface runoff river flow tended to decrease with increasing watershed forest area in wet years, it showed an opposite trend in dry years. As Issyk-Kul region is characterized by continental climate, forest, as opposed to gaps (i.e. non-forest sites), can act as a "moisture evaporator" and a "moisture accumulator" under moisture excess and deficit, respectively, contributing to groundwater-fed rivers, in the latter case. This is clear from the model (Fig. 3). It can thus be concluded that changing background climate and weather parameters induce qualitative changes of forest ecosystem hydrological functions. This confirms a hydrologist's statement that, although hydrological processes are controlled by landscape factors, this control is manifested within limited ranges of geosystem conditions (Mandych, 1987; Onuchin *et al.*, 2008).

Decreasing river flow from forest sites, as compared to non-forest areas, occurring in wet years is assumed to be explained by the following two major reasons. First, forest is capable of intercepting much more rain and snow than other ecosystems; therefore, given the same precipitation amount, soil of non-forest ecosystems will receive more water than forest soil. As a result, heavier river flow is characteristic of non-forest vegetation communities as compared to forest,

other

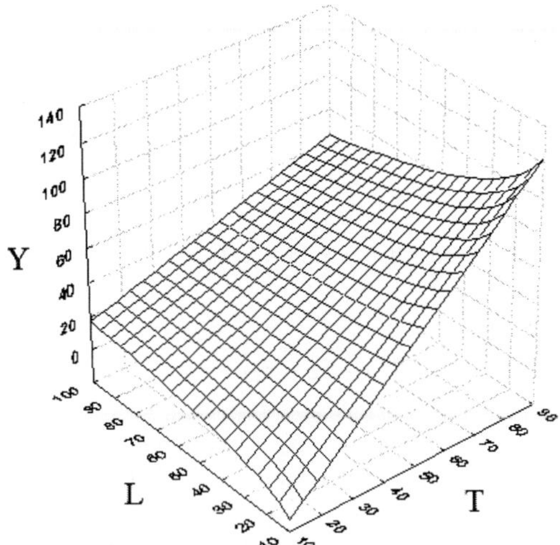

Fig. 3 Annual river flow variability depending on watershed forest area and moisture cycling.

conditions being equal, except where condensation precipitation prevails, and Issyk-Kul region is not among these exceptions.

Precipitation interception by forest canopy and the absolute difference in moisture amount received by soil between forest and non-forest watersheds increases with increasing precipitation (Onuchin 2001, 2003). For this reason, precipitation interception and physical evaporation by forests are greater compared to non-forest areas in wet years. Microclimatic conditions (lower insulation, low wind, radiation and thermal regime characteristics) developing under forest canopy in dry years, reduce the amount of moisture used for physical evaporation and, therefore, heavier river flow from forest sites can exceed that from treeless areas. Secondly, a forest is a more complex barrier than open sites in terms of moisture circulation. As forests are structured systems having great living biomass and detritus loads, they transform soil physical properties and, hence, forest hydrological functions differ greatly from those of other terrestrial ecosystem types. Forest soils characterized by high water permeability and capacity can retain the precipitation received for a long time. These account for the fact that interannual moisture redistribution processes, i.e. involvement of a wet-year precipitation in the next-year river flow are better pronounced in forest than non-forest watersheds.

River flow was found to change most rapidly with forest area changing from 0 to 30% of the total watershed area, both for wet and dry years, whereas forest area increases over 60% showed little effect on the amount of heavier river flow (Fig. 3). It can be concluded, therefore, that a forest ecosystem loses its hydrological and water regulation function with a reduction of a watershed forest area to less than 30%, whereas its increase to over 60% does not yield their adequate improvement. Forest areas making up 30% and 60% of the total watershed area should be thus considered as the lower and upper watershed forest area limits for the Issyk-Kul region. This forest area range enhances moisture interannual redistribution and conservation providing, as a result, temporal smoothing of river flow.

The river flow data we collected for the river basins of the Low Angara region were based upon developing equations describing annual river flow dependence on hydrological and climatic factors, and time since forest logging for:

Taseyeva River:

$$Y = -495.8 + 60.9 \times \ln X_z \times T_{75} + 198.3 \times \ln X_g \times Y_p/(t_5 + t_7 + 5)\ 1.22 \times X_g/(t_g + 11) \quad (5)$$

$R^2 = 0.45$; $G = 24.8$; $F = 15.5$.

Chadobets River:

$$Y = 41.9 + 0.0054 \times X_z \times T_{86} + 0.23X_z \times X_l/(t_5 \times t_7) \quad (6)$$

$R^2 = 0.50$; $G = 26.2$; $F = 20.8$.

Irkineyeva River:

$$Y = 1150.0 + 0.103 \times T_{84} \times (X_z - 11.5) + 0.202 \times X_z \times X_l/(t_5 \times t_7) - 1.03 \times T_{60} - 8.64 \times X_z \quad (7)$$

$R^2 = 0.51$; $G = 25.8$; $F = 9.3$.

Mura River:

$$Y = 91.0 - 1.3 \times T_{67} + 1.3 \times T_{84} - 5.3 \times t_7 + 7.3 \times t_9 + 1.8 \times t_9 \times t_g) + 3.25 \times X_g/(t_g + 11) - 0.35 \times X_l \quad (8)$$

$R^2 = 0.53$; $G = 14.3$; $F = 5.8$.

Karabula River:

$$Y = 58.2 + 0.121 \times X_z \times X_l/(t_5 \times t_7) + 0.54 \times t_9 \times (t_g + 11) - 0.5 \times T_{62} \quad (9)$$

$R^2 = 0.46$; $G = 15.8$; F = 12.1.

where Y is annual runoff, mm; Y_p is previous year runoff, mm; X_z is precipitation as snowfall recorded by the nearest weather station, mm; X_g is annual precipitation recorded by the nearest weather station, mm; $t_{5,7}$ is mean air temperature in May and July, respectively; t_g is average annual air temperature; X_l is precipitation as rainfall recorded by the nearest weather station, mm; t_9 is mean September air temperature; T_{60} is time period between 1960 and 1984: T_{60} = 60const prior to 1960 and T_{60} = 84const after 1984; T_{62} is time period between 1962 and 1980: T_{62} = 62const prior to 1962 and T_{62} = 80const after 1980; T_{67} is time period between 1967 and 1984: T_{67} = 67const prior to 1967 and T_{67} = 84const after 1984; T_{75} is time period between 1975 and 1999: T_{75} = 75const prior to 1975; T_{84} is time period between 1994 and 1999: T_{84} = 84const prior to 1984; T_{86} is time period between 1986 and 1999: T_{86} = 86const prior to 1986; R^2 is multiple determination coefficient; G is standard error; and F is Fisher criterion.

These models show that the annual river flow amounts are closely related to a number of hydro-climatic parameters, as well as time. The annual channel flow of each river appeared to increase with increasing precipitation and decrease with increasing July air temperature, with those of the first four rivers also decreasing with increasing May air temperature. Two rivers, Karabula and Mura, exhibited increases in channel flow with increasing air temperature in September, which indirectly indicates that permafrost water produced by periodical upper soil horizon thawing contributes to channel flows of these rivers.

Channel flows of Karabula, Irkineyeva, and Mura rivers were found to decrease between the early 1960s and 1980s. This might be a result of extending forest logging during that period of time, and wind activity and moisture evaporation are known to increase on logged sites. Channel flows of these rivers were estimated to decrease by about 0.5–1.3 mm annually to total 10–20 mm over the above 20-year period characterized by heavy logging-caused vegetation disturbances. Also, each river channel flow, except for Karabula, has tended to increase since certain points of time. These points appeared to be 1975 for Taseyeva River, 1984 for Irkineyeva and Mura rivers, and 1986 for Chadobets River. These points of time coincide fairly well with when young secondary forests began to occupy the sites logged in each watershed in the 1950–1960s. Snow interception by young secondary deciduous tree crowns appeared to decrease as compared to mature trees, with simultaneously increasing snow accumulating capability of these woody species. This was also confirmed by an identified synergetic effect that indicated that snow contribution to channel flows of Taseyeva, Chadobets, and Irkineyeva rivers increased in time. The channel flows related to increasing capability of young secondary deciduous stands to accumulate snow was estimated to increase by 1.5–3.0 mm/year. Consequently, the initial pre-logging average annual channel flow had increased by 20–40 mm over the first 20 years of intensive logging.

Although the watershed area has no direct influence on river flow occurrence and development, it influences spatial and temporal river flow redistribution patterns (Lvovich, 1963). As watershed area, along with watershed geological structure, small stream talweg and river valley inset depth determines groundwater drainage, river flow decreases with decreasing area. Our investigation showed that river flow increased with increasing watershed total and forested areas in wet years, when forest functioning as a moisture circulation control was most pronounced, and thus confirmed the above dependence for the small rivers of interest. The river flow dependence on watershed area appeared to be especially clear for watersheds with high forest area proportions, in which case groundwater contribution to river flow was considerable. The greatest river flow increase was found with watershed forest area ranging from 0% to 20–40%, and this does not contradict our conclusion about the optimal size of watershed forest area made earlier.

As for moisture cycling, both high- and low-moisture cycles were determined to last 2–4 years and return every six years in the Issyk-Kul region (Fig. 4). The data collected showed the amount of moisture in a given moisture cycle to be controlled, besides the total precipitation, by the amount of moisture and duration of the previous cycle, as well as thermal regime. For example, the longest cycle characterized by the lowest moisture amount (1995–1998) was found to follow the shortest and the highest-moisture cycle (Fig. 4). It is noteworthy that although the last year of this lowest-moisture cycle had the highest precipitation amount recorded for the period of investigation, a considerable river flow increase was only observed in the following year. This indicates that moisture circulation is controlled by inert soil hydrological conditions, and inertia influences future river flow.

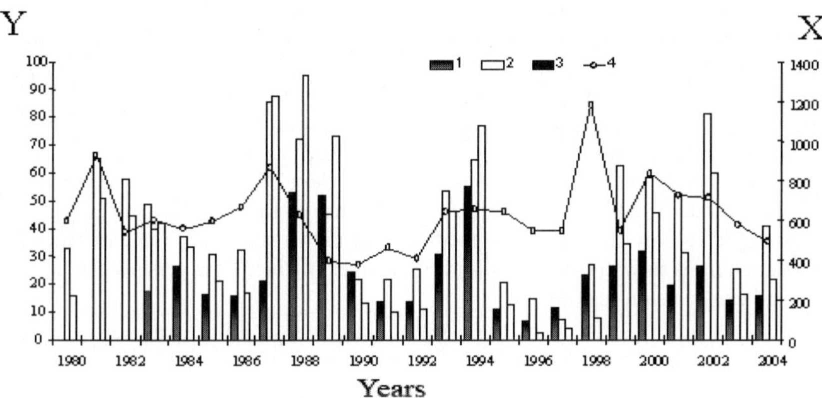

Fig. 4 Periodicity of the river flow cycles for Adybaevo (1), Ak-Tash (2), and Bel-Kara-Su (3) river basins and total precipitation (4). Y, river flow, mm; X, precipitation, mm.

Discrepancies existing nowadays in data on the hydrological role of forests are largely of methodological origin. As moisture circulation processes, river flow development particularly depends on numerous interacting geographical, soil, vegetation, weather, and climatic factors, and a simple statistical approach based on investigating big geographically different data sets does not provide insights into river flow genesis. Analysing long-term river flow data obtained for structurally similar watersheds characterized by fairly similar geographical conditions gives much better results. With this approach, a spatial analysis of river flow dynamics in the context of human and environmental influences ensures better understanding of the cause–effect relations between the processes under study and allows us to make justified conclusions on river flow development mechanisms. However, one should be very careful when extrapolating the results of this approach or making general conclusions from them, as forest ecosystem hydrological functions are determined, among other factors, by the geophysical background. In a study of snow water balance in the boreal forest zone (Onuchin, 2001), for example, snow moisture evaporation in the forest was found to exceed that in open sites in countries characterized by mild and snowy winters and *vice versa* where cold winters are common. This explains the contradictory estimates of the hydrological role of forests in different regions.

Human activity on watersheds has a considerable influence on river flow development. It can result in watershed water balance redistribution, particularly increasing the total evaporation and, hence decreasing surface runoff, or *vice versa*, depending on activity type, geophysical conditions, and time factor. Extensive clear cuts and large fires heavily disturb forest vegetation and can thus become the major river flow control. Analysing annual river flow genesis often helps date human activities in watersheds and changes to this regime can be based upon in optimal land use decision-making (Onuchin *et al.* 2006, 2008). River flow reflects land cover changes occurring across watersheds. These changes can be of opposite signs and can therefore compensate each other. For this reason, the proposed method, based on spatial analysis of river flow changes and dynamics of human and/or environmental factors, should be used primarily for small and moderate-sized watersheds found in fairly similar environments.

CONCLUSIONS

The study found climate to be the major factor controlling annual river flow to the rivers of interest. However, non-climatic factors, such as watershed forest area, soils, total area, and geological characteristics appeared to also contribute to river flow. As all these factors were determined to have a combined influence on river flow, synergetic effects were identified that can be based upon estimating river flow regulation feasibility by conducting appropriate weather-and-climate-based forest treatments. Our study confirmed that assessing ranges of values of weather and climatic factors that induce qualitative changes to the forest moisture regulation function is possible by conducting multi-scenario experiments with river flow models. Changing the sign of the forest water-saving function enabled estimation of climatic parameter values at which forest "works" as a moisture evaporator, or a streamflow feeder, compared to non-forest sites. The river flow development mechanisms revealed in our study allowed us to quantify forest logging ecological effects. The study results can also be used for developing justified regional-scale watershed forest area standards that would ensure sustainable use of forest and water resources.

Analysing annual river flow genesis by the method used in this study helps to date and assess the level of human activities in watersheds. Moreover, changes of river flow genesis can be based upon in optimal land use decision-making.

REFERENCES

Bolshakov, M. N. (1962) Waters and water energy resources. In: *Nature of Kirgizia*, 17–24. Frunze, Bishkek, Kyrgyzstan.

Khamaladze G. N. (1961) Influence of glaciation on annual channel flow of the rivers of Big Caucasian and the method of its calculation. *Proceedings Hidrometeorological Institute* **9**, 148–165. Gidrometeoizdat Publishing, Leningrad, Russia.

Lvovich, M. I. (1963) *Man and the Water. Water Balance and River Flow.* Geografgiz Publications, Moscow, Russia.

Lvovsky, E. N. (1988) *Statistical Methods for Building Empirical Equations.* High School Publications, Moscow, Russia.

Mandych, A. F. (1987) Elementary geosystem water stability and variability. In: *Hydrological Regime Influence on Ecosystem Structure and Functions* (Proceedings of All-Union Conference, Syktyvkar, Russia).

Onuchin, A. A. (2001) General snow accumulation trends in the boreal forest. *Academy News (The Geography Series)* **2**, 80–86.

Onuchin, A. A. (2003) Moisture circulation in mountain forests of Siberia. PhD Thesis 03.00.16, V. N. Sukachev Institute of Forest Publications, Krasnoyarsk, Russia (abstract).

Onuchin, A. A., Balzter, H., Borisova, H. & Blyth, E. (2006) Climatic and geographic patterns of river runoff formation in Northern Eurasia. *Adv. Water Resour.* **29**(9), 1314–1327. doi: 10.1016/ j.advwatres 2005.10.006.

Onuchin, A. A., Gaparov, K. K. & Mikheyeva, N. A. (2008) Influence of watershed forest area and climatic factors on annual channel flow of the rivers of Issyk-Kul region. *J. Forest Sci. (Lesovedenie)* **1**, 45–52.

Puzachenko, G. Yu. (1987) Simplest problems of ecosystem-based water regime management. In: *Hydrological Regime Influence on Ecosystem Structure and Functions*, 5–8. (Proceedings of All-Union Conference, Syktyvkar, Russia).

The USSR Climate Guide (1969) Parts 2 and 3, Issue 21. GUGS Publications, Krasnoyarsk, Russia.

The USSR Terrestrial Water Resources (1973) Vol. 16, Issues 1, 2, and 3. Gidrometeoizdat Publishing, Leningrad, Russia.

2 Hydrological Applications of Hydroinformatics

Keynote Paper

Harnessing hydrological systems knowledge for improved water resource management in water stressed cities such as Singapore

VLADAN BABOVIC & RAGHURAJ RAO

SDWA, Department of Civil Engineering, National University of Singapore, EW1(Annex)-02, Engineering Drive 2, 117576 Singapore

cvebv@nus.edu.sg; cverr@nus.edu.sg

Abstract Many water-stressed countries, such as Singapore, need to efficiently manage their resources. The reservoir system often needs to operate optimally, catering to multiple demands within many operational constraints. Suitable solutions are usually quite complex and call for holistic approaches combining the systems knowledge of the various hydrological and hydrodynamic processes involved. The governing mechanisms and their interactions are often complex and are not easily described by simple numerical models. With the increasing availability of stored data on each of these processes, data-driven techniques provide additional support in understanding these processes. A unified multi-reservoir management framework is suggested and various possible uses of different hydroinformatics techniques are outlined. Specific applications of model emulation and data assimilation using genetic programming techniques are illustrated with representative examples. The advantages of using these computationally efficient techniques are highlighted.

Key words hydroinformatics; data mining; genetic programming; data assimilation

INTRODUCTION

The growing world population and industrialization lead to increasing water demand. However, increasing global warming and changing weather patterns are depleting water resources. Overall, water is becoming a scarce commodity and highly urbanized cities, like Singapore, are already water stressed. As a consequence, efficient management of available water resources is getting serious attention. Reservoir operation and control relies heavily on the support of operational management tools and intelligent Decision Support Systems (DSS). Being one of the highly water-stressed countries, Singapore manages water through a system of 15 reservoirs, which are partly interconnected. Two other new reservoirs are planned, with an increase in the number of connections between these reservoirs. The reservoir operation is governed by the domestic and industrial water demands as well as flood risk management. A very fast runoff process due to the highly paved/concretized land area (runoff concentration time is in the order of 30 minutes), heavy tropical rains, anomalous sea level dynamics, etc. provide further challenges to water quantity and quality regulation. Hence, the multi-reservoir multi-objective system of Singapore belongs to the category of very complex water systems and state-of-the-art technologies have to be investigated to support appropriate water resource management decisions. Numerical models of regional/local hydrological processes can provide major support, but they may lack speed and accuracy in terms of real-time long-term forecasting. Fast developing data assimilation and hydroinformatics techniques and tools can provide additional support and new generation technology to address many problems associated with such complex systems. This contribution explores the utility of different "system knowledge" data-driven hydroinformatics techniques in efficient management of water resources. It outlines different schemes that can be implemented to achieve the optimum real-time reservoir operation.

WATER RESOURCE OPERATIONAL MANAGEMENT FRAMEWORK

Water resource management in water-stressed cities like Singapore could be based upon the following objectives:

- Maximizing catchment yields for the production of drinking water with rainwater being the primary available natural resource.
- Maintaining the water quality standards, overcoming the effects of first flush runoffs and salt water ingression.
- Flood risk management without overfilling the reservoirs.

The main challenge in addressing these objectives is to have a systematic knowledge of the main hydrological processes involved, viz.: rainfall, runoff and the hydrodynamic processes governing reservoir (inflow, outflow, storage and inter-reservoir dynamics) and ocean dynamics (sea level changes – tidal and non-tidal) along with various factors governing the water quality (land area flushing, turbidity, biomass growth and salinity). The water quantity and quality in the reservoirs can be optimally regulated if the outcome of these processes and factors can be predicted well in advance. The "system knowledge", once established, can be harnessed to design a time- and energy-efficient operational procedure for the optimum management of resources. Figure 1 gives a schematic framework to implement this strategy. The strategy involves the following major steps:

- Step (1) Forecast the rainfall in the region of interest with sufficient lead time.
- Step (2) Use this information to predict the runoffs from the catchments, forecasting the future inflows to the reservoir.
- Step (3) Design a control action to optimally operate the reservoir pumps and gates in order to efficiently manage the future inflows by satisfying the constraints on water quantity and quality standards as well as flood risk mitigation.
- Step (4) Update the forecasts and improve the suggested operating procedure, after the models are corrected using the real-time observations, as and when available.

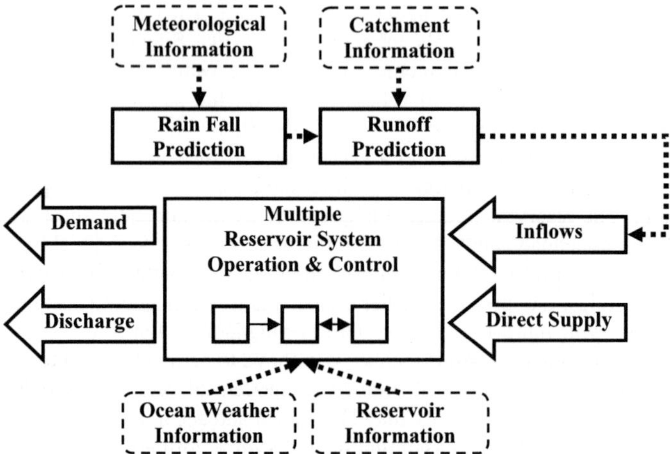

Fig. 1 Different components of a reservoir management framework.

As indicated in the framework (Fig. 1), different components of water management systems require different "system knowledge" in the form of information (data) as inputs. In Step (1), rainfall forecast lead times may be enhanced by combining a high-resolution limited area weather forecasting model with the quantitative precipitation estimates from meteorological radars or satellites. The quality of the rainfall–runoff predictions in Step (2) can be enhanced by using topographical information and also by learning the historical patterns of local discharges into the reservoirs. State updating (in Step (4)) can be carried out using the ground-based rainfall monitoring stations in a real-time set-up.

In Step (3), faster emulation of detailed hydrodynamic models to represent reservoir behaviour is essential for real-time optimization. Such model order reduction based on the sensitivity of the state indicators, will rely heavily on historical data and reservoir information. Since each of the components in this framework largely relies on the data, application of suitable hydroinformatics techniques is inevitable. The major implication here would be to extract the "system knowledge" using the past records by applying data-driven techniques, and to use it in upgrading the performance of the primary models used in Steps (1) to (3).

HYDROINFORMATICS SUPPORT FOR RESERVOIR MANAGEMENT

Different modes and stages of hydrological system contributing to the reservoir set-up are believed to be highly nonlinear, time varying, and spatially distributed. Processes like evapo/convective transports, precipitation, drainage, runoff, inter-reservoir recycling, etc. are complex in their mechanisms and governed by multiple factors. Hence, these processes are often difficult to describe fully by hydrodynamic models. The challenges of holistic analysis of these systems and the limitations of traditional modelling approaches have attracted alternative techniques. With improved data acquisition and processing capabilities in the field, data-driven hydroinformatics techniques have found applications to solve many different problems in hydrological sciences. These applications range from basic understanding of the kinetics and dynamics of the processes (for water quantity and quality) to real-time applications like forecasting, water collection/ distribution system design-optimize-operate-control and water resource management. Recent interest has also been on harnessing the power of techniques such as Artificial Neural Networks (ANN), Genetic Algorithm (GA), Genetic Programming (GP), Chaos theory, and Support Vector Machines (SVM) (Butts *et al.*, 2004). They have been extensively used to optimally tune the parameters of the local model (model calibration to fit the observed water quality/quantity data) and also to learn/predict the nature of model-observation mismatch (model error correction) (Madsen & Skotner, 2005). As a representative case, we present the utility of Genetic Programming, one of the useful data-driven techniques, in addressing various issues pertaining to multi-reservoir management set-up as described above.

Genetic programming – support tool for reservoir management systems

Genetic programming (GP) (Koza, 1992) is a data-driven evolutionary computing method for automatically generating input–output relations. It differs from other black-box type data-driven modelling techniques like Artificial Neural Network (ANN), Fuzzy Logic and Regression Trees, in the sense that it provides mathematically-meaningful structures (with optimum parameters) relating input–output variables of the system (Babovic & Keijzer, 2000). The basic GP algorithm follows the natural evolution principle, which relies on the Darwinian theory of survival of the fittest. Given a set of observed data (training data) on input–output variables, GP generates a population of models with random structure by applying the selective mathematical functions (e.g. sin, cos, log, sqrt, exp) and operations (e.g. +, –, /, *, ^) on the variables and constants. These models are then evaluated using part of the initial data not used during training (validation data) using a suitable fitness measure (e.g. RMS). The probability of a given model surviving during the model evolution process is proportional to how well it predicts the output of the validation data. Components of successful models are continuously recombined with those of others to form new models using genetic operations (mutation, cross-over). In each generation, GP optimizes the model structure, with an underlying nonlinear least-squares algorithm harnessed to estimate the associated model parameters. Over several generations, the GP algorithm evolves the best fit model that indirectly describes the dynamics (relation between input and output variables) of the system from which the data were used during the training (Babovic *et al.*, 1999).

In the present study, we highlight the utility of this unique modelling tool to address two important issues of the multiple reservoir management system.

- GP as a model emulator – representative numerical model.
- GP as a data assimilation tool – correct the numerical model error.

Modelling the model – GP as a model emulator tool

Many of the real-time water resource operational management systems rely on an optimization routine, which requires very fast and efficient internal models to forecast water quantity and quality. This objective is difficult to achieve as most of the real-time forecast models (especially for water quality) involve large-scale numerical simulations which take longer computing time. Hence, there is a need to develop alternate tools that are faster, yet can repeat the performance of original models with sufficient accuracy. Model order reduction and model emulation are some of the options used to achieve this speed up activity. The latter approach has attracted considerable attention in hydrology, mainly using data-driven techniques (Grzeszczuk *et al.*, 1998; Proaño *et al.*, 1998; Schiller, 2007). Though model emulation is not really a new topic in hydrology, applications so far have focused on relatively simpler processes such as rainfall–runoff, with a low dimensional input and output. Emulators are generally regression-based simple transfer functions (in time or space) applied for emulating the state-of-the-art physics-based large-scale numerical models. In this section, we envisage a few strategies for implementing hydroinformatics tools as model emulators. We also provide an illustrative example of one such possibility.

GP as a representative model for model emulation

Numerical model simulation data can be used to design the GP models that can closely emulate the time propagation of water quantity and quality parameters at desired locations (stations of interest) with finer prediction intervals (minutes to hourly scale) for the period during which original model updates are still running. Here the data-driven models act as temporal interpolators and can be used to forecast entities with higher time resolution using the numerical simulation output data available at larger time samples (e.g. twice a day) as initial conditions. Figure 2 depicts the scheme for implementing such a strategy for a real-time forecasting set-up.

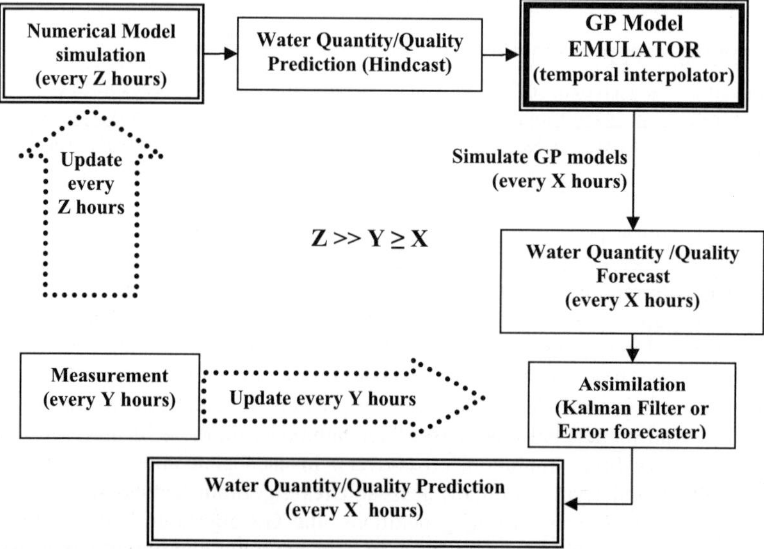

Fig. 2 Model emulation and data assimilation scheme.

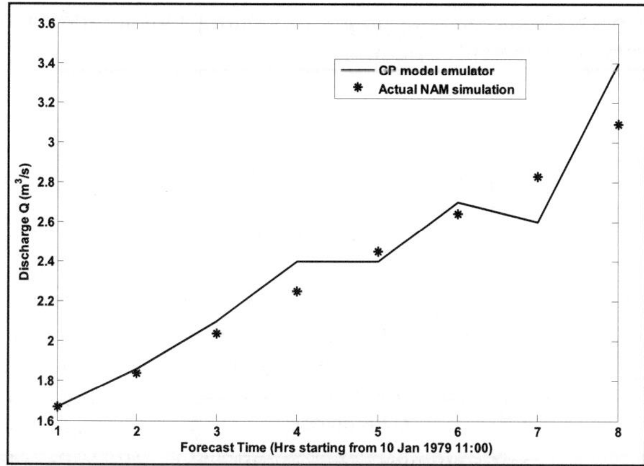

Fig. 3 Iterative forecast using NAM model emulator.

A sample model emulation example is demonstrated in Fig. 3. A representative catchment (Babovic & Keijzer, 2002) is simulated using the NAM simulation package (Nielsen & Madsen, 1973). The simulated data for the runoff at a desired discharge location (hourly time series data) is used as input data during GP training. A one-hour Q_{sim} forecast model is built using $Q_{sim}(t)$ as input and $Q_{sim}(t+1)$ as the output variable. The GP model emulator, with $Q_{sim}(t)$ as input, is then used iteratively to forecast Q_{sim}, k steps ahead in time. A sample forecast is presented in Fig. 3 using a selected time point as the initial condition. It is clearly seen that GP model emulator can predict the actual NAM simulation quite well up to 6 hours ahead in time. During these 6 hours the numerical model can be re-updated or calibrated using observed data (assimilation). At the end of 6 hours the new initial conditions can be taken into the GP model and the next 6 hours of forecast can be obtained. This shows the efficiency of GP model emulators in reproducing the short-term temporal propagation of Q_{sim}. It also establishes the utility of using GP trained model emulators as temporal interpolators for numerical models. Another strategy would be to use emulator models as spatial interpolators which can quickly predict model output on finer grid points using the coarse grid model data as input. The idea here is to obtain real-time parameter (say rainfall forecast, sea levels) using the numerical model run on a coarser grid (hence quicker computationally) and use hydroinformatics tools to improve the spatial resolution. Such model emulation exercises (using them as spatial, temporal interpolators) can significantly contribute to the computational swiftness and efficiency of hydrological forecasting systems. This in turn bridges the gap between the computationally-intense numerical models and real-time decision support systems used for optimized operation and control of water resources.

Modelling the model error – GP as data assimilation tool

The numerical simulation models are based on the fundamental hydrodynamic principles and hence they most realistically represent the hydrological process. Nevertheless, they can potentially be susceptible to errors arising from incorrect parameter settings, incomplete local geometry/ bathymetry and wrong initial states. Hence, they often show deviations in their predictions from the actual observations. Data assimilation procedures are employed to overcome this model–system mismatch and update the model in a real-time set-up. Such procedures either directly update the input variables (model forcing), model states, model parameters or correct the error in the predicted output variables. Output error correction procedures have been applied in hydrology for updating catchment runoff in conceptual rainfall–runoff (R–R) models (Refsgaard, 1997; Khu *et al.*, 2001) and for updating discharge flows at specific locations (Babovic *et al.*, 2002). In these cases, an error correction forecast model is built using the observed model residual errors

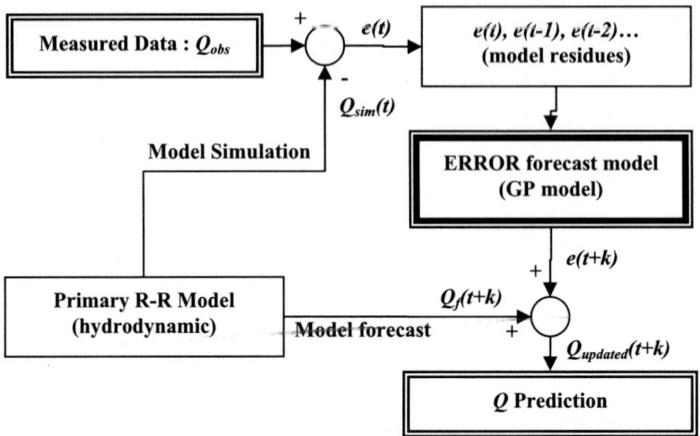

Fig. 4 Implementation scheme and data flow for data assimilation strategy using genetic programming-based error forecast models.

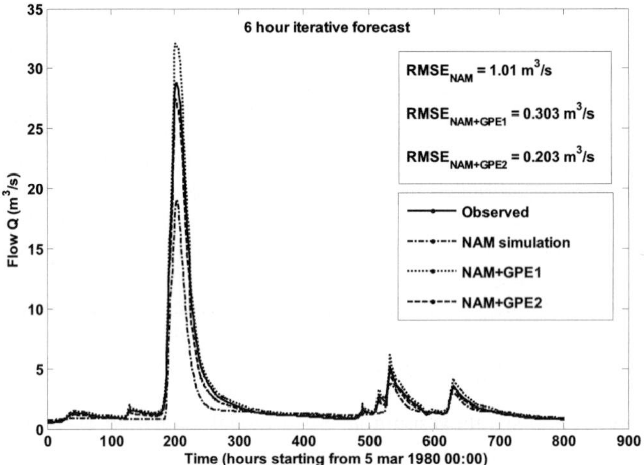

Fig. 5 Data assimilation using GP error forecast models. Comparing GPE1 and GPE2 models.

($Q_{obs} - Q_{sim}$) and then superimposed on the simulation model output. Figure 4 gives a schematic representation of the error correction strategy for real-time forecast systems. Here, we also demonstrate a small example of utilizing GP as a numerical model error learning tool and its utility in updating the rainfall–runoff forecasts given by the NAM model.

Error forecast model (GPE1) obtained using present/past error data:

$$e(t+1) = (3*e(t) + (-2.623*e(t-1)) + 0.16008*e(t-2))$$

Error model (GPE2) obtained using present/past error + Q_{obs} data

$$e(t+1) = (((-1.602*e(t-1)) + 2.255*e(t)) + (-0.1677*((e(t) + (-1.037*e(t)))*Q_{obs}(t)))$$

The updated forecasts are obtained by correcting the simulation model (NAM) forecasts $Q_{sim}(t+1)$ with the error forecast:

$$Q_f(t+1) = Q_{sim}(t+1) + e(t+1)$$

Figure 5 provides the comparison of performances for the above analysis using GP models as data assimilation tools. It is observed from the comparison of Q forecast for 800 hours of simulation that both the error forecast models can improve the prediction performance of the numerical model. GPE2, with terms involving both past errors and Q_{obs}, can provide an improvement of up to 80% in average error of the numerical model. The models GPE1 and GPE2 are designed using default settings of the GP tool and hence have further potential to improve the performance by tuning GP settings and also by incorporating past and predicted R terms into the model structure, as suggested by Babovic & Keijzer (2002). These observations provide further evidence to establish the utility of the GP technique as an effective data assimilation tool.

CONCLUSIONS

Complex water resource systems and associated hydrological processes in water-stressed urban settings calls for intelligent decision support systems which can harness the powers of hydro-informatics tools. The Genetic Programming concept has found special attention as a modelling tool in a multi-process water management framework. Though GP is considered here as an illustrative technique for model emulation and data assimilation, many similar data-driven techniques can be implemented for the same. Hence, hydroinformatics techniques can harness the system knowledge of past hydrological events and vital support for real-time operational management.

Acknowledgements The authors gratefully acknowledge the support and contributions of the Singapore-Delft Water Alliance (SDWA) and, in particular, the advice and guidance of Professor Arthur Mynett of Deltares.

REFERENCES

Babovic, V. & Keijzer, M. (2000) Genetic programming as a model induction engine, *J. Hydroinformatics* **2**(1), 35–60.

Babovic, V. & Keijzer, M. (2002) Rainfall runoff modelling based on genetic programming. *Nordic Hydrology* **33**(5), 331–346.

Babovic, V., Liong, S. Y. & Yu, X. (2002) Hydrologic forecasting with chaotic approach and support vector machine. In: *Proceedings of the XIII IAHR-APD Congress* (January, Singapore).

Babovic, V, Keijzer, M, Daida, J., Eiben, A. E., Garzon, M. H., Honavar, V., Jakiela, M., Smith, R. E. & Banzhaf, W. (1999) Dimensionally aware genetic programming. In: *Proceedings of the Genetic and Evolutionary Computation Conference,* July, Orlando (Florida, USA).

Butts, M. B., Madsen, J. & Refsgaard, J. C. (2004) Hydrologic forecasting, In: *Encyclopaedia of Physical Science and Technology*, 547–566.

Grzeszczuk, R., Terzopoulos, D. & Hinton, G. (1998) NeuroAnimator: fast neural network emulation and control of physics-based models. In: *Proceedings of SIGGRAPH 98*, July, Orlando(Florida, USA).

Khu, S. T., Liong, S., Babovic, V., Madsen, H. & Muttil, N. (2001) Genetic programming and its application in real-time runoff forecasting. *J. Am. Water Res. Assoc.* 37(2), 439-451.

Koza, J. R. (1992) *Genetic Programming: on the Programming of Computers by Means of Natural Selection.* MIT Press, Cambride, Massachusetts, USA.

Madsen, H. & Skotner, C. (2005) Adaptive state updating in real-time river flow forecasting—a combined filtering and error forecasting procedure. *J. Hydrol.* 308(1-4), 302–312.

Nielsen, S. & Hansen, E. (1973) Numerical simulation of rainfall runoff process on daily basis. *Nordic Hydrology* **4**, 171–190.

Proaño, C. O., Verwey, A. van den Boogaard, H. F. P. & Minns, A. W. (1998) Emulation of a sewerage system computational model for the statistical processing of large numbers of simulations. In: *Proceedings of Hydroinformatics '98* (Rotterdam, The Netherlands), vol. 2.

Refsgaard, J. C. (1997) Validation and inter-comparison of different updating procedures for real-time forecasting. *Nordic Hydrology* **28**(2), 65–84.

Schiller, H. (2007) Model inversion by parameter fit using NN emulating the forward model — evaluation of indirect measurements. *Neural Networks* **20**(4), 479–483.

Assessment of future climate change impact on non-point source pollution loads of a small rural watershed using QuickBird high resolution satellite imagery

MI-SEON LEE[1], GEUN-AE PARK[2], MIN-JI PARK[2], JONG-YOON PARK[2] & SEONG-JOON KIM[2]

1 *Dept of Rural Engineering, Konkuk University, 1 Hwayang-dong, Gwangjin-gu, 143-701 Seoul, South Korea*
misun03@konkuk.ac.kr

2 *Dept of Civil & Environmental System Engineering, Konkuk University, 1 Hwayang-dong, Gwangjin-gu, 143-701 Seoul, South Korea*

Abstract This study is to assess the impact on runoff and non-point source pollution of a small rural watershed of future climate change scenarios using Soil and Water Assessment Tool (SWAT) model. The NIES (National Institute for Environmental Studies) MIROC3.2 high-resolution climate data by SRES (Special Report on Emissions Scenarios) A1B and B1 scenarios of the IPCC (Intergovernmental Panel on Climate Change) were adopted, and the data was downscaled by Change Factor method through bias-correction using 30 years (1977–2006) of weather data for three meteorological stations. As a model set-up, the model was calibrated for two years (1999–2000) using daily streamflow and monthly water quality (SS, T-N and T-P) data, and validated for another two years (2001–2002) based on Landsat land use of a 255.4 km^2 watershed. The future assessment was accomplished by preparing QuickBird land use data for a 1.21 km^2 sub-watershed located in the upstream watershed. The future impact on the runoff and non-point source pollution by the climate change was analysed from the simulated results of SWAT for the 2020s (2010–2039), 2050s (2040–2069) and 2080s (2069–2099). The future predicted annual streamflow decreased with a change of –16.5 % (2020s), –9.4 % (2050s) and –13.8 % (2080s) comparing with 2002 streamflow as a base year. For the future water quality, the future sediment load increased depending on surface runoff change in spring and summer seasons.

Key words high resolution satellite image; non-point source; SWAT; future climate change; GCM; downscaling

INTRODUCTION

Recently, the practical use of high spatial resolution imagery such as IKONOS, QuickBird and KOMPSAT (KOrea Multi-Purpose SATellite)-2 has been accommodated in various land-use-related application fields, including agriculture as well as the natural hazard and environmental assessment fields.

The environmental problem of non-point source (NPS) pollution in South Korea is becoming the big issue for the healthy watershed management. Accurate NPS evaluation will improve efficient watershed management which will eventually improve the streamwater quality.

Land use is the essential information in NPS assessment. The land cover affects the hydrological components such as evapotranspiration, infiltration and soil water storage, the dynamics of surface runoff, subsurface flow and groundwater recharge. The effects of land use are directly linked to changes in streamflow and NPS pollution load such as sediment, T-N (total nitrogen) and T-P (total phosphorus). Therefore, the precise and accurate information on land-use data will enhance the evaluation of NPS load from a watershed.

The Intergovernmental Panel on Climate Change (IPCC) report re-affirms that the climate is changing in ways that cannot be accounted for by natural variability and that “global warming” is occurring (IPCC, 2001). This global warming is likely to have significant impacts on the hydrological cycle (Arnell, 1999; IPCC, 2001).

SWAT (Soil Water Assessment Tool) is a continuous, lumped parameter model that uses physically-based and empirical relationships to simulate surface and subsurface flows, tree and crop growth, land management practices, and the fate and transport of water quality constituents (Mishra *et al.*, 2007). Among the various NPS evaluation models, SWAT has been widely applied to evaluate the future hydrological and water quality impacts of land management and agricultural practices (Borah & Bera, 2004; Arnold & Fohrer, 2005; USDA, 2006).

This study aims to assess the impact on runoff and non-point source pollution of a small rural watershed of future climate-change scenarios using the SWAT model with land-use data produced from QuickBird satellite imagery.

STUDY AREA DESCRIPTION

A 255.4 km^2 watershed located in the northwest of South Korea was selected for the SWAT basic set-up using Landsat land use as in Fig. 1(a). The NPS loads evaluation with future climate data was carried out for a 1.21 km^2 catchment located in the upstream of the watershed as in Fig. 1(b). The watershed stream is the tributary of the Han River basin directly linked to Paldang lake. The lake plays an important role for the municipal water supply for Seoul metropolitan city. Thus protection from non-point source pollution by agricultural cultivation becomes more important because of the recently raised consciousness of water quality problem. The watershed average annual temperature is 10.9°C and the average annual precipitation is 1371.1 mm over the last 30 years (1977–2006). Table 1 shows the watershed characteristics.

Table 1 Watershed characteristics.

Watershed	Area (km^2)	Mean elevation (m)	Mean slope (degree)
(a)	255.44	115.0	23.0
(b)	1.21	241.9	52.4

(a); model setup watershed, (b) future NPS loads evaluation watershed.

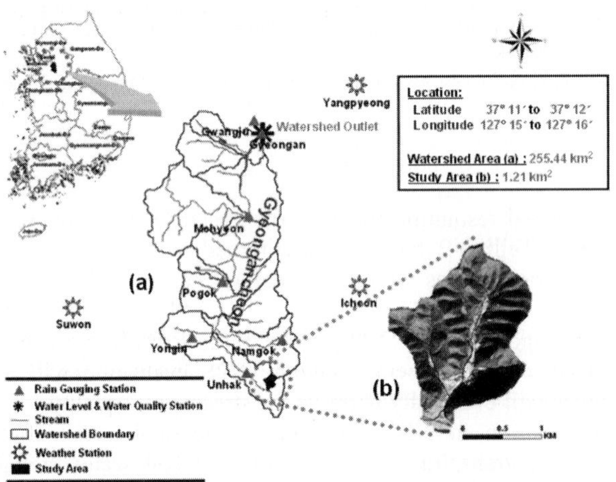

Fig. 1 The study watershed: (a) model setup watershed (255.4 km^2), and (b) climate change impact application watershed (1.21 km^2).

SWAT MODEL DESCRIPTION

The SWAT (Soil and Water Assessment Tool) is a physically-based and continuous, long-term, distributed-parameter model designed to predict the impact of land management practices on the hydrology, sediment and contaminant transport in agricultural watersheds with varying soils, land use, and management conditions (Arnold *et al.*, 1998). SWAT is based on the concept of hydrological response units (HRUs) which are portions of a sub-watershed that possess unique land use/management/soil attributes. The runoff, sediment, and nutrient loadings from each HRU are calculated separately using input data about weather, soil properties, topography, vegetation, and land management practices, and then summed together to determine the total loadings from the sub-watershed.

Soil water balance

To predict the runoff generation, SWAT uses a modified version of the SCS (Soil Conservation Service) CN (Curve Number) method (USDA-SCS, 1972). The hydrological cycle as simulated by SWAT is based on the water balance equation:

$$SW_t = SW_0 + \sum_{i=1}^{t}\left(R_{day} - Q_{surf} - E_a - W_{seep} - Q_{gw}\right) \tag{1}$$

where SW_t is the final soil water content (mm H_2O), SW_0 is the initial soil water content on day i (mm H_2O), t is the time (days), R_{day} is the amount of precipitation on day i (mm H_2O), Q_{surf} is the amount of surface runoff on day i (mm H_2O), E_a is the amount of evapotranspiration on day i (mm H_2O), W_{seep} is the amount of water entering the vadose zone from the soil profile on day i (mm H_2O), and Q_{gw} is the amount of return flow on day i (mm H_2O).

Sediment and nutrient transport by soil erosion

Erosion and sediment yield are estimated by using the Modified Universal Soil Loss Equation (MUSLE) method (Williams & Berndt, 1977), which uses the amount of runoff to simulate erosion and sediment yield (Mishra et al., 2007). USLE predicts average annual gross erosion as a function of rainfall energy. In MUSLE, the rainfall energy factor is replaced with a runoff factor. The modified universal soil equation (Williams, 1995) is:

$$Sed = 11.8(Q_{surf} \cdot q_{peck} \cdot area_{hru})^{0.56} K \cdot C \cdot P \cdot LS \cdot CFRG \tag{2}$$

where Sed is sediment yield on a given day (metric ton), Q_{surf} is surface runoff volume (mm/ha), q_{peak} is peak runoff rate (m^3/s), $area_{hru}$ is area of the HRU (ha), K is ULSE soil erodibility factor, C is USLE cover and management factor, P is USLE support practice factor, LS is USLE topographic factor, $CFRG$ is coarse fragment factor.

The model tracks nutrients dissolved in the stream and nutrients adsorbed to the sediment. Dissolved nutrients are transported with the water while those sorbed to sediment are allowed to be deposited with the sediment on the bed of the channel. Nutrients may be introduced to the main channel and transported downstream through surface runoff and lateral subsurface flow.

SWAT MODEL SET-UP

For the 255.4 km^2 watershed, three spatial input data (elevation, soil, and land use) were prepared (Fig. 2). The 30-m DEM (Digital Elevation Model) was produced using the 1:5000 vector data. The 1:25 000 soil information was obtained from Korea Rural Development Administration. The 30-m land use was produced using Landsat TM (Thematic Mapper) satellite image.

Using the four years (1999–2002) daily weather data from three weather stations (Suwon, Icheon, Yangpyeong) and seven AWSs (Automatic Weather Station) rainfall data, the SWAT model was calibrated for two years (1999–2000) using daily streamflow and monthly water quality (SS, T-N, T-P) records from 1999 to 2000, and validated for another two years (2001–2002).

Figure 3 shows the comparison results of streamflow and NPS loads, and Table 2 and Table 3 show the statistical summaries. For the validation, the Nash and Sutcliffe model efficiency (ME) for streamflow was 0.63, and the coefficients of determination (R^2) for SS, T-N and T-P were 0.88, 0.72 and 0.68, respectively.

FUTURE CLIMATE CHANGE SCENARIOS

The MIROC3.2 hires data and downscaled results

The MIROC3.2 high resolution GCM data by SRES (Special Report on Emissions Scenarios) climate change scenarios (A1B and B1) of the IPCC (Intergovernmental Panel on Climate Change)

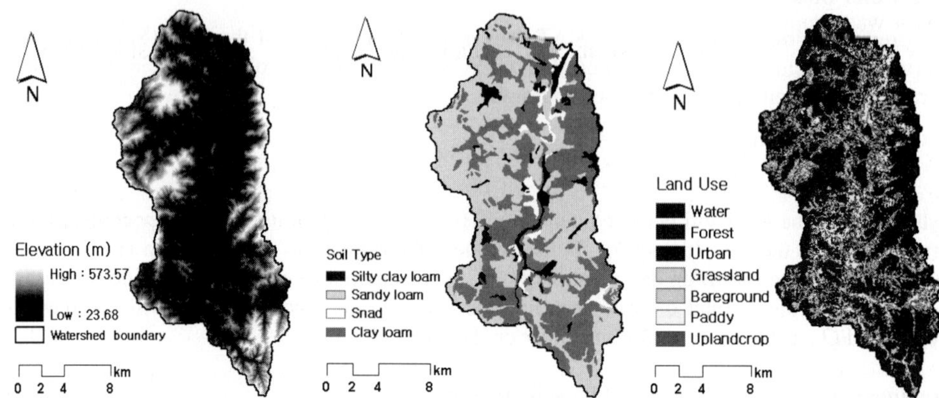

Fig. 2 GIS data: (a) elevation, (b) soil type and (c) land use.

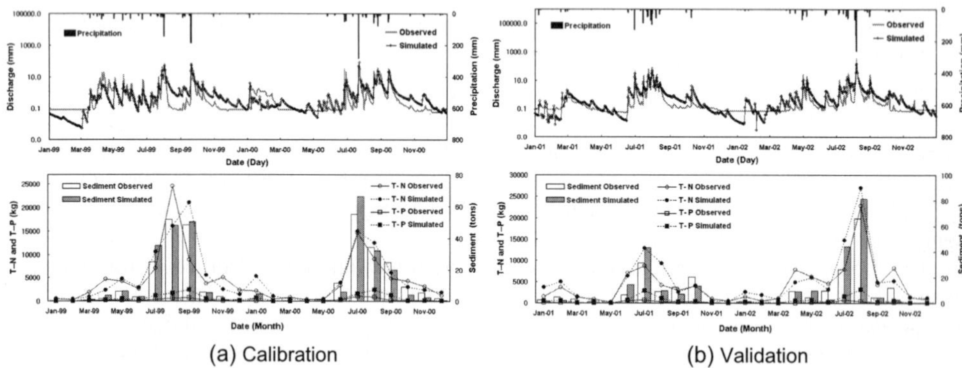

Fig. 3 Comparison between simulated and observed daily streamflow and monthly water quality (sediment, T-N and T-P) values during: (a) calibration, and (b) the validation period.

Table 2 Summary of streamflow calibration and verification.

Period		RMSE (mm)	ME	R^2
Calibration	1999	3.46	0.77	0.80
	2000	4.24	0.45	0.53
Verification	2001	4.52	0.66	0.80
	2002	7.10	0.63	0.71
Mean		4.83	0.63	0.71

RMSE: Root mean square error, ME: Nash-Sutcliffe model efficiency, R^2: Coefficient of determination.

Table 3 Summary of stream water quality (sediment, T-N and T-P) calibration and verification.

Period		R^2			RMSE		
		Sediment	T-N	T-P	Sediment	T-N	T-P
Calibration	1999	0.97	0.55	0.93	3.30	4.57	0.63
	2000	0.95	0.72	0.70	4.98	3.85	0.61
Verification	2001	0.64	0.72	0.63	0.03	5.16	0.69
	2002	0.95	0.89	0.44	0.01	3.60	0.88
Mean		0.88	0.72	0.68	2.08	4.30	0.70

AR4 was adopted. The data has an approximately 1.1° grid cell spatial resolution. The A1B is "middle" GHG (green house gas) emission scenario, and B1 is "low" GHG emission scenario.

The GCM data were downscaled by two steps for the study watershed. First, the GCM data was corrected to ensure that 30 years observed data (1977–2006, baseline period) and GCM model output of the same period have similar statistical properties. Secondly, the future data were downscaled using the Change Factor (CF) method (Diaz-Nieto & Wilby, 2005). The monthly mean changes in equivalent variables from the 30 years data and GCM simulations for three time periods: 2020s (2010–2039), 2050s (2040–2069) and 2080s (2070–2099) were calculated for the GCM grid cell. The percentage changes in monthly mean were applied to each day of the 2002 weather data (base year for future assessment) of each weather station. Table 4 show the future seasonal precipitation by CF downscaling method. The future precipitation shows a tendency of decrease in the summer season. Other seasons showed an increasing tendency on the whole. The biggest change of precipitation was –11.6 % in summer season of 2020s B1 scenario. The winter seasons showed the increase tendency. The biggest change of precipitation was +119.1 % in winter season of 2020s A1B scenario. The results showed a 24.1% precipitation increase in the case of A1B scenario and 15.9% precipitation increase in case of B1 scenario for 2080s.

Table 4 The changes in future seasonal precipitation by applying CF method.

Scenario	Precipitation change (mm)		Precipitation change (%)	
	A1B	B1	A1B	B1
Baseline (2002)	317.4	Spring (March–May)		
2020s	412.1	478.4	+29.8	+50.7
2050s	443.3	477.0	+39.7	+50.3
2080s	517.3	490.2	+63.0	+54.4
Baseline (2002)	957.3	Summer (June –August)		
2020s	881.9	846.1	–7.9	–11.6
2050s	878.9	897.6	–8.2	–6.2
2080s	903.0	894.3	–5.7	–6.6
Baseline (2002)	108.1	Autumn (September–November)		
2020s	146.9	171.5	+35.9	+58.6
2050s	212.8	187.1	+96.9	+73.1
2080s	229.2	170.9	+112.0	+58.1
Baseline (2002)	80.7	Winter (December–February)		
2020s	176.8	155.7	+119.1	+92.9
2050s	152.8	162.9	+89.3	+101.9
2080s	166.4	140.5	+106.2	+74.1

LAND USE PREPARATION USING QUICKBIRD SATELLITE IMAGERY

The QuickBird-2 satellite image has the spatial resolution of 0.61 m at perpendicular in the case of panchromatic and 2.44 m at perpendicular in the case of multi-spectrum. In this study, the QuickBird panchromatic and multi-spectral images of 1st May 2006 were used. It lies between the coordinates of latitude N 37° 11' 5" to N 37° 12' 0" and longitude E 127° 15' 46" to E 127° 16' 36". The image was ortho-rectified and geometrically corrected by using 2 m DEM from NGIS (National Geographic Information System) 1:5000 digital map and 30 GCPs (Ground Control Points) acquired from 1 m accuracy GPS (Global Positioning System) equipment (Fig. 2). The land use was produced by on-screen digitizing method with GPS field investigation data. The land use was classified with 26 categories (Fig. 4).

THE FUTURE CLIMATE CHANGE IMPACT ON HYDROLOGICAL BEHAVIOUR

Using the validated SWAT model, the model was applied to a 1.21 km^2 catchment with the QuickBird land-use data to assess the future hydrological impact first.

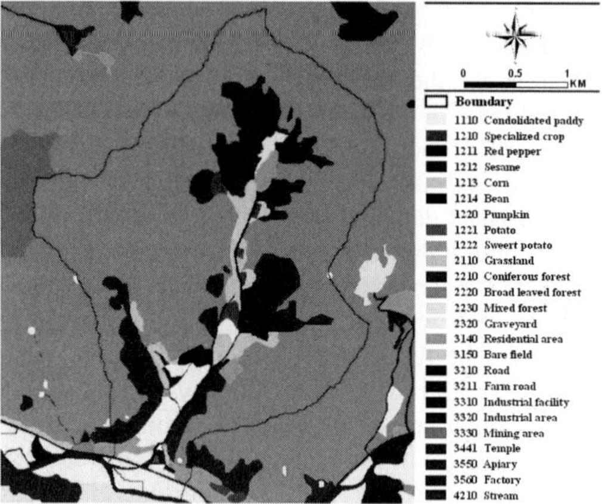

Fig. 4 The Classified land use from QuickBird satellite image.

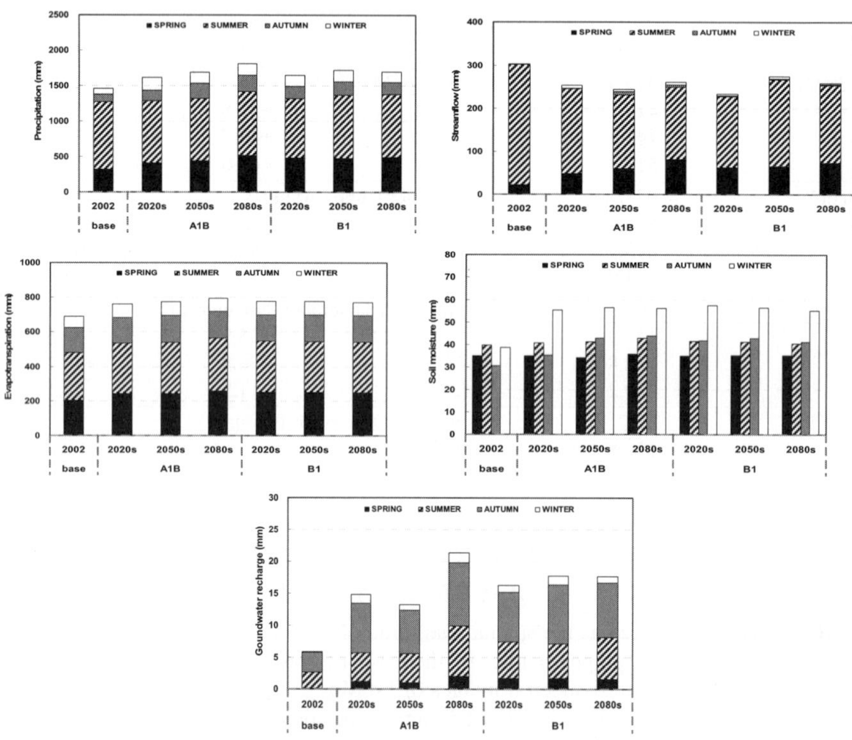

Fig. 5 The future MIROC3.2 hi-res impacts on hydrological components.

Figure 5 shows the predicted streamflow, evapotranspiration, soil moisture and groundwater recharge for the 2020s, 2050s, and 2080s based on 2002 data. Table 5 summarizes the future predicted annual hydrological components for the MIROC3.2 hi-res A1B and B1 scenarios. The 2020s annual streamflow were predicted to change by –16.5% and –23.1% under A1B and B1 scenarios, respectively. The 2080s annual streamflow were predicted to change by –13.8% and –14.6% under A1B and B1 scenarios, respectively. The future ET increased by +10.5 % to +15.5%.

Table 5 Summary of the future predicted hydrological components for MIROC3.2 hi-res A1B and B1 scenarios.

Scenario		P [%]	ET [%]	SW [%]	ST [%]	GW
Baseline	2002	1463.5	688.6	432.5	303.3	5.9
A1B	2020s	1617.7	760.6	499.2	253.2	14.8
		[+10.5]	[+10.5]	[+15.4]	[−16.5]	
	2050s	1687.8	774.7	525.0	244.2	13.2
		[+15.3]	[+12.5]	[+21.4]	[−19.5]	
	2080s	1815.9	795.0	535.7	261.4	21.4
		[+24.1]	[+15.5]	[+23.9]	[−13.8]	
B1	2020s	1651.7	778.0	528.4	233.3	16.3
		[+12.9]	[+13.0]	[+22.2]	[−23.1]	
	2050s	1724.6	777.1	527.9	274.9	17.7
		[+17.8]	[+12.9]	[+22.1]	[−9.4]	
	2080s	1695.9	771.5	516.2	259.0	17.6
		[+15.9]	[+12.0]	[+19.3]	[−14.6]	

P: Precipitation (mm), ET: Evapotranspiration (mm), SW: Soil Water (m^3/m^3), ST: Streamflow (mm), GW: Groundwater recharge (mm).

FUTURE CLIMATE CHANGE IMPACT ON NON-POINT SOURCE POLLUTION LOADS

Figure 6 shows the future monthly changes of sediment, T-N and T-P loads for MIROC3.2 hi-res A1B and B1 scenarios based on 2002 year, and Table 6 summarizes the results. The future sediment load showed the general tendency of increase in A1B scenarios. The nutrient (T-N and T-P) loads are often correlated with surface runoff and sediment transport rates (USDA-SCS, 1972). However, the fugitive sediment from the landscape is carried by overland flow (runoff), but the dominant pathway for nitrate loss is through leaching to groundwater, and then via baseflow or tile drains (Randall & Mulla, 2001). For the A1B scenarios, the annual T-N and T-P loads were predicted to change from +26.9 to +69.8%, and from +28.0 to +72.9%, respectively. For the B1 scenarios, the annual T-N and T-P loads were predicted to change from −62.7 to −68.4%, and from +38.3 to +43.9%, respectively.

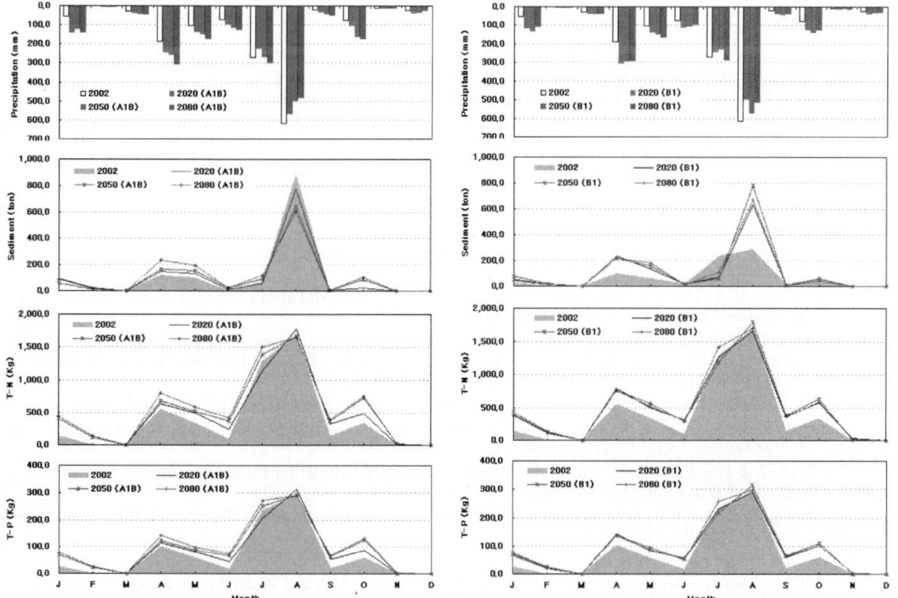

Fig. 6 The future predicted monthly NPS loads for MIROC3.2 hi-res A1B and B1 scenarios.

Table 6 Summary of the future predicted NPS pollution load for MIROC3.2 hires A1B and B1 scenarios.

Scenario		P (mm)	Sediment (ton)	Sediment variation (%)	T-N (kg)	T-N variation (%)	T-P (kg)	T-P variation (%)
Baseline	2002	1463.5	1009.2	-	3026.6	-	527.1	-
A1B	2020s	1617.7	1050.4	+ 4.1	3840.1	+ 26.9	674.8	+ 28.0
	2050s	1687.8	1024.3	+ 1.5	5137.8	+ 69.8	911.4	+ 72.9
	2080s	1815.9	1109.3	+ 9.9	4883.2	+ 61.3	857.9	+ 62.7
B1	2020s	1651.7	955.6	– 5.3	955.6	– 68.4	729.2	+ 38.3
	2050s	1724.6	1128.8	+ 11.8	1128.8	– 62.7	758.6	+ 43.9
	2080s	1695.9	1040.3	+ 3.1	1040.3	– 65.6	747.4	+ 41.8

CONCLUSIONS

This study tried to assess the future climate impacts of runoff and non-point source pollution of a small rural watershed using MIROC3.2 hi-res A1B and B1 scenarios and the SWAT model incorporating QuickBird land use data. For the 1.21 km^2 agricultural watershed, the QuickBird land use composed of 26 categories data were prepared. The impervious area of the watershed detected in QuickBird land use affected runoff volume, sediment, T-N, and T-P. The precise land use and topography can give more detailed information for watershed management.

KOMPSAT (KOrea Multi-Purpose SATellite)-3, that will have spatial resolutions of 0.8 m panchromatic and 2.8 m multi-spectral images, was scheduled to launch in 2008. KOMPSAT-3 imagery can produce USGS (United States Geological Survey) Level IV (0.25–1.0 m spatial resolution) land-use data. It is expected that the data can be used to identify detailed hydrological cycle, soil erosion process, sediment and pollutant transport mechanisms.

Acknowledgements This study was funded by the research project "Development of Extraction and Analysis Technique of Detailed Agricultural Information (M104DA010003-08D0100-00119)" of Ministry of Education Science and Technology remote sensing technology development program.

REFERENCES

Ahn, J. H., Yoo, C. S. & Yoon, Y. N. (2001) An analysis of hydrologic changes in Daechung Dam basin using GCM simulation results due to global warming. *J. Water Resour. Assoc.* **34**(4), 335–345 (in Korean).

Arnell, N. W. (1999) Climate change and global water resources. *Global Environmental Change* **9**, S31–S49.

Arnold, J. G., Srinivasan, R., Muttiah, R. S. & Williams, J. R. (1998) Large area hydrologic modeling and assessment part I: model development. *J. Am. J. Water Resour. Assoc.* **34**(1), 73–89.

Arnold, J. G. & Fohrer, N. (2005) SWAT2000: Current capabilities and research opportunities in applied watershed modelling. *Hydrol. Processes* **19**(3), 563–572.

Borah, D. K. & Bera, M. (2004) Watershed-scale hydrologic and nonpoint-source pollution models: review of applications. *Trans. ASAE.* **47**(3), 789–803.

Caitcheon, G. G. (1998) The significance of various sediment magnetic mineral fractions for tracing sediment sources in Killimicat Creek. *Catena* **32**(2), 131–142.

Diaz-Nieto, J. & Wilby, R. L. (2005) A comparison of statistical downscaling and climate change factor methods: Impacts on low flows in the River Thames, UK. *Climate Change* **69**, 245–268.

IPCC (2001) Climate Change 2001: Scientific Basis. In: *Contribution of Working Group III to the Third Assessment Report of the Intergovernmental Panel on Climate Change* (ed. by B. Metz *et al.*). Published for the Intergovernmental Panel on Climate Change. Cambridge University Press, Cambridge, UK, New York, USA.

IPCC (2006) IPCC Data Distribution Centre. Available at: ipccddc.cptec.inpe.br/ipccddcbr/html/cru_data/support/faqs.html.

Kim, S. H., Lee, M. S., Park, G. A. & Kim, S. J. (2007) Application of QuickBird satellite image to storm runoff modeling. *Korean J. Remote Sensing* **23**(1), 15–20 (in Korean).

Mishra A., Froebrich, J. & Gassman, P. W. (2007) Evaluation of the SWAT Model for assessing sediment control structures in a small watershed in India. *Trans. ASABE.* **50**(2), 469–477.

Randall, G. W. & Mulla, D. J. (2001) Nitrate nitrogen in surface waters as influenced by climatic conditions and agricultural practices. *J. Environ. Quality.* **30**, 337–344.

USDA-SCS. (1972) *National Engineering Handbook*, Section 4 Hydrology 1972 (chapters 4–10).

USDA (2006) SWAT: Peer-reviewed publications. USDA-ARS Grassland, Soil and Water Laboratory, Temple, Texas, USA. Available at: www.brc.tamus.edu/swat/pubs_peerreview.html.

Multi-objective optimization analysis for the Canberra water supply system

LIJIE CUI & GEORGE KUCZERA

School of Engineering, University of Newcastle, Australia

lijie.cui@newcastle.edu.au

Abstract Planning and management of urban water supply headworks systems is a complex and difficult task. Typically, such systems have not only multiple users with different objectives and risk tolerances, but also multiple sources with different levels of quality. This complexity gives rise to an incredibly large number of infrastructure and operating policy options. This is often complicated by the existence of multiple competing objectives, whereby a gain in one particular objective may result in a loss in another objective. The solution to these problems requires simultaneous consideration of conflicting objectives. Multi-objective optimization deals with the process of simultaneously optimizing two or more conflicting objectives subject to certain constraints. The genetic algorithm is a highly suitable technique for solving multi-objective optimization problems. This study demonstrates the applicability of multi-objective optimization method, in particular, the ε-dominance multi-objective evolutionary algorithm (εMOEA) for an urban water supply system.

Key words multi-objective optimization; genetic algorithm; water supply

INTRODUCTION

Most realistic optimization problems, particularly those in design, require the simultaneous optimization of more than one objective function. This is because, in these problems, one is never satisfied by finding one solution that is optimum with respect to a single criterion. Conflicting objectives introduce trade-off solutions and make the task complex yet interesting to execute. Making decisions for management of urban water supply headworks systems is a complex and difficult task. Typically, such systems have not only multiple users (urban, irrigation, and environmental) with different objectives and risk tolerances, but also multiple sources with different levels of quality. This complexity gives rise to an incredibly large number of infrastructure and operating policy options. The solution to these complex decision problems requires the use of mathematical techniques that are formulated to take simultaneous consideration of conflicting objectives. Furthermore, there is, in effect, a large element of uncertainty and risk in all water supply decisions due to hydrological, climatic and anthropogenic uncertainty, and the inability to predict the future with reasonable accuracy.

The classical way of tackling multi-objective problems (MO) is to convert multiple objectives into one objective and solve for the optimal solution. Since multiple objectives are converted into a single objective, the resulting optimal solution is usually subject to the parameter settings used for the conversion. Moreover, since classical single objective optimization methods are used to solve the converted problem, only one solution can be found in one simulation run. Furthermore, the classical methods have been found to be sensitive to the convexity and continuity of the Pareto-optimal region. Most of these drawbacks can be eliminated by using multi-objective genetic algorithms. For a multi-objective problem there exists a set of alternative solutions (Pareto optimal solution set) for which there are no other solutions that are superior when all objectives are simultaneously considered. Hence, optimizing a multi-objective problem is comprised of finding Pareto optimal solutions.

Multi-objective optimization problems have received increased interest from researchers with various backgrounds since early 1960. Over the past years there has been interest in applying genetic algorithms to solving multi-objective optimization problems. One of the striking differences between classical search methods and GAs is that the GA works with populations of solutions instead of only one solution, which makes it naturally well-suited for multi-objective problems, in particular, for searching intractably large, poorly understood problem spaces. The first implementation of a multi-objective evolutionary optimization algorithm (MOEA) dates back to the mid-1980s (Schaffer, 1984). Goldberg (1989) suggested a new non-dominated sorting procedure using the concept of domination to give preference to non-dominated individuals in the population. Since then, research interests in this field have remained strong and a variety of multi-objective optimization techniques have been developed.

This study aims to demonstrate the applicability of multi-objective optimization method for an urban water supply system. It first describes the principles of the multi-objective optimization. It is followed by a case study involving the Canberra system in Australia, demonstrating the application of the multi-objective objection procedure. Finally, the results and discussions are presented and conclusions are drawn.

CANBERRA WATER SUPPLY SYSTEM AND ITS SIMULATION

WATHENT simulation of the Canberra water supply system

A case study of the Canberra headworks water supply system was undertaken to investigate the applicability and efficacy of multi-objective optimisation. Canberra is the capital city of Australia, with a population of over 340 000. ACTEW is the authority to be responsible for Canberra's water supply. This section describes the Canberra system and WATHNET simulation model used to demonstrate the concept. At the regional scale, demand for water within the Canberra precinct can be represented by a single node. In the study, the population of Canberra has been increased by 75% and 100% to enable investigation of a system in a stressed state. Releases from the reservoirs have to meet not only the consumptive needs of the Canberra urban area, but also environmental flow requirements defined in ACTEW's operating licence. Downstream of each reservoir, there are several requirements based on maintenance of base flow, pool and riffle flows. During periods of restriction, these requirements are relaxed, depending on the severity of the drought.

WATHNET (Kuczera, 1992) is an example of a generalised simulation model that departs from the traditional approach to system operation. It uses a network linear program (NetLP) to simulate the operation of a wide range of water supply headworks configurations. Instead of using explicit rules to make water assignments, it uses information about the current state of the system as well as forecasts of streamflow and demand to formulate a network linear program. In a single time step, the NetLP determines the water allocation for given streamflow and demand in accordance with the following hierarchy of objectives:

1. satisfy demand at all demand zones;
2. satisfy all instream flow requirements;
3. ensure that reservoirs are at their end-of-season target volumes;
4. minimise delivery costs; and
5. avoid unnecessary spill from the system.

A WATHNET model of the Canberra system was constructed and is depicted in Fig. 1. The model is based on the REALM model developed by ACTEW, but has only implemented the physical constraints affecting the system. Operating rules were not constrained as in the ACTEW REALM model to allow the optimisation search engine maximum flexibility to search the decision space. The ACTEW REALM model reflects rules that have been tuned to give good performance. In this case study, the objective is to demonstrate the potential of a generic optimization approach to identify decisions (or rules) which lead to good performance.

The Canberra system was simulated with the WATHNET model using monthly streamflow data for the period 1871 to 2008. During this period several major droughts were experienced. These droughts are best appreciated by examining the time series of total storage for a typical simulation shown in Fig. 2. The use of total storage time series is a particularly effective visual tool for highlighting drought because storage is sensitive to the cumulative effect of persistent below-average streamflow conditions. Figure 2 identifies several major droughts, the Federation drought in the early 1900s, 1940 to 1950, 1976 to 1982 (the severest drought) and the current drought commencing around 2000.

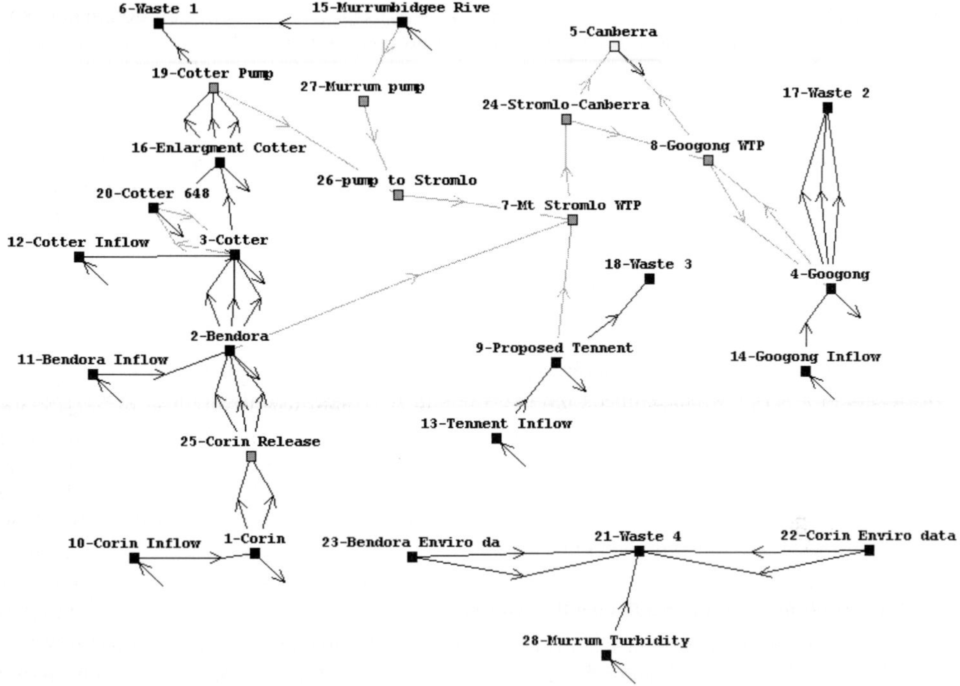

Fig. 1 WATHNET schematic diagram of Canberra water supply system.

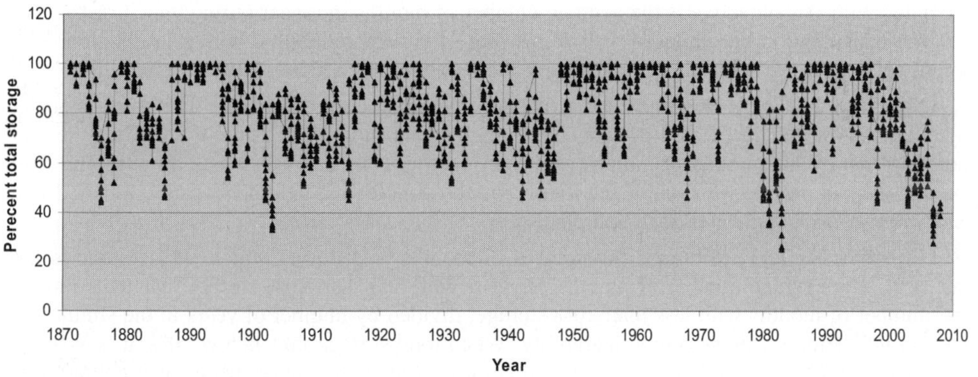

Fig. 2 Total storage (%) time series for a typical WATHNET simulation.

The version of WATHNET used in this case study was adapted from the full version. The adaptation involved consolidating and simplifying WATHNET into two parts: the simulation engine which is data file driven, and the graphical user interface, which allows construction of the network in a controlled manner and simple visualisation of results. This architecture simplified the interfacing of WATHNET with the multi-criterion search engine.

The schematic in Fig. 3 describes the relationship between the WATHNET simulation model and the optimization engine. The WATHNET graphical user interface (GUI) engine is used to create a network which is saved in a WATHNET data file. An important part of this process is to develop scripts which evaluate the performance of the system. It also embeds the number of decisions to be optimised, the number of criteria to be optimised. For example, a script may compute the total operating cost or the frequency of restrictions during a simulation.

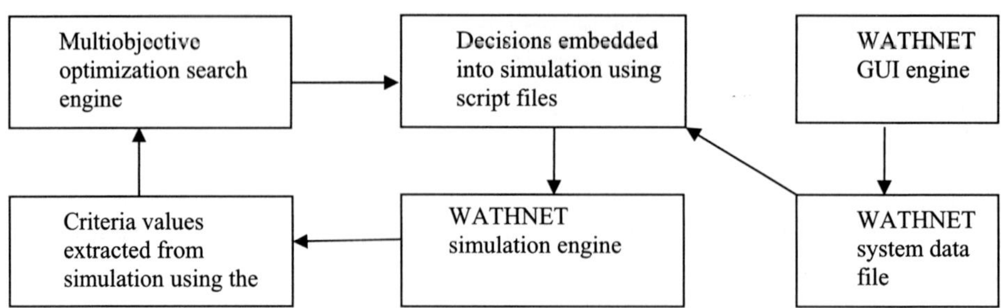

Fig. 3 Interface between WATHNET and the optimization search engine.

The multi-objective optimization engine implements the search for Pareto frontier. It passes decisions to a script which embeds the decisions in the system configuration. WATHNET then performs the simulation. During the simulation, the user-defined script monitors the performance of the system and calculates objective functions values which are passed to the search engine. The search engine continues to iterate through the loop trialling different decision vectors until convergence has been achieved. This represents a generic model-independent protocol for interfacing the search and simulation engines.

Optimization objectives and decision variables

The operation of the Canberra headworks system can be guided by several objectives. In this study, we consider the following three objectives to demonstrate the potential of multi-criterion optimization:

Objective 1 (*Obj1*) Minimize the expected number of restrictions (months/year): The criterion associated with this objective is the average number of months in a year during which restrictions on water consumption are imposed. It is a measure of system reliability, with a lower number of months spent in water restrictions indicating that the system has a higher level of reliability.

Objective 2 (*Obj2*) Minimize the expected operating cost ($/month): The total expected operating cost of the system is the average running cost per month calculated by dividing the total operating cost by the number of months of the simulation. It includes the costs for pumping from Cotter Reservoir and the Murrumbidgee River, and the transfer and treatment costs associated with Bendorra and Googong water treatment plants.

Objective 3 (*Obj3*) Minimize the number of months/year during which total storage is less than 20%. The average number of months per year with less than 20% storage, calculated as the total number of months with less than 20% storage, divided by number of years in the simulation, gives an indication of the system vulnerability. When total storage falls below 20%, it is assumed the system enters crisis mode with the need to employ contingency measures to secure alternative sources of water and to impose more draconian rationing measures. If the system runs out of water and no viable alternative sources are found, the city of Canberra would face economic and social collapse.

There are three decisions variables for this case study:

Decision Variable 1 (*Decision1*) Googong base reservoir gain – The base reservoir gain in the Googong reservoir refers to the negative cost for the first storage carryover arc from Googong reservoir. The storage carryover arcs are used to enable setting of storage targets within the reservoirs. The cost of storage carryover arc j is given by the following equation:

$$Cost(j) = BG + (j-1) * IG, j = 1,\ldots,n \tag{1}$$

where *BG* is the base reservoir gain, *IG* is the incremental reservoir gain, and *n* is the total number of storage carryover arcs. The gain assigned to each carryover arc affects the way water is stored in the reservoirs. For example, if carryover gains for Googong are higher than for Cotter, water will be preferentially stored in Googong reservoir.

Decision Variable 2 (*Decision2*) Googong incremental reservoir gain – The incremental reservoir gain represents the incremental gain added to each successive storage carryover arc. IG in conjunction with BG controls the spatial distribution of target storage.

Decision Variable 3 (*Decision3*) Canberra level 1 trigger $level_1$ – The Canberra level 1 trigger is the combined storage that activates the first level of water restrictions. By imposing restrictions on water consumption, the rate of strong drawdown is reduced during drought. This increases the chance of the system surviving the drought.

In this study the εMOEA implementation described by Jefferson *et al.* (2005) and Cui & Kuczera (2006) was used.

MULTI OBJECTIVE OPTIMIZATION

According to Deb *et al.* (2003), multi-objective optimization has two fundamental goals: guiding the search towards finding a non-dominated set of solutions as close as possible to the Pareto optimal front, and keeping a diverse set of non-dominated solutions. εMOEA was one of the recent advances using the concept of the ε-dominance to achieve the above goals (Laumanns *et al.*, 2002; Kollat & Reed, 2006). The concept of ε-dominance allows the user to specify the precision with which they want to quantify each objective in a multi-objective problem. Figure 4 demonstrates the concept of ε-dominance using a three step approach for a two-objective minimization problem. First, a user specified ε grid is applied to the search space of the problem. Larger ε values result in a coarser grid (and ultimately fewer solutions) while smaller ε values produce a finer grid. Grid blocks containing multiple solutions are then examined and only the solution closest to the bottom left corner of the block is kept (assuming minimization of all objectives). In the second step, non-domination sorting based on the grid blocks is then conducted resulting in a "thinning" of solutions (step 3) and promoting a more even search of the objective space. ε-dominance allows the user to define objective precision requirements that make sense for their particular application. The interested reader can refer to Laumanns *et al.* (2002) and Deb *et al.* (2003) for a more detailed description of ε-dominance.

The basic idea of the εMOEA is to divide the search space into a number of grids (hyperbox). Diversity is maintained by ensuring that a grid can be occupied by only one solution. In εMOEA, there are two co-evolving populations: an EA (evolutionary algorithm) population (Reed *et al.*, 2003) and a ε-dominance archiving population. The εMOEA begins with an initial population P(0). The archived population E(0) is assigned with the ε-dominance solutions of P(0). At generation *t*, parents (designs), one each from both the population and the archive respectively, are chosen for mating. To choose one from the P(*t*–1), three parents are picked randomly and the one which is dominant is selected for mating. The parent from the E(*t*–1) is simply chosen at random

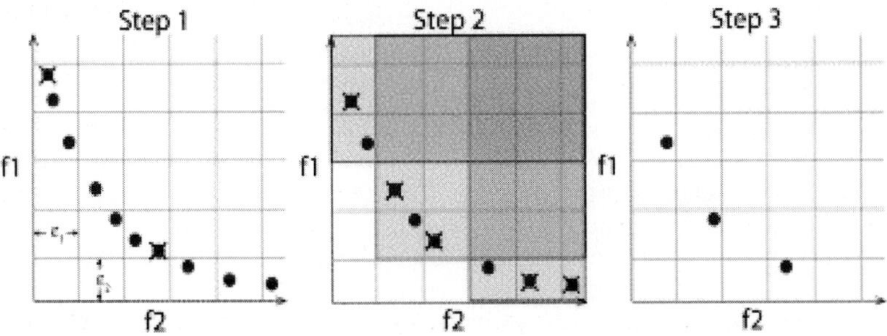

Fig. 4 Example illustrating the concept of ε-dominance (from Kollat & Reed, 2006).

among the archive members. An offspring is then produced and evaluated. For its inclusion in the population, three scenarios exists: (1) if the new solution dominates any designs which already exist, it replaces one at random; (2) if it is dominated by any existing designs, it is rejected; (3) if it is non-dominated with respect to the existing design, it replaces a random member of the population. For its inclusion in the archive, there are also three scenarios: (1) if the new design is ε-dominated by any design in the archive, it is not accepted; (2) if it ε-dominates any member of the design, it randomly replaces a dominated design; (3) if the new design is ε-non-dominated, and if it does not occur within any of the archive design's hyperboxes, it is accepted. Otherwise the two designs occurring in the same hyperbox are compared and the best one with respect to domination in the traditional sense is accepted. The size of the archive is inherently bounded by the user specified ε resolution of the objective. This process is repeated until termination. The archive members at termination are declared to be the final Pareto optimal solutions for the given problem.

RESULTS

In this study, εMOEA is parameterized according to the most commonly recommended settings from the literature (Deb *et al.*, 2003). The probabilities of crossover and mutation were set to $P_{cross} = 0.9$, $P_{mutate} = 0.05$, respectively, and the population size was assigned to 100. The εMOEA stops if it reaches 5000 evaluations. The ε values for the *objF1*, *objF2* and *objF3* were assigned values of 0.05, 50 and 0.01, respectively, for bounding the size of the archive hypercubes. For each case discussed below, two runs were performed with the same GA parameter settings: Scenario 1 involves demand corresponding to a population 175% of the current level, and Scenario 2 involves a demand corresponding to a 200% population.

Results for case 1

Case 1 involves two decisions and two objectives. *Decision* 1 is the Googong base reservoir gain and *Decision* 2 is Googong incremental reservoir gain. These two decisions allow the optimization to identify the best drawdown policy for operating Googong and Corin reservoirs. The base and incremental gain for Corin are fixed. By varying the base and incremental gain for Googong, the optimiser can preferentially draw down either Googong or Corin during drought. This allows assignment of air space to the reservoir most likely to benefit from inflow and hence minimizing spill from the system. Objective 1 minimizes the time spent in restrictions and Objective 2 minimizes the expected operating cost.

Figure 5 shows the εMOEA performance for every 1000 evaluations for Scenario 1 involving 175% population level. It is seen that after the 2000 evaluations, εMOEA is able to rapidly move

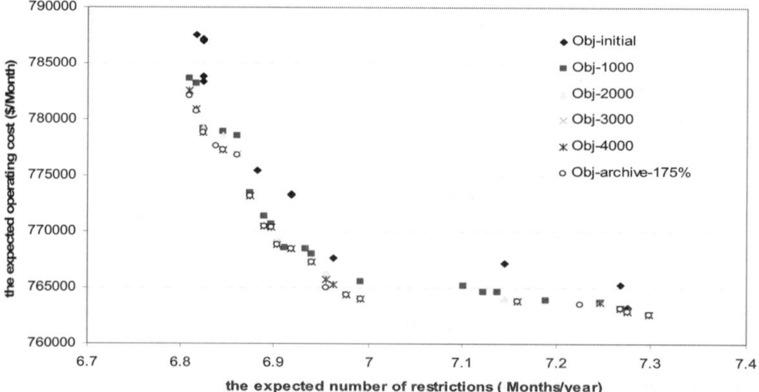

Fig. 5 εMOEA Pareto front for scenario 1 (175% population).

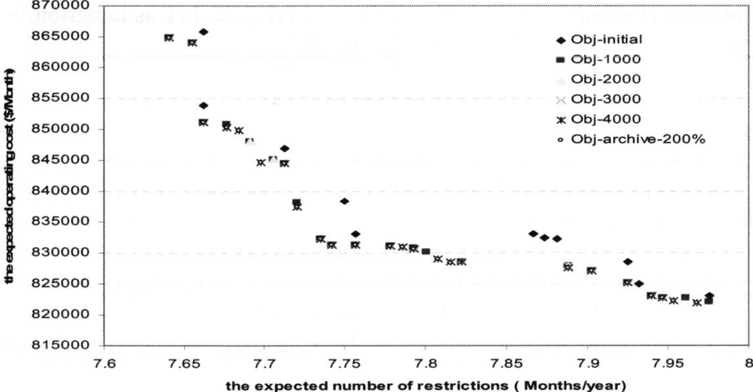

Fig. 6 εMOEA Pareto front for scenario 2 (200% population).

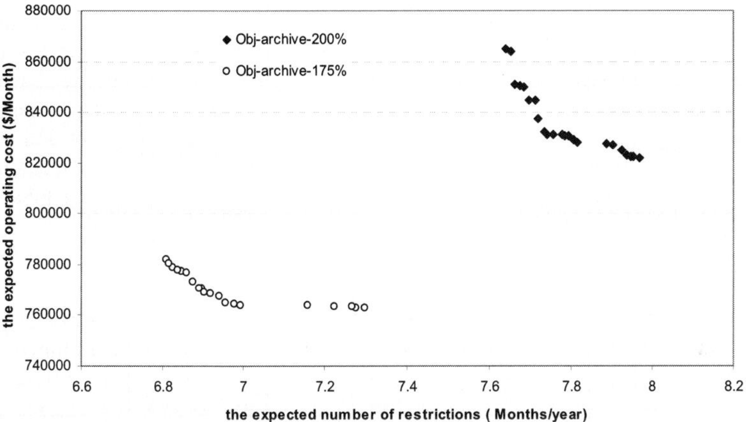

Fig. 7 Comparison of the Pareto fronts for scenarios 1 and 2.

towards the Pareto optimal front. Figure 6 shows similar results for population level 200%. These plots illustrate the trade-off along the Pareto front with higher frequency of restrictions leading to lower operational cost. One striking result is the limited range in restriction frequency and operating cost. It highlights the insensitivity of system performance to variation in storage targets.

Figure 7 compares the Pareto optimal solutions for scenario 1 and 2. For scenario 2 with the higher demand, the Pareto front shifts upward with higher operating costs and imposition of restrictions.

For all the results after each 1000 evaluations, the Euclidean distance between each result and the coordinate origin (which corresponds to the ideal solution of zero cost and zero restrictions) was calculated. Figure 8 presents the histograms for the distribution of Euclidean distances for all non-dominated solutions after every 1000 evaluations for scenario 1. It shows how εMOEA evolves in the search space for scenario 1. It is seen that after 3000 evaluations, the distribution of distances is very similar and hence the Pareto solutions will not improve much.

Results for Case 2

Case 2 involves three decisions and three objectives:
Decision1 – Googong base reservoir gain;
Decision2 – Googong incremental reservoir;
Decision3 – Canberra level 1 restriction trigger level.

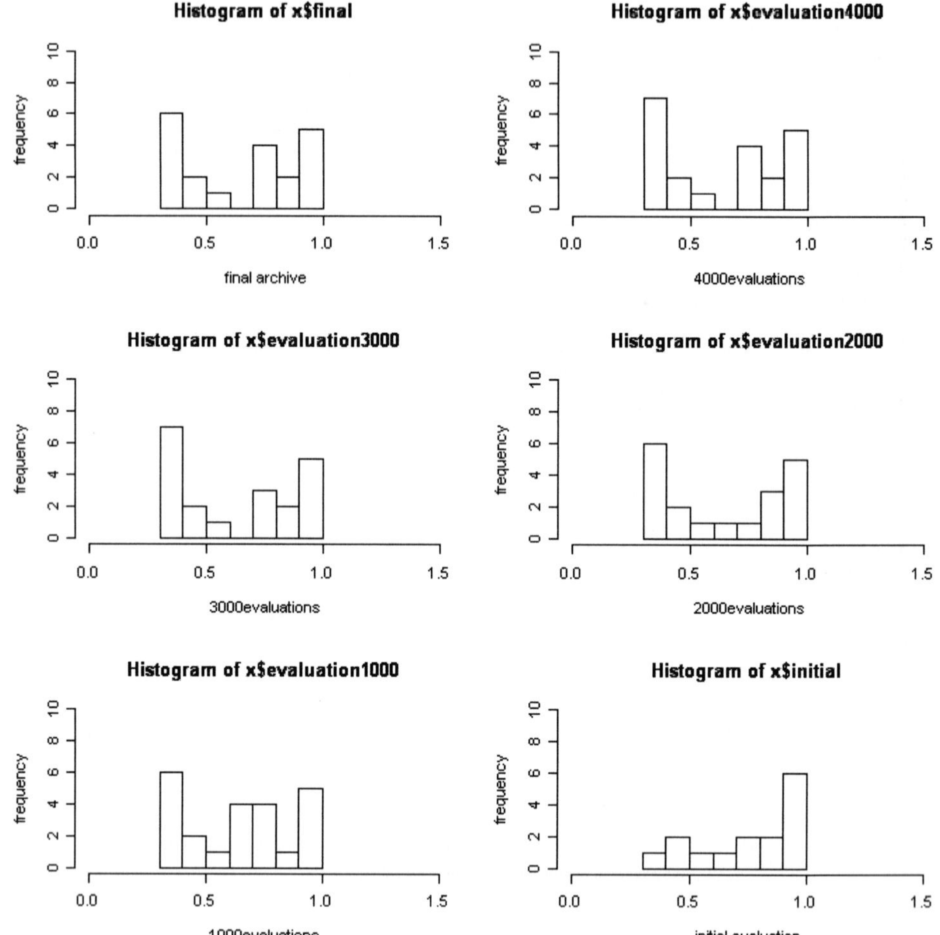

Fig. 8 Histograms of the Euclidean distances for every 1000 evaluations for scenario 1.

Obj1 – Minimize the expected number of restrictions (months/year)
Obj2 – Minimize the expected operating cost ($/month)
Obj3 – Minimize the average number of months/ year that total storage is less than 20%.

To aid visualization, Figs 9 to 11 show the 2-D projections of Pareto fronts. While Fig. 12 presents an isometric perspective of the 3-D Pareto front. Two significant observations can be made:

- The inclusion of the restriction trigger level as a decision has produced a much wider range of restriction frequencies and operating costs. This highlights the fact that performance of the system is quite sensitive to choice of restriction and only mildly sensitive to variations in BG and IG.
- In general, for three objectives, one would expect the Pareto front to be a 2-D surface. However, Figs 9 to 11 show strong trade-offs between all pairs of criteria on the Pareto front. What is of particular interest is that if the system operates with a high restriction trigger keeping the system under restrictions for six or more months every year, the vulnerability of the system is maintained at low levels regardless of the population scenario.

Figure 12 reveals that the Pareto front collapses to a 1-D thread snaking through the 3-D criterion space. This arises because the criteria are complimentary. Increasing frequency of restrictions reduces demand and thus reduces operating costs. Likewise, increasing the frequency

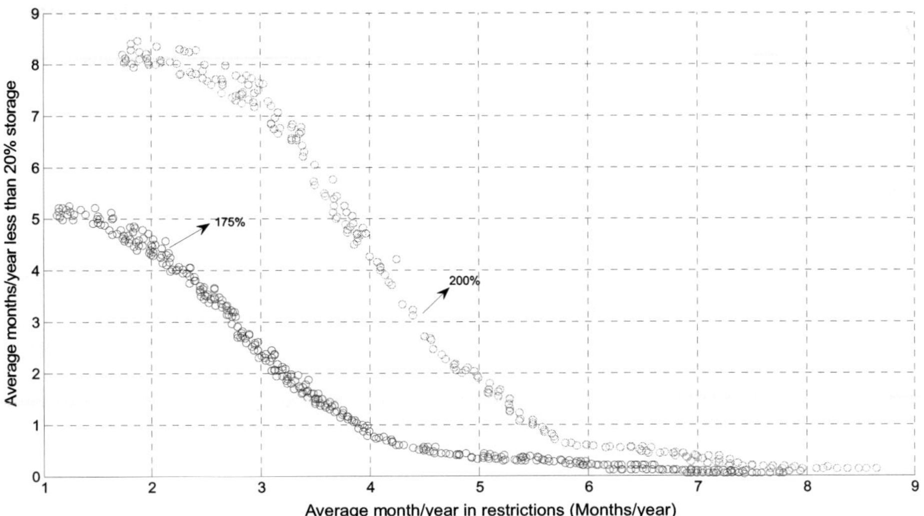

Fig. 9 *Obj1* and *Obj3* projection of Pareto front for scenarios 1 and 2.

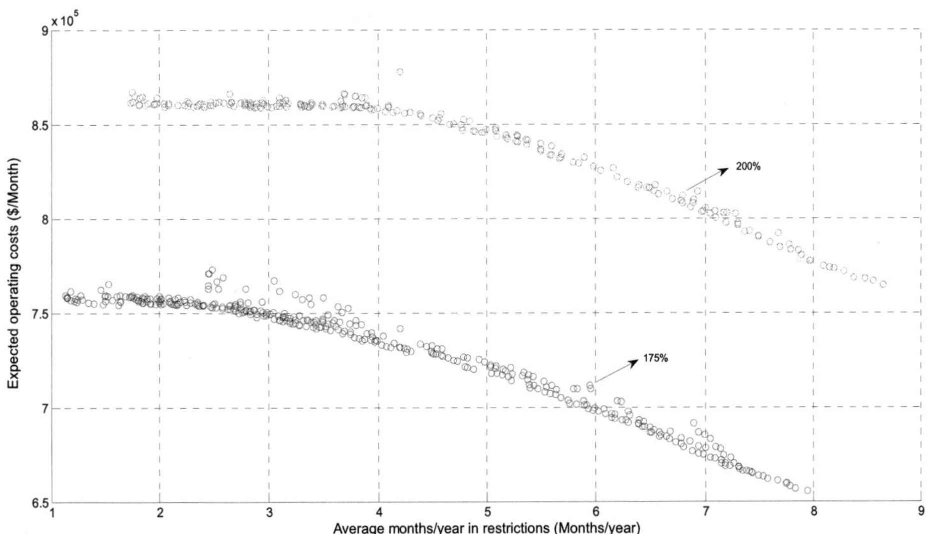

Fig. 10 *Obj1* and *Obj2* projection of Pareto front for scenarios 1 and 2.

of restrictions will reduce the rate of reservoir drawdown and hence reduce the chance of total storage dropping below 20%.

CONCLUDING REMARK

The results reported in this study demonstrate the very significant potential of the generic optimization to optimize the water supply system. The search engine had no knowledge of the way the WATHNET model operated the headworks system. It communicated with WATHNET via

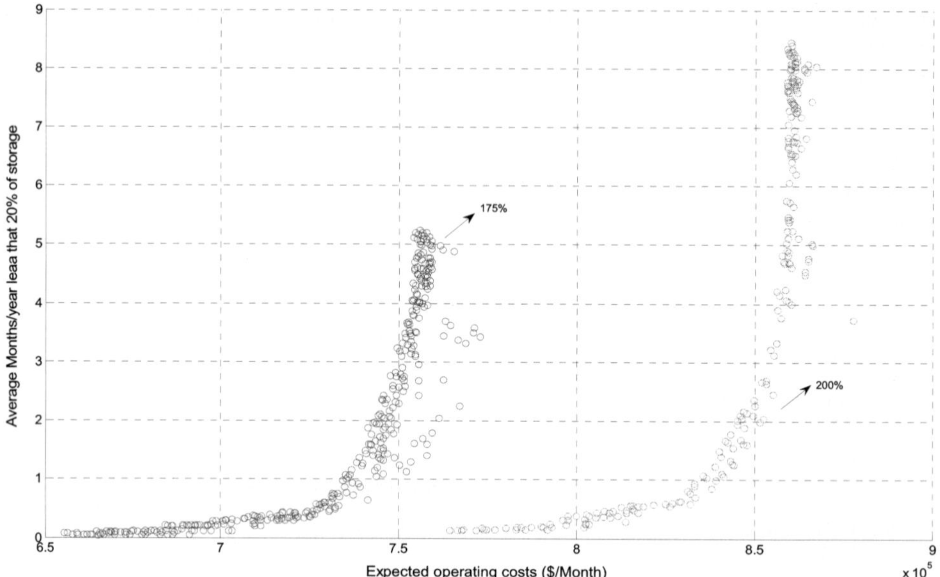

Fig. 11 *Obj2* and *Obj3* projection of Pareto front for scenarios 1 and 2.

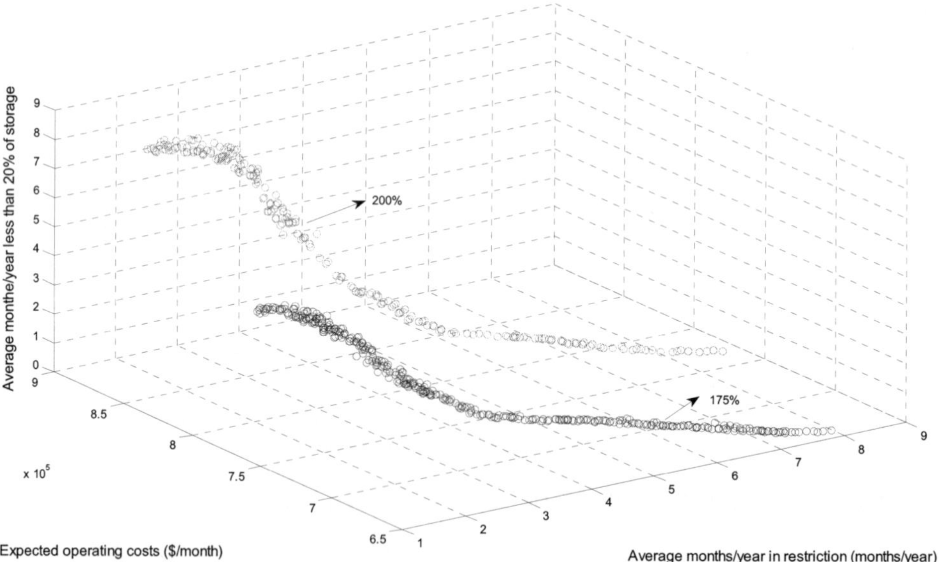

Fig. 12 Pareto front for scenarios 1 and 2.

scripts files. The independence of the search engine from the simulation model means that the optimization framework can be placed over a wide range of simulation models. Provided sensible objectives and decision variables are specified, the search engine will generate Pareto fronts that represent good solutions, which should be the focus of the decision making and trade-off process. Importantly, it will enhance productivity by freeing modellers from the tedious and inefficient trial-and-error search for good decisions.

In addition, future work will include the comparison of two widely-used multi-criterion optimization methods, NSGAII and εMOEA. Finally, two of the three objectives are risk variables associated with drought risk. It will be necessary to use stochastic hydroclimate data to ensure these risks are accurately estimated. This will present major computational challenges.

Acknowledgements The contributions of eWater PhD scholars, Andrew Graddon and Mohammad Mortazavi, are acknowledged. As part of their studies, they assisted in adapting WATHNET for this case study.

REFERENCES

Cui, L. & Kuczera, G. (2006) Application of multi-objective optimization for water supply system decision support using parallel genetic algorithms. *Hydrological Sciences for Managing Water Resources in The Asian Developing World and 2nd GWSP Asia Network Workshop*. Guangzhou, June 8–10, China.

Deb, K., Mohan, M. & Mishra, S. (2003) A Fast multi-objective evolutionary algorithm for finding well-spread Pareto-Optimal solutions. KanGAL Report no. 2003002, Indian Institute of Technology, Kanpur, India.

Goldberg, D. E. (1989) Genetic Algorithms in Search, Optimization and Machine Learning, *Addison-Wesley Longman Publishing Co., Inc.,* Boston, Massachussetts, USA.

Jefferson, C., Holz, L., Hardy, M. & Kuczera, G. (2005) Integrated urban water management: combining multi-criterion optimization and decision analysis. In: *Proceedings of the 2005 ASCE International Conference on Computing in Civil Engineering* (July 2005, Cancun, Mexico), 1765–1776.

Kollat, J. B. & Reed, P. M. (2006) Comparing state-of-the-art evolutionary multi-objective algorithms for long-term groundwater monitoring design. *Adv. Water Resour.* **29**(6), 792–807.

Kuczera, G. (1992) Water supply headworks simulation using network linear programming. *Adv. Engineering Software* **14**. 55–60.

Laumanns, M., Thiele, L., Deb, K. & Zitzler, E. (2002) Combining convergence and diversity in evolutionary multiobjective optimization. *Evol. Comput.* **10**, 263–282.

Reed, P., Minsker, B. S. & Goldberg, D. E. (2001) A multiobjective approach to cost effective long-term groundwater monitoring using an Elitist Nondominated Sorted Genetic Algorithm with historical data. *J. Hydroinform.* **3**, 71–90.

Schaffer, J. D. (1984) Some experiments in machine learning using vector evaluated genetic algorithms. Doctoral Thesis, Vanderbilt University, Nashville, Tennessee, USA.

Development of a decision-support system for enhanced operational management of the reservoirs of Bhakra and Beas dams

CLAUS SKOTNER, LARS-CHRESTEN EKEBJERG & RAJEEV BANSAL

1 *Water Resources Department, DHI, Denmark*
cso@dhigroup.com
2 *DHI Solutions, DHI, Denmark*
3 *BBMB, Chandigarh, India*

Abstract This paper describes the design, development and application of a new type of decision-support system to assist dam operators in their effort to increase hydropower production, ensure water for irrigation, and mitigate flood risk. In a robust and resilient setting, the system applies remotely sensed data on snow cover, point telemetry data, and numerical weather forecast data to drive advanced hydrological and hydraulic forecasting tools, forecast optimization algorithms and uncertainty assessment tools. Decision makers are presented with accurate inflow forecasts, flood forecasts, optimal dam operation strategies and uncertainty estimates for the near term. Complementing these tools, a water allocation and demand module enables operators to identify optimal medium and long-term water use schemes given *a priori* requirements for irrigation, hydropower, flood management and other objectives. The toolsets are embedded in an IT framework designed for easy use by technical as well as managerial staff.

Key words decision-support system; real-time hydropower and irrigation optimization; flood forecasting; hydrological; hydraulic modelling; river diversion; planning

INTRODUCTION

Reservoir operation is a complex problem that involves a number of often conflicting objectives, including flood control, hydropower generation, water supply for various users, navigation control, etc. Traditionally, fixed reservoir rule curves are used for guiding and managing the reservoir operation. These curves specify reservoir releases according to the current reservoir level, hydrological conditions, water demands and time of the year. Established rule curves, however, are often not very efficient for balancing the demands from the different users. Moreover, reservoir operation often includes subjective judgments by the operators. Thus, there is a potential for improving reservoir operating policies and small improvements can lead to large benefits.

For the optimization of water resources systems, procedures are applied that couple simulation models and numerical search methods. Traditionally, the simulation–optimization problem has been solved using mathematical programming techniques such as linear or nonlinear programming. Application of these methods, however, puts severe restrictions on the formulation of the optimization problem with respect to the description of water flow in the system, the definition of control variables to be optimized and the associated optimization criteria. Recently, procedures that directly couple fully dynamic (nonlinear) simulation models with heuristic optimization procedures such as evolutionary algorithms have been developed. These methods have proven to be effective for the optimization of water resources systems.

In this paper, modern hydrological, hydraulic and water allocation models are adopted for simulating reservoir operations and river flows for the short, medium and longer term. The reservoir operation module embedded into the model engines facilitates the implementation of complex control strategies, whereby reservoirs can be operated in accordance with different conditional control operation strategies. Further, any constraints in the operation of the structures can be included. One of the strengths of the chosen modelling system is its capability to capture the short term dynamics of the system, for example river flows resulting from gate operations or sudden reservoir inflows. Moreover, the use of several control strategies makes it possible to simulate multi-purpose reservoirs, taking into account a large number of objectives, including flood protection, hydropower production and water supply for irrigation purposes.

The modelling systems are combined with a numerical optimization tool that is used for optimizing different control variables defined for the reservoir operation strategies. The optimization tool includes a general multi-objective optimization framework that searches for the set of non-dominated or Pareto-optimal solutions according to the trade-offs between the various objectives. For solving the optimization problem, the shuffled complex evolution algorithm (Duan *et al.*, 1992) implemented in the optimization software is applied.

The paper is outlined as follows: the simulation–optimization framework applied in the present project is described, and the next section provides information on the forecast optimization method for short and long term forecast optimization. The procedure used to update the model state at the time of the forecast is then summarised, and the integration of the optimization technology in a real-time data management, forecast modelling and decision-support system is described. The next section describes the application of the new technology to the Sutlej and Beas River Basins in Northern India. Conclusions are then given in the last section.

SIMULATION-OPTIMIZATION FRAMEWORK

The simulation-optimization framework applied in the present project is illustrated in Fig. 1. The optimization criteria (e.g. flood control, hydropower generation, irrigation) are defined as objective functions comprising numerical measures that are used to compare the model output with user specified targets at each time step of the simulation (e.g. flood level, hydropower demand, irrigation demand). Based on the calculated objective functions the optimization algorithm selects new sets of control parameters to be evaluated. The process is repeated a number of times until no further improvement can be made.

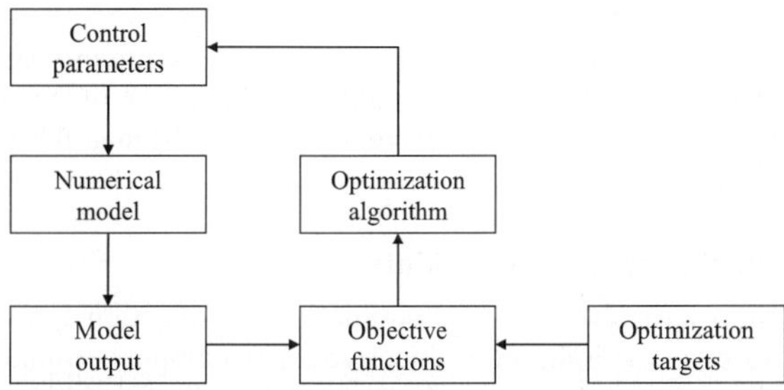

Fig. 1 Simulation-optimization framework.

As opposed to conventional technologies, the simulation-optimization framework is very flexible because it applies to any simulation model and facilitates the definition of any control parameters (both constants and time variables) and objective functions to be optimized. It handles efficiently user-defined constraints such as lower and upper search limits of each control parameter as well as linear or non-linear equality and inequality constraints. In order to facilitate the optimization of computationally demanding models, the optimization framework supports the use of distributed computing solutions such as office grid technology whereby the computational burden is distributed to multiple processors (multiple processors PCs and/or a network of PCs). In effect, this makes it feasible to carry out reservoir optimization in an operational environment.

REAL-TIME OPTIMIZATION OF FORECASTED DAM RELEASES

General

Operation of reservoir systems using optimized rule curves provides a general optimal operation of the system, however it does not make use of information that may be available concerning the current or future state of the system in the short term. Real-time optimization provides additional benefits, whereby real-time and forecast information concerning reservoir levels, reservoir inflows and water/hydropower demands are utilised. In this case, the reservoir system is optimized with respect to the short-term operation, using both short-term and long-term objectives. Often there is a conflict between short-term and long-term benefits, and hence the inclusion of long-term objectives in the optimization is important. The system supports the following methods:

- Combined short-term and long-term simulation-optimization.
- Short-term simulation-optimization with long-term objectives as constraints.

Modelling tools are applied to predict future conditions, given rainfall predictions and other operational forecasts. Real-time data assimilation is applied to update the model system state based on observations. Control strategies are explicitly described in the numerical model of the hydraulic structures. Finally, online optimization tools are implemented to resolve conflicting demands during operation with a dynamic feedback to the control strategies.

Combined short-term and long-term optimization

In this approach the system is optimized over a long period of time, and the optimal solution is implemented for the short-term operation. When new real-time data and forecast information become available, the simulation–optimization is repeated and a new optimal solution is implemented for the short-term operation. Control variables to be optimized include reservoir releases and gate operations from the time of forecast to the various water users.

In the optimization, real-time data, forecast data and historical or synthetically-generated data are combined. Up to the time of forecast real-time data are used to update the simulation model. Forecast data are applied for the first days after the time of forecast, where reliable weather forecasts are available. After this period, data obtained by resampling the historical record or by generating synthetic data according to a given stochastic model are used in the simulation to provide a probabilistic forecast. The combination of forecast data and historical/synthetic data in the optimization is illustrated above. The length of the historical/synthetic series to be used depends on the characteristics of the reservoir system. If the reservoir system is designed for over-year storage, several years of simulation are required.

Short-term optimization with long-term objectives as constraints

One of the drawbacks of the combined short-term and long-term simulation-optimization approach is the use of a long simulation period to evaluate long-term objectives. For complex simulation models involving several reservoirs, such an approach may not be computationally feasible in real time. Instead an approach that includes long-term objectives as constraints in the optimization can be applied.

In this case simulation is only carried out for a short period using the available forecast data the first days after time of forecast. The control variables to be optimized consist of future releases to the various water users in the forecast period. The objective functions to be optimized include short-term objectives (e.g. flood control, hydropower production, irrigation) and an objective function that penalises the deviation of the reservoir level at the end of the forecast period from a general target level. This target level is determined from the optimized rule curves, so that short-term operations that results in large deviations from the rule curves are penalised. The trade-off between short-term operation and long-term penalty constraints can be evaluated using Pareto optimization, which provides the basis for choosing preferred solutions that balance short-term and long-term objectives taking other considerations into account.

UPDATING OF THE MODEL STATE BASED ON LIVE DATA

In order to optimize accurately forecasted dam releases in the short, medium or longer term, the state of the hydrological and hydraulic system needs to reflect the physics prevailing at the time of the forecast. To this end, real-time point measurement time series of lateral inflow, river water level and reservoir operation are required. In the current project, the state of the system at the time of the forecast is updated on the basis of real-time data using a data assimilation procedure embedded into the hydraulic model. The updating procedure, which is based on a predefined time invariant gain (Madsen & Skotner, 2005), ensures that the model state portrays accurately the hydraulics and the water quality of the river system at the time an optimization forecast needs to be issued.

The data assimilation procedure described above can be applied to update the state of the river system regardless of whether measurements are available, up to the very time of the forecast. As such, the updating procedure is insensitive to missing data; a situation commonly encountered in an operational setting. Once the state of the system is known at the time of the forecast, the optimization–simulation framework is applied to compute mathematically optimal forecasts of dam releases. The whole operation thus applies a blend of different data and technologies, all of which are fully integrated into a data management and forecasting modelling framework.

INTEGRATED DATA MANAGEMENT AND FORECAST OPTIMIZATION FRAMEWORK

The system integrates real-time data sources, dam optimization algorithms and dissemination tools in a GIS-based client-server environment. The server side implements standard tools and technologies, while the client side is composed from project to project in order to meet specific project needs in terms of functionality and look and feel.

By its design, the system includes rigid user validation, robust data validation and fall-back strategies, data replication and duty server fall-back procedures. The system can be expanded when and as needed. Interfaces to other components such as real-time data and forecast optimization tools are implemented using industry based formats.

The server side to the system handles: (a) database processes, including data replication and redundancy, (b) system duties, including real-time data interfacing, forecast optimization, dissemination of data and information and task execution, (c) event and alarm handling, and (d) system maintenance duties. As such, all system processes are executed by the duty servers that form the backbone of the system.

The client side to the forecast optimization system includes a Windows-based graphical user interface and a web client composed in collaboration with Bhakra Beas Management Board (BBMB), India, using core components shipped with the system.

APPLICATION TO THE SUTLEJ AND BEAS RIVER BASINS

Description of the system

The northwestern region of the Indian sub-continent is the land of the Indus. Principal tributaries of the Indus River from the west are the Kabul and the Kurrem rivers; the five main tributaries from east are the Jhelum, Chenab, Ravi, Beas and Sutlej rivers. Most of the Indus Basin lies in India and Pakistan, and only about 13% of its total catchment is in Tibet and Afghanistan. Within India, the Indus Basin lies in the States of Jammu and Kashmir, Himachal Pradesh, Punjab, Haryana and Rajasthan (Fig. 2).

This project relates to the Sutlej and Beas river basins that are part of the Indus. The basins are inter-connected and include Bhakra Reservoir on the River Sutlej, the Pong Reservoir on River Beas, Pandoh reservoir for diverting a portion of supplies from Beas River into Bhakra Reservoir (Beas Sutlej Link), Ranjit Sagar Reservoir on Ravi River and Madhopur-Beas Link connecting Ravi and Beas rivers. The Bhakra Dam, the Pong Dam, the Beas Sutlej Link and six power houses with a total installed capacity of about 2866.30 MW, as well as the connected transmission network, are the major components controlled by BBMB.

Large volumes of water are stored in the form of snow in the Himalayas. The seasonal snow accumulation in the winter months and melting of this seasonal snow in the summer months is a

SATLUJ & BEAS BASINS

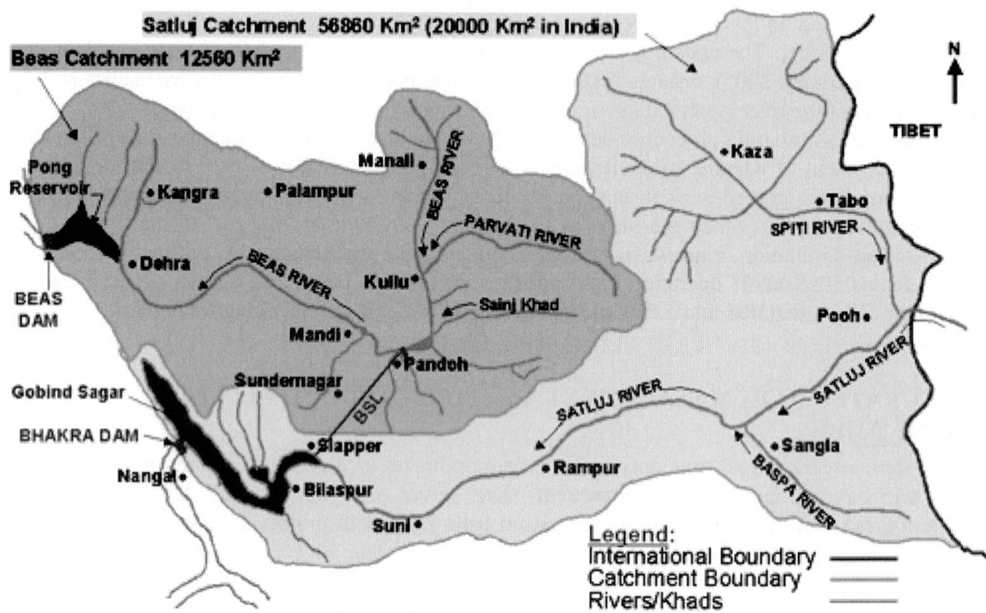

Fig. 2 Outline of the Sutlej and Beas river basins.

regular phenomenon contributing to snowmelt runoff, which constitutes an important part of annual inflows to the Bhakra and Pong Reservoirs. These flows occur during a period of high demand for hydro-electric power generation, irrigation and water supply for northern India.

The sharing of waters of the Indus has been defined in the Indus Waters Treaty. Under this treaty, the water of the three eastern rivers (the Ravi, the Beas and the Sutlej) is defined for the exclusive use of India while the water of the three western rivers (the Indus, the Jhelum and the Chenab) is defined for the exclusive use of Pakistan.

The waters of the Sutlej and Beas basins are stored in the Gobind Sagar and Pong Reservoirs and applied to meet the requirements of the partner states of Punjab, Haryana and Rajasthan, besides drinking water requirements of Chandigarh and Delhi. The sharing of Sutlej, Beas and Ravi waters is conducted in accordance with the various agreements entered into by the partner states and ad hoc arrangements.

The Bhakra and Pong Reservoirs, and the Beas Sutlej Link, are operated in an integrated manner to derive optimal benefits of irrigation and hydropower production. During the summer and rainy season, after meeting the irrigation requirements of the partner states, water is stored in the reservoir and released in a planned manner. Following the construction of Bhakra and Beas dams, the major floods in the rivers have either been absorbed successfully in reservoirs or mitigated through proper flood routing. To this end, a cushion of 5 ft (~1.5 m) in the Bhakra Reservoir and 10 ft (3 m) in the Pong Reservoir is kept for flood routing.

Modelling challenge

A significant part of the catchments of the Bhakra and Pong reservoirs is located in the high Himalayas, hence a significant part of the inflows originates from snow and glacier melt. With the Bhakra headwater located in Tibet (China) with no possibility for obtaining real-time ground data, snow accumulation and snow melt based on remote sensing data and meteorological forecast models will be used to estimate the runoff from the contributing catchments.

CONCLUSIONS

A real-time forecast optimization system has been designed. The system is currently under development and will be installed with BBMB. The system comprises an integrated data management and forecast optimization system, a simulation-optimization framework, and a set of hydrological and hydraulic forecasting models for short, medium and long term operation of the Bhakra and Beas Dams, and the Beas Sutlej Link.

The system implements a client-server based architecture, in which system duties such as real-time data interfacing, model simulations and forecast optimizations are undertaken by backbone servers either upon request by clients, according to task schedules or on an event-driven basis. In an operational environment, the system will be used to optimize dam water releases for hydropower production, flood protection and supply of water for irrigation purposes.

REFERENCES

Duan, Q., Sorooshian, S. & Gupta, V. (1992) Effective and efficient global optimization for conceptual rainfall–runoff models. *Water Resour. Res.* **28**(4), 1015–1031.

Madsen, H. & Skotner, C. (2005) Adaptive state updating in real-time river flow forecasting – a combined filtering and error forecasting procedure. *J. Hydrol.* **308**(1–4), 302–312.

Combining an improved harmony search algorithm with the One Tank Model calibration

MIAO-MIAO MA[1], WEI-MIN BAO[1] & XI-FENG LI[2]

1 *Department of Hydrology and Water Resources, University of Hohai, 210098 Nanjing, China*
wenmiaowu@163.com

2 *The Administration of Yinhuangjiqing Project, Changyi Station, 261300 Weifang, China*

Abstract Conceptual rainfall–runoff models that aim to predict streamflow have been used as a basic tool for hydrological forecasting. The Tank Model, which is a typical deterministic conceptual rainfall–runoff model, can yield good results for catchments of many regions in spite of its simplicity. The parameter calibration of the Tank Model is a very time consuming task and hence the demand for an automatic calibration method has been increasing. In this study, three meta-heuristic algorithms, i.e. Harmony Search, Improved Harmony Search and Global-best Harmony Search, are tested and numerical results reveal that IHS can find better solutions in comparison to HS and GHS. Detailed analysis shows that automatic calibration has the risk of outputting meaningless parameter values despite giving excellent results, if a single objective function is used to evaluate runoff simulation. Thus, multiple criteria are tested and the results show multiple criteria are feasible and easier to apply in practice.

Key words One Tank Model; automatic calibration; Harmony Search; Improved Harmony Search; Global-best Harmony Search; objective function

1. INTRODUCTION

Conceptual rainfall–runoff models (CRRM) that aim at predicting streamflow from the knowledge of precipitation over catchments have been used as a basic tool for hydrological forecasting (Cheng & Zhao, 2006). CRRM generally have many parameters, which cannot be directly obtained from measurable quantities of catchments characteristics (Sorooshian & Dracup, 1980; Sorooshian *et al.*, 1993); some of them are even physically meaningless, hence parameter calibration is needed. Previous studies show that with manual calibration it is not only difficult to assess explicitly the confidence of the model simulation, due to the subjective judgment involved and the need for an experienced hydrologist, but it is also a very time consuming task (Cheng & Ou, 2002). Thus, the demand for an automatic calibration method has been increasing. Moreover, automatic calibration, incorporating new methods of hydrological research, is propitious to the evolution of models. Therefore, how to utilize computer technology and mathematical algorithms for the automatic calibration of CRRM had long been accepted as having a significant influence on the quality of model applications.

There are many traditional optimization methods in solving the automatic calibration of CRRM. However, none of the traditional optimization methods used for calibrating CRRM, even for those with a moderate number of parameters, is robust and efficient in locating or nearly locating the global optima (Wang *et al.*, 1991; Duan & Sorooshian, 1992). Thus, using meta-heuristic algorithms, including genetic algorithm (GA), harmony search algorithm (HS), improved harmony search algorithm (IHS), global-best harmony search algorithm (GHS) and so on, is a new way to find the global optima and has shown its own advantages on calibrating CRRM.

In recent years, the harmony search algorithm (Geem *et al.*, 2001; Lee & Geem, 2004), mimicking the improvisation process of music players, has entered into widespread use. This algorithm generates a new vector, after considering all of the existing vectors, whereas the genetic algorithm (GA) only considers the two parent vectors (Geem *et al.*, 2001, 2002; Kim *et al.*, 2001; Lee & Geem, 2004; Mahdavi & Fesanghary, 2007). Thus, the HS algorithm is accomplished in identifying the high performance regions of the solution space in a reasonable time, but gets into trouble in performing local searches for numerical applications. In order to improve the fine-tuning characteristic of the HS algorithm, the IHS algorithm employs a new method that enhances the fine-tuning characteristic and convergence rate of the HS algorithm (Mahdavi & Fesanghary, 2007). Concerning the global-best harmony search (GHS), a new version of the HS algorithm using concepts from swarm intelligence is proposed, and is used to enhance the performance of the HS algorithm (Omran & Mahdavi, 2008).

In addition to parameter calibration, the objective function to evaluate the results of runoff simulations, which are the main output of the One Tank Model, remains to be improved. Detailed analysis shows that automatic calibration, compared with manual calibration, has a risk of outputting meaningless parameter values despite giving excellent result, if only one criterion is used to evaluate CRRM. Therefore, multiple objective functions that are capable of exploiting all the useful information are needed.

The purpose of this paper is to use mathematical algorithms and computer technology to provide a simple and robust automatic calibration model. So we combine three existing meta-heuristic algorithms (i.e. harmony search algorithm, improved harmony search algorithm, global-best harmony search algorithm) with the One Tank Model in order to test the efficiencies of these three algorithms in CRRM. The comparison proves the superiority of the IHS algorithm over the others in the parameter calibration. The rest of this paper is composed in the following sequence. First, brief reviews of the One Tank Model and applied optimization algorithms are provided. Then, in Section 4, the IHS algorithm is compared with the HS and GHS algorithms both with respect to computational time and multiple objective functions. Finally, the conclusions are given in Section 5.

2. ONE TANK MODEL

The tank models developed by Dr Sugawara are intended to simulate either flood events or long-term runoff by a combination of storage vessels. The model structure is comprised of several tanks in series or parallel connection to achieve the best possible state, according to the demand of different climate zones. From this sense, the basic structure of the tank models is a one-dimensional water tank (i.e. One Tank Model), that is, one-dimensional water tanks are the fundamental component of complex tank models. Similarly, the parameters relating to complex tank models are a combination of several one-dimensional water tank parameters. Therefore, the One Tank Model represents complex tank models well in terms of the similarities in the model structure and parameters. Since the focus of this study is to apply meta-heuristic algorithms to the automatic calibration in CRRM to test the efficiency of meta-heuristic algorithms and discover the problems in an automatic method, choosing the basic and simple One Tank Model as the study object is appropriate. The schematic of the One Tank Model used here is shown in Fig. 1. The parameters of the One Tank Model are the side outlet coefficient (S) and initial storage in tank (Hs), and the possible ranges of these two basic parameters are listed in Table 1.

In addition, the threshold setting for the water tank is that the initial storage at each moment should not be less than zero, in other words, negative storage or negative runoff is not considered in this study.

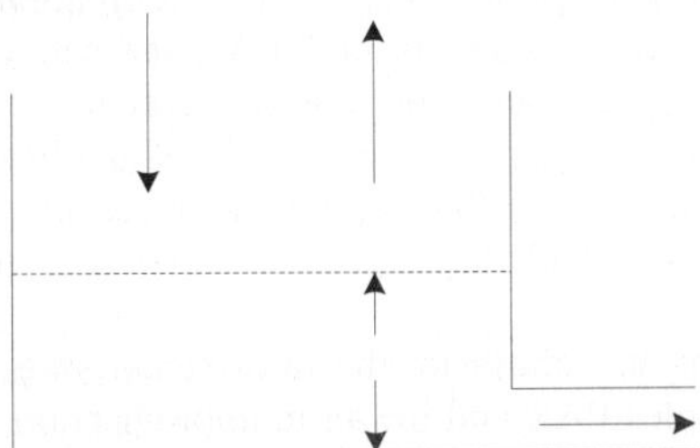

Fig. 1 The structure of the One Tank Model. P, E and R stand for precipitation, evaporation and runoff respectively. Hs is the initial storage of the tank. Simulated runoff is calculated as $Q(t) = A_d \times S(t) \times Hs(t) / 86.4$, where A_d is drainage area and $Hs(t)$, $S(t)$ stand for initial storage in tank and side outlet coefficient of tank at time t, respectively.

Table 1 The possible ranges of One Tank Model parameters.

Parameters	Min value	Max value
Initial storage in tank, Hs (mm)	0	50
Side outlet coefficient of tank, S (1/day)	0	1

3. OPTIMIZATION ALGORITHMS

Generally, the meta-heuristic algorithms work as follows: a population of individuals is randomly initialized where each individual represents a potential solution to the problem. The quality of each solution is evaluated using a fitness function. A selection process is applied during each iteration of algorithms in order to form a new population. The selection process is biased toward the fitter individuals to increase their chances of being included in the new population. Individuals are altered using unitary transformation (mutation) and higher-order transformation (crossover). This procedure is repeated until convergence is reached. The best solution found is expected to be a near-optimum solution (Omran & Mahdavi, 2008). Thus, three meta-heuristic algorithms: HS algorithm, IHS algorithm and GHS algorithm, are presented and compared.

Harmony search algorithm

The harmony search (HS) algorithm is devised in an analogy with the music improvisation process for searching better harmony. Since development of this algorithm, it has come into widespread use in various application areas.

The HS algorithm has three basic parameters: harmony memory (HM) size, harmony memory considering rate (HMCR) and pitch adjusting rate (PAR). HM are used to preserve the better state of experienced harmonies. The HMCR is introduced to escape from local optima and the PAR is adopted to improve the solution by searching adjacent values, thus avoiding being trapped in local optima (Paik & Kim, 2005).

Here several advantages with respect to traditional optimization techniques are presented as follows: (a) the HS algorithm imposes fewer mathematical requirements and does not require initial value settings of the decision variables. (b) As the HS algorithm uses stochastic random searches, derivative information is also unnecessary. (c) The HS algorithm generates a new vector, after considering all of the existing vectors, whereas the genetic algorithm (GA) only considers the two parent vectors (Geem *et al.*, 2001, 2002; Kim *et al.*, 2001; Lee & Geem, 2004; Mahdavi & Fesanghary, 2007). Thus, the HS algorithm is good at identifying the high performance regions of the solution space in a reasonable time compared with GA.

Improved harmony search algorithm

Although HS algorithm has greater flexibility and produces better solutions than GA in several applications (e.g. Geem *et al.*, 2001, 2002; Kim *et al.*, 2001; Lee & Geem, 2004), HS algorithm gets into trouble in performing local searches for numerical applications. In order to improve the fine-tuning characteristic of HS algorithm, IHS uses changed values for both PAR and bandwidth (bw) during new generations to enhance the fine-tuning characteristic and convergence rate of harmony search. Mahdavi & Fesanghary (2007) investigated the performance of PAR and bw with different value combinations, and found that large PAR values with small bw values usually give improvement of the best solutions in final generations which the algorithm converged to an optimal solution vector.

To improve the performance of the HS algorithm and eliminate the drawbacks lying with fixed values of PAR and bw, IHS algorithm uses variables PAR and bw an in improvisation step. PAR and bw change dynamically with generation number and are expressed as follow:

$$PAR(gn) = PAR_{\min} + \frac{(PAR_{\max} - PAR_{\min})}{NI} \times gn \quad (1)$$

where *PAR* is the pitch adjusting rate for each generation, PAR_{min} is the minimum pitch adjusting rate, PAR_{max} is the maximum pitch adjusting rate, *NI* is the number of solution vector generations, and *gn* is the generation number.

In addition, bw is dynamically updated as follows:

$$bw(gn) = bw_{max} \exp\left(\frac{Ln\left(\frac{bw_{min}}{bw_{max}}\right)}{NI} \times gn\right) \tag{2}$$

where *bw*(*gn*) is the bandwidth for each generation, bw_{min} is the minimum bandwidth, and bw_{max} is the maximum bandwidth.

Global-best harmony search algorithm

The new approach, called global-best harmony search (GHS) algorithm, modifies the pitch-adjustment step of the HS algorithm such that this new harmony can mimic the best harmony in the HM. Thus, replacing the bw parameter altogether and adding a social dimension to the HS algorithm is the main purpose of GHS algorithm. The GHS algorithm has exactly the same steps as the IHS algorithm with the exception that the improvisation step is modified to $x = x^{best}$, where best is the index of the best harmony in the HM, if the random value is smaller than *PAR*.

This modification allows the GHS algorithm to work efficiently on both continuous and discrete problems. For further details of the GHS, readers should refer to Omran & Mahdavi (2008).

4. APPLICATIONS

4.1. Study area

The One Tank Model linked to the optimization algorithms was applied to the Shaowu station, located in the northwest of Minjiang basin, drainage area of 2745 km^2 (Fig. 2). The Minjiang River is one of the main rivers in southeast of China. The climate is relatively humid and belongs to the semi-humid region. From the analysis of observed data, 68.9% of the total rainfall falls between March and June, while the evaporation is mostly greatest between May and September. Hereby, the phenomenon of no runoff comes into being, especially in winter. In this study, 10 years of historical data, 1988–1997, including daily precipitation, pan evaporation and observed runoff, are used for parameter calibration whilst the data between 1998– 1999 are used for the parameter validation.

Fig. 2 Location map of Shaowu Station in Minjiang basin, China.

4.2. Setting of optimization parameters

For the HS algorithm, high HMCR, especially from 0.9 to 0.95, contributed to excellent outputs, whereas the HM size and the *PAR* showed little influence on performance improvement compared with HMCR (Paik & Kim, 2005). The HM size and HMCR for the HS algorithm are also used for the IHS and GHS algorithm, but the *PAR* and *bw* values changed during new generations in IHS algorithm. The parameter values determined, except *bw*, for the IHS algorithm are also used for the GHS algorithm, since they share the same basic structure in which the influence of these optimization parameters is limited. The optimization parameter values used in the following analysis are summarized in Table 2.

Table 2 Optimization parameter values determined.

	HM size	HMCR	*PAR*	PAR_{max}	PAR_{min}	bw_{max}	bw_{min}
HS	50	0.90	0.3				
IHS	50	0.90		0.99	0.01	$1/(20 \times (LB - UB))$	0.0001
GHS	50	0.90		0.99	0.01		

UB and LB denotes the upper and lower possible value bound of corresponding parameter, respectively.

4.3. The objective functions

Since these optimization parameters greatly influence the efficiency of the optimization algorithms, their sensitivity was analysed to determine the best value for each optimization parameter. The sensitivity was evaluated by the criterion of the sum of the squares (F^2), which is given as

$$F^2 = \sum_{t=1}^{T} [q(t) - Q(t)]^2 \tag{3}$$

where $q(t)$ (m^3/s) and $Q(t)$ (m^3/s) are the observed and the simulated runoff, respectively, and T is the number of records at evenly spaced time intervals.

In the comparison, minimization of SSQ is adopted as the objective function, while percent error in volume (PEV; equation (4)), is used as an inspector to check the adequacy of the calibrated parameters. This is to reduce the risk of obtaining physically meaningless parameter values despite excellent fitting, the problem of blind automatic calibration having single criterion.

$$PEV(\%) = \left| \frac{q - Q}{q} \right| \times 100 \tag{4}$$

where q and Q (m^3) are the total observed and simulated runoff volume, respectively.

4.4. Comparison of optimization algorithms

To further understand the HS, IHS and GHS algorithms, we use the hydrological data of Minjiang basin in 1991 to show the algorithms behaviour in consecutive generations. The best state of HM in different iterations for the algorithms HS, IHS and GHS are shown in Table 3. All of the algorithms improvised a near optimal solution but the result that is obtained by the IHS is better than the results of the other two algorithms.

The same 12 periods that were mentioned in the previous section and the values of optimization parameters in Table 2, were used for the calibration. Comparison of the three algorithms showed that the IHS algorithm has much smaller F than those obtained using HS and GHS algorithm. The IHS algorithm showed the best results followed by the HS and GHS algorithm after 5000 iterations. The GHS algorithm gave F values slightly bigger than HS algorithm after 5000 iterations. This comparison result was also confirmed by the PEV comparison, since the rank of algorithms based on the PEV agreed exactly with the rank determined by the objective function of minimizing F (Table 4).

Table 3 The best state of HM in different iterations for the algorithms HS, IHS and GHS algorithms.

Algorithm	Iterations	Hs (mm)	S (1/day)	F	PEV (%)
HS	500	2	0.075	800.0271	9.0139
	1000	1.75	0.08	799.4718	9.4955
IHS	500	7.25	0.0646	808.0926	8.4431
	1000	4.5	0.0794	799.6444	9.7889
GHS	500	12	0.0688	805.1587	9.5796
	1000	6	0.0784	799.8886	9.8756

Table 4 Comparison of Optimization algorithms (after 5000 iterations).

	HS		IHS		GHS	
Period	F	PEV(%)	F	PEV(%)	F	PEV(%)
1988	1983.9164	9.5655	1983.8292	9.5008	1984.0195	9.7043
1989	2240.4808	13.5479	2240.452	13.489	2246.7444	13.9174
1990	1197.2315	10.5178	1197.065	10.2784	1200.3299	10.3885
1991	799.4718	9.4955	799.4247	9.3854	799.4999	9.0456
1992	2727.134	5.515	2725.8753	5.6926	2726.6755	5.6667
1993	2861.8452	4.4525	2861.796	4.4159	2863.6321	4.7604
1994	2234.1422	12.1386	2234.1421	12.1332	2254.0528	10.7276
1995	2834.3751	1.6529	2834.2645	1.6667	2838.6741	1.5807
1996	1794.9043	17.4577	1794.5472	17.6941	1794.6253	17.7462
1997	3303.8847	6.1611	3303.5767	6.3428	3304.9492	5.9813
1998	5360.7696	2.1137	5360.7688	2.1047	5360.8503	2.0039
1999	2042.5837	9.3569	2042.5727	9.3694	2046.1617	9.8219
Rank	Good	Good	Best	Best	Fair	Fair

For the evaluation of the heuristic algorithms, the computational time is as equally important as the objective function value. In the case of difficult problems, where the global optimum is seldom expected, if time allows, the heuristic algorithms often keep making progress, although their performance is significantly decelerated over time. Therefore, a fair comparison should imply checking which algorithm yields better results in a limited time (Paik & Kim, 2005).

The actual computational time of 30 000 iterations was measured for the HS algorithm, IHS algorithm and GHS algorithm (Fig. 3). The HS algorithm is the fastest among these due to its

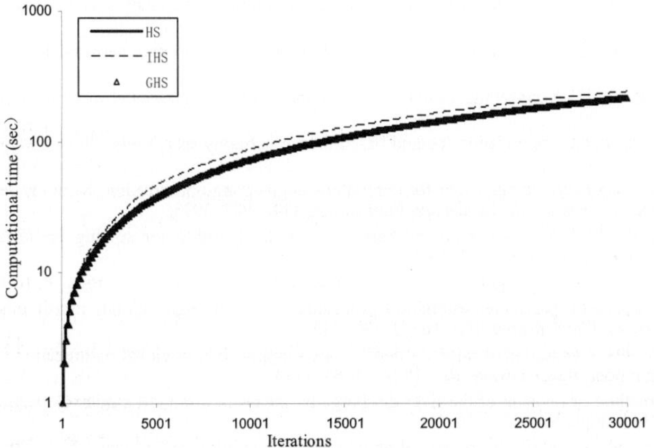

Fig. 3 Logarithm relationship between computational time and numbers of iterations.

simple structure. The strategies of changing values for both PAR and bw during new generations to enhance the fine-tuning characteristic and convergence rate of harmony search, make the IHS algorithm consume more computational time than the HS algorithm for the same iterations. The GHS algorithm, replacing the bw parameter altogether and adding a social dimension to the HS algorithm, uses slightly more computational time than the HS algorithm, but this is still less than the IHS algorithm.

5. CONCLUSIONS

Combining three existing meta-heuristic algorithms (i.e. the harmony search algorithm, improved harmony search algorithm, and the global-best harmony search algorithm) with the One Tank Model in order to test the efficiencies of these three algorithms in CRRM are compared both on computational time and multiple objective functions values in this paper. The comparison proves the superiority of the IHS algorithm over the others in the parameter calibration, in spite of consuming more computational time.

In this study, we provide an automatic calibration model, but a simple and robust calibration model still remains to be investigated given the complication of the automatic calibration task in CRRM. Even though the accuracy of simulated runoff is qualified, but the accuracy is not good enough because of structural problems of the model, so a nonlinear model structure and the diversity of parameters resulting from spatial and temporal distribution are worth considering and exploring. Thus, much effort should be made in work in this field.

The next step of future work aims to examine the most efficient combination of parameters in different climate zones (i.e. humid region and semi-humid region) and explain the trend of some parameters considering climatic variation.

Acknowledgements We are grateful to Dr Kyungrock Paik and our team-mates at Hohai University for their valuable help and suggestions.

REFERENCES

Cheng, C. T., Ou, C. P. & Chau, K. W. (2002) Combining a fuzzy optimal model with a genetic algorithm to solve multi-objective rainfall–runoff model calibration. *J. Hydrol.* **268**, 72–86.

Cheng, C. T., Zhao, M. Y., Chau, K. W. & Wu, X. Y. (2006) Using genetic algorithm and TOPSIS for Xinanjiang model calibration with a single procedure. *J. Hydrol.* **316**, 129–140

Duan, Q., Sorooshian, S. & Gupta, V. (1992) Effective and efficient global optimization for conceptual rainfall–runoff models. *Water Resour. Res.* **28**(4), 1015–1031.

Geem, Z. W., Kim, J. H. & Loganathan, G. V. (2001) A new heuristic optimization algorithm: harmony search. *Simulation* **76**(2), 60–68.

Geem, Z. W., Kim, J. H. & Loganathan, G. V. (2002) Harmony search optimization: application to pipe network design. *Int. J.Modeling & Simulation* **22**(2), 125–133.

Paik, K, Kim, J. H., Kim, H. S. & Lee, D. R. (2005) A conceptual rainfall–runoff model considering seasonal variation. *Hydrol. Processes* **19**, 3837–3850.

Lee, K. S. & Geem, Z. W. (2004) A new structural optimization method based on the harmony search algorithm. *Computers and Structures* **82**(9–10), 781–798.

Lee, K. S. & Geem, Z. W. (2005) A new meta-heuristic algorithm for continuous engineering optimization: harmony search theory and practice. *Computer Methods in Applied Mechanics and Engineering* **194**, 3902–3933.

Mahdavi, M., Fesanghary, M. & Damangir, E. (2007) An improved harmony search algorithm for solving optimization problems. *Applied Mathematics and Computation* **188**, 1567–1579.

Omran, M. G. H. & Mahdavi Mehrdad (2008) Global-best harmony search. *Applied Mathematics and Computation* **10**, 1016.

Sorooshian, S. & Dracup, J. A. (1980) Stochastic parameter estimation procedures for hydrologic rainfall–runoff models: correlated and heteroscedastic error cases. *Water Resour. Res.* **16** (2), 430–442.

Sorooshian, S., Duan, Q. & Gupta, V. K. (1993) Calibration of rainfall–runoff models: application of global optimization to the Sacramento soil moisture accounting model. *Water Resour. Res.* **29** (4), 1185–1194.

Sugawara, M. & Funiyuki, M. (1956) A method of revision of the river discharge by means of a rainfall model. Collection of Research Papers about Forecasting Hydrologic Variables, 14–18.

Wang, Q. J. (1991) The genetic algorithm and its application to calibrating conceptual rainfall–runoff models. *Water Resour. Res.* **27**(9), 2467–2471.

Accessing and sharing data using CUAHSI Water Data Services

DAVID R. MAIDMENT[1], RICHARD P. HOOPER[2], DAVID G. TARBOTON[3] & ILYA ZASLAKSKY[4]

1 *Center for Research in Water Resources, University of Texas, Austin, Texas 78712, USA*

2 *CUAHSI, 2000 Florida Avenue, NW, Washington, DC 20009 USA*
rhooper@cuahsi.org

3 *Utah Water Research Laboratory, Utah State University, 8200 Old Main Hall, Logan, Utah 84322-8200, USA*

4 *San Diego Supercomputing Centre, University of San Diego, 9500 Gilman Drive, La Jolla, California 92093 USA*

Abstract The Hydrologic Information System (HIS) project of the Consortium of Universities for the Advancement of Hydrologic Science, Inc (CUAHSI) has developed Water Data Services (WDS) using a services-oriented architecture. The underlying technological developments include WaterML, an XML-based language for transmission of time-series data, and WaterOneFlow, a set of web services that can provide access to data and metadata using standard web protocols. These technologies form the basis for an easy-to-use data publication system. WDS also includes a registration service for published web services and maintains a metadata catalogue of all services. An ontology of hydrological concepts is included as part of this central service to enable variables to be mapped to a common set of concepts. A map-based discovery tool, Hydroseek (http://www.hydroseek.net/), has been developed using the ontology and metadata catalogue. CUAHSI has been working with US government agencies, such as the US Geological Survey, on providing access to their data holdings using web services and transmitting data using WaterML. Metadata from these agencies has been included in the central metadata catalogue, thereby enabling seamless access to both government and academic environmental data. CUAHSI WDS is an open system in which any group or government agency around the world can participate. All software is freely available at http://his.cuahsi.org.

Key words CUAHSI Water Data Services; data publication; time series; web services; WaterML; ontology

INTRODUCTION: WHAT IS A HYDROLOGICAL INFORMATION SYSTEM?

A Hydrologic Information System (HIS) was conceived initially as an analogy to a geographic information system (GIS), defined by Tomlinsin (2003, p.3) as follows: "a GIS stores spatial data with logically-linked attribute information in a GIS storage database where analytical functions are controlled interactively by a human operator to generate the needed information products". There are key differences between geographic and hydrological information however. First, hydrological data are far more dynamic than geographic data. Secondly, hydrological data come from many owners, each with their own information systems. Indeed, when contemplating the plethora of web sites that present water observation data collected by government agencies and universities, a user is struck by the fact that no two web sites are the same, and there are no standards of consistency in how the information is provided – everyone uses their own data format and method of describing their data. It is almost like a "Tower of Babel" of data languages – everyone speaks their own.

The CUAHSI HIS project (Maidment, 2008) has as a goal the development of standards, systems, and software to overcome these inconsistencies and enhance the interoperability of water information. Ultimately the goal is for water data to be universally accessible and easy to use. The system that we envisage is built on interoperable components connected via the Internet following the services-oriented architecture paradigm (Josuttis, 2007). CUAHSI Water Data Services (WDS) is the first product of HIS and is the subject of this paper. However, achieving the goal described above takes more than technology. HIS must also engage the community through partnerships of data providers, developers and collaborators. HIS must provide leadership in laying the key foundations that others can participate in and build on. In this context of partnering, the mission of HIS is to build a system for accessing and sharing academic and public water observation data and data about the water environment, and to enable the linking of data and models to understand how water systems function.

This paper has seven sections. The first lays out the information model used for the representation of data in WDS. The second gives the system model, describing the conceptual framework for the components of the water data services architecture that underlies the CUAHSI WDS. This system requires that data be stored, transmitted and discovered. The third section describes the framework for persistent storage of water data in a relational model. The fourth section discusses the framework for data transmission using web services. These models for persistence and transmission of water information are core foundational concepts upon which the CUAHSI WDS rests. The fifth section describes the use of a metadata catalogue to support hybrid water data services where data is actually resident on a host agency server. The sixth section describes some of the partnerships CUAHSI WDS have with these disciplines in terms of working towards interoperability among data services across these disciplines. The final section contains some concluding remarks.

INFORMATION MODEL

A model is a simplified view of the real world. A hydrological simulation model is a simplified representation of hydrological processes with equations that attempts to mimic the behaviour of hydrological phenomena. A statistical model of a time series of hydrological observations is a summary using statistical parameters of an assembly of sampled data, sometimes associated with a mathematical probability model describing how the sampled phenomenon varies in nature. A geographic model of a landscape in a GIS is a simplified representation using themes or data layers of different classes of information that describe the characteristics of the landscape. All these concepts are relevant to a hydrological information system – a GIS component that can represent the physical landscape through which water flows; an observational component that describes how the properties of water vary through time and space in the landscape; a simulation component that can mimic and predict the functioning of hydrological phenomena.

A common point of reference among GIS, observations and modelling for hydrology, is that all of these data deal with *variables*, that is, with quantities like the rainfall rate, streamflow discharge or slope of the land surface that vary through *space*, and may also vary through *time*. Indeed, water flow is one of the primary forces in shaping the landscape over millennia, so if we lengthen our time horizon to include geological time scales, the landscape itself is also dynamic. In mathematical terms, suppose there are a set of n variables, denoted by $V_1, V_2, \ldots, V_n$. Each variable may change in space, denoted by s, and in time denoted by t, and at a particular point of space and time, they have *values*, $v_1(s,t), v_2(s,t), \ldots, v_n(s,t)$. The *region of space* over which these variables are defined may be defined by $\mathbf{S}$, and the applicable *time horizon* by $\mathbf{T}$, so, again in mathematical terms, $s \in \mathbf{S}$ and $t \in \mathbf{T}$. As shown in Fig. 1, the value $v_i(s,t)$ denotes the value of the variable V_i at the spatial location s and time t, and one may think of i, s and t as the coordinates in a "data cube" which reference this particular value. A simple way to think about this is that it is a "what-where-when" model – "what" variable (V_i) is represented "where" (s), and "when", (t).

In a perfect world, all the variables would be defined precisely over the whole space–time domain, $\mathbf{V}(\mathbf{S}, \mathbf{T})$, and there would be no uncertainty about hydrological conditions anywhere. In reality, however, we do not live in a perfect world. Observation of water properties can be done at gauges and sampling sites at a few points in space and often irregularly in time, especially for water quality or biological sampling. Remote sensing creates continuous coverages across space at particular points in time, but the connection between remotely-sensed images and hydrological variables may be tenuous. There are many weather and climate grids available through time and across space, but these are produced by computer models of variable accuracy. Even a quantity like land surface elevation, which seems so fixed and unchanging that it must be known without ambiguity, is in fact not so – the root mean square error in elevation of the US National Elevation Dataset when compared to survey benchmarks is 2.34 m (National Research Council, 2007, p.5). Rather than the perfect world of $\mathbf{V}(\mathbf{S}, \mathbf{T})$ without uncertainty, what actually exists is a much

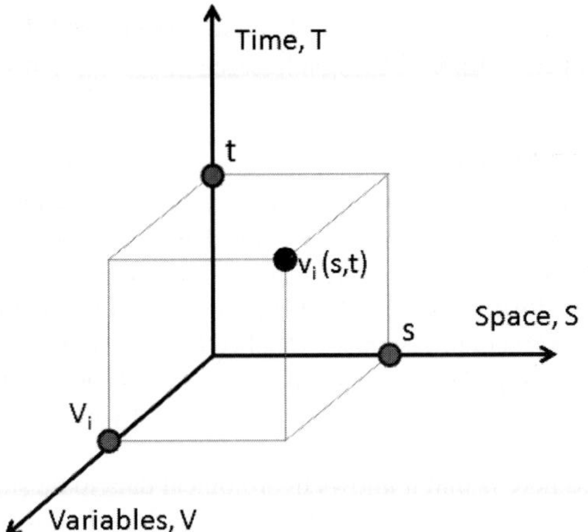

Fig. 1 The "what-where-when" method of representing a value in a data cube.

smaller data collection $\{v(s,t)\}$. Those data are of varying quality, and are produced by many organizations in many formats and descriptions.

Despite these limitations, the fundamental information model used in the CUAHSI Hydrologic Information System is simply this – a collection of variables defined sparsely and with some degree of uncertainty over a domain of space and time. It is important to note that this data model, while seemingly very general, has its limitations – it describes information measured in the natural environment, whose context is defined, in part, by a location measured in geographic coordinates and time referenced to Coordinated Universal Time (UTC). For example, we are not dealing with data sets describing water percolation through soil columns in a laboratory experiment.

In contemplating how to implement this data model it is useful to distinguish between data values, data collections, data series, and data sets. A data *value* is a single entity that exists by itself and represents the value of one variable at one point of space and time. A data *collection* comprises the values of a group of variables all referenced to the same point of time and space. A data *series* is a sequence of data values through time of a particular variable measured regularly or irregularly in time at a particular location in space. Data collections and data series are groups of values where the individual values may have no relation at all with one another. A data *set* is a collective entity where individual values have meaning only in relation to one another. For example, a set of results from a chemical analysis of a water sample is a collection of data values. A multidimensional array of results from a climate simulation model defined over a regular grid in space and time is a data set.

Another useful distinction is the concept of discrete *versus* continuous space. A *discrete space* representation of a spatial region means that the region is represented by a collection of separate points, lines, areas or volumes, as in a GIS vector data set, such as a set of stream gauge location points, river reach lines, watershed polygons or aquifer volumes. A continuous space representation of a region means the domain is covered with a regular mesh of cells or points such that the variables are defined continuously over the whole region. Some examples are digital elevation models of land surface terrain or grids of NEXRAD precipitation maps. In the computational world, discrete and continuous space representations are sometimes called *unstructured* and *structured* grids, respectively.

In a completely configured information model for a hydrological information system, all these different concepts would be included. In what is being presented in this paper, which focuses on observations measured repeatedly at gauges and sampling sites, only data values and data series are being represented, and those only on a discrete space domain of a set of fixed point locations.

SYSTEM MODEL

The problem of sharing information in a consistent fashion is being addressed in the CUAHSI HIS through the services-oriented system model illustrated in Fig. 2. The top three boxes in this figure depict different ways for a user to interact with and access data in the system. The bottom three boxes depict the servers that support the system. The thicker lines between these boxes depict data transfers using WaterML, the XML-based mark-up format for data transmission, described below, that CUAHSI web services use for data transfer. WaterML and the water data web services that serve to link the components are keys to the systems functioning across distributed servers and clients. Other modes of data transfer are depicted using thinner lines. Within the HIS server boxes depicted as the two leftmost bottom boxes are Observations Data Model (ODM) databases and a number of tools for working with ODM. ODM represents the standard for persistence of point observations in the CUAHSI WDS. WaterML and ODM are two key foundational concepts upon which the system is based.

A strength of the system as conceptualized here is that it allows hydrological data to be stored, found, accessed, interpreted and analysed in a uniform way. The system is a distributed network of computers, linked together using the Internet. The core of the system is a group of servers, referred to as HIS Central. In addition to this are numerous servers that are all based on a CUAHSI HIS design, called HIS Servers. These are operated by independent investigators or organizations that wish to easily publish hydrological data. The third group of servers are large third-party data sources, such as the USGS National Water Information System (NWIS), that are of significant value to the hydrological community, that the CUAHSI HIS project has integrated into the system, even though they are not based on the HIS Server design. To access the data on these various servers are a number of client tools, communication protocols and websites that are accessible to anyone with an Internet connection. These various pieces function together as an integrated system, where investigators publish their data and everyone has easy access to both investigator data as well as data in a variety of large third-party data repositories, in a standard way.

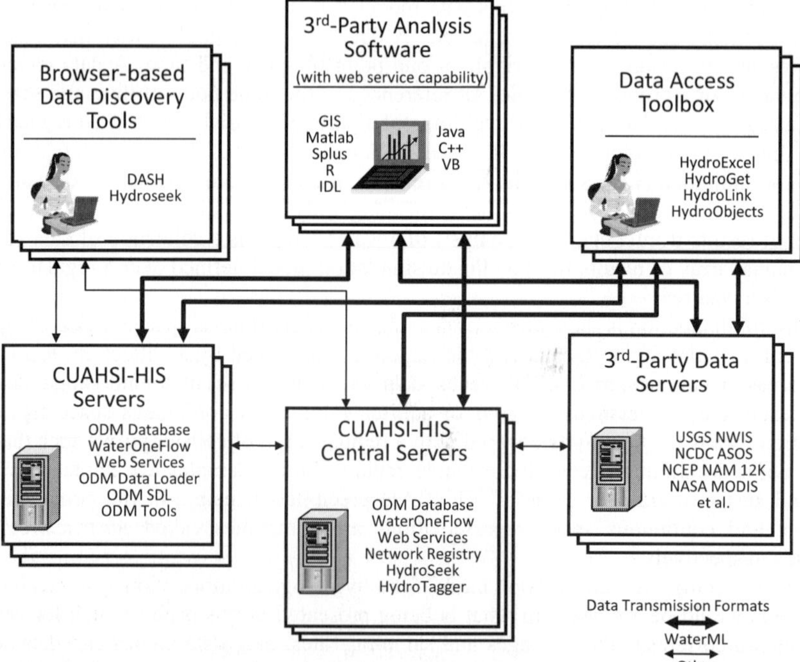

Fig. 2 Services-oriented system model used by CUAHSI Water Data Services.

To get started using the system a user may use one of the browser-based discovery tools. These interact directly with servers, either at HIS Central (middle bottom box) or distributed around the country, e.g. at observatories, (bottom left box) using standard browser http communications. Hydroseek, for example, provides map- and concept-based data discovery capabilities. These are supported by metadata catalogues on the servers.

Once a user has discovered the data they want to use, then it can be accessed directly using water data web services (thicker lines). Web services are machine independent and available from a number of user application environments. This allows HIS users to use application analysis environments of their own choice to work with HIS data. This is depicted in the top middle box. HIS users accessing data this way do not need to learn a new system. Rather they need to learn new data access functions within the working environment of their choice. The system thus has a level of platform and application environment neutrality.

The top right box depicts components that CUAHSI HIS has developed to also provide data access. These include HydroExcel that allows users to directly access data from water data web services from within Microsoft Excel. HydroGet provides data access from within ArcGIS, while HydroLink provides capability to link web services to OpenMI compliant models, linking HIS and hydrological modeling capabilities. OpenMI is the Open Model Interface (www.openmi.org) for linking hydrological models.

Data available through the system includes academic data at CUAHSI HIS servers (bottom left and central boxes) as well as data that has become part of the system through the collaboration of third parties. Third-party data servers are depicted in the bottom right box and comprise servers of various US Federal agencies, where CUAHSI has established data access arrangements. The development and support of water data web services for these third party data sets is a key way that semantic heterogeneity is resolved. While many of these agencies have their own web pages and data sharing systems, each has historically been different, making access tedious and difficult. The WaterML water data services for these data sets provide a consistent access method across multiple agencies. Furthermore, collaboration with third-party agencies has resulted in a number of them moving towards supporting their own WaterML-based web services. This data access mode is depicted by the red lines between the top boxes and third party data servers.

The system also provides the capability to support the point observation needs of hydrological observatories. Software for CUAHSI HIS servers (bottom left) is distributable and may be installed by individual investigators or research groups. Tools included provide capability to load, edit and quality control the data as well as publish it using CUAHSI water data services. Registration of published water data services with the Network/WSDL registry (centre bottom box) connects HIS server with the system allowing it to be discovered and accessed using the centralized discovery tools.

Relational Data Persistence Model

While it is easy to define information models in the abstract, the difficulty comes when you try to implement them. A distinction is made in information technology between a logical data model and a physical data model. A *logical model* is a schema or structure for storing data that defines the various entities, their descriptions or attributes, and their relations with one another. The set of linked relational database tables for the ODM (Tarboton & Horsburgh, 2008) is a logical model. This logical model can be implemented within any relational database system, such as Microsoft SQL/Server, which is the default environment for the CUAHSI HIS, Microsoft Access, the default database within Microsoft Office, or MySQL which is an open source relational database system. The physical data model refers to the actual positioning of the data elements within the data storage device; with relational databases, the user has no control over this and can interact with the data only through the functions that the makers of the relational database allow. In other words, depending on what particular relational database is used the actual physical location of the data elements stored in the computer's memory will be different, and is not accessible to the user. This rigid discipline makes relational databases very robust and indeed we are not aware of any data having been inadvertently lost so far from any implementation of the CUAHSI Observations Data Model. However, the same rigidity can make relational databases difficult to load with data and difficult to work with unless a good set of user tools is provided. We intend that the ODM loading and editing tools provided with CUAHSI HIS will help overcome these difficulties.

Within the CUAHSI Observations Data Model, each data value is treated individually and all data values are stored in one large table, called the DataValues table, as the entity called "DataValue" listed second in the column of entities shown in Fig. 3. This DataValue must be a number. ValueID {PK} is the *Primary Key* or the index number for the values in this table – ValueID is an integer that starts at 1 for the first record in the table and then continues incrementing by one for all subsequent records. The entities shown with the association {FK} are links called *Foreign Keys* to the similarly constructed primary keys in other tables of the database. For example, SiteID links the data value to the observation site at which it was measured, and VariableID links the value to the variable it describes.

Time is actually represented by three entries – LocalDateTime for the time in the time zone where the measurements are made, DateTimeUTC for the corresponding Coordinated Universal Time, and UTCOffset to record the difference in hours between these two times. Other descriptors of the data value are signified by OffsetValue (if measurements are made at various depths within a water body, for example), Qualifier, Method, Source, Sample and Quality Control Level. All these descriptors are applied to every single data value and could be different from one value to the next in the table.

The merit of this approach to storing observations data is its generality – physical, chemical and biological data describing conditions in water systems have all been stored in this structure in the many ODM implementations that have been made at our partner universities. We do not care whether data are recorded regularly or not, since each data value has its own *time stamp* or value in calendar date and time when it applies. It is a relational database convention that all data that represent values over *intervals* of time are time stamped at the instant that the interval begins. This means that annual data, monthly data for January, and daily data for the first of January, are all time stamped at midnight on the beginning of the first day of January in that year, and time support metadata included in the Variables table of the ODM are needed to distinguish what kind of data series this is, if it is regularly recorded.

In relational database design terms, this approach to data structuring is called a *star schema*, because all the actual data values are stored in a single table and other tables arranged around this one in a star configuration contain the metadata or descriptors of these values.

DataValues
ValueID {PK}
DataValue
ValueAccuracy
LocalDateTime
UTCOffset
DateTimeUTC
SiteID {FK}
VariableID {FK}
OffsetValue
OffsetTypeID {FK}
CensorCode
QualifierID {FK}
MethodID {FK}
SourceID {FK}
SampleID {FK}
DerivedFromID
QualityControlLevelID {FK}

Fig. 3 The DataValues table from the CUAHSI Observations Data Model.

WATER DATA WEB SERVICES

The relational database design just described is a structure for internal storage of data in an HIS, much like that envisaged by Tomlinson (2003) for a GIS. If one wishes simply to store observations data and work with them independently, then the Observations Data Model and its attendant ODM Tools function perfectly well as an independent entity for that purpose. Suppose, however, that the hydrological scientist wishes also to publish these data by providing automated external access to the data through the Internet. We are now faced with an entirely different problem – how to communicate data rather than store them. One solution to this problem is to make a copy of the ODM database and put it on an ftp site so that others can download it. This simple, traditional approach has the merit that it completely conveys the entire content of the database. It is, however, greatly limited by the fact that any user of the data has to employ the same relational database system used in creating it – in other words, the physical data model now really matters when the data are moved to another computer.

A more flexible approach is to transform the database so that it becomes a *water data service*. This is accomplished by linking the database to a special set of functions constructed according to the conventions of the Web Services Description Language (WSDL) and a transmission protocol known as Simple Object Access Protocol (SOAP), a set of conventions defined by the World Wide Web Consortium that enable one computer to make legitimate requests of another computer at a remote location, and to receive appropriately constructed responses to those requests (http://www.w3.org/TR/wsdl, http://www.w3.org/TR/soap/). The CUAHSI HIS team has used these conventions to define a specialization of XML (http://www.w3.org/TR/xml/) called *WaterML* for communicating water observations data through the Internet, and a set of functions called *WaterOneFlow* web service methods that operate on the Observations Data Model and produce responses in WaterML to external requests.

Suppose, moreover, that analogous water data services could be constructed for existing national and regional water archives collected by government agencies, even though the database system they are using internally for storing their data may be completely different than the CUAHSI Observations Data Model. In other words, when WaterOneFlow web service requests are made of these government databases, they also produce responses in WaterML. Now, the "Tower of Babel" of a plethora of water web sites and water data formats referred to earlier is vastly transformed – all the water data services have a unique *WSDL address* on the Internet and provide data in WaterML to WaterOneFlow web service requests.

For example, http://his02.usu.edu/littlebearriver/cuahsi_1_0.asmx?WSDL defines a CUAHSI water data service provided by Utah State University, which presents data from observations they are making in the Little Bear River in Utah, and http://ccbay.tamucc.edu/CCBayODWS/cuahsi_1_0.asmx?WSDL defines a similar CUAHSI water data service for measurements being made in Corpus Christi Bay by researchers at Texas A&M University – Corpus Christi. Similarly, http://river.sdsc.edu/wateroneflow/NWIS/DailyValues.asmx?WSDL is the WSDL address for the CUAHSI water data service for all the Daily Values data from more than 30 000 sites in the USGS National Water Information System, and http://cbe.cae.drexel.edu/CIMS/cuahsi_1_0.asmx?WSDL is the WSDL address for water observations for Chesapeake Bay from the Chesapeake Information Management System linked by researchers at Drexel University in Philadelphia.

The point observations information model used to define WaterOneFlow web service requests operates in hierarchical fashion using the following definitions:

a. A **data source** is an organization or individual operating an observation network.
b. A **network** is a set of observation sites.
c. A **site** is a point location where one or more variables are measured, defined geospatially by latitude and longitude coordinates.
d. A **variable** is a property describing the physical, chemical or biological conditions in the water.
e. A **value** is an observation of a variable at a particular time.
f. A **metadata** attribute provides additional information to qualify the value.

There are currently four WaterOneFlow web service methods:

a. **GetSites** – for a given observation network, return the list of observation sites in the network;
b. **GetSiteInfo** – for one site in the network, return the list of variables measured there, the count of the number of values available, the begin date time and end date time of the first and last measurements, and metadata about the site and the variables;
c. **GetVariableInfo** – for a given network, return the list of variables measured at any location on that network (not all variables need be measured at each site but all sites must use the same variable definitions and units), or – if a variable code is specified – return metadata about that variable such as its units and time support;
d. **GetValues** – for one site and one variable, return all data values recorded between a begin date time and an end date time.

WaterML is one example of an eXtensible Markup Language (XML) specification, which is a generalization of the HyperText Markup Language (HTML) used to define normal web pages. In XML specifications, there are a hierarchy of *elements* indicated by <element> … </element> tags, and descriptors or *attributes*. The WaterML response consists of five blocks. The first block describes the query itself – when it was made and for what site, variable, and time period. The second block describes the measurement site: its site name, site code (unique identifier in the network); its location in latitude and longitude coordinates and in local UTM easting and northing coordinates, and some further descriptive information about political units and other comments.

Third, there is a block of WaterML response describing the variable being measured, its name, units, time support and other metadata. The fourth block of WaterML provides the data values themselves and some metadata about them. Finally, there is a block of WaterML response providing information about the method used to collect the data and the researcher to be contacted in the event there are questions about it.

Thus, WaterML provides a complete data package unto itself – it includes the data values and where they were collected, by whom, by what method, and also what data quality you should expect from this information.

METADATA CATALOGUE

A CUAHSI water data service can be built in two ways:

a. A *unified* water data service, where all four WaterOneFlow methods are supported by observations and metadata in a single ODM database – this is the case in nearly all the water data services publishing data from academic data collections;
b. A *hybrid* water data service, where the three metadata functions (GetSites, GetSiteInfo, GetVariableInfo) are supported by information in an ODM database, but the GetValues function which extracts the data goes to the web site or web service of the host water agency – this is the case for all the CUAHSI Water Data Services presently operating for large federal databases, such as those of the US Geological Survey.

In the case of the hybrid water data service, the usual path for building the GetValues function is to have a "web page scraper" that programmatically mimics the action of a human user going to an agency web site and downloading data from it. It turns out that about 50% of the hits on the USGS National Water Information System actually come from automated data mining systems built as web page scrapers. The advantage of a web page scraper is that it accepts the agency's method of publishing its data and does not disrupt its security procedures for protecting its data archive. The disadvantage is that any time the data provider "improves" the format of the web page, the web page scraper breaks, and has to be reprogrammed to operate again.

A more reliable means of supporting the GetValues function is for the data agency to provide a custom-programmed web services function with WaterML as its output. The US Geological Survey has done this for CUAHSI to provide access to its Daily Values database (http://www.cuahsi.org/docs/usgs-cuahsi-webservices.pdf), and is presently programming a similar web service to expose its Instantaneous Data Archive, or historical record of instantaneous measurements within a day.

Whether the GetValues function is supported by a web page scraper or by a custom-programmed web service, there is still the need to harvest the metadata needed to support site and variable metadata function. For CUAHSI's Water Data Services providing access to USGS data, this is done by the USGS doing a metadata transfer from NWIS and providing the resulting files to CUAHSI, where they are reformatted to fit the ODM structure.

Within the Observations Data Model, there is one table, the SeriesCatalog, which is not input by the user, but is constructed using a stored procedure by the database itself, once data loading is complete. Its Primary Key is the SeriesID, which indexes how many series there are, and each series has a set of descriptors listed subsequently. The VariableID and SiteID are the indices i and j identifying variables V_i and sites S_j, and the BeginDateTime, EndDateTime and ValueCount are specified at the end of the table. The data source, measurement method and quality control level are also tabulated here for each series, so that searches can be made for data from particular organizations, or using particular methods, or with defined standards of quality control.

Now, imagine that each of these data series is similarly catalogued over all measurement networks, and all that information is accumulated in a single database. This is what has been done at the HIS Central at the San Diego Supercomputer Center where more than eight million data series are presently catalogued from measurements at nearly two million locations in the nation. We call this the National Water Metadata Catalog and it is a valuable information resource in its own right – it does not store any data, but only metadata about who has measured what and where and when, but that is very valuable when you are searching for particular kinds of information in particular regions and periods of time.

One last major step still remains to render the USA nation's water information scientifically searchable – the variables have to be tagged using a set of scientific concepts so that searches by concept can operate over all the observation networks without regard to the fact that the same variable may be described differently by different organizations. This process is called semantic mediation using an ontology of hydrological concepts described by Piasecki (2008).

INTEROPERABILITY WITH OTHER SCIENTIFIC DISCIPLINES

A key requirement for successful cyberinfrastructure is that it be interoperable among scientific disciplines. Water is important as a physical environment which defines its physical and chemical properties; water is important as a living environment for fish and aquatic organisms, and to support plant life; water is important to the human environment – water accounts for more than half of the weight of a human body. Although water interacts with many disciplines, active cyber-infrastructure efforts in four disciplines are particularly relevant for hydrological science – those in atmospheric science, geological science, ocean science and ecological science. The first three of these are the environments above, below, and surrounding the domain of hydrology of the land surface and near subsurface. The fourth, ecological science, connects hydrology with living things.

Atmospheric science

Unidata is the equivalent organization to CUAHSI HIS within the domain of the Universities Corporation for Atmospheric Research (http://www.unidata.ucar.edu/). Technology developed by Unidata is also being used within the National Weather Service and related agencies to access and transmit atmospheric data. Two elements of Unidata's function are critical for CUAHSI HIS – netCDF and THREDDS. NetCDF is a file format for storing multidimensional arrays of data, where the index dimensions, such as those for time and space are called *coordinate dimensions*, and dimensions that store actual data values like temperature or relative humidity are called *variable dimensions*. NetCDF is a very good format for storing datasets which are documented and interpreted as a whole, rather than individual data values as we are doing with our Observations Data Model and WaterML web services.

THREDDS (Thematic Real-Time Environmental Data Distribution Services) (http://www.unidata.ucar.edu/projects/THREDDS/) is a means like WaterOneFlow web services of publishing multidimensional array information on the Internet without regard to the underlying form of the gridded data. CUAHSI WDS has adopted netCDF as its default file format for multidimensional array data and intends to include publication of such arrays using THREDDS in a future version of its HIS Server. One existing CUAHSI data service, that for the North American Forecast Model (NAM), uses the THREDDS server as a data source, and returns a time series in WaterML of forecast weather conditions for a single geographic point selected anywhere within the spatial domain of the model. CUAHSI data services to other data sources published in THREDDS could similarly be developed, such as for NEXRAD precipitation data published in THREDDS by the National Climatic Data Centre, in Asheville, NC.

Geological science

A large cyberinfrastructure project in geological sciences is GEON (http://www.geongrid.org/), a collaboration between geological scientists and computer scientists. One GEON activity of particular interest to CUAHSI HIS is its capacity to process large point clouds of LIDAR information for land surface terrain collected in NSF projects. We are working with GEON to add processing functions to their terrain processing system that will yield information products of interest to hydrologists, such as watershed and stream network delineation. LIDAR data is much more precise than conventional National Elevation Dataset terrain data but is so voluminous that processing it is beyond the range of desktop computers for significant regional coverage.

Ocean science

The International Ocean Observing System, IOOS, operates an observations registry (http://obsregistry.org/index.php) which catalogues the measurement of water conditions along the coasts of the USA, similar to the approach that we have developed. One of these networks is the Texas Coastal Ocean Observing Network (TCOON) (http://lighthouse.tamucc.edu/TCOON/). As part of building a set of Texas Water Data Services (http://data.crwr.utexas.edu), a project distinct from HIS, we succeeded in building a hybrid water data service for TCOON where the metadata was read from the IOOS XML metadata specification and the GetValues function was made operational using a web scraper on the TCOON web site, so that coastal water observations data are published as a water data service just the same as water data services constructed for inland waters. It may later be possible to accomplish that goal for other networks in the IOOS Observations Registry.

Ecology

The ecological science community and the National Center for Ecological Analysis and Synthesis (NCEAS) have created an Ecological Metadata Language (EML) (http://www.nceas.ucsb.edu/ecoinfo/approach) whose purpose is to describe ecological data sets. These data sets can use any physical format and are not restricted to the rigid structure we have used in our CUAHSI Observations Data Model. The focus of EML is on the digital description of data so that it can be subsequently searched for and discovered, rather than homogeneity of format and meaning of terms, as we are doing with CUAHSI water data services. EML is the standard metadata language used at all the Long Term Ecological Research (LTER) sites, and some discussion has taken place between CUAHSI HIS and the LTER Network Office in Albuquerque about greater interoperability of cyberinfrastructure. The LTER Network has a TRENDS project that documents long-term trends in annual data measured at LTER sites. This would be an excellent candidate to be published as a CUAHSI water data service.

CONCLUDING REMARKS

The concept of water data services for point observations is well understood, and CUAHSI WDS has built and proven in operation an effective services-oriented architecture for this purpose. The support it has received from US federal agencies suggests that it can serve as an emerging standard for access to these governmental data holdings. The need of hydrological scientists to be able to publish and access academically collected data is quite a different task, and one which will require support. CUAHSI will be developing support services for the adoption of WDS and plans to develop a permanent Hydrologic Data Access Centre for maintaining the metadata catalogue. We believe that this lays a foundation for an extensible global network for hydrological and environmental data.

Acknowledgements This material is based upon work supported by the National Science Foundation under Grants EAR 03-26064 and 07-53521 to the Consortium of Universities for the Advancement of Hydrologic Science, Inc. and Grant EAR 06-22374 to the University of Texas.

REFERENCES

Josuttis, N. M. (2007) *SOA in Practice – The Art of Distributed System Design*. O'Reilly Press, Sebastopol, California, USA.

Maidment, D. R. (ed) (2008) *CUAHSI Hydrologic Information System: Overview of version 1.1*. Consortium of Universities for the Advancement of Hydrologic Science, Inc., Washington, DC. Available at http://his./cuahsi.org/documents/HISoverview.pdf.

National Research Council (2007) *Elevation Data for Floodplain Mapping*. Report of the Committee on Floodplain Mapping Technologies, National Academy Press, Washington DC, USA.

Piasecki, M. (2008) "Semantic mediation—linking data with concepts. In: *CUAHSI Hydrologic Information System: Overview of version 1.1* (ed. by D. R. Maidment). Consortium of Universities for the Advancement of Hydrologic Science, Inc., Washington, DC, USA. Available at http://his./cuahsi.org/documents/HISoverview.pdf.

Tarboton, D. M. & Horsburgh, J. S. (2008) Observations data model. In: *CUAHSI Hydrologic Information System: Overview of version 1.1* (ed. by D. R. Maidment). Consortium of Universities for the Advancement of Hydrologic Science, Inc., Washington, DC, USA. Available at http://his./cuahsi.org/documents/HISoverview.pdf.

Tomlinson, R. (2003) *Thinking About GIS*. ESRI Press. Redlands California, USA.

Data mining for hydrological time series analysis

RULIN OUYANG[1,2], LILIANG REN[1] & CHENGHU ZHOU[2]

1 *State Key Laboratory of Hydrology, Water Resources and Hydraulic Engineering, Hohai University, Nanjing 210098, China*
oyrl@hhu.edu.cn

2 *State Key Laboratory of Resources and Environmental Information System, Institute of Geographic Sciences and Natural Resources Research, Chinese Academy of Sciences, Beijing 100101, China*

Abstract The rapid development of data mining supplies a new way for hydrological information analysis and mining, which use artificial intelligence methodology. Applying data mining theory and technology, we analysed hydrological daily discharge time series of Shaligunlanke station in the Tarim River basin in China for the years 1961–2000. First, according to the four monthly statistics, mean monthly discharge, monthly maximum discharge, monthly amplitude and monthly standard deviation, *K*-mean clustering was used to segment the annual process of the daily discharge. The clustering result showed that the annual process of the daily discharge can be divided into five segments: snowmelt period I (April), snowmelt period II (May), rainfall period I (June–August), rainfall period II (September) and dry period (October–December and January–March). Secondly, dynamic time warping (DTW), which is a different distance metric method from the traditional Euclidian distance metric, was used to search for similarity in the discharge process. Based on the similarity matrix, the similar discharge processes can be mined in each period. Finally, the physical causes of the similar processes identified in the previous steps were analysed. It was found that the discharge had a close relationship with the temperature and the precipitation, and the discharge processes were more similar under the same climatic conditions. Our study shows that data mining is a feasible efficient approach to discovering the hidden information in the historical hydrological data and to mining the implicative laws under the hydrological process.

Key words data mining; hydrological time series; dynamic time warping; similarity search

INTRODUCTION

With the development of database technology, the explosive growth in data collected from applications, including business and management government administration, science and engineering, and environmental control, has led to the increasing demand for efficient and effective data analysis and data understanding tools. Data mining, which is also referred to as knowledge discovery in databases, is a process of nontrivial extraction of implicit, previously unknown and potentially useful information from data in databases (Piatetsky-Shapiro & Frawley, 1991). It is a young interdisciplinary field, drawing from areas such as database systems, data warehousing, statistics, machine learning, artificial intelligence, data visualization, information retrieval, and high-performance computing. Other contributing areas include neural networks, pattern recognition, spatial data analysis, signal processing. Nowadays, data mining has been paid great attention in the world and widely applied in many fields, in particular: commerce (Berson *et al.*, 1999; Berry & Linoff, 2004), finance (Benninga, 2000; Higgins, 2001), and biomedicine (Waterman, 1995; Gusfield, 1997; Baldi & Brunak, 2001; Baxevanis & Ouellette, 2001).

In recent years, new emerging applications of data mining in hydrology and water resources have contributed to hydrology science. In the literature (Solomatine, 2002) gives an overview of successful applications of several data mining techniques in the problems of hydrology and water resources, including: using clustering for finding groups of hydrographs on the basis of their shape and magnitude (Hannah *et al.*, 2000), using artificial neural networks (ANN) to model the rainfall–runoff process (Govindaraju & Rao, 2000), using fuzzy rule-based systems to predict precipitation events (Abebe *et al.*, 2000), and using chaos theory in predicting water levels (Solomatine, 2000). More and more hydrologists have begin to notice that tremendous potential of data mining and believe that data mining may be the most appropriate way forward to combine the best of the two approaches: theory-driven, understanding-rich processes with data-driven discovery processes (Babovic, 2005).

Data mining for hydrology depends on the hydrometeorological data which generally take the form of time series. Traditional hydrological time series analysis is established on mathematical theories and some hypotheses by logical deduction method. For example, many hydrological time series regression models are based on the stationarity hypothesis, normal distribution or the linear hypothesis, but hydrological time series always have the characteristics of non-stationary, non-normal distribution and nonlinearity. The hydrological system is so complex, and human knowledge is so finite that it is impossible to establish a reasonable data model by traditional time series analysis approaches. Data mining, which is particularly useful in modelling processes about which adequate knowledge of the physics is limited, is presented as a tool complimentary to hydrological time series analysis.

In principle, data mining can be applicable to any kind of data, such as spatial-temporal data, text, multimedia, and the World Wide Web. Time series, as the most common data set, stores sequences of values that change with time, such as data collected regarding the stock exchange, the amount of sales, temperature changes, earthquake eruption, a patient's heart-rate changes, and so on. Hydrological time series are sets of various hydrological elements' record values that change with time. Though hydrological time series can be expressed in continuous time or at discrete times, they are generally dealt with at discrete times, such as years, months and days. Note that the graphical representation of discrete series is often made as continuous curves just because it is easier to observe the graphical appearance and the overall configuration of the series (Salas *et al.*, 1980).

In this paper, some data mining algorithms are applied to hydrological time series analysis. The purpose of the research is to develop a data mining approach using modern information technology to discover the hidden information in the historical hydrological data and the implicative laws under the hydrological process, and also to look for a new way to meet the requirements of hydrological time series analysis.

The rest of this paper is organized as follows: in the next section, we present the background of two concepts of data mining: time series clustering and similarity search, including their common algorithms. In the following section, a data mining procedure is established for hydrological time series analysis; *K*-means clustering algorithm and dynamic time warping algorithm are utilized to achieve discharge process clustering, hydrological period segmentation and discharge process similarity search. Then, the application of the data mining procedure is described and the results are shown and analysed. The final section presents conclusions and future work directions.

BACKGROUND

Clustering of time series

Clustering of time series has received considerable attention in recent years, due to it being a fundamental task in data mining; it could be defined as the organization of a collection of patterns into groups, based on similarity. There exists a large number of clustering algorithms in the literature. In general, major clustering methods used in time series are: *K*-means clustering (Das *et al.*, 1998; Rafiei & Mendelzon, 1998; Goutte *et al.*, 1999; Knab *et al.*, 2003; Vlachos *et al.*, 2003), *K*-medoids clustering (Kalpakis *et al.*, 2001), nearest neighbour clustering (Zhang *et al.*, 2004), hierarchical clustering (Oates *et al.*, 1999; Van & Van, 1999), self-organizing maps (Fu *et al.*, 2001), and so on.

In all clustering algorithms, *K*-means clustering has become the most well-known and commonly used method because it is relatively scalable and efficient in processing large data sets (Han & Kamber, 2001). The *K*-means clustering algorithm proceeds as follows: first, it randomly selects k of the objects, each of which initially represents a cluster mean or centre. For each of the remaining objects, an object is assigned to the cluster to which it is the most similar, based on the distance between the object and the cluster mean. It then computes the new mean for each cluster. This process iterates until criterion function converges. Typically, the squared-error criterion is used.

Similarity search

Similarity search in time series analysis is one of the rapid development fields in data mining. Unlike normal database queries, which find data that match the given query exactly, a similarity search finds data sequences that differ only slightly from the given query sequence. Similarity search can be classified into two categories (Das *et al.*, 1998):

(a) Whole matching in which it is assumed that all sequences to be compared are the same length.

(b) Subsequence matching in which we have a query sequence *X*, and a longer sequence *Y*. The task is to find the subsequence in *Y*, beginning at Y_i, which best matches *X*, and report its offset within *Y*.

The primary difficulty is defining a similarity measure. For similarity analysis of time series data, Euclidian distance is typically used as a similarity measure. Given two sequences $X = \{x_1, \ldots, x_n\}$ and $Y = \{y_1, \ldots, y_m\}$ with $n = m$, their Euclidean distance is defined as:

$$D(X,Y) = \sqrt{\sum_{i=1}^{n}(x_i - y_i)^2} \quad (1)$$

However, Euclidean distance can be an extremely brittle distance measure (Chu *et al.*, 2002). The reason why Euclidean distance may fail to produce an intuitively correct measure of similarity between two sequences is that it is very sensitive to small distortions in the time axis.

A method that allows this elastic shifting of the *X*-axis is desired in order to detect similar shapes with different phases. The technique, Dynamic Time Warping (DTW), was introduced to the data mining community by Berndt & Clifford (1994). The DTW algorithm was briefly introduced as follows (Keogh, 2002):

Suppose two time series *X* and *Y* which we have given above: $X = \{x_1, \ldots, x_n\}$ and $Y = \{y_1, \ldots, y_m\}$. To align these two sequences using DTW, we construct an *n*-by-*m* matrix where the (*i*th, *j*th) element of the matrix contains the distance $d(x_i, y_j)$ between the two points x_i and y_j. Each matrix element (i, j) corresponds to the alignment between the points x_i and y_j. This is illustrated in Fig. 1. A warping path, *W*, is a contiguous (in the sense stated below) set of matrix elements that defines a mapping between *X* and *Y*. The *l*th element of *W* is defined as $w_l = (i, j)_l$, so we have:

$$W = w_1, w_2, \ldots, w_l, \ldots, w_L \qquad \max(m, n) \le L < m + n - 1 \quad (2)$$

The warping path is typically subjected to several constraints:

Boundary conditions: $w_1 = (1, 1)$ and $w_L = (m, n)$. Simply stated, this requires the warping path to start and finish in diagonally opposite corner cells of the matrix.

Continuity: Given $w_l = (a, b)$ then $w_{l-1} = (a', b')$, where $a - a' \le 1$ and $b - b' \le 1$. This restricts the allowable steps in the warping path to adjacent cells (including diagonally adjacent cells).

Monotonicity: Given $w_l = (a, b)$ then $w_{l-1} = (a', b')$, where $a - a' \ge 0$ and $b - b' \ge 0$. This forces the points in *W* to be monotonically spaced in time.

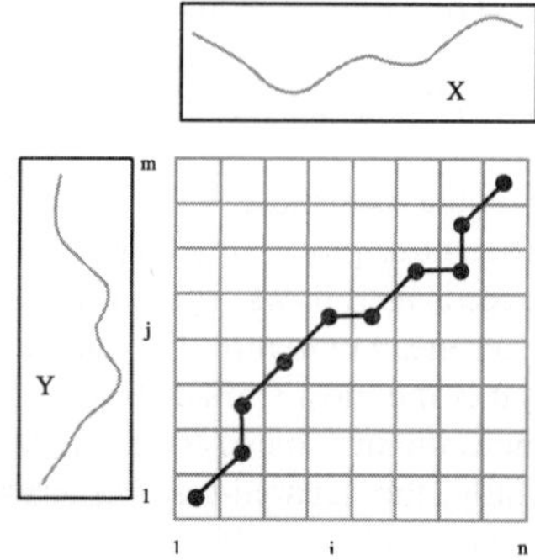

Fig. 1 An example of the warping path.

There are exponentially many warping paths that satisfy the above conditions; however, we are interested only in the path which minimizes the warping cost:

$$\mathrm{DTW}(X,Y) = \min\left\{\frac{1}{L}\sqrt{\sum_{l=1}^{L} w_l}\right\} \tag{3}$$

The L in the denominator is used to compensate for the fact that warping paths may have different lengths. This path can be found very efficiently using dynamic programming to evaluate the following recurrence which defines the cumulative distance $r(i, j)$ as the distance $d(i, j)$ found in the current cell and the minimum of the cumulative distances of the adjacent elements:

$$r(i,j) = d(x_i, y_j) + \min\{r(i-1, j-1), r(i-1, j), r(i, j-1)\} \tag{4}$$

This review of DTW is necessarily brief; we refer the interested readers to Kruskall & Liberman (1983) for a more detailed treatment.

A DATA MINING PROCEDURE FOR HYDROLOGICAL TIME SERIES

The steps of the data mining procedure for hydrological time series analysis used in this work are outlined below (Fig. 2). The procedure mainly involves two data mining algorithms, a clustering algorithm and similarity search, which are discussed above and will be further described with hydrological time series analysis.

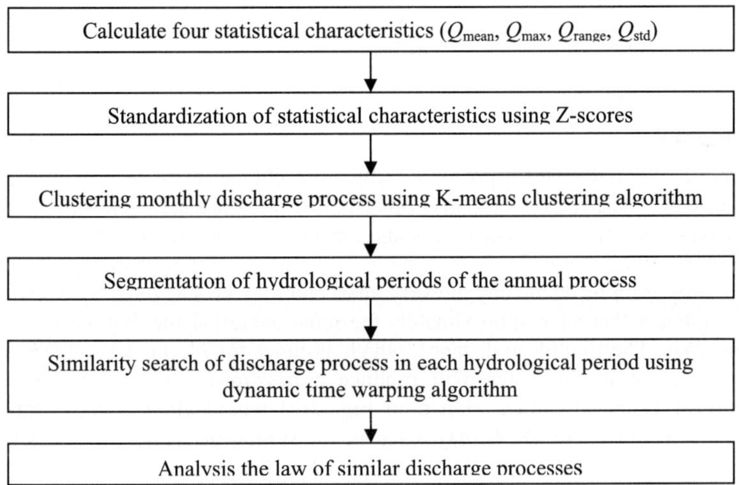

Fig. 2 Flowchart of data mining for hydrological time series analysis.

Clustering analysis

The change of hydrological phenomena is complex, fuzzy and random, but it has its regularity simultaneously. The work of clustering analysis for historical hydrological data makes the advantage of mining the laws in the similar hydrological process and improving the accuracy in hydrological prediction by parameters classification and models classification. The K-means was applied for clustering the processes of monthly discharge based on four statistical characteristics estimated from the discharge process for each month. If the given monthly discharge time series is denoted by $Q = \{q_1, \ldots, q_n\}$, where n is equals the total number of days in the given month, the formulas of these four statistical characteristics can be expressed (Table 1).

All of the monthly statistical characteristics were calculated as the attributes of each month. Then each of the statistical characteristics were standardized using Z-scores (mean = 0; standard deviation = 1) prior to cluster analysis in order to remove major variations that can adversely affect the output from cluster analysis.

The standardized values of these statistical characteristics (N attributes) for each of the month in the discharge time series (M cases) were clustered to achieve the clustering of monthly discharge process using K-means clustering algorithm. Months with a similar discharge process could be grouped together, so that months that are similar to one another within the same cluster and are dissimilar to other months in other clusters. M cases are grouped in k clusters, so 12 months in a year have different distributions in the clusters. If the months can be found with the same distribution in the clusters, they may represent the identical hydrological period. In this case, the annual discharge process can be segmented into some hydrological periods.

Table 1 Statistical characteristics of monthly discharge.

Symbol	Statistical characteristics	Formula
Q_{mean}	Mean monthly discharge	$Q_{mean}=\frac{1}{n}\sum_{i=1}^{n}x_i$
Q_{max}	Monthly maximum discharge	$Q_{max}=\max(x_1,x_2,\cdots,x_n)$
Q_{range}	Monthly discharge range	$Q_{range}=Q_{max}-Q_{min};Q_{min}=\min(x_1,x_2,\cdots,x_n)$
Q_{std}	Standard deviation of discharge	$Q_{std}=\frac{1}{n}\sum_{i=1}^{n}(x_i-Q_{mean})^2$

Similarity search and DTW

As was mentioned above, the hydrological processes which are characterized by seasonal nature always show some similarity. The reason they are similar is that the associated controlling factors of hydrological processes have more consistent seasonal changes. Of course, the changes are not entirely consistent. For example, storm floods in summer are always similar processes in overall shape, but they do not appear on the same day in every year, and may be staggered by a few days. Consider two flood sequences that have approximately the same overall shape, but the shapes are not aligned in the time axis. The Euclidean distance, which assumes the ith point in one sequence is aligned with the ith point in the other, will produce a pessimistic dissimilarity measure (Fig. 3(a)). The nonlinear DTW alignment shown in Fig. 3(b) would allow a more intuitive distance measure to be calculated. The DTW may discover the similar discharge processes which are not aligned on the time axis.

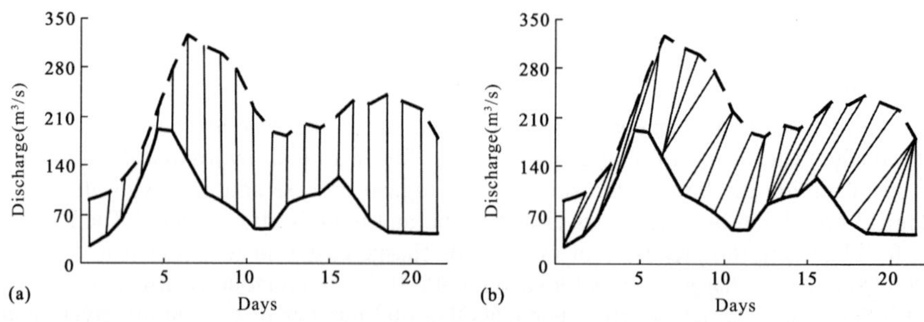

Fig. 3 Different distance measurement methods: (a) Euclidean; and (b) DTW.

Instead of Euclidean distance, similarity search of the discharge time series processes was achieved by calculated DTW distances between every two discharge time series in each hydrological period which has been segmented after the *K*-means clustering algorithm. The lower the DTW distance means, the greater the similarity between the two discharge time series.

APPLICATION OF DATA MINING

Study area and basic data sets

This novel data mining procedure for hydrological time series analysis was applied to the daily discharge time series at Shaligunlanke station, on the Aksu River in the Tarim River basin in the inland arid northwestern part of China, for the years 1961–2000. Shaligunlanke station (78°36′E; 40°57′N) is located at the mountain pass of the Aksu River, the headstream of the Tarim River, and controls a catchment area larger than 1.84×10^4 km^2 with a mean elevation of 3328 m. The vast territory, diverse topography and complex terrain in the basin lead to typical temperate continental dry-climate regional characteristics, such as minimal precipitation, intense evaporation, adequate sunshine and extreme heat (Tang *et al.*, 1992).

Because of the glacier and aged snow in the high mountain zone of the basin, the combination of winter snowmelt in summer and summer precipitation forms the hydrological process of Shaligunlanke station. So, the floods are typical of snowmelt floods or snowmelt mixed with cloudburst floods. Table 2 is indicative of significant seasonal and annual variations of precipitation and discharge: precipitation which is centralized in summer causes uneven distribution of the annual discharge process that accounts for a large portion in summer too.

Table 2 Annual distributions of precipitation and discharge.

	Annual total quantity (10^8 m^3)	Spring (%)	Summer (%)	Autumn (%)	Winter (%)
Precipitation	64.96	20–25	55–60	14–18	3–7
Discharge	26.50	16.8	62.1	16.4	4.7

The other two sets of hydrological time series—precipitation data and temperature data—are also tested to analyse the causes of discharge similarity in hydrological data mining work. Since discharge has roughly the same distribution tendency with the precipitation in the basin, precipitation data are essential to our data mining experiment. However, it is impossible to clearly and completely measure precipitation conditions in such a large basin, of which the elevation range exceeds 5000 m. We collected precipitation data from five rainfall stations, of which only Shaligunlanke station is inside the basin, the others being outside but nearby the basin. We used 500 hPa upper air temperature data for the years 1985–2000 to analyse its relationship with daily discharge. Moreover, this correlation analysis between temperature and discharge may confirm the supply source of the runoff at Shaligunlanke station on the Aksu River. Note that the upper air temperature data cannot match all discharge data, which date from the years 1961–2000, so the correlation analysis was performed on the periods for which the discharge time series and temperature time series overlap.

Clustering analysis

Four statistical characteristics, Q_{mean}, Q_{max}, Q_{range} and Q_{std}, were calculated and standardized in each month from 1961 to 2000; that is, 480 cases (M = 480) were produced from 40 years of discharge data and every case had four attributes (N = 4). Then the *K*-means clustering was performed on 480 cases, setting the number of clusters to k = 4. The result of the clustering analysis is shown in Table 3.

According to the distribution of cases in the clusters, some processes of monthly discharge can be found with five types of distribution in the clusters:

1. all the cases of January, February, March, October, November and December are grouped in Cluster 3;
2. the cases of April account for 65% in Cluster 3, 25% in Cluster 4 and 10% in Cluster 2;
3. the cases of May account for about 20% in Cluster 3, 65% in Cluster 4 and 10% in Cluster 2;
4. the cases of September account for about 70% in Cluster 3 and 25% in Cluster 4;
5. the cases of June, July and August account for about 70% in Cluster 4 and 25% in Cluster 2.

Then the annual process of the daily discharge was correspondingly divided into five hydrological periods, referred to herein as: Snowmelt period I (April), Snowmelt period II (May), Rainfall period I (June–August), Rainfall period II (September) and Dry period (October–December and January–March). Figure 4 shows the five hydrological periods clearly.

Table 3 The result of the processes of monthly discharge clustering – number of cases.

	Cluster°1	Cluster°2	Cluster°3	Cluster°4	Total
January	0	0	40	0	40
February	0	0	40	0	40
March	0	0	40	0	40
April	0	4	26	10	40
May	1	4	8	27	40
June	1	10	1	28	40
July	1	11	0	28	40
August	2	11	0	27	40
September	0	1	29	10	40
October	0	0	40	0	40
November	0	0	40	0	40
December	0	0	40	0	40
Total	5	40	305	130	480

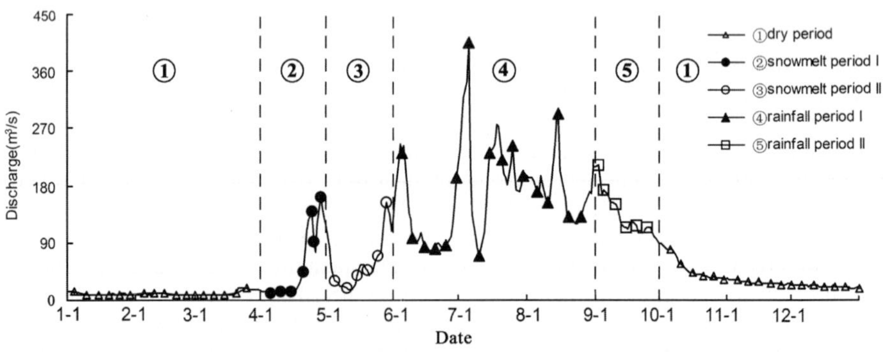

Fig. 4 The five hydrological periods identified.

When the temperature begins to rise in April, snowmelt occurs during the snowmelt season and makes up the major water source for discharge generation. In May, the water quantity of snowmelt is increasing as the temperature continues to rise. From June to August, the summer precipitation has a strong effect upon the generation of discharge with heavy-volume and high peak floods. Discharge decreases gradually with a decrease in temperature in September. In the following months, it enters into the recession season until March the following year. Different climates result in different discharge processes, so it would be reliable to research the hydrological discharge processes under the same formation mechanism.

Similarity search

From 1961 to 2000, there are 40 sets of discharge time series in a certain hydrological period. Suppose two sets of discharge time series Q_u and Q_v, where the subscript denotes the year the discharge time series in. Then, we can construct a 40 × 40 similarity distance matrix, where the (uth, vth) element of the matrix equals the DTW distance DTW(Q_u, Q_v) between the two discharge processes Q_u and Q_v. Table 4 lists a part of the similarity DTW distance matrix in Snowmelt period I and matrices in other periods have the same tabular form. The lower the values of the elements of the similarity distance matrix and the lower the DTW distance the two discharge processes, the more similar the discharge processes are. The diagonal elements all equal 0, which means the DTW distance of one discharge process itself is 0; that is to say, one discharge process is the most similar or the same as itself. That is self-evident but of little practical significance.

Based the similarity distance matrix, the similar discharge processes can be mined in each period. Some examples of similar discharge processes with small values of DTW distance are shown in Fig. 5, including the four hydrological periods, except the Dry period, which is because the discharge process is low and unfluctuating in the Dry period.

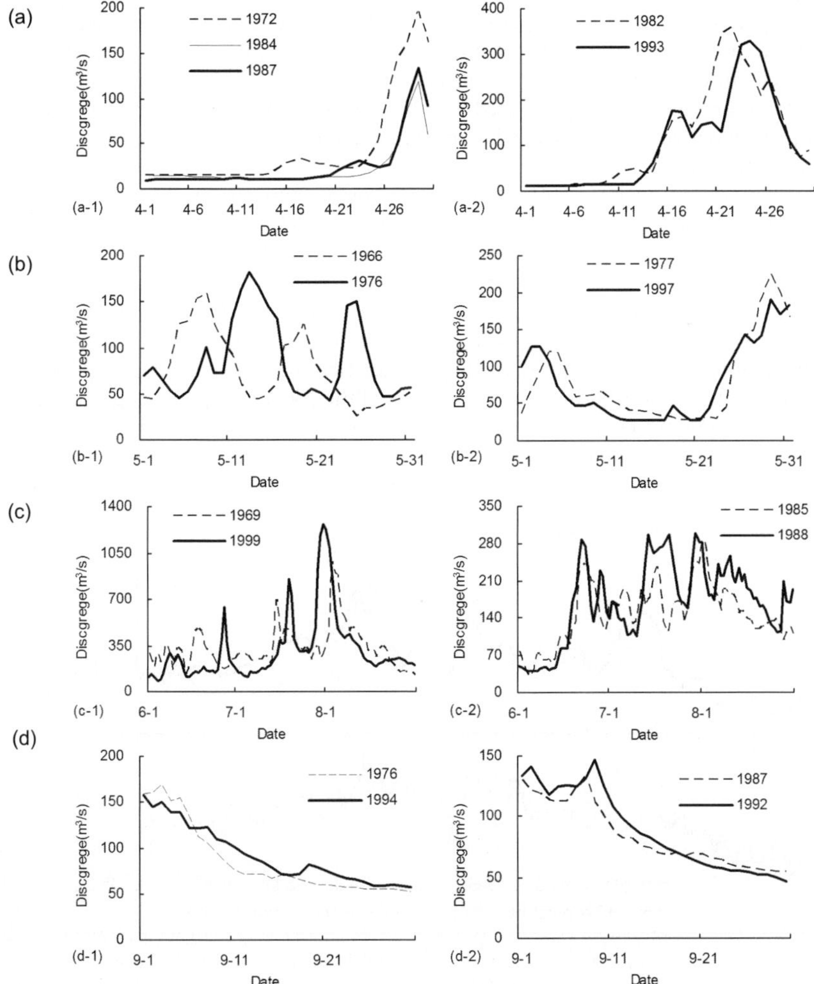

Fig. 5 Examples of similar discharge processes in four hydrological periods: (a) Snowmelt period I; (b) Snowmelt period II; (c) Rainfall period I; and (d) Rainfall period II.

Table 4 Part of the similarity DTW distance matrix in Snowmelt period I (1961–2000).

	Q_{1961}	Q_{1962}	Q_{1963}	…	Q_u	…	Q_{1998}	Q_{1999}	Q_{2000}
Q_{1961}	0.000	3.572	2.996	…	…	…	1.576	3.237	3.066
Q_{1962}	3.572	0.000	5.613	…	…	…	3.626	2.614	4.468
Q_{1963}	2.996	5.613	0.000	…	…	…	2.065	3.798	2.760
…	…	…	…	…	…	…	…	…	…
Q_v	…	…	…	…	…	…	…	…	…
…	…	…	…	…	…	…	…	…	…
Q_{1998}	1.576	3.626	2.065	…	…	…	0.000	3.045	1.932
Q_{1999}	3.237	2.614	3.798	…	…	…	3.045	0.000	2.917
Q_{2000}	3.066	4.468	2.760	…	…	…	1.932	2.917	0.000

Results analysis

The domain knowledge of the data has not been given adequate consideration in the work of data mining before, which may explain why the results of data mining cannot be understood by users. For this reason, an attempt is made to use hydrological knowledge to analyse and interpret the results from the previous data mining experiment for hydrological discharge time series.

Hydrological processes reflect the changes of climate, geography, environment and the underlying surface in the basin, so the seasonal variation of discharge at Shaligunlanke station in an arid mountainous region is mainly affected by air temperature, besides precipitation. Therefore, both the relationship between discharge and temperature and that between discharge and precipitation are analysed below.

From the comparison chart of 500 hPa upper air temperature process curves and discharge process curves from June to August in Rainfall period I in 1985 and 1988 (Fig. 6), it is obvious that the upper air temperature fluctuation process consists of a discharge fluctuation process in some measure. The correlation coefficients between these four time series processes are shown in Table 5. The range of the values of the correlation coefficient is 0.570–0.725. It can be deduced that the discharge had a close relationship with the temperature in the basin.

The combination of the winter snow melting in summer and the summer precipitation forms the hydrological discharge process in the flood seasons of the Aksu River (Ouyang *et al.*, 2007), so precipitation is an important factor in discharge generation. The average precipitation data which originated from the five rainfall stations inside and nearby the basin were calculated to

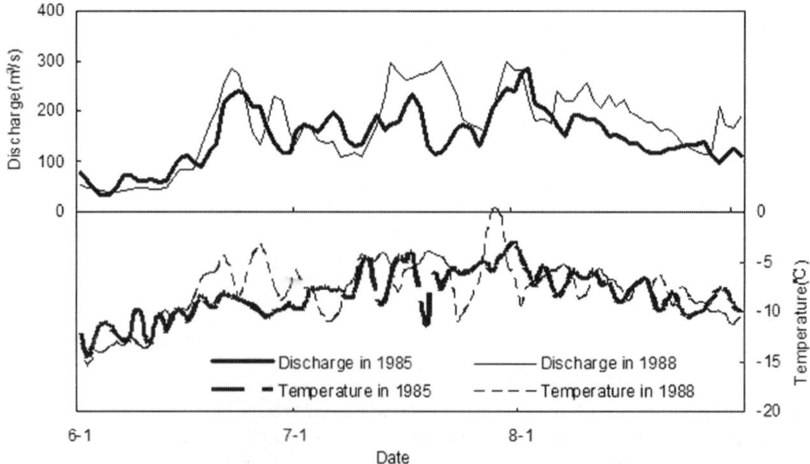

Fig. 6 Comparison of 500 hPa upper air temperature process curves and Shaligunlanke station discharge process curves from June to August, in Rainfall period I: 1985 and 1988.

Table 5 The correlation coefficients of discharge time series and temperature time series from June to August in Rainfall period I: 1985 and 1988.

	Discharge 1985	Discharge 1988	Temperature 1985	Temperature 1988
Discharge 1985	1.000	0.725	0.685	0.616
Discharge 1988	0.725	1.000	0.674	0.658
Temperature 1985	0.685	0.674	1.000	0.570
Temperature 1988	0.616	0.658	0.570	1.000

approximately represent the precipitation in this basin. We selected two similar discharge processes from June to August in Rainfall period I in 1969 and 1999, which mined in the previous work of similarity search. The precipitation processes corresponding to discharge processes in the same periods are also shown in Fig. 7. The precipitation record indicates that heavy rainstorm events are strongly associated with the extreme discharge processes. It is obvious that the three peaks of discharge process are corresponding to three processes of heavy rainstorm. It can be proved that the hydrological discharge process has a strong relationship with the precipitation in summer season in the basin.

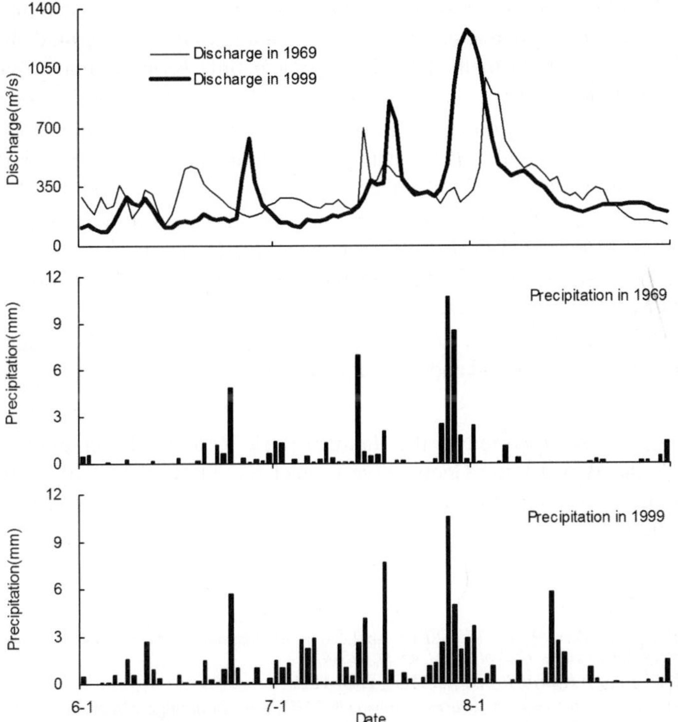

Fig. 7 Shaligunlanke station discharge process curves and precipitation charts from June to August in Rainfall period I: 1969 and 1999.

CONCLUSIONS AND FUTURE WORK

In this paper, data mining algorithms including clustering and similarity search were selected to establish a data mining procedure for hydrological time series analysis. This data mining procedure was applied to daily discharge time series records of the Tarim River basin and some similar hydrological discharge processes were mined successfully. The result shows that this novel data mining procedure can objectively find out the similar hydrological discharge processes. After comparison of precipitation and temperature conditions in the corresponding periods, it can be concluded that the discharge processes were more similar under the same climatic conditions.

Though data mining has many algorithms to solve various problems in our society; not all of the algorithms are suitable for the domain data. Data mining has its practical meaning only when it is combined to the particular fields using the domain knowledge. As this paper showed, data mining is an objective feasible efficient method for hydrological time series analysis, and can be used to discover the hidden information in the historical hydrological data and to mine the implicative laws under the hydrological process. The data mining approach has created a new intelligent tool to extract useful hydrological information and knowledge automatically. It will certainly help us advance hydrological sciences even further. It can be foreseen that data mining will be further combined with hydrology.

At present, the work of data mining for hydrological time series analysis is still in the early stages and mainly focuses on single and local hydrological data; while little work has been done on multi-factors data mining based on the global hydrological data and on hydrological data adaptation. In order to fully utilize the function of the knowledge discovery in the field of data mining, there are several directions in which we intend to extend the work of data mining for hydrological time series analysis:

(a) *Hydrological time series classification, clustering and association rules* In such numerous methods of classification, clustering and association rules in data mining, it is the fundamental issue to search for the most suitable method for hydrological time series characterized by periodicity, tendency and randomness.

(b) *Hydrological time series similarity, periodicity and sequential patterns* Periodicity can be regarded as a similarity problem, and also as frequent patterns in data mining. The methods of hydrological periodicity may be developed in another new way.

(c) *Multi-dimensional hydrological data mining, including discharge, temperature, precipitation and other hydrological data* Because data mining can mine all kinds of data, it can be used in mining the possible relationships of multi-dimensional hydrological data to discover the essential mechanism of hydrological processes.

(d) Hydrological data mining software and data platform This will enable the results of hydrological data mining to be understood and further applied by users.

Acknowledgements This work was supported by the National Basic Research Program of China (Grant no. 2006CB400502), the World Bank Cooperative Project (Grant No. THSD-07) and the 111 Programme of the Ministry of Education and the State Administration of Foreign Expert Affairs, China (Grant no. B08048).

REFERENCES

Abebe, A. J., Solomatine, D. P. & Venneker, R. G. W. (2000) Application of adaptive fuzzy rule-based models for reconstruction of missing precipitation events. *Hydrol. Sci, J.* **45**(3), 425–436.

Babovic, V. (2005) Data mining in hydrology. *Hydrol. Processes.* **19**(7), 1511–1515.

Baldi, P. & Brunak, S. (2001) *Bioinformatics: the Machine Learning Approach.* MIT Press, Cambridge, Massachusetts, USA.

Baxevanis, A. D. & Ouellette, B. F. F. (2001) *Bioinformatics: a Practical Guide to the Analysis of Genes and Proteins.* Wiley-Interscience, New York, USA.

Benninga, S. (2000) *Financial Modeling.* MIT Press, Cambridge, Massachusetts, USA.

Berry, M. J. A. & Linoff, G. S. (2004) *Data Mining Techniques: for Marketing, Sales, and Customer Relationship Management.* Wiley Computer Publishing, USA.

Berson, A., Smith, S. & Thearling, K. (1999) *Building Data Mining Applications for CRM.* McGraw-Hill Professional, New York, USA.

Berndt, D. & Clifford, J. (1994) Using dynamic time warping to find patterns in time series. In: *AAAI-94 Workshop on Knowledge Discovery in Databases* (ed. by U. M. Fayyad & R. Uthurusamy) (Seattle, Washington, USA, 31 July 1994), 359–370. AAAI Press, California, USA.

Chu, S., Keogh, E., Hart, D. & Pazzani, M. (2002) Iterative deepening Dynamic Time Warping for time series. In: *Proc. Second SIAM Int. Conf. on Data Mining* (ICDM'02, Maebashi, Japan, December 2002) (ed. by R. Grossman). SIAM Publ., Philadelphia, USA.

Das, G., Lin, K., Mannila, H., Renganathan, G. & Smyth, P. (1998) Rule discovery from time series. In: *Proc. Fourth Int. Conf. on Knowledge Discovery and Data Mining* (ed. by R. Agrawal, P. E. Stolorz & G. Piatetsky-Shapiro) (KDD'98, New York, USA, August 1998), 16–22. AAAI Press, California, USA.

Fu, T. C., Chung, F. L., Ng, V. & Luk, R. (2001) Pattern discovery from stock time series using self-organizing maps. In: *Proc. Seventh ACM SIGKDD Int. Conf. on Knowledge Discovery and Data Mining* (ed. by D. Lee, M. Schkolnick, F. Provost & R. Srikant) (KDD'01, San Francisco, USA, August 2001), 27–37. ACM Press, New York, USA.

Goutte, C., Toft, P., Rostrup, E., Nielsen, F. & Hansen, L. K. (1999) On clustering fMRI time series. *Neuroimage* **9**(3), 298–310.

Govindaraju, R. S. & Rao, A. R. (2000) *Artificial Neural Networks in Hydrology*. Kluwer Academic, Dordrecht, The Netherlands.

Gusfield, D. (1997) *Algorithms on Strings, Trees, and Sequences*. Cambridge University Press, New York, USA.

Han, J. W. & Kamber, M. (2001) *Data Mining: Concepts and Techniques*. Morgan Kaufmann, San Francisco, California, USA.

Hannah, D. M., Smith, B. P. G., Gurnell, A. M. & McGregor, G. R. (2000) An approach to hydrograph classification. *Hydrol. Processes* **14**(2), 317–338.

Higgins, R. C. (2001) *Analysis for Financial Management*. Irwin/McGraw-Hill, Boston, Massachusetts, USA.

Kalpakis, K., Gada, D. & Puttagunta, V. (2001) Distance measures for effective clustering of ARIMA time series. In: *Proc. of IEEE Int. Conf. on Data Mining* (ed. by V. Kumar & R. Grossman) (ICDM '01, Düsseldorf, Germany, April 2001), 273–280. SIAM Publ., Philadelphia, USA.

Keogh, E. (2002) Exact indexing of dynamic time warping. In: *Proc. 28th Int. Conf. on Very Large Data Bases* (ed. by P. Bernstein, Y. Ioannidis, R. Ramakrishnan & D. Papadias) (VLDB 2002, Hong Kong, China, 20–23 August 2002), 406–417. Morgan Kaufmann, San Francisco, California, USA.

Knab, B., Schliep, A., Steckemetz, B. & Wichern, B. (2003) Model-based clustering with Hidden Markov Models and its application to financial time-series data. In: *Between Data Science and Applied. Data Analysis* (ed. by M. Schader, W. Gaul & M. Vichi), 561–569. Springer, Cologne, Germany.

Kruskal, J. B. & Liberman, M. (1983) The symmetric time warping algorithm: from continuous to discrete. In: *Time Warps, String Edits and Macromolecules: The Theory and Practice of String Comparison*. (ed. by D. Sankoff & J. B. Kruskal), 125–162. Addison-Wesley, New York, USA.

Oates, T., Firoiu, L. & Cohen, P. R. (1999) Clustering time series with Hidden Markov Models and Dynamic Time Warping. In: *Proc. IJCAI-99 Workshop on Neural, Symbolic and Reinforcement Learning Methods for Sequence Learning* (ed. by T. Dean) (IJCAI'99, Stockholm, Sweden, July 1999), 17–21. Morgan Kaufmann Publ., San Francisco, USA.

Ouyang, R. L., Cheng, W. M., Jiang, Y., Zhang Y. C. & Chen, X. (2007) Research on runoff forecast approaches to the Aksu River basin. *Science in China* (Series D: Earth Sciences), **50**(s1):16–25.

Piatetsky-Shapiro, G. & Frawley, W. J. (1991) *Knowledge Discovery in Databases*. AAAI/MIT Press, Boston, USA.

Rafiei, D. & Mendelzon, A. (1998) Efficient retrieval of similar time sequences using DFT. In: *Proc. Fifth Int. Conf. on Foundations of Data Organizations and Algorithms* (ed. by K. Tanaka & S. Ghandeharizadeh) (FODO'98, Kobe, Japan, November 1998), 69–84. Kluwer Academic, Dordrecht, The Netherlands.

Salas, J. D., Delleur J. W., Yevjevich, V. & Lane, W. L. (1980) *Applied Modeling of Hydrologic Time Series*. Water Resources Publications, Littleton, Colorado, USA.

Solomatine, D. P. (2002) Applications of data-driven modeling and machine learning in control of water resources. In: *Computational Intelligence in Control* (ed. by M. Moammadian, R. A. Sarker & X. Yao), 197–217. Idea Group Publishing, USA.

Solomatine, D. P., Rojas, C., Velickov, S. & Wust, H. (2000) Chaos theory in predicting surge water levels in the North Sea. In: *Proc. Fourth Int. Conf. on Hydroinformatics 2000* (Iowa Institute of Hydraulic Research, Iowa City, USA, July 2000), 1–8. University of Iowa College of Engineering, USA.

Tang, Q. C., Qu, Y. G. & Zhou, Y. C. (1992) *Hydrology and Water Resources of Arid Area in China*. Science Press, Beijing, China (in Chinese).

Van, W. J. J. & Van, S. E. R. (1999) Cluster and calendar based visualization of time series data. In: *Proc. IEEE Symp. on Information Visualization* (ed. by P. C. Wong, P. Whitney & J. Thomas) (InfoVis'99, San Francisco, USA, November 1999), 4–9. IEEE Computer Society, Washington, USA.

Vlachos, M., Lin, J., Keogh, E. & Gunopulos, D. (2003) A wavelet-based anytime algorithm for K-means clustering of time series. In: *Int. Conf. on Data Mining, Workshop on Clustering High Dimensionality Data and its Application* (ed. by D. Barbará & C. Kamath) (ICDM' 03, San Francisco, USA, May 2003), 23–30. SIAM Publ., Philadelphia, USA.

Waterman, M. S. (1995) *Introduction to Computational Biology*. Chapman & Hall, New York, USA.

Zhang, H., Ho, T. B. & Lin, M. S. (2004) A non-parametric wavelet feature extractor for time-series classification. In: *Proc, Eighth Pacific-Asia Conf on Knowledge Discovery and Data Mining* (ed. by H. Dai , R. Srikant & C. Zhang) (PAKDD'04, Sydney, Australia, May 2004), 595–603. Springer-Verlag, Berlin, Germany.

Improving the disaggregation of daily rainfall into hourly rainfall using hourly soil moisture

SAT KUMAR[1], M. SEKHAR[1] & D. V. REDDY[2]

1 *Department of Civil Engineering, Indian Institute of Science, Bangalore 560 012, India*
satkumar@civil.iisc.ernet.in

2 *National Geophysical Research Institute, Hyderabad 500 007, India*

Abstract For various applications in hydrology, rainfall in short time steps (hourly) is needed. As direct measurements of sub-hourly rainfall are hardly possible in practice (especially in developing and under-developed countries), a potential alternative may be to try to derive such data from the available daily data through a disaggregation procedure. Random multiplicative cascade (RMC) models appear as an appealing solution to disaggregate the daily rainfall into sub-hourly rainfall. The RMC model provides a number of ensembles of hourly rainfall; taking the ensemble average one can get the hourly rainfall. This model is reasonably good in providing the magnitude of hourly rainfall, but fails to give the exact time of occurrence of hourly rainfall. Improvements can be made in the time of occurrence of hourly rainfall with the help of variables, which are sensitive to hourly rainfall. Soil moisture is one such a variable, which is directly sensitive to hourly rainfall. This paper deals with the potential of using hourly soil moisture in the improvement of hourly rainfall. Using this RMC model, ensembles of time series of hourly rainfall are generated. Among these ensembles of hourly rainfall, best ensembles are identified using generalized likelihood uncertainty estimate (GLUE) approach combined with the soil moisture model. The RMC model is applied and the performance is tested on several station data in the semi-arid and humid regions of South India. The model is able to capture the magnitude of hourly rainfall. Further, the RMC model is combined with a soil moisture data at a station in a semi-arid zone having hourly soil moisture profile data over a monsoon season. A soil moisture model was developed based on the Richards equation, and parameters of the model were calculated through laboratory experiments. The hourly rainfall is modelled through the calibration of soil moisture data at this station. It is shown that using the hourly measured soil moisture, improved estimates of hourly rainfall is feasible.

Key words rainfall disaggregation; soil moisture; multiplicative random cascade; GLUE

INTRODUCTION

In most hydrological applications rainfall data in short time steps is required. However, rainfall data on a short time scale is limited worldwide, mainly because of the high cost and low reliability of monitoring such data. Rainfall data at daily time steps is available worldwide because of the low cost and ease of collection, therefore a possible alternative to provide short time step rainfall data is to use the daily data to retrieve short time scale data. There are various models available for disaggregation of daily rainfall into short time scales, reproducing the statistics such as mean, variance, autocorrelations and the probability of an interval being dry (referred to as dry probability). Such disaggregated rainfall can only be used in hydrological applications where only statistics of the data are used, rather than the data itself. For a wide variety of hydrological applications, where directly short time scale rainfall is required, e.g. in the case of a distributed hydrological model, the model should reproduce the true short time scale rainfall.

There have been various successful attempts to disaggregate the rainfall, e.g. a direct approach (e.g. Arnaud, 2002), Poisson cluster model (e.g. Deidda, 2000), artificial neural network (e.g. Burian & Durrans, 2002), as well as models based on the principle of simulated annealing (e.g. Bardossy, 1998). Among these models, the multiplicative cascade disaggregation model appears to be promising and therefore has received growing attention during the last decade (Menabde *et al.*, 1997; Olsson, 1998; Deidda, 2000). Under certain conditions, these models produce rainfall series exhibiting scaling (multifractal) properties. The invariance scaling behaviour of rainfall has been found over a large area (Schertzer & Lovejoy 1987; Tessier *et al.*, 1993; Olsson & Niemczynowicz, 1996; Over & Gupta 1996; Svensson *et al.*, 1996) and time steps (Olsson *et al.*, 1993; Olsson, 1996; Burlando & Rosso 1996; Tessier *et al.*, 1996). The scaling nature of rainfall is still discussed (Marani, 2003, 2005). Nevertheless, multiplicative cascade models appear as appealing rainfall simulation tools because of their link with the multifractal theory (Gaume, 2007). Moreover, they are equally adapted for the simulation of rainfall in space and time, are parameter parsimonious, and theoretically easy to calibrate and use. Nevertheless, their calibration and validation give rise to some difficulties and questions which have seldom been commented upon.

The main objective of this study is to illustrate the calibration and validation of a random cascade rainfall model and to elaborate their capability to produce the truth. It has been shown that with the help of the variables which are sensitive to hourly rainfall, e.g. hourly soil moisture, improvement can be made to produce the disaggregated rainfall for the simulation of point–rainfall time series to be used in a distributed hydrological model. This paper is organised in three parts. The first part presents the general principles of random multiplicative cascade models, their multifractal properties and the choice of a random cascade model. The issue involved in the calibration and validation, i.e. to produce the truth, is discussed in the second part of the paper. The third part deals with the improvement in the disaggregated rainfall with the help of hourly soil moisture.

THEORY

Firstly, the concept of Random Multiplicative Cascade (RMC) model to disaggregate the rainfall is discussed. Then, the statistical moment scaling function method of scale identification to fit the RMC model is illustrated. Various types of distribution, which can be used to generate the RMCs, are given. Further, the Generalized Likelihood Uncertainty Estimate (GLUE) approach is discussed, which will be used to incorporate the hourly soil moisture into a rainfall disaggregation scheme to improve the disaggregated hourly rainfall. Finally, to use the soil moisture data, a soil moisture model based on the Richards equation is discussed.

Random multiplicative cascade model

Random multiplicative cascade models distribute mass of an interval on successive regular subdivisions in a multiplicative manner. This multiplicative random cascade models of rainfall disaggregation were successfully applied to rainfall modelling (e.g. Schertzer & Lovejoy, 1987; Over & Gupta, 1994, 1996; Olsson, 1998). There are two types of RMC model available, namely, a canonical model which preserves mass on the average in disaggregation, and a microcanonical model which preserves mass exactly in disaggregation. This section summarizes the basic theory behind random cascade models.

The basic cascading structure of the multiplicative random cascade model distributes rainfall on successive regular subdivisions, with b as the branching number. The ith interval after n levels of subdivision is denoted as Δ_n^i (there are i = 1, …, b_n intervals at level n). The dimensionless spatial scale is defined as $\lambda_n = b^{-n}$, i.e. $k_0 = 1$ at the 0th level of subdivision. The distribution of mass occurs via a multiplicative process through all levels n of the cascade, so that the mass in subcube Δ_n^i is:

$$\mu_n(\Delta_n^i) = R_0 \lambda_n \prod_{i=1}^{n} W_j(i), \text{ for } i = 1,2, \dots, b^n \ n>0 \tag{1}$$

Here R_0 is an initial rainfall depth at n = 0 and W is the so-called cascade generator. The basic structure of the discrete multiplicative random cascade model as used in this study is illustrated in Fig. 1, with b = 2 as the branching number and n = 0, 1, and 2.

Scaling identification: statistical moment scaling function method Properties of the cascade generator W can be estimated from the moment scaling behaviour across scales. Sample moments are defined as:

$$M_n(q) = \sum_{i=1}^{b^n} \mu_n^q(\Delta_n^i) \tag{2}$$

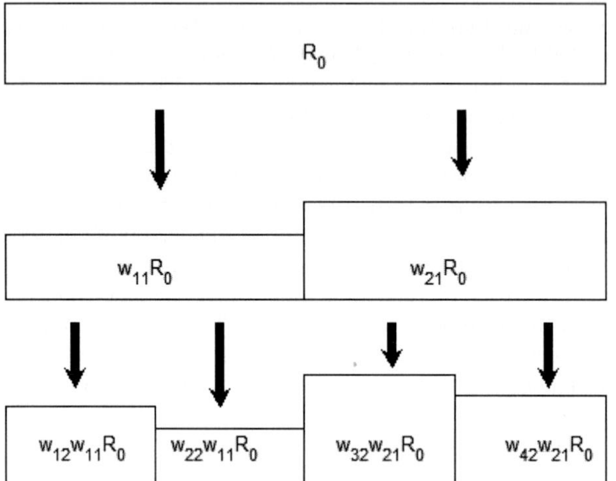

Fig. 1 Graphical representation of a random cascade process with $b = 2$.

here q is the moment order ($q > 0$). For large n, the sample moments should converge to the ensemble moments. However, because the ensemble moments diverge to infinity or converge to zero as $n \to \infty$, instead the (rate of) convergence/divergence of the moments with scale should be considered. In a random cascade, the ensemble moments are shown to be a log–log linear function of the scale of resolution k_n. The slope of this scaling relationship is known as the Mandelbrot–Kahane–Peyriere (MKP) function (Mandelbrot, 1974; Kahane & Peyriere, 1976) expressed as:

$$\chi_b(q) = 1 - q + \log_b E(W^q) \tag{3}$$

The MKP function contains important information about the distribution of the cascade generator W, and thus determines the scaling properties of rainfall. The slope of the sample moment scaling relationship can be found as:

$$\tau(q) = \lim_{\lambda_n \to 0} \frac{\log M_n(q)}{-\log \lambda_n} \tag{4}$$

For large n (as $n \to \infty$) and for a specific range of q, slopes of the moment scaling relationships for sample and ensemble moments converge, i.e.

$$\chi_b(q) = \tau(q) \tag{5}$$

In data analysis, the scaling exponent function $\tau(q)$ will be used to identify the best suited distribution for the random generator W of the cascade and to calibrate the parameters of this distribution. Some examples of PDF candidates for W and the corresponding expression of the $\tau(q)$ for the IID multiplicative cascade are presented in Table 1. The least square method is used to calibrate the parameters of each tested W's PDF on the empirical scaling exponent functions. Once the parameters of the PDF have been found, the cascade approach can be preceded in two ways. If the quantity R_0 is preserved on average, i.e. the expectancies of the W random numbers are equal to 1, then on average, the cascade process is then called canonical. If moreover, the quantity R_0 is preserved on each of the b sub time steps, the cascade process is called micro-canonical.

The Generalized Likelihood Uncertainty Estimate (GLUE)

The Generalised Likelihood Uncertainty Estimation (GLUE) technique was introduced partly to allow for the possible equifinality (non-uniqueness, ambiguity or non-identifiability) of parameter sets during the estimation of model parameters (inverse problem) in the overparameterised models. These problems may arise in the disaggregation of rainfall coupled with soil moisture. Hence in this work, GLUE has been used to identify the optimal rainfall. The basic idea of this approach follows from recognizing the equifinality of models and parameter sets. It is possible to give different degrees of belief to different models or parameter sets, and among them it may be possible to reject many of them because they clearly do not give the right sort of response for an application. The "optimum", given a data set for calibration, will have the highest degree of belief associated with it. The implementation of GLUE requires the likelihood measure of a different ensemble. Once such available likelihood measure is the modelling efficiency E of Nash & Sutcliffe (1970) expressed as:

$$E = 1 - \frac{\sigma_r^2}{\sigma_o^2} \tag{6}$$

where, σ_r^2 is the variance of the observation, and σ_o^2 is the variance of the errors.

Table 1 Expression of Probability density function of W and expression for $\tau(q)$.

Model	Probability density function	$\tau(q)$
β	$P(W = b^c) = b^{-c}, P(\eta = 0) = 1 - b^{-c}$	$\tau(q) = q(1-c) + c$
α	$P(W = b^{\gamma+}) = b^{-c}, P(W = b^{\gamma-}) = 1 - b^{-c}$	$\tau(q) = q - \frac{\ln\left[b^{q\gamma+-c} + \frac{(1-b^{\gamma+,-c})^q}{(1-b^{-c})^{q-1}}\right]}{\ln(b)}$
Log-Poisson	$W = A\beta^Y, P(Y = m) = \frac{c^m \exp(-c)}{m!}$	$q = -c\frac{q(1-\beta) + \beta^q - 1}{\ln(b)}$
Uniform	$P(W) = \frac{1}{b}, b \in [0, b]$	$q - \frac{\ln\left(\frac{b^q}{q+1}\right)}{\ln(b)}$
Log-Normal	$W = \beta^{m+\sigma Y}, P(Y) = \frac{1}{\sqrt{2\pi}}\exp(-Y^2/2)$	$q - \frac{(\sigma \ln \beta)^2 (q^2 - q)}{2\ln(b)}$

Soil moisture modelling

Flow in the unsaturated porous media can be described by the Darcy's law. Insertion of Darcy's law into continuity equation leads to the well known Richards equation.

$$\frac{\partial \theta}{\partial t} = \frac{\partial}{\partial z}\left(K\left(\frac{\partial \psi}{\partial z} - 1\right)\right) \tag{7}$$

where θ is volumetric soil moisture content, z is the elevation relative to a plane (positive downward), K is hydraulic conductivity, ψ is pressure head and t is time. Analytical solution of the Richards' equation for the field boundary conditions is not available. So one has to discretize the equation into space and time using either finite difference or finite element methods. A soil moisture model based on the finite element method has been developed in the MATLAB using the pde toolbox. The model has been tested with the standard numerical solution available for infiltration and exfiltration dynamics in the presence of stratification of stratified soil layers.

STUDY AREA

For testing the model, datasets of four stations located in South India, with extremely different climate conditions were used: Wailapally, Gundelpet, Maddur and Ambalvayal. Wailapally is situated in the western part of Nalgonda district, Andhra Pradesh. Hourly rainfall and hourly soil moisture at three depths 50 cm, 110 cm and 200 cm from the ground surface, are measured at this station. Apart from these, the daily rainfall is also measured separately. Due to battery failure and some other problems, the hourly rainfall data could sometimes not be measured, so only daily rainfall data is available. Since application of the distributed hydrological model requires fine rainfall, there is a need to get the hourly data from measured daily data. Gundelpet and Maddur come under the Chamrajanagar district of Karnataka. At Gundelpet station, 15-min rainfall data is measured, but length of data is very short. Maddur is an experimental watershed and has an automatic raingauge, which can measure hourly rainfall. Ambalvayal is located in the Wayanad district of Kerala and has hourly rainfall measurements. Table 2 shows the location and approximate annual rainfall in all four stations. It can be seen from the table that the approximate annual rainfall of stations varies from 600 mm to 2000 mm, which corresponds to a semi-arid to humid zone.

Table 2 Stations names, their latitude and longitude and approximate annual rainfall.

SN	Station name	Latitude (N)	Longitude (E)	Approximate average annual rainfall (mm)
1	Wailapally	17.095°	78.892°	650
2	Gundelpet	11.808°	76.691°	600
3	Maddur	11.776°	76.558°	950
4	Ambalavayal	11.557°	76.301°	2000

PROCEDURE AND RESULTS

Procedure of the disaggregation of daily rainfall data into hourly data can be described as:

- Firstly, the observed rainfall at available minimum time step has been taken.
- Then this rainfall is aggregate with $b = 2$ up to the higher time of available data, e.g. for station Wailapally the hourly rainfall is aggregated to two hourly and two hourly to four hourly, etc.
- Then the sample moments from equation (2) has been calculated with q varying from 0 to 5 at the interval of 0.5.
- Then the value of the $\tau(q)$ from equation (4), e.g. slope of the line of λ *vs* M for constant q's on a log–log plot has been calculated.
- Then appropriate distribution of W has been found by fitting the curve of $\tau(q)$ as given in Table 2.
- Once the distribution of W has been found out, W can be generated using the PDF of the distribution.
- W can be used in equation (1) to get the ensemble of hourly rainfall from the daily rainfall.
- Using these ensembles of disaggregated hourly rainfall, the ensemble of soil moisture has been generated. GLUE has been used to identify the best ensemble of soil moisture and corresponding rainfall.

Scaling behaviour of the q-moments

Figure 2 shows the empirical q-moments of the rainfall amount series with the time step on log–log plots for all the four stations. Moment of all orders shows linear relation with time step on log–log scale. Although some of the previous studies on rainfall series shows the different linear relationship with different range of scale, the number and the limits of the identified ranges of time scales vary from one study to another, and both are sensitive to rainfall patterns and hence to the climate.

Calibration of a rainfall disaggregation model

The scaling exponent function $\tau(q)$ has been used to identify the best suited distribution for the random generator W of the cascade and to calibrate the parameters of this distribution. The least

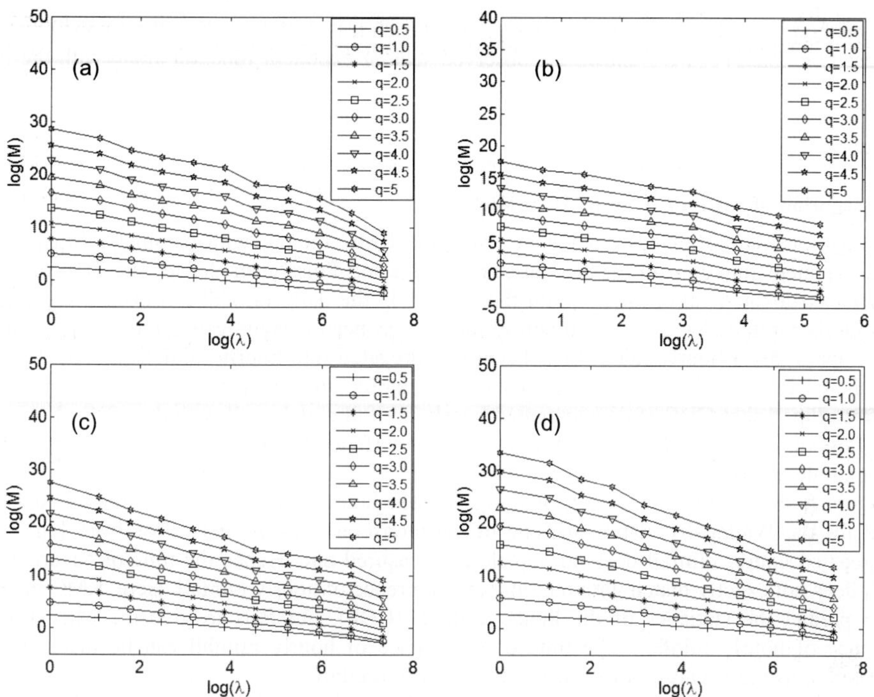

Fig. 2 Log–log plots of the empirical q-moments *versus time* step for all the stations: (a) Wailapally; (b) Gundelpet; (c) Maddur; (d) Ambalavayal.

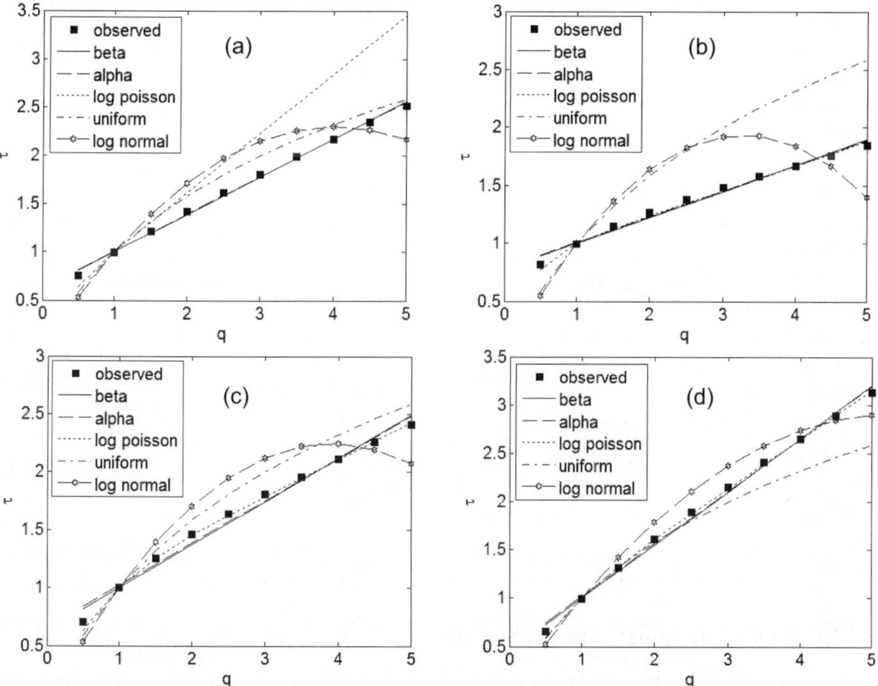

Fig. 3 Calibration of a cascade model on the basis of the scaling exponent function $\tau(q)$ for all the stations: (a) Wailapally; (b) Gundelpet; (c) Maddur; (d) Ambalavayal.

square method is used to calibrate the parameters of each tested PDF of W on the empirical scaling exponent functions. Figure 3 shows the observed scaling exponent function along with the fitted scaling exponent function. In all the four stations the log-Poisson leads to best fits to the empirical scaling exponent functions. However, the parameters of log Poisson cascade model differ for all the stations.

Disaggregation of rainfall from RMC model

Disaggregation of rainfall has been done using the RMC model, with parameter fitted on the scaling exponent function. In all the stations only the log Poisson model for the PDF of W has been selected for generation of W. With the help of W the daily rainfall has been disaggregated into 6-hourly rainfall (b = 2). Then 6-hourly rainfall has been disaggregated into 3-hourly rainfall (b = 2), again this 3-hourly rainfall has been disaggregated into hourly rainfall (b = 3). Figure 4 shows the CDF of the disaggregated hourly rainfall, along with the measured hourly rainfall. It is clear from Fig. 4 that CDF of disaggregated hourly rainfall match well with the CDF of observed hourly rainfall, except for rainfall of lower magnitude. Hence the RMC model is able to simulate the magnitude and frequency of hourly rainfall.

Figure 5 shows the comparison of disaggregated hourly rainfall with the measured hourly rainfall for the Wailapally station for one of the monsoon seasons. It is clear from Fig. 4 that disaggregated hourly rainfall does not match the measured one. Hence the time of occurrence of rainfall does not exactly match with the time of occurrence of hourly rainfall. Here, RMC model is able to predict the statistics of the hourly rainfall (CDF), but fails to give the exact time of occurrence of hourly rainfall. The time of occurrence of hourly rainfall can be made through hourly soil moisture because of its sensitivity to hourly rainfall.

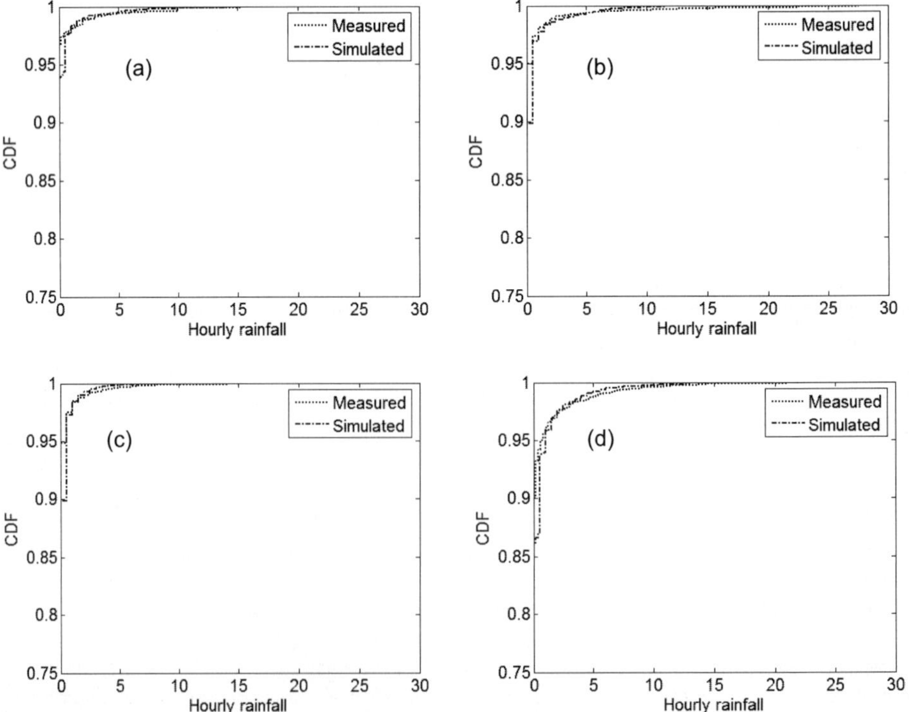

Fig. 4 CDF of measured *vs* simulated hourly rainfall as simulated by RMC for all the four stations: (a) Wailapally; (b) Gundelpet; (c) Maddur; (d) Ambalavayal.

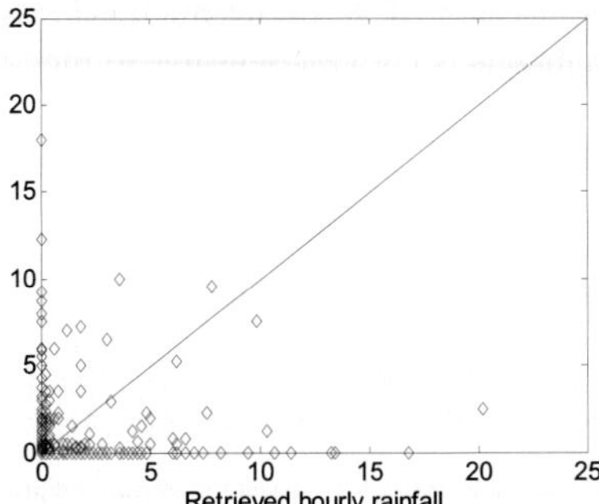

Fig. 5 Relation between measured and simulated hourly rainfall as simulated by RMC.

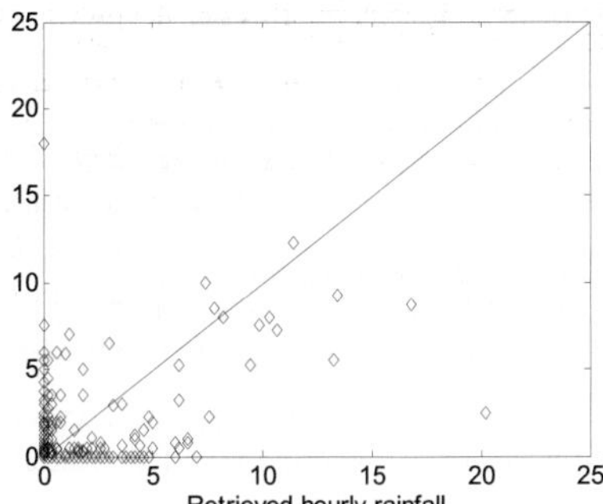

Fig. 6 Relation between measured and simulated hourly rainfall retrieved after combining the hourly soil moisture.

Figure 6 shows the disaggregated hourly rainfall as retrieved by combining the hourly root zone soil moisture with RMC. The combination of root zone soil moisture with the RMC model retrieved hourly rainfall better than the RMC alone. However, the rainfall of lower magnitude is not retrieved well. This may be due to the fact that either the RMC model itself is not giving the hourly rainfall of lower magnitude correctly as shown in Fig. 4, or due to the less sensitivity of root zone soil moisture with the rainfall. The rainfall of lower magnitude goes to atmosphere through evaporation and does not reach up to the root zone. Hence the root zone has no information about the rainfall. Availability of surface soil moisture can be useful in retrieving the hourly rainfall because of its sensitivity to rainfall of lower magnitude. Nevertheless this simulated hourly rainfall from hourly soil moisture can be used to predict the root zone soil moisture and deeper percolation more accurately.

CONCLUSIONS

This study illustrates the usefulness of the RMC approach, which is based on the moment scaling method, for disaggregation of daily rainfall into hourly rainfall. This has been done by disaggregation of rainfall at four stations located in extremely different climate conditions. All the four stations used in this study show the scaling behaviour, irrespective of the climate variability. The RMC model is able to predict the magnitude and frequency of hourly rainfall, but fails to give the exact time of occurrence of hourly rainfall. Further, the hourly measured root zone soil moisture is used to get the exact time of occurrence of hourly rainfall. Use of root zone soil moisture is able to improve the time of occurrence of hourly rainfall. However, the root zone soil moisture fails to predict the time of occurrence of hourly rainfall of lower magnitude because of its decreased sensitivity to rainfall of lower magnitude. Since the rainfall of lower magnitude is not very sensitive to root zone soil moisture, this approach can be used to simulate the root zone soil moisture and deep percolation.

REFERENCES

Arnaud, P. & Lavabre, J. (2002) Coupled rainfall model and discharge model for flood frequency estimation. *Water Resour. Res.* **6**, 1075.

Bardossy, A. (1998) Generating precipitation time series using simulated annealing. *Water Resour. Res.* **7**, 1734–1744.

Burian, S. & Durrans, S. R. (2002) Evaluation of an artificial neural network rainfall disaggregation model. *Water Sci Technol.* **2**, 99–104.

Burlando, P. & Rosso, R. (1996) Scaling and multiscaling models of depth–duration–frequency curves for storm precipitation. *J. Hydrol.* **187**, 45–64.

Deidda, R. (2000) Rainfall downscaling in a space-time multifractal framework. *Water Resour. Res.* **7**, 1779–1794.

Gaume, E., Mouhous, N. & Andrieu, H, (2007) Rainfall stochastic disaggregation models: Calibration and validation of a multiplicative cascade model. *Adv. Water Resour.* **30**, 1301–1319.

Kahane, J. P. & Peyriere, J. (1976) Sur certaines mertingales de Benoit Mandelbrot. *Adv. Math.* **22**, 131–145.

Mandelbrot, B. B. (1974) Intermittent turbulence in self-similar cascades: divergence of high moments and dimension of the carrier. *J Fluid Mech.* **62**, 331–358.

Marani, M. (2003) On the correlation structure of continuous and discrete point rainfall. *Water Resour. Res.* **5**, 1128.

Marani, M. (2005) Non-power-law-scale properties of rainfall in space and time. *Water Resour. Res.* **41**, W08413-10.

Menabde, M., Harris, D., Seed, A., Austin, G. & Stow, D. (1997) Multiscaling properties of rainfall and bounded random cascades. *Water Resour. Res.* **12**, 2823–2830.

Olsson, J. (1998) Evaluation of a cascade model for temporal rainfall disaggregation. *Hydrol. Earth Syst. Sci.* **2**, 19–30.

Olsson, J. & Niemczynowicz, J. (1996) Multifractal analysis of daily spatial rainfall distributions. *J. Hydrol.* **187**, 29–43.

Olsson, J. Niemczynowicz, J. & Berndtsson, R. (1993) Fractal analysis if high-resolution rainfall time series. *J Geophys. Res.* **98**(D12), 23265–23274.

Olsson, J. (1996) Validity and applicability of a scale-independent, multifractal relationship for rainfall. *Atmos. Res.* **42**, 53–65.

Over, T. M. & Gupta, V. K. (1994) Statistical analysis of mesoscale rainfall: dependence of a random cascade generator on large scale forcing. *J. Appl. Meteorol.* **33**, 1526–1542.

Over, T. M. & Gupta, V. K. (1996) A space–time theory of mesoscale rainfall using random cascades. *J. Geophys. Res.* **101**(D21), 26319–26331.

Schertzer, D. & Lovejoy, S. (1987) Physical modeling and analysis of rain and clouds by anisotropic, multiplicative turbulent cascades. *J. Geophys. Res.* **92**(D8), 9693–9714.

Svensson, C., Olsson, J. & Berndtsson, R. (1996) Multifractal properties of daily rainfall in two different climates. *Water Resour. Res.* **8**, 2463–2472.

Tessier, Y., Lovejoy, S., Schertzer, D. & Pecknold, S. (1996) Multifractal analysis and modelling of rainfall and river flows and scaling, causal transfer functions. *J. Geophys. Res.* **101**(D21), 26427–26440.

Tessier, Y., Lovejoy, S. & Schertzer, D. (1993) Universal multifractals: theory and observations for rain and clouds. *J. Appl. Meteorol.* **32**, 223–250.

Towards modelling of streamflow using soft tools

S. N. LONDHE[1] & S. B. CHARHATE[2]

1 *Vishwakarma Institute of Information Technology, Kondhwa (Bk), Pune 411048, India*
2 *Datta Meghe College of Engineering, Airoli, New Mumbai, India*
shreel69@yahoo.com; sbcharate@yahoo.co.in

Abstract Streamflow modelling is perhaps the most sought after research topic for hydrologists all over the world owing to its vital importance in design, construction, operation and maintenance of many hydraulic structures. Accurate forecasting of streamflow well in advance will help in saving human life as well as property damage. This paper presents modelling of streamflow using soft computing tools of Artificial Neural Networks (ANN) and Genetic Programming (GP) at two stations in the Narmada River basin, India. The developed models forecast streamflow one day in advance with reasonable accuracy. The relatively new soft technique of GP seems to work better than already established technique of ANN.

Key words streamflow; soft tools; artificial neural networks; genetic programming

INTRODUCTION

Streamflow modelling is perhaps the most sought after research topic for hydrologists all over the world owing to its vital importance in design, construction, operation and maintenance of many hydraulic structures. The purpose behind developing such models is generally forecasting of streamflow a few hours or days in advance. The models are developed either using the underlying physics of the rainfall–runoff process (mathematical-physical lumped or distributed models) or by simulating the natural physical processes such as infiltration, evaporation (conceptual models). Additionally, traditional statistical methods are also adopted using the principal statistical characteristics of runoff time series. Recently, a technique which has the caught attention of hydrologists is artificial neural networks (ANN), or for short: neural networks (NN), as evident from a plethora of papers in international journals on streamflow prediction using ANN. The ANNs can be grouped under so-called soft computing tools which treat the human brain as their role model and mimic the ability of the human brain to effectively employ modes of reasoning that are approximate rather than exact. The conventional hard computing techniques require a precisely stated analytical model and often a lot of computational time. Premises and guiding principles of hard computing are precision, certainty, and rigour (Zadeh, 1994). Soft computing differs from conventional (hard) computing in that, unlike hard computing, it is tolerant of imprecision, uncertainty, partial truth, and approximation. The guiding principle of soft computing is to exploit the tolerance for imprecision, uncertainty, partial truth, and approximation to achieve tractability, robustness and low solution cost (Zadeh, 1994). The present work deals with modelling of streamflow using two soft tools, namely ANN and genetic programming (GP), and compares them with respect to their accuracy of forecasting. Although ANN is now an established tool in modelling various hydrological parameters, the relatively new technique of GP has yet to be fully explored in terms of its utility in hydrological modelling. This paper is an effort in that direction. The models will be developed to forecast streamflow one day in advance at two stations, namely Rajghat and Mandaleshwar, in the Narmada River basin, India. Daily streamflow data at these two stations for 1987–1997 are used to develop the models after dividing the data into training and testing subsets. The data were obtained from the Central Water Commission Narmada Division, Bhopal, India.

ARTIFICIAL NEURAL NETWORKS

A neural network is a massively parallel distributed processor that has a natural propensity for storing experiential knowledge and making it available for use (Haykin, 1994). A detailed background of ANN may be found in Bose & Liang (1996). In brief, ANNs essentially involve an input layer and an output layer, which are used for supervised training. In between the input and output layers, one or more hidden layers connected by weights, biases, and transfer functions are often used. The input layer data are multiplied by initial trial weights and a bias is added to the product. This weighted sum is then transferred, through either linear or sigmoid transfer functions, to yield an output. This output then becomes the input for the following hidden layers and the procedure is continued until the output layer is reached. The difference between the network output and the target is used and transformed by an error function, and the resulting error is propagated back ("back-propagation") to update the weights and the biases using an optimization technique such as gradient descent, which strives to minimize the error. The entire procedure is repeated for a number of epochs until the desired accuracy in the outputs or other specified condition is achieved ("training"). Once the network is trained, it can be used to validate against unseen data using the trained weights and biases. Figure 1 shows a typical three-layer feed-forward network. Various algorithms, namely Standard Back-Propagation, Conjugate Gradient, Broyden-Fletcher-Goldfarb-Shanno (BFGS) and Levernberg-Marquardt (LM) can be used for training the neural network. Details of application of neural networks in the field of hydrological modelling can be found in ASCE Task Committee (2000), Maier & Dandy (2000) and Dawson & Wilby (2001).

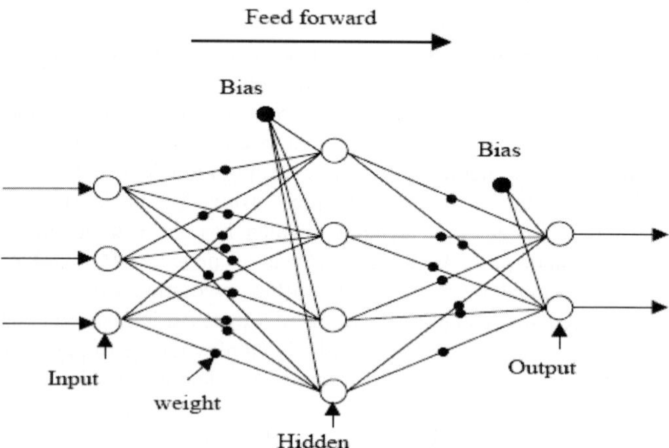

Fig. 1 Typical three-layer feed-forward neural network.

Particularly in the case of streamflow modelling, many research workers have adopted the cause–effect kind of modelling for predicting runoff, in which time series of causative variables such as rainfall (Thirumalaiah & Deo, 2000), temperature (Tokar & Johnson, 1999; Raghuwanshi *et al.*, 2006), and soil moisture content (Anctil *et al.*, 2008), as well as runoff either singly or in combination, are used to predict runoff. However, some have used only previous values of runoff to forecast runoff as in univariate time series modelling (Kisi, 2004; Singh & Deo, 2007; Wang *et al.*, 2007). This paper also uses the previous values of streamflow as the only inputs to forecast streamflow one day in advance. The difference lies in the selection of inputs and model development methodology suggested in our work. Kisi (2004) and Wang *et al.* (2007) have developed runoff models using the data of the entire year, while Singh & Deo (2007) divided the data into four seasons: monsoon, post-monsoon, winter, pre-monsoon, and developed a separate model for each season. Instead of developing a yearly model or seasonal models we have developed separate models for the months of July, August, September and October, and a single model for the non-monsoon months of November to June. Secondly, the previous values of runoff time series to be used as input are selected based on auto-regressive analysis of the data sets. Details of the data analysis are presented in the section on model development.

GENETIC PROGRAMMING

Like genetic algorithm (GA), the concept of genetic programming follows the principle of "survival of the fittest" borrowed from the process of evolution occurring in nature. However, unlike GA, its solution is a computer program or an equation as against a set of numbers in the GA, and hence it is convenient to use the same as a regression tool rather than an optimization one like the GA. A good explanation of various concepts related to GP can be found in Koza (1992). In GP a random population of individuals (equations or computer programs) is created, the fitness of individuals is evaluated and then the "parents" are selected out of these individuals. The parents are then made to yield "offspring" by following the process of reproduction, mutation and crossover. The creation of offspring continues (in an iterative manner) until a specified number of offspring in a generation are produced and further until another specified number of generations are created. The offspring resulting at the end of this process (an equation or a computer program) is the solution of the problem. The GP thus transforms one population of individuals into another one in an iterative manner by following the natural genetic operations like reproduction, mutation and cross-over. Figure 2 shows the genetic operations namely crossover and mutation. Figure 2(a) represents a program $[(-q + (\pi)^{1/2})/ 3p]$ in which p and q are typical input variables in the form of a tree structure. Two random nodes are selected from inside such a program (parents) and thereafter the resultant sub-trees are swapped, generating two new programs as in Fig. 2(b) (crossover). A sub-tree is replaced by another one randomly as in Fig. 2(c) (mutation). Reproduction is an exact duplication of the program if it is found to be acceptable by the fitness criteria.

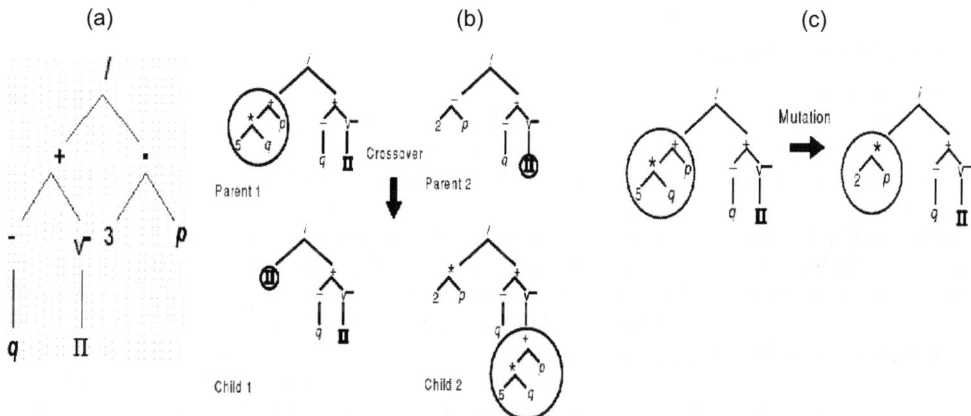

Fig. 2 (a, b, c) Genetic operations.

The application of GP to model water flow (both in hydrology and ocean engineering) is a recent development and authors could find only a handful of paper in this field. Applications of genetic programming to water flow are given by Drecourt (1999), Muttil & Liong (2004) and Drounpob *et al.* (2005) (rainfall–runoff modelling).

STUDY AREA AND DATA

The present study aims to predict average daily flow values one day in advance at two locations: Rajghat and Mandaleshwar, in the Narmada basin, India (Fig. 3). Narmada is the largest west-flowing and seventh largest river in India; it covers a large area of Madhya Pradesh besides some area of Maharashtra and Gujarat between the Vindhya and Satpura hill ranges, before discharging into the Gulf of Camby, Arabian Sea, about 10 km north of Bharuch. The Narmada Basin lies between longitudes 72°32′ to 82°45′E and latitudes 21°20′ to 23°45′N. The total catchment area

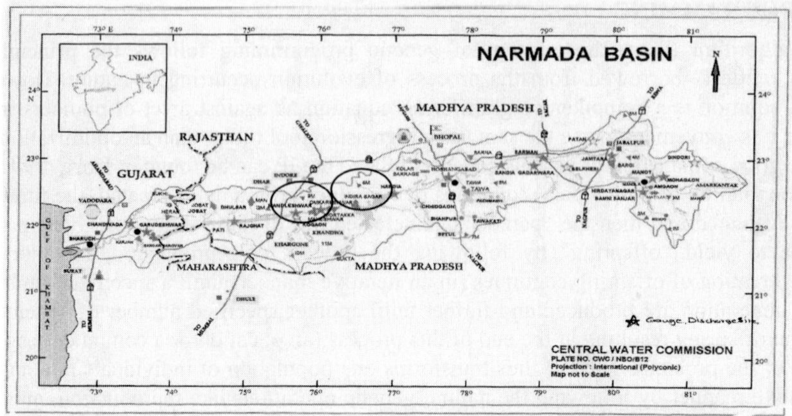

Fig. 3 Narmada basin.

is 98 796 km^2. Observations of daily average streamflow values for the Rajghat and Mandaleshwar stations for the years 1987 to 1997 were obtained from the Central Water Commission, Narmada Division, Bhopal. Rajghat is about 84 km downstream from Mandaleshwar.

MODEL DEVELOPMENT

After examining the data it was found that the average discharge values for each month differed considerably. Figure 4 (a) and (b) shows average daily runoff at Rajghat and Mandaleshwar for 11 years. Considering the variation in daily streamflow across the months and the monsoon and non-monsoon seasons of India, it was decided to develop separate models for each of the monsoon months: July, August, September and October. For non-monsoon months of November to June, it was decided to develop a common model. Thus, in all, five models were developed, i.e. four for monsoon months and one for non-monsoon months, to predict discharge at Rajghat and Mandaleshwar (total 10 models) one day in advance. The models are called: RajJuly, RajAug, RajSept, RajOct, RajNovJune, ManJuly, ManAugust, ManSept, ManOct and ManNovJune hereafter, for the sake of brevity.

The total data available for each model were divided for training and validation or testing purposes. Approximately 70% of data were used in training for each model. The next task was to determine the number of antecedent discharge values to be used for predicting discharge one day in advance. This was done by drawing a correlogram for each data set, which indicated the correlation of previous values with the current value for as many as 20 lags. It was observed that

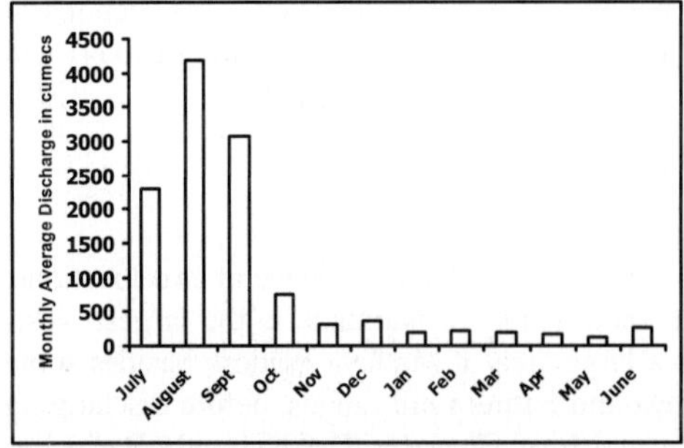

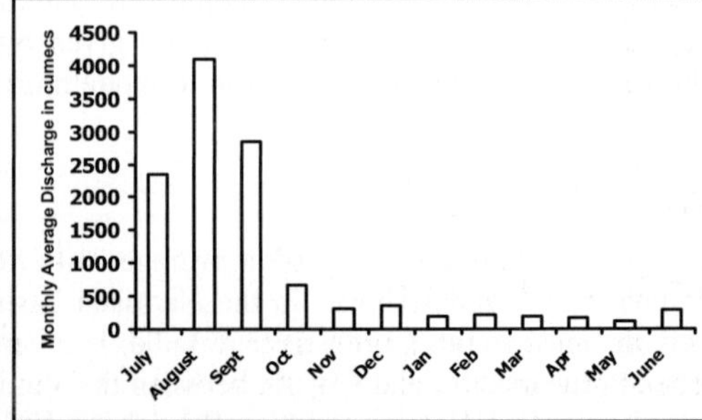

Fig. 4 Daily average discharge at: (a) Rajghat, and (b) Mandaleshwar.

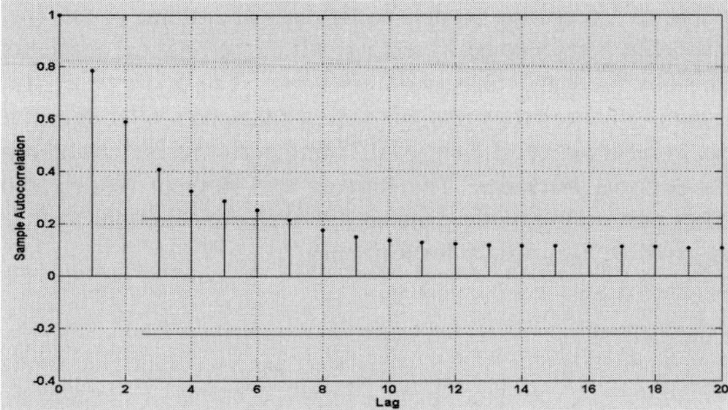

Fig. 5 Auto-correlogram for RajNovJune Model.

the auto-correlation analysis showed considerable influence of the three preceding values on the current value, except for the ManJuly model wherein the influence of only two preceding values was observed. Figure 5 shows a typical correlogram for the non-monsoon model of RajNovJune as an example. The functional relationship for RajNovJune thus can be stated as $Q_{t+1} = f(Q_t, Q_{t-1}, Q_{t-2})$ where Q is streamflow or discharge flowing through the river.

For ANN models the Levernberg-Marquardt algorithm was used. The data were normalized between –1 and 1. The transfer functions of "logsig" and "purelin" were used in all 10 models. The model architecture was 3: Hidden neurons: 1 (except for ManJuly model) representing three previous discharge values: Hidden neurons: next discharge value. The number of neurons in the hidden layer was decided by trial and error. The models were trained till a low error goal was reached and their weights and biases were retained for testing on the remaining data sets. The MATLAB Neural Network Toolbox was used to develop the ANN codes. Table 1 gives details about ANN model architecture and training and testing data sets. The GP models were developed on selection of major control parameters such as fitness function in terms of mean square error, initial population size, 2048, mutation frequency, 95%, and the crossover frequency, 53%. The commercial software Discipulus was used to develop the GP models. The GP models were developed with same data division so that their results can be compared with ANN models.

Table 1 ANN model architecture and data sets.

Model	ANN architecture	Training data sets	Testing data sets
RajJuly	3:3:1	196	112
RajAug	3:3:1	196	112
RajSept	3:3:1	189	108
RajOct	3:7:1	196	84
RajNovJune	3:5:1	1674	957
ManJuly	2:6:1	223	96
ManAug	3:2:1	210	98
ManSept	3:3:1	207	90
ManOct	3:2:1	210	98
ManNovJune	3:4:1	1820	810

RESULTS AND DISCUSSION

All the developed forecasting models were tested for unseen inputs and the qualitative and quantitative performance of models was judged by means of correlation coefficient between the observed and forecasted values or plotting scatter plots between the same. The hydrographs were also plotted to visualize the behaviour of forecasting models particularly at extreme events (peaks). The RajJuly model exhibited reasonable performance in testing, with a correlation coefficient of 0.75 for ANN model between the observed and forecasted discharge. The scatter plot (Fig. 6) between the observed and forecasted discharges confirms this with a balanced scatter except at the peaks. The GP model shows better correlation coefficient of 0.78 and performs better at the peaks though it shows over predictions at some instances. The RajAug and RajSept models showed similar performance with GP models performing better compared to their ANN counterparts (r_{GP} = 0.75, r_{ANN} = 0.7 for RajAug and r_{GP} = 0.79, r_{ANN} = 0.76 for RajSept).

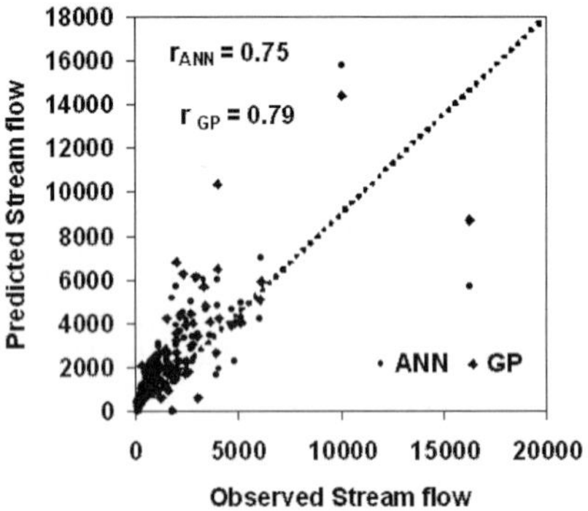

Fig. 6 Scatter plot for RajJuly model.

For the RajOct model the discharge values predicted in testing are highly in agreement with the observed values for both the models as shown by the discharge hydrograph (Fig. 7). It seems that both the soft techniques have understood the underlying phenomenon of rise and fall in the discharge and its dependence on previous discharge values. The results are also supported by a high value of correlation coefficient (r = 0.92) for both ANN and GP models in testing.

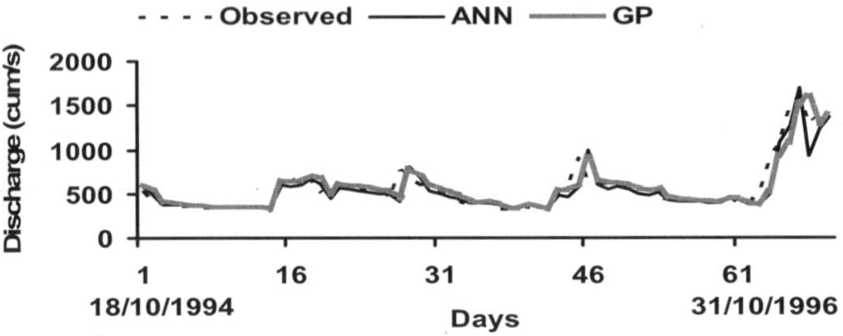

Fig. 7 RajOct model results.

The quantum jump in the prediction accuracy of both models compared to previous RajJuly, RajAug and RajSept may be an effect of the absence of large discharge values as present in the previous three models. The RajJuly model has a maximum discharge of 33 600 and 16 266 m^3/s in training and testing, respectively. There are very few events of such magnitude in the entire data sets, and particularly in testing, all other values except one event of 10 030 m^3/s are less than 6200 m^3/s. The rare occurrence of large events, as well as the large difference between the maximum value and other values in training and testing, may result in over-prediction of many smaller events and, consequently, yields correlation coefficients in the range of 0.7 to 0.8. For the

RajAug model the maximum discharge values in training and testing are 38 102 and 22 850 m^3/s, respectively, and for RajSept model these are 48 745 and 6500 m^3/s, respectively. However for the RajOct model the maximum discharge in training is 8284 m^3/s and in testing 1582 m^3/s. Thus it seems that variability in the data sets plays an important role in prediction capabilities of these soft models. The RajNovJune model supports the above conclusion by depicting high correlation coefficient of 0.92 for ANN model and 0.95 for GP models in testing. The maximum values of discharge are 4159 and 5537 m^3/s in training and testing, respectively. One more observation of the RajNovJune model is that the maximum discharge of 5537 m^3/s is predicted better by the GP model (5555 m^3/s) as compared to the ANN model (2486.8 m^3/s). Thus for all Rajghat models, GP models are working better in terms of accuracy of prediction, and in situations of extreme events. Figure 8 shows the streamflow hydrograph for RajNovJune models in testing.

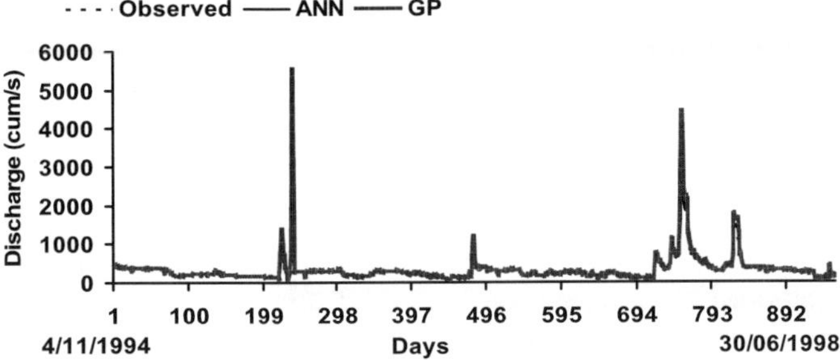

Fig. 8 RajNovJune model results.

The Mandaleshwar models behaved in a similar fashion to the Rajghat models with correlation coefficients of more than 0.73 for both ANN and GP models, except for ManJuly models, which show a considerably low correlation coefficient of 0.63 and 0.65 for ANN and GP, respectively. This may be attributed to the fact that the models were trained with a maximum discharge of 20119 m^3/s and maximum observed discharge in testing was 36045 m^3/s which resulted in large prediction errors at peaks. Similarly, large difference between the discharge values of extreme events and normal events, as experienced in Rajghat models may also have contributed to poor performance of ManJuly models compared to other models. Figure 9 shows a scatter plot of ManJuly models in testing. For ManAug the performance of both ANN and GP models improved considerably with correlation coefficient of 0.74 and 0.78, respectively. For ManSept the ANN results were slightly better than the GP model with a correlation coefficient of 0.9 as compared to 0.87 GP. Figure 10 shows the results of ManSept models in testing.

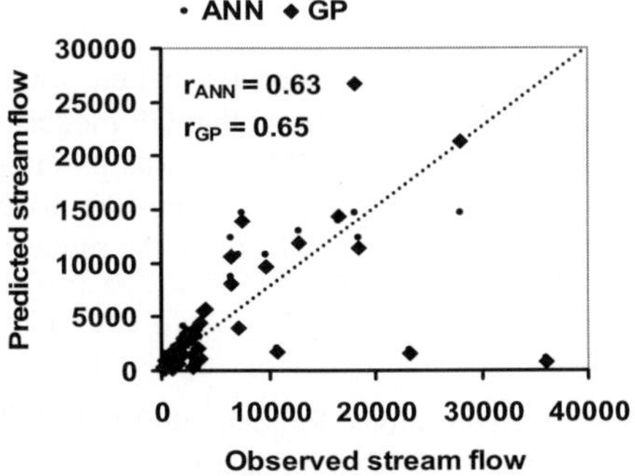

Fig. 9 Scatter plot for ManJuly model.

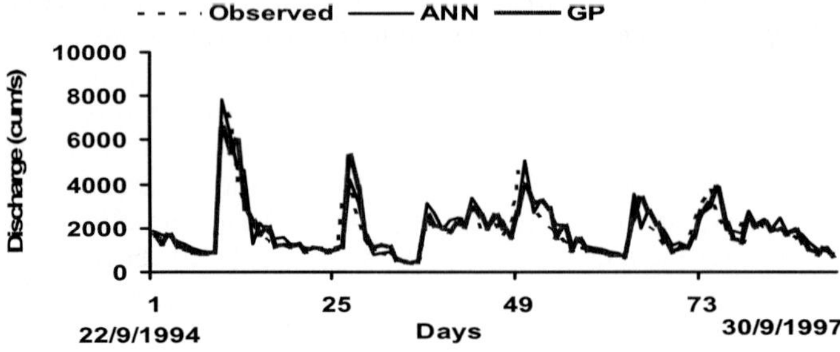

Fig. 10 ManSept model results.

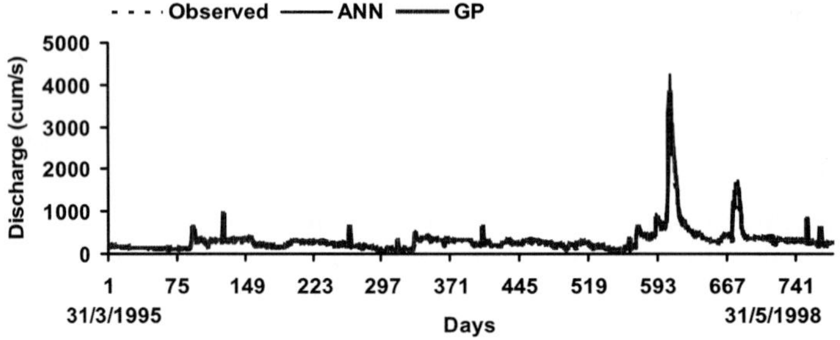

Fig. 11 ManNovJune model results.

Table 2 Results table.

Model	r_{ANN}	r_{GP}
RajJuly	0.75	0.78
RajAug	0.7	0.75
RajSept	0.76	0.79
RajOct	0.92	0.92
RajNovJune	0.92	0.95
ManJuly	0.63	0.65
ManAug	0.74	0.78
ManSept	0.9	0.88
ManOct	0.89	0.89
ManNovJune	0.93	0.96

Note: r = correlation coefficient.

The ManOct and ManNovJune models of both ANN and GP performed better, with high correlation coefficients in testing ($r > 0.89$). However, it was again observed that GP models work better while predicting extreme events. The maximum discharge of 3790 m^3/s was predicted as 1742 m^3/s by the ANN model, and as 3342 m^3/s by the GP model. Figure 11 shows the discharge hydrograph for ManNovJune models. Table 2 shows the consolidated results of all the models.

CONCLUDING REMARKS

The paper presents development of streamflow models at two stations, Rajghat and Mandaleshwar in Narmada basin, in India, using the soft tools of ANN and GP. The models were developed to forecast streamflow one day in advance. All the models performed reasonably well in testing, except the ManJuly model. The GP models performed better than the ANN models in almost all cases and particularly in predicting peak runoff. The results of both models seem to be influenced by variability in the data. Use of such models at operational levels would certainly help the disaster management systems and efficient operation of water impounding structures.

Acknowledgements This study is supported by research funding provided by B.C.U.D. through University of Pune, India.

REFERENCES

Anctil, F., Lauzon, N. & Filion, M. (2008) Added gains of soil moisture content observations for streamflow predictions using neural networks. *J. Hydrol.* **359**, 225–234.

Bose, N. K. & Liang, P. (1998) *Neural Network Fundamentals with Graphs, Algorithms and Applications.* Tata McGraw-Hill Publication, India.

Dawson, C. W. & Wilby, R. L. (2001) Hydrological modeling using artificial neural networks. *Progr. Phys. Geogr.* **25**(1), 80–108.

Drecourt, J. P. (1999) Application of neural networks and genetic programming to rainfall runoff modeling. Danish Hydraulic Institute, HIT.

Drounpob, A, Chang, N. B. & Beaman, M. (2005) Streamflow rate prediction using genetic programming model in a semi-arid coastal water-shed. In: Proc. World Water Congress, Alaska, 2005.

Haykin, S. (1994) *Neural Networks: A Comprehensive Foundation.* Macmillan, New York, USA.

Kisi, O. (2004) River flow modeling using artificial neural networks. *J. Hydrol. Engng* **9**(1), 60–63.

Koza, J. R. (1992) *Genetic Programming on the Programming of Computers by Means of Natural Selection.* A Bradford Book, MIT Press, Cambridge, USA.

Londhe, S. N. (2008) Soft computing approach for real-time estimation of missing wave heights. *Ocean Engng* **35**, 1080–1089.

Maier, H. R. & Dandy, G. C. (2000) Neural networks for prediction and forecasting of water resources variables: a review of modeling issues and applications. *Environ. Modeling Software* **15**, 101–124.

Muttil, N. & Liong, S. Y. (2004) A superior exploration-exploitation balance in shuffled complex evolution. *J. Hydraul.Engng* **130**(12), 1202–1205.

The ASCE Task Committee (2000) Artificial Neural Networks in Hydrology I: Preliminary concepts. *J. Hydrol. Engng* **5**(2), 115–123.

Raghuwanshi, N. S., Singh, R. & Reddy, L. S. (2006) Runoff and sediment yield modeling using artificial neural networks: Upper Siwane River, India. *J. Hydrol. Engng* **11**(1), 71–79.

Singh, P. &, Deo, M. C. (2007) Suitability of different neural networks in daily flow forecasting. *Applied Soft Computing* **7**, 968–978.

Sivapragasam, C., Maheswaran, R. & Veena, V. (2008) Genetic programming approach for flood routing in natural channels. *Hydrol. Processes* **22**, 623–628.

Thirumalaiah, K. & Deo, M.C. (2000) Hydrological forecasting using neural networks. *J. Hydrol. Engng* **5**(2), 180–189

Tokar, A. S. & Johnson, P. A. (1999) Rainfall–runoff modeling using artificial neural networks. *J. Hydrologic Engng* **4**(3), 232–239.

Wang, S. K., Lu, W. Z., Cao, S. Y. & Fang, D. (2007) Using time-delay neural network combined with genetic algorithms to predict runoff level of Linshan watershed, Sichuan, China. *J. Hydrol. Engng* **12**(2), 231–236.

Zadeh, L (1994) Fuzzy logic, neural networks and soft computing. *Communications of the ACM* **37**(3), 77–84.

Geomorphology-structured hydroinformatics for downward basin modelling with flexible accounting for net rainfall variability

C. CUDENNEC[1,2,3], J-C. POUGET[4], S. CHARGUI[1,2,5], H. BOUDHRAÂ[1,2,3,5], A. JAFFREZIC[1,2] & M. SLIMANI[5]

1 *Agrocampus Ouest, UMR1069, Soil Agro and hydroSystem, F-35000 Rennes, France*
cudennec@agrocampus-ouest.fr

2 *INRA, UMR1069, Soil Agro and hydroSystem, F-35000 Rennes, France*

3 *IRD, UMR G-EAU, 1004 Tunis, Tunisia*

4 *IRD, UMR G-EAU, Quito, Ecuador*

5 *INAT, Lab. STE, 1082 Tunis, Tunisia*

Abstract The HydroStruct software application is dedicated to the combined spatial, topological and scaling analysis of the morphometry of river networks. It was also developed to easily link geomorphometric observations with hydrological analysis and modelling approaches, such as geomorphology-based transfer functions and accounting for associated dominant variabilities. Flexibility of the accounting for net rainfall variability is demonstrated in the sense of downward basin rainfall–runoff modelling.

Key words river network; geomorphology-based transfer function; hydrological space-time variability; downward flexibility

INTRODUCTION

The river network is the basic geomorphological structure of a river basin, in both topological and geometric terms. Furthermore, it has fundamental hydrological significance, being the pattern that governs upstream–downstream transfers both from hillslopes and through channelized paths towards the outlet. Moreover, the branched organisation of the river network presents strong scaling properties. The HydroStruct software application is dedicated first to the combined spatial, topological and scaling analysis of the morphometry of river networks. Secondly, major facilities for easily linking geomorphometric observations with hydrological analysis and modelling approaches are developed and explored in order to gain functionality and flexibility, especially through a downward approach in which dominant effects are accounted for, using available data where possible, from a hydro-geomorphological viewpoint. This paper demonstrates this flexibility with a focus on accounting for net rainfall variability.

HYDRO-GEOMORPHOMETRIC METHODOLOGY

From a systemic hydrological point of view, one can focus on a function describing the basin organization in terms of flow paths through the network to the outlet. In this context, some functions have been introduced such as the area function and the width function. These functions are the probability density functions (pdfs) of the basin contributing area and of the number of links in the network, with respect to flow distance to the outlet.

Within the path followed by a water drop to reach the outlet, Cudennec *et al.* (2004a) considered separately the hillslope path and the channelized path through the river network (Fig. 1). The lengths of the hillslope and the channelized paths, l_h and L, respectively, were considered, as well as pdf(l_h) and pdf(L), the latter being a slightly alternative basin-level structural function. Furthermore, the splitting of the channelized path through the Strahler ordering cascade was considered, introducing the "ith order component", which is the part run through successive links of the same Strahler order i (Fig. 1); and scaling effects were identified and analysed, leading to some evidence of an underlying structural pattern (Cudennec *et al.*, 2004a, 2006; Cudennec & Fouad, 2006).

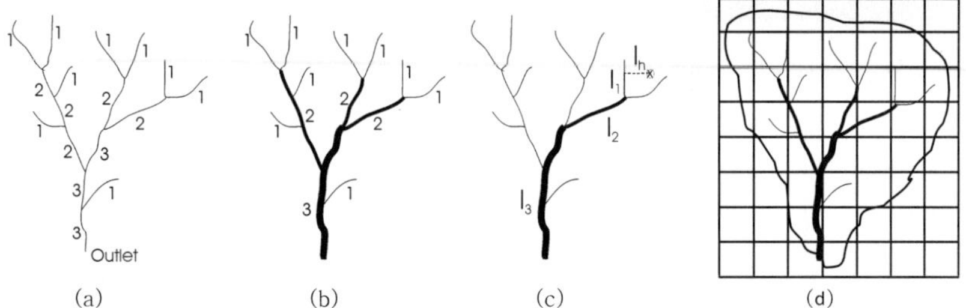

Fig. 1 Schematic network of order $n = 3$. (a) links and Strahler ordering; (b) streams; (c) example of the path taken by a water drop from a point to the outlet, where l_h is the length of the path through the hillslope; and l_1, l_2 and l_3 are the lengths of first-, second- and third-order components, respectively; and (d) sampling grid for spatial analysis (statistics and maps).

HydroStruct: dedicated analysis software

The HydroStruct software application was developed in order to: (a) explore the ubiquity and variability of the identified underlying structural pattern through a great diversity of contexts and constraints (climate, lithology, elevation, etc.), and of scales, together with more classical evidence of scaling and branching; (b) explore the eventual self-organization of networks towards organization rules through their genesis and maturation; (c) observe the metrics and likelihood of network models; (d) feed a downward approach to rainfall–runoff modelling to be adjusted according to hydrological context, degree of knowledge, scales and queries; and (e) prepare coupling with related models, such as of water resources allocation or of spatial analysis of water demand (Pouget *et al.*, 2006a,b).

The software is developed within the object-oriented OdefiX modelling framework (Pouget *et al.*, 2006a), which aims to facilitate the development of software platforms for water resources management decision support, leading to easy-to-use and comprehensive tools, thanks to a user-friendly graphically-oriented interface (map representations, charts, etc.) (Fig. 2), in order to support discussions between specialists and stakeholders. An embracing longer-term aim is to allow the co-construction of dedicated environments in which several models (hydrological, bio-physical, socio-economic, etc.) can cooperate to address river basin management issues.

Rationale for data structure and analysis

The considered input data are georeferenced vector layers of the basin river network and limits, which can be obtained from various kinds of data sources and through diverse human-based and/or automatic procedures. This is illustrated on Fig. 2 for the central-Tunisian semi-arid basin of Skhira (35°44′15″N/9°23′05″E UTM system, 192 km^2; Cudennec *et al.*, 2005, 2006), part of the major central-Tunisian Merguellil basin (Cudennec *et al.*, 2007; Lacombe *et al.*, 2008). Then a topological analysis of the relationships between the vectors allows a branched tree to be built. In addition, a square grid is used as a set of points to sample the basin territory, and for each of these points the water path is identified across the hillslope and the network, i.e. also through the Strahler ordering cascade for hierarchical analyses (Fig. 1). Then geomorphological variables, including the hillslope and the channelized paths lengths, l_h and L, are assessed for each point, and georeferenced raster images of the obtained values are built (Fig. 2(b)). This multi-layer raster mapping of geomorphometric variables allows one to use image processing to both analyse their spatial distribution and extract their statistics (Fig. 2(b) and (c)). Traditional and innovative scaling analyses are further allowed, in particular: (a) through the consideration of the channelized path splitting through the Strahler ordering cascade (Cudennec *et al.*, 2004a, 2006; Cudennec & Fouad, 2006); (b) through nested studies based on sub-settings of the whole-basin tree and territory, as proposed by Cudennec *et al.* (2006) for the Skhira basin; and (c) through considering hydro-geomorphological entities comprising several basins through the spatial and statistical merging of results from neighbouring basins (Cudennec & Fouad, 2006).

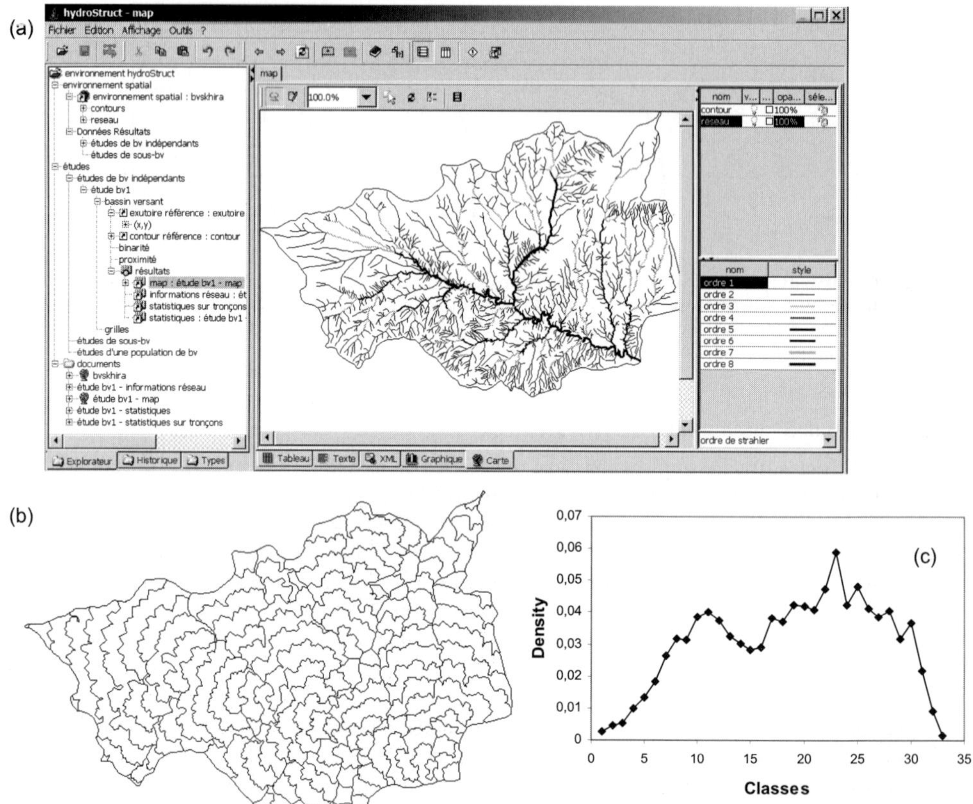

Fig. 2 HydroStruct facilities and outputs illustrated with the Tunisian Skhira basin (Cudennec *et al.*, 2005, 2006): (a) viewing and management interface; (b) iso-L contour mapping, the basis of the isochrone mapping; and (c) pdf(L) (classes of 840 m), the basis of the transfer function TF.

FLEXIBLE ACCOUNTING FOR NET RAINFALL IN RAINFALL–RUNOFF MODELLING

Both hierarchy-based and whole-basin geomorphometric analyses can feed rainfall–runoff modelling approaches (Rodriguez-Iturbe & Rinaldo, 1997; Cudennec, 2007). The explicit observable geomorphometric basis of such approaches is of major interest for hydrological modelling in data-sparse contexts where cumbersome calibration cannot be implemented. Moreover, the conceptualisation of water transfers can be adjusted according to the available data and knowledge related to flow (Rinaldo *et al.*, 1991; Woods & Sivapalan, 1999; Rodriguez *et al.*, 2005), allowing methods to be proposed for the assessment of design floods (Nasri *et al.*, 2004), or for transposing discharge between outlets (Boudhraâ *et al.*, 2006, 2009). Thus, the iso-L map and pdf(L) are the basis of the isochrone map and of the transfer function TF, i.e. pdf(t), respectively (Fig. 2).

When considering basin-level geomorphometric functions, the geometric and functional coupling between the river or drainage network and upslope elements differs. In the global approach, when using the pdf(L) function as the basis for the transfer function TF, the production function PF to be coupled encompasses all the hillslope processes; and the coupling variable, net rainfall, R_n, is understood as the outflow from the hillslope element that enters rivers along the whole river network, and is considered to flow in a conservative way to the outlet (Fig. 3).

The variability of R_n can occur as a consequence of the variability of rainfall input itself over the basin, of hillslope heterogeneity within the basin, or of non-conservative effects along the network (Fig. 4). Such causes of variability can coincide with a variable hierarchy according to the

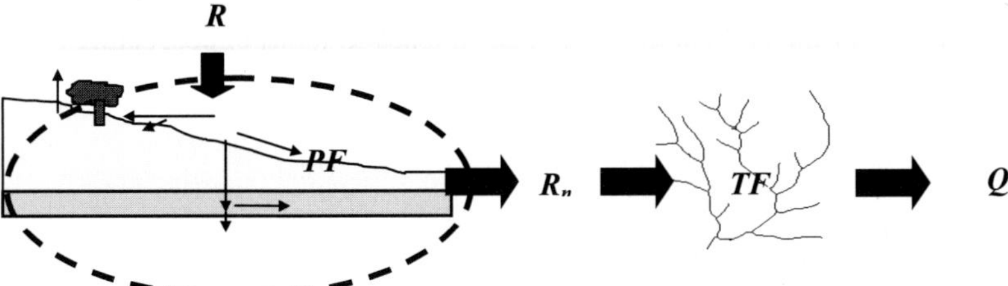

Fig. 3 Geomorphology-based rainfall–runoff (R–Q) modelling: net rainfall, R_n flowing from hillslopes and entering the river network, seen as the coupling variable between a hillslope-level production function, PF, and a network-level transfer function, TF.

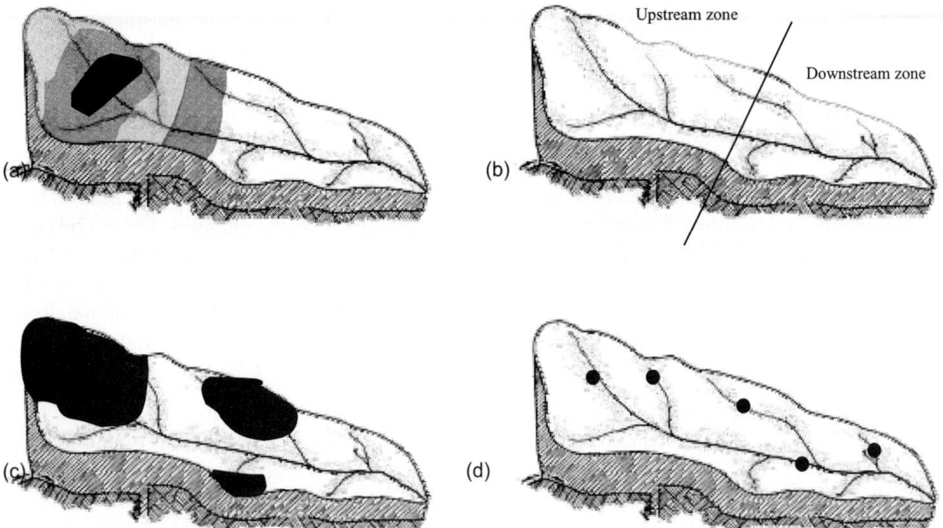

Fig. 4 Types of variability conceptualisations: (a) continuous rainfall input field; (b) semi-distribution zonation; (c) distribution of hillslope types; and (d) distribution of sources of non-conservatism within river networks. (Background drawing inspired from Auzet, 1999.)

particular basin and event. The pdf(L)-based rainfall–runoff modelling allows the study and simulation of such effects thanks to its strong spatial correspondence. New robust and flexible modelling approaches are facilitated by some of the HydroStruct functions, extended by additional rainfall–runoff developments.

Rainfall input variability

The explicit territorial significance of a transfer function based on the whole-basin pdf(L) allows enhancement of the convolution through the accounting for a variability matrix. This variability matrix can be built differently depending on the available raingauge network. This allows one to adapt to evolving network configurations, even in data-sparse contexts, such as through historical changes, breakdown episodes or designed optimization schemes.

As proposed by Cudennec *et al.* (2005), and still applied to the Skhira basin where rainfall variability is strong (Kingumbi *et al.*, 2005; Jebari *et al.*, 2007; Slimani *et al.*, 2007), the classical convolution:

$$Q(t)=\frac{S}{\Delta t}\sum_{\tau=1}^{t}\overline{R_n(t-\tau+1)}\,\mathrm{pdf}(\tau) \qquad (1)$$

where S is the surface of the basin, Δt the time step, $Q(t)$ the discharge at the outlet at time t, $\overline{R_n(t)}$ the average net rainfall over the basin at time t, can be enriched in terms of accounting for the net rainfall space–time variability, as follows:

$$Q(t)=\frac{S}{\Delta t}\sum_{\tau=1}^{\max(K,t)} R_{n\tau}(t-\tau+1)\cdot \mathrm{pdf}(\tau) \qquad (2)$$

where $R_{n\tau}(t)$ is the average net rainfall over the isochrone area τ at time t. This can be turned into:

$$Q(t)=\frac{S}{\Delta t}\sum_{\tau=0}^{\max(K,t)} \overline{R_n(t-\tau+1)}.V_\tau(t-\tau+1)\cdot \mathrm{pdf}(\tau) \qquad (3)$$

where $V_\tau(j)=\dfrac{R_{n\tau}(j)}{\overline{R_n(j)}}$ is a variability factor which forms a hydrologically significant space–time variability matrix when τ ranges over the whole basin up to the ultimate isochrone area K and j for the duration of the whole event. The iso-L map obtained with HydroStruct (Fig. 2) is the foundation for building this variability matrix. This can be implemented by image processing, through the development of step-by-step spatially-interpolated rainfall field raster maps and the successive crossing with the iso-L map, transformed into an isochrone map (Cudennec *et al.*, 2005).

This can further be developed in an improved matrix approach, both to gain calculation functionality and to allow flexibility in terms of raingauge networks changes. For each raingauge, G_i, of a given configuration of n gauges, a matrix of geometric weight is built for each raster element (x,y) of the whole basin, based on the same raster discretization scheme as the geomorphological analysis. Then a combination of these n weight matrices and of the isochrone matrix allows assessment of the weight matrix of each raingauge G_i for each isochrone area τ, α_1^i. At each time step, t, the net rainfall over the isochrone area τ, $R_{n\tau}(t)$ is then assessed as the following linear combination of punctual net rainfalls R_n^i deduced from observations from the raingauges i of the considered configuration and the application of a production function:

$$R_{n\tau}(t)=\sum_{i=1}^{n}\alpha_\tau^i R_n^i(t) \qquad (4)$$

Alongside this, the average net rainfall over the whole basin, $\overline{R_n(t)}$, is obtained by a linear combination, which allows the deduction of any $V_\tau(j)$. The quality of this robustly improved rainfall–runoff modelling is shown by Cudennec *et al.* (2005), and its functionality and flexibility are highly improved as a result of the dedicated HydroStruct software application and the additional matrix tools. For a given basin, any available geomorphological data can be considered, as well as all the existing and historical raingauge configurations. Furthermore, this flexibility can be used in a proactive way, especially in terms of network optimisation and simulation of design events.

In addition, the matrix approach can be simplified in terms of very few zones (for instance an upstream and a downstream zone, Fig. 4(b), leading to a kind of semi-distribution, when the contrast is marked in terms of rainfall input gradient (Cudennec *et al.*, 2004b):

$$Q(t)=\frac{S}{\Delta t}\sum_{\tau=1}^{\max(K,t)}\left[\overline{R_{n\,\mathrm{downstream}}(t-\tau+1)}\cdot \mathrm{pdf}_{\mathrm{downstream}}(\tau)+\overline{R_{n\,\mathrm{upstream}}(t-\tau+1)}\cdot \mathrm{pdf}_{\mathrm{upstream}}(\tau)\right] \qquad (5)$$

Heterogeneity of hillslopes and streams

When part of the R_n variability comes from hillslope heterogeneity within the basin, e.g. because of hillslope harvesting works (Sumarjo Gatot *et al.*, 2001), or drainage density diversity (Cudennec *et al.*, 2006; Cudennec, 2007), the semi-distribution approach corresponding to equation (5) can be set up for contrasting zones. This can eventually be further implemented for hillslope types i, from which outflowing net rainfall, $R_{n,i}$, is assessed through the application of corresponding production functions, PF_i, and computed according to their spatial distribution within the isochrone mapping (Fig. 4(c)), i.e. along pdf(L) (Sumarjo Gatot *et al.*, 2001):

$$Q(t) = \frac{S}{\Delta t} \sum_{\forall i} \sum_{\tau=1}^{\max(K,t)} \left[\overline{R_{n\,i}(t-\tau+1) \cdot \mathrm{pdf}_i(\tau)} \right] \tag{6}$$

When additional variability is introduced to the transferred net rainfall at the outlet because of non-conservative effects along the network transfer, for instance the aggregated impact of multiple small dams or river–aquifer exchanges (Cudennec *et al.*, 2004b; Lacombe *et al.*, 2008), convolution can be enriched through the use of an adapted transfer function, based on the spatial distribution of these artefacts along pdf(L) (Fig. 4(d); Cudennec *et al.*, 2004b):

$$Q(t) = \frac{S}{\Delta t} \sum_{\tau=1}^{\max(K,t)} R_n(t-\tau+1)\mathrm{pdf}_{\mathrm{adapt}}(\tau) \tag{7}$$

CONCLUSION

The dedicated HydroStruct application was developed within the generic Java OdefiX framework. Vector data of river networks and basin boundaries are analysed through both tree formalisation and spatial sampling. Results are extracted as global, classified and scaled statistics; and in terms of spatial distribution through raster images. Raster images of scaled geometric variables can be combined to assess and explore further structural evidences, and be linked with territorial maps of hydrological parameters. Sub-basins and populations of basins can be easily analysed, by crossing, filtering and merging vector, raster and tree data. Major geomorphology-based hydrological developments are further allowed: the main one consisting of crossing hydro-geomorphometric results with spatio-temporal variables in a flexible way through the coupling of the geomorphometric HydroStruct software with additional convolution-based hydrological tools. An overview of modelling options and an associated short review of first applications are provided, focusing on the accounting for net rainfall space–time variability as a result of the variability of rainfall input over the basin, of hillslope heterogeneity within the basin, or of non-conservative effects along the network. Major developments are also allowed in terms of water-related solute and suspended transportation, based on the same concepts: hillslope–network segmentation, iso-L geometry and accounting in-stream non-conservative effects. These modelling options can be standalone or used in a combined manner, depending on the basins and events, in the framework of a downward approach. The geomorphological basis and rationale, together with the flexible hydroinformatics tools, make the approach highly adaptable for either actual or prospective studies.

REFERENCES

Auzet, A. V. (1999) Les cheminements de l'eau naturels et/ou influencés. In: *L'influence humaine dans l'origine des crues* (ed. by E. Leblois), 19–46. CEMAGREF, Paris, France.

Boudhraâ, H., Cudennec, C., Slimani, M. & Andrieu, H. (2006) Inversion d'une modélisation de type hydrogramme unitaire à base géomorphologique: interprétation physique et première mise en œuvre. In: *Predictions in Ungauged Basins: Promises and Progress* (ed. by M. Sivapalan, T. Wagener, S. Uhlenbrook, E. Zehe, V. Lakshmi, X. Liang, Y. Tachikawa & P. Kumar) (Seventh IAHS Scientific Assembly, Foz do Iguaçu, Brazil, April 2005), 391–399. IAHS Publ. 303, IAHS Press, Wallingford, UK.

Boudhraâ, H., Cudennec, C., Slimani, M. & Andrieu, H. (2009) Hydrograph transposition between basins through a geomorphology-based deconvolution–reconvolution approach. In: *New Approaches to Hydrological Prediction in Data-Sparse Regions* (ed. by K. Yilmaz *et al.*) (Proc. Symp. HS.2 at the Joint IAHS & IAH Convention, Hyderabad, India, September 2009). IAHS Publ. 333, IAHS Press, Wallingford, UK.

Cudennec, C. (2007) On width function-based unit hydrographs deduced from separately random self-similar river networks and rainfall variability. *Hydrol. Sci. J.* **52**, 230–237.

Cudennec, C. & Fouad, Y. (2006) Structural patterns in river network organization at both infra- and supra-basin levels: the case of a granitic relief. *Earth Surf. Proc. Landf.* **31**, 369–381.

Cudennec, C., Fouad, Y., Sumarjo Gatot, I. & Duchesne, J. (2004a) A geomorphological explanation of the unit hydrograph concept. *Hydrol. Processes* **18**, 603–621.

Cudennec, C., Sarraza, M. & Nasri, S. (2004b) Modélisation robuste de l'impact agrégé de retenues collinaires sur l'hydrologie de surface. *Rev. Sci. Eau* **17**, 181–194.

Cudennec, C., Slimani, M. & Le Goulven, P. (2005) Accounting for sparsely observed rainfall space–time variability in a rainfall–runoff model of a semiarid Tunisian basin. *Hydrol. Sci. J.* **50**, 617–630.

Cudennec, C., Pouget, J. C., Boudhraâ, H., Slimani, M. & Nasri, S. (2006) A multi-level and multi-scale structure of river network geomorphometry with potential implications towards basin hydrology. In: *Predictions in Ungauged Basins: Promises and Progress* (ed. by M. Sivapalan, T. Wagener, S. Uhlenbrook, E. Zehe, V. Lakshmi, X. Liang, Y. Tachikawa & P. Kumar) (Seventh IAHS Scientific Assembly, Foz do Iguaçu, Brazil, April 2005), 422–430. IAHS Publ. 303. IAHS Press, Wallingford, UK.

Cudennec, C., Leduc, C. & Koutsoyiannis, D. (2007) Dryland hydrology in Mediterranean regions—a review. *Hydrol. Sci. J.* **52**, 1077–1087.

Jebari, S., Berndtsson, R., Uvo, C. & Bahri, A. (2007) Regionalizing short-term rainfall affected by topography in semi-arid Tunisia. *Hydrol. Sci. J.* **52**, 1199–1215.

Kingumbi, A., Bargaoui, Z. & Hubert, P. (2005) Investigation of the rainfall variability in central Tunisia. *Hydrol. Sci. J.* **50**, 493–508.

Lacombe, G., Cappelaere, B. & Leduc, C. (2008) Hydrological impact of water and soil conservation works in the Merguellil catchment of central Tunisia. *J. Hydrol.* **359**, 210–224.

Nasri, S., Cudennec, C., Albergel, J. & Berndtsson, R. (2004) Use of a geomorphological transfer function to model design floods in small hillside catchments in semiarid Tunisia. *J. Hydrol.* **287**, 197–213.

Pouget, J. C., Bousquet, F., Le Goulven, P., Quaranta, D., Rebatel, J. C. & Rolland, D. (2006a) OdefiX Java framework for developing and interfacing hydrological and water management models—generic components and application for water resources allocation. In: *Proc. Seventh Int. Conf. on Hydroinformatics* (ed. by P. Gourbesville, J. Cunge, V. Guinot & S-U. Liong) (Nice, France, September 2006), 2348–2355. Research Publishing Services, Chennai, India.

Pouget, J. C., Poussin, J. C., Pettinotti, B., Quaranta, D. & Rolland, D. (2006b) Regional and prospective analysis of agricultural activities and water demands: the ZonAgri modelling environment within the OdefiX generic Java framework. In: *Proc. Seventh Int. Conf. on Hydroinformatics* (ed. by P. Gourbesville, J. Cunge, V. Guinot & S-U. Liong) (Nice, France, September 2006), 2563–2570. Research Publishing Services, Chennai, India.

Rinaldo, A., Marani, A. & Rigon, R. (1991) Geomorphological dispersion. *Water Resour. Res.* **27**, 513–525.

Rodriguez, F., Cudennec, C. & Andrieu, H. (2005) Application of morphological approaches to determine unit hydrographs of urban catchments. *Hydrol. Processes* **19**, 1021–1035.

Rodriguez-Iturbe, I. & Rinaldo, A. (1997) *Fractal River Basins; Chance and Self-organization.* Cambridge University Press: Cambridge, UK.

Slimani, M., Cudennec, C. & Feki, H. (2007) Structure du gradient pluviométrique de la transition Méditerranée–Sahara en Tunisie: déterminants géographiques et saisonnalité. *Hydrol. Sci. J.* **52**, 1088–1102.

Sumarjo Gatot, I., Pérez, P. & Duchesne, J. (2001) Modelling the influence of irrigated terraces on the hydrological response of a small basin. *Environ. Model. Software* **16**, 31–36.

Woods, R. & Sivapalan, M. (1999) A synthesis of space–time variability in storm response: rainfall, runoff generation, and routing. *Water Resour Res.* **35**, 2469–2485.

Nonlinear dynamic analysis of reservoir inflows: a case study from South India

V. JOTHIPRAKASH[1], MAYANK DOBHAL[1], MAYANK MEHTA[1] & BELLIE SIVAKUMAR[2]

1 *Department of Civil Engineering, Indian Institute of Technology – Bombay, Powai, Mumbai 400076, India*
vprakash@iitb.ac.in

2 *Department of Land, Air and Water Resources, University of California, Davis, California 95616, USA*

Abstract In water resource studies, it is a common practice to use stochastic methods for simulation of future inflows for reservoir operation purposes. However, recent studies have revealed that complex and random-looking streamflows could also be the outcome of simple deterministic systems. The present study investigates the dynamic nature of inflows to the Pechiparrai Reservoir in the state of Tamil Nadu in South India. Inflows over a 32-year period are studied, and the correlation dimension method is employed to identify the nature of the underlying system dynamics. Considering the weekly time scale normally adopted for operation of the Pechiparrai Reservoir, the nature of inflow dynamics at daily, 2-day, 4-day, and 8-day scales is investigated. The low correlation dimension values of 1.98, 2.10, 2.21, and 2.35 estimated for these four series suggest the presence of low-dimensional chaotic dynamics, with the number of variables dominantly governing being in the order of 2 or 3.

Key words reservoir inflows; time series analysis; stochastic; chaos; autocorrelation function; phase space reconstruction; correlation dimension; scale; Kodiyar Reservoir

INTRODUCTION

The nonlinear dynamic nature of hydrological processes has been known for many decades (e.g. Izzard, 1966; Amorocho, 1967). However, it has been a common practice to employ linear stochastic approaches in hydrological and water resource research, especially in the simulation of future inflows for planning, development and operation of reservoirs and other water resource structures (e.g. Thomas & Fiering, 1962; Valencia & Schaake, 1973; Salas *et al.*, 1995). Although the linear approaches continue to be prevalent, advances in computational power and measurement technology during the last two decades or so have facilitated formulation of nonlinear approaches as viable alternatives. The nonlinear approaches that are popular in hydrology include: nonlinear stochastic methods, artificial neural networks, data-based mechanistic models, and deterministic chaos theory.

Among the nonlinear approaches, chaos theory seems to be the simplest, especially in its view of "complex" systems. In the nonlinear science literature, the term "chaos" is normally used to refer to situations where complex and random-looking behaviours arise from simple nonlinear deterministic systems with sensitive dependence on initial conditions (Lorenz, 1963). With the obvious relevance of its fundamental properties (i.e. nonlinear interdependence, hidden determinism and order, and sensitivity to initial conditions) in hydrological systems and processes, chaos theory has been finding increasing applications in hydrology during the last two decades or so. Early studies essentially focused on the investigation and prediction of chaos in rainfall, river flow, temperature, and lake volume data in a purely single-variable data reconstruction sense. Subsequent studies attempted its application on other hydrological problems, including scaling and data disaggregation, missing data estimation, reconstruction of system equations and other processes, such as rainfall–runoff and sediment transport; they also addressed some of the important issues that had been, and continue to be, perceived to significantly influence the outcomes of chaos methods when applied to real hydrological data, including minimum data size, data noise, presence of zeros, selection of optimal parameters, and multi-variable data reconstruct-tion. More recently, studies have applied the ideas of chaos theory to either advance the earlier studies (on scaling, for example) or to address yet other hydrological processes and problems, including groundwater contamination, parameter estimation, and catchment classification, while at the same time also continuing investigations into the potential problems with chaos identification methods. Extensive reviews of these studies are already available in the literature (Sivakumar 2000, 2004, 2009) and, therefore, details are not reported herein.

The outcomes of these applications are very encouraging, especially from the perspectives of model simplification and more accurate short-term predictions. In light of these, an attempt is made in the present study to apply the ideas of chaos theory to investigate the nature of the dynamics of inflows to reservoirs. Data embedding and dimensionality concepts are employed to identify the dynamic nature of inflows, whether stochastic or chaotic. For application, the Pechiparrai Reservoir in the Kodiyar River basin in the state of Tamil Nadu in South India is considered as a case study. The study specifically involves the application of the correlation dimension method to inflow data observed at four different temporal scales (daily, 2-day, 4-day, and 8-day). These four scales are considered in view of the weekly-scale operation of the Pechiparrai Reservoir. It is hoped that the outcomes of this study would be useful towards the selection of appropriate model(s) for simulating/forecasting reservoir inflows and thus eventually in the operation of the reservoir.

The rest of this paper is organized as follows. First, the correlation dimension method, including the phase space reconstruction concept, is described. Next, details of the study area and data considered for analysis are provided. Then, results of the application of the correlation dimension method are presented. Finally, conclusions drawn are highlighted.

CORRELATION DIMENSION METHOD

The dimension of a time series is, in a way, a representation of the number of dominant variables present in the evolution of the corresponding dynamic system. Dimension analysis reveals the extent to which the variations in the time series are concentrated on a subset of the space of all possible variations. Correlation dimension is a measure of the extent to which the presence of a data point affects the position of the other points lying on the attractor (a geometric object which characterizes the long time behaviour of a system) in the phase space (see below). The correlation dimension method uses the correlation integral or function for determining the dimension of the attractor and for distinguishing between stochastic and chaotic behaviours. The concept of the correlation integral is that a time series arising from deterministic dynamics will have a limited number of degrees of freedom equal to the smallest number of first-order differential equations that capture the most important features of the dynamics. Thus, when one constructs phase spaces of increasing dimension, a point will be reached where the dimension equals the number of degrees of freedom, beyond which increasing the phase-space dimension will not have any significant effect on correlation dimension.

Among the methods available for correlation dimension estimation, the Grassberger-Procaccia algorithm (Grassberger & Procaccia, 1983) is the most widely-used one and, therefore, is also employed herein. The algorithm uses the concept of phase-space reconstruction for representing the dynamics of the underlying system from an available single-variable time series. Using past history of such a variable, X_i, $i = 1, 2, ..., N$, and an appropriate delay time τ (an integer multiple of sampling time), the multi-dimensional phase–space can be reconstructed according to (Takens, 1981):

$$Y_j = (X_j, X_{j+\tau}, X_{j+2\tau}, ..., X_{j+(m-1)\tau}) \quad (1)$$

where $j = 1, 2, ..., N - (m - 1)\tau/\Delta t$, m is the dimension of Y_j, called the embedding dimension. A (correct) phase-space reconstruction in a dimension m facilitates an interpretation of the underlying dynamics in the form of an m-dimensional map, f_T, according to:

$$\mathbf{Y}_{j+T} = f_T(\mathbf{Y}_j) \quad (2)$$

where $\mathbf{Y}_j$ and $\mathbf{Y}_{j+T}$ are vectors of dimension m, describing the state of the system at times j (i.e. current state) and $j + T$ (i.e. future state), respectively.

For an m-dimensional phase space, the correlation function $C(r)$ is given by:

$$C(r) = \lim_{n\to\infty} \frac{2}{N(N-1)} \sum_{\substack{i,j \\ (1\le i<j\le N)}} H\left(r - \left|Y_i - Y_j\right|\right) \quad (3)$$

where H is the Heaviside step function, with $H(u) = 1$ for $u > 0$, and $H(u) = 0$ for $u \leq 0$, $u = r - (\boldsymbol{Y}_i - \boldsymbol{Y}_j|$, and r is the radius of the sphere centred on Y_i or Y_j. If the time series is characterized by an attractor, then for positive values of r, the correlation function $C(r)$ is related to the radius r by:

$$\underset{\substack{r \to 0 \\ N \to \infty}}{C(r) \sim \alpha r^{\nu}} \tag{4}$$

where α is a constant and ν is the correlation exponent given by:

$$\nu = \lim_{\substack{r \to 0 \\ n \to \infty}} \frac{\log C(r)}{\log r} \tag{5}$$

which is also the slope of the log $C(r)$ *versus* log r plot.

The slope is generally estimated by a least squares fit of a straight line over a certain range of r, called the scaling region. The distinction between (low-dimensional) determinism and (high-dimensional) stochasticity can be made using the ν *versus* m plot. If ν saturates after a certain m and the saturation value is low, then the system is generally considered to exhibit (low-dimensional) deterministic behaviour. The saturation value of ν is defined as the correlation dimension (d) of the attractor. On the other hand, if ν increases without bound with increase in m, the system under investigation is generally considered to exhibit (high-dimensional) stochastic behaviour.

The reliability of the Grassberger-Procaccia algorithm (or any other algorithm) for estimating correlation dimension and thus for distinguishing between low-dimensional (chaotic) and high-dimensional (stochastic) systems has been under considerable debate, in view of their potential limitations when applied to real data. Some of the relevant issues are: data size, data noise, presence of zeros, delay time selection, and even stochastic processes yielding low correlation dimensions. Many studies (e.g. Lorenz, 1991; Tsonis *et al.*, 1993; Sivakumar *et al.*, 1999, 2002; Sivakumar, 2000, 2005) have extensively responded to these concerns. Addressing these issues is beyond the scope of this paper.

STUDY AREA AND DATA

In the present study, the dynamic nature of inflows to the Pechiparrai Reservoir in the Kodiyar River basin in the state of Tamil Nadu, South India is investigated. The Kodaiyar basin is located between 8°05′N and 8°35′N latitudes and 77°05′E and 77°35′E longitudes (Fig. 1). The area of the basin is 1500 km^2. The capacity of the Pechiparrai Reservoir is 150×10^6 m^3. The average annual inflow to the reservoir is 340×10^6 m^3.

For analysis, inflows observed over a period of 32 years (June 1970–June 2002, consistent with the "water years") are considered. Jothiprakash & Dobhal (2006) present the statistical and frequency analysis of this inflow series. Table 1, for instance, presents some of the important statistical properties of the inflows derived for a monthly scale. Among the observations, the maximum inflow occurs during the month of October and minimum inflow occurs during the month of April.

Since the operation of the Pechiparrai Reservoir is normally carried out at weekly intervals, it would be ideal to study the nature of the inflow dynamics at this scale, rather than monthly. Further, in the spirit of this (chaos) study and with the fact that changes in inflows may sometimes occur quite dramatically at finer scales than weekly, it would be more prudent to study the nature of the inflow dynamics at a few different scales up to about weekly. To this end, and with the available observed inflow data at the daily scale, it is decided to analyse data corresponding to daily, 2-day, 4-day, and 8-day scales. Inflow values corresponding to 2-day, 4-day, and 8-day are obtained by aggregating (i.e. simple addition) the daily values, as deemed appropriate. Figure 2 shows the time series plots of these four series, and Table 2 presents their important statistical properties.

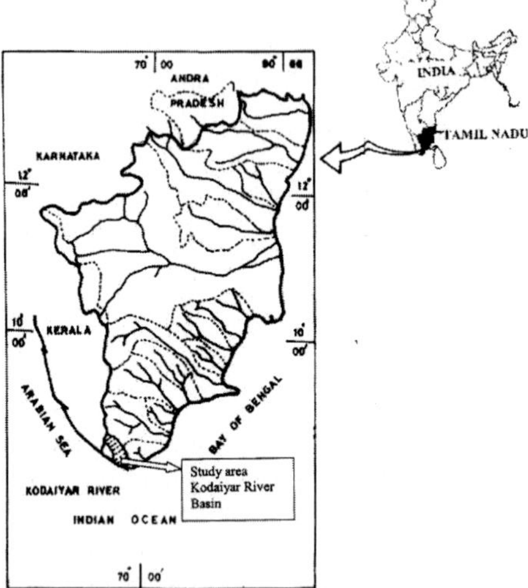

Fig. 1 Location of the Kodaiyar River basin.

Table 1 Statistical properties of the inflow data.

Month	Average ($\times 10^6$ m^3)	Std. dev ($\times 10^6$ m^3)	Coefficient of variation	Skewness	Kurtosis
Jan	25.32	7.90	0.31	0.44	0.42
Feb	23.23	9.88	0.42	0.0351	–0.28
Mar	12.50	5.99	0.47	–0.051	–1.067
Apr	8.24	4.56	0.55	0.84	0.43
May	10.67	6.51	0.60	0.85	–0.26
Jun	36.92	31.40	0.85	3.21	12.98
Jul	37.16	13.64	0.36	0.59	1.49
Aug	33.81	10.42	0.30	0.76	0.27
Sep	36.17	14.77	0.40	1.33	3.75
Oct	44.10	20.70	0.46	0.99	0.68
Nov	41.90	20.19	0.48	1.82	3.31
Dec	30.19	19.11	0.63	3.91	19.11
Annual average	340.21	–	–	–	–

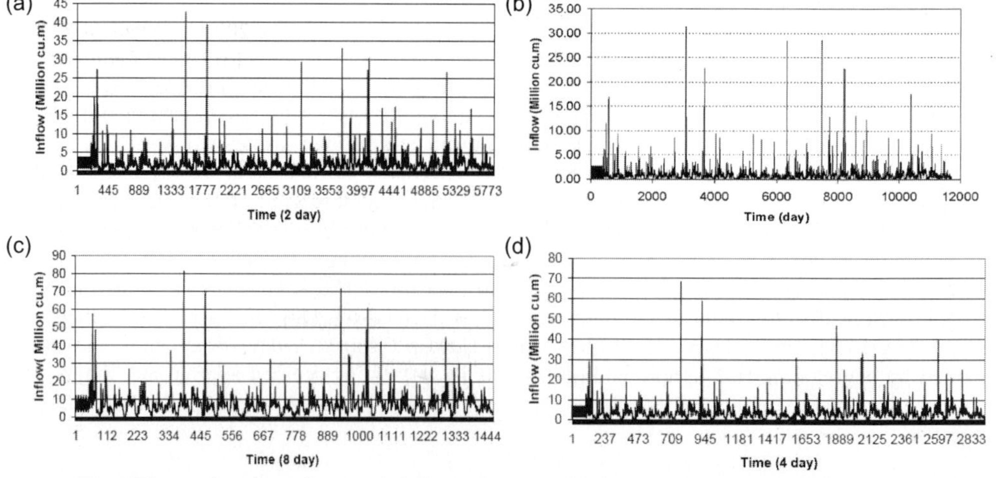

Fig. 2 Time series plots of reservoir inflows: (a) daily, (b) 2-day, (c) 4-day, and (d) 8-day.

Table 2 Statistics of reservoir inflow data of different temporal scales.

Statistic	Daily	2-day	4-day	8-day
No. of data	11712	5856	2928	1464
Mean (Mm^3)	0.93	1.85	3.71	7.42
Standard deviation ($\times 10^6$ m^3)	1.217	2.18	3.90	6.77
Maximum value ($\times 10^6$ m^3)	31.3	42.6	68.44	81.42
Minimum value ($\times 10^6$ m^3)	0	0	0	0.147
Coefficient of variation	1.30	1.17	1.05	0.91
Skewness	9.21	7.18	5.92	4.19
Number of zero values	57	6	1	0

RESULTS AND DISCUSSION

Figure 3(a)–(d) shows the phase space diagrams for the daily, 2-day, 4-day, and 8-day inflow series, respectively. These diagrams correspond to reconstruction of the series in two dimensions (m = 2) with delay time τ = 1, i.e. the projection of the attractor on the plane (X_i, X_{i+1}). The projections yield reasonably good attractors, i.e. no scatter all over the phase space. For each series, the attractor seems to indicate that the dynamics are evolving in some defined "boundary" (around the diagonal 1:1 line) and thus suggests the possibility of some kind of deterministic dynamics. Another notable observation may be that the size of the attractor increases with increase in the scale of observation (i.e. from daily to 8-day), a possible indication that the complexity of the dynamics increases with the scale of aggregation (see below for more details).

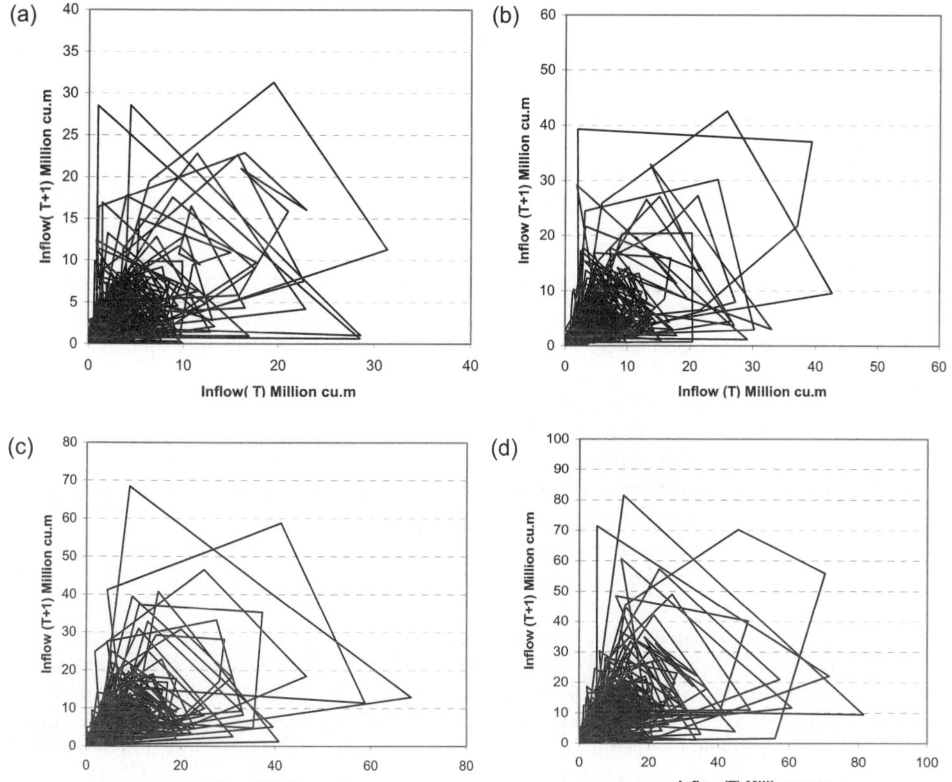

Fig. 3 Phase space diagram of reservoir inflows: (a) daily, (b) 2-day, (c) 4-day, and (d) 8-day.

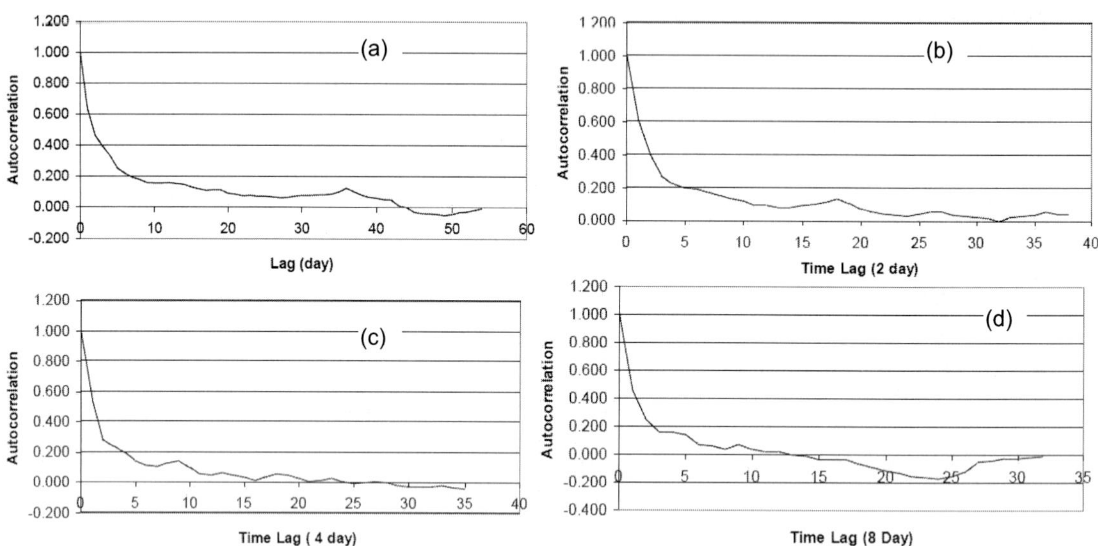

Fig. 4 Autocorrelation function plots for reservoir inflows: (a) daily, (b) 2-day, (c) 4-day, and (d) 8-day.

It may be noted that an appropriate delay time for phase space reconstruction is necessary because only an optimum delay time gives the best separation of neighbouring trajectories within the minimum embedding space, whereas an inappropriate selection of delay time may lead to an underestimation or overestimation of correlation dimension. There are different methods and guidelines to choose the optimum delay time value: the autocorrleation function method, the mutual information method, and the correlation integral method. Each has its own advantages and disadvantages. In this study, the autocorrelation function method is used, and the optimum delay time is taken as the lag time at which the autocorrelation function first crosses the zero line.

Figure 4(a) to (d) presents the autocorrelation function plots for the daily, 2-day, 4-day, and 8-day inflow series, respectively. For all the four series, the autocorrelation functions exhibit relatively slow and exponential decays. These slow decays are indication of temporal persistence of the time series that may be related to determinism in the underlying system dynamics. For the four series, the lag time at which the autocorrelation function first crosses the zero line is 44 (44 days), 32 (64 days), 23 (96 days), and 13 (104) days, respectively, which are subsequently used for phase space reconstruction.

The correlation functions and the exponents are now computed for the four inflow series, with phase spaces reconstructed in embedding dimensions from 1 to 15. Figure 5(a)–(d) presents the relationship between the correlation function $C(r)$ and the radius r in log-log scale. For all the four series, the $\log C(r)$ *versus* $\log r$ plots exhibit reasonably large scaling regions, allowing fairly accurate estimation of the slopes (i.e. correlation exponents).

Figure 6 shows the relationship between the correlation exponent and the embedding dimension for the four series. Saturation of the correlation exponent is observed in all the cases, suggesting that the dynamics of the system that produces the inflows is nonlinear deterministic in nature. The correlation dimension (i.e. saturation value of correlation exponent) are found to be 1.98, 2.10, 2.21, and 2.35, respectively (see Table 3). These low correlation dimension values suggest the presence of low-dimensional chaotic dynamics of the inflows to the reservoir. Since the nearest integer above the correlation dimension value provides the number of variables dominantly governing the dynamics, these results suggest that the reservoir inflow dynamics at the four scales are dominantly governed by just two or three variables, as the case may be. The results also indicate that the complexity of the inflow dynamics changes with aggregation in temporal scale, from daily to 8-day. This may be an indication that certain mechanisms in the river basin take place only at a larger coarser scale.

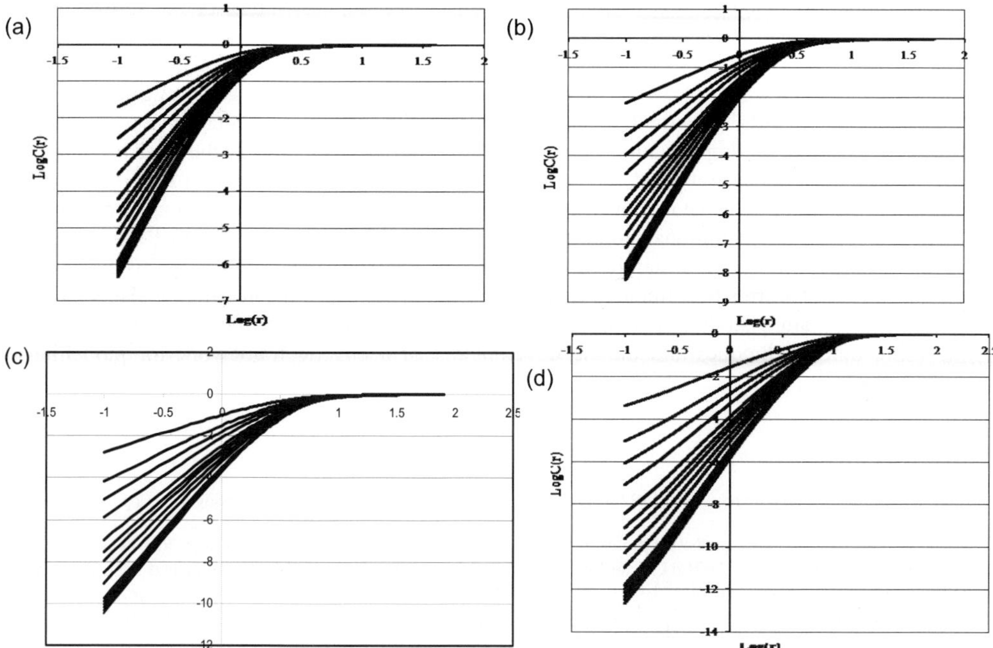

Fig. 5 Correlation function for reservoir inflows: (a) daily, (b) 2-day,(c) 4-day, and (d) 8-day.

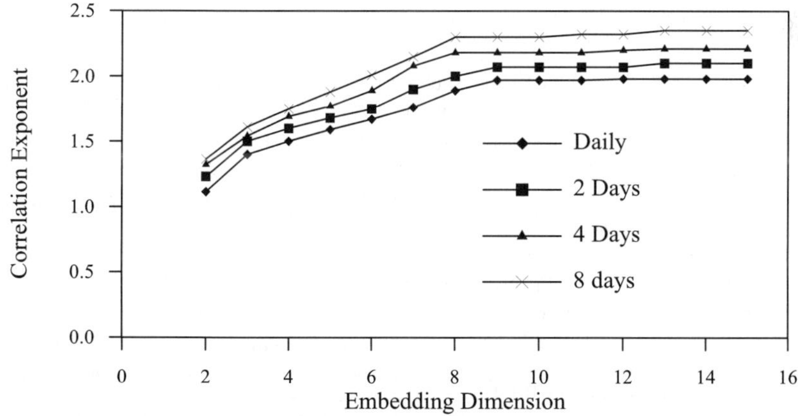

Fig 6 Correlation dimension results for reservoir inflow.

Table 3 Correlation dimension results for reservoir inflow at different temporal scales.

Statistic	Daily	2-day	4-day	8-day
Delay time, τ (in days)	44 (44)	32 (64)	24 (96)	13 (104)
Correlation dimension	1.98	2.10	2.21	2.35
Number of variables	2	3	3	3

CONCLUSIONS AND SCOPE FOR FURTHER STUDY

This study investigated the dynamic nature of inflows to a reservoir, with a case study of the Pechiparrai Reservoir in the Kodiyar River basin in the state of Tamil Nadu in South India. The correlation dimension method was employed to estimate the dimensionality of the inflow time series and to distinguish between chaotic and stochastic dynamics. Results from the analysis of inflows of four different temporal scales (daily, 2-day, 4-day and 8-day) reveal the presence of low-dimensional chaotic nature of the underlying system dynamics. They also suggest that the dynamics are dominantly governed by only two or three variables, which is encouraging, especially from a model simplification perspective. The results also suggest the possibility of a chaotic scaling relationship, at least in the range of the four temporal scales (daily to 8-day) studied. The correlation dimension values for the four time series indicate that data aggregation in temporal scale produces additional complexities. This is probably due to the fact that certain mechanisms only influence the dynamics at coarser scales, linked to the spatial extent of the basin. Results from the autocorrelation function method and the phase space reconstruction also support this observation. Future studies will focus on the verification and confirmation of the present results using other techniques and on the development of a chaotic framework for modelling reservoir inflows.

REFERENCES

Alligood, K. T, Sauer, T. D. & Yorke, J. A. (1997) *Chaos: An Introduction to Dynamical Systems*. Springer-Verlag, New York, USA.

Amorocho, J. (1967) The nonlinear prediction problems in the study of the runoff cycle. *Water Resour. Res.* **3**(3), 861–880.

Bahohui, M., Liang, C. & Zhao, X. (2004a) Chaotic analysis on precipitation time series of Sichuan middle part in upper region of Yangtze. *Nature and Science* **2**(1), 74–78.

Baohui, M., Zhao, X. & Liang, C. (2004b) Chaotic analysis on monthly precipitation on hills region in middle Sichuan of China. *Nature and Science* **2**(2), 45–51.

Casdagli, M. (1989) Nonlinear prediction of chaotic time series. *Physica D* **35**, 335–356.

Casdagli, M. (1992) Chaos and deterministic *versus* stochastic nonlinear modeling. *J. Roy. Statist.. Soc. B* **54**(2), 303–328.

Grassberger, P. & Procaccia, I. (1983) Measuring the strangeness of strange attractors. *Physica D* **9**, 189–208.

Henon, M. (1976) A two-dimensional mapping with a strange attractor. *Commun. Math. Phys.* **50**, 69–77.

Izzard, C. F. (1966) A mathematical model for nonlinear hydrologic systems. *J. Geophys. Res.* **71**(20), 4811–4824.

Jayawardena, A. W. & Lai, F. (1994) Analysis and prediction of chaos in rainfall and stream flow time series. *J. Hydrol.* **153**, 23–52.

Jothiprakash, V. & Dobhal, M. (2006) Probabilistic forecasting of reservoir inflows using frequency analysis: a case study. In: *Proc. National Conference on Hydraulics and Water Resources*, 230–237 (HYDRO-2006, Bharati VidyaPeeth University, 8–9 Dec. 2006).

Lorenz, E. N. (1991) Dimension of weather and climate attractors. *Nature* **353**, 241–244.

Robert, H. C. (1999) *Chaos and Nonlinear Dynamics: An Introduction for Scientists and Engineers*. Oxford University Press, New York, USA.

Salas, J. D., Delleur, J. W., Yevjevich, V. M. & Lane, W. L. (1995) *Applied Modeling of Hydrologic Time Series*. Water Resources Publications, Littleton, Colorado, USA.

Schuster, H. G. (1987) *Deterministic Chaos: An Introduction*. Library of Congress, Weinheim, Germany.

Sivakumar, B. (2000) Chaos theory in hydrology: important issues and interpretations. *J. Hydrol.* **227**(1–4), 1–20.

Sivakumar, B. (2004) Chaos theory in geophysics: past, present and future. *Chaos, Solitons and Fractals* **19**(2), 441–462.

Sivakumar, B. (2005) Correlation dimension estimation of hydrologic series and data size requirement: myth and reality. *Hydrol. Sci. J.* **50**(4), 591–604.

Sivakumar, B., Phoon, K. K., Liong, S. Y. & Liaw, C. Y. (1999) A systematic approach to noise reduction in chaotic hydrological time series. *J. Hydrol.* **219**(3–4), 103–135.

Sivakumar, B., Persson, M., Berndtsson, R. & Uvo, C. B. (2002) Is correlation dimension a reliable indicator of low-dimensional chaos in short hydrological time series? *Water Resour. Res.* **38**, doi:10.1029/2001WR000333.

Takens, F. (1981) Detecting strange attractors in turbulence. In: *Dynamical Systems and Turbulence* (ed. by D. A. Rand & L. S. Young), 366–381. Lecture Notes in Mathematics 898, Springer-Verlag, Berlin, Germany.

Thomas, H. A. & Fiering, M. B. (1962) Mathematical synthesis of streamflow sequences for the analysis of river basins by simulation. In: *Design of Water Resource Systems* (ed. by A. Mass *et al.*), 459–493. Harvard University Press, Cambridge, Massachusetts, USA.

Tsonis, A. A., Elsner, J. B. & Georgakakos, K. P. (1993) Estimating the dimension of weather and climate attractors: important issues about the procedure and interpretation. *J. Atmos. Sci.* **50**, 2549–2555.

Valencia, D. & Schaake, J. C. (1973) Disaggregation process in stochastic hydrology. *Water Resour. Res.* **9**(3), 580–585.

Deriving short-term reservoir inflow forecasts for real-time optimization of reservoir operation to mitigate flood damage

GIOVANNI CARDOSO[1], DAVID MONCUNIL[2], DIRCEU S. REIS Jr[1], EDUARDO MARTINS[4] & LUIZ NASCIMENTO[1]

1 *Department of Water Resources, Research Institute for Meteorology and Water Resources – FUNCEME, Brazil*
giovannibrigido@gmail.com

2 *Department of Meteorology, Research Institute for Meteorology and Water Resources – FUNCEME, Brazil*

3 *Research Institute for Meteorology and Water Resources – FUNCEME, Brazil*

Abstract This research was motivated by very damaging floods that occurred in the State of Ceara, Brazil, in the first weeks of 2004. The strategy adopted here is to use inflow forecasts, based on precipitation forecasts obtained by several atmospheric models and hydrological modelling to perform real-time optimization of reservoir operations. Results based on three years of forecast and observed data (2006–2008) show that the information contained in these precipitation forecasts can be of great value in the short-term reservoir inflow forecasts; comparisons show that the inflow forecast and the inflow estimated by COGERH are very similar, which is useful for the decision maker in a period of a flood event.

Key words forecast; flood; optimization; inflow forecast; damage curve; reservoir operation; Brazil

INTRODUCTION

Flood tragedies are a recurring problem that many governments have faced over the last several decades. An example is the events that occurred in the first two weeks of 2004, in Ceara State, Northeast Region, Brazil. Those events were so serious that the government had to declare a situation of public calamity in many villages and cities that were isolated without water supply. There was much infrastructural damage, such as destruction of small reservoirs, bridges and roads, and a 70% loss of crop yield. These problems occurred mainly because of the lack of information on future reservoir inflows and the lack of knowledge of how the system should be operated in situations like this one.

The goal of this article is the development of reservoir inflow forecasts based on coupling of a hydrological model with different atmospheric conditions. The study presented here focuses on the main reservoirs of the Banabuiu watershed – Banabuiu Reservoir (6269.5 km^2), Fogareiro (5143.5 km^2), Patu (993.3 km^2) and Quixeramubim (1908.1 km^2) (see Fig. 1).

DATA ASSIMILATION

The reservoir inflow forecasts are obtained through the use of atmospheric numerical models, whose precipitation forecasts are used to feed a hydrological model that generates the inflows. In fact, simply using observed precipitation forecasts could generate the inflow forecasts; however, as the times of concentration of these watersheds are around several hours, the incorporation of precipitation forecasts has the objective to increase the forecast horizon.

The idea is to work with a composite of several numerical atmospheric models, run by different weather centres, whose precipitation forecasts are widely available on the internet. This strategy is referred to here as multi-model.

The multi-model strategy is based on the precipitation forecasts of 11 models having different initiation times and different spatial resolution:

(1) CPTEC ETA 40KM 12TMG (2) DHN HRM 30KM 00TMG (3) CPTEC ETA 20KM 12TMG
(4) CPTEC ETA 20KM 00TMG (5) DHN HRM 30KM 12TMG (6) CPTEC ETA 40KM 00TMG
(7) CPTEC GL 60KM 12TMG (8) MARYLAND ETA 80KM 00TMG (9) NCEP AVN 60KM 00TMG
(10) CPTEC GL 100KM 12TMG (11) INMET MBAR 25KM 00TMG

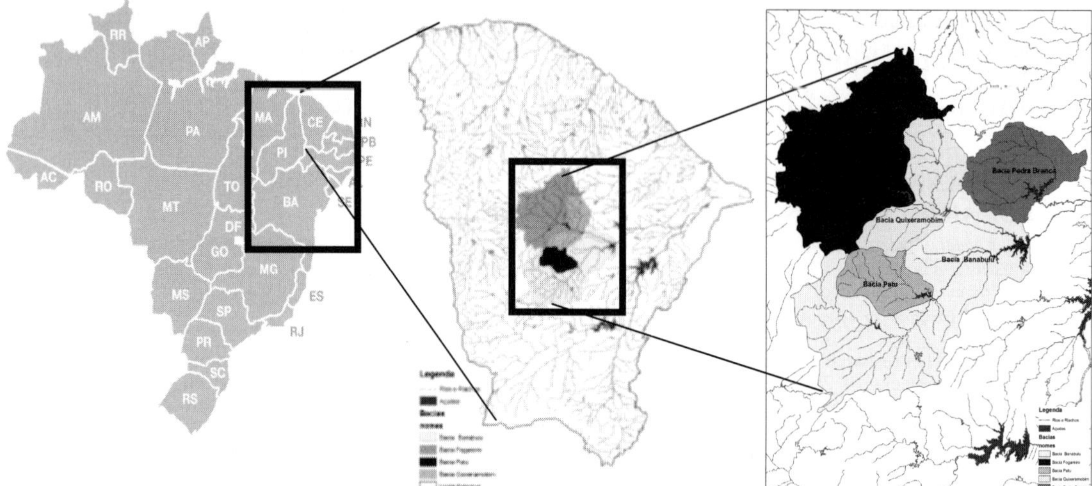

Fig. 1 Brazil, Northeast Region (rectangle to the left), Ceara State (CE) and the Banabuiu watershed with the five drainage areas that contribute to the five main reservoirs in the region.

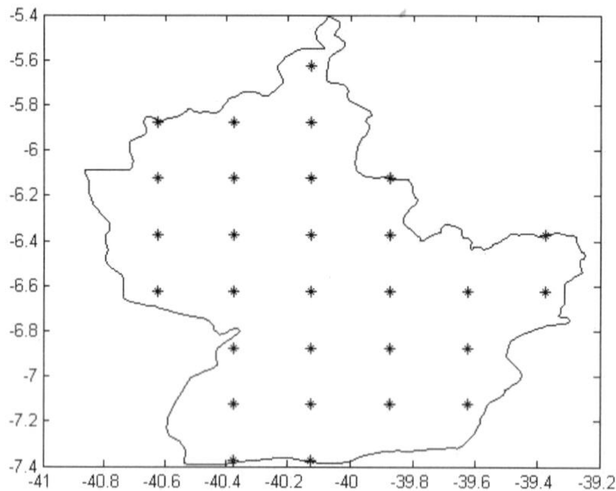

Fig. 2 The common grid mesh for precipitation forecasts over the Patu watershed.

The precipitation forecasts obtained by the multi-model strategy are based on the selection and combination forecasts obtained by a few of the 11 models cited earlier. Details of this selection and combination are given in the methodology section. In order to have an operational system, several scripts were prepared in order to obtain and interpret the images available on the weather centre websites, and to transform all the data into a common spatial grid of ¼ of a degree.

The precipitation forecasts provided by each model at each grid point are used to compute average-basin precipitation to be used by the lumped hydrological model (Fig. 2).

CALIBRATION OF THE HYMOD HYDROLOGICAL MODEL

Historical reservoir inflow series provided by the Water Management Company (COGERH) of the State of Ceara, and computed based on mass balance equations applied for each reservoir (daily storage and release data are available), were used to calibrate the parameters of the hydrological model.

The evolutionary optimization algorithm named Particle Swarm Optimization (PSO) was used to calibrate the model. The objective function employed in the study was the efficiency index of Nash & Sutcliffe (1970):

$$F_1 = \max_{\theta} \left(1 - \frac{\sum_{1}^{n} (I_{obs} - I_{sim})^2}{\sum_{1}^{n} (I_{obs} - \bar{I}_{obs})^2} \right) \quad (1)$$

where θ is a set of model parameters, I_{obs} is a set of observed inflows, I_{sim} is a set of simulated inflows, n is the length of the set in days and $\bar{I}_{obs}$ is the mean of daily observed inflows.

The hydrological model used in the calibration was HYMOD defined by Moore (1985) and used by Barros (2007). The results of the calibration are shown in Table 1.

Table 1 Results of calibration and verification for the reservoirs of Banabuiu watershed.

Reservoirs	Calibration	Verification	Start date	End date	Days of calibration	Days of verification
Banabuiu	60%	52%	26/06/1993	31/05/2008	3000	2454
Fogareiro	81%	54%	29/07/1996	31/05/2008	3000	1325
Patu	67%	64%	28/06/1993	31/05/2008	3000	2452
Quixeramubim	52%	51%	03/06/1996	31/05/2008	1000	3381

METHODOLOGY

This section starts with a description of the methodology used to select and combine results of numerical atmospheric models to generate precipitation forecasts to be used by the hydrological model to provide the reservoir inflow forecasts.

Observed precipitation data and the corresponding inflow simulated series are used to evaluate each model performance. Despite the fact that forecasts of all 11 atmospheric models were freely available on the Internet, these models occasionally may not produce forecasts because of operational features. Therefore, the weighting scheme to be used to select and combine these models must take into account the number of days without failure before the day the forecast is issued. In addition, only a few of the 11 models have good forecasting skill, and as such, only models with good performance were used in the scheme. A sensitivity analysis was carried out on the number of days prior to the forecast day and on the number of models to be used in the weighting scheme. Based on these sensitivity analyses, the best relative performance was identified for a weighting scheme composed of the best two models (among 11) using the 10 days before forecast date to evaluate the weights of each model. It should be clarified that a model with a failure over these 10 days was discarded from the weighting scheme.

The weather forecast estimated by the weighting scheme, or simply "multi-model", is obtained on a daily basis by:

$$P_{multi-model} = \frac{\sum_{i=1}^{2} P_i / AE_i}{\sum_{i=1}^{2} \frac{1}{AE_i}} \quad (2)$$

where P_i is the rainfall and AE_i is the average absolute error evaluated for model i between the observed and modelled rainfall over the 10 previous days. Only the two models with the lowest average absolute error are selected. The multi-model procedure was compared to the alternative models: no-rainfall model, persistent precipitation model, and the 11 individual forecast models.

HYDROLOGICAL MODEL

The hydrological model employed in this study was the daily rainfall–runoff model HYMOD (Moore, 1985). The version used is a relatively simple model, and uses the probability distribution concept to describe the spatial variation of runoff production process parameters. This allows the integration of the flow response over the whole watershed represented by algebraic expressions. The idea behind the model is that the watershed can be viewed as a set of points without interaction among them, while each one has a water storage capacity that, when exceeded, generates runoff.

RESULTS

A comparison between the inflow forecast and the inflow estimated by COGERH is presented in Fig. 3, which represents the Banabuiu watershed; the inflow forecast for 24 hours is reasonable when compared with the information data. In this case, the tendency of the inflow forecast followed the observed inflow.

Table 2 presents the results in terms of Nash-Sutcliff efficiency metric (N-S) for 24-h and 48-h lead times. One can see that, in both cases, the use of multi-model precipitation forecast procedure provided better results compared to individual forecast procedures.

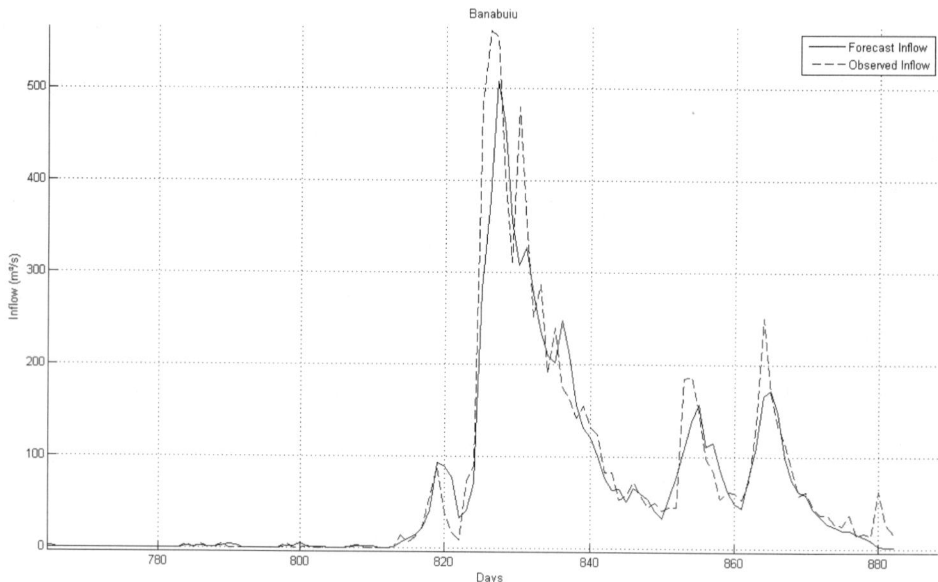

Fig. 3 Comparison between the inflow forecast through the multi-model and observed inflow through COGERH data in the Banabuiu Reservoir 24-h lead time.

Table 2 Performance (Nash-Sutcliffe efficiency, N-S) of models for inflow forecasts for Banabuiu Reservoir.

Inflow (24-h lead time)	N-S	Inflow (48-h lead time)	N-S
Multi-model	0.87	Multi-model	0.70
Persistent model	0.74	Persistent model	0.60
No-rain model	0.66	No-rain model	0.53
Model 10 (best individual model)	0.83	Model 7 (best individual model)	0.66
Model 2 (worst individual model)	0.70	Model 1 (worst individual modelo)	0.17

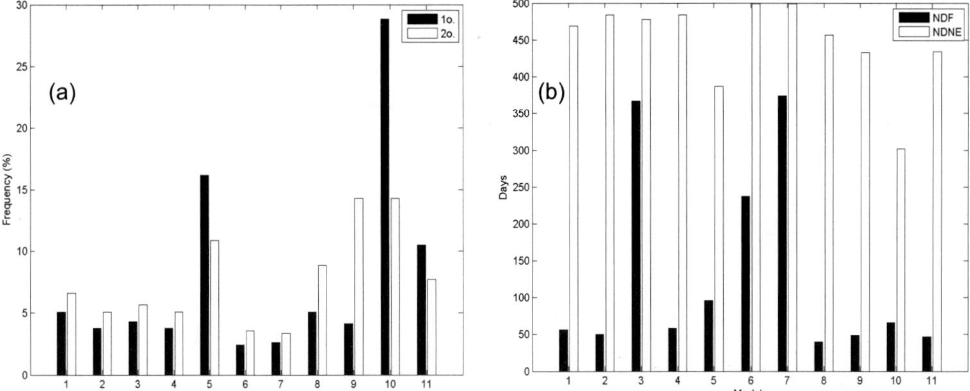

Fig. 4 (a) Percentage of time models is selected in the multi-model procedure (best model in black and second best in white) for 24-h lead time. (b) Number of days that the model failed (in black) and number of days that the models were not chosen (in white) for 24-h lead time.

As the multi-model procedure combines the forecasts of different models on a daily basis, based upon performance on previous days, it is important to evaluate how often each of the 11 models used in the study was in fact employed in the forecast. This analysis may provide some insights for defining a better set of models in the future.

Figure 4(a) shows the percentage of time each of the 11 models employed in the forecast system was selected in the multi-model procedure. There are two reasons for a model not being selected on a given day: (1) poor performance on previous days compared to other models, and (2) unavailable forecast.

Models 5 and 10 were selected try far the most often, whereas models 6 and 7 were selected in very few cases. Models 3, 6 and 7 had the largest proportion of unavailable forecast days, which explains the small selection rate of models 6 and 7. Models 4 and 8 had a small selection rate because they had poor forecast performance (Fig. 4(b)).

Figure 5 presents a comparison of the percentage error with respect to streamflow forecasts obtained with the multi-model precipitation forecast procedure (MM), persistent precipitation (Pr), no-rain scenario (Zr), and precipitation forecasts obtained by 11 individual numerical weather models. Each of these streamflow forecast alternatives has two associated box plots, the first one related to the percentage forecast error for 24-h lead time, and the second representing the percentage forecast error for 48-h lead time. The multi-model presented the more significant results because it had the smallest range of errors.

CONCLUSIONS

All the algorithms developed in this article were implemented in the MATLAB script language along with geographic data processed in the ArcView GIS used to calculate the boundary of each reservoir.

The use of multi-model could be very useful, mainly for the poor countries that do not have the technology to develop a safe forecast.

The results presented here show that it is better to use a couple of models instead of using them individually, as can be seen in Fig. 5 and Table 2 with the use of percentage error and Nash-Sutcliffe index.

Results based on three years of forecast and observed data (2006–2008) show that the information contained in these precipitation forecasts can be of great value in short-term reservoir inflow forecasts; comparison shows that the inflow forecast and the inflow estimated by COGERH is very similar, which must be very useful for the decision maker in a period of a flood event.

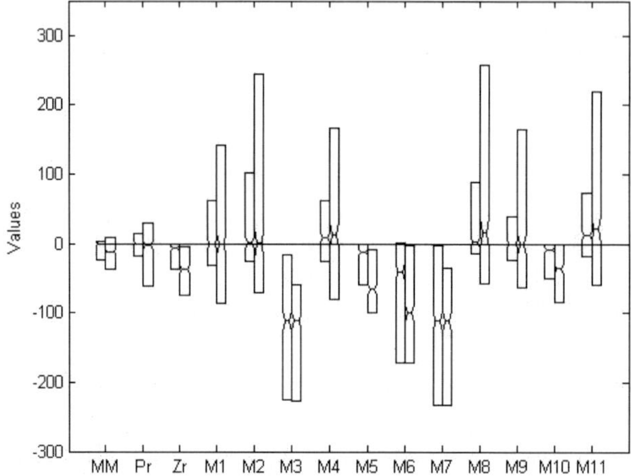

Fig. 5 Boxplots of error (%) in streamflow forecasts (24-h and 48-h lead times) obtained with multi-model precipitation forecast procedure (MM), persistent precipitation model (Pr), no-rainfall model (Zr), and precipitation forecasts obtained by 11 individual models for Banabuiu Reservoir.

Acknowledgements The authors thank CNPq and FUNCAP for providing financial support for this research; and Metereological Institute of Ceara State (FUNCEME) and Water Resources Management Company of Ceara State (COGERH) for providing information about meteorological models and the state's reservoirs data, respectively.

REFERENCES

Al-Futaisi, A. & Stedinger, J. R. (1999) Hydrologic and economic uncertanties and flood risk project design. *J. Water Resour. Plan. Manage.* **125**(6), 314–323.

Barros, F. V. F. (2007) The use of evolutionary algorithms in the calibration of watershed models and in the optimization of reservoirs' system operation. MSc Thesis, Department of Hydraulic and Environmental Engineering, Federal University of Ceara, Fortaleza, CE, Brazil (in Portuguese).

Kennedy, J. & Eberhart, R. C. (1995) Particle swarm optimization. In: *Proc 1995 IEEE International Conference on Neural Networks*, 1942–1948. IEEE Service Center, Piscataway, New Jersey, USA.

Lopes, J. E. G., Braga, B. P. F. & Conejo, J. G. L. (1982) SMAP – A simplified hydrological model. In: *Applied Modelling in Catchment Hydrology* (ed. by V. P. Singh), 167–176. Water Resources Publications, Littleton, Colorado, USA.

Moore, R. J. (1985) The probability-distributed principle and runoff production at point and basin scale. *Hydrol. Sci. J.* **30**(2), 273–297.

Nash, J. E. & Sutcliffe, J. V. (1970) River flow forecasting through conceptual models, Part I – A discussion of principles. *J. Hydrol.* 10(3), 282–290.

Osny, E. da Silva & Pacheco, C. (2004) The control of floods in the Jaguaribe watershed during 2004. Water Resources Management Company – COGERH.

Real-time correction on the river stage forecasting with an equivalent stage approach

XIAOQIN ZHANG[1,2], WEIMIN BAO[1,2], ZHONGBO YU[1,3] & SIMIN QU[1,2]

1 *State Key Laboratory of Hydrology-Water Resources and Hydraulic Engineering, Hohai University, Nanjing 210098, China*
zxqin403@163.com

2 *College of Hydrology and Water Resources, Hohai University, Nanjing, 210098, China*

3 *Department of Geoscience, University of Nevada Las Vegas, Las Vegas, Nevada 89154, USA*

Abstract The auto-regressive model (AR model) has been widely used to correct forecasted river stages. However, the nonlinear relationship between the widths of channel and stages in the cross-sections affects its correction performance. An equivalent stage technique is proposed to reduce this effect, in which the observed and forecasted stages are converted into the observed and forecasted equivalent stages. The errors between these two kinds of equivalent stages are employed in the AR model to forecast equivalent stage errors which are reduced to the forecasted stage errors for correcting the forecasted stages. The real-time correction method that combines the technique with the AR model (EAR method) is applied to the tidal reach in the Caoe River, China, for simulations of river stage. The results indicate that the EAR method can provide better accuracy for stage forecasting than the AR model, and especially has the capability of reducing the errors at simulated peak stages.

Key words river stage forecasting; equivalent technique; flood plain; AR model

INTRODUCTION

Flood forecasting is an important non-structural approach for flood mitigation. River stage is chosen as the variable to be forecast because it is practically useful in the issuance of a flood warning. River stage in a tidal river is especially important for flood prevention, because of the fact that the flood of a tidal river is contributed to not only by the upstream runoff but also by the downstream tide. Generally, stage forecasting is more complicated than discharge forecasting, because it is closely related to factors such as cross-sections, river slope, fluvial erosion deposition, and backwater. The corresponding stage method and the conversion method based on a rating curve have been widely used in stage forecasting. However, their performances are not satisfactory when applied to rivers with complex cross-sections. There exists some uncertainty in the conversion of discharge into stage because the relationship between discharge and stage is imperfect.

Therefore, river stage forecasting has attracted increasing attention. A set of stage forecasting models was proposed and developed, such as hydrodynamic models based on the Saint Venant equations (Lamberti, 1996; He, 2008), a stage model based on the Muskingum method (Franchini, 1994), and stage forecasting with support vector regression (Yu, 2006). These stage forecasting models can obtain satisfactory results when applied to a river with a stable river bed with little variation in the stage. But they cannot perform well in rivers with obvious flood plains with significant changes in stage. It is important to improve the performance of these models. Various adaptive techniques were used to correct forecast stages, including the Kalman filter (Neal, 2007; Wu, 2008), errors regression (Guo, 2002), and comprehensive correction (Qu, 2003). Moreover, some new methods have been developed, such as the real-time stage correction method for flash flood forecasting (Hsu, 2003), the roughness coefficient updating methods (Cheng, 2005; Hsu, 2006), the time-variant parameter considered using a variable weighted forgetting factor (Mu, 2007), and the stage correction considering the relationship between channel width and stage (Hu, 2007).

Hydrological prediction error is of fundamental importance to the feasible operation of a real-time flood forecasting system. The conventional auto-regressive model (AR model) that is based on the errors between observed and calculated stages has been widely used to correct the forecasted stage. However, the forecasted stage corrected with the AR model still cannot achieve a desirable accuracy in a river with complex cross-sections. The main reason is that little attention has been paid to the nonlinear relationship between channel width and stage. Also, river channel morphology has significant impacts on flood routing. The change in water storage caused by a change in stage is different. Even if the changes in stage are the same, there still exist different changes in storage. This difference is significant when flow is over flood plains. Obviously, it is mainly due to the nonlinearity between the channel width and stage. Hence, considering this nonlinearity, an equivalent stage technique was established to reduce this effect. The correction method (EAR method) which combines the equivalent stage technique with the AR model, was developed in this study.

METHODOLOGY

Equivalent stage technique

The equivalent stage technique proposed here focuses on reducing the effect of the nonlinear relationship between channel width and river stage. In a river with obvious flood plains, stages can be classified into several ranges that are dealt with using the equivalent stage technique, respectively. That is, river stage is converted into equivalent flood plain and equivalent main channel stages divided by bankfull water level. Due to the occurrence of small changes in the relation between the widths of the main channels and stages, this study focuses on the equivalent flood plain stages. In other words, the equivalent stage technique was applied when flows over flood plains was calculated. The equivalent stage technique is specified as:

$$Z_{t,d} = Z_{t,o} \times \left(B_t / \overline{B} \right)^a \quad (1)$$

$$ZC_{t,d} = ZC_{t,o} \times \left(B_t / \overline{B} \right)^a \quad (2)$$

where $Z_{t,o}$ and $Z_{t,d}$ are the observed stage and its equivalent stage respectively at time t; $ZC_{t,o}$ and $ZC_{t,d}$ are the calculated stage and its equivalent stage respectively at time t; B_t is the width of channel at the forecasted station according to the stage at time t; $\overline{B}$ is the average width of the whole channel; and a is the equivalent coefficient; $a = 1$ represents the linear equivalent technique.

Error model

In main channels, errors between the observed stage and calculated stage are computed. When the forecasted flows overtop flood plains, the errors between the observed and calculated equivalent stages are computed, in view of the effect of the nonlinear relationship between the width of channel and stage. These error series are used in the AR model. The least-squares method is employed to estimate the AR model parameters. The AR model is specified as:

$$err_t = c_1 e_{t-1} + c_2 e_{t-2} + \cdots + c_p e_{t-p} + x_t \quad (3)$$

where p is the step of auto-regression; ξ_t is white noise at time t; $c_1, c_2, \ldots, c_p$ are the AR model parameters; e_{t-p} is the error between observed and calculated stage or observed and calculated equivalent stage at time $(t - p)$; and err_t is the forecasted error at time t.

Error reduction

The forecasted errors obtained without using the equivalent stage technique are added to the forecasted stages directly. However, the forecasted errors obtained by using the technique should be reduced before they are added to the forecasted stage. The reduction method is specified as:

$$err2_t = err1_t / (B_t / \overline{B})^a \quad (4)$$

$$ZC_{t,2} = ZC_{t,1} + err2_t \quad (5)$$

where $err1_t$ is the forecasted error obtained by using the equivalent stage technique at time t; $err2_t$

are the errors after reduction with equation (4); $ZC_{t,1}$ is the forecasted stage at time t; $ZC_{t,2}$ is the forecasted stage corrected. Equation (5) represents the reduction of the equivalent stage errors.

STUDY CASE

Study area and data

The Caoe River catchment located in southeast China has a drainage area of 6046 km^2, and is one of the eight largest rivers in Zhejiang Province. Originating in the Shengzhou in Zhejiang Province, it discharges into the Hangzhou Bay (Fig. 1). The length from Huashan to Baiguan is about 49 km, and about 20 km from Baiguan to Sangpendian. The flood plain stage is about 6.5 m at Baiguan. The cross-sections between Huashan and Baiguan are of compound nature varying with different main channels and flood plains. River stage forecasting at Baiguan station, especially during a flood, is crucial.

The stage at Baiguan is forecast with the stages at Huashan and Sangpendian using the bi-directional stage routing model (Bao, 1997). Baiguan is the upstream boundary controlled by flood mainly, and Sangpendian is the downstream boundary dominated by tide. Ten flood events (1988–1990) were selected to simulate the stage at Baiguan. Seven events (1988–1989) were used to calibrate the parameters, and three events (1990) were used for validation. The forecasted stage at Baiguan was corrected using the EAR method.

Calculation procedure

The stage routing model (Bao, 1997) was employed to forecast river stage, and then the forecast stage is corrected by the EAR method. Figure 2 presents the scheme of the calculation procedure.

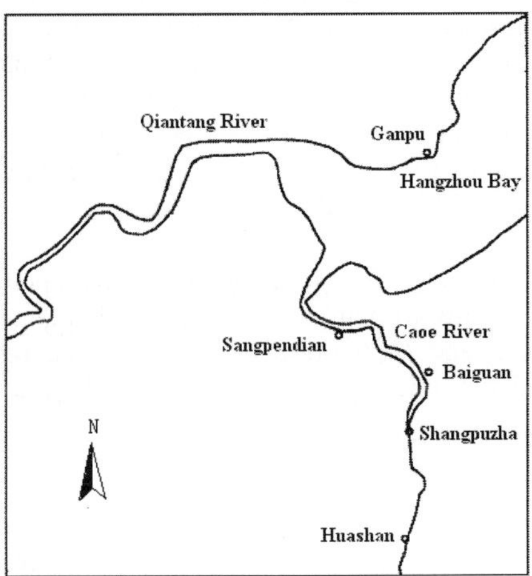

Fig. 1 Location of the study area.

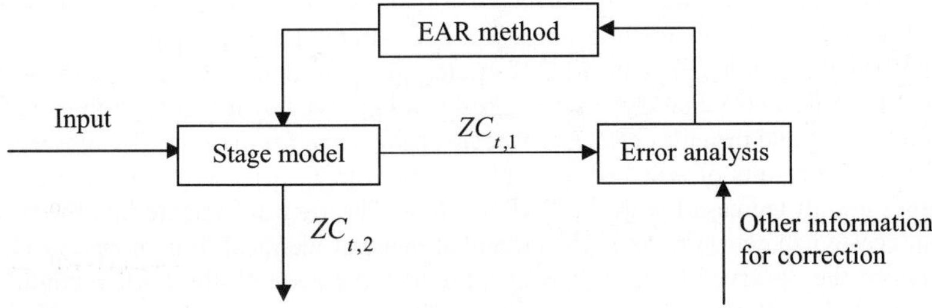

Fig. 2 Schematic flow chart of calculation procedure.

Evaluation criteria

Six statistical indices are selected to evaluate model performance, which are the peak stage error (*PSE*), the relative peak stage error (*RPSE*), the time-at-peak error (*TPE*), the root mean square error (*RMSE*), and the Nash-Sutcliffe coefficient (*CE*). *RMSE* and *CE* are employed to fully indicate how well individual simulated values match observed values.

$$PSE = Z_p - ZC_p \quad (7)$$

$$RPSE = (Z_p - ZC_p) \times 100\% / Z_p \quad (8)$$

$$TPE = OT_p - CT_p \quad (9)$$

$$RMSE = \sqrt{\sum_{i=1}^{n} (Z(i) - ZC(i))^2 / n} \quad (10)$$

$$CE = 1 - \sum_{i=1}^{n} [ZC(i) - Z(i)]^2 / \sum_{i=1}^{n} [Z(i) - \overline{Z}]^2 \quad (11)$$

where $Z(i)$ and $ZC(i)$ represent observed stage and calculated stages at time i respectively; n is the total number of observed stages; Z_p and ZC_p are observed and calculated stage peaks respectively; OT_p is the occurrence time of observed peak stage; CT_p is the occurrence time of forecasted peak stage; and $\overline{Z}$ is the average observed stage.

Model application

The flood plain stage equivalent coefficient a is the only parameter needed to be estimated in the equivalent stage technique. By using a trial-and-error analysis, we found that a is insensitive. Change in a causes little change in the corrected results. Some test results of a are shown in Table 1.

As shown in Table 1, the equivalent stage technique produces a better performance. In general, the variation in parameter a gives rise to little change in the corrected results, moreover nearly no change in the peak stages of these flood events. By comparing these results, the a value is estimated as 0.5 through seven events (1988–1989). The validation results are also shown in Table 1 for three events in 1990.

In order to show the performance of the EAR method, Table 2 lists the comparison results of (1), (2) and (3). Here, (1) represents the forecasted results without any correction; (2) represents the forecasted results corrected with the AR model; and (3) represents the forecasted results corrected with the EAR method. Some observed and simulated hydrographs of these flood events are displayed in Figs 3 to 6.

RESULTS AND DISCUSSION

The results of (2) and (3) in Table 2 show that the EAR method outperforms the AR model. With the EAR method, the *PSE* and *RPSE* of eight flood events (except 880807 and 890616) are obviously reduced. The *PSE* and *RPSE* of three flood events (890521, 890627, and 900531) with the EAR method are zero, whereas the results of nine events (except 880807) with the AR model are more than zero. The *TPE* of these events are not more than 1 h with the EAR method, whereas the *TPE* of flood events 900614 and 900830 with the AR model are more than 1 h (the *TPE* of the event 900614 is up to 3 h). This shows that the EAR method can improve the accuracy in the peak stage. It emphasizes high flows and provides effective forecast peak stage for the river flood prevention.

Compared with the results of AR model (Table 2), the *RMSE* of these events are all reduced and the *CE* values are all increased with the EAR method. The figures indicate that the observed and corrected forecasted stages with the EAR method are almost identical. It demonstrates that the agreements between the observed hydrographs and calculated ones with the EAR method are all better than those with the AR model.

Table 1 Test of *a*.

Flood event	*a*	*PSE* (m)	*RPSE* (%)	*TPE* (h)	*RMSE* (m)	*CE*
880617	0.1	0.01	0.11	0	0.084	0.996
	0.5	0.01	0.11	0	0.085	0.996
	1	0.01	0.11	0	0.086	0.996
880807	0.1	–0.02	–0.20	0	0.145	0.978
	0.5	–0.02	–0.20	0	0.146	0.978
	1	–0.02	–0.20	0	0.149	0.977
890411	0.1	0	0	0	0.024	0.999
	0.5	0	0	0	0.025	0.999
	1	0	0	0	0.026	0.999
890521	0.1	0	0	0	0.046	0.997
	0.5	0	0	0	0.047	0.997
	1	0	0	0	0.048	0.997
890616	0.1	0.02	0.20	–1	0.068	0.994
	0.5	0.02	0.20	–1	0.070	0.994
	1	0.02	0.20	–1	0.072	0.993
890627	0.1	0	0	–1	0.093	0.993
	0.5	0	0	–1	0.092	0.993
	1	0	0	–1	0.091	0.993
890911	0.1	–0.11	–1.08	–1	0.177	0.984
	0.5	–0.10	–1.00	–1	0.174	0.985
	1	–0.09	0.91	–1	0.171	0.985
900531	0.1	0	0	0	0.040	0.990
	0.5	0	0	0	0.041	0.989
	1	0	0	0	0.044	0.988
900614	0.1	0.03	0.40	1	0.036	0.991
	0.5	0.03	0.40	1	0.037	0.991
	1	0.03	0.40	1	0.037	0.990
900830	0.1	–0.09	–0.80	0	0.135	0.990
	0.5	–0.09	–0.80	0	0.133	0.991
	1	–0.09	–0.80	0	0.132	0.991

Note: The minus "—"of the *PSE* and *RPSE* indicates that the stage simulated is higher than the observed, and the plus of the *PSE* and *RPSE* means that the stage simulated is smaller than the observed. The minus "—"of the *TPE* indicates that the peak stage simulated occurs after the observed, and the plus of the *TPE* means that the peak stage simulated occurs before the observed.

Comparing the results of (1) with (2) in Table 2 shows that the EAR method improves the agreement between the forecasted and observed stages greatly. For example, the *PSE* of the flood event 890521 is reduced by 0.23 m, and the *CE* increased from 0.946 to 0.997. Additionally, the values of *CE* without correction range from 0.834 to 0.983, whereas the values with the EAR method range from 0.978 to 0.999. The ranges of the values of *RMSE* fall from 0.100–0.401 m to 0.025–0.174 m when corrected with the EAR method. It indicates that the EAR method is able to correct flood routing processes efficiently.

As shown in Figs 3–6, there is better agreement in the recession limbs for these floods when the EAR method was applied. The agreements in the recession limbs of the flood events 880807, 890616, 890627, and 890911 are not that good when the AR model was applied, but the EAR method simulates the fluctuation in the recession limbs well. In brief, the equivalent stage technique is able to model the regression process effectively.

Table 2 Results of river stage at Baiguan.

Flood event	Correction type	*PSE* (m)	*RPSE* (%)	*TPE* (h)	*RMSE* (m)	CE
880617	(1)	0.21	2.13	2	0.265	0.957
	(2)	0.12	1.21	0	0.149	0.986
	(3)	0.01	0.11	0	0.085	0.996
880807	(1)	0.05	0.56	0	0.401	0.834
	(2)	0	0.00	–1	0.233	0.944
	(3)	–0.02	–0.20	0	0.146	0.978
890411	(1)	0.09	1.05	1	0.100	0.983
	(2)	0.05	0.56	0	0.058	0.994
	(3)	0	0	0	0.025	0.999
890521	(1)	0.23	2.56	0	0.197	0.946
	(2)	0.11	1.27	0	0.109	0.984
	(3)	0	0	0	0.047	0.997
890616	(1)	–0.01	–0.12	3	0.135	0.977
	(2)	0.01	0.12	1	0.088	0.990
	(3)	0.02	0.17	–1	0.070	0.994
890627	(1)	–0.11	–1.19	2	0.241	0.951
	(2)	–0.05	–0.57	0	0.144	0.983
	(3)	0	0	–1	0.092	0.993
890911	(1)	–0.25	–2.56	0	0.353	0.937
	(2)	–0.13	–1.30	–1	0.237	0.972
	(3)	–0.10	–1.00	–1	0.174	0.985
900531	(1)	0.05	0.67	1	0.135	0.885
	(2)	0.03	0.36	0	0.080	0.959
	(3)	0	0	0	0.041	0.989
900614	(1)	0.08	1.09	3	0.101	0.929
	(2)	0.06	0.81	3	0.064	0.971
	(3)	0.03	0.42	1	0.037	0.991
900830	(1)	0.10	0.88	–4	0.280	0.958
	(2)	0.10	0.91	–1	0.179	0.983
	(3)	–0.09	–0.80	0	0.133	0.991

Note: The minus "—"of the *PSE* and *RPSE* represents that the stage simulated is higher than the observed, and the plus of the *PSE* and *RPSE* means that the stage simulated is smaller than the observed. The minus *TPE* values represent that the peak stage simulated occurs after the observed, and the plus of the *TPE* means that the peak stage simulated occurs before the observed.

From the performance of the EAR method in high and low stages divided by the bankfull stage (namely 6.5 m at Baiguan in this case) in Figs 3–6, we found that although only flows over flood plains are dealt with by the equivalent technique, and this technique is not employed in main channels, the accuracy of low stages is also better than these with only the AR model. It shows that the approach of flows over flood plains corrected with the equivalent stage technique can reduce the effect of nonlinearity between the channel width and stage on forecasted stages caused by complex cross-sections.

CONCLUSIONS

Based on the nonlinear relation between channel width and stage, an equivalent stage technique was developed in this study. This technique can reduce the effect of nonlinearity on forecasted stages. There is only one parameter a, which is insensitive in this technique, and its value is easily calibrated. Moreover, the EAR method coupled this technique with the AR model can be easily carried out. It can be used for stage forecasting without any changes in the arithmetic program.

From the study case, the agreement between the forecast stage hydrographs corrected with the EAR method and the observed ones is reasonably good, and the EAR method can obtain a better

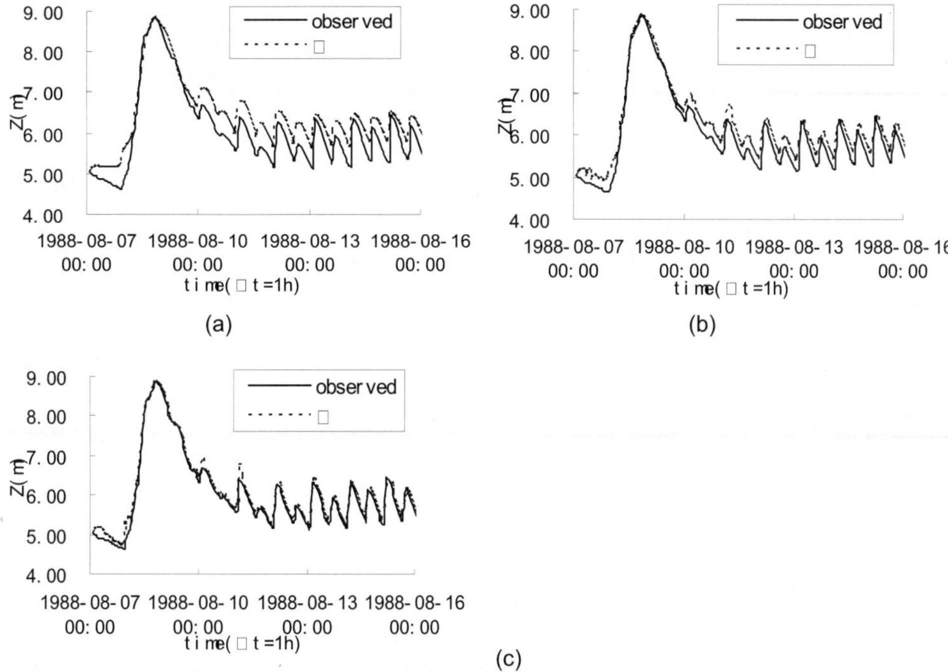

Fig. 3 Stage hydrograph of the flood event 880807 at Baiguan.

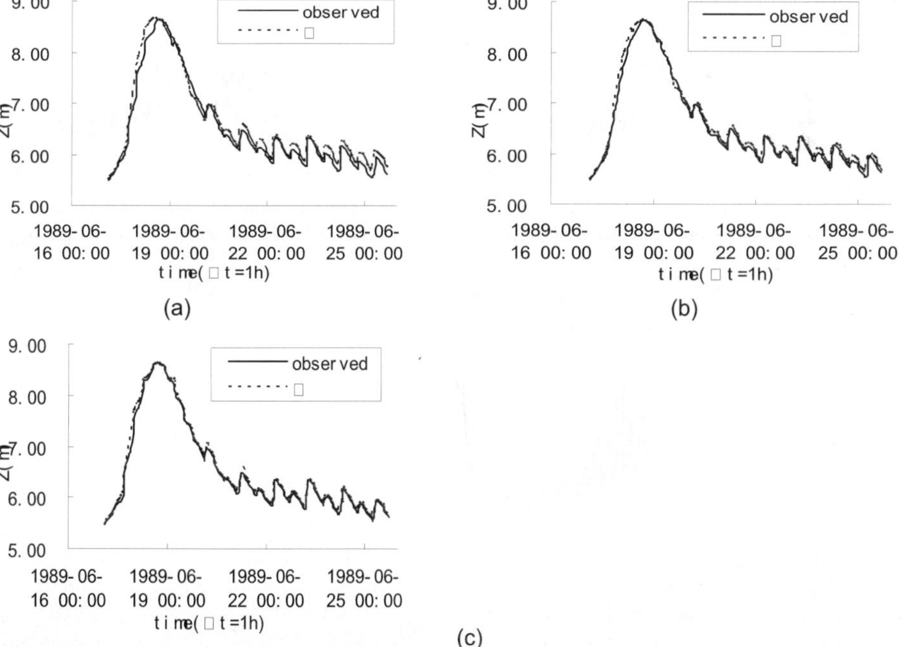

Fig. 4 Stage hydrograph of the flood event 890616 at Baiguan.

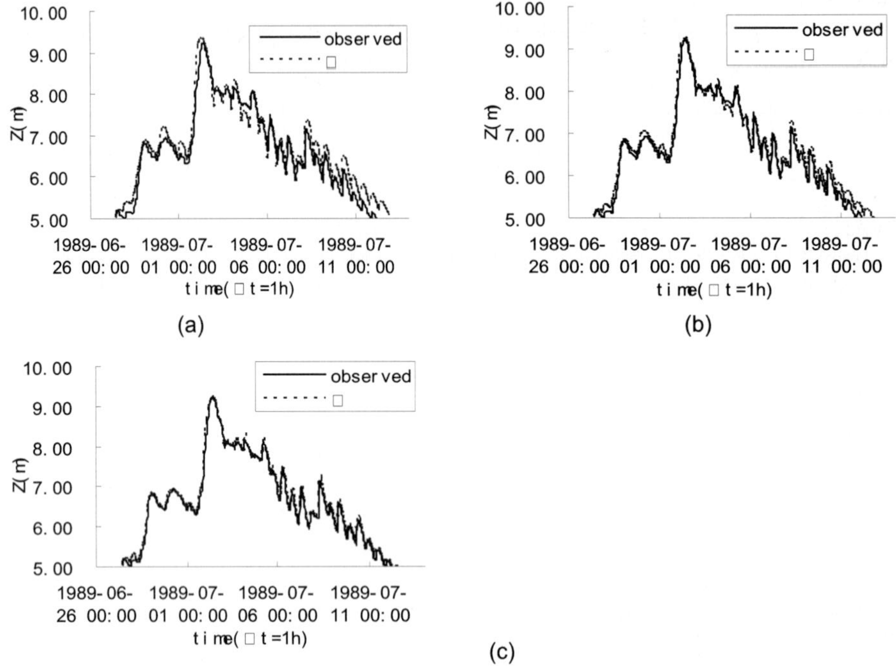

Fig. 5 Stage hydrograph of the flood event 890627 at Baiguan

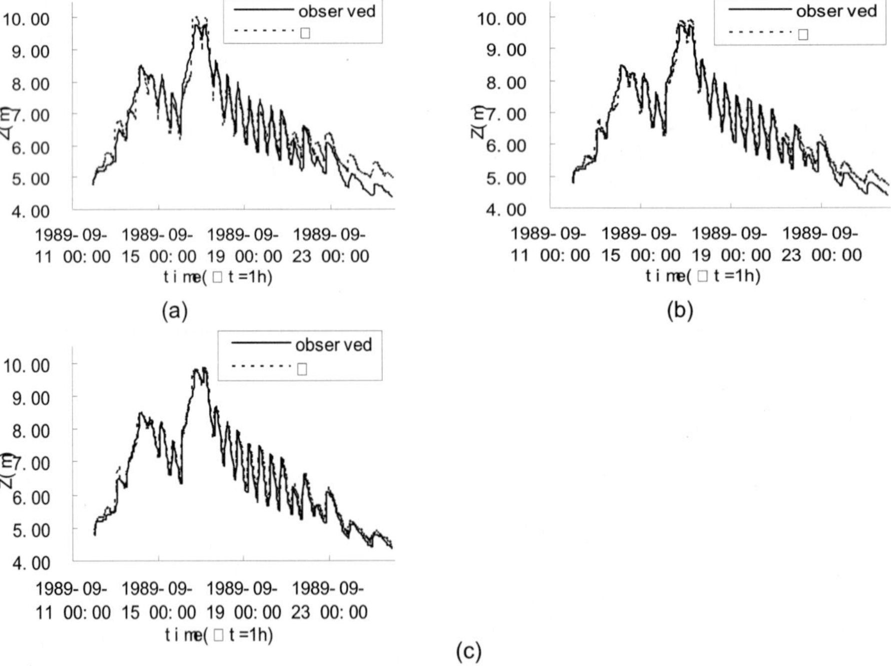

Fig. 6 Stage hydrograph of the flood event 890911 at Baiguan.

accuracy than the AR model. The results also indicate that the EAR method can provide an accurate prediction of the peak stage and peak timing. In addition, the lead time is available, and it can be used in the real-time stage forecasting. Generally, the EAR method provides a tool to assist management decisions for flood mitigation. Hence, a further study is recommended on this method.

Acknowledgements The study was supported by the Natural Science Foundation of China (50679024), Eleven-five Program for Science and Technology of China (2006BAC05B02), Development Program for Changjiang Scholars and Innovation Team (IRT0717), and Program for the Ministry of Education and State Administration of Foreign Experts Affairs, P.R. China (B08408).

REFERENCES

Bao, W. M. & Bian, S. M. (1997) Study on stage routing model of tidal-reach. *J. Hydraul. Eng*ng (Chinese) **11**, 34–38.

Cheng, Y. H., Ge, S. X. & Li, Y. R. (2005) The real-time correction methods of the hydrodynamic model. *Yangtze River* (Chinese) **36**(2), 6–8.

Franchini, M. & Lamberti, P. (1994) A flood routing Muskingum type simulation and forecasting model based on level data alone. *Water Resour. Res.* **30**(7), 2183–2196.

Guo, L. & Zhao, Y. L. (2002) Study on adjustment methods of real-time flood forecasting in view of autoregressive model. *Hydroelectric Energy* (Chinese) **9**, 25–27.

He, H. M., Yu, Q., Zhou, J., Tian, Y. Q. & Chen, R. F. (2008) Modeling complex flood flow evolution in the middle Yellow River basin, China. *J. Hydrol.* **353**, 76–92.

Hsu, M. H., Fu, J. C. & Liu, W. C. (2003) Flood routing with real-time stage correction method for flash flood forecasting in the Tanshui River, Taiwan. *J. Hydrol.* **283**, 267–280.

Hsu, M. H., Fu, J. C. & Liu, W. C. (2006) Dynamic routing model with real-time roughness updating for flood forecasting. *J. Hydraul. Eng.* **132**(6), 605–619.

Hu, L. (2007) The Real-time stage forecasting in the estuarine area of Qiantang River. Master Thesis, HoHai University, Nanjing, China.

Mu, J. B. & Zhang, X. F. (2007) Real-time flood forecasting method with 1-D unsteady flow model. *J. Hydrodynam.* **19**(2), 150–154.

Neal, J. C., Atkinson, P. M. & Hutton, C. W. (2007) Flood inundation model updating using an ensemble Kalman filter and spatially distributed measurements. *J. Hydrol.* **336**, 401–415.

Lamberti, P. & Pilati, S. (1996) Flood propagation models for real-time forecasting. *J. Hydrol.* **175**, 239–265.

Qu, S. M. & Bao, W. M. (2003) Comprehensive correction of real-time flood forecasting. *Adv. Water Sci.* (Chinese) **14**(2), 167–172.

Wu, X. L., Wang, C. H., Chen, X. & Xiang X. H. (2008) Kalman filtering correction in real-time forecasting with hydrodynamic model. *J. Hydrodynam.* **20**(3), 391–397.

Yu, P. S., Chen, S. T. & Chang, I. F. (2006) Support vector regression for real-time flood stage forecasting. *J. Hydrol.* **328**, 704–716.

Long-term reservoir operation optimized by DP models with one-month ensemble forecast of precipitation

DAISUKE NOHARA[1], AMI TSUBOI[2] & TOMOHARU HORI[1]

1 *Water Resources Research Center, Disaster Prevention Research Institute, Kyoto University, Gokasho, Uji, Kyoto 611-0011, Japan*
nohara_d@mbox.kudpc.kyoto-u.ac.jp

2 *Graduate School of Engineering, Kyoto University, Yoshida-Honmachi, Sakyo, Kyoto, 606-8501, Japan*
(Current affiliation: *School of Engineering, The University of Tokyo, 7-3-1, Hongo, Bunkyo, Tokyo, 113-8656, Japan*)

Abstract Optimizing processes of three (dynamic programming) DP-based models are analysed through the application of the models to optimization of long-term reservoir operation. One-month ensemble forecast of precipitation provided by the Japan Meteorological Agency is introduced to long-term reservoir operation focusing on water utilization. Ensemble inflow scenarios are then predicted from estimated precipitation scenarios by use of a regression model. Optimization of long-term reservoir operation which only focuses on water utilization is conducted by use of the predicted inflow scenario. Deterministic DP used with ensemble mean prediction and stochastic DP and sampling stochastic DP with all members of the ensemble forecast were employed for the optimization. By the application to Sameura Reservoir in the Yoshino River basin, Japan, it was shown that distribution characteristics of predicted streamflow, such as the relationship between the normal and median value, can be an index for decision of a DP-model employed, combining with prediction error tendency.

Key words ensemble forecast of precipitation; dynamic programming; reservoir operation; drought management; optimization; real-time operation

INTRODUCTION

Reservoirs, which can diminish seasonal variation of water flow in the downstream by storing stream water, have a significant role in water resources management. It is important to make reservoir management more systematic in developing more efficient water resources management measures. Because it is necessary to consider the future hydrological condition of river basins, it is considered important to include hydrological predictions, such as future precipitation or streamflow in the basin, into reservoir operation.

In Japan, prediction of precipitation or typhoon paths over short periods, such as several hours, has started in some reservoirs for judgment on issuing flood alerts or estimation of abnormal flooding possibility (Wada *et al.*, 2005). The accuracy of prediction for such a short range has improved recently, and it is considered that this brings a situation where these kinds of forecasts start to be accepted by reservoir managers. However, long-term prediction of precipitation or streamflow, especially that of more than several weeks range, is not considered in the actual water resources management. This is because the accuracy of such predictions, focused on longer futures, essentially decline too much to introduce it into water resources management directly due to growth of the observation error and calculation errors through calculation by the prediction model. Therefore, only hydrological observations and statistics are mainly taken into account in actual reservoir management, though some inclusion of measurements of precipitation predictions for the coming several months, derived by long-range weather forecasting, into reservoir operation has been proposed (Ikebuchi *et al.*, 1990; Kojiri *et al.*, 1994).

Ensemble forecasting measures, which synthetically make a prediction with multiple unique prediction sequences, have been developed worldwide to alleviate progressive decrease of accuracy in long-range meteorological or hydrological predictions. An ensemble forecast provides not only a predicted value, but also information about prediction accuracy and the reliable range of the prediction by seeing the behaviour of all prediction sequences. Because it is important to assess prediction accuracy or reliability when using a forecast, the accuracy of which is not considered high, it is considered useful for decision making for reservoir operation to take such an ensemble forecast into account, especially in long-term reservoir management.

Against such a background, some application measures for ensemble hydrological prediction for reservoir operation have been proposed. Faber & Stedinger (2001) developed an optimization method of an actual reservoir system combining weekly updated forecast information from ensemble streamflow prediction (ESP) of the US National Weather Service, with sampling stochastic dynamic programming (DP; SSDP), which is one of stochastic DP (SDP) expanded to reflect the various characteristics of multiple predicted streamflows. They also compared the performance of SSDP with that of deterministic DP (DDP) and SDP, concluding with the advantages of SSDP with ESP. Kim *et al.* (2007) developed an optimizing model for operational policies of Korean multi-reservoir system by use of SSDP with monthly updated ESP. They confirmed that the SSDP model with ESP showed better performance than that without ESP. However, they did not describe what procedures exactly cause those differences among results derived by DDP, SDP and SSDP models in the optimization of reservoir operation. Applying these optimization models to another region, with an ensemble forecast which has different characteristics, is useful to understand the procedures of optimization models with ensemble weather prediction.

Therefore, as a first stage of analysis of the optimization procedure, three different optimization models of reservoir operation considering ensemble forecasts of precipitation are analysed in this paper. One-month ensemble forecast of precipitation provided by the Japan Meteorological Agency (JMA) is employed as the ensemble forecast. Long-term operation, which takes only water use into account, is considered as long-term reservoir operation, and is optimized with the foresight of coming one month by three different DP based models, namely DDP with ensemble mean prediction, SDP and SSDP, with all of 50 ensemble prediction sequences.

ESTIMATION OF ENSEMBLE STREAMFLOW SEQUENCES FROM ONE-MONTH ENSEMBLE FORECAST OF PRECIPITATION

Japan Meteorological Agency (JMA) introduced ensemble measures into mid and long-term weather forecast, and started to provide forecasting results including precipitation prediction for one week, one month, three months, and warm/cold season (six months) from 2001, March 1996, April 2003, and September 2003, respectively (Furukawa & Sakai, 2004). Ensemble sequences of the forecast have also been provided, from March 2001, for the one-month forecast. Because of the accessibility and accumulated amount of forecast data mentioned above, the one-month ensemble forecast is employed in this study, although conditions in the future from one month to six months ahead are necessary to make a decision in long-term reservoir operation.

The one-month ensemble forecast provides 50 sequences of pressures, wind, temperature and relative humidity at several heights, and integrated precipitation at the surface for the coming 34 days at every 2.5° grid point. The one-month ensemble forecast is updated every week, so it is applicable to real-time optimization of reservoir operation. In this study, 50 forecasted sequences of integrated precipitation, which is considered directly relative to reservoir operation, are considered for the optimization of long-term reservoir operation.

At first, integrated precipitation sequences are transferred to daily precipitation sequences: 50 daily precipitation sequences for the coming one month (34 days, strictly), which are averaged over the study basin area, are then estimated from daily precipitation sequences at grid points which cover the study basin, by use of a regression model. Because precipitation sequences are provided as flux at 2.5° (approx. 250 km) grid in a one-month ensemble forecast by JMA, it cannot be said that the provided forecast represents the future precipitation in the study basin. This is the reason why conversion between forecast GPVs and basin mean precipitation is necessary. Finally, 50 streamflow sequences at the dam site and an assessed point are also estimated from precipitation sequences of the basin by use of the regression model. These 50 streamflow sequences for the dam site and the assessed point are used as ensemble streamflow predictions in the proposed optimizing calculation of long-term reservoir operation. An assessed point is usually located downstream of the reservoir, so inflow sequences of the reservoir and runoff sequences between the dam site and the assessed point, respectively, are predicted.

DP MODEL FOR OPTIMIZATION OF LONG-TERM RESERVOIR OPERATION

Outline of DP- based optimization model for long-term reservoir operation

In reservoir operation, release from a reservoir is sequentially decided considering the present and future hydrological conditions in the river basin of the reservoir. From its sequential characteristics, the problem of decision making in reservoir operation can be easily formulated as an optimization problem by use of dynamic programming (Bellman, 1957). Although various DP models for optimization of reservoir operation have been proposed (e.g. Yeh, 1985; Labadie, 2004), the optimization model for long-term reservoir operation focusing on water use employed in this paper is described below.

Focusing on drought management, the objective of optimization can be defined as minimization of damage caused by water deficit. When reservoir operation is optimized taking streamflow predicted for the coming period T into consideration, the objective function of this optimization problem is:

$$\min_{r_t} \sum_{t=1}^{T} H_t \tag{1}$$

where r_t is release amount from reservoir during period t ($t = 0, \ldots, T$), and H_t is a damage function which represents damage suffered in the basin during period t. The damage function for droughts can be generally defined as the product of water deficit and deficit rate against demand, as also used in Ikebuchi *et al.* (1994):

$$H_t = \begin{cases} \dfrac{(d_t - q_t)^2}{d_t} & (d_t > q_t) \\ 0 & (d_t \le q_t) \end{cases} \tag{2}$$

where d_t is water demand at the downstream control point at period t when there is only one control point downstream of the reservoir, q_t is streamflow at the control point during period t. The damage function for droughts can be simply defined as the square of water deficit as used in Kim *et al.* (2007). In either case, it is often seen that a high dimensional function of water deficit or rate is employed for expression of damage suffered by droughts, reflecting the sense that damages of water users would increase when the water deficit increases.

On the other hand, a reservoir has a minimum and maximum value of storage volume and release designed against the physical constraints of the reservoir or for stability of the stream regime downstream. This is defined as a constraint of the optimization problem for reservoir operation, described as follows:

$$S_{\min} \le s_t \le S_{\max} \tag{3}$$

$$R_{\min} \le r_t \le R_{\max} \tag{4}$$

where s_t is storage volume of a focused reservoir at the beginning of period t. Because drought management is the focus in this paper, $S_{\min}$ is defined as storage volume when the water use capacity of the reservoir is 0, and $S_{\max}$ is describing storage volume when the water use capacity is full. In a similar way, $R_{\min}$ is defined as 0, and $R_{\max}$ is defined as maximum release amount designed by the operation rule of the reservoir.

The state transformation equation, defined based on the principle of continuity, is then described as follows:

$$s_{t+1} = s_t + i_t - r_t + \alpha_t \tag{5}$$

where α_t is a variable which represents water loss such as evaporation or spillway water from the reservoir during period t. In this study, water loss is not taken into account to simplify calculation ($\alpha_t = 0$), therefore equation (6) is used instead of equation (5) as the state transformation equation in this study.

$$s_{t+1} = s_t + i_t - r_t \tag{6}$$

The recursive equation is then defined as follows:

$$f_T(s_T) = \min_{r_T} H_T(q_T)$$
$$f_t(s_t) = \min_{r_t}\{H_t(q_t) + f_{t+1}(s_{t+1})\} \qquad (t = 1,\cdots,T-1) \tag{7}$$

where $f_t(s_t)$ is the future damage function (FDF), which represents minimized summation of damage suffered from period t to period T in the case where storage volume at the beginning of period t is s_t. The optimization problem whose objective function is described as equation (1) is solved by sequentially calculating $f_t(s_t)$ from period T to period 1 with the constraint mentioned above. After $f_t(s_t)$ for every period is calculated, optimal s_2 is calculated from initial storage volume s_1 by equation (5), and then optimal release r_1^* is calculated by:

$$r_1^* = \min_{r_1}\{H_1(q_1) + f_2(s_2)\} \tag{8}$$

Optimal release sequence from period 2 to period T are similarly calculated by equation (5) and equation (7). This DP model, which is called as deterministic DP (DDP), is employed for optimization using an ensemble mean forecast of precipitation, which provides only one sequence of prediction, in this paper.

SDP model for long-term reservoir operation with ensemble forecast sequences

Future streamflow in a basin can not be specified when ensemble forecast sequences are used for the prediction of future hydrological condition. The stochastic DP (SDP) approach is often used instead of DDP approach in such a case.

In the SDP algorithm, future streamflow is described stochastically, and the expected value of future damage is minimized. The recursive equation of SDP for reservoir operation is generally described as follows (Faber & Stedinger, 2001):

$$f_t(s_t) = \min_{R^*_{\min} \le r_t \le R^*_{\max}} \mathop{\mathrm{E}}_{q_t}\left\{H_t(q_t) + \mathop{\mathrm{E}}_{i_t}[f_{t+1}(s_{t+1})]\right\}$$
$$(s_{t+1} = s_t + i_t - r_t) \tag{9}$$

where:

$$R^*_{\min} = \max\{R_{\min}, s_t + i_t - S_{\max}\}$$
$$R^*_{\max} = \min\{R_{\max}, s_t + i_t - S_{\min}\} \tag{10}$$

In the case where stochastic streamflow sequences are employed, streamflow at assessed point q_t must be divided into release from a reservoir and runoff water which joins the river channel between the dam site and the assessed point because runoff water is described as a stochastic variable. The objective function and recursive equation for SDP with ensemble streamflow prediction, therefore, are:

$$\min_{R^*_{\min} \le r_t \le R^*_{\max}} \sum_{t=1}^{T}\left\{\sum_{o_t} \mathrm{P}[o_t]\cdot H_t(q_t)\right\} \tag{11}$$

$$f_t(s_t) = \min_{R^*_{\min} \le r_t \le R^*_{\max}} \sum_{o_t} \mathrm{P}[o_t]\cdot\left\{H_t(q_t) + \sum_{i_t} \mathrm{P}[i_t]\cdot f_{t+1}(s_{t+1})\right\}$$
$$(q_t = o_t + r_t) \tag{12}$$

where o_t is runoff amount between a dam site and the assessed point, $\mathrm{P}[o_t]$ is probability that runoff o_t actually occurs, $\mathrm{P}[i_t]$ is probability that inflow i_t actually comes into a reservoir in the future. When Markov characteristics of streamflow are assumed, the recursive equation is defined as follows (Loucks *et al.*, 1981):

$$f_t(s_t,i_t) = \min_{R^*_{\min} \le r_t \le R^*_{\max}} \sum_{o_t} \mathrm{P}[o_{t+1} \mid o_t] \cdot \left\{ H_t(q_t) + \sum_{i_t} \mathrm{P}[i_{t+1} \mid i_t] \cdot f_{t+1}(s_{t+1}, i_{t+1}) \right\} \tag{13}$$

where $\mathrm{P}[o_{t+1}|o_t]$ and $\mathrm{P}[i_{t+1}|i_t]$ are transition probability of runoff and inflow, respectively. Equation (13) can be also described as follows:

$$f_t(s_t,i_t) = \min_{R^*_{\min} \le r_t \le R^*_{\max}} \mathop{\mathrm{E}}_{o_{t+1}|o_t} \left\{ H_t(q_t) + \mathop{\mathrm{E}}_{i_{t+1}|i_t} f_{t+1}(s_{t+1}, i_{t+1}) \right\} \tag{14}$$

In the case where the ensemble streamflow sequences introduced in the previous section is employed with the SDP model, the occurrence probability of inflow i_t^m and runoff o_t^m forecasted by the ensemble scenario m for period t can be defined as follows:

$$\mathrm{P}_t\left[i_t^m\right] = \mathrm{P}_t\left[o_t^m\right] = \frac{1}{M} \qquad (t = 1, \cdots, T) \tag{15}$$

where $\mathrm{P}_t[i_t^m]$, $\mathrm{P}_t[o_t^m]$ are the occurrence probability of inflow i_t^m and runoff o_t^m, M is a number of ensemble scenarios (ensemble sequences), respectively. Because ensemble streamflow sequences are derived values predicted with a numerical forecasting model (by JMA) instead of historical data in this study, it is appropriate to consider no transition probability and to assume equivalent occurrence probability for every streamflow sequence. The objective function and recursive equation are then described as follows:

$$\min_{R^*_{\min} \le r_t \le R^*_{\max}} \sum_{t=1}^{T} \left(\frac{1}{M} \sum_{m=1}^{M} H_t(q_t^m) \right) \tag{16}$$

$$f_t(s_t) = \min_{R^*_{\min} \le r_t \le R^*_{\max}} \frac{1}{M} \sum_{m=1}^{M} \left\{ H_t(q_t^m) + \frac{1}{M} \sum_{m=1}^{M} f_{t+1}(s_{t+1}^m) \right\}$$
$$(q^m{}_t = r_t + o_t^m, \quad s^m{}_{t+1} = s_t + i_t^m - r_t) \tag{17}$$

After calculation of FDF $f_t(s_t)$ for all periods from period T to period 1, optimal release r_t^* is sequentially decided from storage at the beginning of period t by:

$$r_t^* = \min_{R^*_{\min} \le r_t \le R^*_{\max}} \frac{1}{M} \sum_{m=1}^{M} \left\{ H_t(q_t^m) + f_{t+1}(s_{t+1}^m) \right\} \tag{18}$$

SSDP model for long-term reservoir operation with ensemble forecast sequences

SDP can consider stochastic characteristics of multiple streamflow sequences, but cannot consider their temporally sequential characteristics because probabilistic calculation is sequentially conducted at each predicted time in SDP. To express the continuousness of forecast streamflow sequences, optimization of reservoir operation with sampling SDP (SSDP) which considers sequential characteristics of multiple sequences, was proposed (Kelman *et al.*, 1990). In SSDP for optimization of reservoir operation, a future value (FDF in this paper) for each forecast streamflow sequence is calculated at first, and optimization for all streamflow sequences is subsequently conducted by use of future values calculated for each streamflow sequence. In particular, FDF is first calculated by considering each forecast streamflow sequence as a deterministic forecast. Calculation of FDF is conducted following the recursive equations below:

$$f_t(s_t, m) = \min_{R^*_{\min} \le r_t \le R^*_{\max}} \left\{ H_t(q_t^m) + f_{t+1}(s_{t+1}^m, m) \right\}$$
$$(q_t^m = r_t + o_t^m, \quad s_{t+1}^m = s_t + i_t^m - r_t) \tag{19}$$

where $f_t(s_t, m)$ is future damage from period t calculated from forecast streamflow sequence m, o_t^m, i_t^m are runoff joins from dam site to the assessed point and inflow into the study reservoir

Table 1 Details of Sameura Reservoir.

Active storage capacity		289 000 000 m^3
Water use capacity		173 000 000 m^3
Flood control capacity	Flood season (1 July – 10 October)	90 000 000 m^3
	Dry season (11 October – 30 June)	80 000 000 m^3
Power generation capacity	Flood season (1 July – 10 October)	26 000 000 m^3
	Dry season (11 October – 30 June)	36 000 000 m^3
Designed flood inflow		4700 m^3/sec
Designed release discharge		2000 m^3/sec
Maximum release discharge in case of no flood situation		800 m^3/sec

during period *t* calculated from forecast streamflow sequence *m*. The expectation value of future damage from the storage volume at the beginning of the period is subsequently estimated by use of occurrence probabilities of forecasted scenarios. The optimal release amount is then decided so as to minimize the expectation value among the forecasted scenarios:

$$\min_{R^*_{\min} \le r_t \le R^*_{\max}} E\left\{H_t(q_t^m) + f_{t+1}(s_{t+1}^m, m)\right\} \tag{20}$$

Because occurrence probabilities of every ensemble scenario are defined with an identical value (1/*M*) in this study, equation (20) can be transformed to:

$$\min_{R^*_{\min} \le r_t \le R^*_{\max}} \frac{1}{M}\sum_{m=1}^{M}\left\{H_t(q_t^m) + f_{t+1}(s_{t+1}^m, m)\right\} \tag{21}$$

However, equation (16) is also employed for the objective function in SSDP, as in SDP.

APPLICATION

Set-up of optimization problem for applied area

DDP, SDP, SSDP model mentioned above was applied to optimize long-term operation for water use of the Sameura Reservoir in the Yoshino River, Japan, with one-month ensemble precipitation prediction (described as EPP1 below) forecast by JMA from 2007 to 2008.

The Yoshino River basin annually receives an average of more than 3000 mm precipitation, especially in the upstream mountain area (Yoshino Regional Bureau of Japan Water Agency, 2004), droughts have often occurred in this basin because of unstable streamflow regime caused by large variation of precipitation. The details of Sameura Reservoir are shown in Table 1. Sameura Reservoir is operated for multi-purposes such as water use, preventing floods and power generation. Operation for water use only, which is the most important purpose for this basin, is studied for optimization by use of EPP1 in this project to make analysis of the application results easy to understand. Sameura Reservoir and the downstream of Yoshino River basin are modelled as a simple basin shown in Fig. 1 for application of the DP model. The assessed point is located near the upstream of Ikeda Reservoir, which is the lower reservoir in the Yoshino River multi-reservoir system. Runoffs joining the main stream of the Yoshino River between Sameura Reservoir and the assessed point is expressed as one lumped runoff in this study.

GPVs of forecast precipitation at four grids surrounding Sameura reservoir and the downstream were used as predicted information. Details of Sameura Reservoir shown in Table 1 are used for set-up of the constraint conditions. As operation for water use was the focus in this study, only water use capacity was considered in the optimization assuming that operation was conducted following the operation rule of the reservoir in case a flood event occurred. The constraint conditions are defined as follows:

$$0(\mathrm{m}^3) \le s_t \le 173 \times 10^6(\mathrm{m}^3) \tag{22}$$

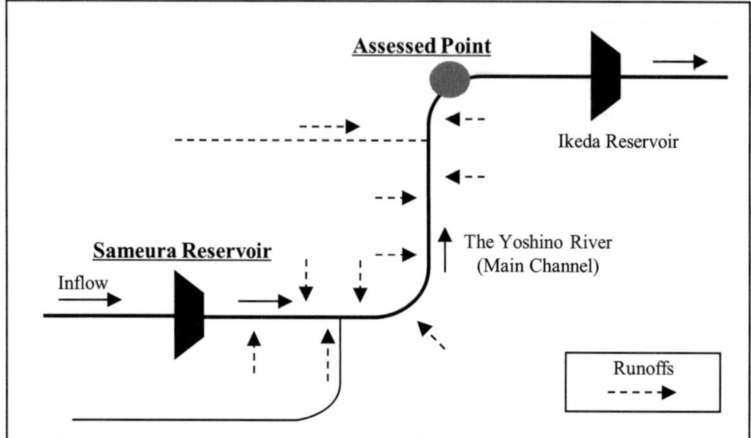

Fig. 1 Simplified model of downstream of Sameura reservoir for application.

$$
\begin{aligned}
&0 \le r_t \le 2000(\mathrm{m}^3/\mathrm{sec}) \qquad (\textit{for flood situation}) \\
&0 \le r_t \le 800(\mathrm{m}^3/\mathrm{sec}) \qquad (\textit{for no flood situation})
\end{aligned} \tag{23}
$$

A flood situation means a period when inflow of more than 800 m^3/sec is observed, and *vice versa*. However, the FDF for the final period in the predicted range (34th day in this paper, i.e. f_{34}) must be given in advance for conducting optimization. In this application, f_{34} was calculated by the equation described below considering the damage suffered for the subsequent one year:

$$
\begin{aligned}
&f_{365}(s_{365}) = \min_{r_{365}} H_{365}(q_{365}) \\
&f_t(s_t) = \min_{r_t} \{H_t(q_t) + f_{t+1}(s_{t+1})\} \qquad (t = 34, \cdots, 364)
\end{aligned} \tag{24}
$$

For the calculation of FDF, normal streamflow was employed for the period after the 35th day, considering no use of forecast information for that period.

Estimation results of ensemble streamflow prediction

The regression model for estimation of basin average precipitation from four GPVs of precipitation mentioned above was developed. The determination coefficient for estimated regression model, however, was not good, and RMSE of calculation by the estimated model was equivalent or large compared with that of the simply precipitation-averaged four GPVs. Therefore, the average of four GPVs was employed as estimation of basin average precipitation.

Subsequently, regression models for estimation of inflow into Sameura reservoir and runoff between the reservoir and assessed point from basin average precipitation were respectively developed. Observational data for 27 years from 1979 to 2005 was used for the development of the two regression models. As a result, the regression equations described below were estimated:

$$
\begin{aligned}
&i(t) = a_1 \cdot p(t) + a_2 \cdot p(t-1) + a_3 \cdot i(t-1) + a_4 \cdot i(t-2) + a_5 \cdot i(t-3) \\
&\quad \begin{cases} a_1 = 0.41476 \quad a_2 = 0.16677 \\ a_3 = 0.11503 \quad a_4 = 0.05309 \quad a_5 = 0.02702 \end{cases}
\end{aligned} \tag{25}
$$

$$
\begin{aligned}
&o(t) = b_1 \cdot p(t) + b_2 \cdot p(t-1) + b_3 \cdot p(t-2) + b_4 \cdot o(t-1) + b_5 \cdot o(t-3) \\
&\quad \begin{cases} b_1 = 0.37581 \quad b_2 = 0.37070 \quad b_3 = -0.07411 \\ b_4 = 0.35023 \quad b_5 = 0.03077 \end{cases}
\end{aligned} \tag{26}
$$

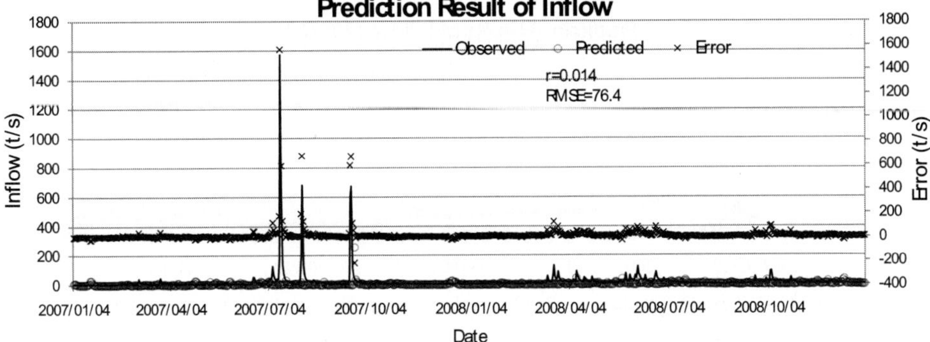

Fig. 2 Prediction result of inflow into Sameura reservoir.

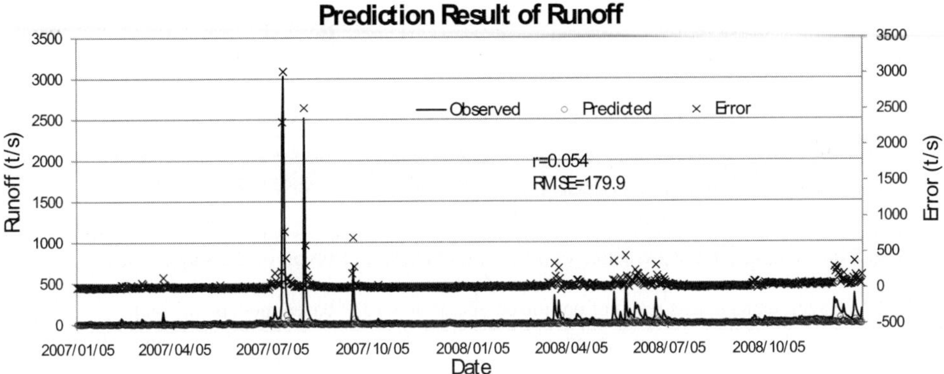

Fig. 3 Prediction result of runoff between Sameura reservoir and the assessed point.

where $i(t)$ is inflow into Sameura Reservoir during period t, $o(t)$ is runoff which joins the main stream of the Yoshino River between the dam site and the assessed point, $p(t)$ is basin average precipitation during period t. The determination coefficient of the regression model estimate of inflow prediction was 0.810, and 0.758 for runoff prediction with more than 99% significance.

Then prediction of inflow and runoff were conducted by use of the developed regression models from basin average precipitation estimated using one-month ensemble precipitation forecasts from 2007 to 2008. Figure 2 shows the prediction result of inflow by the ensemble mean precipitation, and Fig. 3 shows that of runoff. Note that the predicted inflow or runoff sequence in these figures means prediction by the most updated forecast. These sequences are made by connecting all the predicted inflow sequences for seven days from the date of the precipitation forecast was provided. As shown in Fig. 2 and Fig. 3, the accuracies of these predictions are not good, due to the low prediction accuracy of basin average precipitation. Through the applied year, one-month forecasts were likely to underestimate precipitation in total, although some predicted sequences showed greater precipitation than observation. (Similar characteristics of predicted streamflow are shown in Figs 4 and 5.) However, these prediction results were employed for optimization of long-term reservoir operation because it is not essential for analysis of difference among optimizing processes by DP-based models.

Differences among optimization processes by DP-based models

Optimization of reservoir operation for water use was conducted with ensemble streamflow sequences predicted by use of one-month ensemble forecast of precipitation. Three DP-based models, namely DDP with ensemble mean prediction of streamflow (DDP/EPP1), SDP with

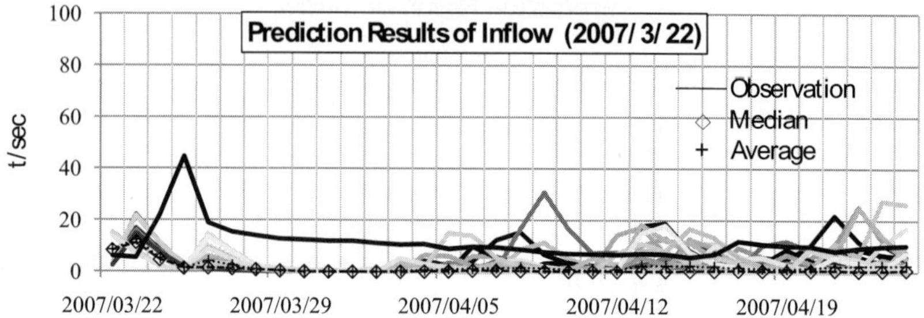

Fig. 4 Result of ensemble prediction of inflow into Sameura Reservoir on 22 March 2007.

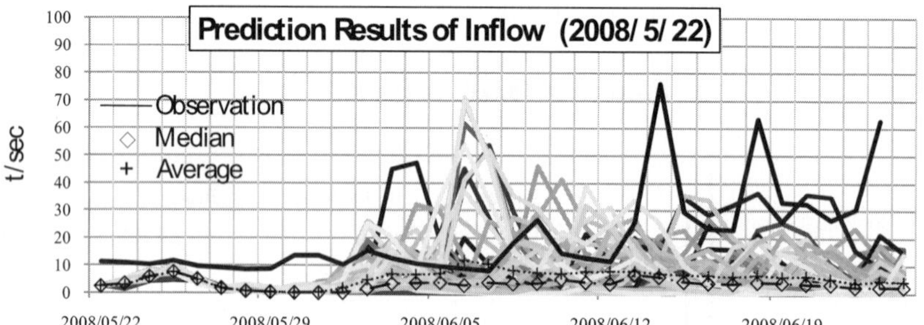

Fig. 5 Result of ensemble prediction of inflow into Sameura reservoir on 22 May 2008.

Table 2 Averaged-damage as a result of optimized operation by DP-based models (m^3/sec).

Model	2007	2008	2007–2008
DDP/Perf	0.760	0.603	0.682
DDP/Ave	5.41	5.18	5.29
DDP/EPP1	3.04	2.77	2.91
SDP/EPP1	2.83	3.74	3.29
SSDP/EPP1	2.68	3.56	3.12
Actual Operation*	0.670	1.30	0.987

* Excluding potential damage caused by alternative release for water use from capacity for power generation into calculation of damage.

ensemble prediction of streamflow (SDP/EPP1) and SSDP with ensemble prediction of streamflow (SSDP/EPP1), were employed for the optimization. Optimization results of these models are shown in Table 2. DDP/Perf and DDP/Ave shown in Table 2 are the optimization result by DDP with perfect foresight (using observed inflow and runoffs) and with normal values of streamflow, respectively.

The DDP model with perfect foresight optimization of reservoir operation gave the smallest damage in the applied models for any period. The damage caused in operation with DDP/Perf was greater than the actual damage. This is because a more severe water demand was set for the application than actual in this study. Although the value of target discharge for design intake flow was used as the demanded water amount in this application, the actual quantity of intake flow was less than that of the designed one in the applied period. DDP with normal streamflow optimization of reservoir operation gave the greatest damage in the simulated models for any period. This is because streamflow in the Yoshino River in 2007 and 2008 was much less than average (Fig. 6).

Any DP models with ensemble streamflow estimated by use of one-month ensemble precipitation prediction conducted operation with greater damage than that of DDP/Perf, and smaller than that of DDP/Ave. This result indicates that forecast is still useful for reservoir operation to an extent, in case accuracy of the forecast is not so good. Concerning the difference

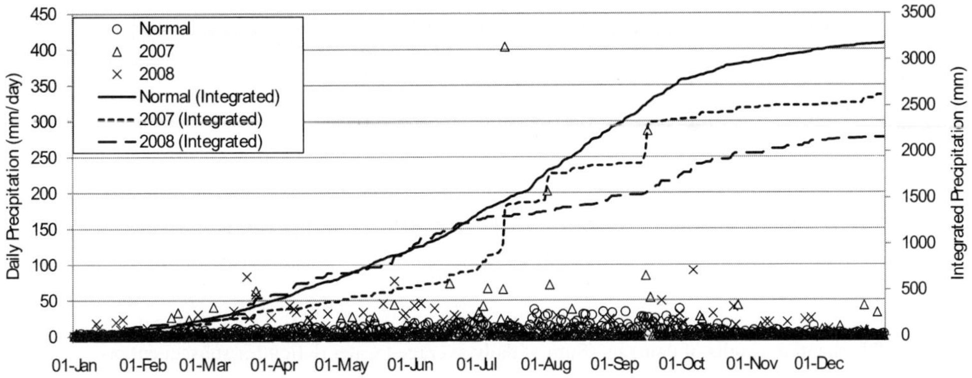

Fig. 6 Observation of precipitation in application period.

between the result by deterministic DP and those by two stochastic DP (SDP and SSDP) models, the DDP model was better than both the two SDP models in 2008, though the opposite was the case in 2007. The reason for this result can be considered as follows.

Figure 6 shows observations of precipitation in the application period. It can be seen here that precipitation in 2007 was consistently smaller than normal precipitation, while in 2008 it was as much as normal precipitation until June, but also smaller after July. The results of ensemble prediction of inflow into Sameura Reservoir for the subsequent 34 days (prediction range) from the date of forecast publication are shown in Figs 4 and 5. It can be seen in these results that the averaged sequence of inflow prediction was consistently much smaller than observed inflow. It can be also seen that the median sequence of these ensemble predictions was even smaller than the averaged sequence. Similar characteristics were seen in the result of runoff prediction. Considering that quadratic function of water deficit was used for estimation of damage, which is the objective function of the optimization, in this study, now it can be considered that the accuracy of optimization was more affected by median values of ensemble predictions than by normal values in SDP and SSDP, because of the difference of calculation algorithm from that of DDP. The future damage value is calculated by first square-summing water deficit then averaging it in SDP and SSDP, while it is calculated first averaging prediction then squaring water deficit calculated by the averaged value in DDP. In this application, median values were consistently less than normal value as mentioned above, so it is considered that the SDP and SSDP models conducted more cautious operations than the DDP model given that smaller streamflow would appear for the coming 34 days. Operation results with DDP model and two SDP based models were especially different from March to October in 2008. In SDP and SSDP models, there was more release, which was often too much for demand as a result, was conducted in the early part of the period, and less release, which was often too little for demand on the contrary, was conducted in the latter half of the period because of the small storage volume existing in that period. This is because SDP and SSDP operated too much release considering lower estimation of streamflow for the predicted period more important than risk in the further future by comparison with DDP. It consequently caused ineffective releases by actually getting much greater amounts of inflow and runoff, which are equivalent to normal values (similar characteristics of precipitation can be seen in Fig. 6). This resulted in rapid decrease of storage water, and finally derived more water deficit in the latter half of the period because of scarcity of storage water. This result means that the DP model considering stochastic characteristics of prediction is not necessarily better than that with no consideration of stochastic characteristics in the case of ensemble prediction with unsatisfactory accuracy. In the optimization of reservoir operation, a high dimension function is often used for evaluation of objective variables, i.e. damage caused by water deficit or flood, or benefit of power generation. And hydrological variables, such as precipitation or streamflow, usually have similar distribution characteristics to the median which is less than normal values, as seen in ensemble streamflow

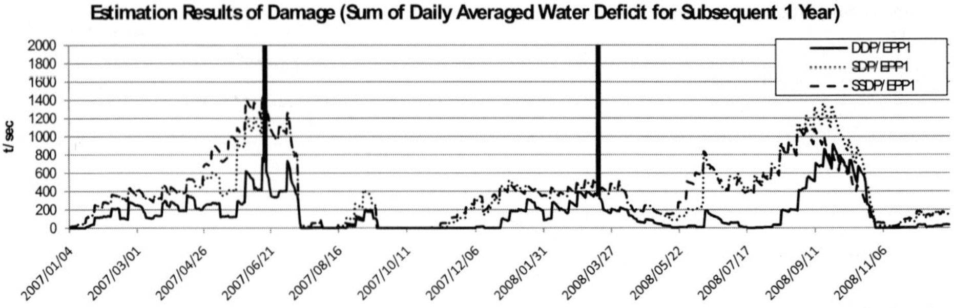

Fig. 7 Estimation results of damage. (Sum of daily averaged water deficit for subsequent one year.)

prediction in this application. Attention is necessary when ensemble prediction is employed, especially for long-term reservoir operation for water use or power generation. It can be considered useful to consider the relationship of median and normal values of the prediction sequences when ensemble prediction is planned to be introduced in reservoir operation.

Estimation results of FDF are shown in Fig. 7. This figure also shows that DDP generally estimated less damage reflecting the estimation of greater streamflow by comparison with SDP and SSDP. Concerning the difference between results of SDP and SSDP, it can be seen that the SDP model had a tendency to estimate FDF tardily, compared with the SSDP model. It is considered that this tendency affected the lower performance of SDP more than that of SSDP. The reason for this, however, did not become apparent by this application: this is an issue in the future.

CONCLUSION

In this paper, optimizing processes of three DP-based models, namely DDP, SDP and SSDP, were analysed through the application of the models to optimization of long-term reservoir operation. Differences in results calculated by deterministic and stochastic DP were revealed. As a result, it was shown that it is necessary to carefully consider the distribution characteristics of the prediction when ensemble prediction is applied to reservoir operation. It is shown that the relationship between normal and median value can be an index for decision of the DP-model to employ, combined with prediction error tendency. However, a reason for the differences between the performance of SDP and SSDP with ensemble prediction, which has unsatisfactory accuracy, was not revealed in this study. Further analysis is necessary in the future.

Acknowledgements Basin data of Sameura Reservoir basin used in this study was offered by Ikeda Operation and Maintenance Office, Yoshino Regional Bureau of Japan Water Agency. Data of one-month ensemble forecast of JMA, is collected by Kyoto University Active Geosphere Investigations for the 21st Century COE Program and is available on the web site of GFD Dennou Club, also used in this study. We would like to express appreciation for all affiliates of them.

REFERENCES

Bellman, R. (1957) *Dynamic Programming*. Princeton University Press., Princeton, New Jersey, USA.

Faber, B. A. & Stedinger, J. (2001) Reservoir optimization using sampling SDP with ensemble streamflow prediction (ESP) forecasts. *J. Hydrol.* **249**, 113–133.

Furukawa, T. & Sakai, S. (2004) *Ensemble Forecast*. Tokyodo Publishing Co. Ltd., Tokyo, Japan (in Japanese)

Ikebuchi, S., Kojiri, T. & Miyakawa, H. (1990) A study on long-term and real-time reservoir operation by using middle and long-term weather forecast. *Ann. Disaster Prevention Res. Inst., Kyoto Univ.* **33**B(2), 167–192. (in Japanese)

Kelman, J. & Stedinger J. R. (1990) Sampling stochastic dynamic programming applied to reservoir operation. *Water Resour. Res.* **26**(3), 447–454.

Kim, Y. O., Eum, H. I., Lee, E. G. & Ko, I. H. (2007) Optimizing operational policies of a Korean multireservoir system using sampling stochastic dynamic programming with ensemble streamflow prediction. *J. Water Resour. Plann. & Manage.* **133**(1), 4–14.

Kojiri, T., Tomosugi, K. & Galvao, C. V. (1994) Knowledge-based decision support system of real-time reservoir operation for drought control. *J. Japan Soc. Hydrol. & Water Resour.* **7**(3), 188–195.

Labadie, J. (2004) Optimal operation of multireservoir systems: state-of-the-art review. *J. Water Resour. Plann. & Manage.* **130**(2), 93–111.

Loucks, D. P., Stedinger, J. R. & Haith, H. A. (1981) *Water Resources Systems Planning and Analysis.* Prentice-Hall, Englewood Cliffs, New Jersey, USA.

Nandaldal, K. D. W. & Bogardi, J. J. (2007) *Dynamic Programming Based on Operation of Reservoirs - Applicability and Limits.* International Hydrological Series, Cambridge University Press, UK.

Wada, K., Kawasaki, M. & Tomizawa, Y. (2005) A study on applicability of precipitation forecasts for river flood management. *J. Japan Soc. Hydrol. &Water Resour.* **18**(6), 703–709 (in Japanese).

Yeh, W. (1985) Reservoir management and operations models: a state-of-the-art review, *Water Resour. Res.* **21**(12), 1797–1818.

Yoshino Regional Bureau of Japan Water Agency (2004) *Life and Water in 21st Century – Water Resources in Shikoku.*

Identification of an appropriate low flow forecast model for the Meuse River

MEHMET C. DEMIREL & MARTIJN J. BOOIJ

Water Engineering and Management, Faculty of Engineering Technology, University of Twente, PO Box 217, 7500 AE Enschede, The Netherlands

m.c.demirel@utwente.nl

Abstract This study investigates the selection of an appropriate low flow forecast model for the Meuse River based on the comparison of output uncertainties of different models. For this purpose, three data driven models have been developed for the Meuse River: a multivariate ARMAX model, a linear regression model and an Artificial Neural Network (ANN) model. The uncertainty in these three models is assumed to be represented by the difference between observed and simulated discharge. The results show that the ANN low flow forecast model with one or two input variables(s) performed slightly better than the other statistical models when forecasting low flows for a lead time of seven days. The approach for the selection of an appropriate low flow forecast model adopted in this study can be used for other lead times and river basins as well.

Key words low flows; linear regression model; ANN; ARMAX; uncertainty; appropriate model; Meuse River

INTRODUCTION

Low flow is defined as a seasonal phenomenon and an integral phase of a flow system of any river (Smakhtin, 2001). However, in Northern European countries high flows are often studied and low flows are generally neglected (Kwadijk & Middelkoop, 1994; Parmet & Burgdorffer, 1995). For the Rhine and Meuse basins, a limited number of studies on low flows have been published in refereed literature (Middelkoop *et al.*, 2001; De Wit *et al.*, 2007; Rutten *et al.*, 2008), and non-refereed literature (Passchier, 2004; Arends, 2005; De Bruijn & Passchier, 2006). This is probably because low flow is a slow process which usually occurs during the dry season, unlike high flow events' fast and eye-catching processes. Low flow events in the Rhine River and Meuse River in dry summers such as in 1921, 1976 and 2003, indicate the importance of considering these events. Moreover, the number of days with low flows in Northern European rivers is expected to increase due to climate change (Middelkoop *et al.*, 2001; De Wit *et al.*, 2007; Te Linde *et al.*, 2008).

Elaborative studies on low flows started in 1976 when a Task Committee on Low Flow was organized by the American Society of Civil Engineers (Riggs, 1980). This committee drew attention to the consequences of hydrological droughts and to the need for further studies using standard low flow indexes. There are several low flow indexes used by different institutes since there are many ways of defining flow conditions as "low flow". One typical way is the lowest flow that has been ever measured in the river. Another is the use of the annual minimum 10-day flow. A more common definition, however, is the flow level being exceeded 95% of the year.

The flow processes are generally represented by different functions embedded into a model. A perfect model including every physical process in a basin may never exist without a certain degree of uncertainty. Therefore, uncertainty analysis in hydrology is necessary, for instance to express the reliability of forecasts. The number of studies applying a systematic quantification of uncertainties has increased rapidly and complementary discussions began to appear in the literature to create consensus in hydrological uncertainty assessment terminology (e.g. Montanari, 2007). Different uncertainty analysis techniques are present for different models (e.g. Monte Carlo simulations, Generalized Likelihood Uncertainty Estimation, etc.).

In this paper, the model output uncertainty is used for the identification of an appropriate low flow forecast model. There have been other studies on model appropriateness (e.g. Booij, 2003; Dong *et al.*, 2005), but the identification of an appropriate low flow forecast model based on uncertainty in predicted low flows has not been done. The objective of this study is therefore to identify an appropriate low flow forecast model for a lead time of seven days by comparing output uncertainties from three different data driven models with different combinations of input variables. The study area is the Meuse basin in Western Europe.

STUDY AREA AND DATA

The Meuse basin covers an area of approximately 33 000 km^2, including parts of France, Luxembourg, Belgium, Germany, and The Netherlands. About 60% of the Meuse basin is used for agricultural purposes (including pastures) and 30% is forested. The average annual precipitation ranges from 1000–1200 mm in the Ardennes to 700–800 mm in the Dutch and Flemish lowlands. The maximum altitude is just below 700 m a.s.l. Snowmelt is not a major factor for the discharge regime of the Meuse. The average discharge at the outlet is approximately 350 $m^3 s^{-1}$, this corresponds with an annual precipitation surplus of almost 400 mm. Precipitation is equally distributed over the year. The seasonal variation in the discharge is a reflection of the variation in evapotranspiration (Booij, 2005).

Daily discharge data at Chooz (upstream area 10 000 km^2) and Monsin (upstream area 21 000 km^2), basin-averaged precipitation data and basin-averaged potential evapotranspiration data for the period 1968–1998 are used; 20 years for calibration and 10 years for validation of the results. Low flows are discharge values measured at Monsin station in the Meuse River of less than 100 $m^3 s^{-1}$ analogously to Booij *et al.* (2006).

METHODOLOGY

The four steps to identify the appropriate low flow forecast model are:

1. Assessment of appropriate temporal input resolutions.
2. Determination of model structures for three data driven models with four different combinations of input variables.
3. Quantification of the model output uncertainty.
4. Identification of an appropriate low flow forecast model.

Following these steps, it is intended to test three different data driven models with different combinations of input variables with appropriate temporal resolutions. Model types used in this study differ in two different dimensions regarding model complexity: the number of input variables and the mathematical description of the models.

Assessment of appropriate temporal input resolutions

The assessment of the appropriate temporal input resolution is based on the determination of cross-correlation coefficients between the input variables at different temporal resolutions and the output variable. The input variables are the discharge at Monsin Q_m, the discharge at Chooz Q_c, the basin averaged precipitation P and the basin averaged potential evapotranspiration PET. The output variable is Q_m seven days ahead of the input variables. For each input variable, the temporal resolution resulting in the largest cross-correlation coefficient is chosen as appropriate temporal resolution in the subsequent modelling steps.

Determination of data driven model structures

Data driven models are built based on input-output relations. Physical processes are generally ignored and model structures are less complex than physically-based or conceptual models. Three different data driven models for 7-day ahead low flow forecasts are compared in this study: a linear regression (LR) model, a multivariate auto regressive moving average model with exogenous inputs (ARMAX) and an artificial neural network (ANN) model. Each model is tested with four different combinations of input variables by adding the input variables according to their cross-correlation with the output variable in a descending order of cross-correlation. The LR model has the simplest structure compared to the other two models as shown in equation (1):

$$Q_{LR}(t+7) = a + b_i x_i(t) \tag{1}$$

where Q_{LR} is the 7-day ahead predicted discharge, a is the intercept, b_i are regression coefficients and x_i are the independent input variables. The objective function in a regression model usually aims to minimize the total sum of squared errors using the ordinary least squares method.

In ARMAX modelling several steps are distinguished: the identification of the model structure, parameter estimation and a diagnostic check to validate the model before using it in forecasting problems. Autocorrelation functions and partial autocorrelation functions are good indicators for univariate AR and MA model orders. The shape of the lagged correlogram generally gives an idea for modellers how to choose the best model order. The Yule-Walker equations are used to estimate the ARMAX multivariate model parameters, see equation (2):

$$Q_{ARMAX}(t+7) = c_i x_i(t) + e(t) \tag{2}$$

where Q_{ARMAX} is the 7-day ahead predicted discharge, c_i are model parameters, x_i are the independent input variables and $e(t)$ is the white noise. Details of ARMAX modelling can be found in core text books such as Ljung (1986).

The ANN model structure has been designed based on the literature, user experience and trial-error processes. A network with one hidden layer with different numbers of hidden nodes has been mostly preferred in hydrological predictions (Raman & Sunilkumar, 1995; Coulibaly *et al.*, 2001; Khan & Coulibaly, 2005; Demirel *et al.*, 2008). More than one hidden layer requires many more parameters to be estimated as many new weights and bias values are necessary in the newly built connections. Following Rumelhart *et al.* (1986) and Govindaraju & Rao (2000) the feed forward ANN model is selected to model daily low flows. The Levenberg-Marquardt algorithm being a fast converging optimisation algorithm for training ANNs, and more efficient than many other present algorithms, is used for optimisation. The hyperbolic tangent transfer function and the logistic sigmoid function were both tested in a trial-and-error process and the former gave better results. The objective function is the Root Mean Squared Error (RMSE) for low flows. The performance goal for this objective function is defined as 1‰ of the observed data variance which is equal to 0.0015. Different guidelines for the number of hidden nodes have been proposed. For example, Ochoa-Rivera (2008) suggests starting with one hidden node and adding new nodes until a significant improvement in performance is achieved. Eberhart & Dobbins (1990) found it useful to commence simulations with a number of hidden nodes equal to half of the number of input nodes. In this study, the number of hidden nodes in the one and two input ANN models is selected by using a trial-and-error procedure following Ochoa-Rivera (2008). For the three and four input ANN models, the number of hidden nodes is assumed to be equal to the number of inputs multiplied with the lead time, hence 21 and 28 hidden neurons are used, respectively, as the lead time is seven days.

Quantification of uncertainty and appropriate low flow model identification

There are many different definitions of uncertainty (e.g. Walker *et al.*, 2003; Refsgaard *et al.*, 2007) Here, uncertainty is assumed to consist of inaccuracy and imprecision following Van der Perk (1997). Inaccuracy is defined as the difference between a simulated value and an observation, while imprecision refers to the possible variation around the average simulated values and observed values. Model inaccuracy can be assessed by, e.g. the RMSE. Obviously, the model inaccuracy does not cover all uncertainties and therefore underestimates the total uncertainty. However, it is expected to give an indication of the trend in uncertainty as a function of model complexity. The appropriate model is selected according to this indicator. An appropriate low flow forecast model is a model that produces output with the smallest uncertainty in low flows. This is quantified by the RMSE of observed and predicted low flows for the validation period.

RESULTS AND DISCUSSION

Assessment of dominant low flow indicators

Figure 1 shows the cross-correlation coefficients between the output variable and input variables as a function of the temporal input resolution. Different temporal resolutions were found to be appropriate for each of the four inputs: seven days for discharge values at Monsin, four days for discharge values at Chooz, and 150 days for basin-averaged precipitation and evapotranspiration.

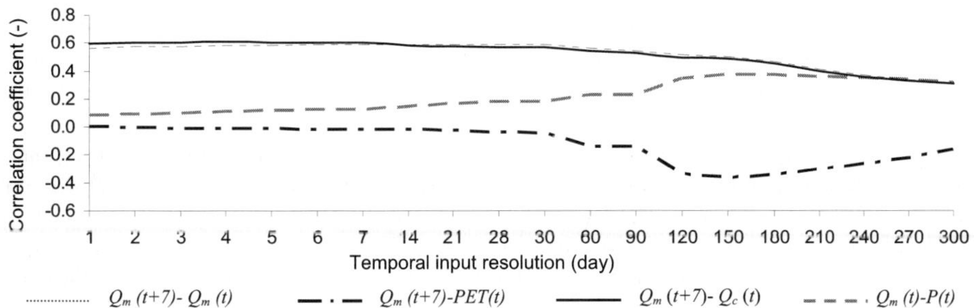

Fig. 1 Cross correlation coefficients between discharge at Monsin $Q_m(t+7)$ and different inputs with a lag time of seven days as a function of temporal input resolution (days) for four different inputs: precipitation $P(t)$, potential evapotranspiration $PET(t)$, discharge at Chooz $Q_c(t)$ and discharge at Monsin $Q_m(t)$.

Determination of data driven model structures

LR model The results of the LR models are presented in Table 1. These results are not promising for low flow predictions and the model order of one can be an important limitation here. The uncertainty (RMSE values) decreased with an increasing number of input variables for both the calibration and validation. The contribution of precipitation to the model is very low. The negative relation between potential evapotranspiration and the discharge at Monsin is apparent in the four input LR model. After the inclusion of the discharge at Chooz (Q_c) as an input to the model, other input variables have marginal effects on the results.

Table 1 Estimated parameter values for LR model and RMSE values in calibration and validation.

Inputs	Estimated parameters					RMSE	
	a	b_1	b_2	b_3	b_4	Calibration	Validation
$Q_m(t)$	0.356	0.680				0.313	0.316
$Q_m(t)$ $Q_c(t)$	0.360	0.141	0.460			0.307	0.292
$Q_m(t)$ $Q_c(t)$ $P(t)$	0.150	0.115	0.458	0.093		0.296	0.278
$Q_m(t)$ $Q_c(t)$ $P(t)$ $PET(t)$	0.519	0.092	0.405	0.095	–0.189	0.246	0.211

Multivariate ARMAX model The results of the ARMAX models are presented in Table 2. The model with two inputs is the better performing model in the ARMAX group. Precipitation and potential evapotranspiration have very low parameter values showing that their contribution to those models is not significant and can be excluded. For that reason, the uncertainty did not decrease when including this meteorological information. In general, the uncertainty is lower than for the LR model.

Table 2 Estimated parameter values for ARMAX model and RMSE values in calibration and validation.

Inputs	Estimated parameters				RMSE	
	c_1	c_2	c_3	c_4	Calibration	Validation
$Q_m(t)$	0.804				0.083	0.094
$Q_m(t)$ $Q_c(t)$	0.366	0.384			0.077	0.081
$Q_m(t)$ $Q_c(t)$ $P(t)$	0.218	0.359	0.079		0.127	0.116
$Q_m(t)$ $Q_c(t)$ $P(t)$ $PET(t)$	0.208	0.336	0.117	–0.038	0.124	0.108

ANN model The results of the ANN models are presented in Table 3. The most promising models were again the one and two input models as for ARMAX. Adding meteorological inputs did not improve the results. This might be due to the large temporal resolution of these inputs. Our aim is to capture daily variations in low flows; however, this can be difficult when using a temporal resolutions of 150 days. The addition of potential evapotranspiration in the other two models had small but positive impacts on the results; however, for the ANN model this is not the case. The possible reason is the model structure which has a very large number of hidden nodes causing difficulty in training and also weakening the effectiveness of the learning cycles. Other forecasting studies also had difficulties in determining the number of hidden nodes (Tingsanchali & Gautam, 2000).

Table 3 ANN model architecture and test scheme and RMSE values in calibration and validation.

Inputs	Outputs	Training	Test	Network structure	Epochs	RMSE	
						Calibration	Validation
$Q_m(t)$	$Q_m(t+7)$	7671	3652	1-10-1	5	0.063	0.062
$Q_m(t)$ $Q_c(t)$	$Q_m(t+7)$	7671	3652	2-20-1	5	0.063	0.061
$Q_m(t)$ $Q_c(t)$ $P(t)$	$Q_m(t+7)$	7671	3652	3-21-1	4	0.091	0.085
$Q_m(t)$ $Q_c(t)$ $P(t)$ $PET(t)$	$Q_m(t+7)$	7671	3652	4-28-1	10	0.098	0.090

Quantification of uncertainty and appropriate low flow forecast model identification

Table 4 and Fig. 2 show the model output uncertainty (RMSE values) as a function of model type and number of inputs. There is a significant decrease in uncertainty with an increasing number of inputs in the LR models in both the calibration and validation, while the ARMAX and ANN models do not show this behaviour.

Table 4 RMSE values for three data driven models (LR model, ARMAX model and ANN model) with four combinations of input variables for calibration and validation.

Inputs	LR		ARMAX		ANN	
	Calibration	Validation	Calibration	Validation	Calibration	Validation
$Q_m(t)$	0.313	0.316	0.083	0.094	0.063	0.062
$Q_m(t)$ $Q_c(t)$	0.307	0.292	0.077	0.081	0.063	0.061
$Q_m(t)$ $Q_c(t)$ $P(t)$	0.296	0.278	0.127	0.116	0.091	0.085
$Q_m(t)$ $Q_c(t)$ $P(t)$ $PET(t)$	0.246	0.211	0.124	0.108	0.098	0.090

The hypothesis is that more inputs to create more predictive accuracy requires more investigation when designing particularly ANN and ARMAX model structures. The possibility of a smaller number of hidden nodes and more efficient training algorithms should be critically tested. Accordingly, the governing trial-and-error processes in ANN modelling should be avoided and more deterministic approaches should be adopted. The improvement in the model results should be observed by the change of training cycles (i.e. epochs) and the number of the hidden

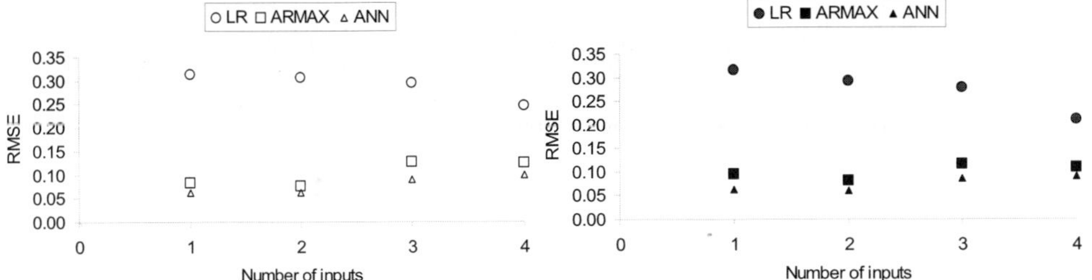

Fig. 2 RMSE values as a function of number of inputs for three data driven models (LR model, ARMAX model and ANN model) for calibration (left) and validation (right).

nodes together. A longer lead time, such as 14 days, should also be tested to understand the behaviour of the models for longer terms. Data averaging for large window sizes should be carefully applied for low flow predictions as it is deteriorating the daily oscillations in low flows while increasing the persistence and correlation coefficient. Hence, we recommend four day to seven day data averaging for predictions with one or two week lead times. The LR and ARMAX models show different behaviour in the validation period, for example the one and two input ARMAX models revealed a higher uncertainty in the validation period than in the calibration period, but the three and four input ARMAX models performed better in the validation period.

The one input and two input ANN models have the smallest uncertainty indicating that they are the most appropriate low flow forecast models in this study. The observed and predicted low flows for these two models are illustrated in Figs 3 and 4. The low flow predictions are in general below the threshold of 100 $m^3 s^{-1}$. The magnitude of the low flows is more successfully captured in the two input ANN model than in the one input model. As shown in Fig. 4 there are very low observed discharges in some validation years, e.g. in 1991, 1992, 1996 and 1997. The two input ANN model only approximated these events in 1992 and 1996. The one input ANN model was successful only for the low flows between 50 and 100 $m^3 s^{-1}$ and not for values below 25 $m^3 s^{-1}$. When the observed low flow is more stationary, as it is in the first and second year of the validation period, the ANN models predict the low flows better. The arbitrary changes in the discharge values are not always well captured due to the learning rate of the networks.

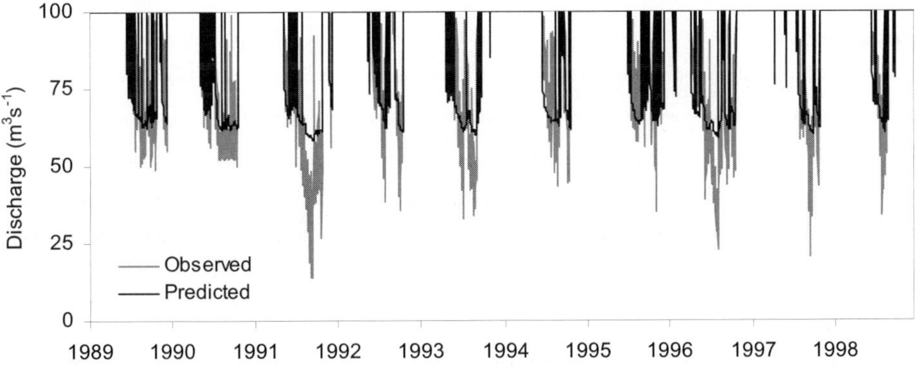

Fig. 3 Observed and predicted low flows at Monsin for one input ANN model in validation period.

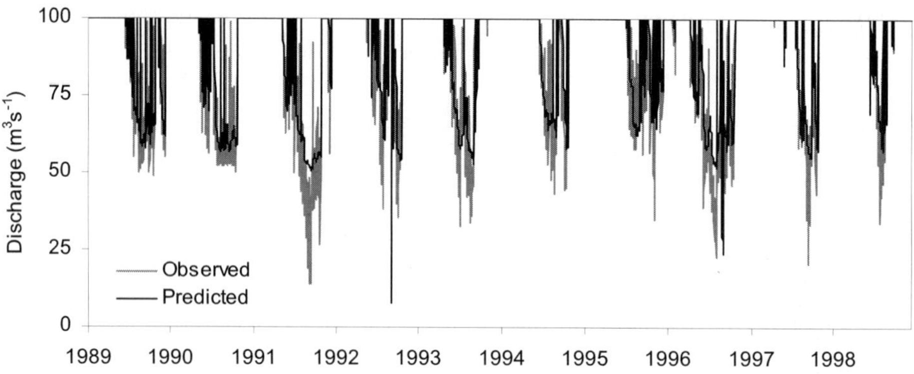

Fig. 4 Observed and predicted low flows at Monsin for two input ANN model in validation period.

CONCLUSIONS

Three data driven models with different combinations of inputs were compared based on modelled output uncertainty. The aim was to identify the most appropriate low flow prediction model for the Meuse River. RMSE is used as an output uncertainty indicator and the ANN models with one and two input variables are found to be most appropriate for predicting low flows, i.e. they have the smallest RMSE values in the validation period. However, the LR model represents the idea that more information should result in less uncertainty very well as additional inputs created smaller RMSE values. This behaviour was not observed for the other two more complicated models, i.e. the ARMAX and ANN models, which might be due to possible obstacles in these two models such as the network structure in the ANN model and the delay factor in the ARMAX model. It can be concluded that low flow characteristics and data scales are the key factors in capturing the daily variations in hydrological low flow events. Futhermore, model inaccuracy can be used as an indicator for model output uncertainty. Finally, the four step methodology introduced in this study can be applied to other river basins and lead times as well.

Acknowledgements We acknowledge the financial support of Dr Ir Cornelis Lely Stichting (CLS), Project no. 20957310.

REFERENCES

Arends, M. (2005) Low flow modelling of the River Meuse: Recalibration of the existing HBV Meuse modelling. MSc Thesis. University of Twente, Enschede, The Netherlands.

Booij, M. J. (2003) Determination and integration of appropriate spatial scales for river basin modelling. *Hydrol. Processes* **17**(13), 2581–2598.

Booij, M. J. (2005) Impact of climate change on river flooding assessed with different spatial model resolutions. *J. Hydrol.* **303**(1-4), 176–198.

Booij, M. J., Huisjes, M. & Hoekstra, A.Y. (2006) Uncertainty in climate change impacts on low flows. In: *Climate Variability and Change – Hydrological Impacts* (ed. by S. Demuth, A. Gustard, E. Planos, F. Scatena & E. Servat) (Proceedings of the Fifth FRIEND World Conference held at Havana, Cuba, November 2006), 401–406. IAHS Publ. 308, IAHS Press, Wallingford, UK.

Coulibaly, P., Bobée, B. & Anctil, F. (2001) Improving extreme hydrologic events forecasting using a new criterion for artificial neural network selection. *Hydrol. Processes* **15**(8), 1533–1536.

De Bruijn, K. M. & Passchier, R. (2006) Predicting low-flows in the Rhine River. Report Q3427, WL | Delft Hydraulics, Delft, The Netherlands.

De Wit, M. J. M., van den Hurk, B., Warmerdam, P. M. M., Torfs, P., Roulin, E. & van Deursen, W. P. A. (2007) Impact of climate change on low-flows in the river Meuse. *Climatic Change* **82**(3), 351–372.

Demirel, M. C., Venancio, A. & Kahya, E. (2009) Flow forecast by SWAT model and ANN in Pracana basin, Portugal. *Adv. Engng Softw.* **40**(7), 467-473.

Dong, X., Dohmen-Janssen, C. M. & Booij, M. J. (2005) Appropriate spatial sampling of rainfall for flow simulation. *Hydrol. Sci. J.* **50**(2), 279–298.

Eberhart, R. C. & Dobbins, R. W. (1990) *Neural Network PC Tools: A Practical Guide*. Academic Press Professional, Inc. San Diego, California, USA.

Govindaraju, R. S. & Rao, A. R. (2000) *Artificial Neural Networks in Hydrology*. Kluwer Academic Publishers, Dordrecht, The Netherlands.

Khan, M. S. & Coulibaly, P. (2005) Streamflow forecasting with uncertainty estimate using Bayesian learning for ANN. In: Proceedings. of IEEE International Joint Conference on Neural Networks held at Montreal, Canada, July 31–August 4, 2005, vol. 5, 2680–2685.

Kwadijk, J. & Middelkoop, H. (1994) Estimation of impact of climate-change on the peak discharge probability of the River Rhine. *Climatic Change* **27**(2), 199–224.

Ljung, L. (1986) *System Identification: Theory for the User*. Prentice-Hall, Inc. Upper Saddle River, NJ, USA.

Middelkoop, H., Daamen, K., Gellens, D., Grabs, W., Kwadijk, J. C. J., Lang, H., Parmet, B., Schadler, B., Schulla, J. & Wilke, K. (2001) Impact of climate change on hydrological regimes and water resources management in the Rhine Basin. *Climatic Change* **49**(1-2), 105–128.

Montanari, A. (2007) What do we mean by "uncertainty"? The need for a consistent wording about uncertainty assessment in hydrology. *Hydrol. Processes* **21**(6), 841–845.

Ochoa-Rivera, J. C. (2008) Prospecting droughts with stochastic artificial neural networks. *J. Hydrol.* **352**(1-2), 174–180.

Parmet, B. & Burgdorffer, M. (1995) Extreme discharges of the Meuse in the Netherlands: 1993, 1995 and 2100—Operational forecasting and long term expectations. *Phys. Chem. Earth* **20**(5-6), 485–489.

Passchier, R. H. (2004) Low flow hydrology. Report Q3427, WL | Delft Hydraulics, Delft, The Netherlands.

Raman, H. & Sunilkumar, N. (1995) Multivariate modelling of water resources time series using artificial neural networks. *Hydrol. Sci. J.* **40**(2), 145–164.

Refsgaard, J. C., van der Sluijs, J. P., Højberg, A. L. & Vanrolleghem, P. A. (2007) Uncertainty in the environmental modelling process-A framework and guidance. *Environmental Modelling and Software* **22**(11), 1543–1556.

Riggs, H. C. (1980) Characteristics of low flows. *J. Hydraul. Div. ASCE* **106**(5), 717–731.

Rumelhart, D. E., Hintont, G. E. & Williams, R. J. (1986) Learning representations by back-propagating errors. *Nature* **323**(6088), 533–536.

Rutten, M., van de Giesen, N., Baptist, M., Icke, J. & Uijttewaal, W. (2008) Seasonal forecast of cooling water problems in the River Rhine. *Hydrol. Processes* **22**(7), 1037–1045.

Smakhtin, V. U. (2001) Low flow hydrology: a review. *J. Hydrol.* **240**(3-4), 147–186.

Te Linde, A. H., Aerts, J., Hurkmans, R. & Eberle, M. (2008) Comparing model performance of two rainfall–runoff models in the Rhine basin using different atmospheric forcing data sets. *Hydrol. Earth Syst. Sci.* **12**(3), 943–957.

Tingsanchali, T. & Gautam, M. R. (2000) Application of tank, NAM, ARMA and neural network models to flood forecasting. *Hydrol. Process.* **14**(14), 2473–2487.

Van der Perk, M. (1997) Effect of model structure on the accuracy and uncertainty of results from water quality models. *Hydrol. Processes* **11**(3), 227–239.

Walker, W. E., Harremoës, P., Rotmans, J., van der Sluijs, J. P., van Asselt, M. B. A., Janssen, P. & von Krauss, M. P. K. (2003) Defining uncertainty: a conceptual basis for uncertainty management in model-based decision support. *Integrated Assessment* **4**(1), 5–17.

Chaos theory for hydrological modelling: a middle-ground between deterministic and stochastic views

BELLIE SIVAKUMAR[1,2] & HUNG SOO KIM[2]

1 *Department of Land, Air & Water Resources, University of California, Davis, California 95616, USA*
sbellie@ucdavis.edu

2 *Department of Civil Engineering, Inha University, Incheon, 402-751, South Korea*

Abstract There have been two approaches prevalent in hydrological modelling: deterministic and stochastic. The use of deterministic approach may be supported on the basis of the "permanent" nature of the Earth, ocean, and the atmosphere and the "cyclical" nature of mechanisms. The stochastic approach may be favoured because of the "highly irregular and complex nature" of hydrological processes and our "limited ability to observe" the details. With these contrasts, the question of whether hydrological processes are better modelled using a deterministic approach or a stochastic approach is meaningless. Indeed, for most hydrological processes, both the deterministic approach and the stochastic approach are complementary to each other, and thus use of an approach that couples these two could be the most appropriate. This paper argues that "chaos theory" can offer such a coupled deterministic–stochastic approach. Support to this argument is provided through study of the dynamic nature of river flow time series.

Key words hydrological modelling; determinism; stochasticity; chaos; linear; nonlinear; autocorrelation function; phase space; correlation dimension

INTRODUCTION

Hydrological processes arise as a result of interactions between climate inputs and landscape characteristics that occur over a wide range of space and time scales. The tremendous changes in climate inputs and heterogeneities in landscape characteristics often give rise to processes that are highly irregular, complex and random. However, the same factors sometimes also give rise to processes that are regular, simple and deterministic. When, where, why and how one or the other of these situations occurs has been (and continues to be) a mystery. To unravel this mystery, two broad approaches have generally been followed: deterministic and stochastic. Each has its own merits, both having solid foundations in scientific principles/philosophies, verifiable assumptions for specific situations, and the ability to provide reliable results. For example, the deterministic approach has merits considering the "permanent" nature of the Earth, ocean, and atmosphere and the "cyclical" nature of mechanisms that happen, whereas the merits of the stochastic approach lie in the facts that hydrological processes exhibit "highly irregular and complex" structures and that we have only "limited ability to observe" their detailed variations.

In light of these, the general question of whether the deterministic or the stochastic approach is better for hydrological modelling is meaningless. Such a question is really a philosophical one that has no general answer, but it is better viewed as a pragmatic one, which has an answer only in terms of specific situations (Gelhar, 1993). These specific situations must be viewed in terms of the process, scale (time and/or space), and the purpose of interest, which collectively form the "hydrological system" (Sivakumar, 2008). For some situations, both the deterministic approach and the stochastic approach may be equally appropriate; for some other situations, the deterministic approach may be more appropriate; and for still others, the stochastic approach may be more appropriate. It is also reasonable to contend that, for most (if not all) hydrological processes, both the deterministic approach and the stochastic approach are actually complementary to each other. This may be supported by our observation of both deterministic and stochastic properties at one or more scales in time and/or space. For example, it is common to observe a significant deterministic nature in river flow in the form of seasonality and annual cycle, while the interactions of the various mechanisms involved and their various degrees of nonlinearity bring stochasticity. The point is that use of a coupled deterministic–stochastic approach, incorporating both the deterministic and the stochastic components, will likely yield a higher "probability of success" compared to either approach when adopted independently. An immediate question is: how to devise such an approach? The answer to this question may well lie in "chaos theory" (e.g. Lorenz, 1963; Sivakumar, 2000, 2004).

The term "chaos" is used to refer to situations where complex and random-looking behaviours arise from simple nonlinear deterministic systems with sensitive dependence on initial conditions (the converse also applies). The three properties inherent in this definition: (a) nonlinear interdependence; (b) hidden determinism and order; and (c) sensitivity to initial conditions, are highly relevant in hydrology. For example: (i) components and mechanisms involved in the hydrological cycle act in a nonlinear manner and are also interdependent; (ii) the daily cycle in temperature and annual cycle in river flow possess determinism and order; and (iii) contaminant transport in surface and sub-surface waters largely depends upon the time at which the contaminants were released. The first property represents the "general" nature of hydrological processes, whereas the second and third represent their "deterministic" and "stochastic" natures, respectively. Further, despite their complexity and random-looking behaviour, hydrological processes may be governed only by a few degrees of freedom (e.g. runoff in a well-developed urban catchment depends essentially on rainfall), another basic idea of chaos theory.

In view of these, this paper argues that chaos theory can bridge the gap between the deterministic approach and the stochastic approach and offer an avenue for a coupled deterministic–stochastic approach. The argument is supported through analysis of four river flow time series, revealing that the system dynamic properties are neither deterministic nor stochastic, but fall somewhere in between. The four river flow series are: daily flows from the Mississippi River and from the Kentucky River in the USA, and monthly flows from the Salmon River in the USA and from the Göta River in Sweden. Both linear and nonlinear tools are employed. To facilitate interpretation of the results, two synthetic time series possessing deterministic and stochastic dynamic properties, respectively, are also generated and analysed.

METHODS

In the analysis of hydrological time series for identification of the nature of dynamic properties, it is customary to use two basic linear tools: autocorrelation function and power spectrum. These tools provide some important information about the system dynamics (including correlation, seasonality, annual cycle, persistence), which can be used to identify whether the system dynamics are deterministic or stochastic. However, since these tools are linearity-based, they are not reliable for identifying nonlinear, and chaotic, signals. Since the dynamic properties of hydrological systems are essentially nonlinear, and possibly chaotic, these linear tools cannot always be relied upon. In view of this, in the present study, both linear and nonlinear tools are employed. The autocorrelation function is employed as a representative linear tool, while the phase space reconstruction and the correlation dimension method are considered for nonlinear tools.

Autocorrelation function

The autocorrelation function (ACF) is a normalized measure of the linear correlation among successive values in a time series. For a discrete time series X_i, where i = 1, 2, ..., N, and for different values of lag time τ, the autocorrelation function $r(\tau)$ is determined according to:

$$r(\tau)=\frac{\sum_{i=1}^{N-\tau} x_i x_{i+\tau} - \frac{1}{N-\tau}\sum_{i=1}^{N-\tau} x_{i+\tau}\sum_{i=1}^{N-\tau} x_i}{\left[\sum_{i=1}^{N-\tau} x_i^2 - \frac{1}{N-\tau}\left(\sum_{i=1}^{N-\tau} x_i\right)^2\right]^{1/2}\left[\sum_{i=1}^{N-\tau} x_{i+\tau}^2 - \frac{1}{N-\tau}\left(\sum_{i=1}^{N-\tau} x_{i+\tau}\right)^2\right]^{1/2}} \tag{1}$$

The use of ACF in characterizing the dynamic properties of a time series lies in its ability to determine the degree of dependence present in the values. For a purely stochastic process, the ACF fluctuates randomly about zero, indicating that the process at any certain instance has no "memory" of the past at all. For a periodic process, the ACF is also periodic, indicating the strong relation between values that repeat over and over again. For signals from a chaotic process, the ACF is expected to decay exponentially with increasing lag, because the states of a chaotic process are neither completely dependent nor completely independent of each other.

Phase space reconstruction

Phase space is a graph or a co-ordinate diagram, whose coordinates represent the variables necessary to describe the state of the system at any moment. The trajectories of the phase space diagram describe the evolution of the system from some initial state and thus represent the history. The "region of attraction" of the trajectories provides qualitative information on the "extent of complexity" of the system. Given a single- (or multi-) variable time series X_i, where $i = 1, 2, ..., N$, a multi-dimensional phase space can be reconstructed as follows (Takens, 1981):

$$\boldsymbol{Y}_j = (X_j, X_{j+\tau}, X_{j+2\tau}, ..., X_{j+(m-1)\tau}) \tag{2}$$

where $j = 1, 2, ..., N - (m - 1)\tau$, m is the dimension of the vector Y_j (embedding dimension), and τ is the delay time. If the "region of attraction" is narrow and clear, then the dynamics are considered as "simple" and the system low dimensional (perhaps deterministic); if it is scattered all over the phase space, then the dynamics are treated as "complex" and the system high dimensional (perhaps stochastic); if it falls in between, then the dynamics are assumed to be of "intermediate complexity" and the system medium dimensional (perhaps chaotic). If not a convincing "simple" *versus* "complex" system identification, the reconstruction at least shows the extent of variability.

Correlation dimension

Correlation dimension is a measure of the extent to which the presence of a data point affects the position of the other points lying on the attractor in phase space. Among the algorithms available for estimation of the correlation dimension of a time series, the Grassberger-Procaccia algorithm (Grassberger & Procaccia, 1983) has been widely used. The algorithm uses phase space reconstruction. For an m-dimensional phase space, the correlation function $C(r)$ is given by:

$$C(r) = \lim_{N\to\infty} \frac{2}{N(N-1)} \sum_{\substack{i,j \\ (1\le i<j\le N)}} H\left(r - \left|\boldsymbol{Y}_i - \boldsymbol{Y}_j\right|\right) \tag{3}$$

where H is the Heaviside step function, with $H(u) = 1$ for $u > 0$, and $H(u) = 0$ for $u \le 0$, where $u = r - \|\boldsymbol{Y}_i - \boldsymbol{Y}_j\|$, r is the vector norm (radius of sphere) centred on $\boldsymbol{Y}_i$ or $\boldsymbol{Y}_j$. If the time series is characterized by an attractor, then $C(r)$ and r are related according to:

$$\underset{\substack{r\to 0 \\ N\to\infty}}{C(r)} \approx \alpha r^{\nu} \tag{4}$$

where α is a constant and ν is the correlation exponent or the slope of the log$C(r)$ *vs* log$r$ plot. The slope is generally estimated by a least square fit of a straight line over a linear range of r or through estimation of local slopes between r values. The distinction between low-dimensional (and perhaps deterministic) and high-dimensional (and perhaps stochastic) systems can be made using the ν *versus* m plot. If ν saturates after a certain m and the saturation value (i.e. correlation dimension, d) is low, then the system is generally considered to exhibit low-dimensional dynamics. On the other hand, if ν increases without bound with increase in m, the system under investigation is generally considered to exhibit high-dimensional dynamics.

DATA SETS

Since the purpose is to identify the nature of dynamic properties of hydrological time series, it is necessary, first of all, to check: (1) what exactly the deterministic and the stochastic time series "look" like; and (2) whether the existing linear and nonlinear tools are capable of identifying them. To this end, two time series that look "similar" to each other and also to "typical" hydrological time series but are outcomes of "systems" possessing deterministic and stochastic dynamic characteristics, respectively, are generated and analysed. The information gained from this step is then used to interpret the results obtained from the analysis of real hydrological time series. To this end, four river flow series are considered: daily flows from the Mississippi and Kentucky rivers, and monthly flows from the Salmon and Göta rivers.

The deterministic series is generated using a simple two-dimensional map (Henon, 1976):

$$X_{i+1} = a - X_i^2 + bY_i \qquad\qquad Y_{i+1} = X_i \tag{5}$$

This map yields irregular solutions for many choices of a and b, but when $a = 1.4$ and $b = 0.3$, a typical sequence of X_i will be chaotic. The initial values of X and Y used are 0.13 and 0.50, respectively. For the present analysis, a total of 5000 values are considered. Figure 1(a) shows a sample of the variation of this time series, over a length of 1000 values. The time series looks highly irregular, complex, and random, although it is the outcome of a deterministic system.

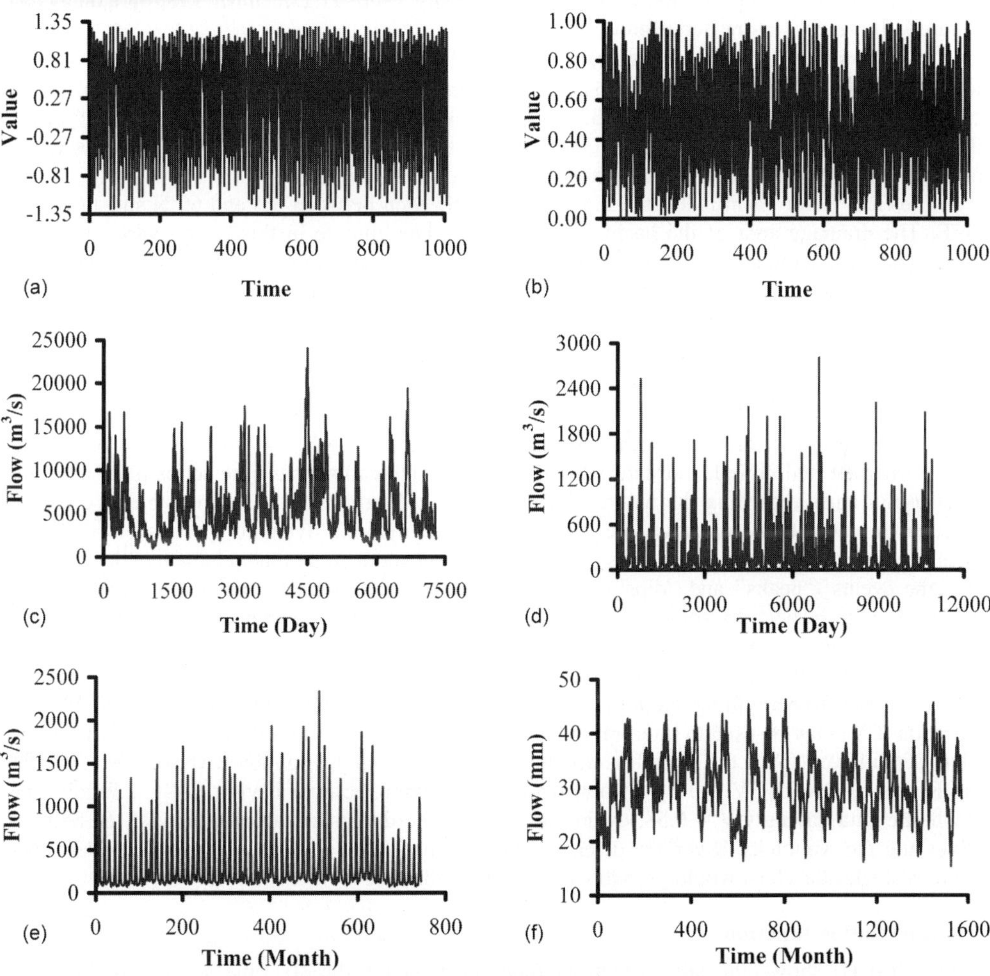

Fig. 1 Time series: (a) synthetic deterministic series; (b) synthetic stochastic series; (c) daily flow series from the Mississippi River; (d) daily flow series from the Kentucky River; (e) monthly flow series from the Salmon River; and (f) monthly flow series from the Göta River.

The synthetic stochastic time series is generated using the random number function:

$$X_i = \text{rand}(\,) \qquad (6)$$

which yields independent and identically distributed (IID) numbers. A total of 5000 values are considered for the present analysis. Figure 1(b) shows a sample of the variation of this time series. As expected, the time series looks highly irregular, complex, and random.

Throughout the Mississippi River, flow data are measured at numerous locations. For the present study, data observed in a sub-basin station at St Louis, Missouri (US Geological Survey station no. 07010000) are considered. The sub-basin is situated at 38°37′03″ latitude and 90°10′58″ longitude, and the drainage area of this sub-basin that falls within the Mississippi River basin is 251 230 km^2. The data used in this study span a period of 20 years (1 January 1961–31 December 1980). The time series plot of this data set is shown in Fig. 1(c).

The Kentucky River is a tributary of the Ohio River, and has a drainage area of about 18 000 km^2. For the present study, daily flow data observed at the gauging station near Winchester, Kentucky (USGS station no. 03284000) are considered. This station is situated at 37°53′41″ latitude and 84°15′44″ longitude. The sub-basin has a drainage area of 10 244 km^2. Flow data observed over a period of 30 years (1 January 1960–31 December 1989) are analysed. Figure 1(d) presents a time series plot of this data set.

The Salmon River basin is situated in the state of Idaho, at 44°59′13″ latitude and 115°43′30″ longitude. The drainage area of this basin is 35 094 km^2. For the present study, monthly flow data collected over a period of 62 years (1932–1993) are considered. Consistent with the "water years", the records used in this study start in October 1931 and end in September 1993 and are average monthly flow values. Figure 1(e) shows the variation of this flow series.

The Göta River basin is located in the south of Sweden between 55° and 60ºN and 12.9° and 16ºE. The drainage area of the basin is 50 132 km^2. The climate in this region varies between boreal and more temperate, without frequently recurring permanent snow cover during winter. For the present study, monthly flow data observed over a period of 131 years (January 1807–December 1937) are considered. The variation of this flow series is shown in Fig. 1(f).

ANALYSIS AND RESULTS

The time series plots for the deterministic and stochastic data look very similar, both exhibiting highly irregular and complex structures. This is a clear indication that time series plots may not offer useful clues regarding the nature of the dynamics. Therefore, it is not possible to construe whether the river flow series are the outcomes of deterministic dynamic systems or stochastic ones. However, these time series plots reveal information about some other characteristics, such as extreme events ("peaks" and "dips") and/or annual cycles. A similar explanation may also be provided with respect to some statistics of the data sets. For example, the mean, standard deviation, maximum, minimum, and number of zeros provide some general idea, but they are in no way reflective of the system dynamic changes. Looking at the statistics alone (Table 1), it is not possible to construe anything about the nature of the system dynamics, although the coefficient of variation (CV) may be used to interpret that the flow data from the Göta River (CV = 0.20) and the Mississippi River (CV = 0.63) are less irregular than those from the Salmon River (CV = 1.14) and the Kentucky River (CV = 1.58). Now, with the deterministic data having CV = 2.78 and the stochastic data having CV = 0.58, could one say that the dynamic nature of flow at the Salmon River and the Kentucky River are deterministic and that at the Göta River and the Mississippi River is stochastic? That would probably be stretching the case a bit too far.

Autocorrelation function

Figure 2(a)–(f) shows the autocorrelation function plots for the six time series. For both the synthetic deterministic and stochastic series, the autocorrelation function fluctuates about zero. This seems only to indicate that the dynamics of the system underlying these two time series are

Table 1 Basic statistics of six different data sets.

Statistic	Deterministic	Stochastic	Mississippi	Kentucky	Salmon	Göta
Basin area*			251 230	10 244	35 094	50 132
Data scale			Day	Day	Month	Month
No. of data	5000	5000	7305	10958	744	1572
Mean†	0.25880	0.49679	5309.97	151.85	317.7	30.53
Std Dev. †	0.71978	0.28949	3333.14	239.96	363.5	6.15
CV	2.78	0.58	0.63	1.58	1.14	0.20
Maximum†	1.27287	0.99989	24100	2806	2338.9	46.30
Minimum†	−1.28450	0.00016	980	3.29	70.4	15.40
No. of zeros	0	0	0	0	0	0

* Units are in km^2

† Units for river flow data are: m^3/s for Mississippi, Kentucky, Salmon; mm for Göta.

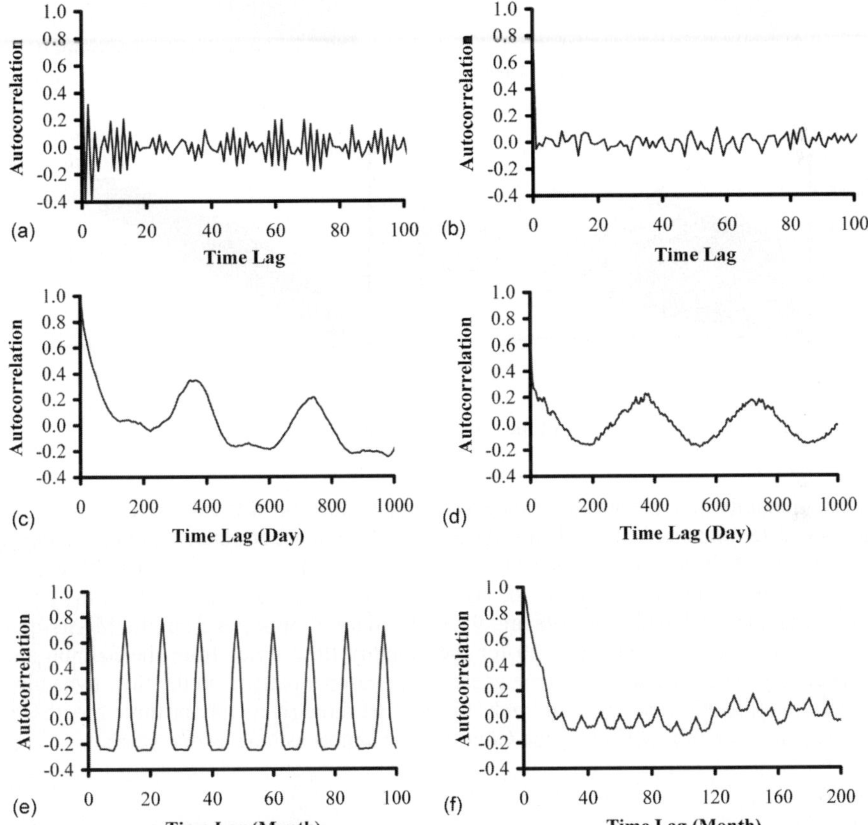

Fig. 2 Autocorrelation function for the six time series shown in Fig. 1.

stochastic in nature. This failure of the ACF to distinguish between the two time series is not just "visual" and "qualitative," but also "quantitative". For instance, for both these series, the time lag at which the ACF first crosses the zero line is 1. These observations suggest that the ACF is not sufficient for identification of the nature of the system dynamic properties.

The ACFs for the river flow series (Fig. 2(c)–(f)) generally indicate slow decays with increasing lag (i.e. temporal persistence), suggesting that the dynamic properties of the underlying

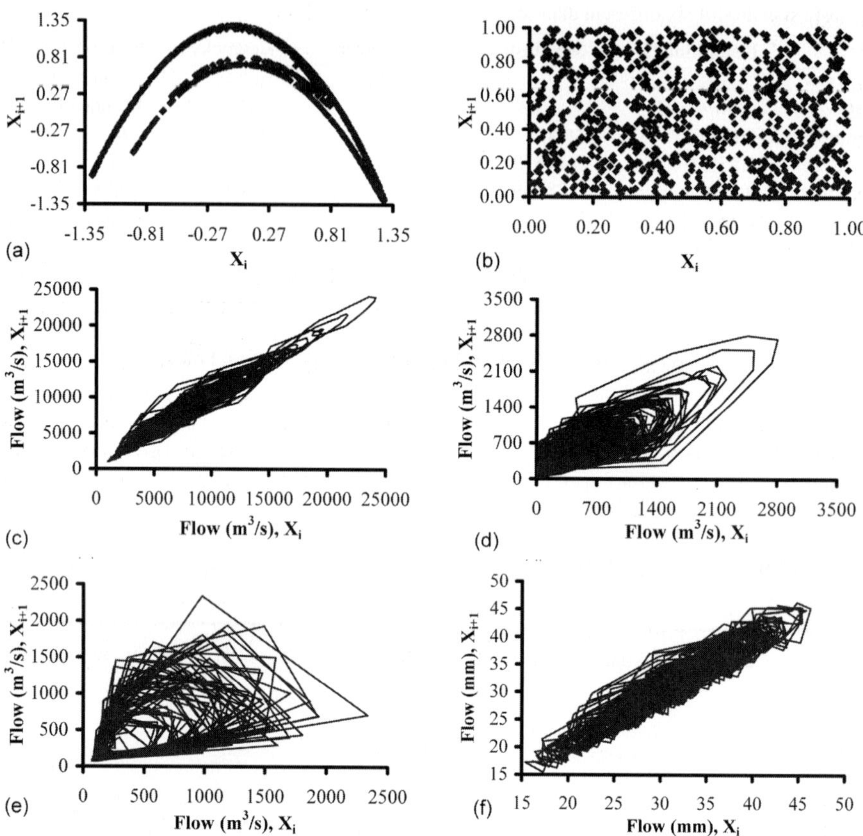

Fig. 3 Phase space diagram for the six time series shown in Fig. 1.

systems are certainly not stochastic. However, it may also be difficult to construe that the dynamics are deterministic. Although the ACFs reveal certain periodicity and/or annual cycle (especially for the Salmon River), such do not provide any clues as to whether the underlying dynamic properties are deterministic or stochastic in nature. The lag times at which the ACF first crosses the zero line are found to be 198 and 95 for the daily flow series from the Mississippi and the Kentucky, respectively, and 3 and 20 for the monthly flow series from the Salmon and the Göta, respectively. These values seem to suggest some seasonal patterns in the flow dynamics, but even such an interpretation may be valid only for the first three rivers (six or three months, as the case may be), since the time period for the Göta River is almost as long as two years.

Phase space

Figure 3(a)–(f) presents the two-dimensional phase space plots for the six time series. For the synthetic deterministic series, the projection yields a very clear attractor (in a well-defined region), indicating a "simple" and "deterministic" nature of the underlying system dynamics. For the synthetic stochastic series, the points are scattered all over the phase space (i.e. absence of an attractor), a clear indication of a "complex" and "random" nature of the dynamic properties of the underlying system. These observations clearly demonstrate the usefulness and ability of phase space reconstruction (and other nonlinear tools) to identify the nature of the underlying dynamics, and more specifically to distinguish between deterministic and stochastic series.

With this, the following observations may be made based on the phase space plots for the river flow series shown in Fig. 3(c)–(f). For the Mississippi River flow series, the phase space diagram exhibits a clear attractor in a well-defined region, suggesting that the system dynamic properties are simple and certainly not stochastic. The phase space diagram for the Kentucky River flow series also shows a reasonably clear attractor, though not as clear as that for the Mississippi River flow series; this seems to suggest that the underlying dynamics may be simple and are certainly not stochastic. The flow series from the Salmon River also seems to exhibit a reasonably clear attractor, although the region of attraction is much larger for the dynamics to be simple; however, it is also difficult to say that the dynamics are stochastic. The phase space diagram for the Göta River flow series shows a clear attractor in a well-defined region, suggesting that the underlying dynamics are simple and certainly not stochastic.

The phase space diagrams for the four river flow series generally indicate the absence of stochastic nature in the underlying system dynamics. They also indicate that the dynamics are "simple" (or less complex) in all four cases, although the "extent of simplicity" certainly varies, with the flow dynamics in the Mississippi River and in the Göta River being simpler than the other two, and with the Salmon River flow dynamics being the most complex of all. These results are indeed encouraging, since they seem to suggest that simpler models may be sufficient, and may be preferred to complex models. At the same time, however, they also do not offer any convincing information to construe that the flow dynamics are deterministic; after all, "simple" does not (necessarily) mean "deterministic."

Correlation dimension

The correlation functions and the exponents are computed. Figure 4(a)–(f) shows the relationship between the correlation exponent and embedding dimension for the six series (embedding dimensions up to 10 or 15 or 20 are chosen for phase space reconstruction, as found appropriate

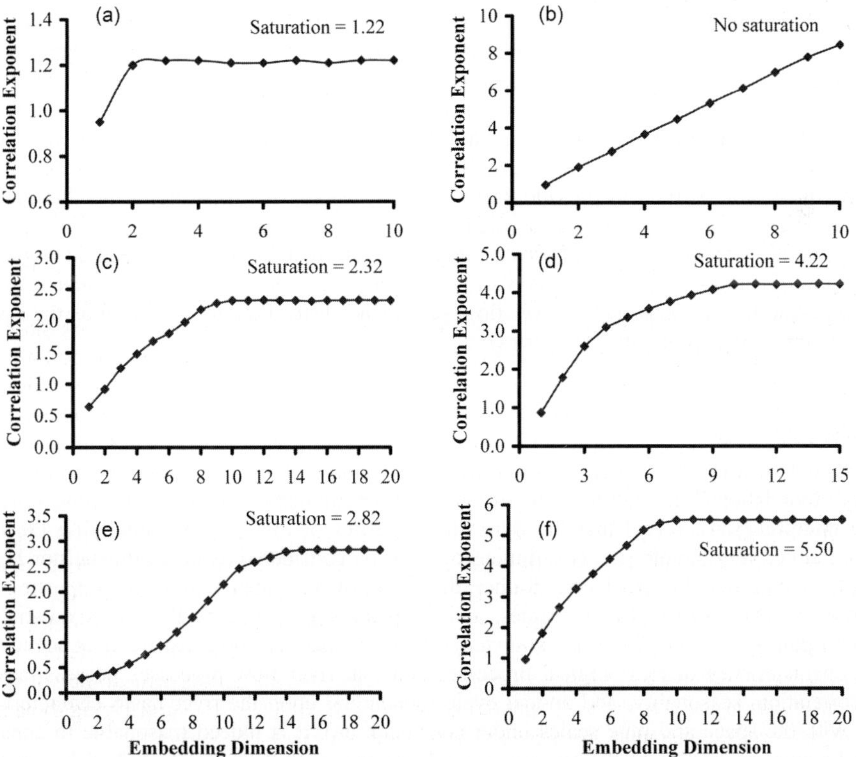

Fig. 4 Correlation dimension results for the six time series shown in Fig. 1.

for the different cases). For the deterministic series, saturation in the correlation exponent is observed when the embedding dimension is increased, suggesting the low-dimensional, and thus simple, nature of the underlying system dynamics. For the stochastic series, no saturation in correlation exponent is observed, suggesting the large- (or infinite-) dimensional, and thus highly complex, nature of the system dynamics. Also, for the deterministic series, the correlation dimension value is found to be 1.22, which means there are two variables dominantly governing the system dynamics. This number is consistent with the number of variables involved in the deterministic system (equation (5)), another indication that the correlation dimension is a reliable tool for identification of "system complexity."

Figure 4(c)–(f) shows that saturation in correlation exponent is observed for all the four flow series, suggesting that the dynamics are certainly not stochastic. Further, the correlation dimension value is found to be small (less than 6) for all the series, suggesting the low-dimensional, and simple to medium-complexity, nature of the underlying dynamics. Looking closely at the correlation dimension values, the Mississippi River flow dynamics (d = 2.32) and the Salmon River flow dynamics (d = 2.82) seem simpler than the others, each dominantly governed by just three variables. The Göta River flow dynamics, on the other hand, seem to have the highest level of complexity (d = 5.50), with the number of dominant governing variables being six. The flow dynamics in the Kentucky River shows a level of complexity somewhere between these two cases (with d = 4.22), with an indication that there are five dominantly governing variables. For any of these four series, while the observation that only a small number of variables dominantly govern the flow dynamics is encouraging (especially from the viewpoint of model complexity), there is no definitive evidence to say that the dynamics are indeed deterministic, since "small number of variables" does not automatically mean deterministic.

Although the phase space reconstruction and correlation dimension results are generally consistent with each other (i.e. absence of stochastic dynamics, simple- to medium-complexity dynamics, no definitive evidence for deterministic dynamics), there are some inconsistencies as well. For example, the phase space reconstruction suggests that the Salmon River flow series is the most complex among the four, but the correlation dimension method indicates that the Göta River flow series has the highest level of complexity. The observations of very low correlation dimension for the Salmon River flow series and the very clear attractor for the Göta River flow series only raise further questions on the ability of the phase space reconstruction and/or the correlation dimension method to provide definitive conclusions. All these point out that caution needs to be exercised in employing these methods and interpreting the outcomes. When such is done, these nonlinear tools can indeed provide important information on the nature and complexity of the system dynamics, as presented above for the synthetic series and also as indicated by the consistent results for the Mississippi River flow series (the clearest attractor as well as the lowest correlation dimension, among the four series).

CONCLUSION

The results obtained for the four river flow time series, using both linear and nonlinear tools, with necessary "foundations" through analysis of synthetic deterministic and stochastic time series to facilitate interpretations, reveal that the dynamic properties of the systems underlying the flow series are neither very simple (i.e. two-dimensional) to be considered as deterministic nor highly complex to be considered as stochastic. Rather, the nature of the systems' dynamic properties falls somewhere in between these two extremes, and dominantly governed by three to six variables, depending upon the system. With our knowledge that nonlinearity and sensitive dependence are inherent characteristics of hydrological processes and that river flow processes possess, among others, correlation, seasonality, and annual cycle (depending upon the river basin characteristics together with the space and time scales under consideration), it is indeed reasonable to construe that the dynamic properties of the systems underlying the river flow series studied herein (and many others) are clearly a combination of deterministic and stochastic.

In view of these observations, the general question of whether the deterministic approach or the stochastic approach is better for hydrological modelling is meaningless. Consequently, any theory that is based purely either on determinism or on stochasticity is probably a misconception of the workings of hydrological processes, and only their combination would be appropriate. Yevjevich (1968) reflected this situation, albeit in his pursuit of the stochastic approach, when he said: "...The misconception that hydrological processes are composed of a limited number of hidden periodicities, and especially the unsuccessful attempts to relate the individual cycles to causal factors, have retarded the use of modern stochastic methods in the analysis of hydrological time sequences, and have tended to perpetuate the deterministic approach to the analysis of hydrological phenomena, instead of combining the deterministic and the stochastic approach to explain the structure of the main hydrological processes."

This is where chaos theory could play a vital role, with its fundamental principles (nonlinear interdependence, hidden determinism and order, and sensitivity to initial conditions) being clearly relevant for hydrological systems and processes and also providing a balancing middle-ground approach between the deterministic and the stochastic views (see also Sivakumar, 2004). Another important thing to note, especially from the viewpoint of model complexity, is that the question is not whether hydrological systems exhibit deterministic dynamics or stochastic dynamics (since they are a combination) but whether a low-dimensional model is sufficient or a high-dimensional model is required. As for the four river flow time series studied herein, the correlation dimension analysis reveals that significant portions of the dynamic complexities of the river systems arise as a result of nonlinear interactions among three to six variables (that are most likely interdependent), depending upon the system. This type of information could play a crucial role in the formulation of a coupled deterministic-stochastic approach.

Acknowledgements This study was supported by the 2007 SOC Project (07-GIBANGUCHUK-D03-01) through the DCRP-AWDP in KICTTEP of MOCT. Bellie Sivakumar acknowledges the support from the Korea Science and Technology Societies and Inha University.

REFERENCES

Gelhar, L. W. (1993) *Stochastic Subsurface Hydrology*. Prentice-Hall, Englewood Cliffs, New Jersey, USA.

Grassberger, P. & Procaccia, I. (1983) Measuring the strangeness of strange attractors. *Physica D* **9**, 189–208.

Henon, M. (1976) A two-dimensional mapping with a strange attractor. *Commun. Math. Phys.* **50**, 69–77.

Lorenz, E. N. (1963) Deterministic nonperiodic flow. *J. Atmos. Sci.* **20**, 130–141.

Sivakumar, B. (2000) Chaos theory in hydrology: important issues and interpretations. *J. Hydrol.* **227**(1/4), 1–20.

Sivakumar, B. (2004) Chaos theory in geophysics: past, present and future. *Chaos, Solitons and Fractals* **19**(2), 441–462.

Sivakumar, B. (2008) Dominant processes concept, model simplification and classification framework in catchment hydrology. *Stoch. Environ. Res. Risk Assess.* **22**(6), 737–748.

Sivakumar, B., Jayawardena, A. W. & Li, W. K. (2007) Hydrologic complexity and classification: a simple data reconstruction approach. *Hydrol. Processes* **21**(20), 2713–2728.

Takens, F. (1981) Detecting strange attractors in turbulence. In: *Dynamical Systems and Turbulence* (ed. by D. A. Rand & L. S. Young), 366–381. Springer-Verlag, Berlin, Germany.

Yevjevich, V. M. (1968) Misconceptions in hydrology and their consequences. *Water Resour. Res.* **4**(2), 225–232.

Proposal of Tank Moisture Index to predict floods and droughts in Peixe River watershed, Brazil

ELFRIDE ANRAIN LINDNER[1] **& MASATO KOBIYAMA**[2]

1 *Department of Civil Engineering, Santa Catarina Western University, Rua Getúlio Vargas 2125, Joaçaba-SC, CEP 89600-000, Brazil*
elfride@cardinal.com.br

2 *Department of Sanitary and Environmental Engineering, Federal University of Santa Catarina, PO Box 476, Florianopólis-SC, CEP 88040-900, Brazil*

Abstract Peixe River watershed (5238 km^2), in southern Brazil, has suffered from natural disasters caused by excess and shortage of rainfall. The mean values (1977–2004) of precipitation (P), potential evapotranspiration (ETP), real evapotranspiration (ETR) and discharge (Q) are 4.95, 2.95, 2.73 and 2.22 mm day^{-1}, respectively. The Tank Model with 4 vertical tanks and 12 parameters was validated for the whole period (Nash, 85% and Nash log, 85%). An index derived from the Tank Model, called the Tank Moisture Index (TMI), is created on a daily basis. It explores the water storage level in tanks 1–4, considering mean and median (md) values. TMI, range 0 to 10, was applied to analyse natural disasters events. The use of TMI_{md} gave a better adjustment than TMI_{mean}: 84% for floods in 161 decrees of water excess; 90% for droughts in 129 decrees of water shortage. TMI_{md} and Tank discharge had 97% correlation with the segmented regression.

Key words Tank Model; floods; droughts; Tank Moisture Index

INTRODUCTION

The Peixe River watershed, southern Brazil, has frequently suffered from hydrological extreme events for a long time. To obtain an adequate watershed management for its sustainable development, the Peixe River Committee requires hydrological models. The Tank Model was applied, with daily monitoring data from 1977 to 2004. A survey of public calamity states associated with the extreme hydrological events was accomplished for the corresponding period. The objective of the present work was to propose the Tank Moisture Index (TMI), on a daily basis, derived from the Tank Model and validated for the studied watershed, which can predict the occurrence of extreme hydrological events (flood and drought).

STUDY AREA AND UTILIZED DATA

Peixe River watershed

The Peixe River watershed (5238 km^2) is a right margin tributary of Uruguay River (ANA, 2006), and is located between the parallels S 26°36′24″ and 27°29′19″ and meridians W 50°48′04″ and 51°53′57″ (Fig. 1). Its main river extension is 307 km. The watershed has the maximum altitude of 1350 m, minimum of 387 m, mean altitude of 876 m and median of 900 m. The basalt rocks are present and the following kinds of soil are found predominantly: Nitosoil (51.5%); Cambisoil (22.3%) and Neosoil (22.4%) (Lindner *et al.*, 2007).

Hydrological data

The hydrological data were processed for the period of 1977–2004: discharge (Q) from one gauging station; rainfall (P) from 19 rainfall stations; and meteorological parameters (temperature, insulation, relative humidity and wind speed) from 4 meteorological stations. In order to calculate the mean rainfall value for the whole watershed, the Thiessen method was utilized.

Lindner *et al.* (2006) estimated the daily potential evapotranspiration (ETP) with the Penman modified method (Doorenbos & Pruit, 1977). Applying the simplified water balance to find the real annual evapotranspiration (ETR), considering that $ETR = P - Q$. The daily ETR values were calculated as $ETP \times kc$. For the whole period (1977–2004) the coefficient (kc) between ETR and

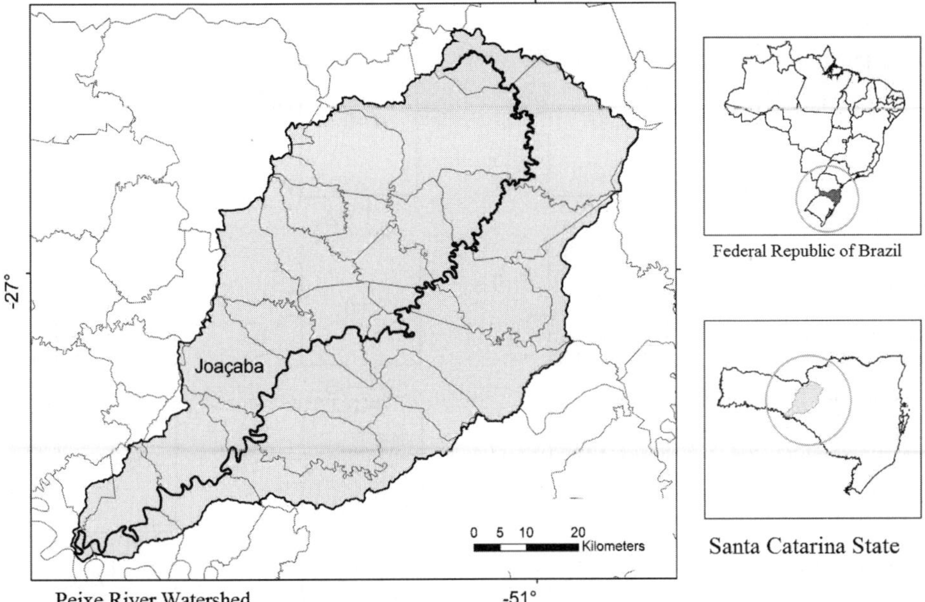

Fig. 1 Location of the Rio do Peixe watershed, Santa Catarina State, Brazil.

ETP, was 0.93. On a daily basis Lindner (2007) showed that the mean values of *P*, *ETP* and *ETR* for the Rio do Peixe watershed are 4.95 mm day^{-1}, 2.95 mm day^{-1} and 2.73 mm day^{-1}, respectively. The aridity index of the watershed by the Thornthwaite and Mather method indicates "little or no water deficiency" and the annual moisture index is 75.90, corresponding to a "humid climate III (B3)".

Natural disasters data

The occurrences of natural disasters have been published in decrees of public calamity state (CP) and emergency situation (SE) signed by mayors and submitted to the National Civil Defense Secretary for recognition (Brazil, 2006). During the period 1977–2004, 330 decrees were established by 25 city halls in Rio do Peixe watershed. Lindner *et al.* (2007) classified them in three categories: excess discharge (161 decrees); drought (129); hail and strong winds (40). The last category would not be considered in the present study. It is observed that the major floods occurred in 1983 (39 decrees), 1990 (28), 1997 (19), and 1992 (18), and that the more severe droughts occurred in 2002 (30), 1991 (27) and 2004 (24).

TANK MODEL APPLICATION

The present study adopted the four tanks in series, based on Sugawara (1995). The Tank Model structure used for the Rio do Peixe Watershed is shown in Fig. 2.

In Excel™, the Tank Model was programmed to generate graphical representations of the hyetograph and hydrographs simultaneously with the change of parameters.

The initial values of parameters suggested by Sugawara (1995) were used for each year of calibration (1977–1990). The automatic calibration proposed by Sugawara (1995) and visual checking were used to optimise the process. For the model performance evaluation multi-objective criteria were applied: coefficient of correlation, coefficient of determination (R^2), and errors indicators such as Relative Error (RE), Volume Standard Error (ΔV), Nash coefficient (NS), Nash Logarithmic coefficient (NSlog), Root Mean Square Error (RMSE), Mean Absolute Error (MAE),

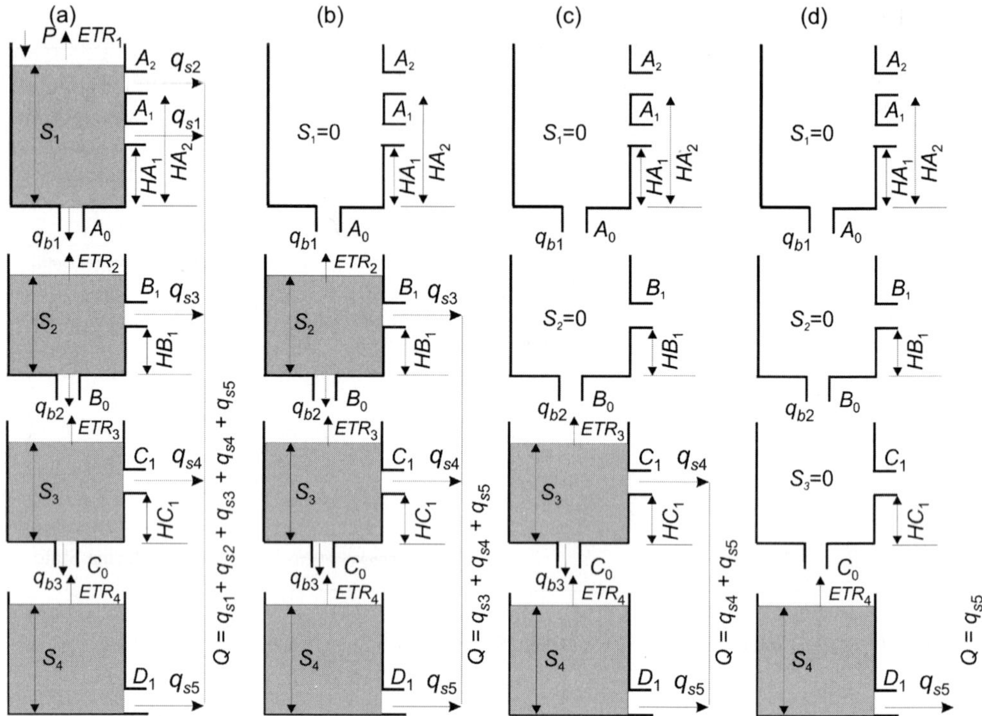

Fig. 2 Change of the storage level in reservoirs 1 to 4 due the precipitation and evapotranspiration, with flow generation: (a) storage in S_1 to S_4; (b) storage in S_2 to S_4; (c) storage in S_3 and S_4; (d) storage in S_4.

Table 1 Tank Model calibrated parameters for Rio do Peixe watershed.

Parameter	Unit	Tank 1	Tank 2	Tank 3	Tank 4
Runoff coefficient	d^{-1}	$A_2 = 0.2800$	$B_1 = 0.0345$	$C_1 = 0.0100$	$D_1 = 0.0010$
	d^{-1}	$A_1 = 0.0579$			
Infiltration coefficient	d^{-1}	$A_0 = 0.0841$	$B_0 = 0.0553$	$C_0 = 0.0081$	
Height of the side outlets	mm	$HA_1 = 13.0$	$HB_1 = 15.0$	$HC_1 = 15.0$	
	mm	$HA_2 = 50.0$			

Logarithmic RMSE (RMSElog) (Setiawan *et al.*, 2003; Fujihara *et al.*, 2004). After calibration, the model validation was carried out for each year from 1991 to 2004. It is noted that the initial storage heights were 0 mm in Tank 1 (S_1), 0 mm in Tank 2 (S_2), 60 mm in Tank 3 (S_3), and 200 mm in Tank 4 (S_4).

The best performance of the Tank Model simulation for the whole period showed the following adjustment: $R^2 = 0.847$, RE = 0.385, $\Delta V = 0.033$, NS = 0.846, Nslog = 0.849, RMSE = 1.363, MAE = 0.656, RMSElog = 0.197. The mean discharge obtained in the simulation was 2.14 mm day^{-1}. The parameters values of the four tanks are shown in Table 1.

TANK MOISTURE INDEX

Observation focused on the hydrograph peaks is needed for the flood studies. The information on the minimum values or base-flow is useful for the water uses planning during the dry periods. As the Tank Model shows, the discharge results from the water quantity stored in the watershed. In other words, the excess and lack in discharge quantity can be associated with the water storage (watershed moisture) condition. Hayes (1999) mentioned that a drought index value is typically a single number, far more useful than raw data for decision making. In this sense, the present study proposes a kind of moisture index derived from the Tank Model which is called Tank Moisture Index (TMI) that can be used for floods and droughts predictions. The TMI considers the daily change in storage values (S) of the Tank Model for runoff generation applied to the Rio do Peixe watershed. TMI aims to represent the hydrological extremes considering the maximum values of storage corresponding to floods and the minimum values to droughts.

Tank Moisture Index (TMI) development

Tank Model with vertical reservoirs represents, schematically, the soil layers from surface to the bottom. Deduced the *ETR*, the exceeded precipitation infiltrates and percolate to reservoirs 1–4.

TMI reaches its maximum when precipitation is occurring, corresponding to water saturation in the superior reservoir (Tank 1), and simultaneously the reservoirs below are filled with water to its maximum capacity (saturation).

The higher the water height, S (mm) stored simultaneously in reservoirs: Tank 1 (S_1); Tank 2 (S_2); Tank 3 (S_3) and Tank 4 (S_4), the higher the index will become. When the reservoirs are losing water through flow or evapotranspiration, without rainfall the index becomes lower. Figure 2 illustrates the TMI concept, considering the multiples combinations of S_1, S_2, S_3 and S_4.

The maximum extreme value of TMI express the situation of maximum S_1 (storage in Tank 1) with the maximum S_4 (storage in Tank 4), combined with the maximum values in reservoir 2 (S_2) and 3 (S_3).

The central tendency indicates the flow situation normality pattern. The mean ($\overline{S_j}$) and the median values (Smd_j) were obtained for the studied period, where the "j" represents the reservoir number, from 1 to 4.

The storage S_1, S_2, S_3 and S_4 (mm) are correlated to one another. The mathematic expression of multiplication was chosen for the present work to relate S_1 and S_4. To moisture indication, from normality to the maximum or to the minimum level, the moisture of day i in Tank 1 is related to the mean (or the median) moisture stored in Tank 4 ($\overline{S_4}$ for the mean or S_{md4} for the median). For Rio do Peixe watershed, during the period 1977–2001, the maximum value corresponds to the day of maximum flow, which occurred on 8 July 1983.

The multiplication result $S_1 \times S_4$ is the most representative in flow generation. To express the transference between the four reservoirs different combinations are used for them. For the daily water balance reservoirs 2 and 3 are considered. The expression $S_1 \times S_4$ corresponds to 78% (mean) and 80% (median) of the total flow, important for high flow (floods). The combination between the storage levels in the others reservoir are important to represent low flow (droughts). Equations (1) and (2) show TMI_1 (step g) by the mean and median based approach, respectively.

$$\mathrm{TMI}_1\ (\mathrm{mean}) = S_{\max 4} \cdot \overline{S_1} + S_{\max 3} \cdot \overline{S_2} + S_{\max 2} \cdot \overline{S_3} + S_{\max 1} \cdot \overline{S_4} \tag{1}$$

where: TMI_1 (mean) is the Tank moisture index (step g), by mean approach; S_{max4}, S_{max3}, S_{max2}, S_{max1} are the maximum storages in reservoirs 4 to 1, respectively; $\overline{S_1}$, $\overline{S_2}$, $\overline{S_3}$ and $\overline{S_4}$ are the mean values of storage in the reservoirs 1 to 4.

$$\mathrm{TMI}_1(\mathrm{median}) = S_{\max 4} \cdot S_{\mathrm{md}1} + S_{\max 3} \cdot S_{\mathrm{md}2} + S_{\max 2} \cdot S_{\mathrm{md}3} + S_{\max 1} \cdot S_{\mathrm{md}4} \tag{2}$$

where: TMI_1 (median) is the Tank Moisture Index (step g), by median approach; S_{md4}, S_{md2}, S_{md3} and S_{md4} are median values of storage in the reservoirs 1 to 4.

The results of equations (1) and (2) are expressed in mm^2. To be a friendly number, the TMI is expressed from zero to 10 and becomes dimensionless with the use of a scale factor. The scale factor initially proposed as F_1 is made equal to the TMI_1 value for the maximum event:

$$\mathrm{TMI}_1 = F_1 = (S_{\max 4} \cdot \overline{S_1} + S_{\max 3} \cdot \overline{S_2} + S_{\max 2} \cdot \overline{S_3} + S_{\max 1} \cdot \overline{S_4})\ \mathrm{mm}^2 \tag{3}$$

where: F_1 is the scale factor (step h).

Dividing TMI_1 by F_1 to get TMI_2, with the value of one unit (TMI, step i):

$$\mathrm{TMI}_2 = \frac{\mathrm{TMI}_1}{F_1} = 1 \quad (4)$$

where: TMI_2 is the Tank moisture index (step i).

Aiming that the result of TMI for the event of maximum is not equal to one unit but equal to 10, the value of TMI_2 is multiplied by 10, getting TMI_3 (TMI, step j). This transformation is transferred to the scale factor, by multiplying 0.1 shown below:

$$\mathrm{TMI}_3 = \frac{10 \cdot \mathrm{TMI}_1}{F_1} \text{ then } \mathrm{TMI}_3 = \frac{\mathrm{TMI}_1}{F_1 \cdot 0.1} \text{ then } \mathrm{TMI}_3 = 10 \quad (5)$$

where TMI_3 is the Tank moisture index (step j).

For the purpose of using TMI in events greater than that registered on 8 July 1983, 90% of the maximum value (TMI, step xiv) is used. Aiming to a good memorization and simplicity, 0.01 is added to the value of 0.1, resulting in 0.11. For the Rio do Peixe watershed data the maximum value of TMI is 9.09, shown in equation (6). A higher level can be expressed in the case of a future catastrophic event:

$$\mathrm{TMI}_4 = \frac{\mathrm{TMI}_1}{F_1 \cdot (0.1 + 0.01)} \text{ then } \mathrm{TMI}_4 = \frac{\mathrm{TMI}_1}{0.11 \cdot F_1} \quad (6)$$

where TMI_4 is the Tank Moisture Index (step k).

The scale factor (F), that is automatically changed by spreadsheet-application used for TMI calculation, is represented by mean and median based approach:

$$F = 0.11\, F_1 \quad (7)$$

where: F is the scale factor (mm^2).

The scale factor (F), by mean and median based approach is shown in equations (8) and (9), respectively:

$$F(\text{mean}) = \max(S_{1_i} \cdot \overline{S_4} + S_{2_i} \cdot \overline{S_3} + S_{3_i} \cdot \overline{S_2} + S_{4_i} \cdot \overline{S_1}) \cdot 0.11 \quad \left| 1 \le i \le n \right. \quad (8)$$

$$F(\text{median}) = \max(S_{1_i} \cdot Smd_4 + S_{2_i} \cdot Smd_3 + S_{3_i} \cdot Smd_2 + S_{4_i} \cdot Smd_1) \cdot 0.11 \quad \left| 1 \le i \le n \right. \quad (9)$$

where: i is the number of the considered day; and n is the number of days of the historical series, such that i is greater or equal to 1 and lower or equal to n.

With the use of mean, for the analysed day, the Tank Moisture Index – $TMI(mean)_i$ is obtained:

$$\mathrm{TMI(mean)}_i = \frac{S_{1_i} \cdot \overline{S_4} + S_{2_i} \cdot \overline{S_3} + S_{3_i} \cdot \overline{S_2} + S_{4_i} \cdot \overline{S_1}}{F(\text{mean})} \quad (10)$$

where $TMI(mean)_i$ is the Tank Moisture Index for the central tendency "mean"; $S_{1i}, S_{2i}, S_{3i}, S_{4i}$ are the water stored in the reservoirs, respectively, Tank 1 to 4; $\overline{S_4}$, $\overline{S_3}$, $\overline{S_2}$, $\overline{S_1}$ are the mean values of storage in the reservoirs, Tanks 4 to 1; F (mean) is the scale factor (mm^2) by mean approach; and i is the variable representing the day of the temporal series.

In the same way, considering the central tendency of median, Tank Moisture Index – $TMI(median)_i$ is obtained:

$$\mathrm{TMI(median)}_i = \frac{S_{1_i} \cdot Smd_4 + S_{2_i} \cdot Smd_3 + S_{3_i} \cdot Smd_2 + S_{4_i} \cdot Smd_1}{F(\text{median})} \quad (11)$$

where $TMI(median)_i$ is the Tank Moisture Index by median approach; S_{md4}, S_{md3}, S_{md2}, S_{md1} are the median values of storage in reservoirs, Tanks 4 to 1; F(median) is a scale factor (mm^2) using the median consideration.

To give flexibility for the equations application, j is considered as the number of the reservoir and m as the number of reservoirs chosen for the Tank Model. Equation (8) for the maximized scale factor (mm^2) becomes:

$$F = \max \left[\sum_{j=1}^{m} S_{j_i} \cdot \overline{S_{(m-j+1)}} \right]_{i=0}^{i=today} \cdot 0.11 \qquad (12)$$

where F is the maximized value of the heights product of storage in the temporal series: $\sum_{j=1}^{m} S_{j_i} \cdot S_{(m-j+1)}$ (mm^2); i is the tested day for maximization; j corresponds to the number of the considered reservoir (in the present example, m = 4); S_{j_i} is the storage in reservoir j in the day i; $S_{(m-j+1)}$ is the storage in the reservoir at the opposite position, that is $(m-j+1)$; $\overline{S_{(m-j+1)}}$ is the mean value of storage in reservoir at the opposite position.

The Tank Moisture Index (TMI), for any number of reservoirs, by mean based approach, is represented by equation (13) for each day i of the analysed temporal series n:

$$\text{TMI(mean)}_i = \frac{1}{F} \sum_{j=1}^{m} S_{j_i} \cdot \overline{S_{(m-j+1)}} \qquad (13)$$

Equation (9), by median approach for any number of reservoirs, is transformed into equation (14):

$$F = \max \left[\sum_{i=1}^{m} S_{j_i} \cdot S_{\text{md}(m-j+1)} \right]_{i=0}^{i=today} \cdot 0.11 \qquad (14)$$

where $S_{\text{md}(m-j+1)}$ is the median value of storage in the reservoir at the opposite position, that is, $(m-j+1)$.

Finally, the Tank Moisture Index (TMI), for any number of reservoirs, by median approach, is represented by equation (15) for each day i of the analysed temporal series n:

$$\text{TMI(median)}_i = \frac{1}{F} \sum_{j=1}^{m} S_{j_i} \cdot S_{\text{md}(m-j+1)} \qquad (15)$$

Tank Moisture Index (TMI) validation procedures

TMI on-site application TMI was applied to the four incremental basins of Rio do Peixe watershed (Lindner, 2007). In the present work the results are presented for the watershed (Pe_4). Using the entire series of daily data, the mean (average), median, maximum and minimum values of the water storage were calculated (Table 2).

Table 2 Mean, median, maximum and minimum values of water storage (S_1, S_2, S_3, S_4) in the Tanks 1, 2, 3, and 4 in mm.

Value/Tank	S_1 (mm)	S_2 (mm)	S_3 (mm)	S_4 (mm)
Mean, $\overline{S_i}$	16.64	16.24	51.14	329.77
Median, Smd_i	12.80	15.77	50.11	322.86
Maximum, S_{max} (8 July 1983)	122.88	54.51	122.70	407.06
Minimum, S_{min} (11 February 1979)	0.00	0.00	0.00	106.25

Exemplifying for the highest flow (day i = 8 July 1983), TMI is equal to 9.09, both, by mean and median based approach. Table 2 data are used in equation (11) corresponding to the median approach resulting in equation (16):

$$\text{TMI} = \frac{122.88 \cdot 322.86 + 54.71 \cdot 50.11 + 122.70 \cdot 15.77 + 407.06 \cdot 12.80}{49549.88 \cdot 0.11} \quad \text{therefore TMI} = 9.09 \tag{16}$$

In the same way, exemplifying for low flow (day i = 11 February 1979), TMI becomes 0.31 by the use of mean approach (equation (10)). Table 2 data are used in equation (11) corresponding to the median approach resulting in equation (17):

$$\text{TMI} = \frac{0 \cdot 322.86 + 0 \cdot 50.11 + 0 \cdot 15.77 + 106.25 \cdot 12.80}{49549.88 \cdot 0.11} \quad \text{therefore TMI} = 0.25 \tag{17}$$

TMI classification Through the data observation as river level (cm), observed discharge (mm day^{-1}), natural disaster decrees of floods and droughts (no.) the TMI is classified in five classes. Table 3 shows the TMI classification, intervals, and on-site application on the Rio do Peixe watershed regarding the observed river water level, h (cm) and discharge (mm day^{-1}) period from 1977 to 2004.

TMI on-site validation The TMI values were checked, day by day, with extreme events registered through the natural disaster decrees published in the Rio do Peixe Watershed (1977–2004). Table 4 shows TMI intervals, and on-site application regarding the number of public decrees of a calamity state and emergency situation due to floods and droughts, for both cases, median-based and mean-based approaches.

Table 3 Tank Moisture Index (TMI) classification, TMI intervals, and on-site application on Rio do Peixe watershed regarding to observed river water level and discharge.

TMI classification	TMI interval	Water level, h (cm)	Discharge, Q (mm day^{-1})
Very wet	TMI > 6	$h > 700$	$Q > 20$
Wet	4 < TMI ≤ 6	$300 < h \leq 700$	$4 < Q \leq 20$
Normal	2 < TMI ≤ 4	$100 < h \leq 300$	$1 < Q \leq 4$
Dry	1 < TMI ≤ 2	$40 < h \leq 100$	$0.4 < Q \leq 1$
Very dry	TMI ≤ 1	$h \leq 40$	$Q \leq 0.4$

Table 4 Tank Moisture Index (TMI) intervals and on-site application on Rio do Peixe watershed with regard to the number of decrees of a public calamity state and emergency situation due floods and droughts.

TMI interval	No. decrees (flood)		No. decrees (drought)	
	Median	Mean	Median	Mean
TMI > 6	72	71	0	0
4 < TMI ≤ 6	63	66	1	2
2 < TMI ≤ 4	22	23	12	21
1 < TMI ≤ 2	4	1	41	54
TMI ≤ 1	0	0	75	52

The adjustment for floods (very wet and wet) of the TMI reached 84% and 85% of adjustment for floods and 90% and 82% for drought (dry and very dry), considering the median and mean approaches, respectively. The mean and median efficiency values are similar for floods, while for droughts the adjustment obtained by median based approach was more favorable. On the whole, the median based approach, compared with the mean based approach, gave better adjustment and was adopted for natural disasters analysis.

The variations of TMI obtained with the use of *Smd* (equation (15)), the observed and calculated discharges are presented for flood (Fig. 3) and drought (Fig. 4) events.

In July 1983 (Fig. 3(a)), at the beginning of the month, TMI was near 3 (normal). TMI increased to 5.1 (wet, day 6); 7.9 (very wet, day 7) and 9.1 on day 8, with the discharges of 67.6 mm day^{-1} (observed) and 51.9 mm day^{-1} (calculated). From 6 to 8 July 1983, 11 municipalities declared "Public Calamity State" (PC) and six municipalities, "Emergency Situation" (ES). In May 1992 (Fig. 3(b)) an episode of gradual flood reached its peak at day 29, TMI increase going above 6, with the discharges of 43.9 mm day^{-1} (observed) and 43.2 mm day^{-1} (calculated) and TMI increasing. In May 1992 two PC decrees and seven ES were recognized.

Drought events are shown in Fig. 4, when TMI decreased going below 1 during the drought of January–February 1979.

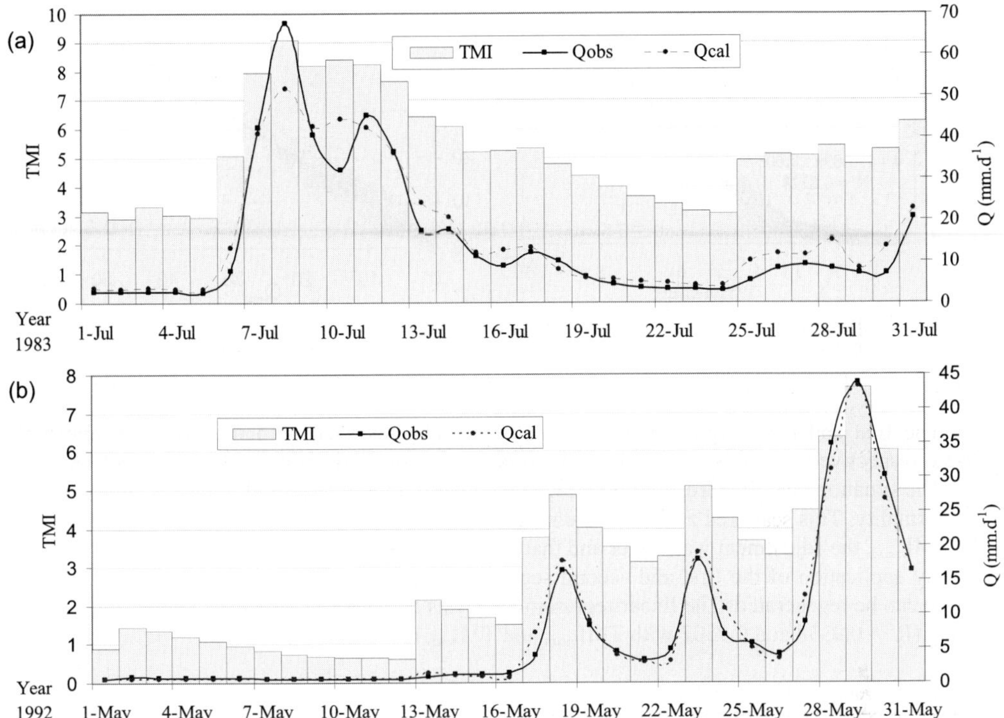

Fig. 3 Variations of TMI, observed (*Qobs*) and calculated (*Qcal*) discharge in Rio do Peixe watershed during flood events: (a) July 1983 (b) May 1992.

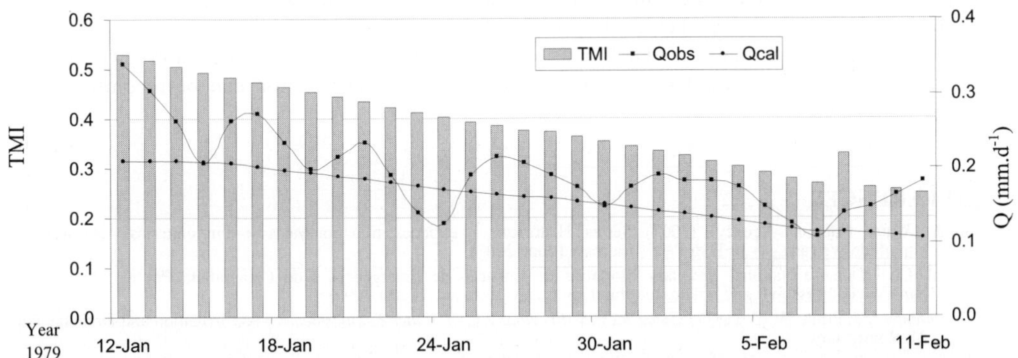

Fig. 4 Variations of TMI, observed (*Qobs*) and calculated (*Qcal*) discharge in Rio do Peixe watershed, drought event of January–February 1979.

TMI and Tank discharge relationship

In general, water level is very stable during the normality and drought periods, while it varies quickly during the flood events. Two broken linear regressions can represent the relationship between daily calculated discharge and the daily TMI_{md} shown in Fig. 4(a).

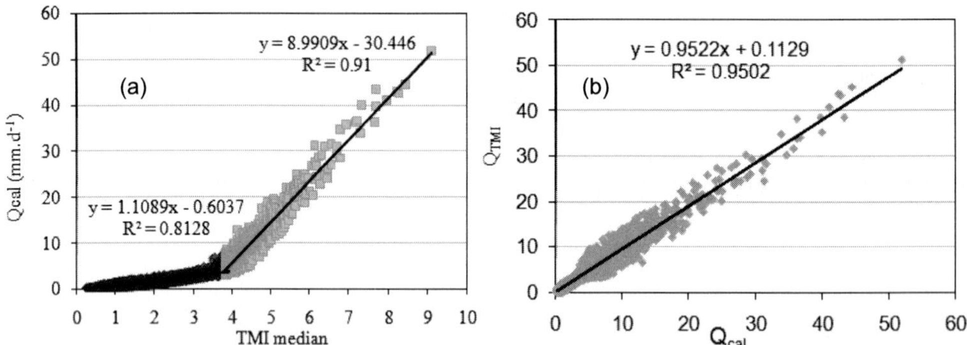

Fig. 4 (a) Segmented linear regression of TMI_{md} and calculated discharge (Q_{cal}); (b) linear regression of calculated discharge and discharged regenerated by TMI for Rio do Peixe watershed.

The first and the second segments are strongly characterized with the low flow and high flows, respectively. A threshold point between two segments is determined to be around TMI_{md} = 3.8. The situation near the threshold point, which contains a lot of scattered points, is considered as the normality. This scattered zone is at the range of 2 to 4. It is noted that in the case of the use of the TMI_{mean}, the adjustment was lower and that the threshold point was also TMI_{mean} = 3.8.

By application of the first and second segmented linear equations in Fig. 4(a), the discharge (Q_{TMI}) can be regenerated. The linear regression analysis between Q_{cal} and Q_{TMI} shows a very good fitting (R^2 = 0.9338 and 0.9502 with TMI_{mean} and TMI_{md}, respectively).

FINAL CONSIDERATIONS

The Tank Model was applied to the Peixe River watershed, southern Brazil, and had a good adjustment. The present study proposed the moisture index derived by Tank Model water storage parameters, and called it Tank Moisture Index (TMI). This index presents daily values with the range 0 to 10. The TMI was validated for both extremes meteorological events (droughts and floods) in the Rio do Peixe watershed for the period of 1977–2004.

It is concluded that the TMI can be a good tool for making decision on watershed management and for natural hazards prevention. The TMI application can be recommended to other watersheds.

REFERENCES

ANA (Water National Agency) (2007) Séries Históricas – estações pluviométricas e fluviométricas. Available at: http://hidroweb.ana.gov.br/HidroWeb/ (accessed June 2007).

Brasil, Civil Defense National Secretary. Portarias de Situação de Emergência e de Calamidade Pública. Available at: http://www.defesacivil.gov.br/situacao/municipios.asp (accessed August 2006).

Doorenbos, J. & Pruitt, W. O. (1977) *Guidelines for Predicting Crop Water Requirements*. FAO Irrigation and Drainage Paper 24, Rome, Italy.

Fujihara, Y., Tamakamaru, H., Hata, T. & Tada, A. (2004) Performance evaluation of rainfall–runoff models using multi-objective optimization approach. Available at:http://www.wrrc.dpri.kyoto-u.ac.jp/~aphw/APHW2004/proceedings/JSC/56-JSC-A603/56-JSC-A603.pdf (accessed in May 2005).

Hayes, M. J. (2002) What is drought? drought indices. NDMC –. National Drought Mitigation Center. Available at: http://drought.unl.edu/whatis/concept.htm (accessed April 2005).

Lindner. E. A., Massignam, A. M., Kobiyama, M. & Zilio, E. (2006) Estimativa da Evapotranspiração Potencial na Bacia Rio do Peixe/SC pelos Métodos de Thornthwaite e Penman Modificado. In: I Simpósio de Recursos Hídricos do Sul-Sudeste, Curitiba. Associação Brasileira de Recursos Hídricos. v. I. p. 125–125.

Lindner, E. A. (2007) Estudo de eventos hidrológicos extremos na Bacia do Rio do Peixe – SC com aplicação de índice de umidade desenvolvido a partir do Tank Model. Florianópolis. Available at: http://biblioteca.universia.net/html_bura/ficha/params/id/31900993.html (accessed January 2009).

Lindner, E. A., Kobiyama, M., Massignam, A. M., Antonello, K. & Canale, D. P. (2007) *Análise dos desastres naturais de excesso e de escassez hídrica decretados na bacia rio do Peixe, SC/Brasil.* In: Jornadas Internacionales sobre Gestión del Riesgo de Inundaciones y Deslizamientos de Laderas. São Carlos/SP.

Nakatsugawa, M. & Hoshi, K. (2004) Long-term runoff calculation considering change of snow pack condition. *J. Hydrosci. Hydraul. Engng.* Available at: http://env-web.ceri.go.jp/houkoku/2004/49.pdf.pdf (accessed May 2005).

Setiawan, B. I., Fukuda, T. & Nakano, Y. (2003) Developing procedures for optimization of Tank Model's parameters. Agricultural Engineering International: the CIGR Journal of Scientific Research and Development. N. LW 01 006. June, 2003. Available at: http://dspace.library.cornell.edu/bitstream/1813/122/42/LW+01+006+Setiawan.pdf (accessed May 2005).

Sugawara, M. (1995) Tank Model. In.: *Computer Models of Watershed Hydrology* (ed. by V. P. Singh), 165–214. Water Resources Publications, Highlands Ranch, Colorado USA p..

Tingsanchali, T. (2001) Application of combined Tank Model and AR Model in flood forecasting. In: *4th DHI Software Conference, Helsingor, Denmark.* Available at: http://www.dhisoftware.com/uc2001/Abstracts_Proceedings/Papers01/057/057.htm (accessed May 2005).

Raingauge siting using the gamma test

MICHAELA BRAY, RENJI REMESAN & DAWEI HAN

Department of Civil Engineering, University of Bristol, Queen's Building, University Walk, Bristol BS8 1TR, UK

michaela.bray@bristol.ac.uk

Abstract The quality of precipitation data can have a significant impact upon rainfall analysis and subsequent hydrological modelling. Many water resource projects are often derived from raingauge networks, but these are only able to measure rainfall at a point location. Although other technologies such as weather radar and weather satellites can tell us about the spatial nature of the rainfall field, many uncertainties remain regarding the reliability of the rainfall estimates they produce under various conditions. Judicious siting of raingauges provides a valuable contribution to spatial rainfall estimations as well as a means of calibration for apparatus which provide rainfall estimations through indirect means. In this paper, the gamma test is investigated as a means for obtaining the most relevant sites for raingauge locations. The gamma test is a nonlinear analysis tool, which can be used for feature selection; it is this characteristic of feature selection which is applied to precipitation data, obtained from a dense raingauge network (49 gauges) across the Brue catchment in southwest England, to single out the most influential raingauge sites. This paper assesses the effectiveness of this use of the gamma test, and considers the impact it has upon improving the quality of rainfall data used in hydrological models.

Key words gamma test; raingauge; feature selection; rainfall analysis

INTRODUCTION

Accurate rainfall data and rainfall forecasting are of great importance and directly affect the performance of hydrological models used for water management and flood forecasting (Larson & Peck, 1974; Peck, 1980), as well as being an important input for climate models. Historically, raingauges have been the key instrument for measuring precipitation and, although current technologies, such as weather radar, weather satellite and numerical weather prediction models, have helped to improve our understanding of precipitation events and provide alternative methods for observing forecasting and measuring rainfall, raingauges are still used as ground truth against which modern equipment is calibrated. However, raingauges measure precipitation at point locations and therefore require some form of areal averaging in order to represent the rainfall occurring at the catchment scale. This, of course, can lead to inaccuracies since typical gauge spacing can range between 10 and 20 km (Jones *et al.*, 2003). Adequate gauge coverage is often lacking; choosing locations for new gauges is difficult but important. This paper explores the use of the gamma test for raingauge siting using the HYREX raingauge network as a test case. In addition, comparisons are made between the gamma test results and areas of similar rainfall using cluster analysis to assess the spatial representation of the gamma test raingauges. The high density raingauge network is only feasible in hydrological experiments since continuing such a network in a catchment is prohibitive in terms of cost and maintenance. In practice, the operational raingauge networks are much less dense and would create uncertainties in the rainfall measured. Nowadays, uncertainty in river flow modelling is very topical (Han *et al.*, 2007) and among the major uncertainty components in the modelling process (data, parameters and system uncertainties); data uncertainty has been an active research in hydrology (e.g. Hamlet & Lettenmaier, 2005).

Data sets

The Brue catchment was chosen as a case study for this research. The choice of catchment was based on the availability of quality data and the representativeness of the catchment, the characteristics of which correspond well with many UK catchments used for rainfall–runoff modelling. Furthermore, during the period starting September 1993 and ending April 2000, a dense raingauge network of 49 raingauges was installed and maintained on the Brue catchment by the National Rivers Authority, UK, as part of the Hydrological Radar Experiment (HYREX). Operationally, most catchments are serviced by only two raingauges, at best. The cost of maintaining a denser network is too prohibitive for it to be feasible on a large scale. Two earlier dense raingauge experimental networks were conducted in Cardington and Winchcombe, UK, between 1957 and 1967 (Holland, 1967; Marshall, 1980). However, these are no longer operational, and the collected data no longer exist (Wheater *et al.*, 2000). Clearly the rich data set provided by the HYREX experiment is rare. With six years of continuous data provided by a 49-dense raingauge network, this catchment is an ideal study area for the analysis of raingauge rainfall estimations.

The Brue catchment (shown in Fig. 1) is located in Somerset, southwest England. It is mainly pasture land with some areas of woodland in the higher eastern half of the domain. It has a drainage area of 135 km^2 and an elevation range between 35 m and 190 m a.s.l. Average annual rainfall for the period 1961–1990 is 867 mm. The raingauges installed on the Brue catchment are typical of those used by the Environmental Agency in the UK; a casella tipping-bucket gauge mounted vertically on a concrete paving slab. The bucket size was 0.2 mm and the gauge aperture was 400 cm^2. The tip time was recorded up to a time resolution of 10 s. The first valid day was considered to be the day after which the gauge records its first value.

Initial quality control of the data was carried out by the former Institute of Hydrology, UK. Cumulative hyetographs using clusters of 10 gauges (all in close proximity) were plotted and were found to be the best of method of identifying anomalies in the gauge behaviour. This was used in combination with the field reports to assess, to the nearest day, when the gauge was last working properly. Met Office Daily Weather summaries were also used to identify unusual weather conditions, since snow and hail would activate the tipping-bucket mechanism in a different way to rain (Wood *et al.*, 2000).

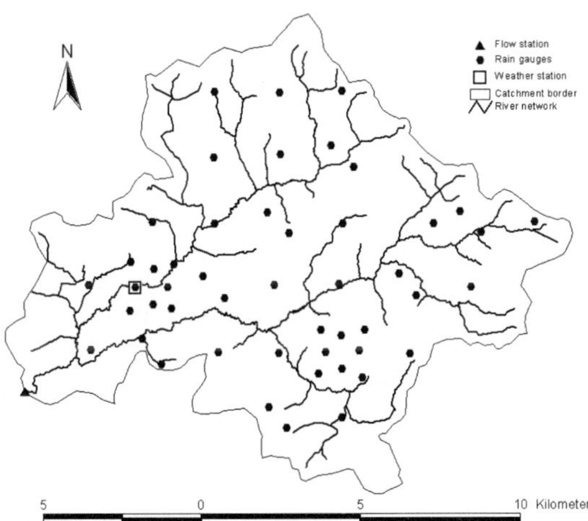

Fig. 1 Brue Catchment, river network and raingauge sites.

Gamma test

Despite an abundance of studies on modelling using nonlinear techniques, such as artificial neural networks (ANN), there are still many questions that need to be answered. For example, the extent to which the inputs determine the outputs and the accuracy output *y*, given an input vector ***x***. So far, these questions have not been addressed adequately by the hydrological community (Han *et al.*, 2007). The advancement of modern computing technology and a new algorithm from the computing science community called the gamma test (Agalbjörn *et al.*, 1997; Končar, 1997), has made it possible to make progress in answering these important questions. The gamma test is used generally for nonlinear modelling, but in our case we are using it for feature selection. The gamma test uses the estimation of variance of noise var(r), computed from the raw data using efficient and scalable algorithms. This novel technique enables us to quickly evaluate and estimate the best mean squared error that can be achieved by a smooth model on unseen data for a given selection of inputs, prior to model construction. This technique can be used to find the best embedding dimensions and time lags for time series analysis. Moreover the gamma test can give an idea of the best combination of inputs for model construction before actual modelling. This is done by using the gamma statistic as a method of selecting the inputs and their relative importance. It is this capability of the gamma test which is used to identify the best input combination of gauges to produce the optimal raingauge network and most effective raingauge position in HYREX network using the catchment average as the target.

Overtraining is considered as one of the serious weaknesses associated with almost all nonlinear modelling techniques, including ANN, which lead to excellent results on the training data but very poor results on the unseen test data. The gamma test is designed to solve this problem efficiently by giving an estimate of how closely any smooth model could fit the unseen data. Thus we can avoid the guesswork associated with the nonlinear curve fitting techniques. A formal proof for the gamma test can be found in Evans (2002) and Evans & Jones (2002).

The gamma test was firstly reported by Konča (1997) and Agalbjörn *et al.* (1997), and later enhanced and discussed in detail by many researchers (Chuzhanova *et al.*, 1998; De Oliveira, 1999; Tsui, 1999; Durrant, 2001; Tsui *et al.*, 2002; Jones *et al.*, 2003). Only a brief introduction on the gamma test is given here and interested readers should consult the aforementioned papers for further details. The basic idea is quite distinct from the earlier attempts with nonlinear analysis.

Suppose we have a set of data observations of the form:

$$\{(\mathbf{x}_i, y_i), 1 \le i \le M\} \tag{1}$$

where the input vectors $\boldsymbol{x}_i \in \Re^m$ are vectors confined to some closed bounded set $C \in \Re^m$ and, without loss of generality, the corresponding outputs $y_i \in \Re$ are scalars. The vectors $\boldsymbol{x}$ contain predicatively useful factors influencing the output y. The only assumption made is that the underlying relationship of the system is of the following form:

$$y = f(\boldsymbol{x}_1, ..., \boldsymbol{x}_m) + r \tag{2}$$

where f is a smooth function and r is a random variable that represents noise. Without loss of generality, it can be assumed that the mean of the distribution of r is zero (since any constant bias can be subsumed into the unknown function f), and that the variance of the noise var(r) is bounded. The domain of a possible model is now restricted to the class of smooth functions which have bounded first partial derivatives. The gamma statistic, Γ, is an estimate of the model's output variance that cannot be accounted for by a smooth data model.

The gamma test is based on $N[i, k]$, which are the kth $(1 \le k \le p)$ nearest neighbours $\boldsymbol{x}_{N[i,k]} (1 \le k \le p)$ for each vector $\boldsymbol{x}_i$ $(1 \le I \le M)$. Specifically, the gamma test is derived from the delta function of the input vectors:

$$\delta_M(k) = \frac{1}{M} \sum_{i=1}^{M} \left| x_{N(i,k)} - \boldsymbol{x}_i \right|^2 \qquad (1 \le k \le p) \tag{3}$$

where $|...|$ denotes Euclidean distance, and the corresponding gamma function of the output values:

$$\gamma_M(k) = \frac{1}{2M} \sum_{i=1}^{M} \left| y_{N(i,k)} - y_i \right|^2 \qquad (1 \le k \le p) \tag{4}$$

where $y_{N[i,k]}$ is the corresponding y value for the kth nearest neighbour of $\boldsymbol{x}_i$ in equation (3). In order to compute Γ a least squares regression line is constructed for the p points $(\delta_M(k), \gamma_M(k))$:

$$\gamma = A\delta + \Gamma \tag{5}$$

The intercept on the vertical axis ($\delta = 0$) is the Γ value, as can be shown:

$$\gamma_M(k) \rightarrow \text{var}(\gamma) \text{ in probability as } \delta_M(k) \rightarrow 0 \quad (6)$$

Calculating the regression line gradient can also provide helpful information on the complexity of the system under investigation. A formal mathematical justification of the method can be found in Evans & Jones (2002).

The graphical output of this regression line (equation (5)) provides very useful information. First, it is remarkable that the vertical intercept Γ of the y (or gamma) axis offers an estimate of the best MSE achievable utilising a modelling technique for unknown smooth functions of continuous variables (Evans & Jones, 2002). Second, the gradient offers an indication of model's complexity (a steeper gradient indicates a model of greater complexity).

The gamma test is a non-parametric method and the results apply regardless of the particular techniques used to subsequently build a model of f. We can standardise the result by considering another term V_{ratio}, which returns a scale invariant noise estimate between zero and one. The V_{ratio} is be defined as:

$$V_{\text{ratio}} = \frac{\Gamma}{\sigma^2(y)} \quad (7)$$

where $\sigma^2(y)$ is the variance of output y, which allows a judgement to be formed, independent of the output range as to how well the output can be modelled by a smooth function. A V_{ratio} close to zero indicates that there is a high degree of predictability of the given output y.

We can also determine the reliability of the Γ statistic by running a series of gamma tests for increasing M, to establish the size of data set required to produce a stable asymptote. This is known as the M test. The M test result would help us to avoid the wasteful attempts of fitting the model beyond the stage where the MSE on the training data is smaller than var(r), which may lead to *overfitting*. The M test also helps us to decide how much data we require to build a model with a mean squared error which approximates the estimated noise variance. In practice, the gamma test can be achieved through the implementation of winGamma™ software (Durrant, 2001). Corcoran *et al.* (2003) applied the gamma test as a method for crime incident forecasting by focusing upon geographical areas of concern that transcend traditional policing boundaries.

Application of the gamma test on HYREX raingauges

The HYREX experiment provides six years of continuous data provided by a 49-dense raingauge network. The date and time of each raingauge tip time was recorded up to a time resolution of 10 s; creating a very large data set. In an effort to make these data manageable, the dense raingauge network data was aggregated into hourly rainfall and monthly rainfall data sets; the gamma test was conducted over a complete year of data for each of the years 1994–1999 for both aggregated data sets. This permitted changes in the most influential sites from year to year to be observed. Moreover, comparisons could also be made between the different temporal resolutions of rainfall times for the same year, since it was not clear at the outset whether or not the preferred raingauge sites were affected by the temporal resolution of rainfall times.

Better model input combination gives the least gamma value. So the removal of the best input data series from the input combination would produce a higher value of the gamma statistic, Γ. In the present study, 49 input combination scenarios were tried; in each case we removed information from a different raingauge, noting the variation in the gamma statistic. The higher gamma static scenario was used to locate the most influencing raingauge point since the "removal" of that data point had produced a higher value of gamma.

Using this method, it was possible to rank each raingauge location by its influence on the target (catchment average). This was done five times in total for each year and each aggregated data set. For the purposes of this study, the five most influential sites were considered to be a reasonable place to stop since, in practice, most catchments would not have more than two raingauges, let alone five. Figure 2 shows the chosen sites for the years 1994 and 1999 using daily raingauge totals. The gamma-test chosen sites varied from year to year for both the daily rainfall totals and the monthly rainfall totals; however, some locations occurred more frequently over the

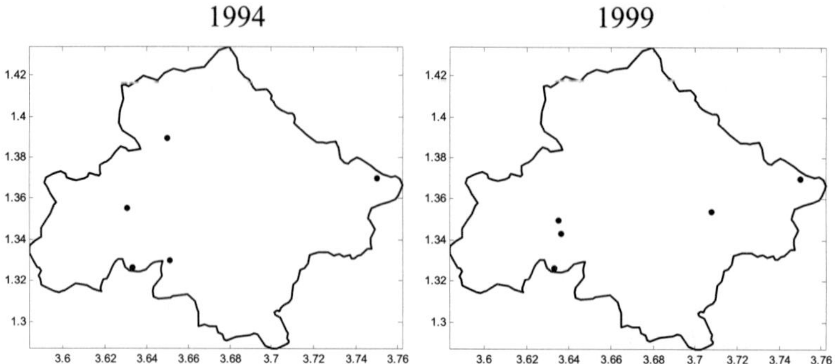

Fig. 2 Gamma test chosen raingauge sites for 1994 and 1999 using daily rainfall totals.

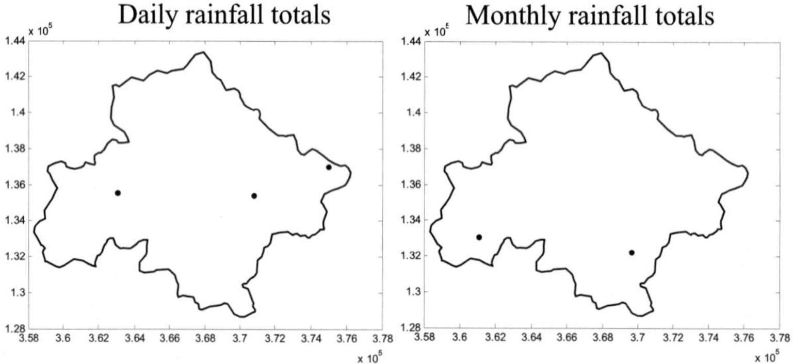

Fig. 3 Raingauge sites which were repeatedly chosen by the gamma test across the years.

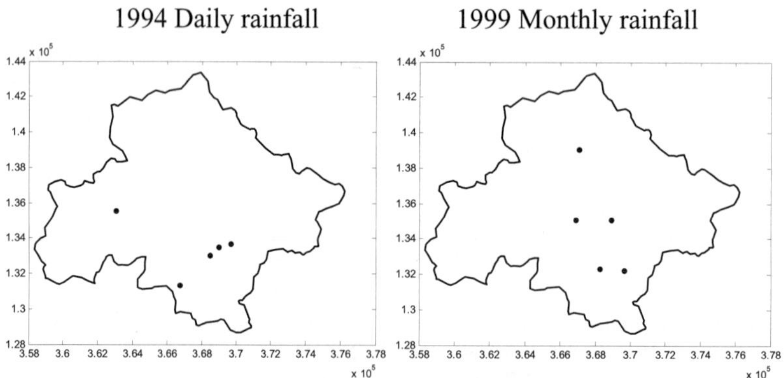

Fig. 4 Gamma test chosen raingauge sites for 1994 and 1999 using monthly rainfall totals.

six years, and could be considered for permanent siting; three such locations were found using daily raingauge totals and two using monthly raingauge totals shown in Fig. 3. It is noteworthy that the suggested permanent raingauge sites differ depending upon whether daily rainfall totals or monthly rainfall totals were used (see Fig. 3). This observation is echoed in comparisons of the same year, but different temporal resolution: generally the sites chosen by the gamma test differ according to whether they are based on the daily or monthly raingauge totals (see Fig. 4).

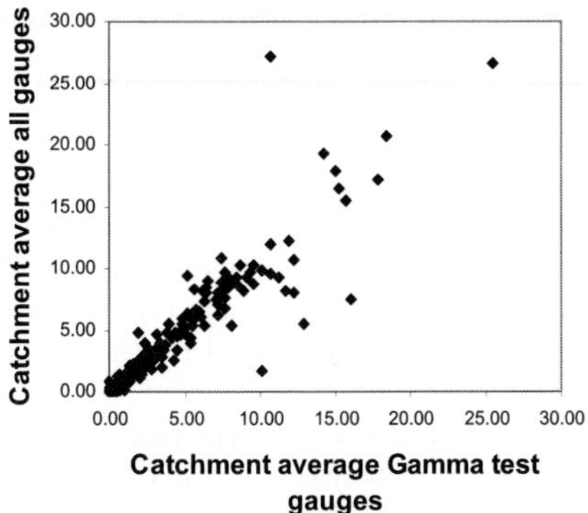

Fig. 5 Comparison of catchment average over 365 days.

Despite the lack of consistency, the chosen raingauge sites performed well within their year and temporal resolution. The catchment average rainfall obtained from the five raingauges compared well to the catchment average produced by the full set of 49 raingauges for all years and both temporal settings. The case of 1994, using daily rainfall totals, is shown in Fig. 5; here we can see that although there is some scatter, the rainfall measured by the chosen sites is positively correlated to that measured by the 49 raingauges for most days. Using the Pearson product moment correlation coefficient (ρ), defined below in equation (8), the correlation between the catchment average produced by the chosen raingauges and the 49 gauges was derived for each of the years; in each case $\rho > 0.85$, and was usually greater than 0.9, confirming strong correlation. In addition, the root mean square error (RMSE), given in equation (9) was produced for each set of chosen raingauges. The RMSE was found to be less than 1 mm using daily rainfall totals, and in the case of monthly raingauge totals the RMSE was less than 5 mm for all years.

$$\rho = \frac{1}{n-1}\sum_{1=1}^{n}\left(\frac{x_i - \bar{x}}{s_x}\right)\left(\frac{y_i - \bar{y}}{s_y}\right) \quad (8)$$

$$\text{RMSE} = \sqrt{\frac{\sum_{i=1}^{n}\left(x_{1,i} - x_{2,i}\right)^2}{n}} \quad (9)$$

The representation of spatial rainfall

Clearly, each set of the five chosen sites could be used to produce as good a catchment average as the complete dense raingauge network; however, it was not known if the chosen raingauges could be used to represent the spatial rainfall over the catchment. In an effort to address this question, the catchment was partitioned into areas of similar rainfall using hierarchical cluster analysis on the full dense raingauge network. The chosen raingauges were then compared to the areas of similar rainfall.

Cluster analysis

Cluster analysis is a popular tool used to group similar data together, allowing trends or patterns in the data to be observed. A formal definition of a cluster, group or class is difficult and is often down to the judgement of the user (Bonner, 1964). Cormack (1971) and Gordon (1980) talk of internal cohesion and external isolation in defining clusters. Although there is no standard definition of a cluster, it is generally felt that it must have something to do with recognition of relative distance between members, and so certain properties are attributed to clusters, such as density, variance, shape and separation. Many applications of cluster analysis can be found in meteorology. In particular, cluster analysis has been put to use for synoptic classification of weather observations. To this end, Kalkstein *et al.* (1987) investigated three common clustering procedures (Ward's method, average linkage and centroid), in order to determine the most meaningful synoptic classifications for use in the development of an objective synoptic methodology. Further uses include the identification of distinct atmospheric flow regimes (Mo & Ghil, 1988; Molteni *et al.*, 1990), as well as the determination of climate regions. Fovell & Fovell (1993) applied hierarchical methods to partition the USA into different climate regions based on temperature and precipitation data. Similar studies have also been carried out by De Gaetano & Shulman (1990) and Gallianin & Filippini (1985).

Clusters are formed by assessing the similarity and dissimilarity of each pair of members. Similarity provides an indication of the strength of association between a pair of objects each having the same p variants. Consider two objects, $\boldsymbol{x}_i$ and $\boldsymbol{x}_j$, with p variables so that $\boldsymbol{x}'_i = [x_{i1}, x_{i2}, x_{i3}, \ldots, x_{ip}]$ and $x'_j = [x_{j1}, x_{j2}, x_{j3}, \ldots, x_{jp}]$, the similarity between them will be some function $s_{il} = f(x_i, x_j)$. For all similarity measures bounded between zero and one, a measure of dissimilarity can be considered as $d_{ij} = 1 - s_{ij}$ (Cormack, 1971; Anderberg, 1973; Sneath & Sokal, 1973; Clifford & Stephenson, 1975; Gower, 1985). An idea of dissimilarity uses the idea of distance or metrics so that membership to a group is excluded if an element does not fall within a given a certain range, i.e. $d_{ij} + d_{ik} \geq d_{jk}$. The most commonly used metric is Euclidean distance (equation (10)); although others such as city block (equation (11)) and Minkowski (equation (12)) are also in common use.

$$d_{ij} = (\sum_{k}^{p} (x_{ij} - x_{jk})^2)^{\frac{1}{2}} \tag{10}$$

$$d_{ij=} \sum_{k}^{p} \left| x_{ik} - x_{jk} \right| \tag{11}$$

$$d_{ij} = (\sum_{k}^{p} (x_{ij} - x_{jk})^q)^{\frac{1}{q}} \tag{12}$$

Hierarchical clustering classifies data by means of a series of partitions which may run from one single cluster of all elements to n clusters each containing one element. Hierarchical classification can be achieved by agglomerative methods, those which fuse n individuals successively into groups, or divisive methods where the n individuals are successively separated into finer groupings.

A number of agglomerative hierarchical clustering methods are in common usage such as single linkage, complete linkage, group average clustering, centroid clustering and Ward's method. Each of these methods differ in the way that similarity or distance between an element and a group of elements are defined and consequently produce different results when using the same data.

Cluster analysis applied to the HYREX raingauges

Hierarchical cluster analysis was used to group the raingauges so that rainfall recorded in one cluster was very similar and the rainfalls recorded in different clusters were quite distinct. Various combinations of linkage and distance were explored to find the best clustering method. Each of these combinations produced a multilevel hierarchical cluster tree from which a level or scale of clustering can be chosen as appropriate since clusters at one level are joined as clusters at the next higher level. A linear correlation coefficient between the original distances (or dissimilarities) used to construct the tree and the cophenetic distances obtained from the tree was used to decide which of the combinations best represented the dissimilarities in the rainfall data. The clustering technique known as Ward's method combined with Euclidean linkage was found to produce the most coherent cluster groups. Ward's method is a popular method which is not based on the distance matrix. The method begins with n single member groups and like other hierarchical methods, merges two groups at each step. All combinations of two groups are considered, the chosen pair is the one which minimises the information loss, namely the sum of the squared distances between the points and the centroids of their respective groups summed over the resulting groups. So for all possible ways of merging two of G+1 groups to make G groups the chosen merge is the one which minimises:

$$W = \sum_{g=1}^{G}\sum_{i=1}^{n_g}\left\|x_i - \bar{x}_g\right\|^2 = \sum_{g=1}^{G}\sum_{i=1}^{n_g}\sum_{k=1}^{K}(x_{i,k} - x_{g,k})^2 \tag{13}$$

Comparison of gamma test raingauges with cluster analysis results

It has already been noted that the locations of the gamma test raingauges varied for each year and temporal resolution. The same inconsistent behaviour was also found in the clusters which would indicate that no repeatable spatial rainfall pattern occurs on a yearly basis, and so it is natural that the gamma test would find different raingauges to be more influential for different years. Figure 6 shows the Brue catchment divided up into three clusters (note two raingauges do not belong to any cluster). We can see that the areas of influence of each raingauge (as defined by Thiessen polygons) are disjointed; nevertheless, in this instance (1999 using daily raingauge totals), at least one of the five gamma test raingauges falls into one of the main cluster groups. Similar correspondence between cluster groups and the chosen raingauges was observed 50% of the time using both the daily raingauge totals and monthly raingauge totals.

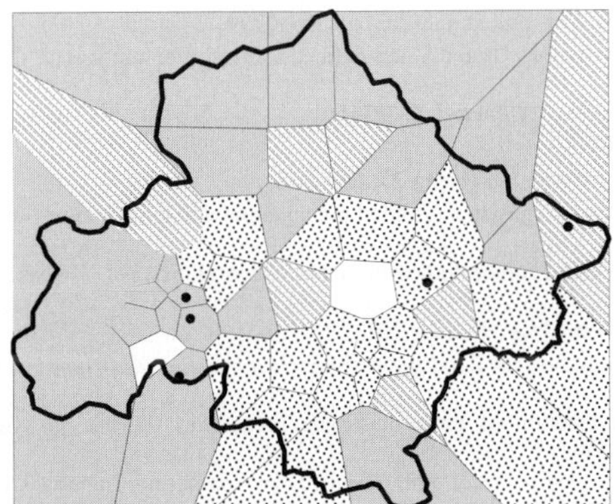

Fig. 6 Brue catchment divided into three raingauge clusters groups with gamma test raingauges. Daily rainfall totals, 1999.

CONCLUSION

Rainfall measurements play a crucial role in understanding and modelling hydrological processes. As such, it is vital that precipitation is accurately measured and recorded whilst reasonably minimising costs. Initial investigations, using the gamma test to select key sites of raingauge location from a dense raingauge, show promise. It has been demonstrated that by using the gamma test, the number of raingauges can be reduced significantly from 49 to five, without any adverse effect upon catchment average rainfall. However, the spatial representation of rainfall based on the same chosen raingauges is variable, and currently no reliable conclusions should be drawn in this respect without further investigation. The chosen raingauge locations varied over the years as did the clusters of similar rainfall, highlighting the variable nature of the rainfall field of the Brue catchment. However, the repeated occurrence of a small number of raingauges in the gamma test results indicates scope for reducing the uncertainty in gauge location for longer time periods of a year, and that with some further analysis, the raingauge locations could be fixed.

Although it is not feasible to run a dense raingauge network, such as HYREX, for every catchment, the methodology described in this piece of work could be transferred to spatial rainfall datasets produced by weather radar. In this case each radar pixel may be considered *in lieu* of a raingauge, the chosen gamma test pixels would indicate the area in which a raingauge should be sited. However substitution of radar pixels for raingauges requires further work in order to verify whether the quality of weather radar information is good enough for this purpose.

REFERENCES

Agalbjörn, S., Končar, N. & Jones, A. J. (1997) A note on the gamma test. *Neural Comput & Applic.* **5**(3), 131–133.

Anderberg, M. R. (1973) *Cluster Analysis for Applications*. Academic Press, New York, USA.

Bonner, R. E. (1964) On some clustering techniques. *I.B.M. J. Res. & Development* **8**, 22–32.

Chuzhanova, N. A., Jones, A. J. & Margetts, S. (1998) Feature selection for genetic sequence classification. *Bioinformatics* **14**(2), 139–143.

Clifford, H. T. & Sephenson, W. (1975) *An Introduction to Numerical Classification*. Academic Press, New York, USA.

Corcoran, J., Wilson, I. & Ware, J. (2003) Predicting the geo-temporal variation of crime and disorder. *Int. J. Forecasting* **19**, 623–634. doi:10Ð1016/S0169-2070(03)00095-5.

Cormack, R. M. (1971) A review of classification. *J. Royal Stat. Soc.* **134**, 321–367.

De Oliveira, A. G. (1999) Synchronisation of chaos and applications to secure communications. PhD Thesis, Dept of Computing, Imperial College of Science, Technology and Medicine, University of London, UK.

DeGaetano, A. T. & Shulman, M. D. (1990) A climatic classification of plant hardiness in the United States and Canada. *Agric. For. Met.* **51**, 333–351.

Durrant, P. J. (2001) winGamma: a non-linear data analysis and modelling tool with applications to flood prediction. PhD Thesis, Dept Computer Sci., Cardiff University, UK.

Evans, D. (2002) Data derived estimates of noise using near neighbour asymptotics. PhD Thesis, Dept Computer Sci., Cardiff University, UK.

Evans, D. & Jones, A. J. (2002) A proof of the gamma test. *Proc. Roy. Soc. A* 458(2027), 2759–2799.

Fovell, R. G. & Fovell, M. Y. (1993) Climate zones of the conterminous United States defined using cluster analysis. *J. Climate* **6**, 2103–2135.

Gallianin, G. & Filippini, F. (1985) Climatic clusters in a small area. *J. Climatol.* **5**, 487–501.

Gordon, A. D. (1980) *Classification*, Chapman & Hall, London, UK.

Gower, J. C. (1967) A comparison of some methods of cluster analysis. *Biometrics* **23**, 623–628.

Hamlet, A. F. & Lettenmaier, D. P. (2005) Production of temporally consistent gridded precipitation and temperature fields for the continental US. *J. Hydromet.* **6**(3), 330–336.

Han, D, Kwong, T. & Li, S. (2007) Uncertainties in real-time flood forecasting with neural networks. *Hydrol. Processes* **21**, 223–228.

Holland, D. J. (1967) The Cardington rainfall experiment. *Meteorol. Mag.* **96**, 193–202.

Jones, A. E., Bell V. A. & Moore, R. J. (2003) Areal rainfall estimation for flood forecasting. *Geophys. Res. Abstracts* **5**, 12164, European Geophysical Society.

Kalkstein, L. S., Tan, G. & Skindlov, J. A. (1987) An evaluation of three clustering procedures for use in synoptic climatological classification. *J. Clim. Appl. Met.* **26**, 717–730.

Končar, N. (1997) Optimisation methodologies for direct inverse neurocontrol. PhD Thesis, Dept Computing, Imperial College of Science, Technology and Medicine, University of London, UK.

Larson, L. W & Peck, E. L. (1974) Accuracy of precipitation measurements for hydrological modelling. *Water Resour. Res.* **10**, 857–863.

Marshall, R. J. (1980) The estimation and distribution of storm movement and structure using a correlation analysis technique and raingauge data. *J. Hydrol.* **48**, 19–39.

Mo, K. C. & Ghil, M. (1988) Cluster analysis of multiple planetary flow regimes. *J. Geophys. Res.* **D93**,10927–10952.

Molteni, F., Buizza, R., Palmer, T. N. & Petroliagis, T. (1996) The new ECMWF ensemble prediction system: methodology and validation. *Quart. J. Roy. Met. Soc.* **122**, 73–119.

Peck, E. L. (1980) Design of precipitation networks. *Bull. Am. Met. Soc.* **61**, 894–902.

Sneath, P. H. A. & Sokal, R. R. (1973) *Numerical Taxonomy*, W.H. Freeman & Co., San Francisco, USA.

Tsui, A. P. M. (1999) Smooth data modelling and stimulus-response via stabilisation of neural chaos. PhD Thesis, Dept of Computing, Imperial College of Science, Technology and Medicine, University of London, UK.

Tsui, A. P. M., Jones, A. J. & de Oliveira, A. G. (2002) The construction of smooth models using irregular embeddings determined by a gamma test analysis. *Neural Comput. Appl.* **10**(4), 318–329. doi:10D1007/s00521020000

Wheater, H. S., Isham, V. S., Cox, D. R., Chandler, R. E., Kakou, A, Northrop, P. J. Oh, L., Onof, C. & Rodriguez-Iturbe, I. (2000) Spatial-temporal rainfall fields: modelling and statistical aspects of Hydrology, *Hydrol. Earth System Sci.* **4**(4), 581–601.

Wood, S. J., Jones, D. A. & Moore, R. J. (2000) Accuracy of rainfall measurement for scales of hydrological interest. *Hydrol. Earth System Sci.* **4**(4), 531–543.

Extracting software component from hydrological information service system

PING AI[1] & YALI CHEN[2]

1 *State Key Laboratory of Hydrology-Water Resources and Hydraulic Engineering, Hohai Univ., Nanjing 210098, China*
aip@hhu.edu.cn

2 *The Bureau of Hydrology, Yangtze River Water Resources Commission, Wuhan 430010, China*

Abstract Because of the extensive application of information technology in hydrology, the scope of the hydrological information service is extended; the frequency of the service's requirement change is also increased. These result in a more and more complex hydrological information service system based on computer and networks. In order to deal with the complexity of the system brought development and maintenance problems, the application of component-based software system technology and evolution mode of system development and maintenance are essential. In the study we analyse the basic features and trends of the hydrological information service system to propose the basic ideas and methods to extract software components from the hydrological information system. We develop the components recognition algorithms (Iterative analysis Algorithms based on the directed weighted graph for the Hydrological Domain software Component Identification). An example of how to extract software components from a real hydrological information system is given to explain the method and condition of application for the algorithm.

Key words hydrology; information service; software component; extraction; system

INTRODUCTION

With the extensive application of the information technology in hydrology and the expansion of the hydrological information service requirements, the complexity and evolution requirements of the hydrological information service system are increasing continuously, the lifecycle of the customization's system is becoming shorter, and system maintenance costs are growing. Therefore, researching and using the software component technology, as well as setting up a mechanism of generating system based on components and a way of system updating and maintenance based on evolution rather than establishing a new system, could be a feasible choice for a complex hydrological information service system to face changing requirements.

From a software architecture point of view, the hydrological information service system could be divided into three parts: data source, data storage and information service (shown in Fig. 1). In these parts, data source mainly includes hydrological data acquisition, transmission, preprocessing and other functions; data storage mainly includes hydrological data organization and storage, as well as integration and consistency management and other functions. The part of information service includes many functions; these functions could be classified into three types: hydrological data service, hydrological information products service, and hydrological information publishing service.

Hydrological data service is a traditional function of the hydrological information service system; it primarily supports a query and access interface for users based on some customized data organization module. Hydrological information products service provides products for users with a particular theme of hydrological information in flood control and drought relief, water resource management, water environmental management and other comprehensive hydrological information requirements. Hydrological information publishing provides various public hydrological information services via a network.

Fig. 1 Basic architecture of Hydrological Information Service System.

To provide hydrological information services based on the hydrological database has become the mainstream. Due especially to the rapid popularization of network, network and database-based comprehensive hydrological information service system, it provides users not only with an electronic information service similar to a traditional service, but also to a customized theme service according to users' demands. As a result of application of the hydrological database, the modern information process has been established. The process includes automatic collecting, transmission, computerized processing, storage and services of hydrological information.

Nowadays, there are several ways of using a hydrological information service based on a hydrological database:

(1) Basic Way – Users access hydrological basic database by network directly, obtaining data and generate data files for application.
(2) Expansion Way – Users obtain customizing data re-extracting and information processing service by the information system based on basic hydrological database.
(3) Senior Way – It can be integrated that user's system and system of hydrological database by interfaces according to information sharing standards.

The senior way is the development of hydrological information service in the future, of which the trend is that: (1) combined with hydrological database system and GIS to improve the capability of information service; (2) establishing network-based distributed hydrological database and its information system; (3) integrating information resource based on heterogeneous database system; (4) inheriting and reconstructing earlier database and information service system.

Because the hydrological information service system is from simple to complex, and from single function to multifunctional integration, so extracting software component from a hydrological information service system and using these components to set up a new system or make an old system evolution is an important technical requirement.

KEY TECHNOLOGIES OF COMPONENT EXTRACTION

Overview

With the development of software technology, software developers have paid more attention to the component extraction and composition for a new system rather than the repeated development of static business logic. Component-based development of the system will gradually replace the old way of development. In the future, software development activities will be to apply software resource composition to form a basic software system (Yang, 2005).

For the extraction or production of reusable components and composition to form a new software system, a basic technology is how to get software components. Because component extraction often relate to characteristics of the application in the domain, so if the component technology is applied to hydrological information service systems, component extraction technology is a key.

There are two main ways to get components: the purposeful production, and extracting from existing systems. Owing to there being a large number of old systems, to extract components from the old systems has become the focus of attention. This paper focused on the second way of components extraction, i.e. analysing reusable component extraction technology from the hydrological information service system according to the requirements in the hydrology domain.

In the hydrological information service system, the component is a fragment of hydrological information service business application logic. As a result, according to the characteristics of the demand of the hydrological information service, components extraction studies how to extract the fragments to repeatedly compose them to form the new hydrological information service function.

How to identify and measure reusable component correctly is very difficult. In order to complete the extraction of components, we need to combine with the domain engineering and software reengineering, and take into account all the relevant factors.

The process of software component extraction from the hydrological information service system can be divided into three phases (see Fig. 2):

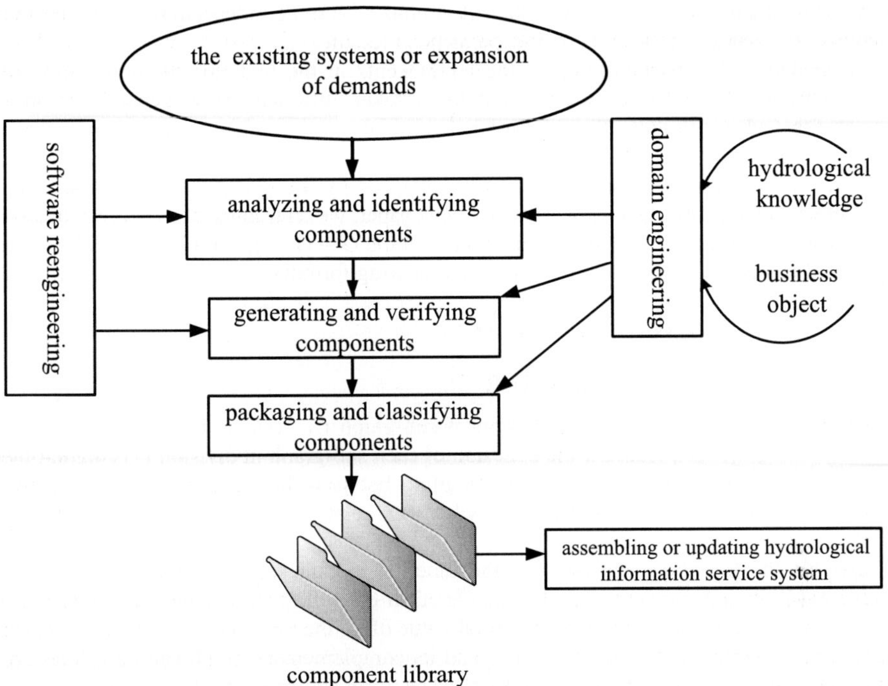

Fig. 2 The basic process of extraction of the components.

Component analysis and identification phase On the basis of the domain engineering and software re-engineering, we understand the current of the hydrological information service software systems, analyse the requirement of hydrological information service, and then identify the relative independent functions to obtain candidate components.

Component generation and verification phase Candidate components are coded or modified, measured and their reusability inspected in order to determine the qualified reusable components.

Component packaging, classification and storage phase The qualified reusable components are going to be packaged and classified, then stored into the component library for assembling or updating the hydrological information service system.

In the above process, the reusability of components is the most important key issue. For the hydrological service information system, the core meaning and main role of reusability is the independence of components in the formation and evolution of the system. Well-independence is the basis of component reusability, and is also the basis of the analysis and identification for components.

Independence measure

In order to identify components, analysing the independence of the system's ingredients is required. At present, object-oriented systems methods for component identification can be divided into two categories: knowledge matching (Pinzger *et al.*, 2002; Zhao *et al.*, 2003; Spinellis *et al.*, 2005) and structure analysis (Mahdavi *et al.*, 2003; Zhou *et al.*, 2003; Canfora *et al.*, 2006). In the knowledge matching, acquiring and representing knowledge is often more difficult in practice. But the structure analysis will abstractly divide the software system into several sub-structures, and each sub-structure corresponds to a candidate component, which is represented by a graph, then using a iterative way to study sub-graphs in the graph with different granularity by some kind of mathematical methods step by step, and the sub-graphs with more independence are determined as the options of reusable components by independence measure.

In considering the independence of the components in the hydrological information service systems, there are some factors which need to be taken into account in general, including the cohesion of the sub-graph itself, the coupling between sub-graphs and its complementary graph, the scales of the sub-graph, etc. The cohesion and coupling between the various types structures in the hydrological information service system can be analysed by the *QMOOD* model (Bansiya *et al.*, 2002; Chiricota *et al.*, 2003; Luo *et al.*, 2003). In this paper, we refer to the above-mentioned idea.

Suppose: $C = (G1, G2, ..., Gn)$ is a division of graph $G = (V, E)$, and $Gi = (Vi, Ei)$ $(1 \leq i \leq n)$ is a sub-graph of G, and its quality is defined by the following formula:

$$MQ(C,G) = \frac{\sum_{i=1}^{n} s(G_i,G_i)}{n} - \frac{\sum_{i=1}^{n-1}\sum_{j=i+1}^{n} s(G_i,G_j)}{n(n-1)/2} \qquad (1)$$

In formula (1), $s(Gi, Gi)$ is the cohesion of sub-graph Gi. $s(Gi, Gj)$ is the coupling between sub-graph Gi and Gj. As a result, if the cohesion of each sub-graph in division C is higher, then the quality of division C is higher, and if the coupling between the sub-graphs is higher, then the quality of division C is lower. The division quality of C relative to the G reflects the independence level of C.

Based on the foregoing analysis, for the function module (sub-graph) Gi in the subsystem indicated by C, its independence can be calculated in the following manner: the cohesion of the sub-graph itself can be calculated from the total value of all the associations weight of edges in the sub-graph; the coupling between sub-graphs and its complementary graph can be calculated from the total value of all the associations weight of edges from sub-graph to its complementary graph; the scales of sub-graph can be determined by the number of vertices in the sub-graph.

So, for directed weighted graph $G = (V, E)$ and its sub-graph $G' = (V', E')$. $O = \{\langle u, v, t\rangle \mid \langle u,v,t\rangle \in E \wedge (u \in V' \wedge v \in V - V')\}$ is the set of edges from G' to its complementary graph. IM – the independence of sub-graph G' is defined as:

$$IM(G',G) = \frac{\sum_{e \in E'} W(e)}{|V'| \sum_{e \in O} W(e)} \qquad (2)$$

In formula (2), $W(e)$ is a weight function of the edge e, associated with the type, direction, and other factors of e. The value of $W(e)$ can be determined through the methods of analysis and experiment in the process of components identification in the hydrological information services system. Formula (2) shows that in a sub-graph, the independence is proportional to the cohesion, and is inversely proportional to the coupling, and is inversely proportional to the scale of the sub-graph.

Identifying algorithm and evaluation criteria

Based on formula (2), combining with object-oriented technology, there are two types of identification algorithms from a system that has been changed into a graph: identifying component algorithms of undirected graph and directed graph. These algorithms are classic mathematical methods. As the relations of the system are directional, directed graph algorithms are more effective for identifying components from the system.

According to object-oriented technology, the directed graph identifying algorithm abstracts the hydrological information service system becomes a weighted-directed graph. A vertex of the graph denotes a class, the edges of the graph denote the relation between classes, and the direction of edges denotes the direction of the relation. According to the different types of the relation, each edge is assigned a different weight. Iterative bottom-up analysis is used when identifying components.

We suppose that the hydrological information service system can be abstracted as a directed graph — $G = (V, E)$, in which V is the set of the graph vertexes, and each vertex represents a class of the system; $E = \{<v1,v2,t> \mid (v1, v2 \in V) \wedge t \in T\}$ is the set of the graph edges, and $T = \{inheritance, whole\text{-}part, instance\text{-}link, message\text{-}link\}$, is the set of relations between classes.

Then the component identifying algorithm is as follows:

- According to the scale of graph to determine the max scale *k* of sub-graph which needs to be iteratively investigated, and then having an iterative analysis for the graph.
- The first iteration input is the graph of whole system. Each iteration inspects input graph's all the connected sub-graphs which scales are smaller than *k*, and using independence measures function (such as formula (2) or formula (1)) calculated for the independence of each sub-graph, and selecting the sub-graph of which the independence is higher than the threshold value – *Wd* as candidate sub-graph (corresponding to the candidate component).
- The candidate sub-graph will be abstracted as a vertex, and using the input graph to form a new graph with smaller scale.
- A new graph can be as the input to the next iteration, repeated iterations until the reconstructing of the new graph smaller than *s*.

Usually the value of *Wd*, *k* and *s* is determined by the function features of the hydrological information services systems.

The pseudo code of identifying algorithm is as follows:

- Input: weighted directed graph – *G*, the scale of maximal sub-graph – *k*, the independence threshold value – *Wd*, and the minimal scale of input graph – *s*;
- Output: the set of candidate sub-graphs – components.

```
//sizeof(G): return |V| — the number of vertexes in G;
//getConnectedComponents(G,k):return all connected sub-graphs in G, and the scale of the sub-
graphs is smaller than k;
//getIndependentComponents(ComponentList,Wd): compute the independence of each sub-graph in
ComponentList, and return all sub-graphs which the independence is higher than Wd;
//constructNewComponent (G , newComponents): contract each candidate sub-graph in
newComponents to a vertex in G; the new graph is the return value when all sub-graphs have been
contracted to a vertex; in this process, if there is intersection between candidate sub-graphs, the
sub-graph which has higher independence should be selected.
Components =∅
While (sizeOf(G)≥s)
{ComponentList=getConnectedComponents(G , k)
  newComponents =getIndependentComponents(ComponentList, Wd)
  Components = Components ∪ newComponents
  G=constructNewComponent(G , newComponents) }
```

Quality evaluation is necessary for the candidate component extracted by recognition algorithms. Traditional component quality evaluation methods have the statistic code lines, the Halstead and the Albrecht function feature. In the objected-oriented system, there is another evaluation method that is the number of classes as factors (Mei *et al.*, 2003). Generally, component evaluation factors have the characteristics of retrieve, usage and modification, and complexity, error rate, coupling degree, as well as the range and the interface of re-use. So, we should consider the following basic technical criteria when extracting the component:

- **Moderate component granularity criteria** The size of component granularity affects the reusability of component to a large extent. Therefore we should adopt the larger granularity component in a relatively stable part of the hydrological information service system. The result of doing so would decrease complexity of the system composition and evolution and improve the stability of the system. In a fickle part of the hydrological information service system, we should adopt a smaller granularity component with a single function.
- **High cohesion criteria** High cohesion means that the component has better independence and reusability, and could be fit for compositions and evolutions of the hydrological information service system, and could effectively reduce the complexity of component composition and replace, as well as affect system stability when the component is changed.

EXAMPLE

An actual hydrological information service system could provide water resource managers (including decision-makers) with several functions, such as water resource information query in the river basin, statistics, reports, monitoring, consultation service support, etc. In this example, the hydrological information service system has graph operation, data editing, browse query, theme monitoring, thematic map making, consultation service support, statistics, report function, etc. Its use case diagram is shown in Fig. 3.

The query statistical analysis module of the hydrological information in the system is chosen as an example of an extracting software component. This module includes a query of the latest and particular hydrological information, river and lake information, sluice dam and tide information and other business information. Not only that, the module is developed by an object-oriented method based on light J2EE. Therefore, it can well demonstrate the component recognition algorithm in this paper.

The relationship within the hydrological information query module is that between classes. A vertex denotes a class, an edge denotes a relationship between classes, and the direction of edge shows the direction of relationship. The figure is constructed in Fig. 4, and the detail information is listed in Table 1.

According to different type of relationship, every edge is given a different weight. In Fig. 4, any edge e is defined as a weight function like $W(e) = f(t)$, in which function f is defined in formula 3.

$$f(t) = \{W_{ih}\,(t = inheritance),\ W_{wp}\,(t = whole\text{–}part),\ W_{il}\,(t = instance\text{–}link),\ W_{ml}\,(t = message\text{–}link)\} \quad (3)$$

There are four types of weight in formula (3) as follows: *Inheritance, Wih*; *Whole and Part, Wwp*; *Instance Link, Wil*; *and Message Link, Wml.*

The value of edge weight in Fig. 4 is shown in Table 2 and the value equals to zero represent that there is no link between one vertex and another.

In this example, according to the features of the system, we determine that $Wih = 0.9$, $Wwp = 0.8$, $Wil = 0.6$, $Wml = 0.6$, parameters of the component recognition algorithm s and Wd is assigned to 3 and 2.6, and k is assigned to the number of vertices: 13.

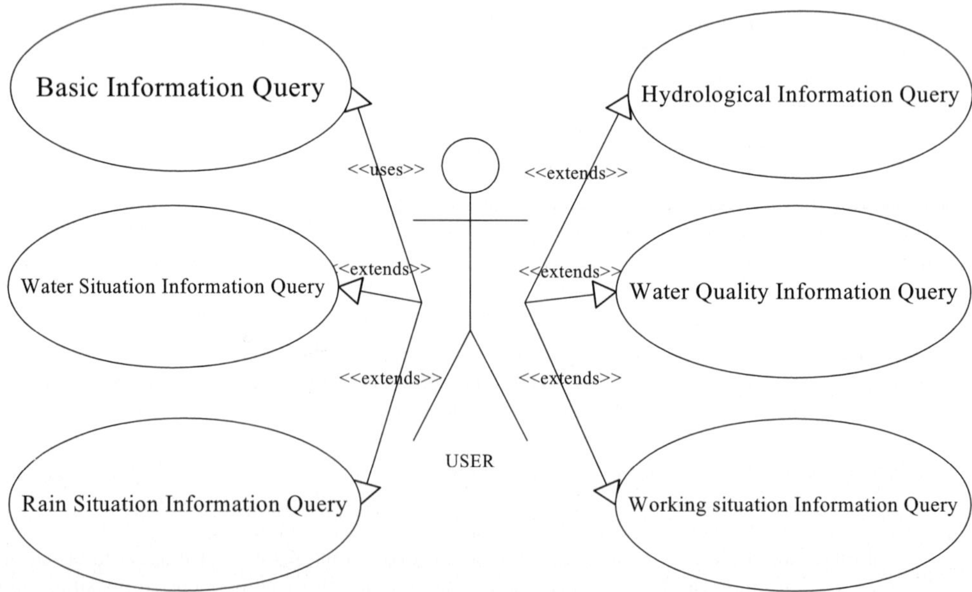

Fig. 3 Use case diagram of the system.

Table 1 Information of water information query directed graph.

Vertex	Class Name	Function
1	com.water.query.form. BaseInfoBean	Agent class of business processing
2	com.water.query.form. WRiverInfoBean	Agent class of hydrological information query business processing, called by Struts Controller
3	com.water.query.input. datainput	Receive input data
4	com.water.query.ouput. formview	Output report
5	com.water.query.service. IRiverQueryServer	Business processing method opened for users
6	com.water.query.service. BaseQueryServer	Implement business processing method in interface
7	*com.water.query.input. datainput*	Receive undealt data
8	com.water.query.service. RiverQueryServer	Implement business processing defined in IRiverQueryServer, access data by the method defined in IRiverInfoDAO
9	com.water.query.dao.iface. IRiverInfoDAO	Provide the method of business processing accepting data for RiverQueryServer, by which access database to receive business data
10	com.water.query.input. pretreat	Preprocessing receiving data
11	com.water.query.ouput. graghview	Output data by graph
12	com.water.query.dao.sqlmapdao RiverInfoDAOImpl	Implement the method of accepting business data in IRiverInfoDAO, query data by data access provided by Ibatis
13	com.water.query.input. datapretreat	Data preprocessing

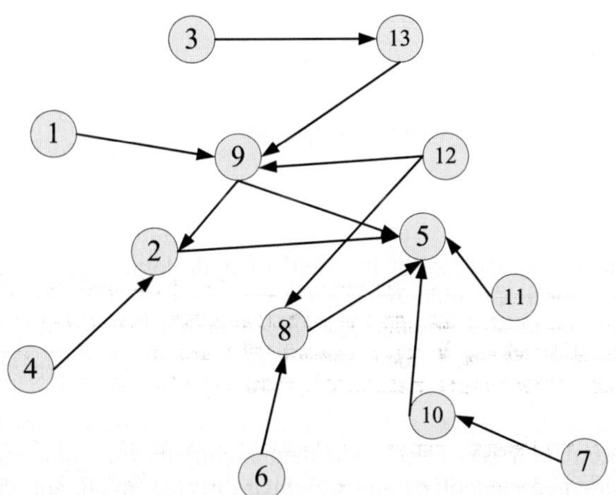

Fig. 4 Hydrological information query directed graph.

As Fig. 4 shows, the hydrological information query module has 13 classes and about 5000 lines of code. Adopting component recognition algorithm based on the directed graph, we could get candidate components to match the premise shown in the sub-graph, which is the connected component constructed by Vertex {2,5,8,9,12}. After evaluating and adjusting this sub-graph, a class graph of candidate component is shown in Fig. 5.

Table 2 Hydrological information query-directed graph weight relationship.

Vertex	1	2	3	4	5	6	7	8	9	10	11	12	13
1	0	0	0	0	0	0	0	0	0.6	0	0	0	0
2	0	0	0	0.6	0.8	0	0	0	0.8	0	0	0	0
3	0	0	0	0	0	0	0	0	0	0	0	0	0.6
4	0	0.6	0	0	0	0	0	0	0	0	0	0	0
5	0	0.8	0	0	0	0	0	0.9	0.8	0.6	0.6	0	0
6	0	0	0	0	0	0	0	0.6	0	0	0	0	0
7	0	0	0	0	0	0	0	0	0	0.6	0	0	0
8	0	0	0	0	0.9	0.6	0	0	0	0	0	0.8	0
9	0.6	0.8	0	0	0.8	0	0	0	0	0	0	0.9	0.6
10	0	0	0	0	0.6	0	0.6	0	0	0	0	0	0
11	0	0	0	0	0.6	0	0	0	0	0	0	0	0
12	0	0	0	0	0	0	0	0.8	0.9	0	0	0	0
13	0	0	0.6	0	0	0	0	0	0.6	0	0	0	0

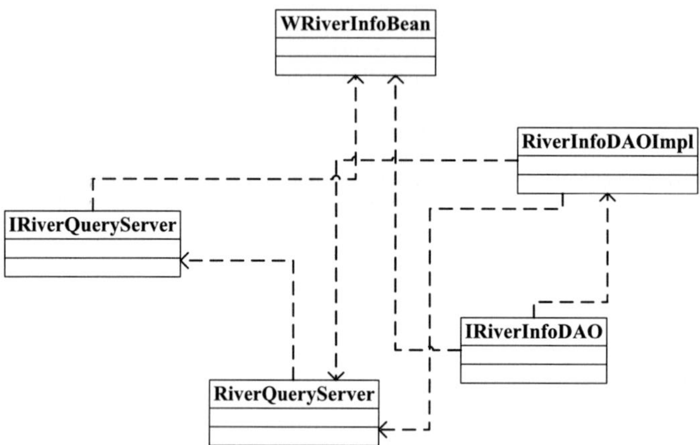

Fig. 5 Class graph of hydrological information query candidate component.

CONCLUSIONS

With extensive application of information technology in the field of hydrological information processing and service, the hydrological information service system is becoming more and more complicated, meaning it will not be suitable for more and more frequent changes in requirements. In this case, introducing software system evolution technology based on components is an effective choice.

In response to the problems above, this paper analyses software component extraction technology from the hydrological information service system, then highlights the components extracting algorithm based on object-oriented technology and weighted directed graph, and gives an example of the algorithm for further description. Practice proves that the algorithm is suitable and could be used as a technical reference for the development of the hydrological information service system.

Acknowledgements This research was supported by the National High-Tech Research and Development Plan (863) of China under Grant no. 2006AA01A126; the National Grand Fundamental Research 973 Program of China under Grant no. 2006CB403204; the Key Project of Science and Technology Research of the Ministry of Education of China under Grant no. 107056.

REFERENCES

Bansiya, J. & Davis, C. G. (2002) A hierarchical model for object-oriented design quality assessment. *IEEE Trans. on Software Engineering* **28**(1), 20–28.

Canfora, G., Czeranski, J. & Koschke, R. (2006) Revisiting the Delta IC approach to component recovery. In: *Proc. 7th Working Conf. on Reverse Engineering*, 140–149.

Chiricota, Y., Jourdan, F. & Melancon, G. (2003) Software components capture using graph clustering. In: *Proc. 11th IEEE International Workshop on Program Comprehension*, 217–226.

Luo, J, Zhao, W., Qin, T., Jiang, R. K., Zhang, L. & Sun, J. S. (2004) A decomposition method for object-oriented systems based on iterative analysis of the directed weighted graph. *J. Software* **15**(9), 1292–1300.

Mahdavi, K., Harman, M. & Hit, R. (2003) A multiple hill climbing approach to software module clustering. In: *Proc. 19th Int. Conf. Software Maintenance*, 315–324.

Mei, H., Xie, T., Yuan, W. & Yang, F. Q. (2000) Component metrics in Jade Bird Component Library System. *J. Software* **11**(5), 634–641.

Pinzger, M. & Gall, H. (2002) Pattern-supported architecture recovery. In: *Proc. l0th Int Workshop on Program Comprehension*, 53–61.

Spinellis, D. & Raptis, K. (2005) Component mining: a process and its pattern language. *Information Software Technol.* **42**(9), 609–617.

Yang, F. Q. (2005) Thinking on the development of software engineering technology. *J. Software* **16**(1), 1–7.

Zhao, W., Zhang, L., Lin, Y. & Luo, J. (2003) Understanding how the requirements are implemented in source code. In: *Proc. of l0th Asia-Pacific Software Engineering Conf.*, 68–77.

Zhou, X., Chen, X. K., Sun, J. S. & Yang, F. Q. (2003) Software measurement based reusable component extraction in object-oriented system. *Acta Electronica Sinica* **31**(5), 649–653.

Isotopic variations of direct runoff

WANG TAO[1,2], BAO WEIMIN[1,2] & HU HAIYING[3]

1 *State Key Laboratory of Hydrology – Water Resources and Hydraulic Engineering, Hohai University, Nanjing 210098, China*
wangtao@hhu.edu.cn

2 *College of Hydrology and Water Resources, Hohai University, Nanjing 210098, China*

3 *School of Civil Engineering and Transportation, South China University of Technology, Guangzhou 510641, China*

Abstract The isotopic variations of direct runoff are studied through laboratory experiments with low initial soil water content and constant isotopic rainfall input situations. The experimental results show that isotopic compositions of direct runoff decrease to a minimum at maximum discharge and increase as discharge decreases; the average values of δD and δ^{18}O are respectively, 5‰ and 0.4‰ lower than those of rainfall. Isotopic compositions of the soil water increase along its pathway; δD and δ^{18}O values of soil water at the end of experiment are respectively, 11‰ and 1.2‰ higher than those of water from surface storage in the trough. Isotopic compositions of rainfall change greatly due to soil regulation. A thorough understanding of the isotopic variation of direct runoff can be useful in providing information for hydrograph separation.

Key words hydrogen and oxygen stable isotopes; direct runoff; isotopic variations

INTRODUCTION

A graphical separation of streamflow components is often used to separate the hydrograph into direct runoff and indirect runoff in rainfall runoff events (Lin, 2001). This empirical method includes a hypothesis which should be studied further. Hydrograph separation is one of the fundamental researches in hydrology, and the calculation of runoff generation and the unit hydrograph both are based on it (Zhang *et al.*, 2006). It is significant to seek a physical hydrograph separation method in order to further develop rainfall–runoff calculation models and investigate the characteristics of runoff generation in catchments. The application of environmental isotopic techniques in hydrology can satisfy these requirements well. Oxygen-18 (^{18}O) and deuterium (D) are environmental isotopes and are frequently used to trace the sources of runoff in rainfall–runoff events. Isotope hydrograph separation based on mass balance usually separates runoff into event water (surface flow generated by the rainfall event of interest) and pre-event water (underground flow storage in soil before the rainfall event of interest). Proportions of event and pre-event water reflect hydrological processes and runoff pathways; thus further understanding of the sources of runoff and their pathways will be useful in developing conceptual hydrological model. Numerous studies about hydrograph separation have been done (Moore, 1989; McDonnell *et al.*, 1990; Sklash, 1990; Kendall & McDonnell, 1998, 2001); however, comparatively less research on this has been carried out in China (Gu, 1992, 1995; Gu & Xie, 1997; Qu *et al.*, 2006). A series of representative studies was made by Gu *et al.* (1992, 1995) at the artificially constructed Hydrohill catchment of the Chuzhou Hydrology Laboratory located in northeastern China near Nanjing. Isotope hydrograph separation is based on several assumptions which have been summarized by Moore (1989), Sklash (1990) and Gu (1996); these are:

(a) The isotope signatures of base flow and groundwater are constant in space and time, and any variations can be accounted for.
(b) The isotope signature of rainfall is constant in space and time, and any variations can be accounted for.
(c) There is a significant difference between the isotopic compositions of the rainfall and base flow or groundwater.
(d) Contributions of soil water must be negligible, or the isotopic composition of soil water must be similar to that of groundwater.
(e) Contributions of water from surface storage to streamflow are negligible.
(f) The isotopic composition of surface flow must be similar to that of rainfall.
(g) Isotopic fractionation is negligible in runoff concentration.
(h) Precipitation and runoff are based on classical simplified runoff mechanisms.

Buttle (1994) and Gu (1996) discussed the reliability of these assumptions in practical application in detail.

The variation of stable isotopic composition in nature is slight, and expression of isotopic abundance and isotopic ratio cannot show this slight difference. So δ-values are used to represent the isotopic composition of a substance. The δ-value for isotopic species *i* is defined relative to the isotopic composition of Standard Mean Ocean Water (SMOW) by:

$$1+\delta_i = R_i / R_i^{SMOW} \quad (1)$$

It is common to separate hydrographs into direct runoff and indirect runoff. Fewer studies are concerned with isotopic variations of a single runoff component such as direct runoff. This paper studies the variations of hydrogen and oxygen stable isotope concentrations in direct runoff through a rainfall–runoff laboratory experiment under constant isotopic rainfall input and low initial soil moisture conditions; this is aimed at providing more information for hydrograph separation.

MATERIALS AND METHODS

The experiment was conducted in the rainfall hall of the State Key Laboratory of Hydrology-Water Resources and Hydraulic Engineering of Hohai University. Experiment time is from 09:00 29 June to 23:00 on 30 June 2007. Daily mean temperature and average relative humidity in the laboratory were 28.0°C and 78%, respectively. The experimental trough (Fig. 1) is 12.1 m in length, 3.35 m in width, and 1.2 m in height. It is made up of two impermeable concrete walls, dividing the whole trough into three equal parts; the two walls are 15 cm in height with a water hole at each bottom. The bottom and sidewalls of the trough are constructed of concrete to prevent water infiltrating, and the three equal parts of the whole trough are marked as troughs 1, 2 and 3 for the purpose of easy identification. Water flowed out at the outlet of trough 3 where sediment in runoff percolates through fine stones and earthwork cloth. The slope of the whole trough is 6°. Uniform soil particles were added to the three troughs after having been naturally air dried and because direct runoff contacts with the surface soil layer during the process of direct runoff concentration, especially the layer of soil between 0 and 15 cm. The thickness of the soil layer in the troughs was designed to be around 15 cm. To this end too, the soil volume water content (3.2%) is measured by Time Domain Reflector (TDR). The soil used for the test was obtained from the hillside of Tangshan twon, in Nanjing city; the clay content of the soil is high.

The rubber tube (2 cm) shown in Fig. 1, serves as the inflow line to direct the "rainfall". One side of the rubber tube is occluded, while the other is connected to a tap water tube with a series of small holes; the rubber tube used in this application is 3.35 m. For the experiment proper, the flow rate of the rainfall in the tube is 139.8 cm^3/s; this represents a rainfall intensity of 12.4 mm/h over

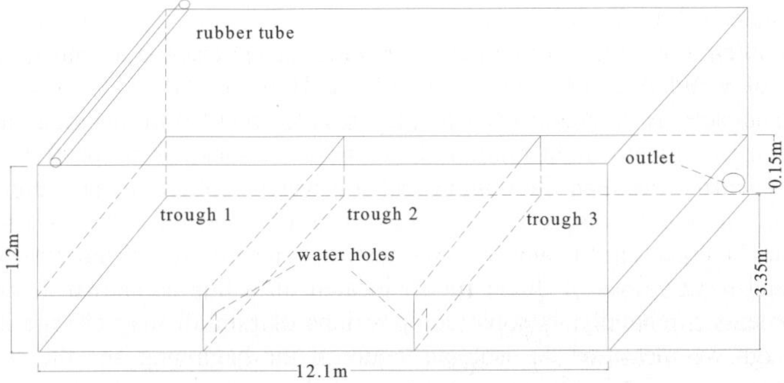

Fig. 1 Trough of rainfall–runoff experiment.

the whole trough. Rainfall intensity was kept constant and rain falls uniformly in the upper section of trough 1. Rainfall was started at 09:40 h on 29 June and ended at 18:30 h. Water flowed out of trough 3 at 18:20, while the discharge began to decrease when the water holes of troughs 1 and 2 were occluded at 18:45 h. Samples of rainfall were taken at the beginning and the end of the experiment; the δ^{18}O and δD values of rainfall were –7.7‰ and –51‰ at the beginning, and –7.7‰ and –49‰ at the end, respectively. So the isotopic composition of the rainfall could be regarded as constant.

The surface storage recession experiment was carried out from 09:00 to 22:30 h on 30 June. The main objective of this experiment is to study the isotopic variations between soil water and surface storage water in troughs. Samples of water from surface storage were collected as the water holes of troughs 1 and 2 were opened at the beginning of the experiment. Discharge measurements and samples were taken at the outlet of trough 3. There was a significant amount of surface storage in troughs 1 and 2 where the soil was soaked over 12 hours.

A polymethyl methacrylate triangular weir was designed to measure discharge at the outlet of trough 3. Water samples were taken at the triangular weir using 30 ml plastic bottles. The sample bottles were stored in a cold environment and sealed tightly with wax to prevent evaporation. Hydrogen and oxygen isotopic compositions were measured using a MAT-253 mass spectrometer in the isotopic laboratory of Ministry of Land and Resources in Beijing. The analytical precisions are ±2‰ and ±0.2‰, both hydrogen and oxygen isotope analyses, respectively.

RESULTS AND DISCUSSION

Isotopic variations of direct runoff

Rain infiltrates into the soil and generates runoff. Runoff is hindered by soil during water flow to the outlet of the trough. The infiltrated rainfall is always in direct contact with the soil, resulting in a great change in the isotopic composition. Because the experiment was carried out with low initial soil moisture, the measured water could be considered as direct runoff measured at 15 cm depth including surface runoff and interflow. Figures 2 and 3 show the variations of hydrogen and oxygen stable isotopes and discharge of direct runoff in trough 3. The observed δ^{18}O values of direct runoff ranged from –8.3 to –6.3‰, while the average value is –8.1‰ and fluctuation amplitude 2‰; the observed δD values of direct runoff ranged from –59 to –44‰, with an average value of –55‰; the fluctuation amplitude is 14‰. δ^{18}O values of direct runoff decreased as discharge increased and reached a minimum δ^{18}O value at maximum discharge, 0.6‰ lower than that of rainfall. Interestingly however, δ^{18}O values of direct runoff increased as discharge decreased, a maximum δ^{18}O value at minimum discharge, 1.4‰ higher than that of rainfall, resulting in an increasing amplitude 2.3 times greater than the amplitude reduction. The trend of δD values varying with time is the same as δ^{18}O values with the minimum value, 7‰ lower than that of rainfall. The maximum value stands at 7‰ higher with increasing amplitude equal to decreasing amplitude.

As shown in Figs 2 and 3, isotopic compositions of direct runoff decreased to a minimum at maximum discharge and increased as discharge decreases under constant isotopic rainfall input and low initial soil water content conditions. Experimental results are different from those of the natural basin, e.g. results of Kendall *et al.* (2001). Stichler & Herrmann(1978) observed the isotopic compositions of Lainbach creek (Bavarian Alps); the results obtained in this study agree with their findings, but, with isotopically light rainfall. Three basic reasons can be proffered for this; first, isotopic mixing occurred between the rainfall and soil water. Isotopic compositions of mixtures are intermediate between the isotopic compositions of rainfall and the initial soil water and this will produce a "line" connecting the isotopic compositions of rainfall and the initial soil water. By extension, δD and δ^{18}O values of direct runoff located on a line as shown in Fig. 4 further explain this phenomenon. Secondly, isotopic compositions of rainfall may change in an experimental set up. Although we measured the isotopic values at the beginning and the end of experiment, isotopic compositions of rainfall may change, and the values may be lower than measured values. Unfortunately, samples were not taken during the whole rainfall process. And, thirdly, evaporation fractionation of soil water and rainfall would result in isotopically enriching the direct runoff.

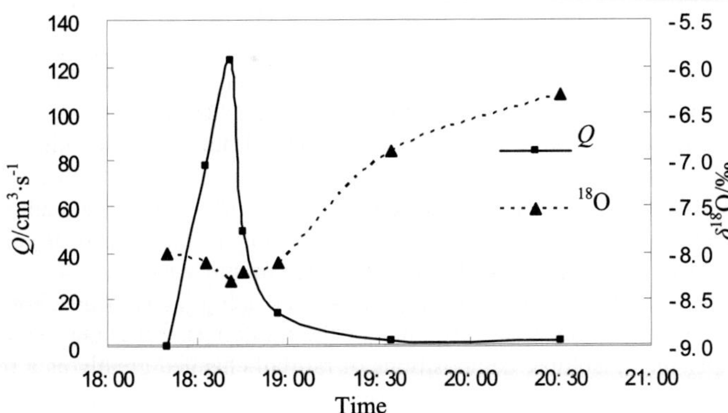

Fig. 2 $\delta^{18}O$ values and discharge of runoff at the outlet of trough 3.

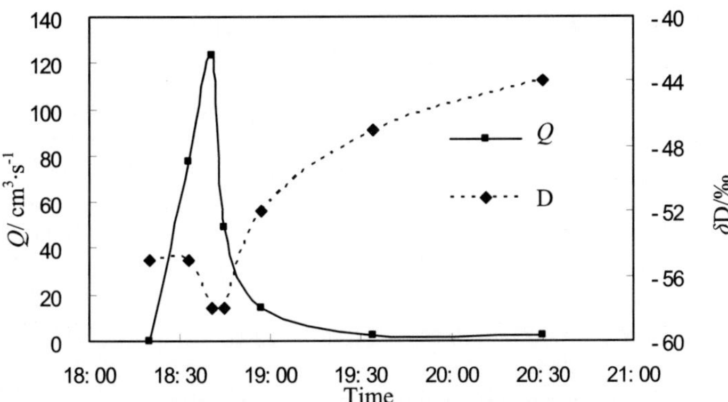

Fig. 3 δD values and discharge of runoff at the outlet of trough 3.

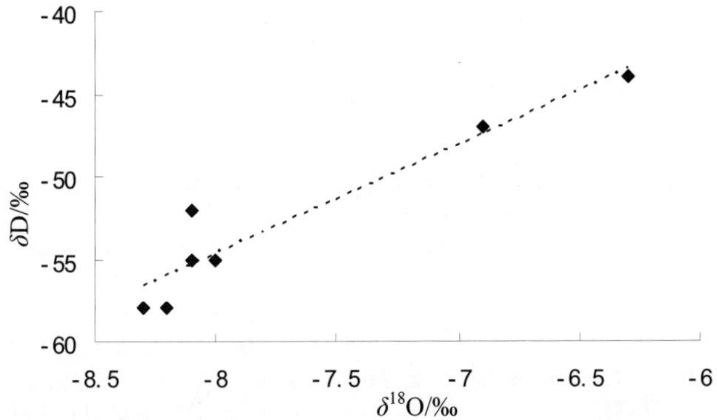

Fig.4 Relationship between δD and $\delta^{18}O$ values of direct runoff.

Isotopic variations of surface storage recession experiment

Because the water holes of troughs 1 and 2 were closed on 29 June, there was plenty of surface storage in troughs 1 and 2 while soil was soaked over 12 hours. Isotopic variations of surface storage and soil water in the troughs were studied by measurements of discharge and samples at the outlet of trough 3. Part of the surface water in trough 2 flowed out at the beginning of experiment, after several minutes, while the mixing water of troughs 1 and 2 also flowed out. Figures 5 and 6 show the variations of isotopic compositions and discharge of runoff during the surface storage recession experiment. They indicate that the isotopic compositions of surface storage in trough 2 were higher than those of mixing water of troughs 1 and 2. On the other hand, the isotopic compositions of surface storage in trough 1 were lower than those in trough 2. The isotopic compositions of runoff increased when no surface storage was detected. At the end of the experiment, $\delta^{18}O$ and δD values of soil water are 1.2‰ and 11‰ higher than those of surface storage in trough 2, respectively. The result indicated that isotopes of soil water diffused slowly into the surface storage. Isotopic compositions of soil water increased along its pathway. From the relationship between hydrogen and oxygen stable isotopes of the rainfall–runoff experiment and surface storage recession experiment, isotopic compositions of rainfall changed greatly as a result of soil regulation.

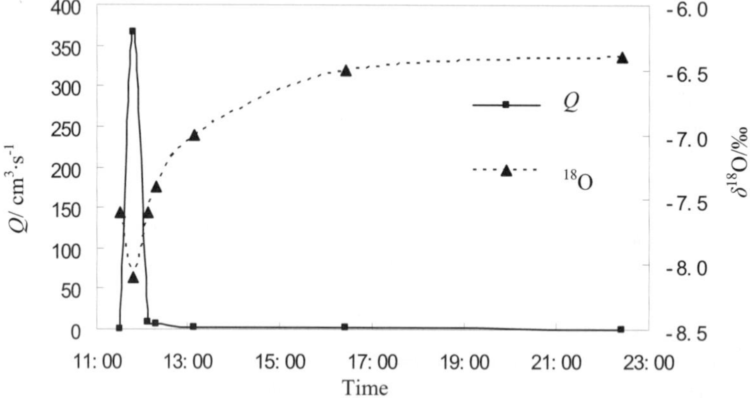

Fig. 5 $\delta^{18}O$ values and discharge of runoff in surface storage recession experiment.

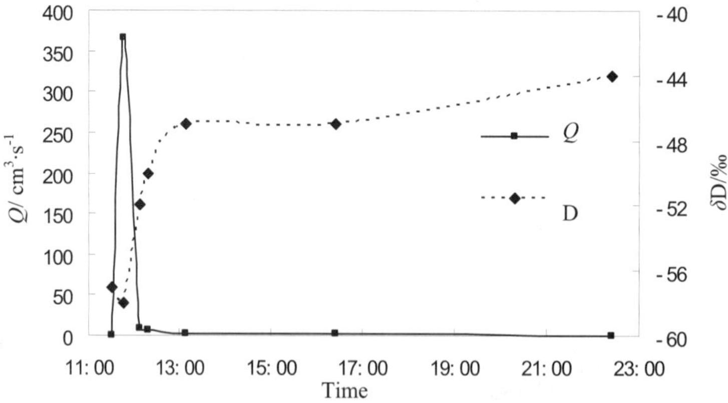

Fig. 6 δD values and discharge of runoff in surface storage recession experiment.

CONCLUSIONS

This paper presents variations of hydrogen and oxygen stable isotopes in rainfall–runoff events through laboratory experiments under constant isotopic rainfall input and low initial soil water content conditions, in order to provide more information for hydrograph separation. The main conclusions of the study are as follows:

(a) Isotopic compositions decreased to a minimum at maximum discharge and increased as discharge decreased under constant isotopic rainfall input and low initial soil water content conditions.
(b) Isotopic compositions of soil water increased along its pathway; δ^{18}O and δD values of soil water at end of surface storage recession experiment are 1.2‰ and 11‰ higher than those of surface storage in trough 2, respectively.
(c) Isotopic compositions of rainfall changed greatly as a result of soil regulation.

Acknowledgements This study is financially supported by the National Natural Science Foundation of China (50679024)

REFERENCES

Buttle, J. M. (1994) Isotope hydrograph separations and rapid delivery of pre-event water from drainage basins. *Prog. Physical Geography* **18**, 16–41.

Gu, W. Z. (1992) Experimental research on catchment runoff responses traced by environmental isotopes. *Adv. Water Sci.* **3**(4), 246–254.

Gu, W. Z. (1995) Various patterns of basin runoff generation identified by hydrological experiment and water tracing using environmental isotopes. *J. Hydraul. Engng* **5**, 9–17.

Gu, W. Z. (1996) On the hydrograph separation traced by environmental isotopes. *Adv. Water Sci.* **7**(2), 105–111.

Gu, W. Z. & Xie, M. (1997) Experimental research on Tenqiao catchment isotope hydrograph separation. *J. China Hydrology* **1**, 29–32.

Kendall, C. & McDonnell, J. J. (1998) *Isotope Tracers in Catchment Hydrology*. Elsevier Science B V, Amsterdam, The Netherlands.

Kendall, C., McDonnell, J. J. & Gu, W. Z. (2001) A look inside 'black box' hydrograph separation models:a study at the Hydrohill catchment. *Hydrol. Processes* **15**, 1877–1902.

Lin, S. Y. (2001) *Hydrological Forecasting*. China Water Power Press, Beijing, China.

McDonnell, J. J., Bonell, M., Stewart, M. K. & Pearce, A. J. (1990) Deuterium variations in storm rainfall: implications for stream hydrograph separation. *Water Resour. Res.* **26**, 455–458.

Moore, R. D. (1989) Tracing runoff sources with deuterium and oxygen-18 during spring melt in a headwater catchment, southern Laurentians, Quebec. *J. Hydrol.* **112**, 135–48.

Qu, S. M., Bao, W. M., Shi, P. & Hu, H. Y. (2006) Review on isotopic hydrograph separation methods. *Water Resour. & Power* **24**(1), 80–83.

Sklash, M. G. (1990) Environmental isotope studies of storm and snowmelt runoff generation. In: *Process Studies in Hillslope Hydrology* (ed. by M. G. Anderson & T. P. Burt), 401–435. Wiley, Chichester, UK.

Stichler, W. & Herrmann, A. (1978) Verwendung von Sauerstoff-18-Messungen fur hydrologische Bilanzierungen. *Deutsche Gewasserkundliche Mitt.* **22**(1), 9–13.

Zhang, Y. H., Wu, Y. Q., Wen, X. H. & Su, J. P.(2006) Application of environmental isotopes in water cycle. *Adv. Water Sci.* **17**(5), 738–747.

Small-scale rainfall–runoff experiments and numerical simulation in a typical small karst basin of Houzhai, Guizhou Province, China

ZHANG RONG-RONG, SHU LONG-CANG, DONG GUI-MING, LU CHENG-PENG & LIU LI-HONG

State Key Laboratory of Hydrology-Water Resource and Hydraulic Engineering, Hohai University, Nanjing 210098, China

lcshu@hhu.edu.cn

Abstract Although the runoff plot experiment is one of the important ways to research the rainfall–runoff process under different conditions, few runoff plots have been built in karst basins due to the complex hydrogeological conditions of karst aquifers. According to the characteristics of large surface leakage coefficient and lack of surface runoff in the typical small karst basin of Houzhai, China, field rainfall–runoff experiments were done in three locations from upstream to downstream with different landforms, soil conditions, and rock fracture development degree in both bare and covered carbonate rock area in Guizhou Province, southwestern China. The runoff components proportion was analysed based on the preliminary law of rainfall–runoff process gained from the observation data. Moreover, a numerical model to simulate soil water motion under experimental conditions was established using HYDRUS-1D. The simulated proportion of runoff components is basically consistent with the field-measured results and the relative error is less than 3.5%.

Key words Houzhai karst basin; bare area; covered area; rainfall–runoff experiments; HYDRUS-1D; numerical simulation

INTRODUCTION

Houzhai basin is located in southwestern China where the largest subtropical karst mountainous area of piece-link bare carbonate rocks is located. Due to a large number of uneven distributed karst fractures and conduits, strong karst development in both surface and underground, and maldistribution of covered soil thickness, this basin differs from common basins in its flow pathway, and hydraulic–hydrological characteristics. As a result of the uneven temporal–spatial distribution of rainfall, in addition to the specific underlying surface conditions in this region (thin surface soil layer, severe soil erosion and large surface leakage coefficient; Chen *et al.*, 2001), and although there is a considerable amount of rainfall, the surface water recharges rapidly to the groundwater by way of injection. Thus there is a severe lack of surface runoff and low-flow state during winter and spring in this region; this is referred to as a "drought and water shortage region in humid climate" (Wang & Shi, 2006). The hydrological model may be supported physically by in-depth study of the rainfall–runoff process, with a view to understanding the runoff generation process in this region. Moreover, it is important for the rational exploitation of water resources to analyse the composition of runoff and the conversion mechanism about PSSK (precipitation–surface water–soil water–karst water).

In this study, the response of surface runoff, interflow and underground runoff to the rainfall process was analysed in small-scale experiments based on the measurement of rainfall, surface runoff, interflow, underground runoff and soil water in the typical small karst basin of Houzhai. Then, the preliminary law of runoff generation was obtained to support the karst distributing hydrological model based on physical data.

EXPERIMENTAL PROCESS

As there is little soil cover in the bare carbonate rock area, rainfall flows into underground karst fractures, caves and rivers through the surface fractured zone, and quickly contributes to quick flow recession and weak water storage function. However, there is a dual structure of rock and soil in the covered carbonate rock area; that is: the upper layer, covered with unequal-thickness soil and the underlying layer, composed of fractured rock and conduits. This kind of vertical variety leads to different responses of surface runoff, interflow and underground runoff to rainfall, and to different runoff generation processes (White, 2002).

Three typical experimental sites were chosen in the upper, middle and lower reaches of the Houzhai basin. The basin shows a transition from bare to covered carbonate rock area from upstream to downstream. Plot 1 is Puding station (Fig. 1) in the upper reaches, which belongs to bare carbonate rock area; Plot 2 is Laoheitan station in the middle reaches, which belongs to both bare and covered carbonate rock area; and Plot 3 is Maoshuikeng station in the lower reaches, which belongs to covered carbonate rock area. On these three plots, typical profiles were dug into the fractured hard rock to measure surface runoff, interflow and underground runoff. In Plot 1, there is no interflow due to absence of soil cover.

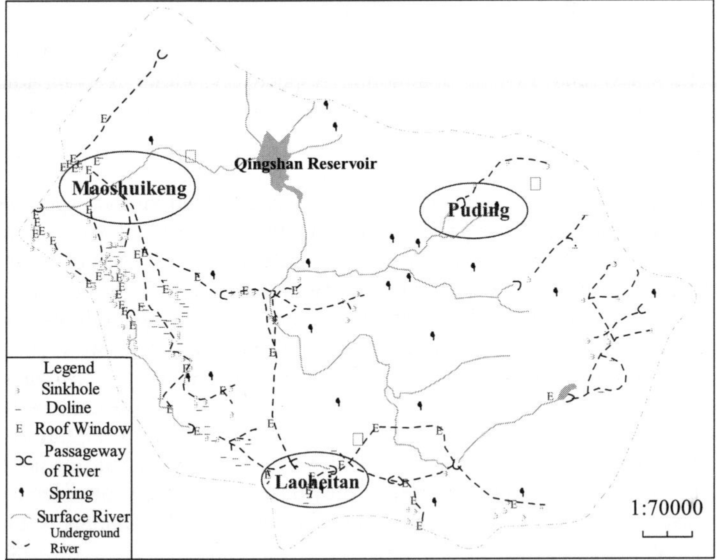

Fig. 1 Schematic map of experimental sites in the small karst basin of Houzhai.

The vertical profiles dug into the fractured hard rock were lined with plastic film, and then filled with clay to create flow-intercepting walls that cut off the water exchange between the sample plots and the surrounding soil on the selected experimental sites (Cao *et al.*, 2005a,b). In addition, considering the fact that the overlying soil of the study area is very thin (more than 97% of which is less than 40 cm), and the soil thickness is generally about 10 cm, the overlying soil was not stratified during the experiments. At a depth of 10 cm below the surface, basically at the interface of the soil layer and the fractured rock, TDR (Time Domain Reflectometry) was installed to measure the soil water content changes of the whole soil layer. The outlets on the slope of the experimental sites were connected with plastic pipes to transmit slope runoff and interflow into the storage buckets that were sealed by plastic rubberized fabric to prevent the influence of rainfall and evaporation.

RESULTS AND ANALYSIS OF EXPERIMENTS

Analysis of the rainfall process

Because of the small scale of the experiments, artificial rainfall was used with a plastic watering pot, the volume of which was determined before the experiments (Chen *et al.*, 2005). During the

Table 1 Relative information of rainfall processes in the experimental area.

Experimental plots	Latitude-Longitude	Situation	Length × width (mm)	Soil layer depth / slope	Rainfall events	Rainfall duration (s)	Accumulative rainfall (mL)	Average rainfall intensity (mm/min)
Upper reaches	26°18′08″N 105°43′59″E	Bare carbonate rock, Puding	310 × 230	None	7	1156	4270	5.38
Middle reaches	26°13′01″N 105°44′00″E	Covered carbonate rock, Laoheitan	900 × 750	10 cm; 20°	8	4956	49008	5.08
Lower reaches	26°16′27″N 105°14′36″E	Covered carbonate rock, Maoshuikeng	1000 × 800	12.25 cm; 11°	7 (1)	5120 (312)	40369 (5969)	2.85 (1.43)

Note: Rainfall conditions in the bare carbonate rock area are shown in the lower reaches plot.

rainfall processes of the three experiments in the upper, middle and lower reaches, the rainfall intensity is 3.27–6.75, 1.04–6.57 and 1.64–3.53 mm/min, respectively. Other relevant rainfall information (average rainfall intensity, duration, cumulative rainfall, etc.) is given in Table 1.

Analysis of soil water changes

There is no soil cover on the upper reaches, so soil water was only analysed on the experimental plots of the middle and lower reaches. However, considering the similarity of the soil water changes law between the middle and lower reaches plots, the middle reaches plot was chosen to be analysed alone. Some results are shown in Fig. 2, which shows a correlation between soil water changes and accumulative rainfall on Laoheitan station of middle reaches as follows: (a) the process of soil water changes responds to the process of rainfall quickly. There are eight peak values of soil water which are consistent with eight rainfall events. (b) When the rainfall stops every time, the increment of soil water in covered layer decreases accordingly, even becomes negative. However, the negative increment last a short time during the first two rainfall events. After the third rainfall event, the negative increment last much longer and the sum of negative increment has exceeded the sum of soil water increment in previous. That is to say, after the end of rainfall, soil water decreases but becomes saturated, so interflow forms and recharges to underground runoff.

Analysis of surface runoff, interflow and underground runoff process

According to the water balance principle, when the whole underlying surface above the hard rock of a closed basin is treated as the study object, the water balance equation of the basin would be

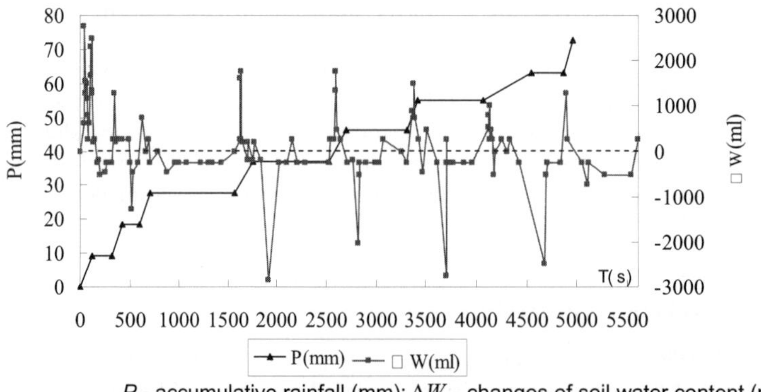

P - accumulative rainfall (mm); ΔW - changes of soil water content (mL)

Fig. 2 The correlation between soil water changes and accumulative rainfall at Laoheitan station.

expressed as follows:

$$\Delta W = W_2 - W_1 = P - E - R_s - R_i - R_g \quad (1)$$

where ΔW is the change of soil water content; W_1 and W_2 are soil water contents before and after rainfall events, respectively; P is rainfall; E is evaporation which can be neglected in this experiment as a result of short experimental duration (4956 s) and intensive rainfall events, so $E = 0$; and R_s, R_i and R_g are surface runoff, interflow and underground runoff, respectively. Each variable is measured during periods of experimental time.

Analysis of rainfall–runoff components in the bare carbonate rock area There is a fracture 13 cm in length, 1.4 cm in width and 12 cm away from the left side of the experimental rock in Puding County (Fig. 3). As there is no soil layer cover here, $R_i = 0$. As the data of rainfall and surface runoff can be obtained accurately, underground runoff, R_g can be got by the water balance.

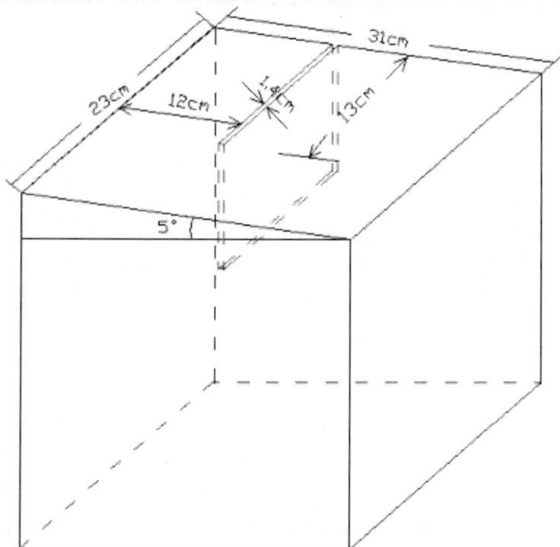

Fig. 3 Sketch map of the experimental fracturing of rock in Puding County.

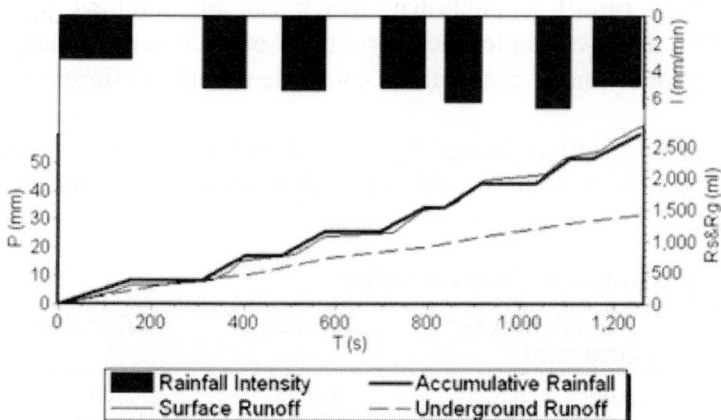

Fig. 4 The hydrograph of accumulative rainfall, surface runoff and underground runoff.

Figure 4 displays the hydrograph of accumulative rainfall, surface runoff and underground runoff. From Fig. 4 we can see surface runoff responds to rainfall significantly. The time of surface runoff outflow lags about 20 s behind the rainfall events. Surface runoff significantly

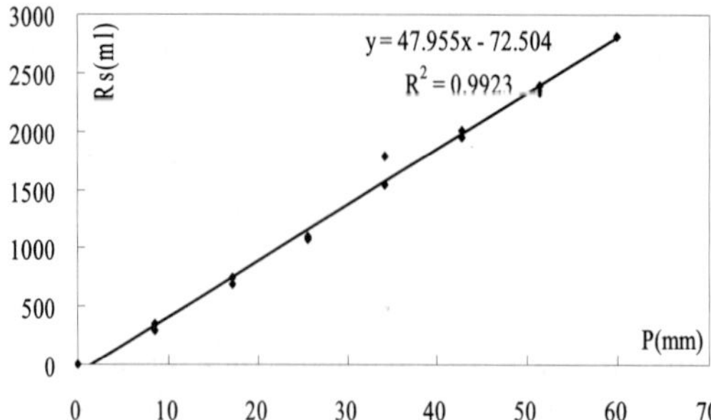

Fig. 5 The correlation between surface runoff and accumulative rainfall.

correlates to accumulative rainfall with a high deterministic coefficient of 0.9923 (Fig. 5). Because there is no soil layer cover to regulate and store the runoff, the rainfall can recharge the groundwater only through the 1.4 × 13 cm fracture of the experimental rock whose hydraulic conductivity is limited, and most rainfall directly transmits to surface runoff. The underground runoff through the penetrated recharge of fracture increases linearly with a high deterministic coefficient of 0.9970.

The experiment was carried out under the following conditions: 1256 s duration of rainfall, 4279 ml accumulative rainfall and 5.38 mm/min average rainfall intensity. The total surface runoff of Puding County formed by rainfall recharge in bare carbonate rock area is 2835 ml, accounting for 66.39% of the total rainfall. The other 33.61% of the rainfall is underground runoff recharged by fracture, the amount of which is 1425 ml. Applying $\alpha = P_r/P = 0.3361$, the rainfall infiltration coefficient in the Puding County plot is 0.3361, where P is the total rainfall and P_r is the underground recharge from rainfall.

Analysis of rainfall–runoff components in the covered carbonate rock area Besides the upper reaches plot which located in the bare carbonate rock area, the other two plots both have soil layer cover of 10 cm depth. After completing the rainfall–runoff experiments, we removed the soil layer from the lower reaches plot to keep rain falling in order to observe the response of fracturing rock under the soil layer to rainfall. Because the soil layer is very thin (only 10 cm) the distribution of soil water content is considered uniform. And the data measured by TDR is taken as the water content of the whole soil layer. The data of rainfall, surface runoff and interflow can be obtained accurately, and then only underground runoff is unknown, which can be obtained by water balance. Also taking Laoheitan station as an example, the proportions of each runoff component can be checked in Table 2, and the proportion of each runoff component under different rainfall intensities is shown in Fig. 6.

The total surface runoff of Laoheitan station formed by rainfall recharge in bare carbonate rock area is 2950 mL under the condition of 4956 s rainfall duration, 49 008 mL accumulative

Table 2 The rainfall–runoff components proportions of Laoheitan station.

Rainfall order	Start (s)	End (s)	Rainfall intensity (mm/min)	R_s/P (%)	R_i/P (%)
1	0	126	4.43	1.03	0.00
2	319	421	5.34	4.65	0.00
3	597	703	5.27	9.39	8.04
4	1575	1752	3.16	4.77	28.81
5	2522	2698	3.09	4.57	38.03
6	3314	3416	5.20	0.67	42.39
7	4066	4549	1.04	0.00	24.23
8	4871	4956	6.57	22.12	49.18

rainfall and 5.08 mm/min average rainfall intensity, accounting for 6.02% of the total rainfall. The interflow accounts for 23.74% (11 635 mL) of total rainfall. The other 34.98% (17 143 mL) of total rainfall is underground runoff recharged by fractures. Therefore, the rainfall infiltration coefficient in Laoheitan plot (Plot 2) is 0.3498.

Similarly, the total surface runoff of Maoshuikeng station (Plot 3) formed by rainfall recharge in covered carbonate rock area is 19 620 mL under the conditions of: 5120 s duration of rainfall, 40 369 mL accumulative rainfall and 2.85 mm/min average rainfall intensity, accounting for 47.71% of the total rainfall. The interflow accounts for 3.37% (1362 mL) of total rainfall, and the other 2.55% (1029 mL) of total rainfall is underground runoff recharged by fractures. Therefore, the rainfall infiltration coefficient in Plot 3 is merely 0.0255, which is out of line with the Laoheitan (Plot 2) rainfall infiltration coefficient. Two possible reasons are: (a) due to the high heterogeneity of the karst aquifer medium, the development of rock fractures is very weak in Plot 3; hence, the rainfall infiltration recharge to the groundwater is very low. (b) Compared to Plot 2, the rainfall area is larger and the soil layer is thicker in Plot 3, so recharge from rainfall to soil water deficiency is more; in consequence, recharge from rainfall to underground runoff is comparatively low. In order to prove the more likely reason is the weak penetrability in the Maoshuikeng plot, we removed the soil layer cover and carried out a group of experiments directly on the bare rock. Although the result is not representative, so it is not analysed in detail, it proved rainfall recharges to groundwater are very low with a very low value of α, 0.0467. The experimental results reflect the fact that the rock fracturing development in this area is very weak.

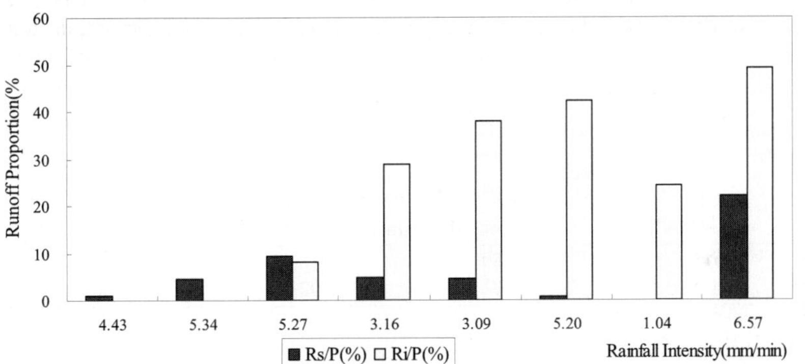

Fig. 6 The runoff components proportions of Laoheitan station under different rainfall intensities.

From Fig. 6, we can see that interflow occurs during the third rainfall process (the rainfall intensity is 5.27 mm/min). This is consistent with conclusion (b) above; that is: the soil layer begins to recharge groundwater when it becomes saturated. After the soil layer cover has been saturated, the rainfall intensity is stronger, the surface runoff proportion larger and the groundwater recharge lower.

NUMERICAL SIMULATION OF THE RAINFALL INFILTRATION PROCESS

Model foundation

Owing to the limited conditions of rainfall–runoff field experiments, the underground runoff cannot be observed, but can be calculated by water balance law in the covered carbonate area. In order to further study the conversion relationship among rainfall, soil water and groundwater, a numerical model to simulate soil water motion under saturated–unsaturated conditions was established using HYDRUS-1D that can be used to quantitatively research the function of the soil layer in the groundwater recharge process. The HYDRUS-1D software package is a finite element model for simulating the one-dimensional movement of water, heat and multiple solutes in variably saturated media, which is published by the International Groundwater Modeling Centre (Šimunek *et al.*, 2008). It is applicable to variably stable boundary conditions, and provided with flexible functions of input and output. The model is described by the Richards equation:

$$\frac{\partial \theta}{\partial t} = \frac{\partial}{\partial z}\left[K(\theta)\frac{\partial h}{\partial z}\right] + \frac{\partial K(\theta)}{\partial z} \tag{2}$$

The equation is solved by the Galerkin linear finite element method (Wang *et al.*, 2005; Hilten *et al.*, 2008).

Model discretization Take the soil layer thickness of Laoheitan plot as the model calculated depth. The space interval is set as 1 mm. The simulated period is from the beginning of rainfall until the end of rainfall with 10-s time interval.

Initial and boundary conditions The initial condition is set as the observed soil water content before rainfall events, that is $\theta|_{t=0} = \theta_0(z)$. Value each node according to linear interpolation. For Laoheitan plot, $\theta_0 = 14.6\%$.

The soil layer of upper boundary receives recharge directly from rainfall and evaporation is neglected, so the upper boundary condition can be set as constant flux boundary. The observed rainfall amount can be valued to the top cells directly in HYDRUS-1D. The lower boundary is set as free drainage boundary (Hu *et al.*, 2006; Wang *et al.*, 2007).

Soil water content parameters The soil particle composition of the Laoheitan and Maoshuikeng plots is shown in Table 3, according to the particle size analysis of the field soil samples.

Table 3 Soil particle compositions of Laoheitan and Maoshuikeng plots.

Location	Soil particle size (%): Clay (< 0.002 mm)	Silt (0.002–0.05 mm)	Sand (0.05–2.0 mm)	UNSODA Name
Laoheitan	7.50	53.59	38.91	Sandy loam
Maoshuikeng	4.09	3.37	92.54	Sand

Based on the observed data of particle composition, dry density, soil water characteristic curve, saturated hydraulic conductivity of 1913 kinds of different rocks, USSL (United States Salinity Laboratory) establishes a function relationship of soil water characteristic parameters with saturated hydraulic conductivity, particle composition and dry density by use of neural network technology (Rosetta software). According to the observed soil particle composition (Table 3), and in addition with the observed saturated soil water content, the soil water characteristic parameters is predicted by Rosetta software, and shown in Table 4.

Table 4 Initial values of soil water parameters by Rosetta software.

Location	θ_r	θ_s	α	n	k_s (mm/s)	l
Laoheitan	0.065	0.41	0.0075	1.89	0.0122801	0.5

4.2. Model identification

Both α and n are empirical parameters. The value of α varies from 0.05 to 0.2, and is not sensitive to the change of soil water content. The parameter n is an exponent power index which is the most sensitive parameter to the change of soil water content. Resident water content θ_r, saturated water content θ_s and saturated hydraulic conductivity k_s do not vary too much. So n should be identified mainly in model identification. By analysing the fitting degree between calculated and observed soil water content, the soil water parameters are modified repeatedly. When their error achieves the precision, the model parameters can represent the soil water parameters. The objective error function of calculated and observed soil water content can be expressed as follows:

$$E = \sum_{i=1}^{m}\sum_{j=1}^{n} W_j(\theta_{ij}^{e} - \theta_{ij}^{0}) \tag{3}$$

Where m is the total numbers of periods; n is the total numbers of observed points; W_j is the weight; θ_{ij}^{e} and θ_{ij}^{0} are the calculated and observed soil water content of the jth point in the ith period, respectively. The parameters of the minimum objective function are optimum.

A comparatively idea model identification was gained by adjusting the parameters. Figure 7 shows the fitting curve of calculated and observed soil water content in the Laoheitan plot. The change tendency fits very well, and the errors are less than ±5%. The optimum parameters are shown in Table 5.

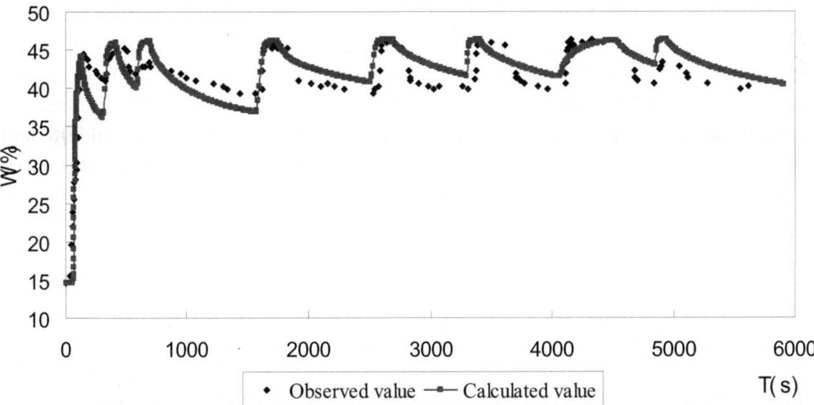

Fig. 7 The fitting curve of calculated and observed soil water content of Laoheitan station.

Table 5 The optimum soil water parameters.

Location	θ_r	θ_s	α	n	k_s (mm/s)	l
Laoheitan	0.087	0.463	0.0036	1.56	0.026	0.5

Table 6 Numerical simulated water balance of Laoheitan station.

Recharge		Discharge		Proportion of rainfall (%)
Equilibrium terms	Amount (mm)	Equilibrium terms	Amount (mm)	
Rainfall infiltration	74.74	Evaporation	0.152	
		Surface runoff	4.357	5.83
		Interflow	18.327	24.52
		Infiltration to groundwater	27.032	36.17
Storage variable of soil water			26.337	35.24
Total	74.74		76.205	
Relative error (%)	1.96			

Model application and analysis of results

The water balance under the numerical simulated conditions was found by the Mass Balance Information of HYDRUS-1D (Table 6). From the water balance table, the recharge from rainfall to soil water is 74.74 mm and evaporation is neglected because of the short simulated period. The results show: 35.24% of rainfall turns to the soil water storage variable; 5.83% of rainfall turns to surface runoff; 24.52% of rainfall turns to interflow; and 36.17% of rainfall turns to underground runoff. The proportions are basically consistent with the experimental result (6.02%, 23.74% and 34.98%). So the numerical model we built can be accepted, meanwhile the model can further prove the veracity of our experimental results.

CONCLUSIONS AND SUGGESTIONS

(a) The degree of rock fracture development influences rainfall infiltration significantly, and strong fracturing rocks benefit rainfall infiltration. The rainfall infiltration coefficient of the Laoheitan plot is 0.3498, much larger than the rainfall infiltration of Maoshuikeng plot (0.0255). It is mainly due to the poor rock-fracture-development degree in Maoshuikeng plot.

(b) The accumulative surface runoff has significant linear correlation to the accumulative rainfall, and the underground runoff from fracture recharge increases linearly with the time for the rock mass with good fractural network. However, the experimental condition is set as only one large fracture in Puding County plot. Therefore, the runoff generation law of the fracture network needs to be validated further.

(c) The numerical model in Laoheitan plot is built using HYDRUS-1D and the model results are rational and reliable. Meanwhile, it supports the experiments we did theoretically. However, there are still some disadvantages. The biggest problem is that the comparatively sensitive parameter is empirical. It can contribute to subjective deviation in the process of man-made adjusting parameters.

(d) The runoff generation process in karst area is much more complicated due to the different natural characteristics of underlying surface and the different runoff generation mechanism. Therefore, we should enforce the mechanism research of hydrological process. By increasing observation intensity and improving observation means, we should recognize and understand the rainfall-infiltration process physically. Moreover, the scale of experiments we have done is comparatively small, the temporal-spatial scale recognition of rainfall-infiltration process should be extended and deepened.

Acknowledgements The research work was funded by the National Development Programme for Basic Key Science Research entitled "The fundamental theoretical problems concerning the processes of rocky desertification in the mountainous areas of Southwest China and adaptive ecological rehabilitation" (2006CB403200), the project "Study on Coupled Processes of Atmosphere–Land Hydrology" (IRT0717), supported by the Programme for Changjiang Scholars and Innovative Research Team in University (PCSIRT), the "111" Project under Grant B08048 supported by the Ministry of Education and State Administration of Foreign Experts Affairs, China and the National Natural Science Foundation of China (70711120407). The authors convey their deepest appreciation for this support.

REFERENCES

Cao Jian-sheng, Liu Chang-ming, Zhang Wan-jun & Yang Yong-hui (2005a) Field experiment on discharge of rain seepage from fissured rock in a catchment. *Shuili Xuebao* **36**(11), 1335–1340 (in Chinese).

Cao Jian-sheng, Liu Chang-ming & Zhang Wan-jun (2005b) Research on the process of rain infiltration to fissured rock in a catchment with a dual structure of rock and soil. *Adv. in Natural Science* **15**(6), 759–763 (in Chinese).

Chen Hong-song, Shao Ming-an, Zhang Xing-chang & Wang Ke-lin (2005) Field experiment on hill slope rainfall infiltration and runoff under simulated rainfall conditions. *J. Soil & Water Conserv.* **19**(2), 5–8 (in Chinese).

Chen Hong-yuan, Hu Xing-hua, Yang Yong & Chen Bang-yu (2001) Study of structures and water resources development patterns in karst watershed. *Carsologica Sinica* **20**(1), 21–26 (in Chinese).

Hilten, R. N., Lawrence, T. M. & Tollner, E. W. (2008) Modeling stormwater runoff from green roofs with HYDRUS-1D. *J. Hydrol.* **358**, 288–293.

Hu Ke-lin, Xiao Xin-hua & Li Bao-guo (2006) Numerical analysis of the effect of the different lower boundary conditions on water drainage in irrigated field. *Adv. in Water Science* **17**(5), 665–670 (in Chinese).

Šimůnek, J., Šejna, M., Saito, H., Sakai, M. & van Genuchten, M. Th. (2008) Hydrus-1D manual. Version 4.0, 1–281.

Wang La-chun & Shi Yun-liang (2006) The availability of rain water resources at karst mountains in southwest China. *Guizhou Science* **24**(1), 8–13 (in Chinese).

Wang Shui-xian, Zhou Jin-long, Yu Fang & Dong Xin-guang (2005) Application of HYDRUS-1D model to evaluating soil water resource. *Research of Soil and Water Conservation* **12**(2), 36–38 (in Chinese).

Wang Xiao-feng, Liu Guang-yan & Wang Tao (2007) Numerical simulation of overall process of rainfall infiltration in unsaturated soil. *J. China Hydrol.* **27**(1), 30–32 (in Chinese).

White, W. B. (2002) Karst hydrology: recent developments and open questions. *Engng Geol.* **65**, 85–105.

Reflections about water quality based upon conceptual aspects of monitoring and modelling of organic content

HELOISE GARCIA KNAPIK[1], MARIANNE SCHAEFER FRANÇA[1], CRISTOVAO VICENTE SCAPULATEMPO FERNANDES[1], JULIO CÉSAR RODRIGUES DE AZEVEDO[2] & MONICA FERREIRA DO AMARAL PORTO[3]

1 *Federal University of Parana, Hydraulic and Sanitary Department, Polytechnic Center s/n, Block 5, PO Box 19011, Curitiba PR, 81531-990, Brazil*
cris.dhs@ufpr.br

2 *Federal Technological University of Parana, Academic Chemistry and Biology Department, Curitiba PR, Brazil*

3 *University of São Paulo, Hydraulic and Sanitary Department, São Paulo SP, Brazil*

Abstract This paper presents reflections about the water quality conditions of the Iguaçu River in the Metropolitan Area of Curitiba, with special attention to the dynamic of the organic content in the overall watershed. The reason for this analysis is to provide consistent water resources planning and management plan for this complex basin in the context of producing reliable and feasible strategies for water quality recovery. The methodology used in this paper relies on a consolidation of hydrological and water quality data base that allowed the implementation, calibration and validation of a water quality model to achieve the main goal. Additionally, a complementary monitoring plan is proposed to better assess the dynamic of organic content for the river system in analysis. These results summarise the potential positive implications of a strong decision support system to this basin that will certainly induce strategies for the water quality recovery of the river system. Additionally, this paper highlights the main issues of the concepts defined that can be reproduced in other critical basins with strong organic water quality pollution.

Key words Upper Iguaçu Basin; monitoring; water quality modelling

INTRODUCTION

Many researchers are motivated by the prospect of preventing the impact of control measures. It is therefore very important to attempt to establish elements for a careful analysis of water clean-up goals based on measures to be implemented in a basin like the Iguaçu River in the Metropolitan Region of Curitiba-Brazil, focusing not only on impact in terms of improving environmental quality, but also on its economic and financial implications. Obviously, from this standpoint, mathematical modelling of water quality in rivers, reservoirs and lakes should be considered a major tool to support the decision-making process, especially when the water resources management tools are implemented.

However, the dynamics of organic matter is slightly complex, especially in the basin where fast urban development occurs, as in the case of the Upper Iguaçu Basin. In some cases, biological activity, responsible for the degradation of organic matter, may not keep up with the degree to which the water body is compromised, either because of chemical interference or due to a characteristic of the ecosystem itself. Therefore, the manager must identify the potentials and weaknesses of each basin with a view to preserving the water body as well as developing society.

Thus, the main contribution of this article is to promote an assessment of water quality concerning the input of organic matter into the Upper Iguaçu Basin, by monitoring water quality based on a qualitative and quantitative study of the organic matter dynamics.

WATER QUALITY MONITORING CHALLENGES

Water quality monitoring is a stage of special interest, and essential to establish a solid foundation of knowledge about the basin studied. More importantly, it provides the manager with a broad view of the degree to which the water body has been affected, and also enables the identification of sources and types of pollution.

However, the monitoring network that currently prevails in most of the main Brazilian hydrographic basins is lagging behind, sometimes as spatial amplitude, with few sampling points, others as temporal amplitude, with extensive periods without information. This failure to perform monitoring, and the way it is done, ultimately has an effect on water resources management, with late or inefficient actions in given regions.

Another problem is the lack of joint water quality and quantity data. It is difficult to estimate the self-deputation or assimilation capacity of the water body without looking quantitatively at flow available to dilute a given pollutant. This lack, as can be seen in the fact that the long river gauging series of qualitative data have fallen behind, prevents their use, for instance in mathematical modelling of the water quality, which in certain procedures such as calibration, requires joint data to build the simulation scenarios.

In qualitative terms, the current base of standards established, according to CONAMA Resolution 357/05, also does not include the present water quality indicators efficiently. Assays that improve understanding of the dynamics of organic matter did not come into standardized use. An example of this is the determination of dissolved organic carbon, the main means of transporting metals and other pollutants in the water column. It is crucially important to study them, as a complement to the classical laboratory procedures, since they include all the organic components of a water sample, independent of its state of oxidation, besides not measuring inorganic compounds which can contribute to the oxygen demand in BOD and COD analyses (Apha, 1998; Thomas, 1999; Villa, 2005). However, determining dissolved organic carbon does not supply data on the nature of the organic substances, because the natural organic matter is constituted both by aquagenic organic compounds, and by pedogenic ones (Buffle *et al.*, 1987). Thus, besides quantitative data, it is necessary to use qualitative assays which will help understand pollution, its source and state of degradation.

Both spectroscopy in the visible ultraviolet region, and fluorescence spectroscopy, are being used by several researchers to characterize the structural composition of dissolved organic carbon and identify its possible sources in the systems studied (Cabaniss & Shuman, 1987; Senesi *et al.*, 1989; Ahmad & Reynolds, 1995; Korshin *et al.*, 1997; Peuravuori *et al.*, 1997; Frimmel, 1998; Artinger *et al.*, 2000; Westerhoft & Anning, 2000; Chen *et al.*, 2002; Pons *et al.*, 2004; Villa, 2005; Azevedo *et al.*, 2006; Spencer *et al.*, 2007). Thomas *et al.* (1999) mention the UV-Vis spectroscopy assay as a complement to the determination of total organic carbon (TOC), since it provides qualitative information about organic matter. Westerhoft & Anning (2000) indicate the fluorescence assay to identify allochthonous (outside the system) and autochthonous (inside the system) sources of organic matter. In their studies, authors such as Ahmad & Reynolds (1995), Galapate *et al.* (1998), Chen *et al.* (2002) and Pons *et al.* (2004) utilized the determination of fluorescence to distinguish between organic matter from natural sources and dissolved organic matter from domestic effluents.

It is therefore important for water quality monitoring to include both conventional assays, such as BOD, COD, DO and TOC, and UV-Vis spectroscopy and fluorescence assays, and other parameters, since they provide a complement that enables the researcher to improve the characterization of organic matter, an activity which is relevant in different situations, such as in the context of the mathematical modelling of water quality, because generally organic matter from aquagenic and pedogenic sources have distinct hydrodynamic properties, and also influence the organic carbon cycle (Zumstein & Buffle, 1989; Villa, 2005).

MODELLING – FROM PHYSICAL REALITY TO NUMERICAL REPRESENTATION

Modelling consists of a simplified representation of different systems and interactions that occur in reality, through hypotheses regarding the structure or the behaviour of a physical medium. It is based on two fundamental components: equations to represent flow and mass transport equations which show the variation of the water quality variable concentration. An example of this representation may be seen in Figs 1 and 2. In this context, it is important to know the self-depuration capacity of the receiving body, as well as the sedimentation and capacity to remove the pollutant mass in the system in order to improve the technical bases of pollutants in the receiving body.

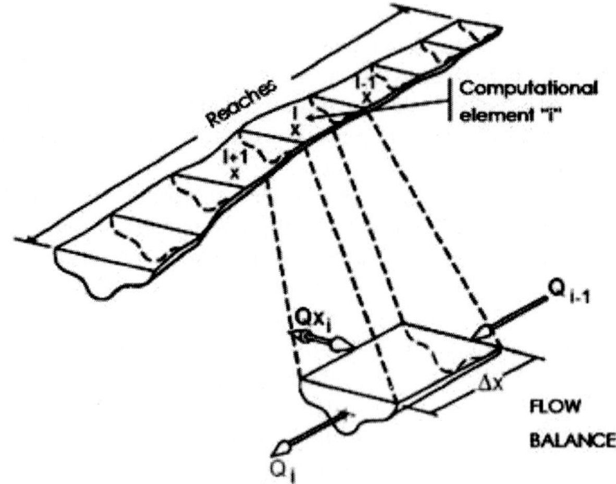

Fig. 1 Schematic representation of a river reach subdivision. Adapted from Brown & Barnwell (1987).

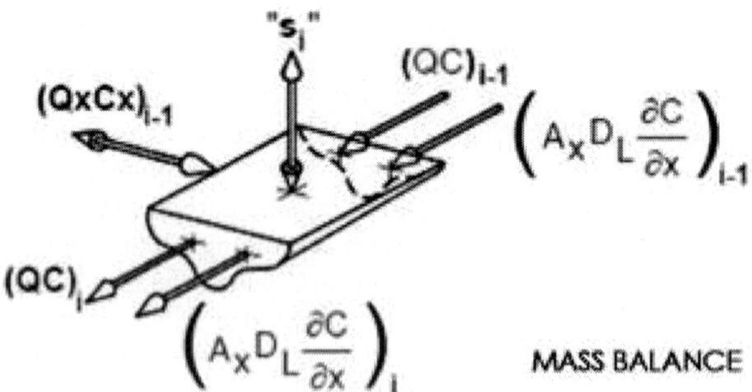

Fig. 2 Schematic representation of the model mass balance. Adapted from Brown & Barnwell (1987).

Water quality simulation should be seen as a major stage to support the implementation process of water resources management instruments. In this context, a mathematical model of water quality is a basic methodological tool, since it allows identification of the current behaviour of the dynamics of different constituents in the water body, and also evaluation of the different impacts as regards improving environmental quality (Porto *et al.*, 2007).

Several researchers have developed or improved water quality simulation models in the last few decades, especially for the DO and BOD variables. The 1925 Streeter-Phelps model, which proposes a first order decay for BOD and DO, pioneered the representation of the combined processes. The most recent models, including QUAL2E (Brown & Barnwell, 1987), enable the simulation of more variables, with details on the physical, chemical and biological processes that interact in the water body.

However, the use of more complex and excessively simplified models requires care and much theoretical grounding. Little further study has been done in recent years, and possibly an important fraction of dissolved oxygen consumption is neglected when one only simulates the contribution of organic matter, usually DO and BOD (Palmieri, 2003; Bäumle, 2005; Rodrigues, 2005; Knapik, 2006). The insertion of contributions of nitrogen, phosphorous and algae may significantly influence decision-making, especially when progressive water quality goals are implemented, which require a good diagnosis of water quality. However, studies that discuss these aspects have not been properly reported in the Brazilian literature. According to Chapra (1997), limiting oneself to studying the problem in a traditional manner is like seeing only what one can reach and considering that only this parcel is significant. This way of seeing things deserves re-evaluation in water resources management.

Figure 3 shows an approach, according to Chapra (1997), to the simulation of the behaviour of dissolved oxygen together with other processes, including reactions with nitrogen (organic, ammoniacal, nitrite and nitrate), with phosphorus (organic and dissolved), chlorophyll A, organic matter (BOD), and oxygen losses and gains by sedimentation and re-aeration, respectively. Parameters F, K, α, β, σ, ρ and μ (Fig. 3) refer to the calibration of the different processes. These parameters need to be adjusted to the physical, chemical and biological conditions of the basins studied.

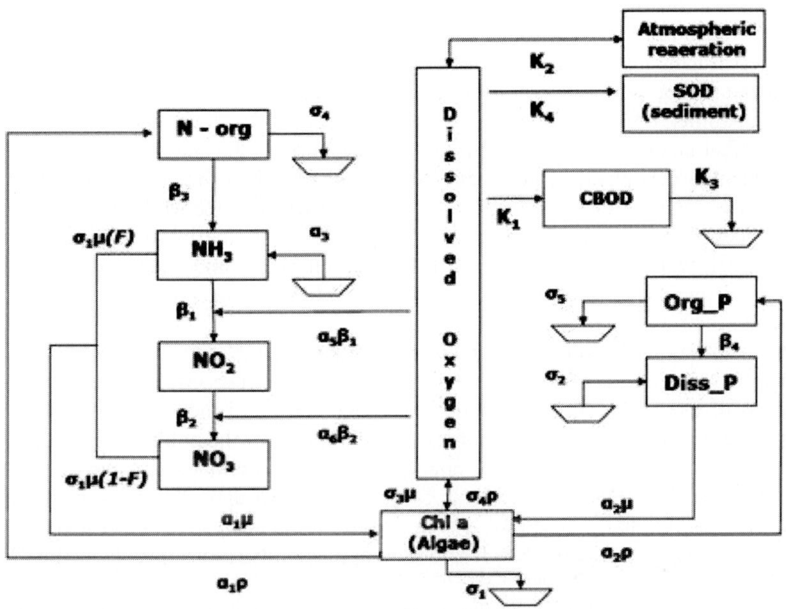

Fig. 3 Scheme of interrelationships in the dissolved oxygen balance. Adapted from Brown & Barnwell, (1987).

Another important aspect in the water quality simulation process is model calibration. In this stage the parameters of hydrological, hydraulic and water quality aspects are defined, reproducing the scenario of information obtained in the field. This consists of varying water quality model parameters to obtain an optimum result among the results of simulation by a model and the field data scenarios. The calibration data are established with a view to the greatest possible similarity for the project condition that is being studied. Despite the importance of the model calibration process, a large proportion of similar studies do not present the methodology clearly, and field data to establish the relationship between model calculations and data scenarios are generally lacking. Furthermore, calibration is done with a view to getting things right with the laboratory data, producing artificial changes in the constants that represent the physical, chemical and biological realities. The calibration methodology is time-consuming (Knapik, 2006; Kondageski, 2008). This means computational costs, which are frequently high. Thus, often not much is invested in calibration and in training people to perform such tasks with a view to minimizing the expenditure during this phase of the project. In such cases, a decision which will produce a mathematical representation is not very representative of the physical, chemical and biological reality of a river.

THE EXAMPLE OF THE UPPER IGUAÇU BASIN

The water quality evaluation presented in the following research is connected to Project "Critical Basins: technical bases for the definition of Progressive Goals for their Classification and Integration with the other Management Instruments", developed in a partnership between UFPR (Federal University of Paraná) and USP (São Paulo University) from 2005 to 2007. This project evaluated the development of methodologies that would allow the basins to establish progressive goals to remove pollution and implement the classification plan, working on their conceptual, institutional and technical aspects applied in the Upper Iguaçu Basin. Two other studies were also used as a database, Project "Analysis of Economic and Environmental Sustainability of Water Clean-Up Goals – Case Study: Upper Iguaçu Basin", the predecessor of Project Critical Basins, developed by the same institution in the years from 2003 to 2005, and the "Pollution Clean-Up Plan for the Upper Iguaçu Basin " developed by SUDERHSA in 2000.

Project Critical Basins integrated concepts such as definition of reference flow, water quality parameters, understanding of issues concerning the calibration of mathematical tools and establishing sustainability criteria to evaluate the executability of water clean-up measures and their consequent effect on classification conditions. For this purpose the activities performed in the project were segmented into physical goals directed at understanding the concept of progressive goals, supported on a database to be consolidated for Iguaçu River in the Metropolitan Region of Curitiba. The project phases involved hydrological studies, with the definition of a hydrological regionalization model, water quality studies, with field monitoring, simulation and calibration of model QUAL2E (Brown & Barnwell, 1987), and application of the concepts in a model of sustainability analyses of water clean-up goals.

The technical base tools considered hydraulic characteristics of the rivers, their self-depuration capacity, water availability, effluent discharge flow, the respective concentration of the pollutant, the class of use of the receiving body, the concentration of the pollutant in the system according to the flow to which it is submitted, besides the removal of the pollutant from the system through the intake points. As to the water quality parameters simulated in model QUAL2E, emphasis was given to the BOD (Biochemical Oxygen Demand) and DO (Dissolved Oxygen) concentrations. The institutionally-based tools were based on the analysis of legal texts available for the classification (CONAMA and CNRH) and their applicability, indicating the difficulties for their implementation.

Technical and institutional tools were integrated by constructing water quality scenarios for the river basin, showing the interference of the reference flow, of the different levels of investment and of removal of pollutant loads. The analysis of scenarios allowed showing the importance of the integrated use of tools for water use concessions, both for intakes and for discharges, and the potential reduction of a pollutant load achieved with the implementation of billing for water use. The concept of water quality permanence curve with a tool for a strategy of classification with progressive goals was also consolidated (Porto *et al.*, 2007).

The Iguaçu River at the Upper Iguaçu Basin serves as an important water resource for the southeastern area of Parana State, Brazil. In the Upper Iguaçu Basin, the river drains an area of about 2800 km^2 and length of the main stream is over 100 km. Currently, about 3 million people reside within the basin, and an additional number intake drinking water from the river. During the last several decades, the combination of rapid population growth coupled with industrial and urban development, mainly with irregular occupation of flood plains, has resulted in a serious deterioration of water quality. Major pollution sources include domestic sewage, industrial wastewater, and urban and agricultural runoff. As a consequence of this process, problems have been encountered in the water supply, sanitary sewage treatment and urban drainage systems, which did not keep up with the growth of cities, negatively affecting the environment and people's quality of life.

Table 1 shows a summary of the mean concentrations obtained for the main variables monitored during the period from June 2005 to July 2006, in six points located in the Iguaçu River, and whether they were covered or not by the water quality classes according to CONAMA Brazilian Resolution 357/05. A more detailed study of the results of monitoring can be found in Knapik (2006) and Porto *et al.* (2007).

Table 1 Summary of the main variables analysed during the monitoring period.

Variables		P1	P2	P3	P4	P5	P6
BOD (mg/L)	Mean	9,30 [(c)]	17,90 [(d)]	20,60 [(d)]	18,40 [(d)]	14,52 [(d)]	9,81 [(c)]
	Standard Deviation	7.46	9.68	11.29	8.19	9.68	4.20
	Samples	19	19	19	19	18	15
COD (mg/L)	Mean	18.06	30.67	27.79	33.18	23.58	16.95
	Standard Deviation	12.79	15.53	7.82	14.83	10.35	5.40
	Samples	18	18	18	18	17	15
DO (mg/L)	Mean	6,10 [(a)]	2,73 [(c)]	2,12 [(c)]	1,51 [(d)]	1,77 [(d)]	2,65 [(c)]
	Standard Deviation	1.26	1.68	1.95	1.38	0.83	0.98
	Samples	15	15	15	15	14	10
TOC (mg/L)	Mean	7.80	14.51	13.57	13.38	10.45	9.71
	Standard Deviation	5.24	9.68	8.88	7.17	5.88	7.18
	Samples	19	19	19	19	18	15
Turbidity (NTU)	Mean	13.35 [(a)]	26.28 [(a)]	20.50 [(a)]	37.01 [(a)]	28.85 [(a)]	21.73 [(a)]
	Standard Deviation	5.93	38.71	16.43	35.39	32.03	20.48
	Samples	18	18	18	18	17	14
Conductivity (μS/cm)	Mean	20.64	123.72	110.33	117.41	104.55	91.55
	Standard Deviation	5.78	41.31	39.56	40.57	33.57	33.20
	Samples	19	19	19	19	18	15
Total Dissolved Solids (mg/L)[(e)]	Mean	75,23 [(a)]	151,67 [(a)]	147,71 [(a)]	176,54 [(a)]	151,78 [(a)]	154,71 [(a)]
	Standard Deviation	47.16	54.32	70.42	96.06	46.52	63.74
	Samples	17	19	19	18	18	15
Total Suspended Solids (mg/L)	Mean	17.24	25.45	28.87	58.98	41.15	34.87
	Standard Deviation	7.92	16.57	10.43	65.66	46.60	24.74
	Samples	19	19	19	19	18	15
pH[(e)]	Mean	6.40 [(a)]	6.94 [(a)]	6.85 [(a)]	6.94 [(a)]	6.84 [(a)]	6.95 [(a)]
	Standard Deviation	0.43	0.39	0.38	0.35	0.35	0.38
	Samples	19	19	19	19	17	15
Temperature (°C)	Mean	17.14	17.61	18.12	18.04	18.22	19.07
	Standard Deviation	2.78	3.15	3.15	3.14	3.18	3.38
	Samples	19	19	19	19	18	15
Secchi depth (cm)	Mean	66.00	35.88	34.38	24.17	46.56	39.58
	Standard Deviation	21.15	20.78	13.15	10.62	15.99	15.59
	Samples	15	17	16	12	16	12
Flow (m^3/s)	Mean	5.11	19.19	28.80	52.95	55.99	67.12
	Standard Deviation	4.32	23.40	29.56	48.51	47.59	61.60
	Samples	17	19	19	17	17	15

[(a)]Class 1, [(b)]Class 2, [(c)]Class 3, [(d)]Class 4, [(e)]standard value for all classes.

As could be observed, for the variables monitored with standards established by CONAMA 357/05, the mean concentration of BOD and DO remained in classes 3 and 4 for most of the points monitored, except for DO concentration at point P1, located in the Iguaçu River headwater, with levels compatible with class 1. At this point, in contrast to the other points monitored along the Iguaçu River, there is no strong influence of pollution by domestic sewage. The turbidity, total dissolved solids and pH variables were highlighted as class 1, although the values established were the same for all classes (except turbidity).

However, the values presented are only a result of the monitoring campaigns performed during a 1-year period, and alone they do not allow a conclusion about the classification of water bodies. Within the scope of the Project Critical Basins, a distinct approach was taken, more detailed regarding classification methodologies. One of the tools used was mathematical modelling, in which model QUAL2E (Brown & Barnwell, 1987) was used to simulate water quality in the Iguaçu River. The results of the simulations are presented in detail in Porto *et al.* (2007).

As presented in Knapik *et al.* (2006), the challenge of calibrating model QUAL2E (Brown & Barnwell, 1987) for BOD and DO was based on solving the equations of quantity of movement, conservation of energy integrated with the pollutant transport equations, assuming as representation of the cycle of organic matter represented only by the coefficients of deoxygenation (K_1), reaeration (K_2), sedimentation (K_3) and sediment oxygen demand (K_4) (Chapra, 1997). Although it means a strong simplification, it is the most convenient one for the field resources available.

Figure 4 presents the model calibration results for BOD_5 and DO in the main stream and the observed profiles of 20 field measurements, with 5000 km travelled and intense laboratory activity. Curve 1 indicates calibration considering values of K_1 and K_3, respectively, 0.1 day^{-1} and 1.2 day^{-1}. Curve 2 is a more critical look at the same parameters based on data from the literature (Sperling, 2006), with a reflex for a variation from 0.09 to 0.45 day^{-1} for K_1 and K_3 depending on depth (Chapra, 1997). Clearly, there is a significant difference, especially in the reach between kilometers 26 and 51, for BOD (Fig. 4(a)), which deserves highlighting and attention. For DO (Fig. 4(b)) the difference is not relevant in the reach involved, but there is a marked difference beginning at km 91.

However, the use only of data from the literature to calibrate the parameters can cover up peculiar characteristics of the intrinsic physical, chemical and biological aspects of the basin, in this case the Upper Iguaçu Basin. Complementary studies, which are currently being developed in the same basin, show that the behaviour of the parameter of carbonaceous deoxygenation, K_1, determined in a laboratory, is below the values used in calibration, indicating low biological activity in the water body, which consequently implies a low rate of organic matter removal. This result, although simple and not yet conclusive, is a warning about the considerations and simplifications usually adopted in mathematical modelling, whose chemical and biological significance is often neglected.

FINAL CONSIDERATIONS

This article essentially compiles the main activities needed to consolidate technical bases for the implementation of the water resources management instruments, providing a view of the results for the basin of the Upper Iguaçu in the Metropolitan Region of Curitiba. In this context the need to integrate different technical strategies that are absolutely crucial for an integrated view toward a better water resources planning and management strategies should be highlighted. This focus is not clearly shown in the Brazilian literature, and that is why this reflection is so important.

Therefore it is necessary to improve the qualitative and quantitative monitoring strategies, considering new non-traditional water parameters like those incorporated here; review the water quality model calibration strategies; gain further understanding of the boundary conditions needed for the better use of statistical methods, and especially integration with strategies to support the decision-making process, in order to really consolidate the implementation of Brazilian Law 9.433/97, particularly as regards the implementation of water resources management instruments.

Acknowledgements This research has the financial support of CNPq, FINEP/CNPq/CT-HIDRO and the PPGERHA–Graduate Program in Water Resources and Environmental Engineering (Programa de Pós-Graduação em Engenharia de Recursos Hídricos e Ambiental) at the Federal University of Paraná.

REFERENCES

Ahmad, S. R. & Reynolds, D. M. (1995) Synchronous fluorescence spectroscopy of wastewater and some potential constituents. *Water Res.* **29**(6), 1599–1602.

Artinger, R., Buckau, G., Geyer, S., Fritz, P., Wolf, M. & Kim, J. I. (2000). Characterization of groundwater humic substances: influence of sedimentary organic carbon. *Appl. Geochemistry* **15**, 97–116.

Azevedo, J. C., Teixeira, M. C. & Nozaki, J. (2006). Estudo espectroscópico de substâncias húmicas extraídas da água, solos e sedimentos da Lagoa dos Patos – MS, Brasil. *Revista Saúde e Biologia* **1**(2), 59–71.

Bäumle, A. M. B. (2005) Avaliação de Benefícios Econômicos da Despoluição Hídrica: Efeitos de Erros de Calibração de Modelos de Qualidade da Água. MSc Thesis, Federal University of Paraná, Curitiba, Brazil.

Brown, L, C. & Barnwell, T. O. Jr (1987) The Enhanced Stream Water Quality Model QUAL2E and QUAL2E-UNCAS: Computer Program Documentation and User Manual. United States Environmental Protection Agency, Athens, Greece.

Buffle, J., Zali, O., Zumstein, J. & Vitre, R. (1987) Analytical methods for the direct determination of inorganic and organic species: seasonal changes of iron, sulfur, and pedogenic and aquagenic organic constituents in the eutrophic Lake Bret, Switzerland. *Sci. Total Environ.* **64**, 41–59.

Cabaniss, S. E. & Shuman, M. S. (1987) Synchronous fluorescence spectra of natural waters: Tracing sources of dissolved organic matter. *Marine Chem.* **21**, 37–50.

Chapra, S. C. (1997) *Surface Water Quality Modeling*. McGraw-Hill, New York, USA.

Chen, J., Gu, B., Leboeuf, E. J., Pan, H. & Dai, S. (2002) Spectroscopic characterization of the structural and functional properties of natural organic matter fractions. *Chemosphere* **48**, 59–68.

Frimmell, F. H. (1998) Characterization of natural organic matter as major constituents in aquatic systems. *J. Contaminant Hydrol.* **35**, 201–216.

Galapate, R. P., Baes, A. U., Ito, K., Mukai, T., Shoto, E. & Okada, M. (1998) Detection of domestic wastes in Kurose River using synchronous fluorescence spectroscopy. *Water Res.* **32**(7), 2232–2239.

Knapik, H. G. (2006) Modelagem da Qualidade da Água da Bacia do Alto Iguaçu: Monitoramento e Calibração. Graduation Project, Federal University of Paraná, Curitiba, Brazil.

Knapik, H. G., França, M. S., Fernandes, C. V. S., Masini, L., Marin, M. C. F. C., Porto, M. F. A. (2006) Análise crítica da calibração de modelos de qualidade de água em rios – Estudo de caso da Bacia do Alto Iguaçu, Proc. I Workshop sobre Gestão Estratégica de Recursos Hídricos, Brasília, Brazil.

Kondageski, J. H. (2008) Calibração de modelo de qualidade de água para rio utilizando algoritmo genético. MSc Thesis, Federal University of Paraná, Curitiba, Brazil.

Korshin, G. V., Li, C.W. & Benjamim, M. M. (1997) Monitoring the properties of natural organic matter through UV spectroscopy: a Consistent Theory. *Water Res.* **31**(7), 1787–1795.

Oliveira, J. L., Boroski, M., Azevedo, J. C. R. & Nozaki, J. (2006) Spectroscopic investigation of humic substances in a tropical lake during a complete hydrological cycle. *Acta Hydrochimica et Hidrobiologica* **34**, 608–617.

Palmieri, V. (2003) Calibração do Modelo QUAL2E para o Rio Corumbataí (SP) MSc Thesis, Catholic University of Rio de Janeiro, Rio de Janeiro, Brazil.

Peuravuori, J. & Pihlaja, K. (2002) Molecular size distribution and spectroscopic properties of aquatic humic substances. *Analytica Chimica Acta* **337**, 133–149.

Pons, M., Bonté, S. L. & Potier, O. (2004) Spectral analysis and fingerprinting for biomedia characterization. *J. Biotech.* **113**, 211–230.

Porto, M. F. A., Fernandes, C. V. S., Knapik, H. G., França, M. S., Brites, A. P. Z., Marin, M. C., Machado, F. W., Chella, M. R., Sá, J. F. & Masini, L. (2007) Critical Basins: Technical bases for the definition of Progressive Goals for their Classification and Integration with the other Management Instruments. Federal University of Paraná – Hydraulic and Sanitary Department. (FINEP/ CT-HIDRO), Curitiba, Brazil.

Rodrigues, R. B. (2005) SSD RB – Sistema de Suporte a Decisão proposto para a Gestão Quali–Quantitativa dos Processos de Outorga e Cobrança pelo Uso da Água. PhD Thesis, University of São Paulo, São Paulo, Brazil.

Senesi, N., Miano T. M., Provenzano, M. R. & Brunetti, G. (1989) Spectroscopy and compositional comparative characterization of i.h.s.s. reference and standard fulvic and humic acids of various origins. *Sci. Total Environ.* **81/82**, 143–156.

Spencer, R. G. M., Baker, A., Ahad, J. M. E, Cowie, G. L., Ganeshram, R., Upstill-Goddard, R. C. & Uher, G. (2007) Discriminatory classification of natural and anthropogenic waters in two U. K. estuaries. *Sci. Total Environ.* **373**, 305–323.

Sperling, M. V. (2006) *Introdução à Qualidade das Águas e ao Tratamento de Esgotos*, 3rd edn. DESA/UFMG, Minas Gerais, Brazil.

Thomas, O., Khorassani, H. El, Touraud. E. & Bitar, H. (1999) TOC versus UV spectrophotometry for wastewater quality monitoring. *Talanta* **50**, 743–749.

Villa, A. T. (2005) Avaliação Ambiental de Qualidade da Água do Lago do Parque Barigüi: Potencial de Poluição Orgânica. MSc Thesis, Federal University of Paraná, Curitiba, Brazil.

Westerhoff, P. & Anning, D. (2000) Concentrations and characteristics of organic carbon in surface water in Arizona: influence of urbanization. *J. Hydrol.* **236**, 202–222.

Zumstein, J. & Buffle, J. (1987) Circulation of pedogenic and aquagenic organic matter in an eutrophic lake. *Water Res.* **64**, 41–59.

Development of canal design software as an educational tool

M. S. AYYANAGOWDAR[1], S. SANTHANA BASU[2] & S. S. SHIRAHATTI[1]

1 *College of Agricultural Engineering, UAS, Raichur 584102, India*
mallikarjunrcr@gmail.com

2 *Agricultural Engineering College and Research Institute, Tamil Nadu Agricultural University, Coimbatore 641 003, India*

Abstract One of the tools frequently used for rehabilitation and modernisation of irrigation projects is the canal flow model. Various models do exist with almost similar accuracy, which can simulate the actual flow conditions, although very few models are user-friendly. In view of this, a computer application tool for canal design was developed, which would empower the irrigation officials/managers to redesign the existing canals, and construct newer stretches of the canal, to enhance their carrying capacities by way of improving the discharge capability within the available cross-section of the canal by taking up lining works under modernization and renovation of existing irrigation projects.

Key words canal design; simulation; irrigation training tools

INTRODUCTION

The conveyance system is an integral part of any irrigation system. It plays a vital role in the overall efficiency of the system. The efficiency of an irrigation system depends on the right planning, execution and regular maintenance. A well-planned conveyance system would deliver the right quantity of water to each and every remote corner of the command area. A canal, in due time, deteriorates; its shape changes and eventually the carrying capacity would be altered. There are various natural and man-made causes for such a change. Hence, regular maintenance is absolutely essential for the overall functioning of the system in the long term. In view of this, a computer application tool for the canal design was developed, which would empower the irrigation officials/managers to redesign the existing canals, and construct newer stretches of the canal, to enhance their carrying capacities by way of improving the discharge capability within the available cross-section of the canal by taking up lining works under modernization and renovation of existing irrigation projects.

STUDY AREA

The computer application tool was tested for the Tungabhadra Canal command area in the northern part of Karnataka, India. Tungabhadra Project is located at 76°21′E longitude and 15°16′N latitude near Mallapur village in Hospet Taluka of Karnataka, having a catchment area of 28 180 km^2, with an annual yield of 423 TMC. Four canals are fed from the reservoir, two on the left bank and two on the right bank of the river. The fifth canal on the right bank serves the Basasvanna and Raya system, an ancient system dating from the Vijayanagar Empire (16th century) and still in use to irrigate 2960 ha. The Left Bank Canal, in which the study was conducted, is wholly situated in Raichur District. The Left bank low level main canal is 227 km long, supplying water to 86 distributaries.

REVIEW

Even though some irrigation canals have been operating for hundreds of years, available canal control methods have changed dramatically during the last few decades. The introduction of sensors, motorized gates, SCADA systems, and computers, combined with advances in hydraulics and control engineering, has allowed new automatic control techniques such as distant downstream controllers and centralized controllers (Vion *et al.*, 2007). In addition, classical local upstream controllers can be improved to become more efficient and flexible using digital technologies. Most new large irrigation canals are now designed and built using modern technologies allowing advanced control procedures (Narmada Canal in India, South–North Water Transfer Canals in China, Toshka Canal in Egypt, etc.). But more than 90% of the irrigation canals in the world are still using traditional technologies and simple manual operation principles. Reasons for this discrepancy are the lack of engineers skilled in this new domain, and the reluctance of canal managers to change to an unfamiliar system with new technologies and new operation and maintenance procedures.

Burt & Gartell (1993) discussed the various irrigation canal simulation models, their detailed working principles, etc. They are of the opinion that canal simulation models can be valuable aids in development/design, but they require a firm and considerable commitment of time and personnel, yet they are not considered as user-friendly by an average engineer. Cemagref has designed an interactive computer game allowing students to manipulate several hydraulic gates in order to stabilize water levels in an irrigation canal following what is called a distant downstream regulation framework (Malaterre *et al.*, 1998).

RESULTS

Canal Design – a computer application tool was developed in Visual Basic on the basis of hydraulic principles for an optimal and efficient trapezoidal canal section. It consists of two screens: Main Screen and Sub Screen (Fig. 1 and Fig. 2). Figure 1 depicts the Main Screen, wherein three inputs are to be entered. They are: (a) discharge requirement of the canal to serve a particular command area; (b) bed slope in concurrence with the existing topographic conditions of the proposed alignment of the canal; and (c) side slope of the canal with respect to the subgrade/terrain characteristics. Figure 2 depicts the Sub Screen for the selection of the lining material, it consists of the six options *viz.*, (a) artificially-constructed earthen channel; (b) natural streams on plains; (c) flood plains; (d) lined concrete channels; (e) masonry channels; (f) channels lined with miscellaneous materials. Further, each one of them has various surface conditions for which standard Manning's roughness coefficient ranges are provided. Ultimately the user can make his selection and return to the Main Screen. The selected Manning's roughness coefficient "n" is used for computation of canal parameters. A run command would lead to the design of the canal for a set of conditions. Figure 1 includes a pictorial representation of the design details.

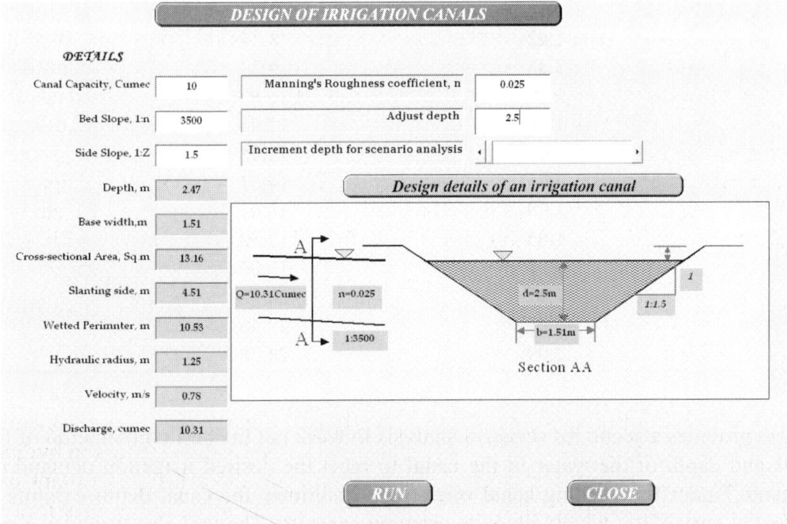

Fig. 1 Main screen of the Canal Design Software.

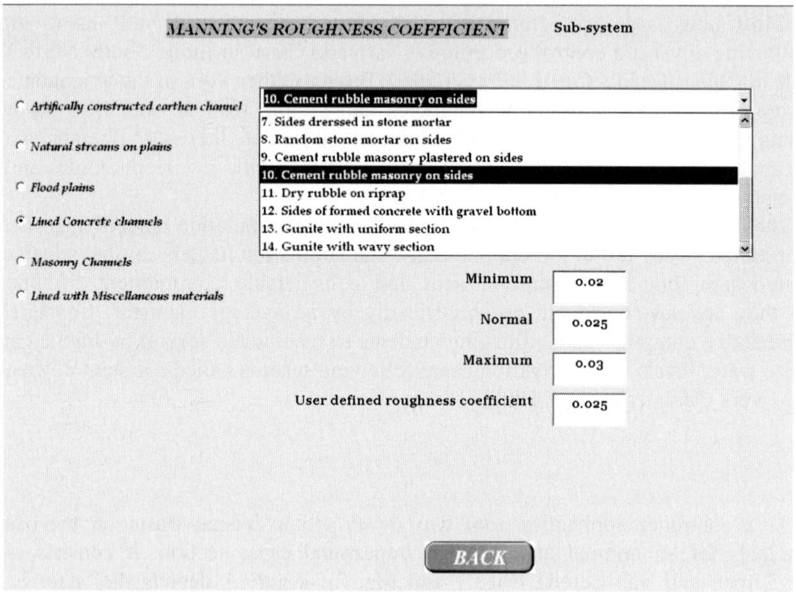

Fig. 2 Sub-screen of the Canal Design Software.

Table 1 Sensitivity analysis of water depth in canal to its carrying capacity.

Sl. no.	Depth (m)	Bed width (m)	Discharge (Cumec)
1	1.55	0.94	2.89
2	1.65	1.00	3.41
3	1.75	1.06	3.99
4	1.85	1.12	4.63
5	1.95	1.18	5.32
6	2.05	1.24	6.08
7	2.15	1.30	6.90
8	2.25	1.36	7.79
9	2.35	1.42	8.74
10	2.45	1.48	9.77
11	2.55	1.54	10.87
12	2.65	1.60	12.04
13	2.75	1.67	13.28
14	2.85	1.73	14.61
15	2.95	1.79	16.01
16	3.05	1.85	17.50
17	3.15	1.91	19.06
18	3.25	1.97	20.72
19	3.35	2.07	22.46
20	3.45	2.09	24.28

The tool also provides a scope for scenario analysis to work out the best combination of lining materials, slope and depth of the water in the canal to meet the desired irrigation demand of an existing command. Under the existing canal operating conditions, the canal depth exhibits both temporal and spatial variability, which alters its carrying capacity. The tool also provides a means for sensitivity analysis with respect to water depths in the canal and its carrying capacity.

The utility of the computer application tool is demonstrated for a set of design considerations: discharge = 10 cumec, bed slope = 1:3500, side slope 1.5:1 and Manning's roughness coefficient for such a condition n = 0.025 for a lined concrete channel with cemented rubble masonry on sides. The results indicate an optimal section with a depth of 2.47 m, bed width of 1.51 m having a discharge capability of 10.31 cumec. The sensitivity analysis was carried out and the result of the analysis is presented in Table 1.

Table 1 reveals the results of the sensitivity analysis, which was carried out by varying the depth in the range of 1.55 to 3.45 m, resulting in the variation of base width from 0.94 m to 2.09 m and correspondingly discharge varies from 2.89 to 24.28 cumec. The scenario analysis can be carried out for most economical section.

SUMMARY

The canal design software is very simple and user friendly and requires minimum inputs for the computation. The irrigation manger schedule can use this software to compute the existing carrying capacity in his routine maintenance, and during the modernization and renovation process, it can be used to design the new canal reach to confirm to the demand considerations.

It can also be used like a training tool for the budding irrigation managers. It can be a vital educational tool for undergraduate students to understand the steps involved in the design of major irrigation projects and canal design in particular. It encourages the active participation of the students with "hands-on" activities, rather than passive listening. The objective is capacity building of the stake holders in the domain of irrigation and command area development, hydraulics and control of irrigation canals. This will facilitate the introduction of new methods and technologies to operate the projects.

Acknowledgements This study was undertaken as a part of the PhD work on " Development of Management Information System (MIS) and Performance Analysis of Tungabhadra Project " under a joint arrangement between UAS Dharwad and TNAU Coimbatore. The facilitation extended by the authorities of both the Universities is appreciated. The help rendered by the technical staff of Irrigation Department, Tungabhadra Board, Command Area Development Authority is gratefully acknowledged.

REFERENCES

Burt, C. M. & Gartell, G. (1993) Irrigation canal-simulation model usage. *J. Irrig. Drain. Engng* **119**(4), 631–636.

Malaterre, P.-O. (1998) PILOTE: linear quadratic optimal controller for irrigation canals. *J. Irrig. Drain. Engng* **124**, 187–194.

Vion, P.-Y., Kulesza, V. & Malaterre, P.-O. (2007) Modernization project and scientific platform on the Gignac canal. In: *Second Conference on SCADA and Related Technologies for Irrigation System Modernization - A USCID Water Management Conference* (6–9 June 2007, Denver, Colorado, USA).

Comparison and analysis of monthly runoff predicted by four different techniques in a rainfall–runoff forecasting system (RRFS)

WOOCHANG JEONG[1], MANHA HWANG[2], TAEMIN HA[3] & YONGSIK CHO[3]

1 *Department of Civil Engineering, Kyungnam University, 449 Woryeong-dong, Masan, Gyeongnam 631-701, Korea*
jeongwc@kyungnam.ac.kr

2 *Water Resources and Environment Research Center, Korea Water Resources Corporation, 462-1 Jeonmin-dong, Yuseung-gu, Daejeon, 305-730, Korea*

3 *Department of Civil Engineering, Hanyang University, 417 Haengdang-dong Seongdong-gu, Seoul 133-791, Korea*

Abstract A rainfall–runoff forecasting system (RRFS) was developed as a toolkit for quantitative and qualitative analysis and prediction of a basin-wide rainfall–runoff relationship. The main technique to predict the long-term runoff of more than one month is the ensemble streamflow prediction (ESP) technique, which is currently integrated as a module in the RRFS. In order to improve the accuracy of runoff predicted by the pre-existing ESP technique, the weather outlook provided by the Korea Meteorological Agency (KMA) was added to that module. The verification of monthly runoff predicted by both techniques was performed for the 12-month period January–December 2007. Predicted monthly runoff values were compared with those from two other methods: mean monthly runoff measured during the 24-year period 1983–2006, and numerical precipitation data provided by the KMA.

Key words monthly runoff prediction; water resources management; Geum River basin; ESP

INTRODUCTION

During the last decade, problems related to water, such as more frequent and severe droughts and floods, conflicts between stakeholders for water use, and degradation of ecosystems due to water pollution, have created a serious burden for governments and local residents in Korea. In order to resolve these problems, the need of advanced tools for basin-wide integrated water resources management (IWRM), permitting more efficient and sustainable operation of existing water resources facilities, is urgently required (Ko & Chung, 2002). For the purpose of meeting this requirement, the new basin-wide IWRM concept seeks environmental sustainability, equality, and economic efficiency in water allocation and operation through integrated considerations of the mutual interests of water stakeholders, the relationship of water quantity and quality, and the relationship of water and adjacent land use.

Since 2001, as a part of "21st Century New Frontier National R&D Projects" which have been planned to overcome the water shortage problem in Korea, the Integrated Real-time Water Management System (IRWMS) has been developed. The ultimate goal of the IRWMS is to acquire and develop technology for basin-wide integrated water management, while taking into account both surface water and sub-surface water, as well as water quantity and quality. This system consists of four different sub-systems: (1) the basin-wide rainfall–runoff forecasting system; (2) the real-time integrated water resources information system; (3) the simulation and optimization system for reservoirs system operation; and (4) the steady and unsteady water quality prediction system for decision-making support in water management. In this paper, among these four sub-systems, the basin-wide rainfall–runoff forecasting system (RRFS) will be introduced.

The major functions of the RRFS are: to classify spatial and temporal runoff generated in a basin into hydrological components; to establish a short-term water supply plan; and to estimate quantitatively the available water resources in a basin by reflecting various patterns of water demand, such as domestic, industrial and agricultural. Another major function is to provide runoff data predicted more than 1 month ahead to forecast the future water supply in reservoir operations.

DEVELOPMENT OF THE RRFS

The schematic procedure of the development of the rainfall–runoff forecasting system (RRFS) is illustrated in Fig. 1. This system will be used to compute monthly and daily runoff components, including surface water, sub-surface water and return water, at key operating stations in a basin. The RRFS can be used for predicting monthly streamflow in conjunction with hydro-meteorological forecasting information. In this system, a short-term water demand projection scheme is developed that can take into account the patterns of municipal, industrial and agricultural water uses. The amount of water available in each sub-basin can be assessed based on the water budget analysis between supplied and projected water demands.

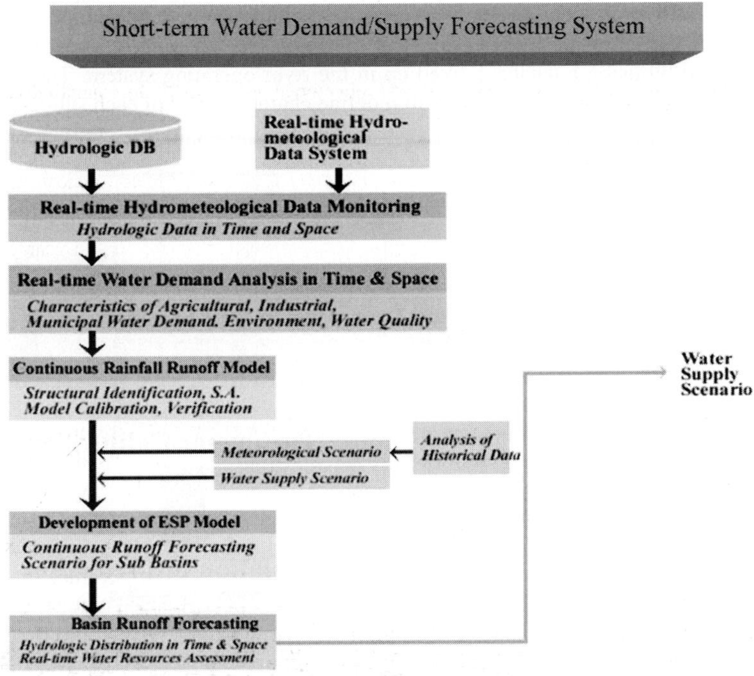

Fig. 1 Development framework of the RRFS.

The model will accommodate efficient uses of water resources by providing short-term water demands and supplies over 10 days. For sustainable execution of the runoff analysis model, a real-time hydro-meteorological data monitoring system is developed to acquire basin-wide hydrological information, such as the amounts of water intake and river discharge at key stations. Moreover, the ensemble streamflow prediction (ESP) process is employed to take into account the uncertainty of the long-term hydro-meteorological information and the risk in decision making.

For the development and utilization of the runoff analysis model, relevant basin information, including historical precipitation and river water stages, geophysical basin characteristics, and water withdrawal and consumption, needs to be collected and stored in the hydrological database with the IRWMS. The RRFS in conjunction with the ESP process can serve to predict monthly runoff in each sub-basin and key station.

The SSARR (Streamflow Synthesis and Reservoir Regulation, US Army Corps of Engineers, 1991) model was adopted as the main engine for calculating the continuous daily runoff. In particular, in order to overcome the uncertainty of the SMI (soil moisture index) parameter, which is used to define soil moisture condition of a basin of interest, the real-time soil moisture monitoring technology by the analysis of meteorological data acquired from a satellite such as NOAA is also developed in this sub-project. The RRFS is designed to be operated easily by users and on the web-based system with graphic user interface.

SYSTEM CONFIGURATIONS OF THE RRFS

The RRFS consists of: a real-time, hydrological input data preparation module for obtaining rainfall, temperature, observed flows at key points, dam release, etc.; a runoff simulation and forecasting module; an output analysis and treatment module; and the ESP runoff prediction module.

The RRFS is directly linked to the data base management system (DBMS) of IRWMS including: configuration of hydrological system, real-time data acquisition and its calibration, real-time inflow computation, operating plan of hydrological facilities, etc. The hydrological analysis in the RRFS is conducted by using a database fixed up in the river operating system. These data are defined only once, but can be corrected by users to redefine characteristics of each sub-basin.

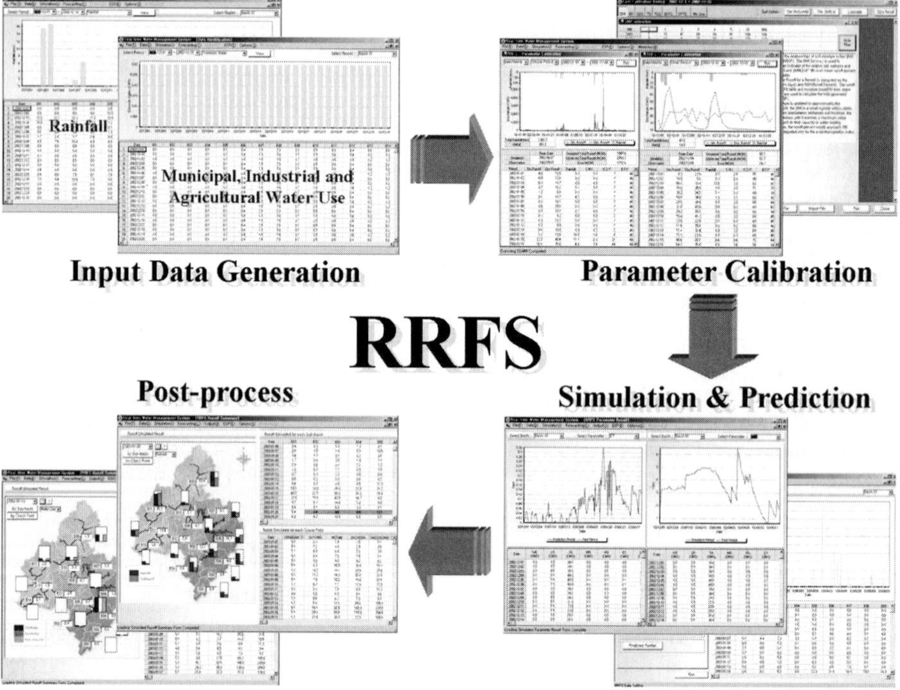

Fig. 2 System configuration of the RRFS and graphic user interfaces.

Figure 2 shows the system configuration of the RRFS and its operating procedure. In the first step it needs to generate the various hydrological input data, including rainfall, temperature, municipal, industrial and agricultural water use, intake, and dam release. After generating the input data, parameters such as SMI, BII (baseflow infiltration index) etc., used to characterize the rainfall–runoff relationship in a basin of interest, are established and calibrated, identifying the comparison between calculated and measured runoff results. Once the set-up of parameters is terminated, the runoff simulation and prediction to investigate the future situation of water resources according to various hydrological scenarios can be switched by users. The runoff results calculated from simulation and prediction process can be identified in the post-process which shows graphically the spatial and temporal distribution of water resources in sub-basins and key operating stations and statistically treated outputs, etc.

APPLICATION

Brief description of Geum River basin

The Geum River basin is the third largest river basin in Korea, situated on the west of the central part of the Korean peninsula (Fig. 3). It occupies an area of 9835.3 km^2, almost one tenth of our country, and extends east to west for about 130 km and north to south for about 160 km. At the north and east, the Geum basin borders on the Han and Nakdong river basins. Its southern part borders on the Mangyeong and Seomjin river basins. At the west of the basin lies the Yellow Sea. The main stream of the Geum River has a length of 395.9 km.

There are two multi-purpose dams in this river basin (Fig. 4): Daecheong Dam, which is located at 150 km upstream from the estuary of the Geum River; and Yondam Dam which is located upstream of the Daecheong Dam.

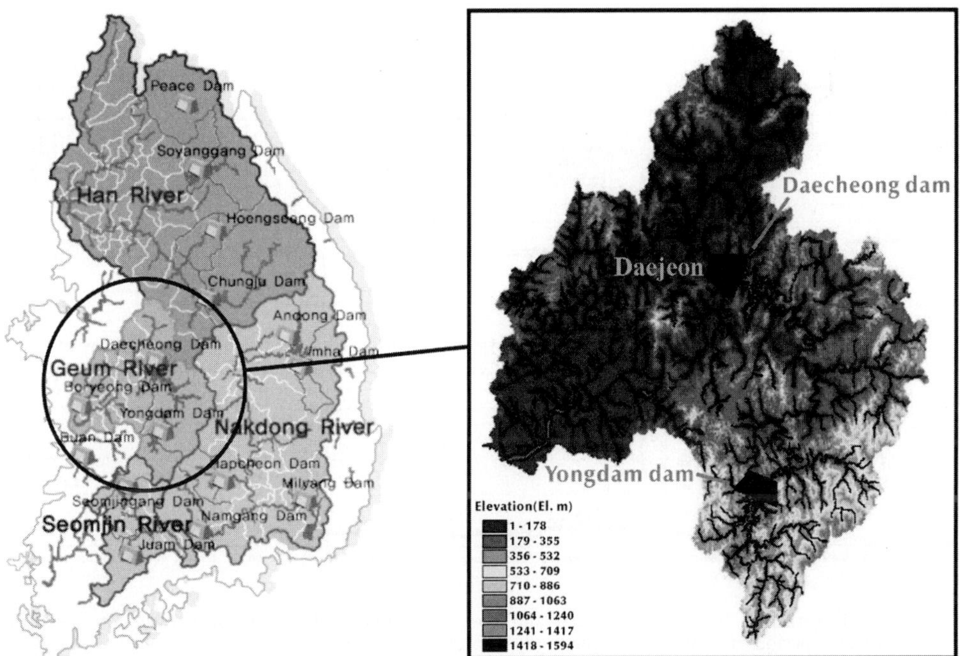

Fig. 3 Layout of the Geum River basin, Korea.

Calibration and verification of RRFS

The calibration and verification of the RRFS were performed by comparing with field discharge data which had been measured at three key points: Yondam Dam, Daecheong Dam and Gongju. The calibration of parameters used in the RRFS was carried out by setting up the test period of 21 years from 1 January 1983 to 31 December 2005. The calibrated RRFS was verified with hydrological data set obtained in 2006.

Figure 4 show the comparison between simulated and observed daily discharge at two key points during the verification period of 2006: Daecheong Dam and Gongju. Table 1 represents the summary of results obtained from the error and statistical analyses.

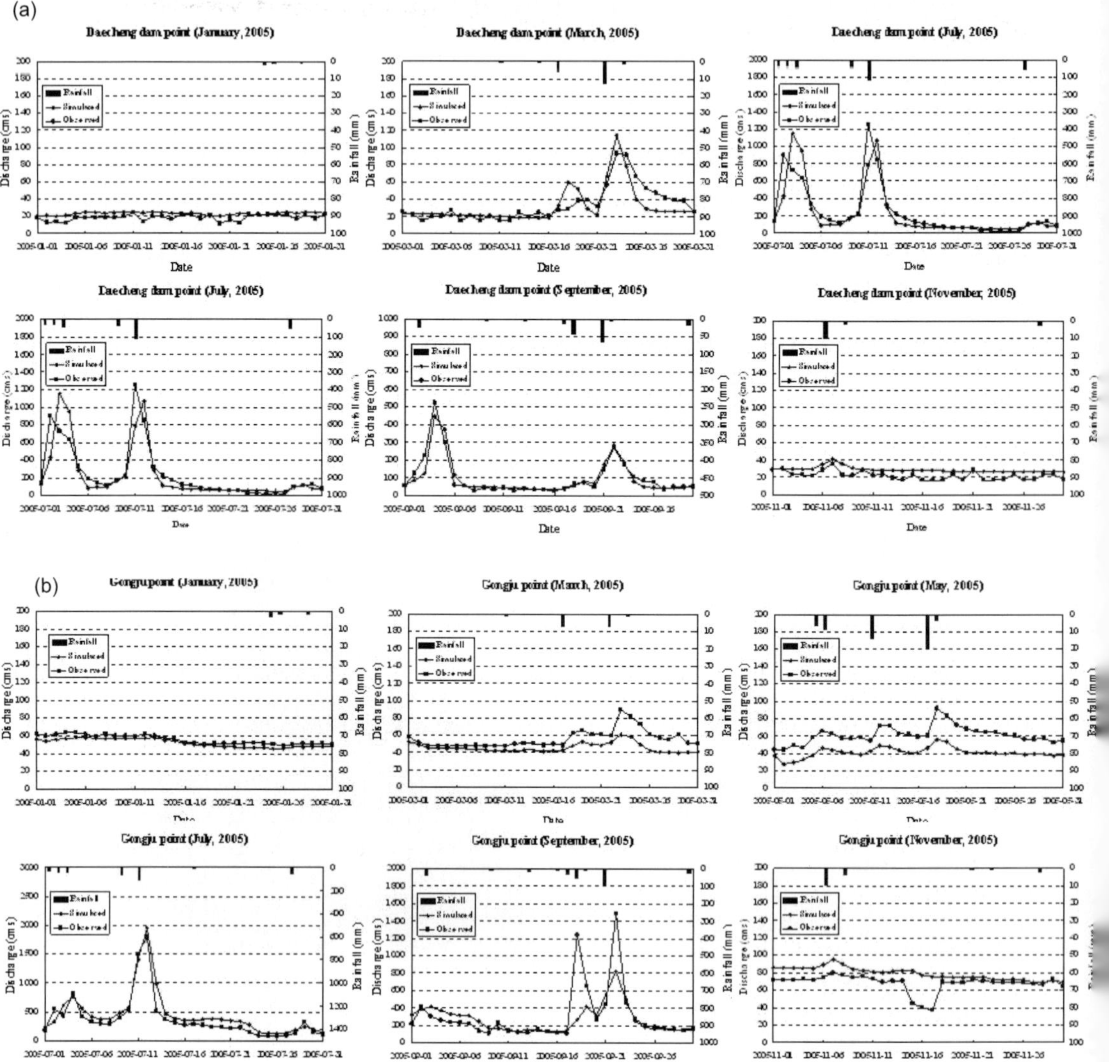

Fig. 4 Comparison between simulated and observed monthly discharge at two key points during verification period of 2006: (a) Daecheong Dam and (b) Gongju.

As shown in Fig. 5 and Table 1, the simulated discharge for three key points generally compare well with observed values, and the parameters used in this verification are thus satisfactorily calibrated. The mean values of monthly root mean square error (RMSE) between simulated and observed discharges are 34.9 for Yongdam Dam, 44.1 for Daecheong Dam, and 50.1 for Gongju, and discharge has a tendency to increase gradually from the upstream to the downstream part of the basin.

However, for the case of Yondam and Daecheong dams, the discharge simulated during the dry season (October–January) is more or less overestimated compared to the observed values. In contrast, the discharge simulated for April and May is underestimated and this is due to the overestimation of agricultural water use in that period.

Table 1 Error analysis of simulated and observed discharge for three key points.

Month		Yongdam Dam		Daecheong Dam		Gongju	
		RMSE	Total discharge ($m^3 s^{-1}$)	RMSE	Total discharge ($m^3 s^{-1}$)	RMSE	Total discharge ($m^3 s^{-1}$)
January	Observed	3.8	34.9	5.6	574.8	4.3	1733.3
	Simulated		137.3		717.9		1609.4
February	Observed	8.4	119.6	19.0	610.4	7.0	1525.9
	Simulated		226.5		848.3		1449.7
March	Observed	7.2	283.4	12.4	1054.6	12.2	1737.2
	Simulated		338.4		1002.7		1420.6
April	Observed	3.2	242.6	9.6	998.8	10.4	1760.1
	Simulated		207.7		806.9		1473.4
May	Observed	2.3	122.8	8.4	691.0	20.7	1895.0
	Simulated		100.3		474.2		1276.1
June	Observed	12.8	463.7	16.9	1309.3	30.1	2919.9
	Simulated		591.5		1495.6		2267.5
July	Observed	120.2	3966.7	161.6	7608.8	130.9	11864.9
	Simulated		3797.7		7096.5		14139.0
August	Observed	238.4	3658.8	231.8	11156.1	100.7	12374.0
	Simulated		3101.4		11182.9		13393.0
September	Observed	12.5	396.6	32.9	2953.0	228.5	9188.6
	Simulated		294.6		2911.2		8188.0
October	Observed	3.6	74.3	16.9	1106.2	35.1	3261.4
	Simulated		176.2		1381.7		3920.3
November	Observed	3.7	30	7.8	661.5	14.8	2045.4
	Simulated		135.3		865.3		2369.7
December	Observed	3.7	18.8	6.6	523.7	5.9	2029.7
	Simulated		129.5		678.0		2041.6
Mean		34.9		44.1		50.1	

Runoff predictions

The main technique used to predict the long-term runoff of more than 1 month is the ESP (ensemble streamflow prediction) technique, which is currently integrated as a module in the RRFS. The ESP uses conceptual hydrological models and historical data to generate a set, or ensemble, of possible streamflow scenarios conditioned on the initial states (e.g. soil moisture, etc.) of a basin of interest. This technique is well known as a useful tool to predict the river discharge and has been used in the NWSRFS (National Weather Service River Forecast System) of the US National Weather Service (Anderson, 1973). The value of ESP probability used to predict the runoff is determined by statistical analysis, with monthly runoff data measured during the period 2003–2006.

In order to improve the accuracy of runoff predicted by the ESP technique, the weather outlook provided by the Korea Meteorological Agency (KMA) is added to that technique. The weather outlook is provided in three levels: AN – Above Normal, N – Normal, and BN – Below Normal. According to the level of weather outlook, the ESP probability is newly determined.

The verification of monthly runoff predicted from the techniques above is performed during the 8-month period from January to August 2006. Predicted monthly runoff data are compared with those from two other methods: mean of monthly runoffs measured during the 23-year period

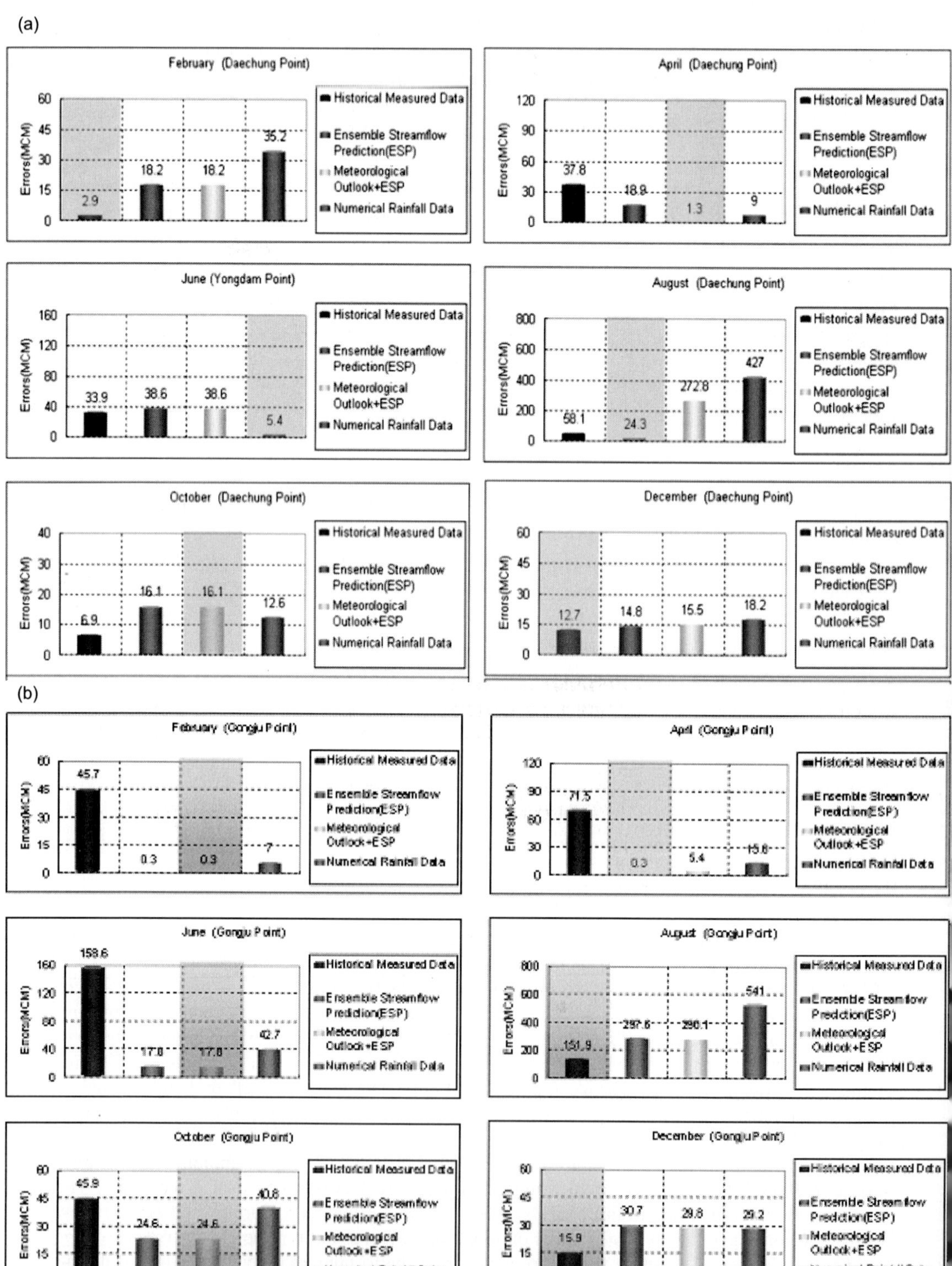

Fig. 5 Comparison of monthly runoff predicted by four methods: (a) Daecheong Dam, and (b) Gongju.

1983–2006 and numerical precipitation data calculated from the Quantitative Precipitation Model (QPM).

Figure 5 graphically compares the error between monthly runoffs predicted by the four techniques and measured ones for two key points: Daecheong Dam and Gongju. For the case of Daecheong Dam, except January, ESP + Weather outlook technique is generally better than the other techniques. This tendency is also similarly found for the case of Gongju. It was found from this comparison that this technique can give runoff results more than 50% improved compared to the traditional technique (that is, mean of monthly measured runoffs).

CONCLUSIONS

In this paper, the rainfall–runoff forecasting system (RRFS), which was designed to perform effectively the rainfall–runoff analyses for the real-time or short-term (<10 days) water demand/supply and the long-term runoff prediction by using ESP and ESP + Weather outlook techniques, and developed with web-based technology, is introduced. The RRFS was calibrated and verified by applying to the Geum River basin in Korea and it was found that this system can be efficiently and sufficiently used to analyse the rainfall–runoff relationship in a basin of interest and to establish the outlook of water supply in reservoir operations.

As the results of comparison of monthly runoff outlooks from four different techniques, it can be shown that the ESP technique coupled with weather outlook gives comparatively more accurate runoff results with respect to three other techniques.

Currently, the RRFS is using in working level at Water Resources Operation Center of Korea Water Resources Corporation and applied to different large river basins in Korea, such as the Han and the Nakdong river basins.

Aknowledgements This research was supported by a grant (code 1-6-3) from Sustainable Water Resources Research Center of "21st Century Frontier Research & Development Program", Korea.

REFERENCES

Anderson, E. A. (1973) National Weather Service River Forecast System—Snow Accumulation and Ablation Model, *NOAA Tech. Memo. NWS HYDRO-17*, Office of Hydrology, National Weather Service, NOAA, Silver Spring, Maryland, USA.

Ko, I. H. & Chung, S. W. (2002) Strategy for developing base technology for integrated water resources management—the trend of technology development for IWRM. *J. Korea Water Resour. Assoc.* **35**(6), 61–70.

US Army Corps of Engineers (1991) *SSARR User's Manual*, North Pacific Division, Portland, Oregon, USA.

[illegible] and monthly precipitation data calculated from the Quantitative Precipitation Model (QPM).

Figure 3 graphically compares the error between monthly runoff predicted by the four techniques and measured data for two key points: Hae-Soong Dam and Gangju. For the case of Dae-cheong Dam, except January, July and October, outputs technique is generally better than the other techniques. This tendency is also similarly found on the case of Gangju. It was found from this comparison that [illegible] the additional technique, mean of monthly measured runoff.

CONCLUSIONS

In this paper, the runoff results of forecasting system (KRFS), which was designed to perform effectively the runoff analysis [illegible] techniques and [illegible] The KRFS was [illegible] to the Geum River basin in Korea and it was found [illegible] and to demonstrate the ability of [illegible] conditions.

[illegible] that the ESP technique [illegible] with respect to [illegible] techniques.

[illegible] the KRFS [illegible] Water Resources Corporation, Korea Water Resources Corporation [illegible] basins in Korea such as the Han and the Nakdong river basins.

Acknowledgements This research was supported by a grant (code [illegible]) from Sustainable Water Resources Research Center of 21st Century Frontier Research Program, Korea.

REFERENCES

[illegible]

3 Hydrogeological Applications and Modelling Large Systems

Using RS/GIS to explore the impact of land-use changes on evapotranspiration and runoff in a semi-arid watershed

XIAOLI YANG, LILIANG REN, XIAOFAN LIU, FEI YUAN, BIN YONG & HONG WANG

State Key Laboratory of Hydrology, Water Resources and Hydraulic Engineering, Hohai University, No. 1 Xikang Road, Nanjing 210098, China

y_xiaoer@163.com

Abstract The paper aims to assess the effects of land-use and land-cover changes (LUCC) on hydrological variables such as evapotranspiration and runoff in the Shalamulun watershed, China. First, land-use types were interpreted from the TM and ETM$^+$ remote-sensed images via the knowledge-based decision tree (K-DT) classification method and LUCC was analysed through the post-classification comparison method. Subsequently, the two-source potential evapotranspiration (PET) model was used to estimate the potential evapotranspiration responses to LUCC. Finally, the influence of LUCC on annual runoff was determined statistically. The results show that in the period of 2001–2007, the grassland and forest had decreased by 85.36 km^2 and 3.92 km^2, respectively; both farmland and residential land have a distinct increasing tendency, increasing by 70.65 km^2 and 24.78 km^2, respectively. This change potentially leads to decrease in the annual PET and runoff. Meanwhile, the land-use types result in spatio–temporal variations of monthly PET in the growing seasons (May–September).

Key words land-use and land-cover change; potential evapotranspiration; headwater; knowledge-based decision tree

INTRODUCTION

In semi-arid regions, land-cover modifications, i.e. subtle changes that affect the characteristics of the land cover without changing its overall classification, are common (Diouf & Lambin, 2001). This change will affect the ecological environment of the whole region. Semi-arid regions play an important role in the system of global climate change. Much research shows that land use and land cover (LULC) influences runoff production and runoff volume (Lørup *et al.*, 1998; Karvonen *et al.*, 1999; Wei *et al.*, 2007). Evapotranspiration (ET) and runoff are the most important key components in the hydrological cycle, which are closely related to LUCC. Potential evapotranspiration (PET) is generally considered to be the amount of water that is lost into the atmosphere from a land surface with ample water supply. It is one of the most complicated variables in the coupled eco-hydrological system, mainly due to its high spatio-temporal variability (Zhao *et al.*, 2004). Land-use change impacts hydrological processes directly through its link with the evapotranspiration regime on one hand, and on the other hand the degree and type of ground cover has an enormous impact on the initiation of surface runoff (Fohrer, 2001). Particularly in semi-arid areas, the impact of land-use changes on basin runoff is of major interest to water resources planners, managers and local authorities (Helmscrot & Flügel, 2002). Furthermore, rainfall patterns in semi-arid areas are unpredictable, both in amount and time. Consequently, the ability to successfully manage the resulting runoff is extremely important (Bellot *et al.*, 2001; Winnaar *et al.*, 2007). Thus in terms of the planning and management of land use and water resources, it is essential to develop an approach for simulating and assessing land-use changes and their effects on hydrological processes at the watershed scale (Lin, 2007). Since the characteristics of hydrological processes are both temporally and spatially variable, it necessary to use remote sensing (RS) and geographical information system (GIS) tools in hydrological science.

In this study, the influence of LUCC on PET and annual runoff was analysed through the physically-based two-source potential evapotranspiration (PET) model and the statistical analysis method, supported by RS and GIS techniques in the Shalamulun watershed, China.

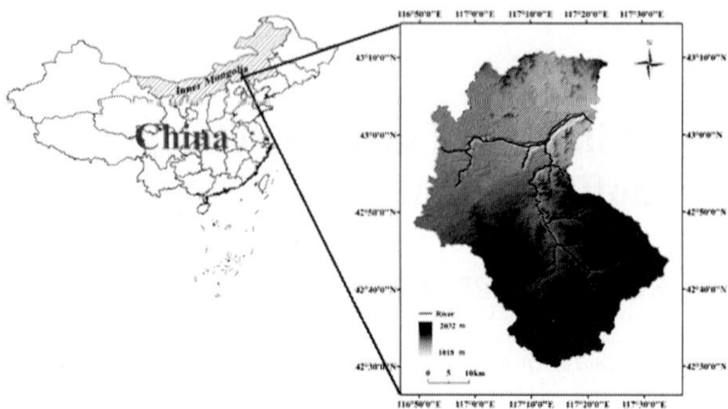

Fig. 1 Location map of the Shalamulun watershed.

STUDY AREA

The Shalamulun River is the headwater of the West Liao River, extending between 42°30′ to 43°15′N latitude and 116°40′ to 117°35′E longitude, in the southeast part of the Inner Mongolia Autonomic Region of China (Fig. 1). The drainage area of the Shalamulun River is 2453 km^2. Many land cover types appear in the watershed: grassland, forest, farmland, and barren land. The majority vegetation type is grassland, about 70% of the whole basin. In this area, the landforms range from grass plains to hills enclosing valleys, uplands ranging from 1000 m a.s.l. near Henanyingzi station, to in excess of 2000 m on the eastern border of the watershed.

The watershed has a semi-arid continental climate with strong climatic gradients. Approximately 90% of annual rainfall occurs during the wet season, which starts from mid to late April, reaches a peak in July to August, and ends in September. The driest months are from October to March, with monthly precipitation usually being below 10 mm. The streamflow peak usually occurs in April when plenty of melted snow water contributes to the streamflow. Mean monthly rainfall varies from 2 mm in January to 108 mm in July, with a mean annual rainfall of 360 mm. There is a small rainfall gradient across the watershed increasing from south to north, reflecting the higher topography in the south. Annual average maximum (minimum) temperature is 7°C ranging from –13°C in January to 25°C in July.

LAND-USE AND LAND-COVER CHANGE IN THE STUDY AREA

On the basis of the Chinese LULC classification system, developed by the Institute of Geographic Science and Natural Resources Research CAS (Liu, 1996), and the specific land use of the Shalamulun watershed, a classification scheme was proposed with six types: water body, residential land, forest, farmland, grassland, and barren land.

Images of two different days (6 July 2001 and 7 October 2007) were used in this study, which were in the Mercator projection (UTM) zone 50, GRS 1984. The geometric correction was carried out using ground control points established with global positioning system (GPS) units on site from land-use survey data in 2008. The 2007 image was registered with the ETM^+ image of 2001, with the root mean square errors of less than 0.45 pixels (12.8 m). The processing of these images was performed with the ENVI 4.5 remote sensing preprocessing software. Then the knowledge-based decision tree (K-DT) classification technique was used to detect the land-use types.

The decision tree (DT) technique is suitable for remote sensing classification problems because of its flexibility, intuitive simplicity and computational efficiency, which leads to increased acceptance (Pal & Mather, 2003). The DT classifiers utilize conditional relationships between vegetation types, spectral and geospatial information such as elevation, slope, aspect and other environmental variables to enhance forest classifications (Gislason *et al.*, 2006). Many researchers have demonstrated that a decision tree is an accurate and efficient methodology for land-use classification in remote sensing (e.g. Hansen *et al.*, 2000; Kandrika *et al.*, 2008).

In this study, by incorporation of other data sources, the knowledge-based decision tree (K-DT) technique was built to detect the land-use types based on the spectral characteristics of the samples and the spatial patterns of the six classes. The K-DT is a kind of decision tree classification incorporating multisource information and expert knowledge to compute the accumulated evidential support for each inferred class at each image pixel. It can integrate geographical knowledge including spectral and geospatial information, such as elevation, slope and aspect, and has been demonstrated by other experts to be an accurate and efficient methodology for land-cover classification in remote sensing (Kandrika *et al.*, 2008). In classification procedures, elevation, percent slope, normal difference vegetation index (NDVI), image texture, and previous land-use data were used, as well as imagery, and this knowledge was represented by rules. Classification maps for 2001 and 2007 (Fig. 2) were proposed with all the user's and producer's accuracies respectively above 85.7% and 87.9%, and the Kappa statistics were above 89.4%. These indices can meet the lowest demand for change detection (Lucas *et al.*, 1989).

A multi-date post-classification comparison change detection algorithm was used to determine the land-cover changes in the period 2001–2007. LULC data generated with K-DT classifiers from TM and ETM+ images were used to quantify changes using a pixel-by-pixel post-classification comparison supported by the ArcGIS 9.2 software. This is perhaps the most common approach to change detection (Jensen, 2004). The results of land-use and land-cover classification and land-use and land-cover changes from 2001 to 2007 are summarized in Table 1. Comparing the areas of the aggregated LULC classes in 2001 and 2007, shows that grassland occupies about 70% of the whole watershed. This is typical of semi-arid regions. The results suggest that the land use changes intensively in the period 2001–2007. The grassland is reduced by 85.36 km^2 (3.47%). Farmland increases by 70.65 km^2 (2.87%), and residential land increases by 24.78 km^2 (1.01%). The forest land, barren land and water body areas are reduced by 3.92 km^2 (0.16%), 5.15 km^2 (0.21%), and 0.98 km^2 (0.04%), respectively.

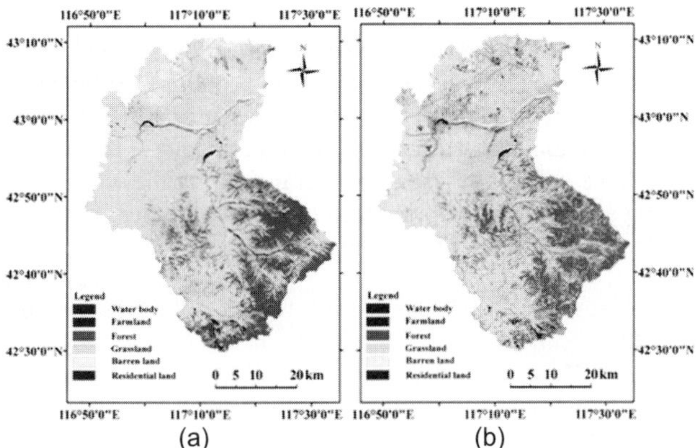

Fig. 2 Land-use and land-cover classification maps of Shalamulum watershed in (a): 2001; and (b) 2007.

Table 1 Results of LULC classification for 2001 and 2007 images showing the area of each category, class percentage and area changed.

Land-use class	2001 Area (km^2)	%	2007 Area (km^2)	%	2001–2007 Area (km^2)	%
Grassland	1722.26	70.21	1636.89	66.73	85.37	3.47
Forest	509.73	20.78	505.81	20.62	3.92	0.16
Barren land	160.43	6.54	155.27	6.33	5.15	0.21
Farmland	48.32	1.97	118.97	4.85	–70.65	–2.87
Water body	6.62	0.27	5.64	0.23	0.98	0.04
Residential land	5.64	0.23	30.42	1.24	–24.78	–1.01

MODEL AND DATA PREPARATION

The physically-based two-source PET model was modified by Yuan (2008) based on the two-source ET model (Mo *et al.*, 2004). It is a distributed physically-based model that is able to calculate PET on each grid cell considering the LULC characteristics. This method was applied successfully in the Laohahe watershed that is close to the Shalamulun watershed (Ren *et al.*, 2009).

Leaf area index (LAI) data

In our study, 8-day LAI production of MODIS data at 1000 m resolution for 2001 and 2007 were acquired from the EOS data gateway (https://wist.echo.nasa.gov/apimalized).

Meteorological data

The meteorological data from 2001 to 2007 at the Chifeng and Weichang meteorological stations were obtained from the China Meteorological Administration. The data include the daily readings of mean, maximum and minimum temperatures, sunshine hours, wind speed, and air water vapour pressure each 6-hour.

Digital Elevation Model (DEM) data

The Shuttle Radar Topography Mission (SRTM) 3-second digital elevation model (DEM) data (http://lpdaac.usgs.gov/gtopo30/hydro/index.html) were used to represent the topography of the study area. The basin part of the DEM is clipped out in ArcGIS 9.2 software using the basin boundary, which is acquired from the Rivertools 2.0 software and then averaged into 300 m × 300 m spatial resolution (the same resolution of the other data).

ANALYSIS OF RESULTS

Influence of LUCC on PET

The mean annual PET from 2001 to 2007 were estimated by the two-source PET model. As shown in Fig. 3, the spatial variation of mean annual PET over the watershed is non-uniform, from 624 to 1422 mm. Higher PET values occur in the river valley covered with water and the mountainous region with forest cover. Low PET values mainly appear in the middle-north part of the watershed.

Most vegetation is considered to have only slight canopy coverage during the winter months, when it has died back, with the coverage increasing in spring through to the summer. Therefore we compared the monthly PET of the Shalamulun watershed during growing seasons (i.e. from May

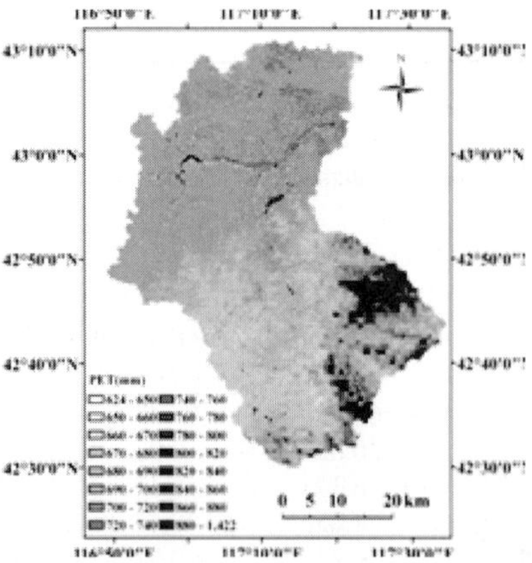

Fig. 3 Spatial distribution of mean annual PET from 2001 to 2007 in the Shalamulun watershed.

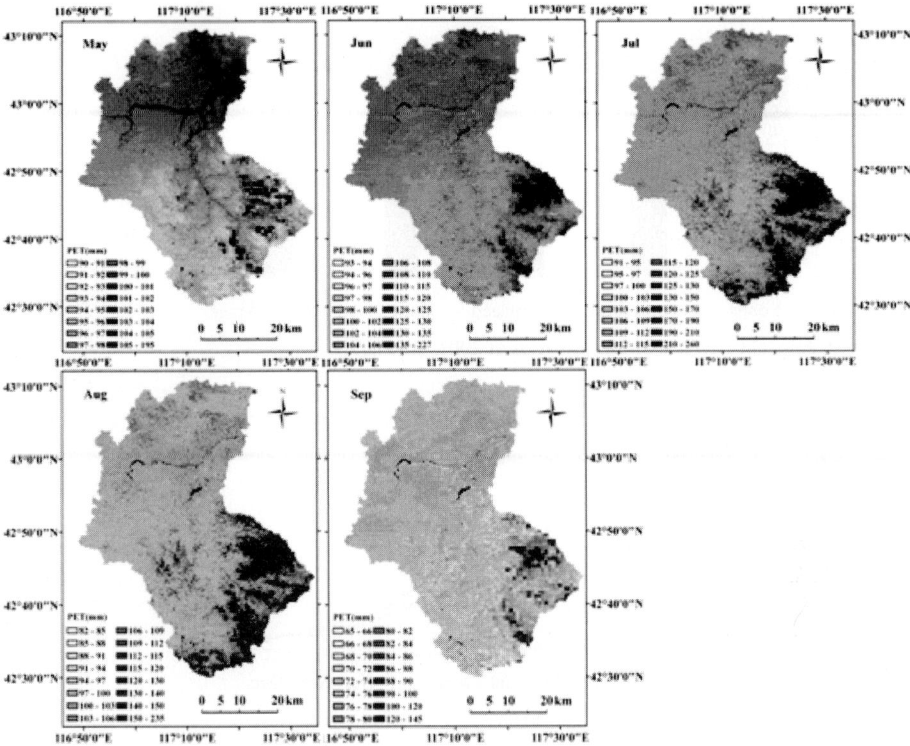

Fig. 4 Spatial distribution of PET during the May–September growing season in Shalamulun watershed.

to September) (Fig. 4). Figure 4 shows that PET value has temporal variation during growing seasons, the highest PET values, ranging from 92 to 257 mm, appearing in July, and the lowest PET, from 65 to 145 mm, being seen in September; the mean value of PET is 82–233 mm in June and August, and lower, 90–195 mm in May. It can also be seen that the monthly PET is spatially non-uniform in different land-use types. Water bodies have much higher evapotranspiration losses than other land covers throughout the growing season. The monthly PET of forest is higher than other vegetation covers. The lowest monthly PET is the value for grassland. It reflects that the land-use and land-cover classes have a significant influence on monthly PET.

Both climate change and land-use and land-cover changes can lead to the variations of PET. To quantify the influence of LUCC on PET, PET was evaluated under the land-use scenarios in 2001 and 2007, with the same meteorological record in 2001. The total annual evapotranspiration losses for various land cover scenarios are given in Fig. 5. It shows that the annual PET has a relatively decrease of 5 mm (from 726 to 721 mm). It also can be found that the annual PET is spatially heterogeneous in different land-use types. The annual PET of farmland changes more than other land-use types, with a decrease of 8 mm as the farmland area increases by about 70.65 km^2. The annual PET of grassland has decreased slightly (3 mm), while the area of grassland reduced significantly (3.47%). In the study area, maize and helianthus are the main crop types, which have higher LAI than grassland during the growing months (Fig. 6). It has been confirmed by other experts that the great change of PET over vegetation with higher LAI is due to the evaporative coefficient increasing exponentially with LAI (Dunin & Mackay, 1982; Zhang, 1999). In other words, forests are more sensitive to changes in canopy resistance than grass. Furthermore, forests have higher LAI than grass (Dunin & MacKay, 1982; Zhang, 1999). Hence, the great change of PET did not occur on the grassland, which has undergone a great change of area (85.36 km^2) during 2001–2007, but on the farmland (70.65 km^2). As a key land surface biophysical

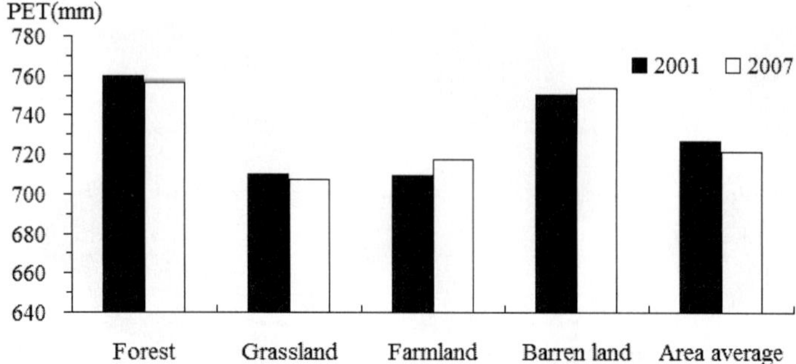

Fig. 5 Annual PET on land-use scenarios of 2001 and 2007.

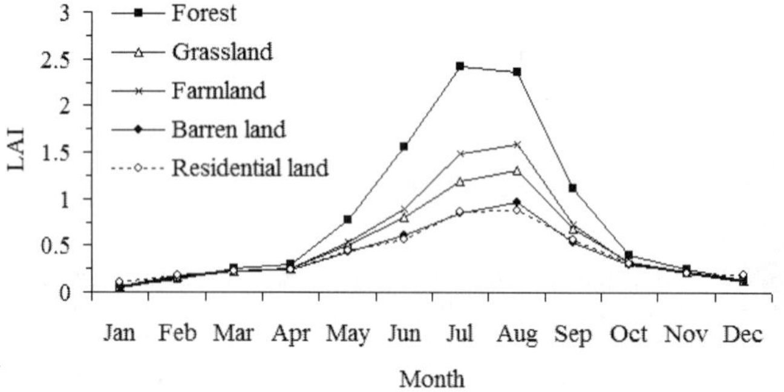

Fig. 6 Mean monthly LAI for various land-use classes.

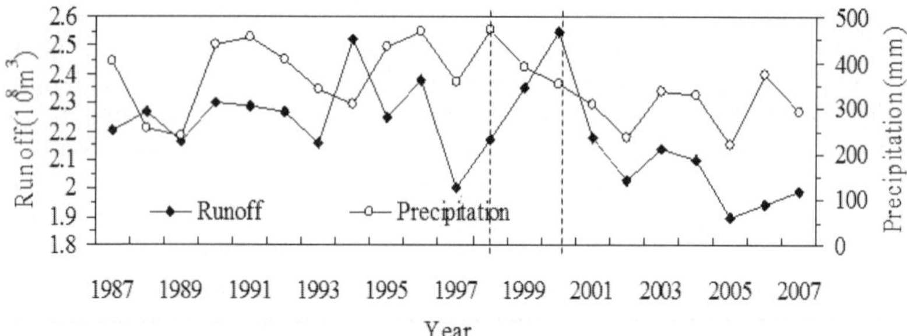

Fig. 7 Variations in annual precipitation and observed annual runoff at Henanyingzi station.

parameter, LAI can reflect the land-use and land-cover class, especially in the growing season (Turner *et al.*, 1999). It reveals that LUCC has a significant effect on the changes of PET.

Influence of LUCC on runoff

Both climate change and LUCC can lead to variations in annual runoff (Lu *et al.*, 2008); thus the variations of the annual precipitation series and observed annual runoff series were analysed. Figure 7 shows that both annual precipitation and runoff tend to increase, with a good correlation relationship before 1998. However, after the year 2000, the annual runoff decreases, while the annual precipitation continues to increase. It reveals that land-use and land-cover changes may be one of the influencing factors on annual runoff decrease after 2000.

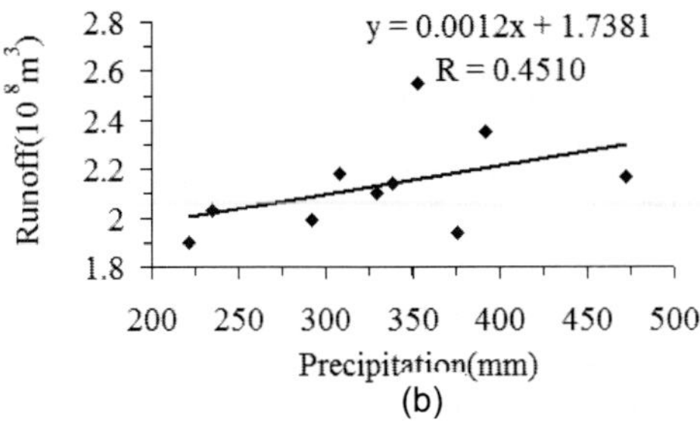

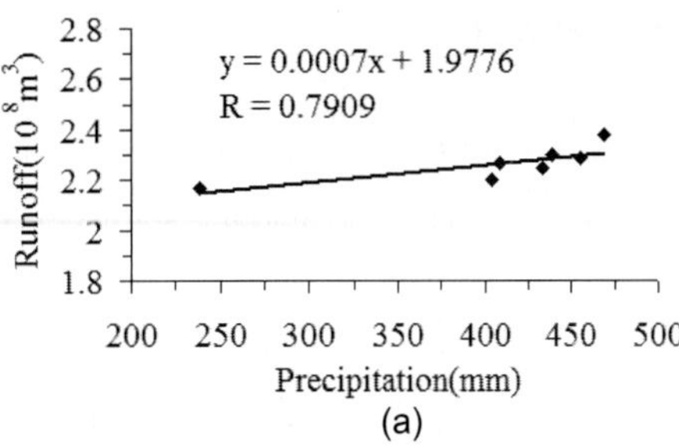

Fig. 8 The relationships between runoff and precipitation: (a) time period 1987–1997; (b) time period 1998–2007.

The correlation analyses of annual precipitation and observed runoff in the years 1987–1997 and 1998–2007 are shown in Fig. 8. It is seen that the correlation between annual runoff and precipitation is better from 1987 to 1997 (R = 0.7909) than that from 1998 to 2007 (R = 0.4510). It also justifies the conclusion that LUCC has a significant effect on the annual runoff.

CONCLUSIONS

The purpose of this study was to assess the quantitative effect of LUCC on PET and runoff in the Shalamulun watershed by the two-source PET model and statistical analysis supported by RS and GIS techniques. Land-use types were interpreted from TM (2001) and ETM^+ (2007) images by the K-DT method in ENVI 4.5. Meanwhile, the spatio-temporal variations of the land-use types were analysed using the spatial analysis algorithm in ArcGIS 9.2. The results show that land-use types of the Shalamulun watershed have undergone remarkable changes. For the whole, the grassland and forest decreased from 2001 to 2007, while the farmland and residential land area show a distinct increasing tendency. The driving forces of land-use and land-cover changes include biophysical and human factors. Based on the field survey and literature, a scenario of possible LUCC was generated taking into account the human growth and economic development. The declining trend of grassland area will certainly cause soil loss. The ecological system in semi-arid regions is more sensitive to any changes, such as land-use and land-cover changes, and climate change.

In the study area, the annual PET has a spatial variation characteristic in different types of land use and land cover. The monthly PET value has spatio-temporal variation during growing seasons (i.e. from May to September). These results reveal that the land-use and land-cover classes and their change obviously influence PET. Meanwhile, both the annual PET and runoff decreased with the decrease of grassland and the increase of farmland area. It reveals that land-use and land-cover changes have a significant impact on the hydrological cycle. These have implications for the watershed water balance in terms of multi-purpose land-use management and rehabilitation strategies.

Acknowledgements This study was jointly funded by National Key Basic Research Program of China under Grant 2006CB400502, the 111 Project under Grant B08048, Ministry of Education and State Administration of Foreign Experts Affairs, China, the Program for Changjiang Scholars and Innovative Research Team in University under Grant IRT0717, and the Grant Sci-Tech Research Project of Ministry of Education under Grant 308012. The authors would like to express their appreciation to the reviewers for their constructive remarks by which this paper has been improved.

REFERENCES

Bellot, J., Bonet, A., Sanchez, J. R. & Chirino, E. (2001) Likely effect of land use changes on the runoff and aquifer recharge in a semiarid landscape using a hydrological model. *Landscape Urban Plan.* **55**, 41–53.

Cunge, K. A. (1969) On the subject of a flood propagation method (Muskingum method). *J. Hydraul. Res.* **7**(2), 205–230.

Diouf, A. & Lambin, E. F. (2001) Monitoring land-cover changes in semiarid regions: remote sensing data and field observations in the Ferlo, Senegal. *J. Arid Environ.***48**, 129–148.

Dunin, F. X. & Mackay, S. M. (1982) Evaporation of eucalypt and coniferous forest communities. First National Symposium on Forest Hydrology, 18–25.

Fohrer, N., Haverkamp, S., Eckhardt, K. & Frede, H. G. (2001) Hydrologic response to land use changes on the catchment scale. *Phys. Chem. Earth (B)* **26**(7-8), 577–582.

Gislason, P. O., Benediktsson, J. A. & & Sveinsson, J. R. (2006) Random forests for land cover classification. *Pattern Recogn. Lett.* **27**, 294–300.

Helmschrot, J. & Flügel, W. A. (2002) Land use characterisation and change detection analysis for hydrological model parameterisation of large scale afforested areas using remote sensing. *Phys. Chem. Earth.* **27**, 711–718.

Hundecha, Y. & Bardossy, A. (2004) Modeling of the effect of land use changes on the runoff generation of a river basin through parameter regionalization of a watershed model. *J. Hydrol.* **292**, 281–295.

Jensen, J. R. (2004) Digital change detection. In: *Introductory Digital Image Processing: A Remote Sensing Perspective*, 467–494. Prentice-Hall, New Jersey, USA.

Kandrika, S. & Roy, P. S. (2008) Land use land cover classification of Orissa using multi-temporal IRS-P6 awifs data: a decision tree approach. *Int. J. Appl. Earth Observ. Geoinform.* **10**, 186–193.

Karvonen, T., Koivusalo, H. & Jauhiainen, M. (1999) A hydrological model for predicting runoff from different land use areas. *J. Hydrol.* **217**, 253–265.

Legesse, D., Vallet-Coulomb, C. & Gasse, F. (2003) Hydrological response of a catchment to climate and land use changes in tropical Africa: case study south central Ethiopia. *J. Hydrol.* **275**, 67–85.

Liu, J. Y. (1996) *The Macro Investigation and Dynamic Research of the Resource and Environment.* China Science and Technology Press, Beijing, China.

Lin, Y. P., Hong, N. M. & Wu, P. J. (2007) Impacts of land use change scenarios on hydrology and land use patterns in the Wu-Tu watershed in Northern Taiwan. *Landscape Urban Plan.* **80**, 111–126.

Liu, Y. S. & Chen, B. (2002) The study framework of land use/cover change based on sustainable development in China. *Geogr. Res.* **21**(3), 324–330.

Lørup, J. K., Refsgaard, J. C. & Mazvimavi, D. (1997) Assessing the effect of land use change on catchment runoff by combined use of statistical tests and hydrological modelling: Case studies from Zimbabwe. *J. Hydrol.* **205**, 147–163.

Lu, G. B., Li, Q. F., Zou, Z. H. Wang, H. J., Xia, Z. Q. & Ma, X. R. (2008) Impact of human activities on the flow regime of the Yellow River. In: *Hydrological Sciences for Managing Water Resources in the Asian Developing World* (ed. by X. H. Chen, Y. Q. David Chen, J. Xia & H. L. Zhang) (Proc. Symp. in Guangzhou, China, June 2006), 184–191. IAHS Publ. 319. IAHS Press, Wallingford, UK.

Mo, X., Liu, S., Liu, Z. & Zhao, W. (2004) Simulating temporal and spatial variation of evapotranspiraton over the Lushi basin. *J. Hydrol.* **285**, 125–142.

Pal, M. & Mather, P. M. (2003) An assessment of the effectiveness of decision tree methods for land cover classification. *Remote Sens. Environ.* **86**, 554–565.

Ren, L. L., Liu, X. F., Yuan, F., Singh, V. P., Fang, X. Q., Yu, Z. B. & Zhang, W. (2009) Quantitative effect of land use and land cover change on green water and blue water in northern part of China. In: *Hydrological Changes and Watershed Management from Headwaters to the Ocean* (ed. by M. Taniguchi *et al.*), 187–193. Taylor & Francis Group, London, UK. ISBN 978-0-415-47279-1.

Turner, D. P., Cohen, W. B., Kennedy, R. E., Fassnacht, K. S. & Briggs, J. M. (1999) Relationships between Leaf Area Index and Landsat TM Spectral Vegetation Indices across three temperate zone sites. *Remote Sens. Environ.* **70**, 52–68.

Wei, W., Chen, L. D., Fu, B. J., Huang, Z. L., Wu, D. P. & Gui, L. D. (2006) The effect of land uses and rainfall regimes on runoff and soil erosion in the semiarid loess hilly area, China. *J. Hydrol.* **335**, 247–258.

Winnaar, G. D., Jewitt, G. P. W. & Horan, M. (2007) A GIS-based approach for identifying potential runoff harvesting sites in the Thukela River basin, South Africa. *Phys. Chem. Earth.* **32**, 1058–1067.

Yuan, F., Ren, L. L., Yu, Z. B. & Xu, J. (2008) Computation of potential evapotranspiration using a two-source method for the Xin'anjiang Hydrological model. *J. Hydrol. Engng ASCE.* **13**(5), 305–316.

Yu, F. M. (2007) The character of land use/cover change in Chifeng and physical factor influences analyse. MSc Thesis, Inner Mongolia Normal University, Inner Mongolia, China.

Zhang, L., Dawes, W. R. & Walker, G. R. (1999) Predicting the effect of vegetation change on catchment average water balance. *Cooperative Research Centre for Catchment Hydrology, Technical Report. 1-35.*

Zhao, R. J. (1992) The Xinanjiang model applied in China. *J. Hydrol.* **135**, 371–381.

Zhao, C. Y., Nan Z. R. & Feng Z. D. (2004) GIS-assisted spatially distributed modeling of the potential evapotranspiration in semiarid climate of the Chinese Loess Plateau. *J. Arid Environ.* **58**, 387–403.

GIS and a remote sensing based approach for urban flood-plain mapping for the Tapi catchment, India

ANUPAM K. SINGH[1] & ARUN K. SHARMA[2]

1 *Department of Civil Engineering, Nirma University of Science and Technology, Ahmedabad 382481, India*
anupam.singh@gmx.net

2 *Division of Marine and Earth Sciences, Space Application Centre (ISRO), Ahmedabad 380015, India*

Abstract In India, floods typically occur during the monsoon season due to heavy tropical storm downpours and unregulated urban development. The floods during August 2006 in Tapi catchment caused great damage to people and property, resulting in 300 people being killed and US$ 4.5 billion worth of property damage. In this paper, geospatial technologies such as remote sensing, GIS, and GPS have been utilised to prepare urban flood hazard maps and to handle entity-specific query and analysis. The research methodology employed is based on statistical probabilities of flood frequency, maximum discharge carrying capacity at river cross-section, mapping of inhabited areas based on high-resolution images, and terrain mapping using global position system. It is estimated that for a mean flood height of 10 m (35-year return period), more than 80% of land in the west and southwest zone will be under flood against 40% in the central, 33% in the northern and 15% in the eastern zones.

Key words flood; hazard mapping; hydraulic remote sensing; Tapi River, India

INTRODUCTION

There was considerable increase in the occurrence of floods and flood-related damage globally during 2006. The flood events accounted for nearly 55% of all disasters registered and approx. 72.5% of total economic losses worldwide. Flooding is considered as the world's most costly type of natural disaster in terms of both human causalities and property damage (ESA, 2004). The annual disaster review indicates that flood occurrence has increased almost 10-fold during the last 45 years, from just 20 events in 1960, to 190 events in 2005 (Scheuren *et al.*, 2007). These extreme flood events have caused major damage to property, agricultural productivity, industrial production, communication networks, and infrastructure, mainly in the downstream parts of catchments. Therefore, floods are posing a great concern and challenge to design engineers, re-insurance industries, policy makers and to the government.

In India, about 40 million ha of land is flood prone, which is about 12% of the total geographical area (328 million ha) of the country. The flooding occurs typically during the monsoon season (July–September), caused by the formation of heavy tropical storms, ever decreasing channel capacity due to encroachments on river beds, and sometime due to tidal back-water effects from the sea. The Indian sub-continent in general, and the western peninsula in particular, experienced heavy floods during August 2006 that caused great damage to personal and property. The arid regions in Rajasthan and Gujarat to humid regions in the northeast were caught by surprise. It is estimated that a single flood event in the lower Tapi basin, a river stretch between the Ukai Dam and Arabian Sea, during 7–14 August 2006 resulted in 300 people being killed and approx. Rs 20 000 Crore (US$4.5 billion) of property damage. Human life came to a stand-still for almost two weeks in Surat and Hazira twin cities, as well as tens of rural villages along the lower Tapi basin. The Tapi River in Surat recorded the highest water levels of the last 35 years. Therefore, it is of prime importance to minimise the property damage, reduce infrastructure disturbances, and identify zones and building structures having greater flood hazard and flood risk.

It is at the strategic planning level that the possibility of utilising information available in cartographic forms assumes significant importance for urban flood mapping. However, a major contribution to flood mapping and planning is derived from the information made available by the use of remote sensing technologies and its integration with geographical information systems, GIS (Prasad, 2006). The high resolution satellite imagery, relief and land-use maps, hydraulic characteristics of river channels and flood-plain surveys, and probable water levels can be used for predictive flood hazard mapping. The evaluation of flood risk is generally based on a two-stage procedure: In the first stage, the statistical probabilities of stage–discharge characteristics of river sections are calculated; thereby the over-bank flow and the river sections at which the flow exceeds the carrying capacity of the river channel are determined. In addition, measurements of critical channel sections are undertaken to assess the hydraulically determined characteristics of the sections of the river course. In the second stage, the inhabited areas falling in the greater or lesser flood risk zone are evaluated based on the relief map and high resolution remote sensing imagery. Research on flood risk mapping (Miwa *et al.*, 2003; Singh *et al.*, 2005) reveals that three elements: maximum water level, the velocity of water flow, and the amount of time flood remains in a given land area, are essential to evaluate possible damage. The described requirements are considered to be essential for preparation of urban flood management and risk plans.

STUDY AREA

Tapi is the second largest westward-draining inter-state river in India after the mighty Narmada River. It covers approximately 51 504 km^2 (79%) of Maharashtra state, 9804 km^2 (15%) of Madhya Pradesh and 3837 km^2 (5%) of Gujarat state. The basin finds its outlet in the Arabian Sea and is bounded on three sides by ranges of hills. The Tapi River and its tributaries flow over the plains of Vidharbha, Khandesh and later to Gujarat, and can be divided into three zones, viz. Upper Tapi basin, Middle Tapi Basin, and Lower Tapi Basin (LTB). The portion between Ukai Dam to the Arabian Sea is considered as LTB, mainly occupying the Surat and Hazira twin cities along with tens of small towns and villages by the river. The Surat and Hazira twin cities are almost 100 km downstream of Ukai Dam and are affected by the recurrence of floods at regular intervals.

The LTB receives an average annual rainfall of 1376 mm, and these heavy downpours result in devastating floods and water logging downstream. The LTB contains the Ukai and Kakrapar reservoirs and part of the flow is diverted for irrigation from Kakrapar weir. The major crops grown are cotton and maize, followed by soybean. The prevailing land use is mixed forest, agricultural land, rural and urban settlements. The topography in LTB comprises narrow valleys and gently sloping ground. Figure 1 shows the location of the study area, Surat City, along with the relief and administrative zones. The main reasons for flooding in LTB are the heavy rainfall and discharge due to high water levels from Ukai Dam. Therefore, the flood problems of the river system are inundation due to over flowing of the banks, inadequate drainage capacity of the river, congestion at the point of confluence, and an excessive silt load factor.

In order to assess the geographical impact of floods, IRS-1D LISS IV with PAN satellite data at 5.8 m resolution has been procured. Besides IRS-1D PAN data, high resolution Google-earth data at the sub-metre scale were downloaded. The hydraulic data on river cross-sections including channel levels and hourly gauge discharge measurements, for three stations in and around Surat City, were collected from several sources. GIS analysis was conducted for all seven zones of Surat Municipal Corporation.

The data used for conducting the present research include geo-coded Indian Remote Sensing (IRS-1D) satellite imagery of April 2005, several Survey of India topo-sheets at 1:50 000 scale, high resolution Google-earth images and physical measurements for river hydraulic parameters obtained at the end of 2006. Contour maps for various city zones at 0.5 m interval were collected from Surat Municipal Corporation (SMC). More than 200 river channel sections at a mean distance of 150 to 200 m were measured by the Irrigation Department for LTB between Ukai Dam and Magdala weir. The water level and river discharge data from hourly to daily scales for discharge stations: Ghala, Mandavi, Ukai Dam, Kakarpar weir, Singanpur weir, and Hope (Nehru) Bridge were collected from the Central Water Commission, State Water Data Centre (SWDC), and Irrigation Department, respectively. All the data were attributed ready to put into a digital database management system (DBMS) using GIS software. Time series data on stage and discharge that are stored as *.ASCII format have been linked with GIS software-based analysis tools. In this paper

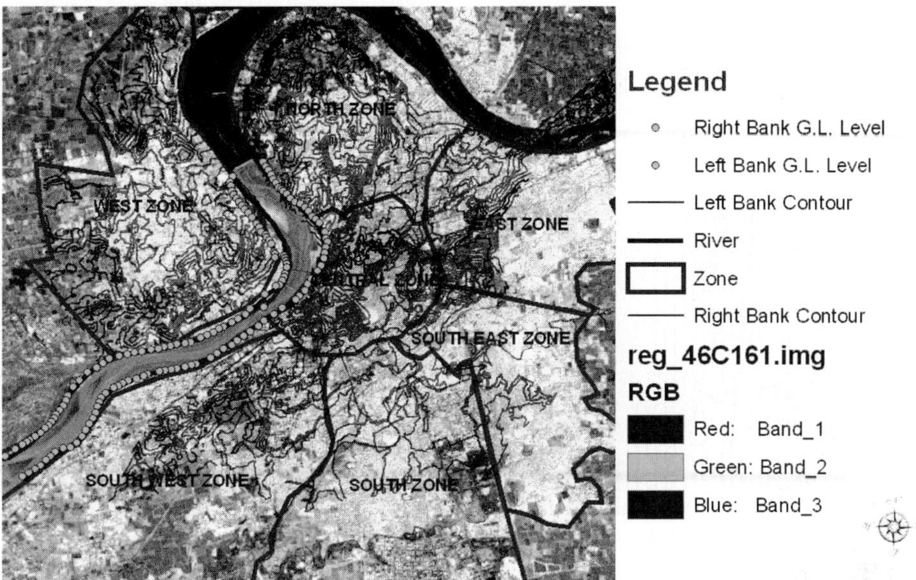

Fig. 1 Satellite image of IRS-1D LISSIV superimposed with municipal zones, contour lines and river channel elevations.

high-resolution remote sensing images from Google-Earth and IRS-1D are combined with river hydraulic analysis and digital elevation model (DEM) to identify the flood susceptible areas.

The step-wise methodology adopted for generation of flood simulation, flood mapping and zone-level flood hazard assessment is described below:

1 Collection of high-resolution remote sensing images for IRS-1D and Google-earth.
2 Collation of topographical features such as contours, river channel sections, and water level and discharge data.
3 Inter-linking of spatial and temporal data using GIS software and customised DBMS tools.
4 Generation of thematic maps.
5 The analysis of results and delineation of areas under various degrees of flood.

This methodology has been applied to prepare a flood potential and risk map for Surat city.

RESULTS

River discharge capacity using hydraulic data

In the study area, Ukai is a multi-purpose reservoir and is designed to cope with projected floods of 49 470 m^3/s (17.48 Lac cusecs) and a probable maximum flood of 59 880 m^3/s (21.16 Lac cusecs). The provided spillway has a capacity of 16.34 Lac cusecs. However, the safe carrying capacity of the river channel below Ukai Dam is considered to be 8.5 Lac cusecs. The river channel carrying capacities have been calculated from river section data collected after the August 2006 flood using the bathymetric method in GIS, as shown in Fig. 2.

The field data on river cross-sections were collected immediately after the August 2006 flood event and we calculated the river carrying capacity. Therefore, this section increases the flood risk in and around neighbouring areas. In our analysis on river channel sections it is found that the peak discharge capacity has been reduced to as low as 1.2 Lac cusecs at Viveknand Bridge. Hence, discharge in the river channel greater than 1.2 Lac cusecs will lead to flooding and this flood volume will lead to submergence of the west and central zones of Surat City.

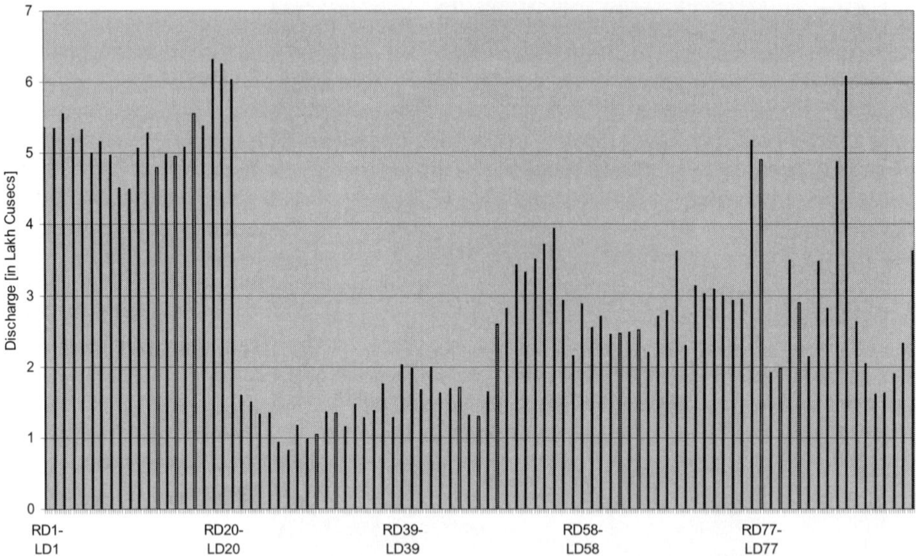

Fig. 2 Analysis of discharge carrying capacity at different river channel sections based on hydraulic modelling between the weir and the Arabian Sea. Surat City lies between sections RD22-LD22 to RD54-LD54.

The safe water carrying capacity of the Tapi River near Surat is reported to have been significantly reduced due to: encroachment in the flood plain areas, silting in the river-bed, and afflux caused by the Singanpore weir constructed on the river very close to the city. It is assessed that the river at Surat can carry only between 2 to 4 Lac cusecs without causing significant damage to urban dwellings and infrastructures.

Analysis of flood frequency at selected stations

The flood frequency analysis of the water level and discharge data shows that there was a likelihood of a flood hazard once in eight years till 1998, but that has increased during the recent decade to one in five years. Figure 3 shows the flood frequency analysis at Ghala discharge station for 5, 10, 15, 20 and 25-year return periods. It is estimated that discharge will exceed or equal 1.0 Lac cusecs every 17 years, 2.0 Lac cusecs every 21 years and 4.0 Lac cusecs every 25 years. The 25-year flood will result in several areas being under flood and river bank breaching at several locations. For quantification of the flood-prone area, the thematic map based on Google-Earth and IRS-1D data reveals that more than 80% of the urban area in Surat City could be flooded by an event of 50-year return period.

Flood hazard mapping using remote sensing data

Based on the contour levels supplied by SMC, a digital elevation model (DEM) for the west zone was developed. After combining the DEM with river bank levels, a flood risk map for various water-level scenarios at 0.5 m intervals was prepared, as shown in Fig. 5. The sample flood mapping potential areas for the west zone have been demarcated. The possible areas under each water level height are depicted with different colours in the flood hazard map. After generating the flood risk map, the water levels of the 2006 flood were compared with Hope (Nehru) Bridge. It was found that Hope Bridge has a bank level of 4.1 m; therefore there will be about 3–4 m water over the right-bank area. Hence, the major parts of the Adajan area will be submerged. The Rander area will also be submerged by 1–2 m depth of water, as was experienced during the August 2006 flood. This shows the accuracy of our hydraulic and GIS model for flood mapping.

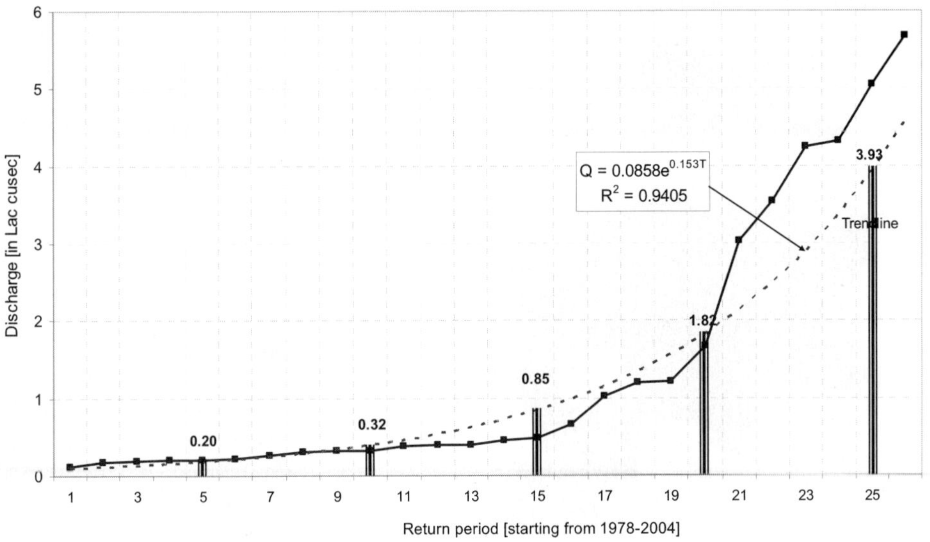

Fig. 3 Probability analysis of flood frequency and flood occurrence at the Ghala River section upstream of Surat City.

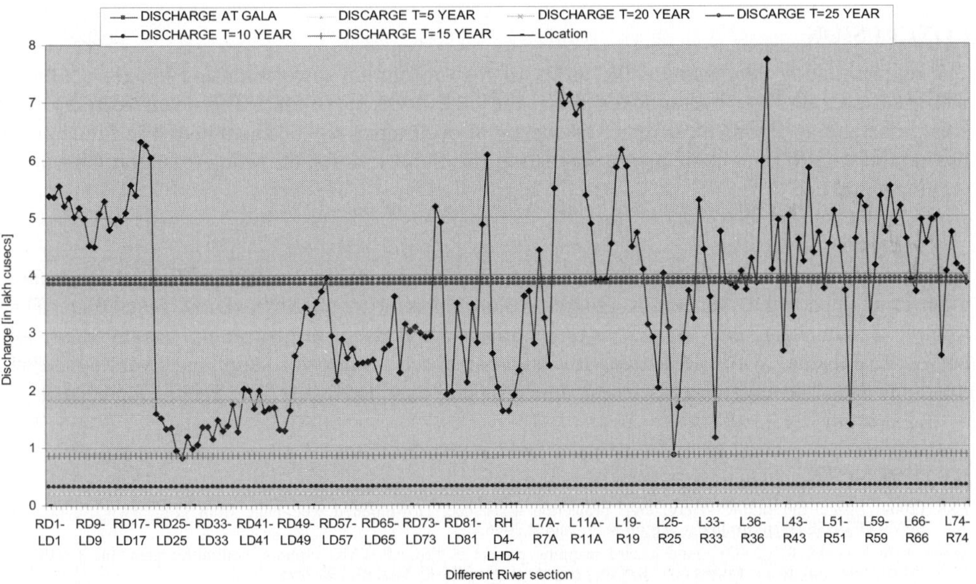

Fig. 4 Safe carrying capacity at river sections, superimposed with flood frequency analysis for all river sections.

In addition to flood mapping, the flood hazard map can also be prepared. This can be done by multiplying by the *a priori* probability map, which indicates the degree of danger to each area of the map derived from the analysis of historical events. The field data for August 2006 and September 1998 are under progress and we expect to get good results in a few months.

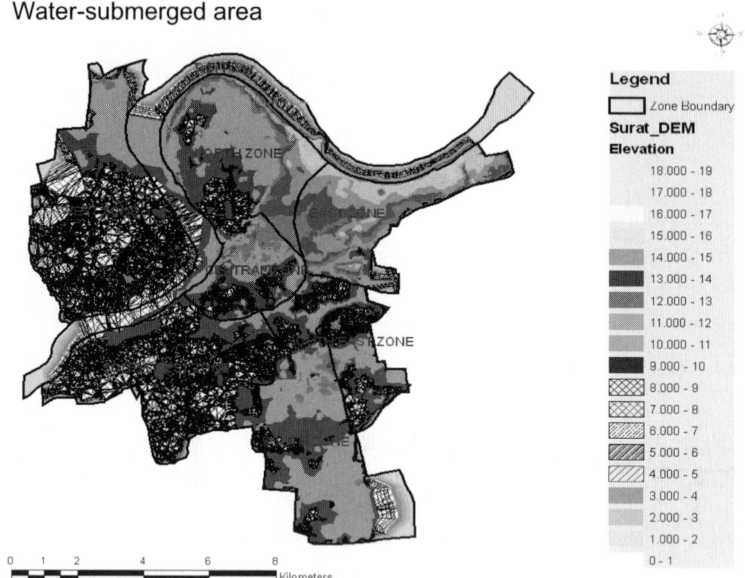

Fig. 5 Estimated flood hazard map of Surat City showing various zones and probable flooding potential.

CONCLUSION

The research study demonstrates the utility of high-resolution remote sensing images combined with field data on river hydraulics in delineating the flood prone area. The quantification of the area vulnerable to floods of different frequency of occurrence has been studied. The future scope lies in applying the methodology for preparing the flood risk for the whole of Surat City under varying scenarios.

Acknowledgements This research work is being carried out under ISRO-RESPOND research grant sanctioned to Dr Anupam K. Singh. We are thankful to Mr M. K. Dixit, State Water Data Centre, Gandhinagar; Mr M. K. Singh, Narmada-Tapi Basin Organisation, Gandhinagar; and Mr K. B. Rabadia, Water Resources Investigation Circle, Surat, for supplying hydrological and hydraulic data. The research assistance and field support provided by Mr D. P. Patel is appreciated.

REFERENCES

ESA (2004) Space and public safety, civil protection assistance. Downloaded web page from http://www.esa.int/esaeo/SEMQFF3VQUD_environment_0.html on 5 August.2006.

Miwa, J. & Kikuchi, R. (2003) Flood hazard mapping project in ESCAP/WMO typhoon committee area. In: *APHW2003 Proceedings* (ed. by K. Takara & T. Kojima) (Kyoto, Japan, 13–15 March), 792–795.

Prasad, A. K. (2006) Potentiality of multi-sensor satellite data in mapping flood hazard. *J. India Soc. Remote Sensing* **34**(3), 219–231.

Scheuren, J.-M., de Waroux, O., Below, R., Guha-Saphir, D. & Ponserre S (2008) Annual Disaster Statistical Review. CRED Brussels, Belgium.

Singh, A. K., Eldho, T. I. & Lindenmaier, F. (2005) GIS, remote sensing and computer models for water resources management. Proceedings of Training Course, Department of Civil Engineering, Nirma University Ahmedabad, India.

Singh, A. K., Shah, R., Desai, S. & Patel, D. P. (2007) High resolution remote sensing and field measurements for urban flood mapping. *High Resolution Remote Sensing and Thematic Application Abstracts* (18–20 December 2007, Calcutta, Indian Society of Remote Sensing), 108–109.

A new GIS-based algorithm for computing the TOPMODEL topographic index

BIN YONG[1], LILIANG REN[1], XIAOLI YANG[1], WANCHANG ZHANG[3], XI CHEN[1], LIHUA XIONG[2] & SHANHU JIANG[1]

1 *State Key Laboratory of Hydrology, Water Resources and Hydraulic Engineering, Hohai University, Nanjing 210098, China*
yongbin_hhu@yahoo.cn

2 *State Key Laboratory of Water Resources and Hydropower Engineering Science, Wuhan University, Wuhan 430072, China*

3 *Key Laboratory of Regional Climate-Environment Research for Temperate East Asia, Institute of Atmospheric Physics, CAS, Beijing 100029, China*

Abstract The TOPMODEL topographic index (TI), frequently used in approximately characterizing the spatial distribution of soil moisture, surface saturation, and runoff generation processes within a watershed, has been widely used in some topography-related geographical processes and hydrological models. However, it is still questionable whether the current GIS-based algorithms of TI can effectively control the local topographic characteristics and correctly indicate the spatial distributions of soil moisture and surface saturation. Based on the commonly used multiple flow direction (MFD) approach, we proposed a new improved multiple flow direction (IMFD) algorithm for accurately computing the TI distribution from DEM in this study. Then the new algorithm is applied to a real catchment located upstream of the Hanjiang River in China. Subsequently, the IMFD algorithm is quantitatively evaluated by using four types of artificial mathematical surfaces. Assessing the results shows that the error generated by IMFD is lower than that computed by the commonly used MFD algorithm, and the IMFD algorithm can more correctly express the relationship of upslope contribution area and soil water content and more accurately reflect the hydrological similarity of watersheds in TOPMODEL.

Key words topographic index; TOPMODEL; DEM; GIS; multiple flow direction algorithm

INTRODUCTION

The topographic index (TI, also called wetness index), originally defined in TOPMODEL (Beven & Kirkby, 1979) to approximate the likely spatial distribution of variable source areas (Quinn *et al.*, 1995) and predict local variations in water table depths within a watershed, has been widely used to study the effects of topography on hydrological processes at the basin scale. The TI concept, owing to its advantages of a simple and physically-based nature as well as its potential to couple with groundwater variations in time and space, has been frequently implemented in several ecological–atmosphere models and some topography-based land-surface process schemes.

The topographic index is defined as $\ln(\alpha/\tan\beta)$ in TOPMODEL, where α is the accumulative upslope area per unit contour length (or specific catchment area) and β is the local slope angle. Current GIS-based computation methods of TI mainly include algorithms of single flow direction (SFD) and multiple flow direction (MFD). Many studies have shown that MFD is obviously better than SFD, especially when the spatial pattern of upslope contributing area needs to be computed (Wolock & McCabe, 1995; Pan *et al.*, 2004).

BASIC ALGORITHM THEORY FOR COMPUTING TI SPATIAL DISTRIBUTIONS

The MFD algorithm, originally proposed by Quinn *et al.* (1991) (referred to as "MFD-Quinn" in this paper), is now recognized as the most accurate. On the basis of MFD-Quinn, we develop an improved multiple flow direction algorithm (IMFD) for the more accurate computation of the TI distribution.

Commonly used multiple flow direction algorithm (MFD-Quinn)

Different from SFD, the MFD algorithm assumes that flow from a current grid cell could drain into more than one downslope neighbouring grid cell. The fraction of the area draining through each current cell to each downslope direction is proportional to the slope gradient of each downhill flow path, so that steeper gradients will attract more water from the upslope contributing area (Quinn *et al.*, 1991).

In MFD-Quinn, the calculation of accumulative upslope area (i.e. α) can be concisely summarized by the following four important equations (i.e. equations (1)–(4)). According to Quinn *et al.* (1991), α was defined as:

$$\alpha = A / \sum_{j=1}^{n} L_j \tag{1}$$

where A is the total area draining into each cell, and L_j is the effective contour length of the cell boundary between current cell and its jth downslope neighbouring cell. The value of L_j is set to a half of grid size for cells in cardinal directions as illustrated as L_1 and L_3 (Fig. 1), and 0.354 for cells in diagonal directions, such as L_2 and L_4 (Fig. 1).

The local slope angle in the downslope direction, $\tan\beta$, is computed as:

$$\tan\beta = \sum_{j=1}^{n} (\tan\beta_j L_j) / \sum_{j=1}^{n} L_j \tag{2}$$

where $\tan\beta_j$ is the slope gradient between the current cell and its jth downslope neighbouring cell. Combining equations (1) and (2), we can obtain the $\ln(\alpha/\tan\beta)$ expression:

$$\ln(\alpha / \tan\beta) = \ln(A / \sum_{j=1}^{n} (\tan\beta_j L_j)) \tag{3}$$

The total upslope area of the current cell is distributed to its downslope neighbouring cells as:

$$\Delta A_i = A(\tan\beta_i L_i) / (\sum_{j=1}^{n} (\tan\beta_j L_j) \tag{4}$$

where ΔA_i is the drainage area passed from the current cell onto its downslope neighbouring cell i.

Improved multiple flow direction algorithm (IMFD)

In MFD-Quinn, α is the effective cumulative area draining from the upslope neighbouring cells to the current cell, which reflects the tendency of upslope water to accumulate at the current cell in the catchment. While $\tan\beta$ is the local slope angle of the current cell, which reflects the tendency for gravitational forces to move upslope water from the current cell to downslope neighbouring cells. Equations (1) and (2) suggest that the effective contour lengths of the grid cell boundaries between current cell and its downslope neighbouring cells (e.g. L_1, L_2, L_3 and L_4 in Fig. 1) are used to determine both α and $\tan\beta$.

However, in IMFD, we consider that the contour lengths used to determine α should be the effective contour lengths of the grid cell boundaries between upslope cells and current cell (e.g. K_1, K_2 and K_3 in Fig. 1) rather than that boundaries between current cell and downslope cells. Thus equation (1) should be amended as follows:

$$\alpha = A / \sum_{i=1}^{m} K_i \tag{5}$$

where K_i is the effective contour length of the grid-cell boundary between the current cell and its ith upslope neighbouring cell. Using equation (5) to substitute (1), equation (3) then becomes:

$$\ln(\alpha / \tan\beta) = \ln(A / \sum_{j=1}^{n} (\tan\beta_j L_j)) + \ln(\sum_{j=1}^{n} L_j / \sum_{i=1}^{m} K_i) \tag{6}$$

Additionally, the calculated total drainage area passed onto the downslope cells does not include the area of the current cell in MFD-Quinn. This shortcoming will bring about some errors in the iterative programming for the computation of TI. Thus it is necessary to modify equation (4):

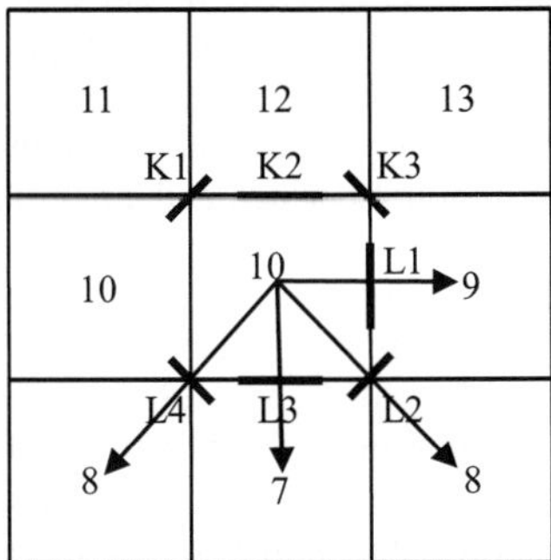

Fig. 1 An example of a simulated flow partition on DEM using the MFD-Quinn algorithm.

$$\Delta A_i = (A + \Delta x^2)(\tan \beta_i L_i) / (\sum_{j=1}^{n} (\tan \beta_j L_j) \quad (7)$$

where Δx is the grid size of the DEM, and the definition of other symbols is the same as that in equation (4).

Although some computing deficiencies of the effective cumulative area have been remedied through the above amendments, another important problem still remains unresolved, i.e. the fixed-exponent strategy to model the flow-partition proportion used in MFD-Quinn. In practice, the fixed-exponent strategy is unsuitable when the complex terrain conditions of a real watershed include both convergence and divergence, because the exponent cannot be altered in response to local terrain conditions. In this study, we adopt a simple but adaptive scheme proposed by Qin *et al.* (2007) to dynamically determine the downslope flow-partition exponent and accurately compute the TI distribution under various complex terrain conditions. According to Qin *et al.*'s (2007) approach, equation (7) can be substituted to the following expression for computing ΔA_i:

$$\Delta A_i = (A + \Delta x^2)((\tan \beta_i)^{f(e)} L_i) / (\sum_{j=1}^{n} (\tan \beta_j)^{f(e)} L_j) \quad (8)$$

Under MFD-Quinn which sets $f(e) = 1$, the partition of downslope flow cannot adapt to varying local terrain conditions. Here, e is the tangent value of the maximum downslope angle; $f(e)$ is a linear function of e, rather than a fixed constant value. More details about how to construct the function $f(e)$ are given in Qin *et al.* (2007).

Because "real-world" DEMs often contain depressions or flat areas where the TI cannot be directly calculated, a pre-processing procedure for raw DEM data is necessary for various TI computation algorithms. IMFD uses the simple but operative method of Jenson & Domingue (1988) to deal with depressions in the DEM. With regard to the computation of slopes in flat areas, Wolock & McCabe (1995) suggested that the slope gradient of a current flat cell equals to (0.5 × vertical resolution)/(horizontal resolution). This method is typical and simple enough, but not too reasonable in some very low relief areas. In IMFD, we use the tracking flow direction (TFD) method recommended by Pan *et al.* (2004) to process the flat cells in the DEM. It has been tested that TFD is more accurate than Wolock & McCabe's method under most terrain conditions because TFD can produce smaller slope value of flat area. IMFD should compute the spatial distribution of TI more accurately than MFD-Quinn, because the above improved equations and approaches can compensate for some serious deficiencies in MFD-Quinn.

Comparison of MFD-Quinn and IMFD computed in the real watershed DEM

To analyse the difference between IMFD and MFD-Quinn, we selected the Baohe catchment (about 2500 km^2) in the Chinese Hanjiang River basin as the experimental area (Fig. 2). This catchment is located in the upstream of the Hanjiang River, extending between 33°38′03″ to 34°1′08″N latitude and 106°48′15″ to 107°25′34″E longitude, and it has complicated topography, including various terrain conditions. The grid size of the DEM is 60 m (Fig. 2).

Based on the above basic theories, we realize the two multiple flow direction algorithms of MFD-Quinn and IMFD by using FORTRAN90 language programming to design and compute the TI spatial distribution values of the Baohe catchment (Fig. 3(a) and (b)). The spatial distributions of TI generated by IMFD and MFD-Quinn look somewhat different. The IMFD produces a spatial distribution with a smoother pattern and higher TI values. Especially in the footslopes and narrower channel and near the initial channel, the TIs computed by IMFD have obvious higher values than those by the MFD-Quinn approach. The spatial distribution of the TI values can be described by their maximum, minimum, mean, variance and skew values (Table 1). Compared to MFD-Quinn, IMFD results in TI distributions for Baohe catchment with a higher mean value and lower variance and skew values. And this conclusion is consistent with the finding of Wolock (1995) that the mean TI values computed by MFD are larger than SFD, while the variance and skew value are lower. The statistical results of the TI distributions of Baohe catchment show that MFD-Quinn is inferior to IMFD for deriving the TI spatial distributions from a real-world DEM and use of IMFD may obtain more accurate TI expressions.

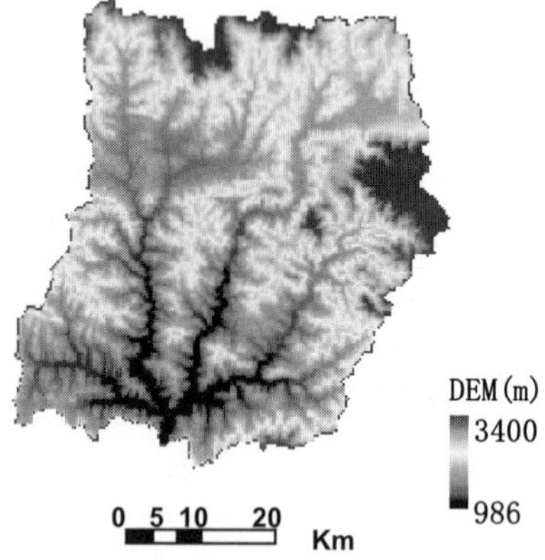

Fig. 2 DEM (60 m × 60 m) of the Baohe Catchment in Hanjiang River basin.

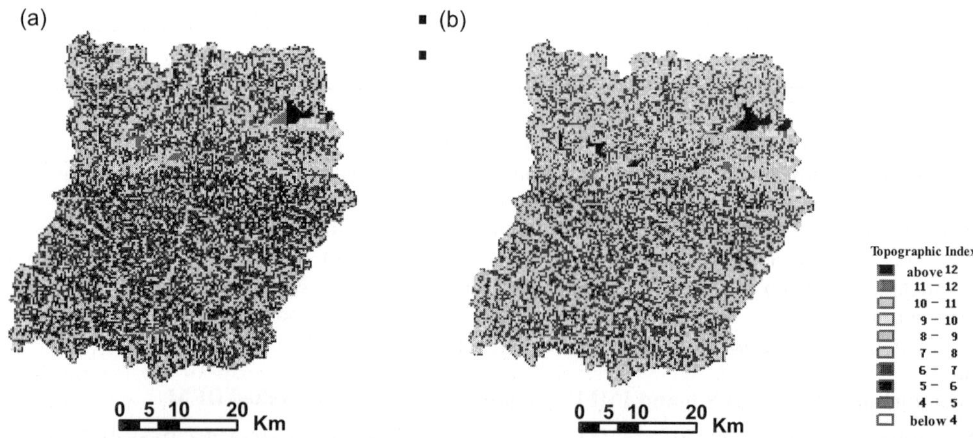

Fig. 3 Spatial distributions of TI computed by: (a) the MFD-Quinn algorithm, and (b) the IMFD algorithm, in Baohe catchment.

Table 1 Statistical results of the $\ln(\alpha/\tan\beta)$ distribution computed using MFD-Quinn and IMFD.

Algorithm	Min	Max	Mean	Variance	Skew
MFD-Quinn	3.06	15.54	7.15	2.90	1.70
IMFD	2.95	14.37	7.59	2.69	1.64
Difference	0.11	1.17	–0.44	0.21	0.06

QUANTITATIVE EVALUATION ON IMFD

To further investigate the accuracy and detect the effectiveness of IMFD, we should quantitatively evaluate and compare the errors generated by MFD-Quinn and IMFD. Using a "real-world" DEM to compare algorithms is data-dependent and difficult to quantify (Qin *et al.*, 2007). By using a selection of mathematical models with controlled parameters, digital terrain analysis (DTA) algorithms can be objectively compared and evaluated independently from data and human analyst's bias (Zhou & Liu, 2004). For these reasons, we adopted some standard and artificial mathematical surfaces to compare different TI algorithms quantitatively. In this paper, we use the evaluation method developed by Zhou & Liu (2002) to assess IMFD. The kernel of Zhou & Liu's (2002) approach is to construct four types of artificial surfaces (i.e. ellipsoid, inverse ellipsoid, saddle and plane) which can be represented by mathematical models, thus the theoretical "true" value of a specific catchment area (i.e. α in this paper) on any point of each surface can be inferred from mathematical formulations. Comparing the theoretical values and the computed values, the errors at each grid cell caused by different algorithms will be estimated easily. This evaluation method can simulate various terrain conditions including convergent, divergent, ridge, planar, and so on. Furthermore, its four types of artificial DEM surfaces can be regarded as a good surrogate for real topography. The approach is described in more detail in Zhou & Liu (2002).

Figure 4 gives an example of the four mathematical models. Based on the artificial DEM surfaces and the theoretical α and $\ln(\alpha/\tan\beta)$ distributions (see Fig. 4), we can compute the root-mean-square errors (RMSEs) of α and $\ln(\alpha/\tan\beta)$ for IMFD and MFD-Quinn, respectively. The root-mean-square error of α ($RMSE_\alpha$) can be computed as:

$$RMSE_\alpha = \sqrt{\frac{\sum_{i=1}^{n}(\alpha_{Ti} - \alpha_{Ci})^2}{n}} \tag{9}$$

where subscripts *Ti* and *Ci* stand for theoretical and computed α at grid *i*, and *n* is the number of grid cells for evaluation. The RMSE of the calculated topographic index ($RMSE_{TI}$) is:

$$RMSE_{TI} = \sqrt{\frac{\sum_{i=1}^{n}\left[\ln(\alpha/\tan\beta)_{Ti} - \ln(\alpha/\tan\beta)_{Ci}\right]^2}{n}} \tag{10}$$

Table 2 lists $RMSE_\alpha$ and $RMSE_{TI}$ computed for the four mathematical surfaces using MFD-Quinn and IMFD. The RMSEs of α show that IMFD yields lower errors than MFD-Quinn under all tested artificial terrain conditions; while for the RMSEs of $\ln(\alpha/\tan\beta)$, except for the inverse ellipsoid surface, IMFD produces greater accuracy for other three kinds of terrain conditions than MFD-Quinn does. We conclude that IMFD performs reasonably well and produces better results in most cases.

In summary, the quantitative evaluation suggests that the specific catchment area values and the topographic index values computed by IMFD are closer to the theoretical true values than those computed by MFD-Quinn. Compared with MFD-Quinn, IMFD can more accurately express the spatial distribution of the depth to the water table and more reasonably reflect the hydrological similarity within a watershed.

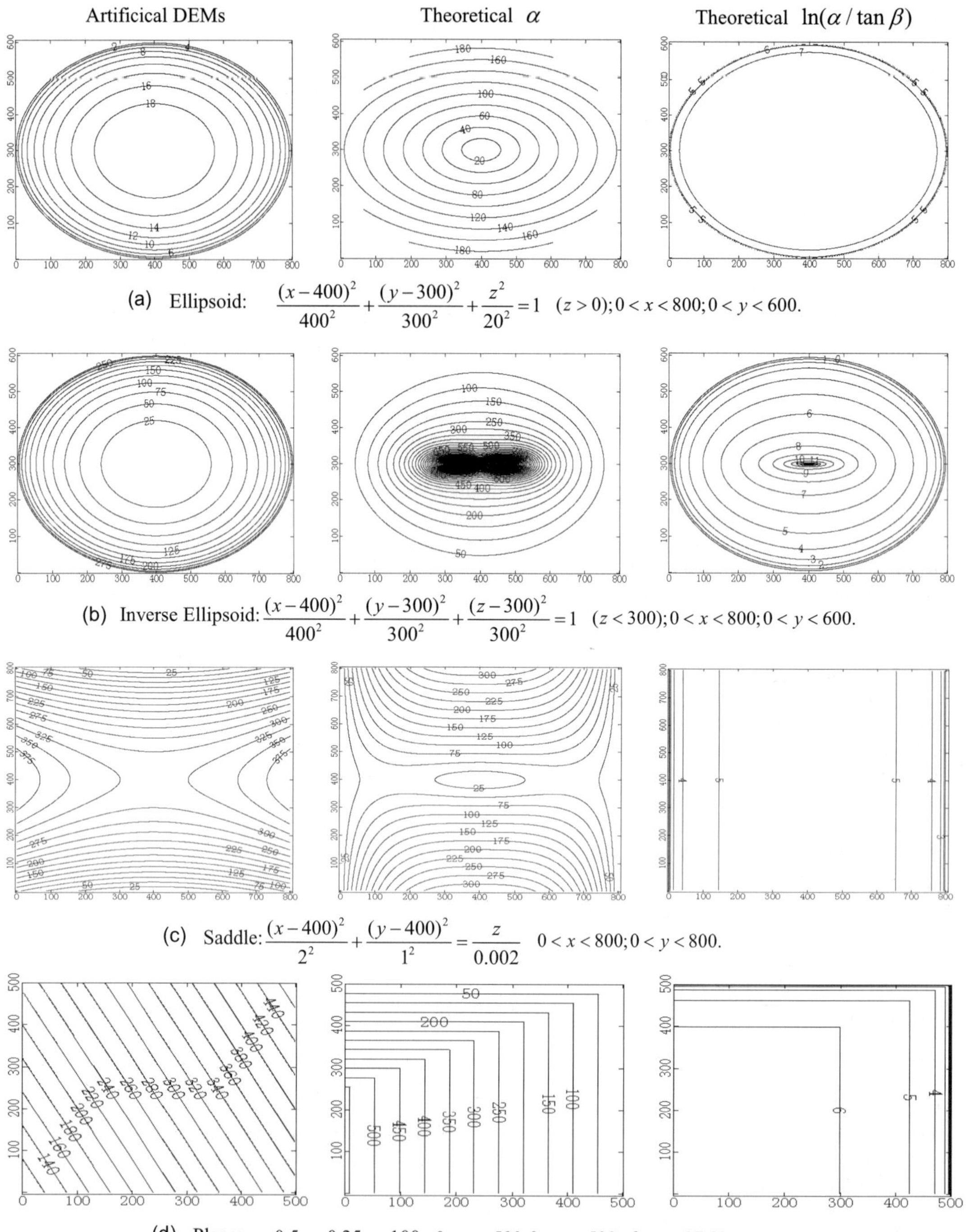

Fig. 4 The contour maps of artificial surfaces, the theoretical α distributions and the theoretical $\ln(\alpha/\tan\beta)$ distributions of four typical mathematical surfaces: (a) ellipsoid, (b) inverse ellipsoid, (c) saddle, and (d) plane.

Table 2 RMSEs for α and $\ln(\alpha/\tan\beta)$ computed from four mathematical surfaces by using the MFD-Quinn and IMFD algorithms.

Surface	$RMSE_\alpha$		$RMSE_{TI}$	
	MFD-Quinn	IMFD	MFD-Quinn	IMFD
Ellipsoid	6.602	5.892	0.063	0.048
Inverse ellipsoid	500.513	394.534	0.081	0.096
Saddle	109.779	102.817	0.573	0.459
Plane	412.448	32.564	0.176	0.158

CONCLUSIONS

Based on the MFD-Quinn algorithm (Quinn *et al.*, 1991), we proposed a new improved scheme (i.e. IMFD) for more accurate computation of the TI distribution. The improvements mainly include:

1. amending the equations of the classical MFD algorithm for reasonably computing the specific catchment area α,
2. adopting a varying exponent strategy for accurately modelling of flow partition, and
3. using a flow-direction tracking method (i.e. TFD) to meticulously deal with flat grid cells.

We quantitatively evaluated IMFD for four types of mathematical surfaces against their theoretical values of the specific catchment area and the topographic index. Assessment of the results indicates that the spatial distribution of TI computed by IMFD is more accurate than the result of MFD-Quinn.

Acknowledgements This study is financially supported by the National Key Basic Research Program of China (Grant 2006CB400502). Also the paper is a part of results supported from the 111 Project (Grant B08048), the Program for the Grant Sci-Tech Research Project of Ministry of Education (Grant 308012) and the open study funds of State Key Laboratory of Water Resources and Hydropower Engineering Science, Wuhan University (Grant 2008B039).

REFERENCES

Beven, K. J. & Kirkby, M. J. (1979) A physically based variable contributing area model of basin hydrology. *Hydrol. Sci. Bull.*, **24**(1), 43–69.

Jenson, S. K. & Domingue, J. O. (1988) Extracting topographic structure from digital elevation data for geographic information system analysis. *Photogramm. Engng Remote Sens.* **54**, 1953–1600.

Pan, F., Peters-Lindard, C. D., Sale, M. J. & King, A. W. (2004) A comparison of geographical information systems-based algorithms for computing the TOPMODEL topographic index. *Water Resour. Res.* **40**, W06303, doi:10.1029/2004WR003069.

Qin, C., Zhu, A.X., Pei, T., Li, B., Zhou, C. & Yang, L. (2007) An adaptive approach to selecting a flow-partition exponent for a multiple-flow-direction algorithm. *Int. J. Geogr. Inf. Sci.* **21**(4), 443–458, doi:10.1080/13658810601073240.

Quinn, P., Beven, K. J. & Lamb, R. (1995) The $\ln(\alpha/\tan\beta)$ index: how to calculate it and how to use it in the TOPMODEL framework. *Hydrol. Processes* **9**,161–185.

Quinn, P., Beven, K. J. & Planchon, O. (1991) The prediction of hillslope flow paths for distributed hydrological modeling using digital terrain models. *Hydrol. Processes* **5**, 59–79.

Wolock, D. M. & McCabe, G. J. (1995) Comparison of single and multiple flow direction algorithms for computing topographic parameters in TOPMODEL. *Water Resour. Res.* **31**(5), 1315–1324.

Zhou, Q. & Liu, X. (2002) Error assessment of grid-based flow routing algorithms used in hydrological models. *Int. J. Geogr. Inf. Sci.* **16**, 819–842.

Zhou, Q. & Liu, X. (2004) Analysis of errors of derived slope and aspect related to DEM data properties. *Comput. & Geosci.* **30**, 369–378.

Analysis of bathymetry and spatial changes of Vembanad Lake and terrain characteristics of Vembanad Wetlands using GIS

R. GOPAKUMAR[1] **& KAORU TAKARA**[2]

1 *Centre for Water Resources Development and Management, Kozhikode 673571, Kerala, India*
gopan65@hotmail.com

2 *Disaster Prevention Research Institute, Kyoto University, Uji, Kyoto 611-0011, Japan*

Abstract A number of artificial interventions, such as large-scale reclamation of water bodies to form polders for rice cultivation, and unscientific construction activities have led to the degradation of the environmental status of the Vembanad Wetlands of Kerala State in India. Increased flood proneness is one of the many major environmental issues faced in the region during the recent decades. A dynamic view of spatial changes in the wetland and knowledge on its relation to the various environmental issues are essential for scientifically planning the land and water resources management of the wetlands. In this study, through the development of geographic data sets and their analysis in GIS, benchmarks are established for the extent of Vembanad Lake and its changes due to reclamation during the 20th century, the bathymetry of the lake and its area–elevation–capacity relationship, the wetland topography and its water holding capacity at different elevations.

Key words Vembanad Wetlands; bathymetry; spatial changes; elevation–area–capacity relationship

INTRODUCTION

As the transitional zones between terrestrial and aquatic ecosystems, wetlands perform a variety of functions that are of vital importance to the environment and to society. They filter contaminants out of surface waters, and attenuate floods and tidal surges, and thus protect coastal settlements. Wetlands are also important as the feeding, breeding, and resting grounds for resident and migratory fish and waterfowl, and host unique vegetative and microbial communities. It is widely accepted that hydrology is one of the most important factors affecting the structure and functions of wetlands (Mitsch & Gosselink, 1986; Brooks, 2005). When human intervention, through activities like drainage, filling, dam construction, water diversions and dredging, alter hydrological conditions of wetlands, profound changes result in the wetlands and their unique functions.

The Vembanad Wetlands of Kerala State in India (Fig. 1) is a tidal wetland which includes the lower deltaic regions of the Achencoil, Pamba, Manimala, Meenachil, and Muvattupuzha rivers and the Vembanad backwater lake connected to the Arabian Sea at Cochin. This wetland has a complex aquatic system of coastal backwaters, lagoons, marshes, mangroves and reclaimed lands, inlaid with intricate networks of natural and man-made channels. Large-scale land reclamation started at the beginning of the 19th century to form polders for rice cultivation, and the closing off of tidal action in the system by the construction of Thanneermukkom (TM) salinity barrage across Vembanad Lake in 1975, were two of the major human interventions which adversely affected the water system of Vembanad Wetlands, which have many water and land-use functions such as flood control, pollution control, inland navigation, tourism, biodiversity, agriculture. In recognition of the urgent need for its conservation, the Vembanad Wetlands, together with the contiguous Kol Wetlands, were designated as a Wetland of International Importance (Ramsar Site) in 2002.

Addressing the various problems related to wetland management requires a dynamic view on the spatial changes in the system and linkages of these changes with the unique functions like flood control and the status of the ecosystem in general. Although the pollution of Vembanad Lake and its impacts on the flora and fauna have often been reported (e.g. Pradeep & Samy 1986; Padmakumar *et al.* 1988, 2002; Ambat, 1992), only a few studies report on the changes in the water system of the wetland. In a case study on the environmental consequences of water resources mismanagement, Balchand (1983) highlighted the adverse impacts of human interventions to moderate floods and to regulate salinity intrusion in the Vembanad system. In another case study on the Vembanad-Kol wetland system, James *et al.* (1997) comprehensively discussed

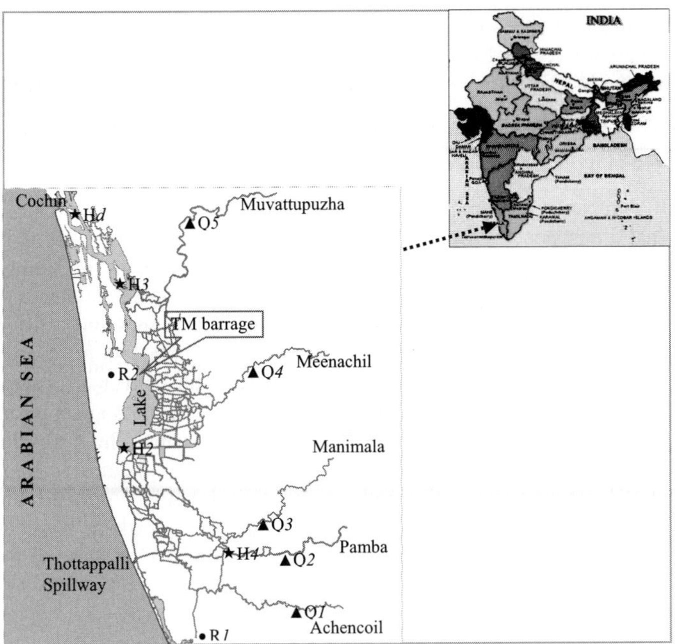

Fig. 1 Location map of Vembanad Wetlands with hydrologic monitoring stations (Rn: Raingauge. Qn: Discharge monitoring station. Hn: Water level monitoring station).

the environmental issues faced in the Vembanad Wetlands and highlighted the importance of an integrated management approach covering the wetland and its river basins for the optimal utilization of wetland benefits. This study highlighted the shrinkage in the area of Vembanad Lake, and changes in the flood carrying capacity of wetlands due to the large-scale reclamation.

Although it has been generally reported that there are major spatial changes to the water system of the wetlands, so far, there has not been any attempt to develop a spatial information system for the water resources of Vembanad Wetlands and its river basins to establish benchmarks and to monitor the spatial changes. This paper presents a study on the development and analysis of geographic information data sets on the physical features of the Vembanad Lake and the wetlands using the available data and maps in a geographic information system (GIS). The study area and its associated river basins are delineated from the Survey of India (SOI) topographic maps in GIS. The geographic information data sets generated on the bottom elevation and water spread area of the Vembanad Lake, and elevation and land use of the wetlands are used for analysing the bathymetry and changes in the extent of the lake, and physical characteristics of the wetlands.

DATA USED

The 1:50 000 scale SOI topographic maps are used for delineating the Vembanad Wetlands and its associated drainage basins. The references used for mapping the past scenarios of Vembanad Lake are: (i) the 1:250 000 topographic map no. NC 43-11 and 43-12 from the U502 map series (1953) of the Army Map Service (AMS) of US Army Corps of Engineers, Washington; (ii) the 1:50 000 scale SOI topographic map (1970); and (iii) the N-43-05 Landsat image mosaic (MDA Federal, 2004) for the year 1990 obtained from the Global Land Cover Facility (GLCF) of NASA Stennis Space Center. Three data sets of spot elevations are used for generating the wetland elevation model: spot heights available from the SOI topographic map, and spot elevations of the wetland

and polder region as obtained from the maps prepared by the Water Resources Department (WRD) of Kerala based on a survey conducted in 1987. Bathymetric data of the Vembanad Lake as on 1987 were also obtained from the WRD.

DELINEATION OF STUDY AREA AND ASSOCIATED RIVER BASINS

The surface water system of Vembanad Wetlands consists of the discharges from five major rivers (Fig. 1) and the semi-diurnal tides penetrating from Cochin estuary, downstream. Within the deltaic region, the rivers join, branch and finally drain to the Arabian Sea through the Vembanad Lake. Basin boundaries of the Achencoil, Pamba, Manimala, Meenachil and Muvattupuzha rivers and a stream named Kariar that could be delineated based on the drainage pattern and contours in the 1:50 000 scale SOI topographic maps are digitized in GIS as vector data sets in geographic coordinate system (GCS), and the maps are subsequently projected to the Universal Transverse Mercator (UTM) coordinate system (Zone 43 N) with WGS84 as datum. The drainage areas of the rivers as defined by the digitized boundaries are referred to as the upper basins. The lower deltaic region of the rivers, where individual catchments cannot be distinguished is delineated into one single non-basin unit titled "Vembanad wetland", with its western boundary defined by the shoreline of the Arabian Sea demarcated in the SOI topographic map. The northern boundary is extended up to the Cochin estuary and the southern boundary to 25 km south of TP flood spillway.

Results

Areas of the drainage units delineated as the Vembanad Wetland and the six river basins are given in Table 1. The upper basins of the rivers cover a total area of 6126.48 km^2 and the non-basin area delineated as the study area, i.e. the Vembanad Wetlands, covers 2033.02 km^2. Thus, the total area covered by the Vembanad wetland and its associated drainage units is 8159.5 km^2.

Table 1 Areas of the Vembanad Wetlands and its associated river basins.

No.	River basin/ Unit	Area (km^2)
1	Achencoil	1013.16
2	Pamba	1705.34
3	Manimala	793.79
4	Meenachil	1030.94
5	Kariar	94.74
6	Muvattupuzha	1488.52
7	Wetland region	2033.01
	Total	8159.50

ASSESSMENT OF CHANGES IN THE AREA OF VEMBANAD LAKE

Although some of the case studies (e.g. James *et al.*, 1997) generally highlighted shrinkage of Vembanad Lake and subsequent reduction in the water carrying capacity as the major spatial changes in the system, so far no change detection studies are reported for Vembanad Lake. In this study, to establish benchmarks for the area of Vembanad Lake and detect the changes in the lake's extent during the past century, digital maps showing three scenarios of the water area of the lake for different time periods during 1917–1990 were prepared in GIS. By comparing the three scenarios, spatial changes to the Vembanad Lake during the 20th century are assessed.

The water area of the lake in 1917 (Scenario-1) is mapped using the 1:250 000 scale U502 series historical topographic map collection of the US Army Map Service (1955). The U502 topographic map series was prepared by compiling the information from the 1:126 720 scale SOI maps of 1917, and British Admiralty Charts. For mapping the scenario of the lake as in 1970 (Scenario-2), the 1:50 000 scale SOI topographic maps are used as reference. The N-43-05

Landsat image mosaic (MDA Federal, 2004) of the NASA Stennis Space Center, USA, is used for mapping of the lake extent as in 1990 (Scenario-3). The Landsat mosaics available in MrSID file format use Universal Transverse Mercator (UTM) projection with World Geodetic System 1984 (WGS84) Datum. Pixel size is 28.5 m and absolute positional accuracy is 50 m root mean square error. In each of the above cases, vector data sets of polygon geometry for the outer boundary of the lake and the interior islands are digitized in GIS and by geometrically intersecting these data sets, a polygon map containing the water spread areas of Vembanad Lake is generated in Universal Transverse Mercator coordinate system with WGS84 as datum. For each scenario, the water spread area of the lake is obtained from the geometric properties of the data set.

Results and discussions

The water spread area of Vembanad Lake under the three different scenarios is presented in Table 2. Scenario-1 is defined as the baseline scenario; in 1917 the water spread area of the lake was 290.85 km^2. As per Scenario-2, by the year 1970, the water spread area of the Vembanad Lake had reduced to 227.23 km^2. Although land reclamation for polder formation started in the Vembanad Wetlands during the 19th century, major land reclamation schemes were completed after 1917. Therefore, during the period from 1917–1970, reduction in the lake area was 63.62 km^2, i.e. 21.88% of area as on 1917. Polders located on the southern sector of Vembanad Lake and expansion of the area of the Wellington Island located at Cochin estuary are the major interventions in the water system after 1917. In Scenario-3 (1990), the lake area has further reduced to 213.28 km^2, a change of 13.95 km^2 (4.80%), which includes relatively small-scale reclamations for construction of the TM barrage and other purposes. Thus the total reduction in areal extent of the Vembanad Lake during the period from 1917 to 1970 when maximum land reclamation was completed is 77.57 km^2, which is 26.67% of the area of the lake at the beginning of the 20th century.

Table 2 Area of the Vembanad Lake under three different scenarios.

Scenario	Year	Source map	Area (km^2)
Scenario-1	1917	Map nos NC 43-11 and 43-12, AMS, US Army, 1957.	290.855
Scenario-2	1970	Map no. 58 C/6, Survey of India, 1970.	227.226
Scenario-3	1990	Landsat mosaic N-43-05, 1990 Edition, MDA Federal	213.282

ANALYSIS OF THE BATHYMETRY OF VEMBANAD LAKE

Water levels of Vembanad Lake, which extends about 60 km from Alleppey in the south to the Cochin estuary in the north, vary according to the tidal conditions at Cochin and the river flows from the upper basins. Benchmark studies have not been conducted to analyse the lake bathymetry and to establish its elevation–area–capacity relationship. Bathymetric data sets of different time periods required to monitor the spatial and temporal changes in the depth of the lake are not available, but to establish a benchmark for the bottom levels of the lake, data from a bathymetric survey conducted in 1987 by the Water Resources Department (WRD) of Kerala State was available in the from of a data book containing cross-sections of Vembanad Lake for its entire water spread area. This data set was digitized and analysed in GIS to generate an elevation model of the lake-bottom and establish the relationship of water spread area and storage volume of the lake with the elevation of lake-bottom.

Processing of bathymetric data and generation of TIN data set

The available bathymetric survey data included cross-sections measured using an echo-sounder and optical range finder at an average interval of about 2 km for the entire length of Vembanad Lake and maps showing the locations of the cross-section measurements. To reduce the echo soundings to mean sea level (MSL), water levels obtained by fixing a network of staff gauges are used. The bathymetry of the lake is analysed by generating an elevation model of the lake in GIS using the available data. Since the bathymetric data were available in hardcopy format, a considerable amount of pre-processing was done to generate a digital database with spot elevations of the lake, which was subsequently used for generating the elevation model of the lake bottom surface in GIS.

At first, a number of points were selected along each cross-sectional profile and for each point, the elevation of lake bottom (Z) is recorded and the geographic location (X,Y) coordinates are computed using the maps showing the location of the cross-sections. An elevation data set was generated with the (X, Y, Z) coordinates of a total of 10 976 points locations selected from the cross-sectional profiles of the lake. Using the elevation data set, a shapefile of point geometry is generated in GIS in GCS, which is exported to a geodatabase as feature class and projected to UTM coordinate system with WGS84 as datum (WGS84_UTM Zone 43N).

The elevation of the bottom surface of Vembanad Lake for its water spread area is modelled and analysed by generating a triangulated irregular network (TIN) data set using the spot elevation data set of the lake in the 3-D analysis module of GIS. TIN data sets contain irregularly spaced points that have X, Y coordinates describing their location and a Z value that describes the surface at that point. A series of edges join the points to form triangles and the resulting triangular mosaic forms a continuous faceted surface (Chang, 2006). To define the interruptions in surface behaviour and continuity at the bank of the lake and due to the islands located within the lake body, the boundaries of the lake and islands in polyline geometry are defined in the TIN data set as breaklines.

Calculation of surface area and storage volume

The elevation model of Vembanad Lake shows that the bottom elevation of the lake varies from –9.9 m MSL to +2 m MSL. At Alleppey on the southern end, bank level is 1 m above MSL and lake bottom elevation is 1.5 m below MSL. At the Cochin estuary, bank level is 2 m to 5 m above MSL and lake bottom elevation varies from 2 m to 6 m below MSL. A profile of the lake bottom elevation from Alleppey to Cochin generated using the TIN data set is shown in Fig. 2. For the wider portion of the lake from Alleppey to Thanneermukkom, variations in lake bottom level are marginal. Near the TM barrage the lake is narrow and comparatively shallow on the upstream side, whereas, on the downstream side, depth varies up to 7 m below MSL. From the TM barrage to Cochin, variations in the altitude of the lake are large and there are many narrow stretches where the lake bottom elevation falls to a depth of 10 m below MSL. The sudden variations observed in the elevation profile are mainly due to the exaggerated vertical scale compared to the horizontal scale.

To determine the elevation–area–capacity relationship, area and storage volume of the Lake below different reference elevations are computed (Table 3) from the TIN data set using the

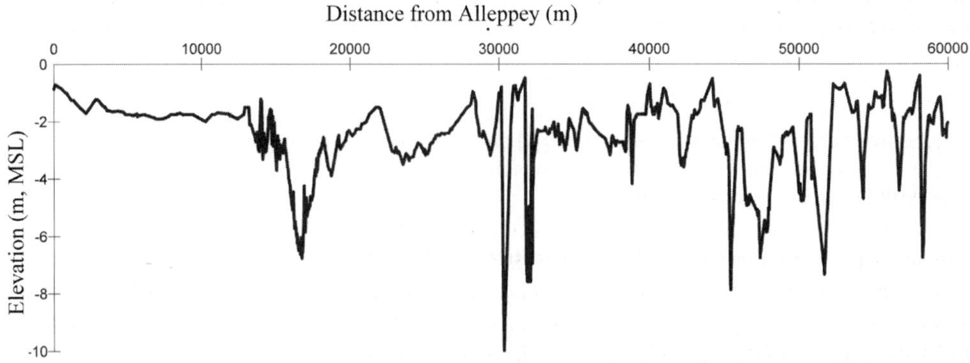

Fig. 2 Depth profile of the Vembanad Lake from Alleppey to Cochin.

Table 3 Area and volume of Vembanad Lake at different elevations.

Elevation, m MSL	Area of lake (km^2)	Volume ($\times 10^6$ m^3)
–6	1.54	1.40
–4	10.69	11.46
–3	26.28	28.49
–2	67.63	72.99
–1	171.79	199.29
0	211.06	393.94
1	221.28	611.47
2	227.50	834.18

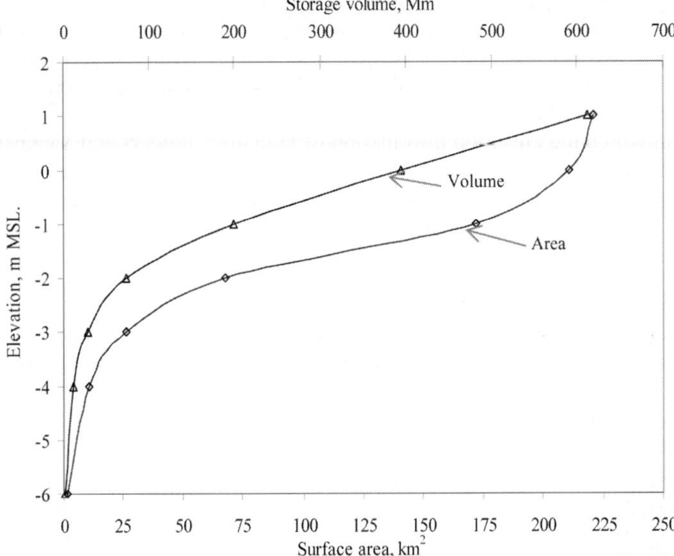

Fig. 3 Elevation–area–capacity curves of the Vembanad Lake.

3-D surface-analysis software of GIS. Variation in the area and volume of the lake is significant between 6 m below MSL to 1 m above MSL. From the TIN data set, mean elevation of the lake is computed as 1.725 m below MSL. At 1 m MSL, the area of lake is 216.53 km^2 and its volume is 611.47×10^6 m^3. The elevation–area–capacity curve of the lake is shown in Fig. 3. The elevation–capacity relationship shows a logarithmic trend with the best-fit line indicating an R^2 value of 0.988. A fifth-order polynomial was found to be the best to indicate the elevation–area relationship with an R^2 value of 0.996.

ANALYSIS OF THE TOPOGRAPHY AND LAND USE OF WETLANDS

Development of elevation model

A digital elevation model (DEM) is an essential feature for many applications including the flood management in Vembanad Wetlands. A DEM can be constructed in GIS using the different topographic information available either in the form of digitized contour lines or spot elevations of selected area or in combinations of the elevation data from different sources. Depending on the availability and accuracy of topographic data, parts of the area considered or certain features of it may be very accurate while others may be quite basic. However, the DEM can be improved as more topographic data become available. Since a DEM is not available for the Vembanad Wetlands, the digital landscape of the wetland is created as a TIN data set in this study using four sets of elevation data. The surface analysis module in GIS was used to combine the different data sets into one single topographic surface. Since the major part of the terrain of Vembanad Wetlands has a ground elevation below 20 m, contour lines are not shown on the SOI topographic maps of the study region, which has a contour interval of 20 m. However, a large number of spot elevations are marked in SOI maps and this information is used to generate the TIN data set of the wetland together with spot elevations obtained from the maps of the Water Resources Department (WRD), and the bathymetric data of the lake.

From the available maps, spot elevations of the land area and the polders are digitized into two shapefiles in GCS, which are later converted into feature classes of the geodatabase in UTM coordinate system. These geographic data sets and bathymetric survey data set were used for generating the TIN data set of the Vembanad Wetland in the following sequence. First, the TIN data set of the wetland for its defined boundary is generated in GIS by combining all the spot elevation data for the land area. In the generated TIN data set, elevations of the polder region and lake water area are modified using the corresponding elevation data sets. The finalized TIN data set of Vembanad Wetland, in which the elevation information of land area, polders and Vembanad Lake are merged together into a single elevation map of land surface, is shown in Fig. 4. The TIN data set is used to analyse the terrain characteristics of the wetlands. For the purposes of other analyses requiring elevation data sets of Vembanad Wetland in raster format, the TIN data set is transformed into a digital elevation model (DEM) with a resolution of 25-m cell size.

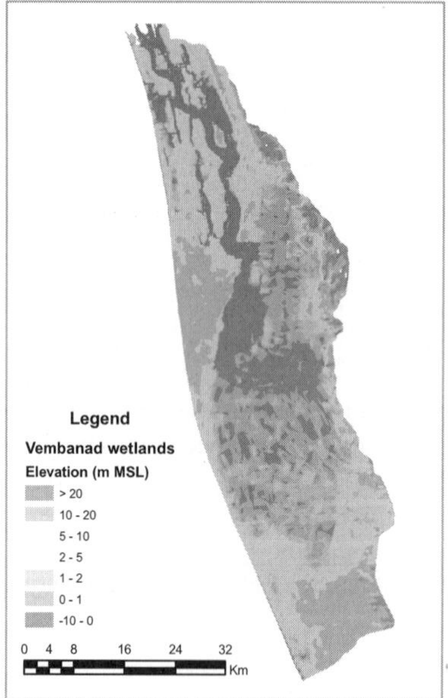

Fig. 4 TIN model of the terrain of Vembanad Wetlands.

Topography of the wetlands

A major part of Vembanad Wetland consists of flat, low-lying plains formed by the deposition of fluvial sediments. The coastal area is covered by alluvium composed of sand, formed mainly as the sandbars along the coast (Chattopadhyay & Sidharthan, 1985). From the TIN data set of the terrain classified into different elevation classes, it is found that the major part of Vembanad Wetland lies at between –6 m and +20 m MSL and the mean elevation of the wetland is 4.64 m

MSL. In the Kuttanad region of the wetlands, elevation of the terrain gradually drops towards the west to below mean sea level, and then gradually rises to above mean sea level at the west coast. Using the 3-D surface analysis tool of GIS, areas of the wetland located below different reference plane elevations from mean sea level to 20 m above MSL, are calculated from the TIN data set. Areas of wetland falling under the different elevation classes are shown in Table 4. Out of a total area of the wetland of 2033 km^2, 19.59% (398.12 km^2) is located below the mean sea level. The total area of the wetland located below 1 m MSL and 2 m MSL are 763.23 km^2 (37.55%) and 976.28 km^2 (48.03%), respectively. Volumes of the wetland at reference elevations MSL, 1 m MSL and 2 m MSL are calculated as 525.08 × 10^6 m^3, 1079.35 × 10^6 m^3 and 1970.65 × 10^6 m^3, respectively.

Table 4 Area of Vembanad wetland covered under different ranges of elevation.

No.	Elevation range (m MSL)	Area (km^2)	% area	Cumulative area (km^2)
1	below MSL	398.12	19.59	398.12
2	0 – 1m	365.11	17.96	763.23
3	1 – 2	213.05	10.48	976.28
4	2 – 5	265.11	13.04	1241.39
5	5 – 10	402.80	19.82	1644.19
6	10 – 20	340.59	16.73	1984.78
7	above 20 m	48.23	2.37	2033.01
	Total	2033.01	100.00	

Land use

Cultivated wetlands within the selected study area are the low formations reclaimed from the surrounding backwaters and divided into polders and enclosed by embankments. Polders digitised in GIS from a map prepared by the Indo-Dutch Mission (1989) show that the total area covered by the polders of Vembanad Wetland is 441.85 km^2. The area of water bodies in 1989 was 236.45 km^2. Using the elevation model, it is found that the polders are located at altitudes which vary from 2.5 m below to 10 m above MSL. A land-use map prepared for the carrying capacity based study of greater Cochin region (NEERI, 2003) also covered the present study area. The land-use classification was done using the LISS III sensor data of Indian remote sensing satellites, IRS 1C and IRS 1D for February 1999. Land-use information of the Vembanad Wetlands as extracted from the above land use map is given in Table 5. During the period 1989–1999, area of cultivated rice lands reduced to 337.28 km^2, with a decline in area of 104.57 km^2. Area of water bodies including lake and unused polders is 341.02 km^2, and mixed crops cover the greatest area of 1283.44 km^2.

Table 5 Land use in the Vembanad Wetlands (Data source: NEERI, 2003).

No.	Land use	Area (km^2)
1	Built-up area	46.828
2	Mixed Crops	1283.44
3	Rice	337.281
4	Rubber	23.786
5	Water body	341.016
	Total	2033.06

CONCLUSIONS

In this study, geographic information data sets of the physical features of the Vembanad Wetlands are generated in GIS using the available data and maps, and the data sets are used for analysing the bathymetry and changes in the extent of Vembanad Lake, and physical characteristics of the wetland. The following conclusions are drawn from the study:

1. Using the 1:50 000 scale SOI topographic map, which is the standard geographic reference map in India, drainage boundaries of the Vembanad Wetlands and the Achencoil, Pamba, Manimala, Meenachil, Muvattupuzha and Kariar rivers draining to it are delineated. The non-basin area delineated as Vembanad Wetlands covers an area of 2033.01 km^2 and the drainage basins cover a total area of 6126.48 km^2.
2. Benchmarks for the area of the Vembanad Lake are established for the time period 1917–1990 using authentic references and the extent of land reclamation quantified for the same period. The change detection study shows that the water spread area of Vembanad Lake in the years 1917, 1970 and 1990 were: 290.85 km^2, 227.23 km^2 and 213.28 km^2, respectively. Land reclamation of 63.62 km^2 of lake area during the period 1917–1970 was mainly for the formation of polders located on the southern sector of the lake, and to enlarge Wellington Island. Total change in the extent of the Vembanad Lake during the period 1917–1970 is 77.57 km^2, i.e. 26.67% of the lake area as on 1917.
3. By generating and analysing the elevation model for the bottom surface of lake, a benchmark is established for the bathymetry of the lake as on 1987. Lake bottom elevation varies from –9.9 m MSL to 2 m MSL. Depth variations are marginal for the wider portion of the lake from Alleppey to Thanneermukkom, and the variations are large from TM barrage to Cochin estuary. Variation in the area and volume of the lake is significant from –6 m MSL to 1 m MSL. At 1 m MSL, the area of lake is 216.53 km^2 and its volume is 611.47 m^3.
4. The elevation model of the terrain shows that a major part of Vembanad Wetland lies at an altitude between –6 m MSL and 20 m MSL, with a mean elevation of 4.64 m MSL. Out of the total area of the wetland, 398.12 km^2 (19.59%) is located below the MSL and 763.23 km^2 area is located below the reference elevation of 1 m MSL. The volume of the wetland at reference elevations of MSL, and 1 m MSL are 525.08×10^6 m^3, and 1079.35×10^6 m^3 respectively. The large storage capacity available in the Vembanad Wetlands, even after the land reclamation, indicates that flood control is an important function of the wetlands which receives the runoff from a total catchment area of 6126.5 km^2.
5. The digital elevation model fulfils an essential requirement for the applications including emergency flood risk management in the Vembanad Wetlands. The elevation models can be improved further when more detailed topographic data are available.
6. In 1987, the total area covered by the polders in the Vembanad Wetland was 441.85 km^2. As per the land-use information of 1999, area of polders used for rice cultivation reduced to 337.28 km^2. During the period 1989–1999, with the reduction in the cultivated area of polders, the total area of water bodies including lake and unused polders increased from 236.45 km^2 to 341.02 km^2. This increase in storage will benefit to control the floods, which occur during the northeast monsoon season.

In short, the study achieved its basic objective of establishing an information base to provide a dynamic view of the changes in spatial developments in the system, which is essential to tackle the various land and water management issues of the wetland system.

Acknowledgements The first author acknowledges the help and support of the Executive Director, Centre for Water Resources Development and Management (CWRDM), Kozhikode, Kerala for conducting the study presented in this paper.

REFERENCES

Ambat, B. (1992) *Kuttanad – Facts and Fallacy*. Kerala Sastra, Sahitya Parishad, Calicut, Kerala.

Army Map Service (AMS) (1953) US Army Corps of Engineers, Washington, USA.

Balchand, A. N. (1983) Kuttanad: A case study on environmental consequences of water resources mismanagement. *Water International* **8**, 35–41

Brooks, R. T. (2005) A review of basin morphology and pool hydrology of isolated ponded wetlands: implications for seasonal forest pools of the northeastern United States. *Wetlands Ecology and Management* **13**, 335–348.

Chang, K. T. (2006) *Introduction to Geographical Information Systems*. Tata McGraw-Hill Publishing Company Limited, New Delhi, India.

Chattopadhyay, S. & Sidhardhan, S. (1985) Regional analysis of the greater Kuttand, Kerala. Technical Report 43, Centre for Earth Science Studies, Trivandrum, Kerala, India.

Indo-Dutch Mission (1989) Kuttanad water balance study (draft final report), vol. IV, Government of Kerala, India.

James, E. J., Anitha, A. B., Joseph, E. J., Nandeshwar, M. D., Nirmala, E., Padmini, V. & Unni, P. N. (1997) Case Study – I. Vembanad Kole Wetland system in relation to drainage basin management. In: *Wetlands and Integrated River Basin Management: Experiences in Asia and Pacific*. UNEP/Wetlands International-Asia Pacific, Kuala Lumpur.

MDA Federal (2004) Landsat GeoCover ETM+ 1990 Edition Mosaics Tile N-03-05.ETM-EarthSat-MrSID, 1.0, USGS, Sioux Falls, South Dakota, USA.

Mitsch, W. J. & Gosselink, J. (1986) *Wetlands*. Van Nostrand Reinhold, New York, USA.

NEERI (2003) Carrying capacity based developmental planning for Greater Kochi Region. National Environmental Engineering Research Institute, Nagpur, India.

Padmakumar, K. G., Anuradha, K., Radhika, R., Manu, P. S. & Shiny, C. K. (2002) Thanneermukkom barrage and fishery decline in Vembanad wetlands, Kerala. *Proc. Fourteenth Kerala Science Congress.*

Padmakumar, K. G., Nair, R. J. & Mohamed Kunju, U. (1988) Observation on the scope of rice-cum-fish culture in the rice fields of Kuttanad, Kerala. *Aquatic Biology* **7**, 161–166.

Pradeep, R. & Samy L. P. (1986) Distribution of faecal indicator bacteria in Cochin backwaters. *Indian J. Mar. Sci.* **15**, 99–101.

Groundwater management in the Darb El Arbaein, Southwestern Desert, Egypt

MOHAMED EL KASHOUTY[1] & A. ABDEL-LATTIF[2]

1 *Cairo University, Faculty of Science, Geology Department, Cairo, Egypt*
mohamedkashouty@yahoo.com

2 *National Research Institute of Astronomy and Geophysics, Helwan, Cairo, Egypt*

Abstract Groundwater is the only water resource in the Darb El Arbaein area of Egypt. The aquifer system extends from Palaeozoic-Mesozoic to Upper Cretaceous sandstone rocks. These overlay the basement rocks and the aquifer is confined. The hydrogeological and hydrogeochemical data of 94 groundwater samples were obtained from the GARPAD authority and used in this paper. In the northern and southern Darb El Arbaein, the hydrogeology did not match with hydrogeochemistry, due to structural elements. Processing using the Modflow Windows (PMWIN) program one can estimate the future out-coming of the hydrogeological environment of Nubian sandstone aquifer, which is a non-renewable groundwater resource in the arid zone. The calibration was made and adjusted to within acceptable hydrogeological characteristics ranges. The model was used to study the water balance elements and to forecast the effect of low to high pumping rates (increase by 25, 50, 100, 200, and 300%) on the water balance and on hydrogeological features. In the northern Darb El Arbaein, the maximum drop of hydraulic head by pumping was in the northwestern part, due to a maximum initial water depth. The hydraulic head dropping was high in the southeastern part, attributed to the lowest transmissivity of the aquifer system. The minimum potentiometric level drop was in the southwestern and northeastern part, because of geographic position relative to regional recharge and local recharge from the surrounding aquifers. In the southern Darb El Arbaein, the minimum hydraulic head dropping was in the southern part, due to the increased aquifer thickness and hydrogeological characteristics values. There is also minimum drop in potentiometric level in the central-northern part, caused by high concentration of faults that receive more groundwater recharge through high hydraulic conductivity. The proposed locations of pumping wells for more irrigation and urbanization were established and concentrated in the minimum potentiometric level drop areas.

Key words Nubian sandstone aquifer; statistical; AquaChem; PMWIN programs

INTRODUCTION

The development in Darb El Arbaein, Egypt, was determined by GARPAD (General Authority for Rehabilitation Projects and Agricultural Development) in 1997. The groundwater is considered as the only water resource for domestic and irrigation uses. They reclaimed 12 456 acres and used these to construct 16 villages for farmers along 400 km from the town of Paris towards the Egypt–Sudan border. The pumping wells are located in three parts, the northern, the central, and the southern part of Darb El Arbaein (Fig. 1). The northern part extends 90 km south from Paris town and has a 90 km^2 area. The central part extends 80 km south of the northern part and is 120 km^2. The southern part extends 200 km south of the central part and is 170 km^2. The Darb El Arbaein is located in an arid region; absolute maximum and minimum air temperatures are 48.6°C in May and –2°C in February, respectively (Galal *et al.*, 2002). The annual rainfall is <1.1 mm, and the total monthly rainfall is 0.3 mm. The annual evaporation is about 172 mm, the maximum potential evaporation rate in June is about 21.32 mm.

Geomorphology and geological setting

Issawi (1971) subdivided the Darb El Arbaein into three geomorphologic units: the southern Naklai-Sheb peneplain; the western Atmur peneplain; and plateau surface. Geologically, the CONOCO map (1989) showed that the exposed rocks ranged from Pre-Cambrian to Quaternary sediments. The lithostratigraphic successions are divided into seven units, from base to top (CONOCO, 1989; Fathy *et al.*, 2001; Korany *et al.*, 2002): Pre-Cambrian basement, Palaeozoic-Mesozoic sandstone, Lower Cretaceous, Upper Cretaceous, Palaeocene, Eocene, and Quaternary. Issawi (1973) subdivided the Nubian sandstone into Shab, Qusier, and Taref members.

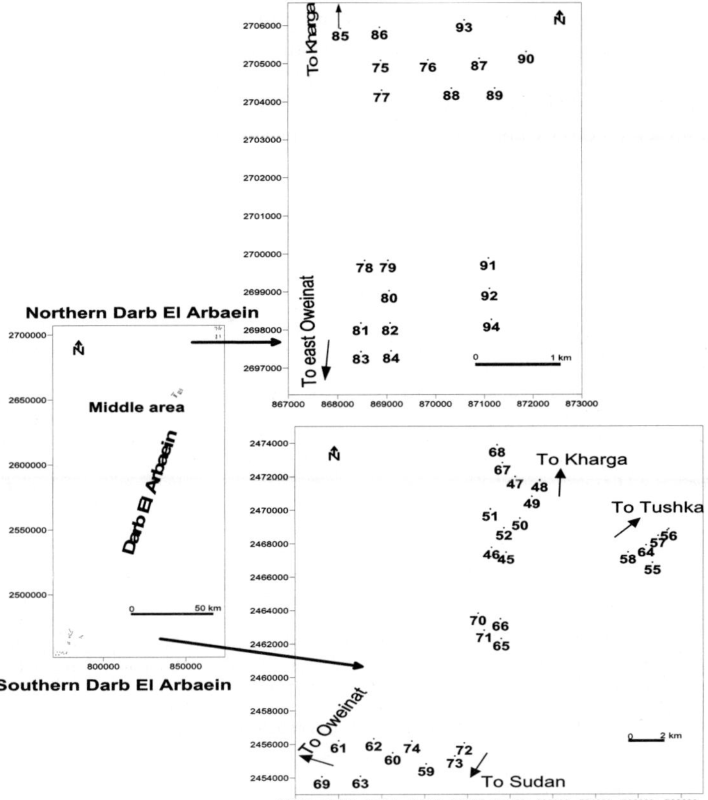

Fig. 1 Location map of Darb El Arbaein.

CONOCO (1989) classified the Kisieba Formation as equivalent to both Shab and Qusier members and Taref is equivalent to Taref member. The Dakhla shale represents the capping aquifer only in the central Darb El Arbaein. The Darb El Arbaein area is related structurally to the Red Sea and southwestern regions (EGSMA, 1987a,b). Issawi (1971) has identified the faults in E–W, NE–SW and NW–SE directions and three anticlines (Bir Kiseiba, Rage, and Shirshir). The uparching of basement rocks is also encountered (Fathy *et al.*, 2002).

Hydrogeological conditions

Eighty wells are drilled by GARPAD in 1998–2001, with depths of 130–535 m. They penetrated the Nubian sandstone successions (Palaeozoic–Upper Cretaceous), which overlay Pre-Cambrian basement rocks. The generalized hydrogeological section indicates the area dissected by 18 normal exposed faults and concealed faults (Fathy *et al.*, 2002). The natural recharge of the aquifer system in the Western Desert was supposed as about 1.5 billion m^3/year (Ambroggi, 1966), while groundwater flow from Libya to Egypt was about 3.782 million m^3/year (Ezzat & Abu Atta, 1974). It is believed that the groundwater has accumulated and been preserved in the Nubian sandstone aquifer since 10 000 BC, or even earlier, as revealed by C_{14} and H_3 isotopes (German Water Group, 1977). The stored water in the Nubian sandstone is mainly fossilized water and ranges from 20 000 to 40 000 years in age (Shata *et al.*, 1962; Shata, 1982).

Northern Darb El Arbaein The aquifer is composed of sandstone of the Qusier and Taref Formations and Paleozoic-Mesozoic sandstone layers. The uppermost layer of Qusier Fm. acts as a confining layer. The aquifer thickness increases in the northern part. The aquifer is underlain by basement rocks. The latter become deeper northward indicating the influence of concealed faults (NW–SE) and the increase of sedimentary cover due north (Fathy *et al.*, 2002). The groundwater flow is directed towards the eastern, northern and northeastern parts (Fig. 2(a)). The water depth increases in the western, southwestern and northwestern parts (Fig. 2(c) and (d)), following the topography. The hydraulic conductivity and transmissivity increase in the northern part (768 m^2/d and 2.9 m/d) and decrease in the southeastern part (300 m^2/d and 1.5 m/d), respectively (Fig. 2(e) and (f)).

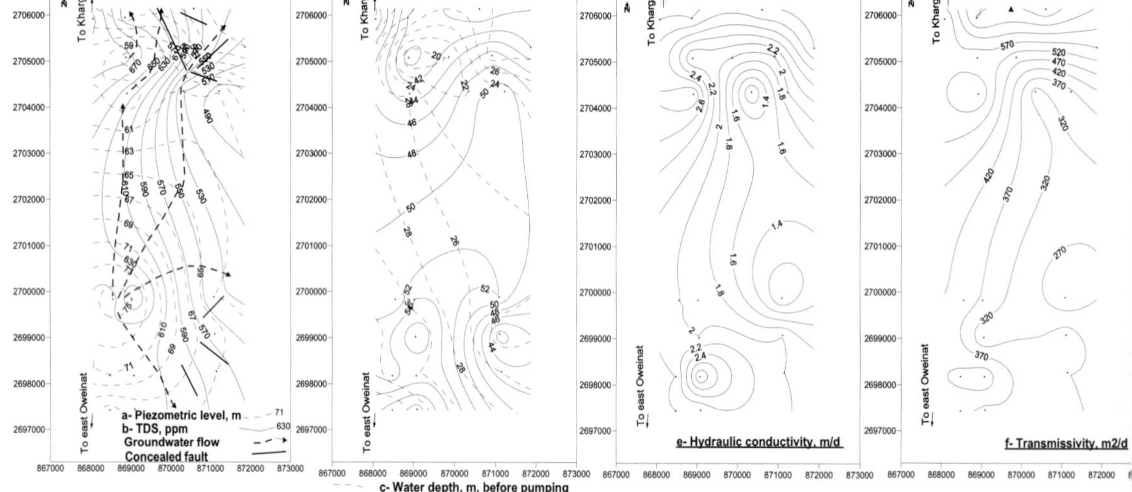

Fig. 2 Piezometric level (a), TDS content (b), water depth before (c) and after (d) pumping, hydraulic conductivity (e) and transmissivity (f) of the Nubian sandstone aquifer in northern Darb El Arbaein.

The storage coefficient ranges between 1.2×10^{-4} to 2.3×10^{-4}. The calculated formation loss varied from 1.09×10^{-3} to 2.05×10^{-3} m^{-2}/d, and the well loss from 1.82×10^{-7} to $2.39\ 10^{-7}$ m^{-5}/d (Fathy *et al.*, 2002). The well efficiency ranged from 61 to 71% at a discharge of 150 m^3/h.

Southern Darb El Arbaein The aquifer is built of Kiseiba and Taref Formations of the Upper Cretaceous (Fathy *et al.*, 2001). The upper most layer of Kiseiba Fm. act as confining layer. The aquifer thickness increases in the southern part, attributed to a high concentration of faults due north. The aquifer is composed of coarse grained sandstone at the base, changing gradually to fine and medium upward. It directly overlays the basement rocks. The groundwater flow was in the northeastern part (Fig. 3(a)). The initial water depth increases in the southwestern part (Fig. 3(d) and (e)). Hydraulic conductivity and transmissivity are greater in the southern part (1400 m^2/d and 12 m/d, respectively), and less in the northern and eastern part (500 m^2/d and 2.5 m/d) (Fig. 3(f) and (g)), attributed to fault planes and an increase in clay content (Fathy *et al.*, 2002). The storage coefficient varied from 2.2×10^{-4} to 3.22×10^{-4}, reflecting the confined conditions. The formation loss ranged from 2.36×10^{-3} to 2.51×10^{-3} m^{-2}/d, while well loss varied from 0.93×10^{-7} to 1.45×10^{-7} m^{-5}/d.

MATERIALS AND METHODS

The hydrochemical and hydrogeological data (2002–2004) obtained, after organization by GARPAD, were analysed using statistical, AquaChem (PHREEQC), and groundwater flow model (PMWIN) techniques.

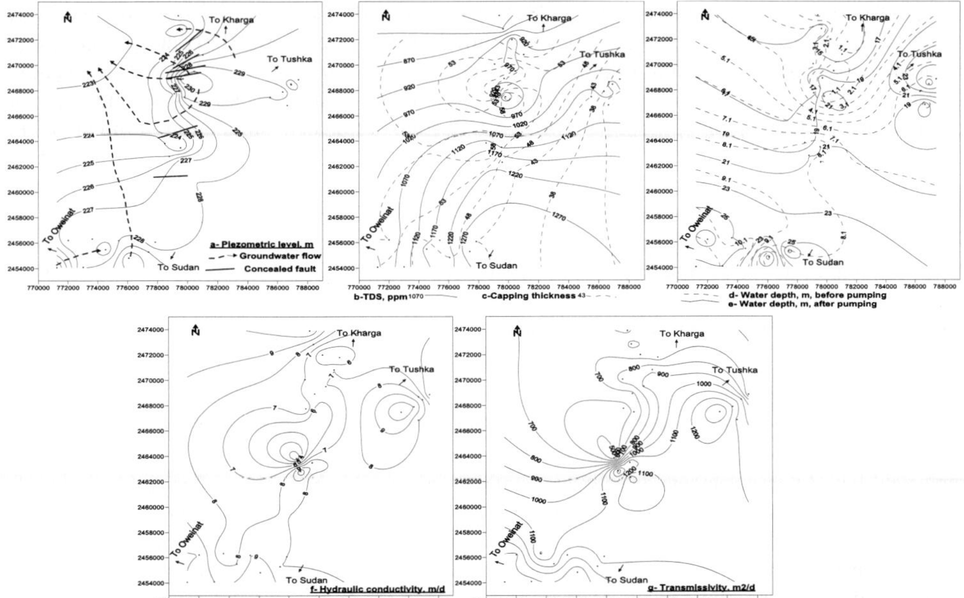

Fig. 3 Groundwater flow (a), TDS content (b), capping thickness (c), water depth before (d), and after (e) pumping, hydraulic conductivity (f) and transmissivity (g) of the Nubian sandstone aquifer in southern Darb El Arbaein.

IMPACT OF GEOLOGY, HYDROGEOLOGY AND HUMAN ACTIVITY ON AQUIFER QUALITY

In northern Darb El Arbaein, the water depth before and after pumping was high (Fig. 2(c) and (d)), indicating a low evaporation rate. The TDS concentration values are lowest in the Darb El Arbaein area, attributed to the increased water depth and therefore low evaporation process. The TDS concentration increased in the northwestern part (Fig. 2(b)), but has a narrow range (450–720 ppm), caused by low evaporation and low evaporite minerals in the aquifer sediments. The TDS concentration decreased in the eastern part with groundwater flow (Fig. 2(a)), caused by presence of concealed faults in the eastern part. The change in hydraulic conductivity with depth enhances the groundwater flow from the surroundings and below into the Nubian aquifer.

In southern Darb El Arbaein, the TDS concentration increased to the southeastern part, coinciding with a low thickness of capping shale (Fig. 3(b) and (c)). It is caused partially by anthropogenic infiltration enhanced by low capping thickness. The wide range of TDS concentration (750–1350 ppm) confirms the impact of non-point source inputs (agricultural activity). The water depth is generally very shallow before pumping (Fig. 3(d) and (e)), which contributed the TDS concentration by evaporation process. After pumping, the water depth is much greater due to the low recharge rate and non-renewable aquifer system. The groundwater flow (Fig. 3(a)) does not match with the TDS concentration (Fig. 3(b)). The TDS concentration decreased in the northern part, in the same groundwater flow. This is attributed to a vertical change; hydraulic conductivity may increase with depth. Such conditions enhance the upward flow of the freshwater into the groundwater flow path. The concealed faults in the northern part (Fig. 3(a)) facilitate the upward flow and flow also from the surrounding aquifers.

GROUNDWATER MANAGEMENT

Processing Modflow Windows (PMWIN) is a simulation system for modelling groundwater flow and transport processes with the modular three-dimensional finite difference groundwater model

MODFLOW of the US Geological Survey (McDonald *et al.*, 1988); the particle tracking model or MODPATH (Pollock, 1988, 1989, 1994), the solute transport model MT3D (Zheng, 1990), and the parameter estimation program, PEST (Doherty *et al.*, 1994). Simulation by PMWIN (Chiang & Kinzelback 1996) was conducted in three steps. First; a steady-state calibration was conducted for the entire area to develop an optimal parameter set. Second; a transient simulation of the model to predict the future development and the outcome was obtained. The cell spacing was chosen to allow the curvature in the regional potentiometric surface map of the aquifers. The chosen values of constant hydraulic head are those at the present time. When we want to use these kind of boundary conditions, they must be imposed on the study area so as not to influence the local aquifer behaviour, as far as possible. When several scenarios are considered, the results obtained provide water resources managers with the means to evaluate and reconcile competing management alternatives (Senthilkumar & Elango, 2004).

Calibration techniques

Calibration of a numerical groundwater flow model is the process of finding a set of boundary conditions, parameter values and stresses that produce simulated groundwater levels and flows that match field-based measurements or estimates within a pre-established range of error (Anderson & Woessner, 1991). Model calibration is performed manually and automatically (Doherty, 2000a,b, 2001). Estimated parameters in the final calibration model are ordered from highest to lowest in terms of sensitivity; they include: hydraulic conductivity, transmissivity, recharge and potential evapotranspiration. Any adjustments to these values were made by trial-and-error and the PEST method. The purpose of PEST is to assist in data interpretation and in model calibration.

Model geometry, boundary conditions and calibration techniques

This study was carried out using PMWIN, which has enhanced capabilities for pre-and post-processing input and output data. PMWIN has been applied extensively for simulating groundwater dynamics. The aquifer was discretized into two model layers of variable thickness for northern and southern Darb El Arbaein. The aquifer is confined with varied transmissivity. The effective porosity is used by PMPATH, MODPATH and MT3D to calculate the average velocity of the flow through the porous medium. Aquifer top and bottom elevations and hydrogeological information were derived from the available data, well-completion reports and logs obtained from published works. To incorporate the spatial distribution of the hydraulic conductivity (K) and the transmissivity (T) over the aquifer, active cells in model layers were initially grouped into zones based on the data. Several boundaries were simulated in the models according to the geology, groundwater flow net, neighbouring aquifers and structural patterns. Recharge to the aquifers was applied to the top model layer. Evapotranspiration was simulated in the models. Groundwater pumping was specified in the model at each well.

Northern Darb El Arbaein model

The grid information is composed of 71 rows and 41 columns with a total grid area of 144 km^2 (Fig. 4(a)). The boundary conditions in the upper and lower zones of the western part are no-flow, derived from the groundwater flow net and geological maps. A constant head boundary condition is imposed on the far northeastern and southwestern parts, where the boundaries are coincident with the 47 and 77-m equipotential lines on the potentiometric map (Fig. 4(a)), respectively. The top and bottom of the Nubian sandstone aquifer were input to the program. During a flow simulation, hydraulic conductivity and transmissivity of each cell (fed manually) varies with the saturated thickness of the aquifer. The grid is based on the three hydrogeological characteristics of the aquifer system (Fig. 4(c)). Estimated values for the optimal parameter set from the calibrated final model (Fig. 4(b)) are tabulated in Table 1.

The model evaluates the response of the Nubian sandstone aquifer system to withdrawals for irrigation. The drop in potentiometric level under different policies of groundwater extraction, after 10 years, is given in Fig. 5(a)–(e). These predictive maps match with current potentiometric

Table 1 Final calibrated of hydrogeological parameters of the Nubian sandstone aquifer; manual and PEST.

Parameter	Estimated value	95% confidence Limits	
		Lower limit	Upper limit
Recharge (m/d)	5.2654×10^{-6}	5.2651×10^{-6}	5.28457×10^{-6}
Evapotranspiration (m/d)	0.22	0.2189	0.220021
Hydraulic conductivity (m/d)	1.3–2.79	1–2.51	1.5–3
Transmissivity (m^2/d)	304–768	304–765	308–770

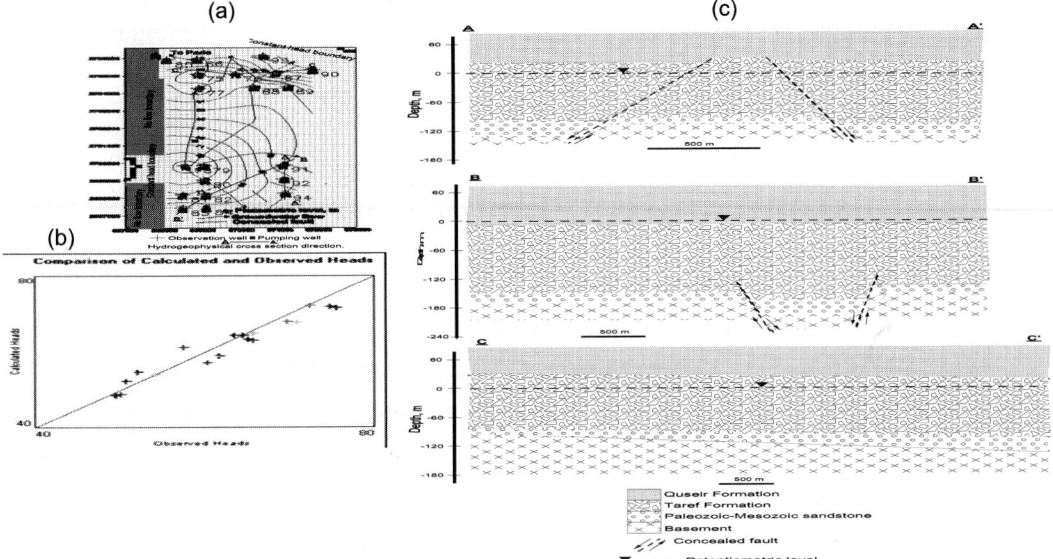

Fig. 4 The grid of the Nubian sandstone aquifer system include the potentiometric levels (a), and the match between observed and calculated heads (b), and (c) hydrophysical cross-section along the three sections in Fig. 4(a), in the northern Darb El Arbaein.

levels to construct the potentiometric drop due to pumping after 10 years (Fig. 5(f), (h)). The maximum potentiometric drop was in the northwestern part, attributed to the maximum initial water depth (Fig. 2(c)) and high topography. The drop was high in the southeastern part (Fig. 5(f), (h)), caused by the lowest transmissivity and hydraulic head of the aquifer system. The minimum hydraulic head drop was in the southwestern part (Fig. 5(f), (h)), because of the geographic position of regional recharge from the surrounding aquifers. The model outputs indicate that increasing the pumping rate by 25, 50 and 100% does not produce much decline in potentiometric levels. The regional recharge from the surrounding aquifers, the vertical change in hydraulic conductivity, upward flow of freshwater, the geology, and faults, can compensate the drop in potentiometric head. If there is a demand to increase well pumping for more irrigation, the selected locations (Fig. 6(a) and (b)) can provide the development. Resource sustainability has proven to be an elusive concept to define in a precise manner and with universal applicability. Groundwater development in the study area should allow long-term sustainability; i.e. development and use of groundwater in a manner that can be maintained for an indefinite time without causing unacceptable environmental, economic or social consequences. The selected promising area was chosen in the lowest potentiometric drop areas (northeast and southwest) (Fig. 6(a), (b)). The hydrogeological conditions of the aquifer mean that the proposed additional pumping wells will not produce much potentiometric level drop (Fig. 6(c)–(f)), with same current fixed pumping rate.

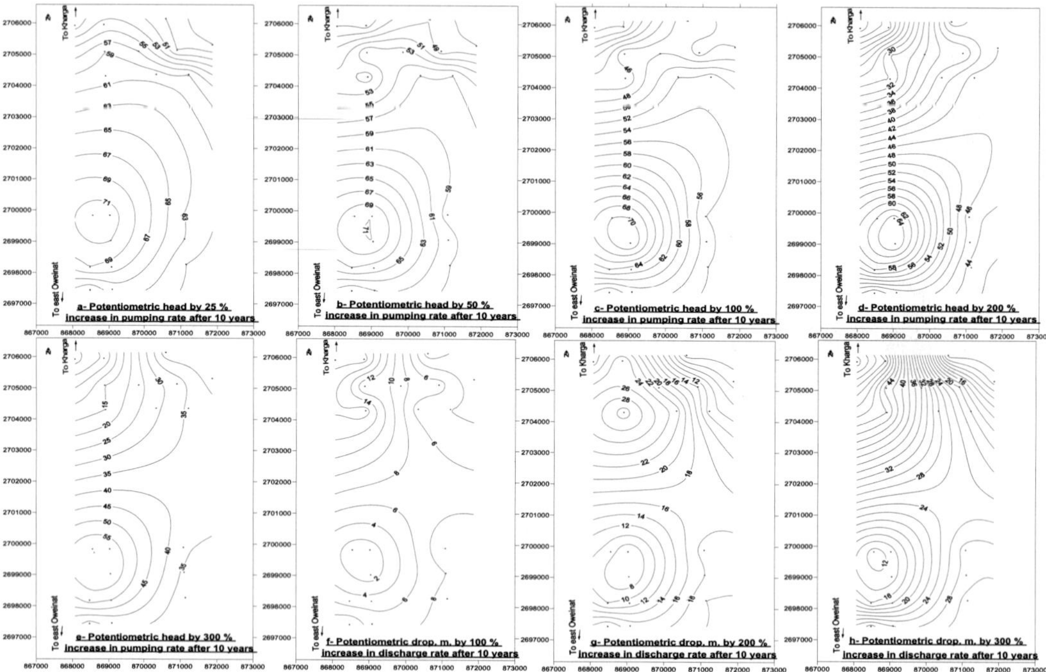

Fig. 5 Simulated scenarios (a)–(e) of the Nubian sandstone aquifer after 10 years and the corresponding potentiometric drops (f)–(h) in northern Darb El Arbaein.

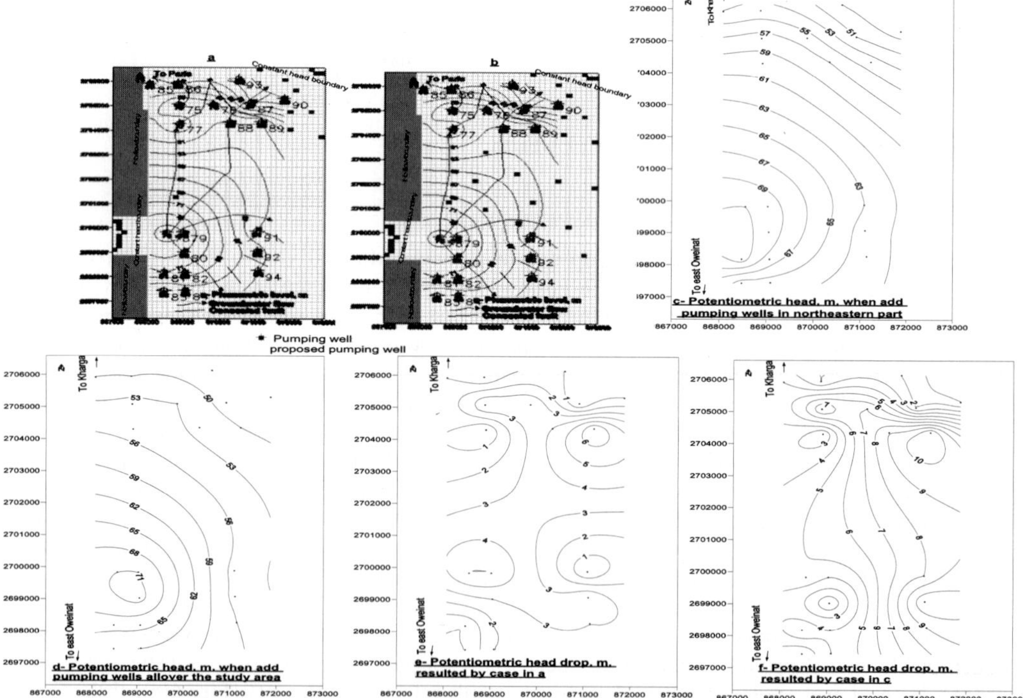

Fig. 6 The proposed add pumping wells in northeastern (a) and all over the study area (b), potentiometric head in case a (c) and in case b (d), potentiometric drop in case a (e), and drop in case b (f) in northern Darb El Arbaein.

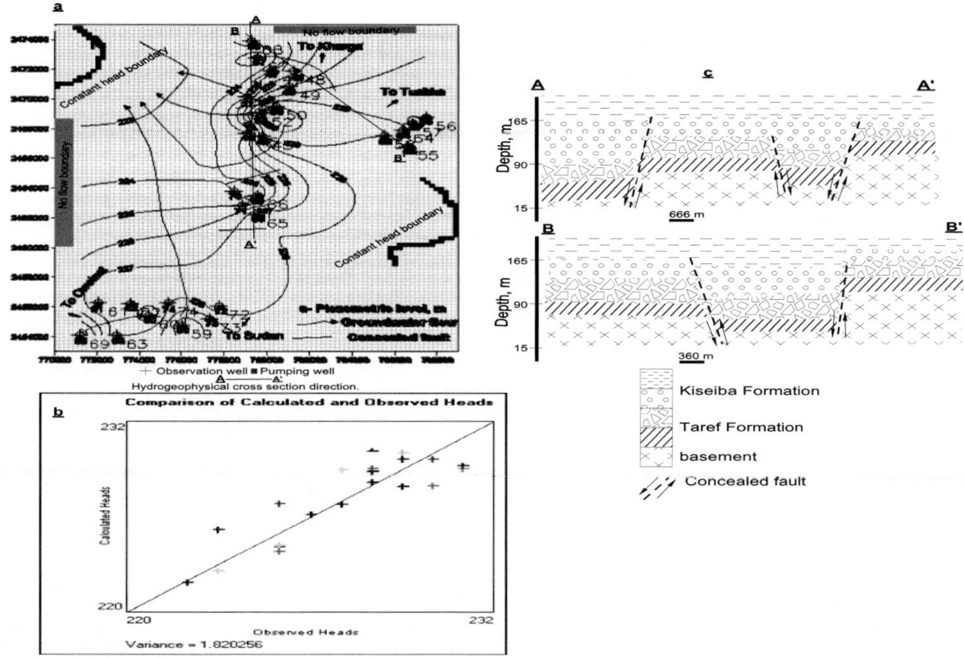

Fig. 7 The grid of the Nubian sandstone aquifer system including the potentiometric levels (a) and the match between observed and calculated heads (b), and hydrophysical cross-section (c) in the southern Darb El Arbaein.

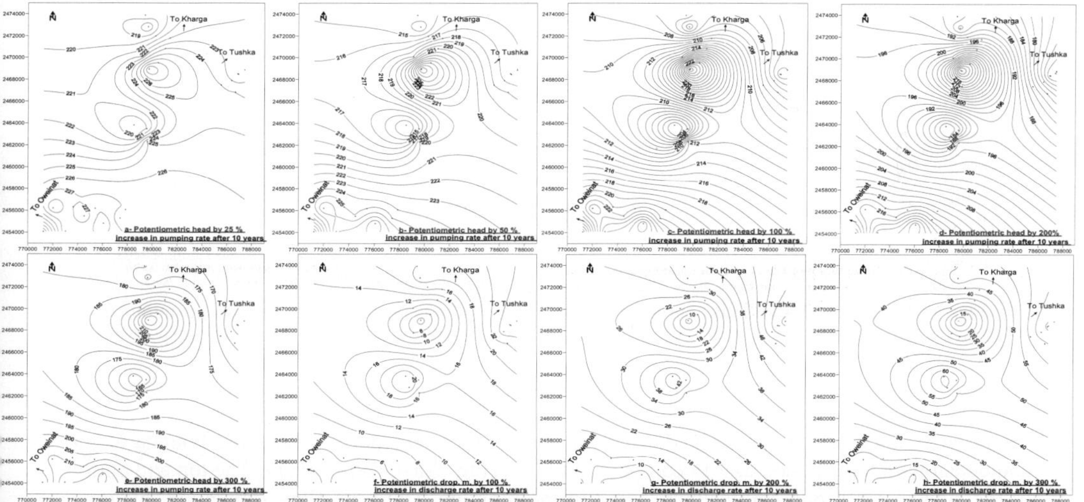

Fig. 8 Simulated scenarios (a)–(e) of the Nubian sandstone aquifer after 10 years and the corresponding potentiometric drops (f)–(h) in northern Darb El Arbaein.

Southern Darb El Arbaein model

The grid information is composed of 99 rows and 86 columns with a total grid area of 230 km^2 (Fig. 7(a)). The boundary conditions in the upper western part and northeastern part are no-flow, as derived from the groundwater flow net and geological maps. A constant-head boundary

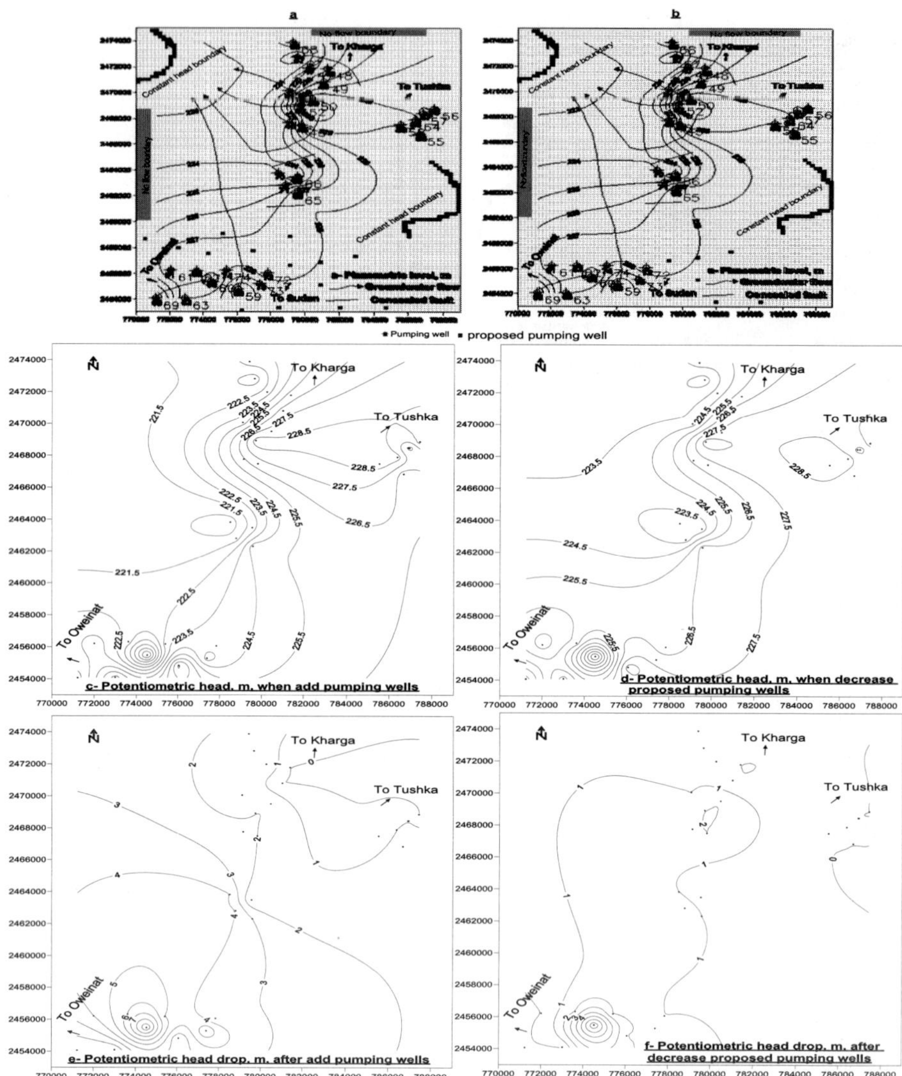

Fig. 9 The proposed additional pumping wells in the southern area (a), and decrease in numbers (b), and corresponding potentiometric heads (c) and (d), and potentiometric drop (e) and (f), respectively, in the southern study area.

condition is imposed on the far northwestern and southeastern parts, where the boundaries are coincident with the 222 and 229-m equipotential lines (Fig. 7(a)), respectively. The top and bottom of the Nubian sandstone aquifer were input to the program. During a flow simulation, hydraulic conductivity and transmissivity of each cell varies with the saturated thickness of the aquifer. They were fed into the grid cell by cell. The grid is based on the two hydrogeological characteristics of the aquifer system (Fig. 7(c)). Estimated values for the optimal parameter set from the calibrated final model (Fig. 7(b)) were within the permissible ranges.

The hydraulic heads predicted for an increase in pumping rate after 10 years are shown in Fig. 8. The maximum potentiometric level drop occurs in the northern part (Fig. 8(f)–(h)), and is attributed to the decrease in aquifer thickness, while in the southern part the drop was minimum due to greater thickness. The minimum drop was also in the north-central part (Fig. 8(f)–(h)), caused by the high concentration of concealed faults that enhance the regional recharge from the surrounding and lower aquifers. The maximum potentiometric level drop was in the south-central part (Fig. 8(f)–(h)), due to low hydrogeological characteristics. This new development area needs investment for more irrigation in time. The proposed new pumping wells are in the lowest potentiometric drop areas (Fig. 9(a) and (b)). The resulting potentiometric level fall (Fig. 9(c) and (e)) was large and may impact negatively on aquifer sustainability. Therefore, the alternative proposal, to decrease the number of proposed pumping wells (Fig. 9(b)) reduces the drop (Fig. 9(d) and (f)) and can be applied. The groundwater is the only water resource, so unacceptable impacts should be avoided.

CONCLUSION AND RECOMMENDATIONS

The geology, hydrogeology, dissolution and recharge of an aquifer influence its quality. The impact of increasing pumping rates by 25, 50, 100, 200 and 300%, on the fall of the potentiometric level after 10 years was estimated by modelling. The first three scenarios can be applied with few unacceptable consequences. The proposed additional pumping wells required for future agricultural investment were concentrated in the best hydrogeological areas. However, if a model is used to address questions about the future response of a groundwater system that are of continuing significance to society, then field monitoring of the groundwater system should continue and the model should be re-evaluated periodically to incorporate new information and new insights (Konikow & Reillow, 1999). Strategies for sustainability include the use of sources of water other than local groundwater, changing rates or spatial patterns of groundwater pumpage, and increasing recharge to the aquifer.

Acknowledgement The authors are greatly thanks the GARPAD organization for supplying data.

REFERENCES

Al Rahman, A. (2003) Hydrogeological aspects of sandstone aquifer in Darb El Arbaein area, Southwestern Desert. Egypt. Master Science, Geology Department, Cairo University, Egypt.

Ambroggi, R. (1966) Water under the Sahara. *Scientific American* **214**(5).

Andersson, M. & Woessner, W. (1992) *Applied Groundwater Modeling: Simulation of Flow and Advective Transport.* Academic Press, Inc., New York, USA.

Chiang, W. & Kinzelbach, W. (1996) User's Manual, processing MODFLOW, a simulation system for modeling groundwater flow and pollution. Scientific Software Group. Washington DC, USA.

CONOCO (Continental Oil Company) (1989) Geologic map of Egypt, scale 1:500 000.

Doherty, J. (2000a) PEST model independent parameter estimation. Watermark Numerical Computing, S.S. Papadopoulos & Associates Inc., Bethesda, Maryland, USA.

Doherty, J. (2000b) PEST utilities for MODFLOW and MT3D parameter estimation. Watermark Numerical Computing, S.S. Papadopoulos & Associates Inc., Bethesda, Maryland, USA.

Fathy, R. G., El Nagaty, M., Atef, A. & El Gammal, N. (2002) Contribution to the hydrogeological and hydrochemical characteristics of Nubian sandstone aquifer in Darb El Arbaein, Southwestern Desert, Egypt. *Al Azhar Bull. Sci.* **13**(2), 69–100.

Fathy, R., Hefnawy, M. & Abdel Hamid, A (2001) Contributions to the hydrogeology aspects of the groundwater aquifer in southern part of Darb El Arbaein area, southern portion of Western Desert, Egypt. *Al Azhar Bull. Sci.* **12**(2), 175–194.

El Gammal, N. (2004). Hydrogeological studies in Darb El Arbaein area, south Egypt. Master Science. Geology Department., Cairo University, Egypt.

Ezzat m; Abu Atta (1974). Exploration of groundwater in El Wadi El Gedid project area (New valley), Part II, Hydrogeological conditions. Dakhla-Kharaga area.Ministry of Agriculture and Land Reclamation, Cairo, Egypt.

General Authority for Rehabilitation Projects and Agricultural Development (GARPAD) (1998–2000) Lithologic logs and geophysical logs of water wells, Darb El Arbaein Area, Southwestern Desert, Egypt.

General Survey of Egypt (EGMSA) (1987a) Geology and geomorphology of the Egyptian Component Transitional Sandstone Project. Report to groundwater research Institute, Egypt.

General Survey of Egypt (EGMSA) (1987b) Geophysical investigation of the Egyptian Component Transitional Sandstone Project. Report to groundwater research Institute, Egypt.

German Water Group (1977) Hydrogeological study of groundwater resources in the Kufra area. German Water Engineers, GB, v 5.

Ghazal, A (2002) Hydrogeological studies in Darb El Arbaein area, Southwestern Desert, Egypt. Master Science. Geology Department, Cairo University, Egypt.

Issawi, B. (1971) *Geology of Darb El Arbaein, Western Desert, Egypt.* Ann. G. S. E. 46, Cairo, Egypt

Issawi, B. (1973) *Geology of Southeastern corner of Western Desert.* Ann. G. S. E. 8. Cairo, Egypt

Konikow, L. & Reillow, T (1999) Groundwater modeling. In: *The Handbook of Groundwater Engineering* (ed. by J. W. Delleur), 20-1–20-40. CRC Press, Boca Raton, Florida, USA.

Korany, E, Fathy, G. & El Nagaty, M. (2002). Contributions to the hydrogeology of Nubian sandstone aquifer in the middle part of Darb El Arbaein, southwestern Desert, Egypt. Sedimentology of Egypt, ISSN 1110-2527, *J. Sediment. Soc. Egypt* **10**, 119–143.

Pollock, D. W. (1988) Semianalytical computation of path lines for finite difference models. *Ground Water* **26**(6), 743–750.

Pollock, D. W. (1989) MODPATH (version 1x) – Documentation of computer programs to compute and display path lines using results from the USGS modular three dimensional finite difference groundwater model. *USGS Open File Report 89-381.*

Pollock, D. W. (1994) User's guide MODPATH/MODPATH-PLOT, version 3: A particle tracking post-processing package for MODFLOW the USGS finite difference groundwater flow model. USGS, Reston, Virginia, USA.

Shata, A., Knetsch, G. & El Shazly, M. (1962) The geology, origin, and age of the groundwater supplies in some desert areas of UAR. *Bull. Inst Desert, El Matariya, Cairo* **12**(2), 61–124.

Shata, A. (1982) Hydrogeology of the great Nubian sandstone basin. *Quatern. J. Engng London* **15**, 127–133.

Senthilkumar, M. & Elango, L. (2004) Three-dimensional mathematical model to simulate groundwater flow in the lower Palar River basin; southern India. *Hydrogeol. J.* **12**(2), 197–208.

Zheng, C. (1990) MT3D, a modular three dimensional transport model. S.S. Papadopulos and Associates, Inc., Rockville, Maryland, USA.

Management Support System for wetlands protection: Red Bog and Lower Biebrza Valley case study

JAROSLAW CHORMANSKI, IGNACY KARDEL, DOROTA SWIATEK, MATEUSZ GRYGORUK & TOMASZ OKRUSZKO

Division of Hydrology and Water Resources, Warsaw University of Life Sciences, Poland

j.chormanski@levis.sggw.pl

Abstract The Biebrza Wetlands (northeast Poland) belongs to the biggest protected areas in Europe aiming at conservation of wetland ecosystems. In order to optimize the expensive and time-consuming protection and restoration measures a Management Support System (MSS) was developed. The MSS consists of following three elements: a data catalogue combined with a Geographic Information System (GIS); a hydrological module, which simulates quantitative and qualitative variability of both ground- and surface waters dynamics; and an ecological module, which predicts directions of changes for particular wetlands' vegetation types as an effect of hydrological condition variability. MSS, as a complex tool for wetlands analysis, indicates the reaction of Biebrza Valley and Red Bog to different scenarios of hydrological conditions – as an effect of the present status (i.e. for various scenarios of drainage network function) and as possible consequences that would occur if all the conservation activities and climate changes were given up.

Key words Management Support System; GIS; the Biebrza wetlands; flood plain flow modelling; groundwater modelling

INTRODUCTION

The Biebrza National Park (BNP) (northeast Poland), which covers an area of 60 000 ha, belongs to the biggest protected areas in Europe aiming at conservation of wetland ecosystems. However, the environmental changes that occurred in the valley of the Biebrza River during the last two centuries enforced the necessity of carrying out conservation activities in order to protect the most valuable areas of wetlands vegetation including *Magnocaricion* vegetation, sedge-moss habitats of rich fen, and pine-birch forest of transition bogs. In order to optimize the expense and time-consuming protection and restoration measures, the Management Support System (MSS) was developed. It is an analysis tool, which is used in a description of relations between particular types of land use, management techniques, surface- and groundwater resources, conservation measures, and the current state of wetland ecosystems.

In order to find feasible solutions for mentioned activities, the Management Support System (MSS) was developed (Chormanski & Wassen, 2005). The MSS was aimed to assist in the following management tasks: operational data acquisition from different monitoring networks, sharing the data sets with scientific and administration institutions, preparing annual environmental reports, planning of land exchanges with private land owners, planning of conservation measures, and selecting specific sites for restoration measures.

The MSS presented herein had to be specifically designed for the Biebrza region regarding its unique characteristics, such as relatively large area of the park and presence of many different wetland habitats. Typical Management Support Systems that function in water management institutions are strongly oriented towards water balances and/or water allocation issues (Gromiec, 2006), as in Biebrza, where the water quality and ecological services of both surface water and groundwater are very important. A "tailor made" approach was essential in order to work out the MSS which can address the specific problems of BNP, and which can be used for multi-user demand for data processing and the need for sustainable development of the river valley and valuable wetland habitats such as the Red Bog, which is used not only by conservationists or resource managers, but also by researchers, farmers and tourists (Mahesh *et al.*, 2007).

STUDY AREA

In particular, two main parts of the BNP are involved in MSS activities. The first one, "Red Bog", is one of the oldest nature reserves in Poland and the biggest raised bog of the European lowlands. The second, the Lower Biebrza Valley, is noted as a unique ecosystem where the flood phenomena that occur regularly after spring thaw, are the most important factor in the ecosystem development. The choice of study area was based on its history and land-use tradition. Once the Lower Biebrza Valley was extensively used for hay-making and pasture; the Red Bog developed almost completely naturally over the last 200 years. Such a comparison allowed the evaluation of the presented MSS as a tool for human activity-caused processes, as well as for the further development of strictly protected areas.

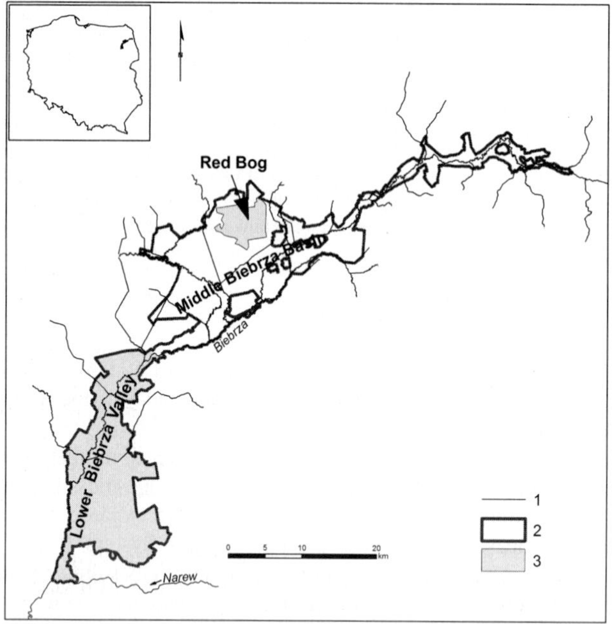

Fig. 1 Study area: Biebrza National Park (1, hydrography network; 2, boundary of Biebrza National Park; 3, areas of MSS implementation).

The Biebrza River Valley is situated in northeast Poland (Fig. 1) in the geographical region of the Podlaskie Lowlands. On almost its entire length, the river course is protected by the Biebrza National Park, which covers the most valuable parts of the river valley that contain non-drained flood plains, marshes and fens, surrounded by a post-glacial landscape with ice-pushed hills, moraines and outwash plains and eolic dunes (Okruszko, 1990). The almost natural characteristic of the Biebrza mires is reflected in a regular pattern of peat-forming plant communities, which run the length and breadth of the valley. The river is not regulated, meandering in the whole course of the valley, and it has many oxbow lakes. However, a number of its tributaries has been drained or turned into drainage canals (Okruszko, 2005). Extensively-used areas are flooded by the river in spring. Further away from the river, at the edge of the valley, the dominant habitats are groundwater-fed rich fens. The Middle Biebrza basin's peatlands are exceptional due to their ecohydrological features. The Red Bog, as a relic of large mid-European peatlands, develops constantly without significant human pressure.

The flood plain contains highly-productive rich fen types which belong taxonomically to *Glycerietum maximae, Caricetum gracilis* and *Caricetum elatae* (Palczynski, 1984). These are tall sedge, grass and herb vegetation, relatively poor in species. Typical associations of the occasionally flooded belt are *Caricetum caespitosae* and *Peucedano–Caricetum appropinquatae.* Transitional fen is found in an intermediate belt, outside the reach of the seasonal river floods in places where the calcareous groundwater from the moraines does not reach the fen surface. It is fed mainly by rainwater (Wassen *et al.*, 1990) and belongs taxonomically to *Betuletum humilis* with affinity to *Caricetum rostrato-diandrae.* It is thin dwarf-shrub vegetation with low sedges and occasionally some *Sphagnum* hummocks. The moss layer has a fairly high standing crop. Low sedge-rich fen types are abundant in a belt along the moraines, but also further away from the moraines, provided that the calcareous groundwater still reaches the fen surface. In the Biebrza Valley several species-rich associations of this fen type (the *Caricion diandrae*) are present (Palczynski, 1984). These are low-productivity sedge and herb vegetation with a well-developed moss layer of *Hypnaceae.* The area of Red Bog, in contrast to open areas of the Lower Biebrza Valley, is a mosaic of unique spots of boreal pine forest *Carici chordorrhizae–pinetum,* alder communities of *Sphagno-squarrosi alnetum* and some mix of *Ribeso-nigri alnetum.* The most important factor of the Red Bog's habitat development is groundwater supply. Spatial diversity of water recharge systems makes development of Red Bog's ecosystems hard to foresee.

The immense ecological value of the Biebrza Wetlands was recognized in a number of publications, e.g. Bootsma (2000), Wassen *et al.* (2006), as well as in international and national protection measures. The Biebrza River Valley is designated as a wetland site of global significance in the frame of the Ramsar Convention; it belongs to the NATURA 2000 EU Ecological Network.

Protection of this unique area requires specific management activities in more than 166 000 parcels, which are either privately owned (40% of the park area), or belong to the state and are managed by BNP. Conservation activities are mainly focused on halting the scrub encroachment process, which decreases the area of open spaces needed for bird breeding and endangers the sedge-moss communities. Therefore, one of the major BNP maintenance aims is to save traditional ways of agriculture by mowing wetland meadows and control hydrological features of the river and its valley. Another management challenge lies in decreasing peat mineralization in areas affected by drainage. Forecasts for potential threats for the future development of peat habitats are detailed based on groundwater and surface water monitoring and modelling combined with plant cover change analyses. Results of scientific research led to the conclusion that to conserve the ecosystems of Biebrza Valley, restoration of the natural hydrographical network is required. Thus, management strategies should lead to decrease or halt the outflow through artificial canals, which is especially relevant in the Middle Biebrza Basin (Mioduszewski & Wassen, 2000).

METHOD

Description of the MSS

The MSS was developed as a GIS database which links point data gathered in the base or spatial data on maps with output calculated by models. The MSS consists of three parts: the Data Catalogue and GIS, Hydrological Module and Ecological Module (Fig. 2). The first part, the Data Catalogue and GIS, aims at an inventory of all important characteristics of BNP and contains several thematic databases (Fauna, Flora, Soils, Hydrology and Meteorology, Water Quality, Pollution Sources, Land Ownerships, Fire Events, Forest Management, Non-Forest Management Activities) mostly related to GIS by SQL language. The spatial part of the database developed in a GIS environment was included as ArcGIS themes. The thematic basis contains information gathered from different research activities conducted since the establishment of BNP in 1993, a biota and abiotic resources inventory, which was done during BNP management plan development in the late 1990s, as well as historical and current management practices recorded by the park staff.

The Hydrological Module predicts the changes in water conditions as a function of climate change and different water and habitat management strategies. The hydrological module consists

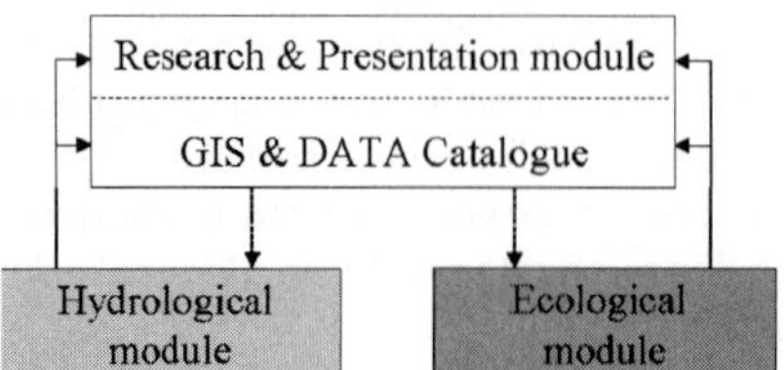

Fig. 2 Scheme of the logical structure of the MSS.

of results of existing mathematical and statistical models of surface and groundwater developed in the last decade in the frame of the different research programmes. The regional groundwater model was the basic tool for the simulation of the impact of planned infrastructure modifications (e.g. blocking the outflow from drainage canals, partial restoration of the historical river network, increasing the water level in the tributaries) on groundwater level in the phreatic zone. The model has been developed using the SIMGRO programme (Querner, 1993), which simulates groundwater flow as well as the flow of surface water. Examples of simulation scenarios have been produced separately for the Middle (Mioduszewski & Wassen, 2000) and the Lower Basin (Slesicka *et al.* 2002) for an historical hydro-meteorological data set of 1990–1995. Exceptional habitats of the Red Bog area, where the balance between groundwater and precipitation water feeding is crucial for precise description of hydrological dimension of ecosystem development, a separate groundwater model is being created. Simulation constructed with MODFLOW treats Biebrza Middle Basin as a dynamic system, where changes of drainage network (straightening of rivers and canal construction in the late 1800s) seem to be the key aspect of its existence. Calibration of this model is based on groundwater monitoring, which consists of 17 measurement locations with 31 piezometers supplied with 17 Divers®, which record groundwater level and additional parameters (t, EC, pH) in 6-hour time steps.

The predictions of the river flow were based again on two separate models: the model of the Middle Basin and the model of the Lower Basin. Models are based on the 1-D St. Venant equation. The model of the Middle Basin uses the specially developed calculation programme (Kubrak & Okruszko, 2000) and the flow simulation in the Lower Basin was calculated using HECRAS-UNET software (HECRAS, 2001). The steady state hydraulic model was used for water level calculation. A digital elevation model of the flood plain constructed with the use of GIS techniques, was combined with calculated water levels for spatial flood extent in the valley. The model was successfully verified with the help of processed Landsat satellite images and field measurements (Swiatek *et al.*, 2006).

RESULTS AND DISCUSSION

Both models of the surface water flow were used for generating hydrological data for developing thematic maps showing the extent of floods for different hydrological conditions.

The Ecological Module aims to predict the direction of changes of plant communities and presence of "Red List" plant species. It also predicts suitability of specific habitats for breeding birds as a function of changes in abiotic conditions (soil, water) and conservation measures (hay-making, grazing, burning). The Ecological Module uses statistical correlations and expert knowledge. Expert knowledge was used for formulating decision rules about ecological relations. Additionally, expert judgement was used in order to replace the empirical model, which could not be calibrated satisfactorily (Chormanski & Wassen, 2005).

The usability of the MSS was already demonstrated soon after its completion, when it was used repeatedly to assist in the planning of management activities, predicting the habitat status for different scenarios or actions in particular sections of the park. In most cases the MSS was used in order to plan small scale management activities (mowing, shrub removal, blocking of drainage ditches, etc.) on different land parcels, taking into account the ecological value of the area and the ownership status. For this purpose, mainly the database and the GIS module were used. However, for the strategic middle- to long-term planning of the natural resources such as the mire's and transitional bog's ecosystems, some kind of scenario studies were performed aiming at larger scale changes. Such a study involved hydrological calculation of flooding phenomena in the Lower Basin of the Biebrza River Valley, and analysis of hydrological parameters of various habitats of the Red Bog in reference to degraded peatlands.

EXAMPLES OF THE APPLICATION OF THE MSS

Flooding analysis in the Lower Basin

The developed system was used as a tool for evaluation of the impact of different land-use management practices on the flood extent in the BNP area in the Lower Biebrza basin (Fig. 3(a)). Vegetation in the valley was determined from the MSS databases for four different land use scenarios. The first (Scenario 1) presents the current state of agricultural and protection activities. The second (Scenario 2) takes into consideration a situation when tall sedge, grass and herb vegetation are used much more intensively for hay-making and pastures compared to the current land use, resulting in short vegetation of wet meadows. Natural succession in the area, leads to bush encroachment, resulting in decreasing biodiversity – including dramatic changes in bird populations. That is why the third scenario (Scenario 3) takes into account removal of willow shrubs and birch trees from the areas currently covered by this type of vegetation. The fourth scenario (Scenario 4) allows for the natural succession of willow and birch, resulting in almost complete disappearance of non-forest ecosystems from the area.

The water profile in steady gradually-varied flow was calculated for all four land-use scenarios. Data which were needed for the hydraulic model were obtained automatically from the MSS database. The water surface calculation was performed for the two following cases: absolute maximum (MAX Q = 229.2 m^3/s) and mean of maximum flows (AVG Q = 70.5 m^3/s) of annual

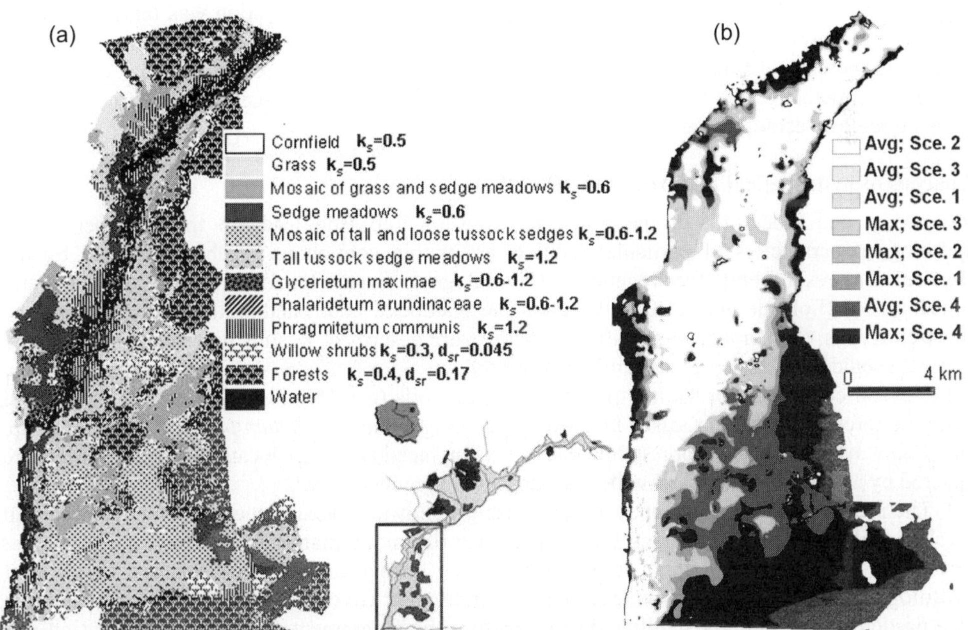

Fig. 3 Location of the Lower Biebrza Basin: (a) land use considered in this study, (b) flooded area calculated for different management scenarios (k_s, roughness height; d_s, diameter plans).

Table 1 Variation of the flooded area and the water depth on the flood plain for different land-use scenarios (MAX Q = 229.20 m3/s, AVG Q = 70.51 m^3/s).

Scenario	Flooded area (km^2)	Average depth (m)	Flow condition
1	93.29	0.65	MAX
2	83.84	0.61	MAX
3	83.21	0.60	MAX
4	179.55	1.44	MAX
1	61.35	0.49	AVG
2	56.54	0.46	AVG
3	56.27	0.45	AVG
4	113.74	0.68	AVG

discharges for historical records 1965–1996. The flood extent and average flood depth were determined by GIS analysis using the MSS database.

Table 1 and Fig. 3(b) show the results of these calculations compared to the current state (Scenario 1). Both management activities, grass mowing (Scenario 2) and cutting of shrubs and trees (Scenario 3) decreased the flood extent by about 8% for average flood conditions and 10%, for high flood conditions. The average water depth decreased by 0.03 m and 0.05 m, respectively. However, the difference between these two management scenarios is not significant. Scenario 4 (natural succession), results in very significant differences in both flood extents and water depths. The flood extent increased by more than 100% in both average and high flood conditions, while the simulated water depth increased by 0.2 m for average and 0.80 m for high flood conditions. In practical terms, this means that regular flooding would occur in both zones currently occupied by vegetation of occasional floods, as well as in groundwater-fed fens which, in the present situation are very rarely flooded.

The results obtained show that the variation of the flood extent is related to the vegetation structure of the flood plain. The most significant difference in flood extent was in the scenario in which human management was absent and natural succession of vegetation was taking place. In this scenario, birch forest and willow shrubs occurred over large areas of wetland resulting in a significant increase of flood extent and water depth. The ecological consequences, i.e. changes of vegetation structure due to the modified flood characteristics, were not assessed due to the lack of decision rules on ecological relations in such a case.

Analysis of hydrological features of the Red Bog

Monitoring and modelling of the transition bogs of Red Bog leads to the conclusion that hydrological parameters of particular habitats are within continuous feedback induced by the physical features of both the vegetation (diurnal groundwater level fluctuation due to water consumption) (Loheide *et al.*, 2005; Banaszuk & Kamocki, 2008) and water feeding. Using the research results, the number of statistical indexes were described. Results presented in Fig. 4 and Table 2 should be regarded as an equilibrium to reference for future modelling scenarios. Average hydrological conditions for habitat of forest and peat soils (e.g. location G2, where the habitat of bog-pine forest is typical for raised bog with *Oxycoccus palustris, Ledum palustre, Sphagnum* sp. and *Pinus sylvestris*) are more stable than in open meadows (e.g. location RB21), which are covered by a mosaic of birch and willow shrubs and sedge meadows.

The presented index of number of days with groundwater occurrence above/beneath ground level describe essential conditions for the presence of unique plant communities, such as alder forest, and become "hard-points" due to the Red Bog further development forecast. This sort of hydrological model of MSS, combined with accurate plant cover analysis, allows the evaluation of a particular ecosystem's stability. In this matter, the assessment of potential restoration and conservation activities will establish proper direction of management of Red Bog's wetland areas.

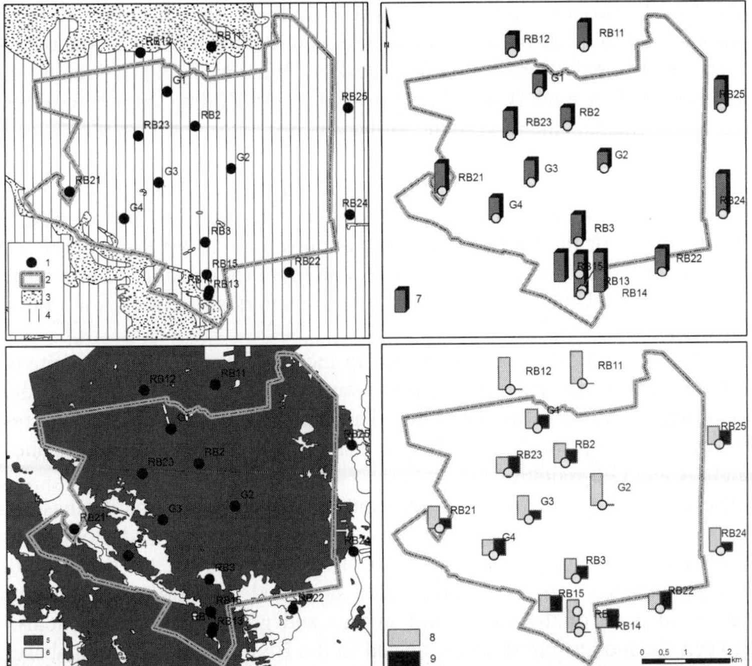

Fig. 4 Hydrological equilibrium of the Red Bog (1, measurement locations; 2, boundary of "Red Bog" Strict Protected Area; 3, mineral soils; 4, peat; 5, forests; 6, open areas (grass and sedge meadows); 7, magnitude of groundwater level fluctuation; 8, number of days with water level above ground level; 9, number of days with water level beneath ground level) – refer to Table 2.

Table 2 Hydrological statistics of particular measurement locations in the Red Bog, example of hydrological year 2008.

Location	Magnitude of water level fluctuation (m)	Number of days with water level below ground level	Number of days with water level above ground level
G1	0.49	218	148
G2	0.46	358	8
G3	0.59	266	100
G4	0.57	177	189
RB11	0.67	366	0
RB12	0.49	362	4
RB13	1.09	366	0
RB14	1.01	163	203
RB15	0.74	188	178
RB2	0.51	218	148
RB21	0.75	251	115
RB22	0.64	170	196
RB23	0.69	181	185
RB24	1.07	264	102
RB25	0.76	207	159
RB3	0.73	225	141

CONCLUSIONS

The local branch of WWF was raised to facilitate the communication process on integration and dissemination of knowledge between the park authority and other relevant stakeholders, such as water boards, farmers and tourism organisations. The results of the different surface- and groundwater scenarios were published on the Internet and were communicated to stakeholders in the area and local and national policy makers (Chormanski & Wassen, 2005). It seemed that the hydrological and hydrogeological models had their function in pinpointing important aspects for optimization of the water management system. However, for successful implementation of measures for restoring hydrology, communication was the key. Since then, the Management Support System is being used as a tool facilitating continuous dialogue and discussion.

The communication process, as well as decision making, is constrained by the huge number of land owners and the small size of the land parcels. This implies that the MSS should be used for information sharing in an almost automatic way as there is no room for personal communication in the case of bigger-scale projects. Updating of the land owners map is also critical for a number of conservation measures.

Establishing the MSS at the BNP headquarters led to increased awareness of the importance of information and the value of this in preparing management plans and application of different financial grants supporting the conservation measures. This has built up some pressure on research groups to share the results of their work in such a way that it can be incorporated in the MSS, especially in the database and GIS module.

However, we also observed that there is a significant bias in the use of the database module compared to the hydrological and ecological modules. In the case of the hydrological module significant organisational (and financial) efforts are needed in order to couple different hydrological models developed in several projects into one final operational unit resulting in coherent data for different parts of the valley and surface and groundwater sub-systems. If an operational link between the monitoring devices located in the BNP and the MSS database could be designed, the process of effective and immediate data use could be enhanced. For the ecological module, it seems that the knowledge base is still a major constraint to producing the required quantitative predictions for different management and climate scenarios. This implies that basic research on ecological relations in wetland ecosystems has to be continued at full speed.

Acknowledgements This study was undertaken thanks to funding of the EEA, Grant number PL 0082.

REFERENCES

Banaszuk, P. & Kamocki, A. (2008) Effects of climatic fluctuations and land-use changes on the hydrology of temperate fluviogenous mire. *Ecol. Engng* **32**, 133–146.

Bootsma, M. C. (2000) Stress and recovery in wetland ecosystems. Thesis, Utrecht University, The Netherlands.

Chormanski, J. & Wassen, M. J. (eds) (2005) Man and nature at Biebrza; integration and dissemination of knowledge for sustainable nature management. PIN-MATRA Final Report 2001/039, Warsaw University of Life Sciences / Utrecht University.

Gromiec, M. (2006) *Decision Support System in Water Management.* IMGW Press, Warsaw, Poland.

HECRAS, (2001) *One-Dimensional Unsteady Flow Through a Full Network of Open Channels.* User's Manual, US Army Corps of Engineers Institute for Water Resources Hydrologic Engineering Center.

Kubrak, J. & Okruszko, T. (2000) Hydraulic model of the surface water system for Central Biebrza Basin. In: *Hydrological System Analysis in the Valley of Biebrza River* (ed. by W. Mioduszewski & E. P. Querner). IMUZ Press, Falenty, Poland.

Loheide, S. P., Butler, J. J., Jr & Gorelick, S. M. (2005) Estimation of groundwater consumption by phreatophytes using diurnal water table fluctuations: a saturated–unsaturated flow assessment. *Water Resour. Res.* **41**.

Mahesh, R., Guoliang, F., Johnson, T., Ginto, C., Varun, C. & Muheeb A. (2007) A web-based GIS Decision Support System for managing and planning USDA's CPR. *Environ. Modelling Software* **22**(9), 1270–1280.

Mioduszewski, W. & Wassen, M. (2000) *Some Aspects of Water Management in the Valley of Biebrza River.* IMUZ Press, Falenty, Poland.

Okruszko, H. (1990) *Wetlands of the Biebrza Valley; their Value and Future Management.* Polish Academy of Sciences Press, Warsaw, Poland.

Okruszko, T. (2005) *Hydrological Criteria in Wetland Protection.* SGGW Press, Warsaw, Poland (in Polish).

Palczynski, A. (1984) Natural differentiation of plant communities in relation to hydrological conditions of the Biebrza Valley. *Pol. Ecol. Studies* **10**, 347–385.

Querner, E. P. (1993) Aquatic weed control within an integrated water management framework, Report 67. DLO Wind Staring Centre, Wageningen, The Netherlands.

Swiatek, D., Kubrak, J. & Chormanski, J. (2006) Steady 1-D water surface model of natural rivers with vegetated floodplain: an application to the Lower Biebrza. In: *Int. Conf. on Fluvial Hydraulics River Flow*, vol. 1, 545–553.

Slesicka, A., Querner, E. P. & Mioduszewski, W. (2002) The assessment of regional groundwater modelling. In: *Hydrological System Analysis in the Valley of Biebrza River* (ed. by W. Mioduszewski & E. P. Querner). IMUZ Press, Falenty, Poland.

Wassen, M., Barendregt, J. A., Palczynski, A., De Smidt, J. T. & De Mars, H. (1990) The relationship between fen vegetation gradients, groundwater flow and flooding in an undrained valley mire at Biebrza. *Pol. J. Ecol.* **78**, 1106–1122.

Wassen, M., Okruszko, T., Kardel, I., Chormanski, J., Światek, D., Mioduszewski, W., Bleuten, W., Querner, E., El Kahloun, M., Batelaan, O. & Meire, P. (2006) Eco-hydrological functioning of Biebrza Wetlands: lessons for the conservation and restoration of deteriorated wetlands. *Ecological Studies* **191**, 285–310.

Groundwater potential zoning by remote sensing, GIS and MCDM techniques: a case study of eastern India

MADAN K. JHA[1], GANESH M. BONGANE[1] & V. M. CHOWDARY[2]

1 *AgFE Department, Indian Institute of Technology, Kharagpur 721 302, West Bengal, India*
madan@agfe.iitkgp.ernet.in; gbongane@gmail.com

2 *Regional Remote Sensing Service Center, Indian Space Research Organization, Indian Institute of Technology, Kharagpur Campus, Kharagpur 721 302, West Bengal, India*

Abstract The main intent of the present study is to highlight the role of RS (remote sensing), GIS (geographic information system) and MCDM (multicriteria decision making) in identifying groundwater potential zones in the Bankura district of West Bengal, eastern India. Remote sensing data and available conventional maps have been used to generate thematic layers for: geology, geomorphology, land use/land cover, drainage density, soil, slope, lineament density and proximity to surface water bodies, using GIS software. All these thematic layers were standardized using fuzzy logic, and weights were assigned to thematic layers according to their relative influence on groundwater occurrence. The assigned weights were normalized using Saaty's analytic hierarchy process (AHP). Finally, all thematic layers were integrated in a GIS environment to generate a groundwater potential map. Thus, four groundwater potential zones were identified, viz., "good" (23% of study area), "moderate" (29%), "poor" (20%) and "very poor" (28%). It is concluded that RS, GIS and MCDM are very useful tools for delineating groundwater potential zones in an area/basin, especially under data-scarce conditions.

Key words groundwater potential; remote sensing; GIS; multicriteria decision making; eastern India

INTRODUCTION

Groundwater is one of the most valuable natural resources, which supports human health, economic development and ecological diversity. Around two-fifths of India's agricultural output is contributed from areas irrigated by groundwater. The contribution from groundwater to India's GDP (Gross Domestic Product) is estimated at 9%. It has also been estimated that 70–80% of the value of irrigated production in India comes from groundwater irrigation (Mall *et al.*, 2006). It is now well-recognized that water is a finite and vulnerable resource, and it must be used efficiently and in an ecologically sound manner for present and future generations. At the beginning of the 21st century, groundwater has become very important worldwide, and, hence, proper monitoring and conservation, as well as exploration of groundwater resources are essential.

Remote sensing (RS) technology, with its advantages of spatial, spectral and temporal availability of data covering large and inaccessible areas within a short time, has emerged as a very useful tool for the assessment, monitoring and management of groundwater resources (Engman & Gurney, 1991; Jha *et al.*, 2007). Since the delineation of groundwater prospect zones involves a large volume of multidisciplinary data from various thematic sources, it is necessary to use geographic information system (GIS), which can provide an ideal platform for convergent analysis of diverse data sets for decision making in groundwater planning and management. Many researchers, such as Krishnamurthy *et al.* (1996), Saraf & Choudhury (1998), Shahid & Nath (2002), Solomon & Quiel (2006), Srivastava & Bhattacharya (2006) and Madrucci *et al.* (2008) have applied integrated remote sensing and GIS techniques for the delineation of groundwater potential zones with successful results. Shahid *et al.* (2002) used the GIS approach integrated with remote sensing and fuzzy logic for the assessment of groundwater potential. A comprehensive review on the applications of remote sensing and GIS techniques in groundwater hydrology, including groundwater prospecting, can be found in Jha & Peiffer (2006) and Jha *et al.* (2007).

The objective of present study was to assess groundwater potential in Bankura district of West Bengal, eastern India by considering suitable thematic layers that have direct or indirect control over groundwater occurrence.

STUDY AREA

Bankura district, located in the western part of West Bengal state, India, was used as the study area to demonstrate the capabilities of integrated RS and GIS techniques in delineating groundwater potential/prospect zones (Fig. 1). Geographically, the study area is located between 22°38′–23°38'N latitude and 86°36'–87°46'E longitude, with a total geographical area of about 6781.22 km^2. The drainage network of the district is mainly controlled by the principal rivers Damodar, Dwarakeswar and Kangsabati and their tributaries. Climatologically, the study area falls in the Indo-Gangetic West Bengal region, with an annual average rainfall of 1420 mm, more than 80% of which falls from June to September. The net cultivable area of the Bankura district is 4300 km^2. A vast area of Bankura is not cultivable, due to undulation of the land and the *morum* soil. The district has a unique geomorphological setting with hard rock upland, laterite covered fringe areas and flat alluvial plains. The district may be divided physiographically into three distinct parts: (a) the hilly terrain to the west, (b) the adjoining undulating tract in the centre, and (c) the alluvial plain to the east. In the extreme southwestern part, undulations are more pronounced than in the Chhotanagpur table land.

METHODOLOGY

Preparation of thematic layers

In order to demarcate the groundwater potential zones in the study area, a multi-parametric data set comprising satellite data and other conventional maps, including Survey of India (SOI) topographic sheets, was used. In the present study, eight themes were evaluated on the raster GIS platform: (i) geomorphology (GM); (ii) geology (GG); (iii) land use/land cover (LU); (iv) soil (ST); (v) slope (SL); (vi) drainage density (DD); (vii) lineament density (LD); and (viii) proximity to surface water bodies (PW). The IRS-1D LISS-III data collected from the National Remote Sensing Agency (NRSA), Hyderabad were used for preparation of thematic maps of land use/land cover, drainage density and proximity to surface water bodies. Toposheets at 1:50 000 scale were obtained from the Survey of India (SOI), Kolkata. All the toposheets covering the study area were scanned separately and mosaicked to form a single image. Thereafter, thematic layers of study area boundary and digital elevation model (DEM) were prepared using ArcView 3.2 software. The DEM was prepared by digitizing contours at 20-m intervals. Further, using the DEM of the study area, 677 micro-watersheds were delineated using the ArcView Soil and Water Assessment Tool (AVSWAT) model. These micro-watersheds were used for the generation of thematic layers of drainage density and lineament density. The procedures followed for the preparation of each thematic layer are presented below.

Geomorphology In the present study, the geomorphologic map available in paper format from State Water Investigation Department (SWID), Government of West Bengal, Bankura was scanned rectified and digitized in ArcView 3.2. On the basis of the physiographic characteristics, the landforms of the study area were classified into six different units: (i) floodplain and alluvial deposit; (ii) upper undulating alluvial fill; (iii) dissected pediment; (iv) mounds and valleys; (v) severely gullied area; and (vi) isolated hills.

Geology The thematic layer on geology was prepared by digitizing a map available at the SWID of Bankura district. In the study area, five types of geology class were found: (i) alluvium; (ii) lateritic; (iii) phyllite and micaschist; (iv) granite gneiss; and (v) gondwana.

Land use/land cover The land-use/land cover classification of the study area was performed using satellite imagery. IRS-1D LISS-III data at 23.5-m spatial resolution were used for preparation of the land-use/land cover thematic map. The raw satellite images were digitally processed in a series of image processing operations: geometric rectification, image enhancement, image interpretation and multispectral classification. Further, supervised image classification was

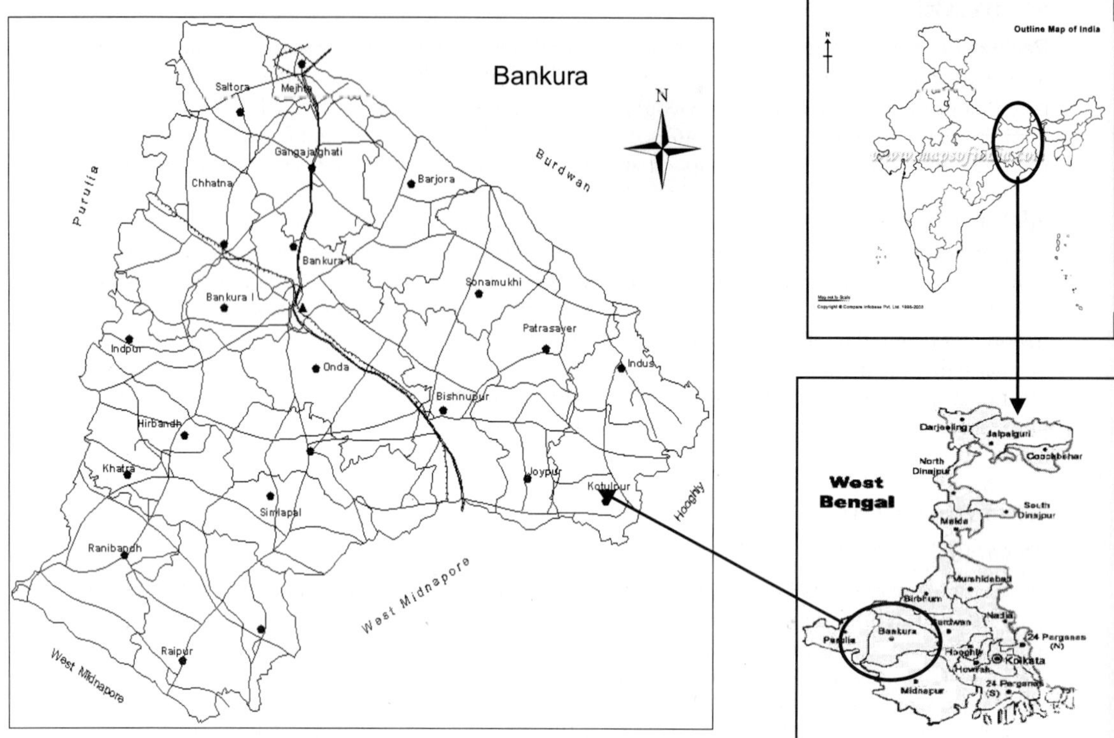

Fig. 1 Location map of the study area.

carried out to classify land-use classes. A land-use/land cover thematic map was distinguished into seven different classes: (i) agricultural land; (ii) dense forest; (iii) degraded forest; (iv) waste lands; (v) settlements; (vi) rivers; and (vii) tanks/ponds.

Drainage density The drainage network of the study area was digitized from SOI toposheets and updated with IRS-1D LISS-III satellite imagery data. The micro-watersheds delineated from the DEM were used to calculate drainage density at a micro-watershed level. The drainage density in each micro-watershed was calculated as the "total length of all streams in the unit micro-watershed area" and expressed as km km^{-2}. The minimum drainage density was found to be 0.05 km km^{-2}, and the maximum was 1.60 km km^{-2}.

Slope The thematic layer of slope was generated from the DEM of the study area, from which, slope in percentage was obtained. The entire range of slope varies between 0 and 73%, but most of the area has a slope of less than 2%.

Soil The soil layer was prepared by digitizing the soil map available from the NBSS&LUP, Government of India, Nagpur at 1:250 000 scale. The thematic layer on soil for the study area reveals six main soil texture classes, viz.: gravelly sandy soil; sandy to loamy sand; sandy loam to loamy sand; gravelly sandy clay to clay loam; sandy clay to clay loam; and clay loam to clay soil. The majority of the study area is dominated by clay loam to gravelly sandy soil, while other soil types cover relatively small areas.

Lineament density The lineament map available at the SWID of Bankura district was digitized in ArcView 3.2. The digitized lineament map of the study area was intersected with a delineated micro-watershed to find the lineament density, which was calculated as the total length of lineaments in a micro-watershed. The lineament density range varies from 0.0 to 1.10 km km^{-2}.

Proximity to surface water bodies The surface water bodies were identified from the satellite imagery of the study area. The water bodies are small in areal extent and are distributed sporadically all over the area. The thematic layer on surface water bodies was prepared by considering the groundwater prospect *vs* distance from the water body. Although there is no yardstick as to what extent the surface water bodies can recharge in the immediate vicinity, two buffer zones with radii of 25 and 50 m were chosen.

Standardization of thematic layers by fuzzy logic

Application of a fuzzy set membership in criteria standardization is highly appealing, because it provides a very strong logic for the process of criteria standardization (Jiang & Eastman, 2000). In this study, fuzzy factor standardization of thematic layers was performed by using the "Fuzzy" module of the IDRISI Andes GIS software.

Initially all the themes were standardized to a byte-level range of 0–255, which provides the maximum differentiation possible while analysing data in bytes (Eastman, 2003). Zero is assigned to the least groundwater influencing features of the thematic layer, and 255 to the most groundwater influencing features, transforming the different measurement units of the factor images, which served as GIS map layers, into comparable values using fuzzy membership functions. In this process, sigmoidal (s-shaped) fuzzy membership functions, specified for each factor, were used, i.e. monotonically increasing (Fig. 2(a)) and monotonically decreasing (Fig. 2(b)). The sigmoidal membership function is perhaps the most commonly used function in fuzzy set theory Eastman (2003). Gemitzi *et al.* (2006) offer a gradual variation from non-membership (i.e. 0) to complete membership (i.e. 1). The sigmoidal membership function can be specified by four parameters: *a* (membership rises above 0); *b* (membership becomes 1); *c* (membership falls below 1); and *d* (membership becomes 0) (Fig. 2). This is expressed mathematically as follows:

$$\mu(x) = \cos^2 a \tag{1}$$

where, in the case of a monotonically decreasing function:

$$a = \frac{x-c}{d-c}\frac{\pi}{2} \qquad x < c,\ \mu(x) = 1 \tag{2a}$$

and, in the case of a monotonically increasing function:

$$\alpha = \frac{1-(x-a)}{b-a}\frac{\pi}{2} \qquad x > b,\ \mu(x) = 1 \tag{2b}$$

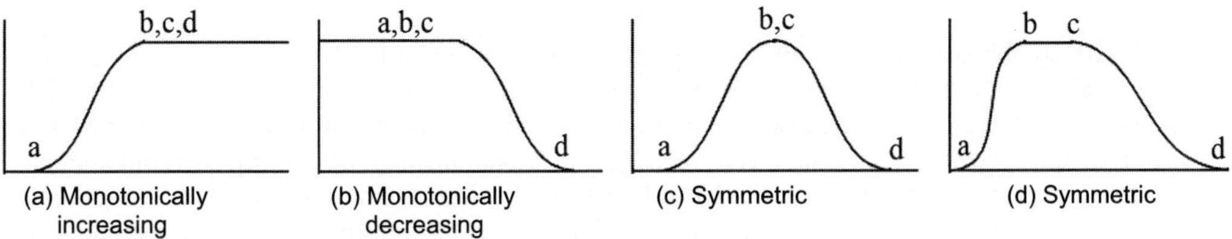

Fig. 2 Sigmoidal fuzzy membership functions (Eastman, 2003).

The thematic layers geomorphology, geology, land-use/land cover, soil type and proximity to surface water bodies with discrete classes require giving a rating to each class based on the influence of each feature class in groundwater potential occurrences. Each feature of an individual theme was rated out of 10 in ascending order of hydrogeological significance. A value of 10 was assigned to features of highest influence, and 1 for lowest influence of groundwater occurrence. The assigned ratings are presented in Table 1. These assigned ratings were used for further defining control points (*a*, *b*, *c*, *d*) of the function in the process of fuzzy factor standardization.

The thematic layers of slope, drainage density and lineament density with continuous value range were also standardized using a suitable fuzzy membership function. The method of standardization of each thematic layer is described below.

Geomorphology Geomorphological feature classes are arranged in decreasing order according to the assigned rating. Flood plain is the youngest geomorphic unit and covers the banks of the rivers Damodar, Dwarkeshwar and Kangsabati. It consists of recent alluvium deposited by these three rivers and, hence, groundwater conditions are very good in this area. The landforms "flood plain and alluvial deposit", and "upper undulating alluvial fill" are considered as most suitable for groundwater recharge and were assigned the highest rating. The feature "isolated hill" was assigned the lowest rating because of its very small groundwater recharge capability, and the other landform types were assigned ratings in between. A monotonically decreasing sigmoidal fuzzy membership function was used for rescaling rating values on a 0–255 scale. The flood plain and alluvial deposit received the highest values (255) and isolated hill received 0; the other feature classes: upper undulating alluvial fill, dissected pediments, mounds and valleys, and severely gullied area, received values in between.

Geology Geology affects the groundwater recharge by controlling the percolation of water. The geological formations, such as alluvium and laterite, were considered best for controlling recharge. Granite gneiss and gondwana were considered the most unsuitable for recharge because of their hard rock structure. Accordingly, the highest rating was assigned to alluvium, followed by laterite, phyllite and micaschist, granite gneiss and gondwana. A monotonically decreasing sigmoidal fuzzy membership function was applied to transform a geology rating to a 0–255 scale. The control points were rating 9 – membership function becomes 1 (i.e. highest groundwater potential), and 1, where membership function becomes 0 (i.e. lowest groundwater potential).

Land use/Land cover Different land-use/land cover classes define the rate of infiltration of rainfall and water to the ground. Water bodies were given the highest rating over other land-use features because of their continuous recharge to ground, followed by agricultural land, which requires more water for irrigation, resulting in groundwater recharge. Dense forest was assigned a higher rating than degraded forest, because vegetation prevents direct evaporation of water from the soil, and the roots of a plant can absorb water, thus preventing water loss. Waste land and settlement were assigned the lowest rating, because they allow less time for the infiltration of water. Different assigned ratings were transformed by a monotonically decreasing sigmoidal fuzzy membership function to a 0–255 scale. The control points were set to 9 – membership function becomes 1, and rating 1 for which the membership function becomes 0. The feature class "river" received the highest value of 255, and settlement received value of 0; tanks/ponds, agricultural land, dense forest, degraded forest and waste land received values in between.

Drainage density The drainage density, expressed in terms of km km^{-2} is an inverse function of permeability. Actually, the less permeable the rock, the lower the infiltration of rainfall, which conversely tends to be concentrated in surface runoff. This gives origin to a well-developed and fine drainage system, which cause less recharge. A monotonically decreasing sigmoidal fuzzy membership function was applied in order to transform the drainage density range to a continuous set of values in the range 0–255. The control points were 0.25 km km^{-2}, where the membership function becomes 1 (i.e. highest groundwater potential) and 1.60, where the membership function becomes 0 (i.e. lowest groundwater potential). Drainage density values <0.25 km km^{-2} receive the same membership function 1.

Slope Rainfall is the main source of groundwater recharge in tropic and subtropic regions. Slope has a direct control on the runoff and therefore on infiltration and finally on recharge. Larger slopes produce a smaller recharge because water runs rapidly off the surface of a steep slope during rainfall, not having sufficient time to infiltrate the surface and recharge the saturated zone. Most of the study area, having 0–1% slope, falls under the nearly flat terrain and has optimal infiltration rate. A monotonically decreasing sigmoidal fuzzy membership function was applied in order to transform the slope values to a continuous set of values ranging from 0 to 255. The

Table 1 Rating assigned to features of individual thematic layers.

Sr. no.	Theme	Feature class	Rating
1	Geomorphology	Flood plain and alluvial deposit	9
		Upper undulating alluvial fill	7
		Dissected pediments	4
		Mounds and valleys	3
		Severely gullied area	2
		Isolated hills	1
2	Geology	Alluvium	9
		Laterite	7
		Phyllite and miscaschist	4
		Granite gneiss	2
		Gondwana	1
3	Land use/land cover	River	8
		Tank/pond	7
		Agricultural land	5
		Dense forest	5
		Degraded forest	3
		Waste land	2
		Settlements	1
4	Soil type	Gravelly sandy soil	9
		Sandy to loamy sand soil	7
		Sandy loam to loam sand soil	5
		Gravelly sandy clay to clay loam	3
		Sandy clay to clay loam	2
		Clay loam soil	1
5	Proximity to surface water body	0 m (rivers and water body)	9
		0–25 m	7
		25–50 m	5
		>50 m	1

control points were 1%, where membership function becomes 1 (i.e. highest groundwater potential) and 30%, where membership function becomes 0 (i.e. lowest groundwater potential). Slope of <1% received a value of 255, while a slope >30% received 0 values.

Soil Soil texture also plays an important role in the infiltration of rainfall. Gravelly and sandy soils are highly permeable, light textured and well drained, and the rate of infiltration is excellent. Loamy soils are moderately well drained and consist of fine loams. The soil moisture content of such soil is very good due to its fine loamy texture. Clay soil is less permeable and high in water holding capacity. The ratings were assigned to each textural class of soil, according to the infiltration characteristics of the soil. The highest rating were assigned to gravelly sandy soil followed by sandy to loamy sand, sandy loam to loamy sand, gravelly sandy clay to clay loam, sandy clay to clay loam, and clay loam to clay soil. A monotonically decreasing sigmoidal fuzzy membership function was applied to transform the soil texture ratings to a 0–255 scale. The control points were 9 – membership function becomes 1 (i.e. highest groundwater potential), and 1 – membership function becomes 0 (i.e. lowest groundwater potential).

Lineament density The lineament density map is a measure of quantitative length of linear feature expressed in a km km^{-2}. Lineament density of an area can indirectly reveal the groundwater potential of that area, since the presence of lineaments usually denotes a permeable zone. Areas with high lineament are good for groundwater development (Edet *et al.*, 1998). A monotonically increasing sigmoidal fuzzy membership function was applied in order to transform the continuous range of lineament density values of micro-watersheds to a continuous set of values in the range

0–255. The control points were 0.0 km km^{-2}, where membership function becomes 0 (i.e. lowest groundwater potential) and 1.11 km km^{-2}, where membership function becomes 1 (i.e. highest groundwater potential).

Proximity to surface water body Surface water bodies, such as rivers, small ponds, lakes, etc., indicate the presence of groundwater in an area as they enhance the recharge procedure. Surface water bodies act as a natural source of recharge and considerably influence the potential of groundwater in its vicinity (Todd, 1980). A surface water body itself is assigned the highest rating, followed by a 25-m buffer zone, and a 50-m buffer zone. Different ratings assigned were transformed by a monotonically decreasing sigmoidal fuzzy membership function to a 0–255 scale. The control points were set to 9 – membership function becomes 1 (i.e. highest groundwater potential) and 1 – membership function becomes 0 (i.e. lowest groundwater potential).

Weight assignment

After preparing and standardizing eight thematic layers, the behaviour and the features of different themes with respect to groundwater potential in the study area were studied and, accordingly, suitable weights were assigned to each thematic layer. For the assignment of weight values, the comparison judgements scale from Saaty (1980) was used, as given in Table 2 with their corresponding meanings. The weights assigned to each thematic layer are presented in Table 3. To make the process of assigning weights more objective, a pairwise comparison was applied: in assigning the weights of the layers, only two layers were considered at a time. The pairwise comparison matrix of assigning weights to thematic layers is presented in Table 4. Following this method, one is more likely to produce a more robust set of criteria weights (Eastman, 2003). The pairwise comparison was developed by Saaty (1980) in the context of a decision-making process known as the Analytic Hierarchy Process (AHP).

Table 2 Comparison judgements from a fundamental scale of absolute numbers for assigning weight values (from Saaty, 1980).

Serial no.	Scale value	Meaning
1	1	Equal importance
2	3	Moderate importance one over another
3	5	Essential or strong important
4	7	Very strong importance
5	9	Extreme importance
6	2, 4, 6, 8	Intermediate values
7	Reciprocals	For inverse judgement

Table 3 Weights assigned and normalized weights of thematic layers.

Serial no.	Thematic layer	Weight assigned
1	Geomorphology	7
2	Geology	9
3	Land use/land cover	6
4	Drainage density (km km^{-2})	5
5	Slope (%)	3
6	Soil type	4
7	Lineament density (km km^{-2})	4
8	Proximity to surface water body	2

Calculation of normalized weight

In Saaty's (1980) technique, factor weights can be derived by taking the principal eigenvector of a square reciprocal matrix of pairwise comparisons between the criteria. The comparisons concern

Table 4 The pairwise comparison table for assigning weight values for thematic layers.

	GM	GG	LU	ST	SL	DD	LD	PW	EV	NW
GM	7/7	7/9	7/6	7/5	7/3	7/4	7/4	7/2	0.77	0.17
GG	9/7	9/9	9/6	9/5	9/3	9/4	9/4	9/2	1.00	0.23
LU	6/7	6/9	6/6	6/5	6/3	6/4	6/4	6/2	0.66	0.15
DD	5/7	5/9	5/6	5/5	5/3	5/4	5/4	5/2	0.55	0.10
SL	3/7	3/9	3/6	3/5	3/3	3/4	3/4	3/2	0.33	0.07
ST	4/7	4/9	4/6	4/5	4/3	4/4	4/4	4/2	0.44	0.12
LD	4/7	4/9	4/6	4/5	4/3	4/4	4/4	4/2	0.44	0.10
PW	2/7	2/9	2/6	2/5	2/3	2/4	2/4	2/2	0.22	0.06
Column total									4.41	1.00

GM: geomorphology; GG: geology; LU: land use/land cover; DD: drainage density; SL: slope; ST: soil; LD: lineament density; PW: proximity to water body; EV: eigenvector; NW: normalized weight.

the relative importance of the two criteria involved in determining suitability for the stated objective Eastman (2003). The "weight" module of the IDRISI Andes software package was used for calculation of normalized weight. Thus the sum of the normalized factor weight values was equal to 1. Table 4 presents normalized weights derived from the pairwise comparison technique of AHP for each thematic layer.

Consistency ratio

An index of consistency, known as the consistency ratio (CR) (Saaty, 1980), was calculated in order to determine whether the pairwise comparisons were consistent or not. The consistency ratio point outs the bias and inconsistency of a decision maker by a weight assignment process Saaty (1980). The consistency ratio (CR) is designed in such a way that if $CR < 0.10$, it indicates a reasonable level of consistency in the pairwise comparisons; if $CR \geq 0.10$ it is indicative of inconsistent judgements (Malczewski, 1999); and pairwise comparison matrices with $CR > 0.10$ should be re-evaluated.

Weight aggregation and integration of thematic layers

All the eight fuzzy standardized thematic layers (GM, GG, LU, ST, SL, DD, LD and PW) were resampled to a pixel size of the IRS-1D LISS-III image at 23.5 m × 23.5 m resolution. These resampled thematic layers were integrated using the "Multicriteria Evaluation" module of IDRISI Andes GIS software. The weighted linear combination (WLC) method of weight aggregation was used for the calculation of groundwater potential index, GWPI, as follows:

$$GWPI = GG_w GG_{wi} + GM_w GM_{wi} + LU_w LU_{wi} + DD_w DD_{wi} + ST_w ST_{wi} + SL_w SL_{wi} + LD_w LD_{wi} + PW_w PW_{wi} \quad (2)$$

where subscript w is the normalized weight of a theme, and wi is the fuzzy membership standardized weight of the individual features of a theme. GWPI is a dimensionless quantity that helps in indexing probable groundwater potential zones in the area. The range of GWPI values were divided into four equal classes (zones) and the GWPI of different pixels falling under different range were also grouped into one class, as described below.

GROUNDWATER POTENTIAL MAP

Integration of the eight fuzzy standardized thematic layers resulted in a final groundwater potential map. The integrated layer has a GWPI value varying from 45.0 to 201.5. This range of GWPI values was divided into four equal classes: 45.0–80.0, 80.1–125.0, 125.1–160.0 and 160.1–201.5. These ranges of GWPI values revealed four distinct groundwater potential zones representing groundwater potential in the study area of: "very poor" (GWPI: 45.0–80.0), "poor" (GWPI: 80.1–125.0), "moderate" (GWPI: 125.1–160.0) and "good" (GWPI: 160.1–201.5). Thus, the entire

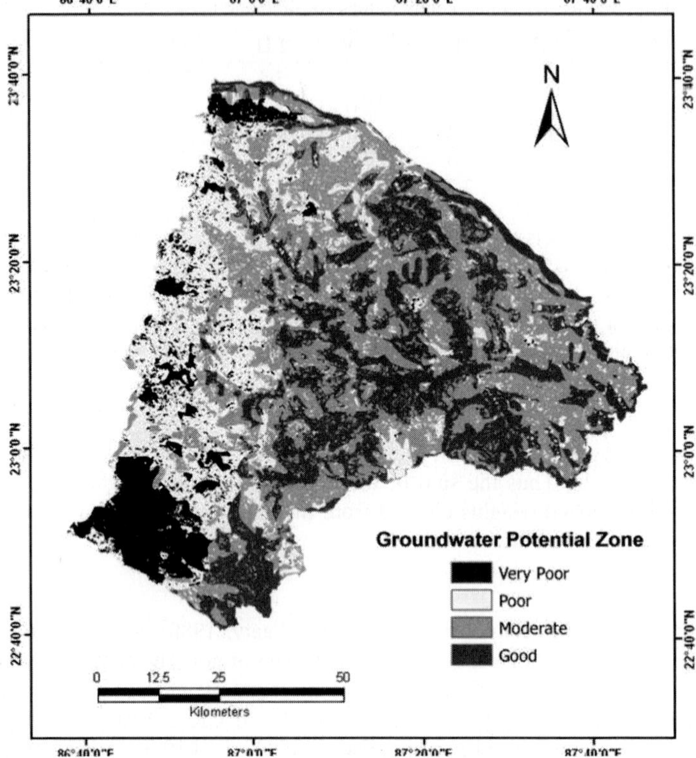

Fig. 3 Groundwater potential zone map of Bankura district.

study area was qualitatively divided into four groundwater potential zones; Fig. 3 shows the groundwater potential map of Bankura district. The "good" groundwater potential zone mainly encompasses the alluvium and flood-plain zones around the major river systems. It demarcates the areas where the terrain is mostly suitable for groundwater storage, and also indicates the availability of water below the ground. The area covered by the "good" groundwater potential zone is about 1559.54 km^2 (23%). The eastern portion and some patches in the central portion of the study area fall under the "moderate" groundwater potential zone, which is about 29% of the total area. The hydrogeomorphic feature available in this portion is alluvial fill with alluvium and laterite geology, which also suggests moderate potential of groundwater storage. However, the groundwater potential in the northwestern and southwestern parts of the study area falls in the "poor" groundwater potential zone, covering an area of about 1369.15 km^2 (20%). Geomorphology of mounds and valley type and geology classes granite gneiss, phylite and micaschist are found in "poor" groundwater potential zones. The extreme western part of the study area was rated "very poor" in groundwater potential and covers about 1960 km^2 (28%). The "very poor" groundwater potential is due to the higher slope, and unfavourable geology and geomorphology in this zone. These prospective groundwater zones can form a basis for the detailed hydrological and/or geophysical investigations required for well siting and proper management of vital groundwater resources.

CONCLUSIONS

This study has successfully used remote sensing data on the GIS platform coupled with multi-criteria decision making to obtain a detailed evaluation of the groundwater conditions in Bankura district, West Bengal, India. To delineate prospective groundwater potential zones of the study area, eight thematic layers, namely, geology, geomorphology, land use/land cover, soil, slope, lineament density, drainage density and proximity to surface water body, were considered in this study. The GIS was used for the generation and integration of various thematic maps to demarcate and identify different groundwater potential zones. The groundwater potential zones map was generated by weighted linear combination method, which revealed four distinct zones in the study area: "good", "moderate", "poor" and "very poor". The central and eastern part of the study area falls in the "good" groundwater potential zone, which is approx. 23% of total study area. The eastern portion and some patches in the central portions of the study area fall under "moderate" groundwater potential zone, which is approx. 29% of the total study area. However, the groundwater potential in the northwestern and southwestern parts of the study area is "poor", which is approx. 20% of the total study area. The extreme western part of the study area is dominated by "very poor" groundwater potential, which is approx. 28% of total study area.

Overall, the results of this study demonstrated that the integrated RS and GIS-based approach coupled with multicriteria analysis is a powerful tool for assessing groundwater potential, based on which suitable locations for groundwater withdrawals could be identified. It is concluded that RS, GIS and MCDM are very useful tools for delineating groundwater potential zones in a basin/area, especially in the areas where field data are very limited.

REFERENCES

Eastman, J. R. (2003) *IDRISI Kilimanjaro: Guide to GIS and Image Processing*, 145–177. Clark Laboratories, Clark University, Worcester, USA.

Edet, A. E., Okereke, C. S., Teme, S. C. & Esu, E. O. (1998) Application of remote sensing data to groundwater exploration: a case study of the cross-river state, Southeastern Nigeria. *Hydrogeol. J.* **6**, 394–404.

Engman, E. T. & Gurney, R. J. (1991) *Remote Sensing in Hydrology*. Chapman and Hall, London, UK.

Gemitzi, A., Petalas, C., Tsihrintzis, V. A. & Pisinaras, V. (2006) Assessment of groundwater vulnerability to pollution: a combination of GIS, fuzzy logic and decision making techniques. *Environ. Geol.* **49**, 653–673.

Jha, M. K. & Peiffer, S. (2006) *Applications of Remote Sensing and GIS Technologies in Groundwater Hydrology: Past, Present and Future.* BayCEER, Bayreuth, Germany.

Jha, M. K., Chowdhury, A., Chowdary, V. M. & Peiffer, S. (2007) Groundwater management and development by integrated remote sensing and geographic information systems: prospects and constraints. *Water Resour. Manage.* **21**(2), 427–467.

Jiang, H. & Eastman, J. R. (2000) Application of fuzzy measures in multi-criteria evaluation in GIS. *Int. J. Geogr. Inf. Sci.* **14**(2), 173–184.

Krishnamurthy, J., Kumar, N. V., Jayaraman, V. & Manivel, M. (1996) An approach to demarcate groundwater potential zones through remote sensing and a geographic information system. *Int. J. Remote Sens.* **17**(10), 1867–1884.

Madrucci, V., Taioli, F. & Carlos, C. A. (2008) Groundwater favorability map using GIS multi-criteria data analysis on crystalline terrain, São Paulo State, Brazil. *J. Hydrol.* **357**, 153–173.

Malczewski, J. (1999) *GIS and Multicriteria Decision Analysis*. John Wiley & Sons, Inc., New York, USA.

Mall, R. K., Gupta, A., Singh, R., Singh, R. S. and Rathore, L. S. (2006) Water resources and climatic change: an Indian prospective. *Current Sci.* **90**(12), 1610–1626.

Saaty, T. L. (1980). *The Analytic Hierarchy Process: Planning, Priority Setting, Resource Allocation.* McGraw-Hill, New York, USA.

Sahid, S., Nath, S. K. & Kamal, A. S. M. M. (2002) GIS integration of remote sensing and topographic data using fuzzy logic for ground water assessment in Midnapur district, India. *Geocarto International* **17**(3), 1752–1762.

Saraf, A. K. & Choudhury, P. R. (1998) Integrated remote sensing and GIS for groundwater exploration and identification of artificial recharge sites. *Int. J. Remote Sens.* **19**(10), 1825–1841.

Shahid, S. & Nath, S. K. (2002) GIS integration of remote sensing and electrical sounding data for hydrogeological exploration. *J. Spatial Hydrol.* **2**(1), 1–12.

Solomon, S. & Quiel, F. (2006) Groundwater study using remote sensing and geographic information system (GIS) in the central highlands of Eritrea. *Hydrogeol. J.* **14**(5), 729–741.

Srivastva, P. K. & Bhattacharya, A. K. (2006) Groundwater assessment through an integrated approach using remote sensing, GIS and resistivity techniques: a case study from a hard rock terrain. *Int. J. Remote Sens.* **27**(20), 4599–4620.

Todd, D. K. (1980) *Groundwater Hydrology*, 111–163. John Wiley & Sons, Inc., New York, USA.

Challenges for three-dimensional hydrogeological modelling of an LNG mined underground storage during construction

ALEXANDRA GOLDSCHNEIDER, ERIC AMANTINI & BLANCA VAN HASSELT

Géostock, 7 rue E. et A. Peugeot, F-92563 Rueil Malmaison Cedex, France

ago@geostock.fr

Abstract Hydrogeological modelling of a membrane-lined liquid natural gas (LNG) underground storage has different aims at different stages of a project: design, construction or operation. Two-dimensional (2-D) modelling focusing on local scale for cavern design is well adapted for optimising cavern depth, as well as for determining the spacing and location of the boreholes of the drainage system. Evaluation of desaturation timing and efficiency as well as impact on hydrogeological environment requires models of larger extent performed in two or three dimensions. A 3-D model created by finite element software FEFLOW® was used to simulate the progress of excavation and its impact on the hydrogeological environment. Challenges for 3-D modelling cover the representation of the storage geometry, the modelling of unsaturated zones and the determination of the most significant parameters influencing the results.

Key words underground storage; mined cavern; membrane lining; design; construction; desaturation; salt intrusion; hydrogeology; modelling; LNG

OPERATION PRINCIPLES OF MEMBRANE-LINED ROCK CAVERNS FOR UNDERGROUND STORAGE OF LIQUEFIED NATURAL GAS (LNG)

Mined caverns for underground storage of LNG are lined caverns where the product is stored at a temperature of –162°C. To ensure the LNG containment and to protect the rockmass against the extremely low LNG temperature, the storage cavern is lined by a combination of corrugated stainless steel membrane and polyurethane foam insulation panels. The concept of a membrane-lined rock cavern for underground storage of LNG is developed in Claude & Londe (1994) and Amantini & Chanfreau (2004).

During the construction and operation periods of an LNG rock cavern, the insulation membrane should not be in contact with water. In order to comply with this requirement, the rockmass needs to be desaturated during storage construction. A drainage system is therefore implemented. This drainage system is composed of a drainage gallery from which drainage boreholes are drilled. This system is located below and at the periphery of the storage galleries. Depending on the rainfall intensity and/or variability, an additional drainage system may be implemented above the storage galleries to avoid rainfall infiltrating down to the storage.

The cooling process of the rockmass begins with the storage cavern operation start-up. During the first years of storage cavern operation, the drainage system is maintained active in order to control ice formation in the rockmass. This corresponds to the time required to freeze the rock faces to a thickness of a few metres. Once the frozen thickness is considered to be sufficient, drainage is stopped and the water table builds up to its hydrostatic level.

DIFFERENT AIMS, DIFFERENT ISSUES: WHEN TO PERFORM A 3-D MODEL?

Designing the storage

The geological formations that are preferred for implementing such caverns are low permeable hard rock type, such as granite, gneiss, volcanic rocks and limestone. Weak materials, such as marls or clays, or porous materials, such as chalk or sandstone, are second choice, since they would require design adaptations.

In order to ensure that the rockmass will be desaturated at the end of construction, the main parameters to be determined during the design stage are:

Depth, size and layout of the storage cavern galleries These are mainly determined by the geological and hydrogeological characteristics of the site and should be optimized to avoid:

- unacceptable thermo-mechanical and hydrogeological environmental impact;
- excessive construction cost; and
- excessive water seepage and drainage quantities during construction and cooling down periods.

Characteristics of the drainage system(s) This refers to the distance between the drainage system and the caverns, as well as spacing of the drainage boreholes: a sufficient number of boreholes should be implemented to allow the desaturation in due time, but drilling boreholes that are not necessary to achieve desaturation will lead to additional construction costs and time for completion.

Need for an upper drainage system Depending on the hydrogeological conditions, a box-type upper drainage system might be required:

- to speed up the drainage of the water table and allow complete desaturation of the rockmass before the installation of the containment system; and
- to control the possible rockmass imbibition effect due to the infiltration of heavy rainfalls.

Specific monitoring and efficiency test procedures and guidelines (Amantini *et al.*, 2005) These are required for identifying unavoidable drainage system design adaptation to effective geological and hydrogeological conditions which are observed during the excavation works progress.

Seepage and drainage flow rates This figure is necessary to determine the characteristics of the water pumps.

Grouting works procedures and guidelines for storage galleries Such works could be necessary for improving the desaturation of cavern walls by the drainage system in areas exhibiting high permeability.

At the stage of running sensitivity studies, 2-D hydrogeological modelling was carried out as it is easier and faster. However, in 2-D models it is not always possible to take into account the exact direction of the boreholes for the drainage system. These boreholes are directed such as to intersect the main permeable and water-bearing joints, in order to constitute an interconnected network of joints and boreholes allowing an optimized drainage process. Consequently, they are not always located in the plane of the 2-D section used for the different sensitivity studies of the storage.

Moreover, 2-D models are not adapted for reliable evaluation of the hydrogeological impact of the underground works, which necessitates integrating complex hydrological boundary conditions and lateral variations of the hydrogeological context.

Confirming the design

However, once the definitive geometry of the storage has been determined, the use of 3-D modelling is particularly appropriate for:

Modelling hydraulic interactions between the different parts of the underground works The main advantage of 3-D modelling consists in the ability of 3-D models to represent exactly the shape of the underground storage. This enables one to consider tunnels and galleries and their various intersections, bends and down slopes, and to take into account their mutual hydraulic effects.

Confirming the drainage system design obtained from 2-D modelling 3-D models allow exact representation of the drainage system with all its boreholes. Transient simulation at the end of construction provides the shape of the water table and indicates whether the rockmass is maintained under stable desaturated conditions before the installation of the membrane lining containment system. If this is not the case, the efficiency of the drainage system is to be improved and the design is to be adapted.

Although 3-D models do not appear as the most appropriate tool for performing sensitivity study, the mesh of the model can include boreholes that do not appear in the design but may be activated if the design proves not to be adapted and therefore allows to perform a sensitivity study. For example, if the design spacing is 20 m in between successive drainage boreholes, the mesh can include boreholes every 10 m, so that if 20 m spacing is not appropriate, the efficiency of a higher density of drainage boreholes can be tested. This type of alternative solution has to be decided prior to creating the mesh.

PARTICULAR BOUNDARY CONDITIONS

As the membrane has to be installed in a dry environment, the main aim of modelling is to determine whether the design of the drainage system is adapted to reach, in due time, a desaturated rockmass in the vicinity of the storage. The consequences of the drainage on the nearby environment are also to be assessed. This implies creation of a transient unsaturated model that simulates the storage construction progress.

The mesh of the model shall include all the different parts of the storage, i.e. access tunnel, upper and lower drainage systems, storage galleries and maintenance chambers. The simulation is based on a weekly or monthly schedule, covering the construction period for the excavation of the different tunnels and galleries, but also the commissioning of both the upper and the lower drainage systems. With the progress of excavation works, atmospheric pressure boundary condition is applied to the excavated parts of the storage. With the installation of the containment system, a no flow condition is progressively applied along completed walls of the storage galleries.

CASE STUDY: EXAMPLE MODEL FOR LNG STORAGE

Hypotheses and challenges

For a theoretical case of a membrane-lined cavern for underground storage of LNG, the consecutive construction steps of the storage have been modelled using FEFLOW® (DHI-WASY GmbH, Germany). So far, only the construction phase, from the excavation to the set-up of the concrete lining, has been modelled. The construction duration was reckoned to be four years, and a monthly schedule was established as the basis of changing boundary conditions. A specific module was developed by DHI-WASY GmbH to ease the change of boundary conditions (Schätzl *et al.*, 2008).

The storage has a complex geometry including: an access tunnel, three parallel storage units with their operation shaft and maintenance chambers, operation tunnels located above the storage galleries and from which boreholes are drilled for the upper drainage system, and the lower drainage gallery with associated drainage boreholes located below the storage galleries for drainage purposes (Fig. 1(a)). The mesh was created to take into account all the different elements of the storage (Fig. 2).

One of the challenges for representing the storage exactly was to represent an inclined structure (downsloping access tunnel) in a model with horizontal layers. Horizontal layering had been chosen for developing this 3-D model because the surface topography is pronounced and horizontal layers were considered most appropriate for this purpose. The inclined access tunnel was implemented by adapting the elevation of the horizontal layers along the tunnel. The tunnel has been implemented with a shape as close as possible to the designed geometry and with only a few recesses due to the horizontal layering (Fig. 1(b)).

In order to evaluate the time required to reach desaturation of the rockmass during the storage construction and active drainage phase, two different values were taken on both sides of the model for hydraulic conductivity of the rockmass (10^{-9} m/s and 10^{-7} m/s).

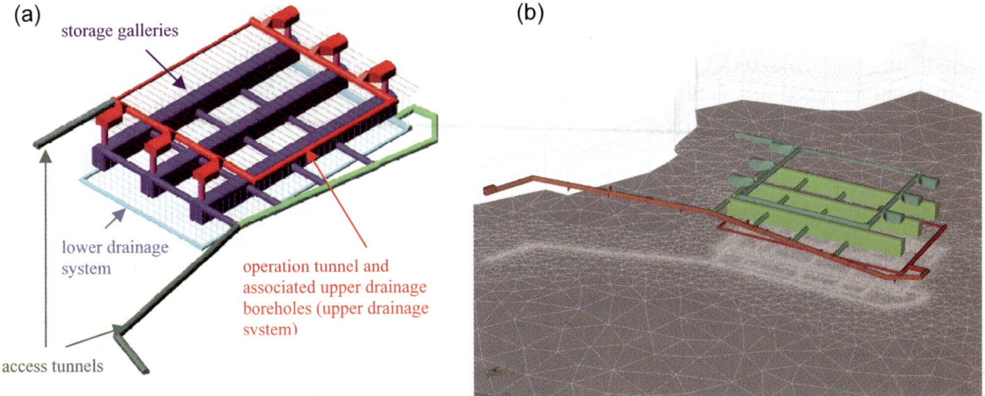

Fig. 1 Geometry of the considered LNG storage cavern: (a) theoretical geometry represented using computer-aided design software; and (b) geometry of the model as represented in the model (drainage boreholes have been implemented but are not shown).

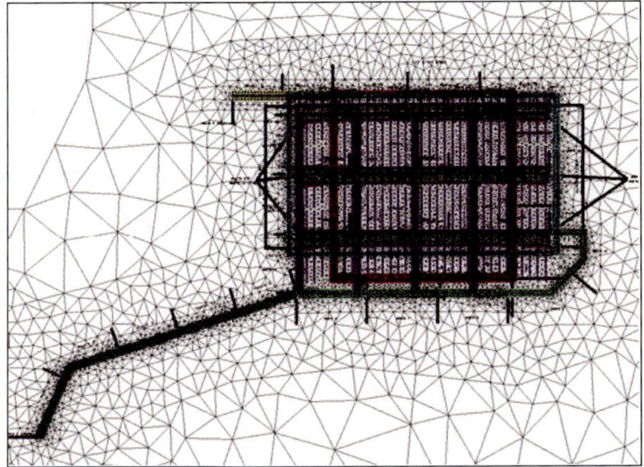

Fig. 2 View of the mesh showing the different parts of the storage as well as the boreholes of the lower drainage system (in blue).

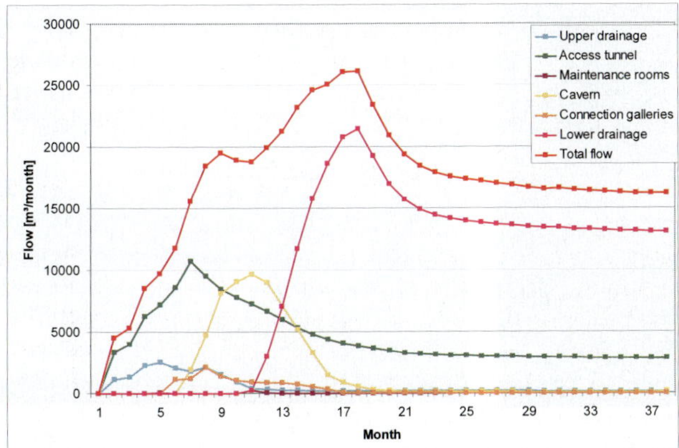

Fig. 3 Evolution of seepage water quantities for different parts of the lined storage cavern during excavation and construction works.

Another important challenge was to perform unsaturated transient modelling to represent the rock desaturation process during the construction of the membrane-lined cavern. The main aim for the modelling was to determine whether the rockmass was sufficiently desaturated by the end of construction, since this condition is required for the installation of the membrane (Amantini & Chanfreau, 2004). Figure 3 represents the water table before and after construction around the storage. A compromise had to be found between mesh size, simulation time and input parameters in order to optimize the runs of the model.

The projected LNG underground storage is located close to the seashore as it will be loaded by LNG tankers. The model was also run as a transport model in order to assess the impact of the storage construction on the salinity of the underground water and namely sea water intrusion resulting from the progressive drainage.

Results

Infiltrating seepage water was computed in the different parts of the storage (Fig. 3) to determine the exact role played by the drainage gallery and associated boreholes and ,thus, to evaluate their efficiency. Logically, the lower drainage system contributes the most to the drainage, which confirms that this lower drainage system is a key point of the design. Both the access tunnel and the cavern itself play a significant role, particularly in the early time of the excavation process, but the access tunnel continues to play an important role throughout the construction.

The role of the upper drainage system does not appear as significant; its main use would be in the case of heavy rainfall, but this case has not been modelled as the rainfall infiltration at this stage of the modelling was considered as constant in time. A sensitivity analysis performed on the 2-D model, preliminary to the 3-D model, showed that both the recharge rate and the characteristics of the upper weathered zone are sensitive parameters. As a consequence, for properly assessing the need of an upper drainage system, it will be necessary to collect very accurate information on the distribution of efficient rainfall, but also to have a good knowledge of the rockmass characteristics.

The evolution of the water table (Fig. 4) showed that the desaturation was feasible within the given time span for the part of the model where the storage is located in the rockmass with hydraulic conductivity of 10^{-7} m/s. However, in the part of the model where the storage is located in the rockmass with a hydraulic conductivity of 10^{-9} m/s, rockmass desaturation was not achieved by the end of the construction. This indicates that the drainage system needs to be adapted to attain a higher efficiency in zones of lower hydraulic conductivity. Furthermore, these results show it will be in the interest of the project, when possible, to start by excavating parts of the storage that are located in the least permeable zones.

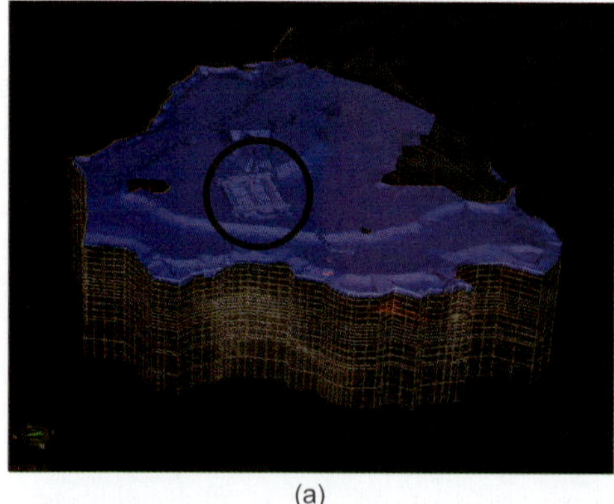

(a)

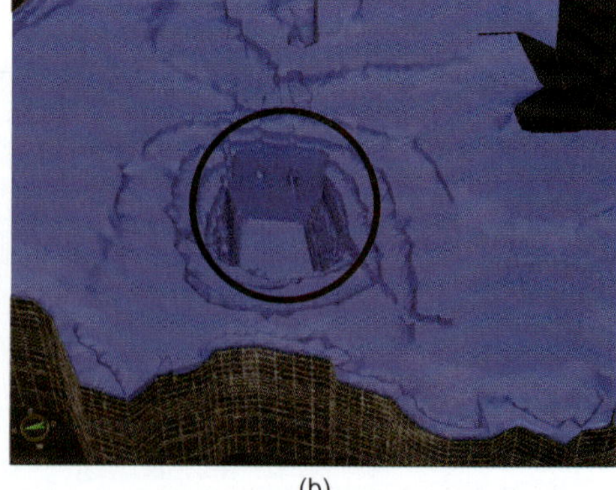

(b)

Fig. 4 Water-table level (in blue): (a) before LNG storage construction; and (b) after storage construction before drainage is released. The storage location is circled.

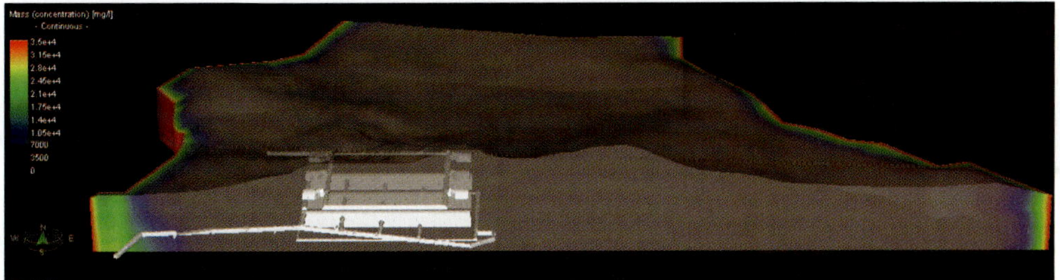

Fig. 5 Distribution of salt concentration at the end of the simulation.

The cone of desaturation does not extend very far beyond the caverns, which means that there will be no significant effects in terms of water-table level far away from the storage caverns. In the natural state, salt water intrusion is not much developed due to the low conductivity of the rockmass and the recharge from freshwater due to rainfall. However, during storage construction, salt intrusion occurs along the coast only at the locations close to the storage (Fig. 5); however, this intrusion remains limited. Results of the simulation also show a salt intrusion into the weathered zone. However it is believed that this intrusion results from the mesh refining techniques used to improve the numerical stability and are not representative of a physical phenomenon.

CONCLUSION

Each stage of a project has specific needs for modelling which can be fulfilled by 2-D or 3-D finite element modelling. Two-dimensional models are very convenient for most calculations in the design stage, and have the advantage of being easy to build and fast to run. Consequently they will be used when sensitivity analyses are required.

Three-dimensional models are more difficult to set up, but are very important to confirm the design obtained in 2-D. They allow representation of the storage in full, including the total complex geometry and the spatial variations of hydraulic conductivity related to various types of rock formations. Therefore, they allow consideration of the hydraulic interactions between the different parts of the storage facility and ancillary galleries.

The 3-D model developed for LNG storage presented numerous challenges regarding the representation of the inclined access tunnel, the simulation of the excavation progress with changes in boundary conditions, and the calculation of the rockmass desaturation.

A next step for the model will be to simulate the formation of the cooled zone around the storage units and to determine how the water table will react during this cooling. In a subsequent step, after drainage has been released, the formation of the ice ring is to be simulated (spatial and time-dependent expansion of a no-flow boundary condition), in order to evaluate the impact on the water table and environment.

Acknowledgements This study was undertaken with the technical help of DHI-WASY GmbH, which was in charge of implementing and running the 3-D models under the supervision of Géostock.

REFERENCES

Amantini, E. & Chanfreau, E. (2004) Development and construction of a pilot lined cavern for LNG underground storage. In: *14th International Conferences and Exhibition on Liquefied Natural Gas* (Doha, Quatar), PO-33. Gas Technology Institute.

Amantini, E., Cabon, F. & Moretto, A. (2005) Groundwater management during the construction of underground hydrocarbon storage in rock caverns. In: *Mine Water 2005 – Mine Closure* (ed. by J. Loredo & F. Pendás) (University of Oviedo, Spain), 311–315.

Claude, J. & Londe, L. (1994) A new concept for LNG storage in lined rock caverns. In: *Gastech 1994* (Kuala Lumpur), 434–443. Gastech RAI, London, UK.

Schätzl, P., Clausnitzer, V. & Diersch, H.-J. (2008) Groundwater modelling for mining and underground construction – challenges and solutions. In: *Mine Water and the Environment* (Proc. 10th IMWA Congress, VŠB – Tech. Univ. of Ostrava, Czech Republic) (ed. by N. Rapantova & Z. Hrkal), 473–476. VŠB – Technical University of Ostrava, Faculty of Mining and Geology, Czech Republic.

Coupled geological and hydrogeological models in fractured systems: understanding interactions between underground storages and their rockmass

ALEXANDRA GOLDSCHNEIDER, JUSTINE MORRUZZI, JEAN-LUC BODIN, ERIC AMANTINI & PHILIPPE VASKOU
Géostock, 7 rue E. et A. Peugeot, F-92563 Rueil Malmaison Cedex, France
ago@geostock.fr

Abstract Underground unlined mined storage caverns are a technique used worldwide for storing LPG and liquid hydrocarbons products. The knowledge of the structural conditions and the hydrogeological behaviour of the rockmass are essential for the elaboration of an optimized and safe design of the underground storage project. For each underground mined storage, and at each project stage, a visual geological and hydrogeological model integrating all observations from the investigation and construction phases is developed. It gives the best representation of the geological and hydrogeological characteristics of the host rockmass with the available data. At the design stage the model is used to optimize the position of the underground works and especially the orientation of the boreholes of the water curtain system. During the construction and operation phases, it is used to interpret the measurements recorded by the monitoring network and to understand water flow paths through the fracture network.

Key words geology; hydrogeology; model; visualization; underground storage; fissured rockmass

1 INTRODUCTION

For more than half a century, underground unlined mined storage caverns for storing liquid petroleum gas (LPG) and liquid hydrocarbon products, proved a reliable technology, and experienced a strong development related to its intrinsic qualities, but also to the innovations and technical progress from which it regularly benefited. Currently, large capacity storage cavern units (with more than several hundred of thousands of cubic metres) for a broad range of hydrocarbon products are successfully operated.

The selection of the site and the location and orientation of the storage caverns are crucial preliminary steps before designing the underground facilities. Most storage caverns are implemented in fissured hard rockmasses likely to exhibit a large range of permeability values. Product containment is ensured by maintaining at all times conditions of water flow from the host rockmass into the cavern (Amantini *et al.*, 2005). In certain cases the hydrogeological conditions are naturally sufficient to maintain acceptable hydrodynamic containment conditions. Nevertheless, in most cases and especially in a discontinuous medium, adding specific artificial water supply systems is required. However, if the groundwater flow gradient towards the cavern is essential for the containment of the stored product, for environmental and economic reasons the resulting water seepage must remain limited. Consequently the knowledge of the structural conditions (mainly the 3-D geometry of layers, folds, faults and joints) and the hydrogeological characteristics of the rockmass are essential for the elaboration of an optimized and safe design of the underground storage project based on representative hydrogeological models.

Unlined mined storage caverns are often located in fissured hard rockmasses that by definition cannot be considered as continuum media and are likely to exhibit significant anisotropic hydraulic behaviour. In many cases it is impossible to define an equivalent porous continuum media around the underground works, owing to the fact that a hypothetical representative elementary volume (REV) is either too large, compared to the storage cavern size, or even non-existent. This scale effect problem is commonly observed for underground works in fissured rockmass engineering, and especially when designing the hydraulic containment conditions for a storage cavern. Therefore, it is necessary:

– to focus on the sensitive areas of the underground works, such as the crown of the storage

galleries, the inverts of the connection galleries, the pillars and the water curtain boreholes system,
- to identify the characteristics of the intersections of these sensitive areas with permeable discontinuities of the rockmass,
- to highlight possible critical flow patterns as far as the hydraulic conditions necessary for the product containment in the cavern are concerned.

Joints and faults are the preferential pathway for water and the distribution of hydraulic conductivities is generally complex but needs to be characterized as precisely as possible to:
- locate the storage where it is the most appropriate,
- optimize the design of the storage to ensure its tightness during operation,
- interpret the monitoring data collected during the construction phase and adjust the design to the actual hydrogeological conditions encountered,
- carry out an analysis of the monitoring data collected during the whole life of the storage and interpret observed tendencies,
- be able to propose efficient remedial actions if a degradation of the initial hydraulic containment conditions of the storage appears in the course of operation.

Although information needs to be collected on a large scale (about a few hundreds of thousands of square metres) to have a good knowledge of the site, borehole drilling and testing is costly and consequently data are only available locally. Geological and hydrogeological visualization models are a tool able to represent the correlations between data gathered on several boreholes and enable the representation of several pieces of information at the same time.

2. DATA COLLECTION DURING SITE INVESTIGATION AND CONSTRUCTION

2.1 Site investigation

At the feasibility stage, one to three boreholes are core-drilled to determine if the geology of the site is appropriate for an underground storage. If the site is found to be suitable, additional boreholes are performed to enhance the knowledge of the rockmass. Depending on the complexity of the site geology, the number of storage units and the volume of the storage, a total of five to twenty boreholes are fully core-drilled to collect information on the rockmass. These boreholes are usually between 100 m and 200 m long to reach the estimated depth of the storage. In practice data are collected over several kilometres of cores, but only a few cubic metres of rock are brought to surface and analysed to understand the behaviour of a rockmass composed of the matrix plus discontinuities that will host a storage of a few hundreds of thousands of cubic metres.

The boreholes are not implemented all at the same time. Geological information is analysed continuously during the investigation and the results obtained in the first boreholes are used to locate and orientate the next boreholes. This methodology allows optimizing of the location of the cored holes to characterize the major geological structures that will affect the construction and operation of the storage, as well as confirming or invalidating and completing the first conclusions. Several analyses and tests are performed, but only those whose results are used in the geological and hydrogeological model are mentioned in the present paper.

A visual description of the cores is performed by an engineering geologist in a first step to describe the lithology and determine the different rock types encountered. All discontinuities are recorded in terms of geometry, i.e. depth and dip angle, and described in terms of thickness and filling. However the strike cannot be assessed on unoriented cores and acoustic borehole televiewer (BHTV) logging is often used to get an image of the borehole wall and to orientate the joints that have been identified during core inspection. Combining visual inspection of the cores and BHTV allows determining the different joint sets and their characteristics.

From the geological survey, the rock quality designation index (RQD; Deere, 1963) is calculated, giving a first quantification of the overall quality of the rockmass, where the underground storage is implemented.

A geophysical survey, especially refraction seismic, is performed over the investigated area to get large-scale data and to make correlations between boreholes. The three main results provided by the geophysical survey are first the thickness of the weathered zone and consequently the depth of the fresh rockmass, second the location of low velocity zones such as fractured and faulted zones, and third the bottom velocity of the fresh rockmass.

Hydraulic tests are performed in the boreholes to get a profile of hydraulic conductivity *versus* depth. In each borehole, intervals about 10-m long are individually tested and a hydraulic conductivity is evaluated for each interval.

Data collected during site investigation and that will be used for elaborating the visualisation model are summarized in Table 1.

Table 1 Summary of data collected during site investigation and construction and used as input data in the geological and hydrogeological model.

	Site investigation	Storage construction
Geology	surface topography of site thickness of weathered zone topography of fresh rock roof lithology oriented discontinuities RQD	oriented joints in galleries structural relationships between discontinuities
Hydrogeology	profile of hydraulic conductivity *versus* depth	hydraulic conductivity in water curtain and monitoring boreholes

2.2 Construction

The most complete knowledge of the rockmass is obtained at the end of the storage construction because the rockmass has been observed from the inside when excavating galleries with a section of several hundred square metres. For each excavated gallery or tunnel, geological mapping is performed by surveying precisely all the joints that have been crossed during excavation and characterizing them in terms of location, filling and water ingress. The scale of the mapping is generally about 1:200.

Discontinuities are locally measured in 3-D during the mapping with their strike and dip. In addition, the existence of tunnels or galleries, such as water curtain galleries, above or below the storage galleries, allows larger correlations of joints identified at different levels and provides an image of their persistence in space. Once all joints have been mapped they are organized into a hierarchy depending on their impact with regards to stability of the works and water ingress (Giafferi *et al.*, 2003). Only the joints having significant impact will be input into the model.

Hydraulic testing is performed (Amantini *et al.*, 2005):

- in the boreholes drilled from the water curtain gallery, which will be used to inject water and enhance the distribution of hydraulic potentials, and
- in the boreholes that will be used for underground monitoring and equipped with a pressure cell or a manometer.

This testing provides a map of the distribution of hydraulic conductivities into the rockmass. Data collected during construction that will be used for elaborating the visualisation model are summarized in Table 1.

3 INTEGRATING THE DATA IN THE VISUALIZATION MODEL

The aim of the 3-D visualization model is to present in 3-D the major geological features of the rockmass, essentially the structure and the tectonic defects, as well as the results of the hydrogeological tests. The method for creating the model is based on observations and manual correlations and the elaborated model is neither a numerical analysis model nor the result of automatic interpolation by a software system. It is a conceptual model established on a fully deterministic approach which requires classification of observations and/or measurements in order to clearly distinguish facts from interpretations.

3.1 Creating a database

Upstream, the software Drill&Log® of Pöyry Infra Corporation (Austria) is used. It is a georeferenced database used to store general (coordinates, length, inclination, deflection) geological (lithology, RQD, weathering grades, discontinuities), hydrogeological (permeability, static pressures, piezometric completion) and geotechnical data. The Drill&Log® database allows a representation of 2-D as well as 3-D geological drillings in order to facilitate visualisation and therefore geological correlations and interpretations. Drill&Log® also outputs graphics in the AutoCAD® 2-D/3-D format, which is very convenient for further data treatment and project integration.

3.2 Creating a visualization of the data

The procedure for building the structural 3-D model is centred on AutoCAD® software (Fig. 1), which is a universal and powerful software used world-wide, allowing easy integration of data or pre-existing maps. This software has advanced 3-D functions allowing the creation of very realistic 3-D models of the storages. It is possible to design structures and smooth the plans, providing a visual rendering of excellent quality. In addition, to get a better visual rendering of 3-D representations (caverns, faults, layers, etc.), 3ds Max®, which is also developed by Autodesk® and fully compatible with AutoCAD®, is used.

The output graphics from Drill&Log® are imported into AutoCAD®. The method for creating the model being manual, the correlations between the faults and the faults' movements are issued on the interpretation and hypothesis of the engineering geologist and represented by him in AutoCAD®.

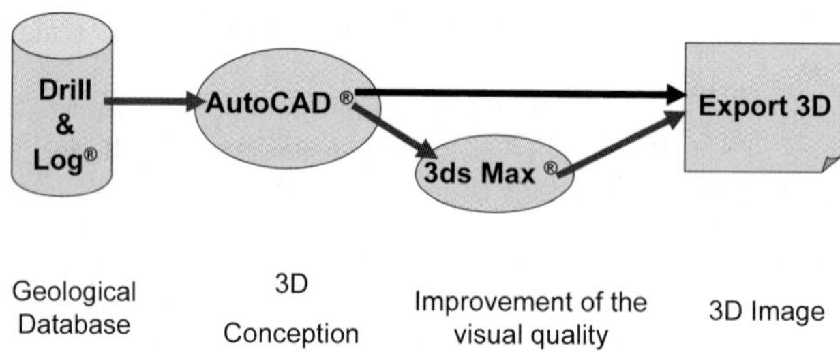

Fig. 1 Modelling procedure.

3.3 Visualizing the model

The geological 3-D model, when generated, is finally exported in VRML 97 format (Virtual Reality Modelling Language) for representing 3-D interactive vector graphics. The export VRML model has been chosen for several reasons:

- the 3-D image is relatively small (a few Mb). The image file can be sent easily over the Internet or carried on a USB key and can be opened on any type of computer as no sophisticated computer configuration is required;
- the VRML export file provides a 3-D static image which cannot be changed by a third party;
- VRML files do not have any loss of accuracy depending on the zooming. There is no pixelization when zooming in or out since the image is recalculated automatically to fit the screen;

- the VRML image supports structuring elements in groups and subgroups. This type of export gives the choice to display desired elements (for example, boreholes, faults, layers).
- VRML files can be opened by free software, and Drill&Log®, AutoCAD® or 3ds Max® is not needed to visualize the model.

4 USING THE MODEL TO UNDERSTAND THE HYDROGEOLOGICAL BEHAVIOUR OF THE ROCKMASS: TWO EXAMPLES

4.1 Use of the model

The first model for a site is built by taking into account all data from the site investigation. This model is a visualization of the geological and hydrogeological understanding of the site at the design stage. Once the shape of the underground works is determined to provide the volume required for a specific project and to ensure stability of the storage, the storage geometry can be input into the model, which is used to optimize the storage depth as well as its orientation with regards to geological structures. For example, it is in the interest of any project to avoid as much as possible locating tunnels or galleries in horizons showing high hydraulic conductivities or to follow sub-vertical discontinuities with high water ingress or instable rock conditions over the whole length of a gallery.

During construction, the model can be updated with the data collected during geological mapping and hydrogeological tests (see Section 2.2). It is therefore possible to confront the interpretation made at the design stage with the reality and to determine whether the design choices that have been made are adapted to the site. In some cases they may not be optimal and the design might require to be adapted during construction. Adaptation can include shortening or lengthening some galleries, modifying the orientation of water curtain boreholes, adding water curtain boreholes or modifying the provisional location of underground monitoring devices. The model is in this respect a useful tool to adapt the design to the geological and hydrogeological characteristics of the site.

In addition, the model is used to interpret the results of the monitoring performed during construction and operation phases. An extensive network of piezometers and underground pore pressure cells is installed to monitor how the hydraulic potential distribution in the rockmass is reacting, and to ensure that the water table is not significantly disturbed (van Hasselt *et al.*, 2003). In fissured hard rockmasses, interpreting hydraulic head measurements requires one to understand how fissures are hydraulically interconnected and what are the pathways water uses. The two examples developed in the next section illustrate how the model can be used to interpret hydrogeological observations. The first example illustrates how the model was used on a site to interpret a piezometric drop. The second example illustrates how the model was used on another site to confirm the design of the hydrogeological support from surface.

4.2 Example of interaction between piezometric levels and water curtain boreholes supply pressure

The studied underground storage is an LPG storage, excavated in gneiss, consisting of two parallel galleries with a roof elevation of –162 m a.m.s.l. and a horizontal water curtain. Figure 2 shows the storage galleries and the two shafts (in red and green) with the water curtain gallery (in blue) and horizontal water curtain boreholes located 15 m above the storage galleries. Water curtain boreholes in pink are the boreholes with the highest hydraulic conductivity (higher than 10^{-8} m/s) while boreholes in blue have a hydraulic conductivity lower than 10^{-8} m/s. Four piezometers belonging to the piezometric network and measuring the hydraulic head at the depth of the cavern are represented in Fig. 2, with their observation interval indicated in purple. Underground manometer holes are represented in red as the inclined boreholes drilled from the water curtain gallery. The line in purple above the storage represents the storage perimeter.

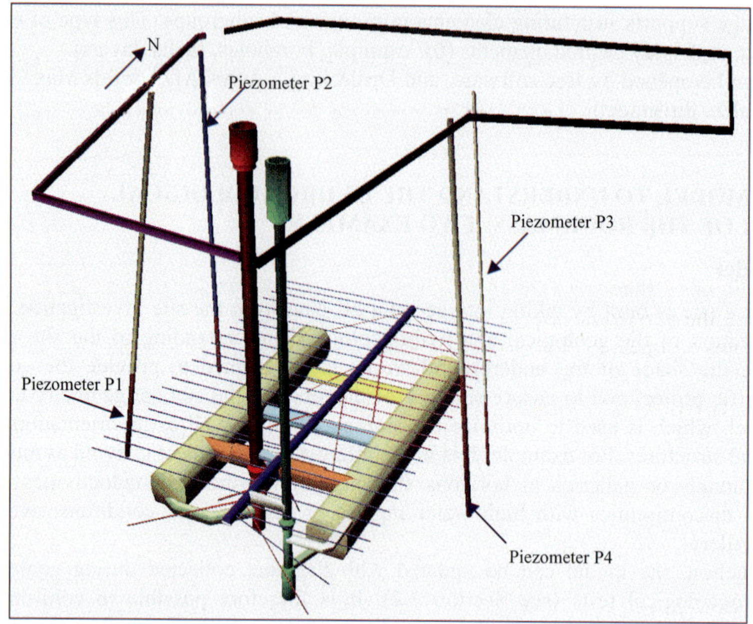

Fig. 2 Representation of the studied LPG storage showing some of the deep piezometers belonging to the monitoring network as well as the water curtain system implemented above the storage cavern.

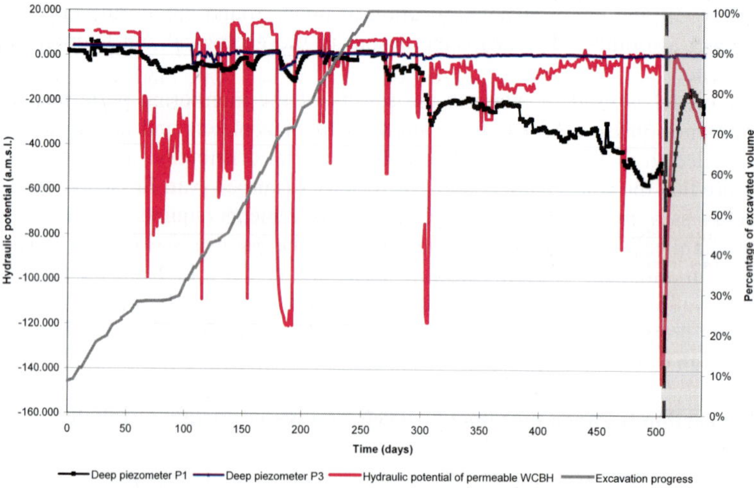

Fig. 3 Evolution of P1 and P3 piezometric level and of the mean hydraulic potential of permeable water curtain boreholes together with excavation progress. Day 0 corresponds to the start of P1 monitoring. Period in white corresponds to the period when the water curtain gallery is at atmospheric pressure and only water curtain boreholes are supplied while the period shaded in grey corresponds to the time span when the water curtain system has been water filled.

During construction the hydraulic potential of piezometers P1 and P2 showed an important decrease: sixteen months after the start of the monitoring the water level had dropped 64 m below its initial monitoring level. Figure 3 shows the hydraulic head measured only for piezometer P1. The hydraulic head for piezometer P2 is not represented since it is very similar to the hydraulic head evolution of piezometer P1. As the observation interval of piezometer P1 is located at the depth of the cavern, this decrease took place in the immediate zone of interest for the cavern. At the same time, the hydraulic head for piezometers P3 and P4 remained relatively stable. Similarly, Fig. 3 only shows the hydraulic head of piezometer P3 since the hydraulic head of piezometer P4 is similar to that of P3.

An analysis was performed to understand the reasons for the decrease of hydraulic head in piezometers P1 and P2 and to recommend remedial actions. Figure 3 shows in addition to the hydraulic head of piezometers P1 and P3, the average hydraulic head in the permeable water curtain boreholes (in pink on Fig. 2) and the progress of excavation shown as the percentage of excavated volume. Figure 3 shows that the hydraulic head of piezometer P1 was influenced by the hydraulic potential in the permeable water curtain boreholes: each time the hydraulic head in the water curtain boreholes dropped, the hydraulic head of piezometer P1 also dropped. Increasing the water supply pressure in the boreholes leads to a recovery of the piezometric level for P1 and P2, which stayed relatively stable until the storage was fully excavated.

At day 300 we observe a strong decrease in the mean hydraulic potential of the water curtain boreholes, which never recovered to its initial potential of about +7 a.m.s.l. At this stage the storage was fully excavated but not yet in operation, which implies that it was acting as a draining hydraulic boundary condition. From this point, the hydraulic potential of piezometer P1 decreased continuously. Increasing the mean hydraulic potential from day 400 to day 450 lead to a temporary stabilization of P1 hydraulic potential at about –34 a.m.s.l. but was not enough to have it stabilize.

From the observations, it appeared that the hydraulic potential of piezometers P1 and P2 was correlated to the injection pressure in the permeable boreholes of the water curtain, although P1 and P2 were located opposite the permeable water curtain boreholes. The model was then used to understand the phenomena involved in this unexpected lateral hydraulic interference. The combination of several joints observed during storage excavation and when drilling piezometers provided a communication between the permeable boreholes and piezometers P1 and P2 (see Fig. 4). It also appeared that P3 and P4 were not connected to the permeable water curtain boreholes: joint 1_7 is not intercepting the piezometers in their observation interval and joint H1 had not been observed on the cores, which is the reason why it does not intercept the piezometers.

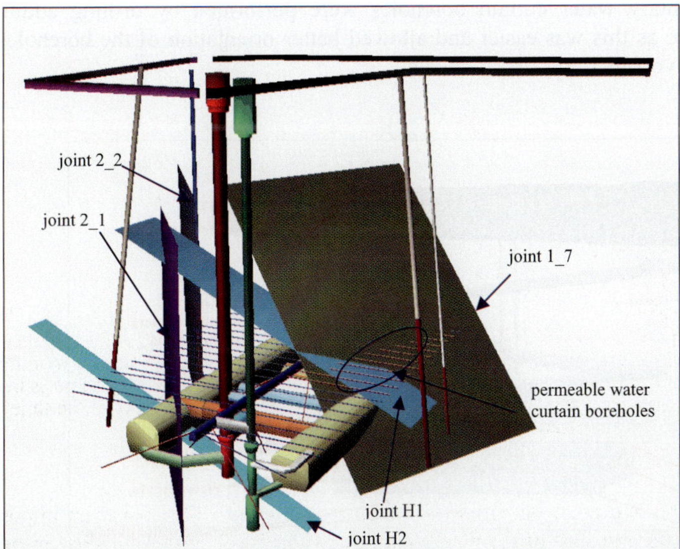

Fig. 4 Network of joints connecting piezometers P1 and P2 to the permeable water curtain boreholes (circled).

Geological structures as represented in the model, mainly joints and locally foliation, were consistent with hydrogeological observations and confirmed the hydraulic connection between piezometers P1 and P2 and the permeable part of the water curtain. The risk of having a hydraulic potential drop in the temporary supply lines used during construction being unavoidable, it was necessary to water fill the water curtain gallery and the shaft to ensure a stable potential that would allow the hydraulic potential of piezometers P1 and P2 to recover. Water filling was performed on day 506 (Fig. 4) and the hydraulic potential of piezometer P1 recovered to –14 a.m.s.l. for a hydraulic head of the water curtain of 0 a.m.s.l. As the water curtain was not supplied after its initial filling, we observe on Fig. 3 a decrease in water curtain potential which leads to an expected decrease in hydraulic heads of P1 and P2.

The visualisation model proved a very useful tool to analyse the hydrogeological monitoring measurements and to understand the hydrogeological connections between different parts of the rockmass surrounding the storage.

4.3 Example of correlations between the geological structures and the hydrogeological support network

The studied underground storage is a propane storage, excavated in gabbro, consisting of three parallel galleries with a roof elevation at –105 m a.m.s.l. and a horizontal and inclined water curtain. Figure 5 shows the storage galleries, the access tunnel with the above water curtain gallery located at –85 m a.m.s.l. and horizontal and inclined water curtain boreholes.

During construction, the existence of a major fault was observed in one of the three galleries on the western side of the investigated area. In gallery C and in the upper transverse connection gallery between galleries B and C, important water ingresses were detected (Fig. 5). Due to the existence of this unfavourable major faulted zone, it was essential to adapt the initial designed horizontal water curtain boreholes system to the geological conditions encountered. New water curtain boreholes were drilled during construction of the water curtain gallery, on both sides of the western part of gallery C, where the major fault is acting as an impervious barrier.

This first set of sub-vertical additional boreholes was necessary to improve the hydraulic conditions at the level of the storage cavern, but was inappropriate to enhance the hydraulic efficiency of the horizontal water curtain boreholes, above the southern part of gallery C. Consequently, complementary water curtain boreholes were performed by drilling additional boreholes from the surface as this was easier and allowed better orientation of the boreholes for intersecting the features involved.

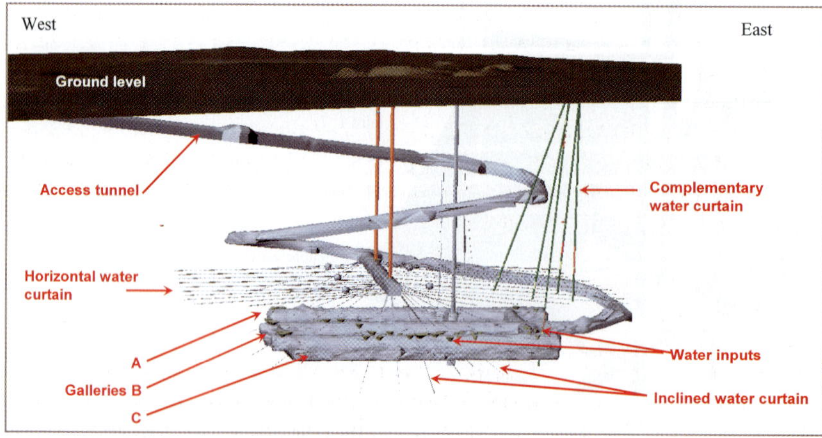

Fig. 5 Representation of studied propane storage.

A coupled geological and hydrogeological 3-D model was recently established to confirm the initial general interpretation of the structural characteristics of the site and to outline geological and hydrogeological conditions of the most sensitive area of the storage cavern. The 3-D model and its interface with the geometric characteristics of the underground works (storage cavern, water curtain systems and monitoring equipment network) provide a valuable support to the interpretation of the hydrogeological monitoring of the storage cavern.

From a geological point of view, the 3-D model has highlighted two perpendicular sets of subvertical major faults on the site. These different sets of faults create the presence of a fractured band bounded by the east–west major faults (close to gallery C and the upper connection between galleries B and C), that had not be identified during construction (Fig. 6).

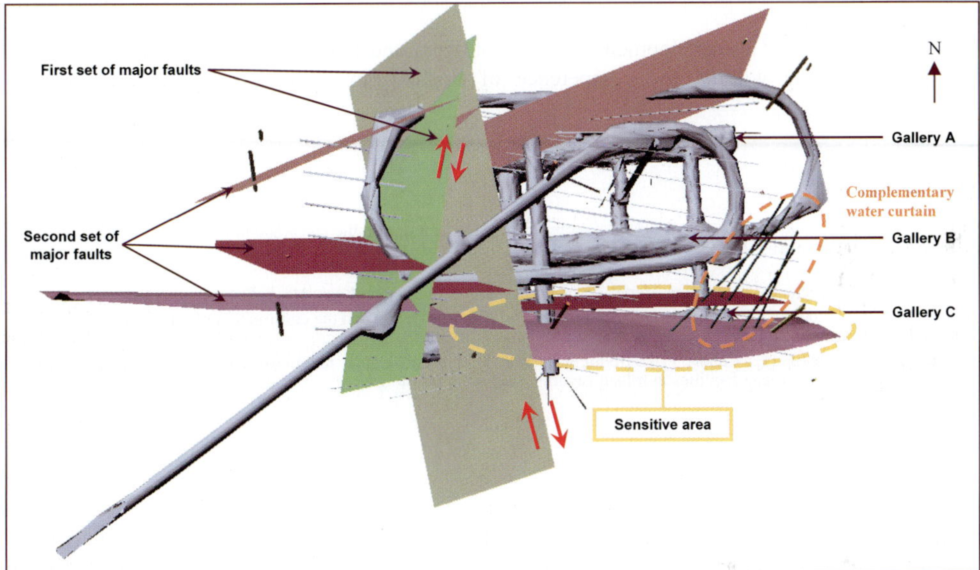

Fig. 6 3-D geological model showing the geological structure and their interaction with the galleries.

The combined geological and hydrogeological 3-D model allows exact visualization of the location of the major faults, but also of some minor faults that have had important consequences during construction. It also allows understanding of the interaction between the geological structures and the hydrogeological monitoring network in place, composed of piezometers and pressure cells. For example, the model highlights the parallelism between the water curtain and fractured band and indeed the rare horizontal water curtain cross the sensitive area (Fig. 6). Therefore the model has confirmed the necessity of the complementary water curtain boreholes drilled from the surface to locally improve the water recharge, and it can therefore be used to design additional water curtains if they appear to be necessary.

5 CONCLUSIONS

Geological and hydrogeological visualization models are developed using a series of software that enable a good picture quality as well as easy access to the model. Due to the export format the models can be easily used as a work tool. The two examples developed showed how the models can be used to understand the hydrogeological behaviour of the rockmass, interpret the data collected during monitoring and take the most adapted remedial actions when necessary.

Models are continuously evolving and representing observations more and more accurately. For the more recent models, joints and faults are represented with their measured thickness and with their observed shape, while in earlier projects they were only represented as planes. Our aim is to be able to compile complete sets of geological, hydrogeological and geomechanical data.

The models allow production of an image of the understanding of the hydrogeological behaviour of the site and this image can be used easily by all people involved in a project independently of their abilities in computer science, geology or hydrogeology. It therefore appears to be an essential tool to carry out the cross-analysis and the synthesis of numerous data of various sources, and draw up a clear representation of the main characteristics of the investigated site. It is also a decision tool that can be used on site for critical issues such as decision to grout, implementation of monitoring network and layout adaptation during construction.

Acknowledgements The development of the geological and hydrogeological models would not have been possible without the competence of Franck Pinbouen in computer-aided design software.

REFERENCES

Amantini, E., Cabon, F. & Moretto, A. (2005) Groundwater management during the construction of underground hydrocarbon storage in rock caverns. In: *Mine Water 2005 – Mine Closure* (ed. by J. Loredo & F. Pendás) (University of Oviedo), 311–315.

Deere, D. U. (1963) Technical description of rock cores for engineering purposes. *Rock. Mech. Engng. Geol.* **1**, 16–22.

Giafferi, J.-L. *et al.* (2003) Characterisation of rock masses useful for the design and the construction of underground structures. *AFTES Guidelines; Tunnels et Ouvrages souterrains*, no. 177, 2–49.

van Hasselt, B., Amantini. E., Cabon, F. & Bodin, J.-L. (2003) Hydrogeological monitoring construction and operation of underground LPG storage facilities in mined caverns. *International Congress Groundwater in Fractured Rocks* (Prague), 389–390.

Inverse modelling approach in a spatial environment to study the dynamics of groundwater flow in the main aquifer of the Doon Valley, India

K. H. V. DURGA RAO[1], **A. M. J. MEIJERINK**[2], **V. VENKATESWARA RAO**[3] **& P. S. ROY**[1]

1 *National Remote Sensing Centre (NRSC), Indian Space Research Organisation, Department of Space, Balanagar, Hyderabad 500 625, India*
durgarao_khv@nrsc.gov.in; khvdurgarao@yahoo.com

2 *Water Resources Surveys and Management, International Institute for Geo-Information Science and Earth Observation (ITC), The Netherlands*

3 *Department of Geo-engineering, College of Engineering, Andhra University, India*

Abstract Study of the dynamics of groundwater flow where the maximum water demands are met from groundwater supplies is a challenging task for water resource planners. Dehradun city, India, is almost entirely dependent on groundwater resources for its domestic and irrigation needs. Irrigation water is being tapped from the downstream baseflows of the main aquifer. High rates of pumping in the main aquifer some times create shortages in the downstream baseflows. In this study special emphasis has been given to demonstrating the inverse groundwater modelling technique in a spatial environment to study the dynamics of groundwater flow in the areas where sufficient field data are not available. Spatio-temporal variabilities of groundwater recharge flux, aquifer parameters, and groundwater flow dynamics in the main aquifer of the Doon Valley in India have been studied based on integration of various data sets and methods within the numerical groundwater model. The data sets and methods include satellite remote sensing solution of the mass water balance, GIS modelling, baseflow measurements, depth to groundwater table, lithological information, pumping data, and river flow simulation. Groundwater recharge has been estimated using a distributed hydrological model and fed into the groundwater model as an input. Average recharge rate in the study area varies from 1.86 to 6.77 mm/day during the monsoon period. Spatial variabilities of hydraulic conductivities have been assessed using an inverse modelling technique. The water budget of the aquifer system for the two stress periods (wet and dry) were computed in the calibration process. Model results are validated with the field observations. Computed and observed depth to groundwater table are found to have very good correlation. The calibrated model can be used to study the dynamics of groundwater flow in the region. As the drinking water and irrigation requirements in the study area are met from the groundwater resources, the optimum pumping rates can be decided by restricting the pumping to the productive aquifer with a minimum drawdown of the water table and minimal influence on the downstream irrigation canals.

Key words groundwater modelling; satellite remote sensing; geographic information system; groundwater recharge; groundwater budget

INTRODUCTION

Groundwater simulation models are important tools in water resources planning and management. Groundwater flow models have been applied to the study of regional steady-state flow in aquifer systems; regional changes in hydraulic head caused by change in discharge or recharge; change in a head near a well field, de-watering well systems, injection wells or infiltration basins, and surface water–groundwater interactions (Fetter, 1994). For regional groundwater flow studies, numerical models (such as finite-difference or finite-element models) are often used to represent heterogeneous hydrogeological parameters and boundary conditions. These numerical models, however, often require large data sets, which are difficult to manage. Assembling data assigning parameters for each model cell, and checking for errors are time-consuming tasks. As a result, once a model grid or mesh is designed and parameters assigned, there is generally no consideration of model sensitivity to discretisation. Moreover, the number of simulations performed during model calibration and verification is often limited. Finally, the visualization and translation of model results can be a tedious and time-consuming task.

Calibration of a flow model refers to a demonstration that the model is capable of producing field-measured heads and flows, which are the calibration values. Calibration is accomplished by finding a set of parameters, boundary conditions and stresses that produce simulated heads and fluxes that match field-measured values within a pre-established range of error. Finding the set of values amounts to solving what is known as the inverse problem. In an inverse problem the objective is to determine values of parameters and hydrological stresses from information about heads, where as in the forward problem system parameters such as hydraulic conductivity, specific storage, and hydrological stresses such as recharge rate are specified and the model calculates heads. Most classroom problems are formulated as forward problems but most field problems require solving an inverse problem.

Geographic Information System (GIS) can greatly facilitate the development, calibration, and verification of groundwater models, as well as the display of groundwater model parameters and results. Through the linking of a digital mapping system to a database, GIS has the ability to integrate data layers and perform spatial operations on data. Thus, GIS can automate many of the data compilation and management duties in groundwater modelling. Spatial statistics and grid design capabilities of GIS can improve the modelling effort and aid in reliability assessment. Furthermore, GIS has many visual display capabilities that can aid in calibration, verification, and the production of final results. Interfacing GIS and groundwater models, however, can be a substantial undertaking.

Lubczynski & Gurwin (2005) have evaluated the complexity of spatio-temporal variable fluxes and generic methodology of modern data acquisition and data integration in transient groundwater modelling for spatio-temporal groundwater balancing. Vazeer & Durga Rao (2001) have worked on remote sensing and GIS applications in groundwater modelling of Visakhapatnam city, India. They studied the groundwater dynamics, with some assumptions. More emphasis was given to groundwater budgeting to study the present situation and for future planning for optimum utilisation of groundwater resources. Chansheng He (1999) has used a hydrological budget method and particle chemical data to evaluate the availability of groundwater for irrigation supply in a humid region through a case study of the Saginaw Bay Basin, Michigan, USA. The study demonstrates that the integration of a hydrological budget and water quality data is an effective approach in the analysis of the limit of groundwater for irrigation development in humid regions when intensive hydrological monitoring and modelling are unaffordable. Tain-Shing Ma *et al.* (1999) emphasised the supportive role of geo-statistics in applying groundwater models. Field data of 1994 groundwater level, bedrock and saltwater–freshwater interface elevations in south central Kansas (USA) were collected and analysed using the geo-statistical approach. The geo-statistically analysed data were employed in a numerical model of the Siefkes site in the project area. Ijsbrand & Remco (1998) studied the interaction of groundwater and surface water to develop a concept for a decision support system that can be used by water managers. Watkins *et al.* (1996) studied the use of GIS in groundwater flow modelling. They studied three main aspects: linking a GIS to a groundwater model through data transfer programs, integrating a model with a GIS database, and embedding model capabilities within a GIS and suggested future work. Nachtnebel *et al.* (1993) addressed the user requirements of GIS to support regional groundwater modelling. In their paper five issues are discussed related to exchange of information with GIS and model design. Olsthoorn *et al.* (1993) discussed the data handling performed for the groundwater model. The groundwater model results were analysed by GIS and further used via a DTM in ecological models. In the present research, dynamics of groundwater flow within the main aquifer of Dehradun Valley has been studied using the inverse modelling approach.

As per Waterworks Department of Dehradun, nearly 80% of the domestic water requirements of the city are met from groundwater through tube wells located at different parts of the city. Decline in the groundwater table has been observed over two decades due to excessive drafts. Due to urban growth, dependability on groundwater resources is ever increasing and it is raising serious concern for the water resources planners. There are some irrigation canals on the downstream side of the area drawing water from the baseflows. Excessive drafts on the upstream side may cause drawdown of the water table in the downstream side also and it may affect the discharges available for canals and hamper the irrigation. Hence, it is not sufficient just to study the groundwater potential zones in the study area; there is a need to study the dynamics of the groundwater flow. Groundwater modelling will help in studying the groundwater table and its fluctuations with drafts; it will help in creation of various scenarios and to study the optimum rate of pumping. Groundwater modelling in conjunction with GIS can help in the decision-making process to identify potential zones for pumping and its effect on the aquifer system. In this study groundwater modelling has been done using the inverse modelling approach for the main aquifer of the study area to understand the groundwater dynamics and to plan for its optimum usage.

MODELLING APPROACH

Different types of modelling approaches exist to study the dynamics of the groundwater flow; some of these approaches are: analog models, scale models, mathematical models, analytical models, numerical models, and stochastic models. With the widespread availability of digital computers, development of mathematical models for aquifer simulation has become a relatively easy task. Applications are expanding, programming techniques are steadily improving, and computer capabilities are growing so that it is safe to say that almost any type of groundwater situation can be studied by means of a digital computer model (Todd, 1980). Finite-difference methods, similar to those for electric analog simulation, are well developed; more recently, finite-element methods have emerged as promising alternative techniques. Finally, hybrid computer models combine a resistance network with a digital computer.

The finite-difference method is a computational procedure based on dividing an aquifer into a grid and analysing the flows associated within a single zone of the aquifer. The flow equation is based on the equation of continuity:

$$\text{Inflow} - \text{outflow} = \text{change of storage} \tag{1}$$

which for a small portion of an aquifer can be restated as:

$$\text{Sum of subsurface flow} + \text{net flow to or from surface} = \text{change in storage} \tag{2}$$

This relation, plus Darcy's law for the equation of motion yields the following equation in two dimensional groundwater flow:

$$\frac{\partial}{\partial x}(T\frac{\partial h}{\partial x}) + \frac{\partial}{\partial y}(T\frac{\partial h}{\partial y}) - Q = S\frac{\partial h}{\partial t} \tag{3}$$

where T and S are the aquifer transmissivity and storage coefficient, respectively; Q is the net external inflow, h is head, and t is time.

In finite-difference form, equation (3) can be expressed as:

$$\sum_i \frac{W_{iB}T_{iB}}{L_{iB}}(h_i^{j+1} - h_B^{j+1}) - A_B Q_B^{j+1} = \frac{A_B S_B}{dt}(h_B^{j+1} - h_B^j) \tag{4}$$

where W, T, and L are the zonal boundary width, transmissivity, and flow path length, respectively (Fig. 1); A is the area of a single zone; the superscript j denotes points along the time co-ordinate with Δt being one time step; the subscripts I and B refer to a contiguous zone and the zone in question, respectively.

The quantity Q represents the algebraic sum of extraction flows (pumping) and replenishment flows (including precipitation, excess irrigation, imported water, stream percolation, and artificial recharge). With the zonal configurations defining values of W, L, and A, and estimates from hydrogeological data for S, T, and Q, time variations of h over the aquifer can then be computed from solution of the system of simultaneous equations. For verification of the model, periods of past records of groundwater levels are required. Adjustments of the physical constants S, T, and perhaps Q are made, as needed, until a satisfactory agreement is reached between the computed water-level responses and the historical data. Once the model has been calibrated, it can be applied to study the dynamic behaviour of the basin for a variety of alternative future operational conditions. Applications of numerical modelling employing finite-difference methods cover a wide range of groundwater topics: groundwater and well flow, unsaturated flow, flow with surface water bodies, dispersion, saltwater intrusion, land subsidence, mass transport (quality models), and management.

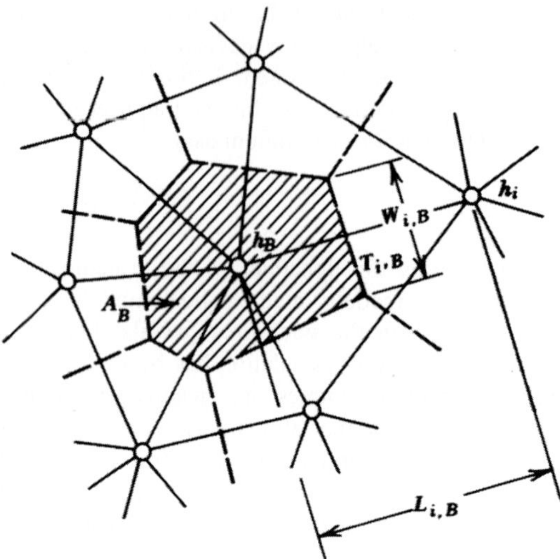

Fig. 1 Geometry of an aquifer zone for a finite-difference solution to flow within an aquifer (source: Todd, 1980).

MODFLOW96, which is a block-centred finite difference code that can simulate all types of aquifers, has been used in the study as a platform for modelling. The sophistication present in MODFLOW means that the input assembly is complex. Pre-processors are available to help with data assembly and post-processors can assist in viewing the output. Various input layers that are required for the model can be created in a GIS domain and can be easily imported to MODFLOW for the modelling; outputs generated in MODFLOW can be exported to GIS as well.

MODEL DESIGN

In order to successfully transform a conceptual model into a mathematical/analog/scale-model/numerical model, it is necessary to have a perfect model database and design that provides adequate information to apply to requisite equations. All the models start with a groundwater flow model; for this, one needs to know the physical configuration of the aquifer. This includes the location, aerial extent, boundary conditions, and thickness of aquifer layers. Hydraulic properties that need to be known include the variation of transmissivity or permeability, and the storage coefficient of the aquifers, the variations of permeability and specific storage of the confining layers, and the hydraulic connection between the aquifers and surface-water bodies. Hydraulic energy, as indicated by water table or potentionmetric-surface maps, and the amounts of natural-aquifer recharge and natural streamflow are also needed.

The design of a groundwater model of any aquifer system requires some functions like: grid design under consideration of the observed network, definition of the boundaries and their hydrological characters, identification of sources and sink terms, interpolation and assignment of hydrological data to the grid cells, selecting time steps, initial conditions and preliminary selection of values for aquifer parameters and hydrological stresses (Nachtnebel, 1993; Anderson & Woessner, 1991).

Grid Size and its Orientation The accuracy of the model, time for calibration, and quantity of data handling depend on the grid size of the model. In this study modelling has been done with different grid sizes: e.g. 500 m, 200 m and 100 m separately. Finally a grid size of 200 m was selected because, the difference between output results with 200 m and 100 m grids is negligible, computer space and processing time required for a 200 m grid structure is less than for a 100 m grid.

Dimensions of the aquifer in both x and y directions has been taken into consideration while finding the grid size to have uniform cell size for the entire area. The grid has been oriented such that all the layers that are created in GIS environment with UTM projection system can be imported directly to MODFLOW without any errors.

Boundary Conditions Boundary conditions are mathematical statements specifying the dependent variable (head) or the derivative of the dependent variable (flux) at the boundaries of the problem domain. Correct selection of boundary conditions is a critical step in model design. In steady-state simulations, the boundaries largely determine the flow pattern. Boundary conditions influence transient solutions when the effects of the transient stress reach the boundary. The boundaries must be selected so that the simulated effect is realistic.

The boundary of the aquifer has been identified using the geological, hydrological information of the study area. The groundwater divide in the western side of the study area has been taken as the no-flow boundary. The main impermeable fault that is passing along Song River has been taken as the eastern boundary of the aquifer. The groundwater divide in the southern part of the area has been considered as a no-flow boundary of the aquifer. Groundwater comes out at two places: Lachiwala village (along Song River) and Clement Town (along Rispana, Baddrao rivers). Cells covering these areas have been considered as the flow boundary. The northern side has been considered as a no-flow boundary because of water divide and fault zone. The aquifer boundary has been verified with the aquifer boundary of the entire Doon Valley delineated by Meijerink (1974) and Bartarya (1995b), and has been digitised, polygonised and rasterised at 200 m grid size in GIS environment and imported to the model. Subsequently boundary conditions were assigned.

Stress Periods and Time Step Most flow codes allow the user the option of dividing the simulation period into blocks of time of variable lengths. These smaller blocks of time are known as stress periods. In the model all important variables such as evapotranspiration, recharge, pumping, river leakage, etc., have to be given as an input for each stress period.

In the present study, to improve the model accuracy, the total simulation period has been divided into two stress periods: the wet period (rainy season) during which maximum recharge is taking place, lasts 122 days; and the remaining 243 days of the dry period are considered as the second stress period.

Just as it is desirable to use small nodal spacing, ideally one would like to use small time steps to obtain an accurate solution. However, it is usually impractical to use extremely small time steps (Anderson, 1991). An optimum time step should be chosen such that it should not affect the solution. In the present study "day" has been considered as the time step, so all time units in the modelling are "day". The simulation period of the model has been taken as 365 days.

Initial Hydraulic Heads The model requires initial heads in a steady-state simulation. In the modelling initial values assigned to different cells are starting values to initiate the computations. Static water level data of 51 tube wells have been obtained from groundwater authorities and water levels were measured in the open wells at some locations in the southern part of the area. Part of this data has been considered as an input to the groundwater model for the model calibration purpose. Remaining water level data has been used for model validation. These initial hydraulic heads have been given as an input at different locations covering the entire area uniformly.

Aquifer Parameters Layers may be designated as always confined, always unconfined, or capable of being either confined or unconfined (convertible). Heads in the layer are calculated under the Dupuit assumptions. From the lithological information at different tube well locations obtained from the *Garhwal Jal Sansathan* and from previous studies (Meijerink, 1974; Bartarya, 1995) in the Doon Valley, the layer type has been considered as unconfined. Though there are few poorly sorted mixtures of clay deposits exists at some pockets in the study area, the layer has been assumed as homogeneous and isotropic throughout its extent.

To compute transmissivities from the given hydraulic conductivity values, saturated layer thickness is an important factor; this can be computed from the bottom elevation of the aquifer and the hydraulic heads (for unconfined aquifer). However, to compute water table depth below ground level, the digital elevation model (DEM) of the aquifer area has been prepared using topographic information. The DEM has been imported to the model as the top of the layers; this represents the elevation of each cell of the aquifer with respect to mean sea level.

The bottom of the layer has been prepared from the lithological information obtained from the tube-well data and from resistivity studies at a few locations (resisitivity data obtained from the Geo-science division of IIRS, India). A point map has been created with this information and interpolated at 200-m cell size using kriging techniques in the GIS domain. This layer has been given as input to the groundwater model as bottom of the aquifer.

Injection and Pumping Wells An injection or pumping well is a point source or sink and is represented in a model by a node. The user typically specifies an injection or pumping rate in units of volume of water per time for each node to designate.

There are nearly 51 pumping wells (tube-wells) within the aquifer area that are pumping water for drinking purposes. The pumping rate of each well in m^3/day has been computed from rates of pumping (L/min) and duration of pumping in a day. A point raster map showing the well locations has been prepared and attributed with the corresponding pumping rates. This has been given as an input to the model using the well package.

In fact, no recharge well exists in the area; however, in the upstream side, baseflows from the hills after passing through some distance infiltrate and can be noticed clearly during dry periods. These flows were measured in the field and an approximate average rate (m^3/day) has been computed and given as an input to the model.

There are a few irrigation wells in the study area. As the rainfall in this area is high, canal discharges are sufficient during the maximum demand period; the irrigation wells operate during the lean period. Compared to tube well (drinking wells) discharges, these irrigation discharges are much less hence; they have been ignored in the model.

Recharge Recharge refers to the volume of infiltrated water that reaches the water table and becomes part of the groundwater flow system. Discharge refers to groundwater that moves upward across the water table and discharge directly to the surface or to the unsaturated zone. No one has yet devised a universally applicable method for estimating groundwater recharge. Numerous methods have been proposed but most have met with limited success (Anderson & Woessner, 1991). Due to lack of a way to quantify the spatial distribution of recharge, models have traditionally assumed a spatially-uniform recharge rate across the water table equal to some percentage of average annual precipitation.

Agnese *et al.* (1996) aimed to study the recharge and discharge components of the Death Valley (USA) regional groundwater flow system by remote sensing and GIS techniques. Elevation map, slope aspect, soil permeability and vegetation maps were reclassified and integrated to identify recharge potential zones and subsequently a recharge rates map was prepared using average annual precipitation.

Burke *et al.* (1998) studied the relationship between rainfall and recharge and the variables affecting the relationship. They found that the sensitivity of recharge to climate and land surface parameters could be grouped into two spatial scales. At the regional scale, recharge responds predominantly to temporal variation in climatic variables. At the site scale, soil and vegetation conditions will affect recharge response within the context of these landscape and regional responses.

Another popular method of estimating recharge is the mass balance method that calculates the recharge in the spatial domain. The mass balance method can be expressed as:

$$\mathrm{P} = \mathrm{R} + \mathrm{ET} + \mathrm{Re} \pm \Delta SM + \mathrm{L} \tag{5}$$

where, P is the precipitation, R is runoff, ET is evapotranspiration, Re is recharge, ΔSM is the change in soil moisture and L is other minor losses (interception, etc.)

By re-arranging the terms, equation (5) can be expressed as:

$$\mathrm{Re} = \mathrm{P} - \mathrm{R} - \mathrm{ET} \pm \Delta SM - \mathrm{L} \qquad 6$$

A rainfall (P) map showing the spatial variation of rainfall has been prepared through Thiessen polygon technique using the rainfall data of 1997–98 obtained from three stations within the study area. Surface runoff (R) has been computed using the distributed modified curve number technique. A land-use/land-cover map was prepared using satellite data, and a soil texture map of the study area has been used in runoff computation. Evapotranspiration (ET) has been estimated through the Penman-Monteith method using meteorological data obtained from the three stations. A spatial ET map was prepared using the Thiessen polygon method. Interception losses were approximated as 10% and 5% of the rainfall in the forest areas and in other vegetated areas, respectively Other minor losses and ΔSM have been ignored in calculating recharge (as these losses are less than 2% of average annual rainfall).

Recharge in the spatial domain has been computed in GIS environment using equation (6) and the above derived layers. Recharge rate (m/day) for both the stress periods (wet and dry season) has been computed separately. These two recharge layers have been imported to the model using the recharge package of MODFLOW. All the prepared GIS layers have been imported to the model through an ASCII matrix format.

River Flow Simulation The river package is used to simulate the flow of water between an aquifer and an overlaying (or underlying) source reservoir, which is usually a river or lake. The model allows water to flow from the aquifer to the source river, thereby removing water from the model by seepage to gaining stream reaches. Water can also flow out of the stream into the aquifer but the seepage out of the stream is independent of the stream discharge (Anderson & Woessner, 1991).

The river package uses the streambed conductance (CRIV) to account for the length (L) and width (W) of the river channel in the cell, the thickness of riverbed sediments (M), and their vertical hydraulic conductivity (Ky):

$$\mathrm{CRIV} = \mathrm{KyLW/M} \tag{7}$$

The rate of leakage between the river and the aquifer (QRIV) is calculated from:

$$\mathrm{QRIV} = \mathrm{CRIV}\ (\mathrm{HRIV} - \mathrm{h}) \quad \text{if } \mathrm{h} > \mathrm{RBOT} \tag{8}$$

where, HRIV is the head in the source reservoir and h is the head in the aquifer directly below the source reservoir. When the water table falls below the bottom of the streambed (RBOT), leakage stabilises and QRIV is calculated from:

$$\mathrm{QRIV} = \mathrm{CRIV}\ (\mathrm{HRIV} - \mathrm{RBOT}) \quad \text{if } \mathrm{h} \leq \mathrm{RBOT} \tag{9}$$

In the field it has been observed that groundwater flow comes out at two places, Lachiwala and Clementown. During the dry season at these locations flow width, length, water depth and discharges were measured at different cross-sections of the flow at regular time intervals. The averages of these measurements (flow width, length, depth) have been computed and given as input to the model using the river package.

Hydraulic Conductivity Hydraulic conductivity is a measure of the permeability of the porous medium. Hydraulic conductivity of a soil or rock depends on a variety of physical factors, including porosity, particle size, distribution, shape, arrangement of particles, and other factors (Todd, 1980). Hydraulic conductivities can be measured in the field by pumping tests. For unconsolidated sediments, hydraulic conductivity may also be obtained from laboratory grain-size analysis. Laboratory measurements of hydraulic conductivity suffer from the scale problems and are not accurate compared to field values (Aunderson, 1991). Since the hydraulic conductivity data of the study area is not available, it has been taken as unknown variable and computed by applying the inverse modelling technique (see next section).

A GIS layer having spatial porosity information has been prepared based on the soil textural information, its average particle size, and its distribution and exported to the groundwater model.

MODEL CALIBRATION

Most computer models are not subjected to field verification, as this is a very time-consuming and expensive step. The more accurate the data that initially go into a model, and the more detailed the data against which it is verified, the more confidence the modeller can have in the model results. The process of model calibration and verification frequently requires many changes in the data parameters that make up the model. In the inverse modelling approach, the unknown variable will be assumed based on other hydrogeological parameters, and subsequently the model will be calibrated by changing the unknown variable until the end results match the ground reality.

The model has been calibrated for steady-state conditions to compute hydraulic conductivities for the known heads at different locations. In spatial modelling, assuming initial values of hydraulic conductivities in a spatial domain is very difficult. A wrong assumption may increase the calibration time. Small changes in the hydraulic conductivity will affect the upstream hydraulic heads and *vice versa*. Hence, in this study a realistic approach has been adopted in assuming the initial values of hydraulic heads. As the hydraulic conductivity is one of the main parameters that control the slope of the groundwater profile, the slope of the water profile has been taken as a basis for assumption. The DEM of the groundwater table has been prepared using the static water levels at different tube well locations by the kriging technique. Slopes of the groundwater table have been computed using the water table DEM and subsequently classified into suitable classes. A zonal map that indicates the boundaries of hydraulic conductivity has been prepared using the classified slope map of the water table. Assumed approximate initial hydraulic conductivity values (Todd, 1980) have been assigned to the polygons of the zonal map in proportion to its water surface slope. Soils information has been taken into consideration while assigning the initial values. The model was run for the initial value of hydraulic conductivity and using the other input layers. Computed heads at different locations were compared with the measured heads. The model was run several times by changing the hydraulic conductivity values of each polygon by trial and error until the measured and computed heads are matched. Finally, the model has been calibrated using the known heads at different locations. Figure 2 shows the calibration procedure in the inverse modelling approach.

For checking the simulation results, the model calculates a volumetric water budget for the entire model at the end of each time step, and saves it in the record file. A water budget provides an indication of the overall acceptability of the numerical solution. In numerical solution techniques, the system of equations solved by a model actually consists of a flow continuity statement for each model cell. Continuity should also exist for the total flows into and out of the entire model.

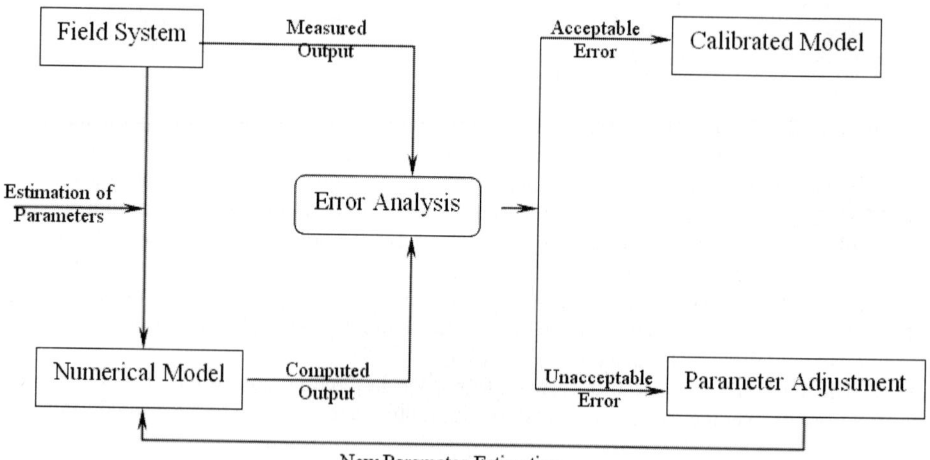

Fig. 2 Calibration procedure in the inverse modelling approach.

RESULTS AND DISCUSSIONS:

Recharge rate, runoff, and evapotranspiration estimated for the year 1997-98 are shown in Fig. 3. From this it can be inferred that almost 95% of the recharge is takes place in the months June–September. Spatial variations of recharge rate (m/day) during the wet season (during which maximum recharge is taking place) are shown in Fig. 4. From the recharge computations it is found that the recharge rate is high in the study area and it ranges from 0.00186 to 0.00677 m/day; a large part of the area has a recharge rate of 0.0064 m/day during monsoon period. This higher rate of recharge may be due to the coarse-grained soils and high rainfall in the study area.

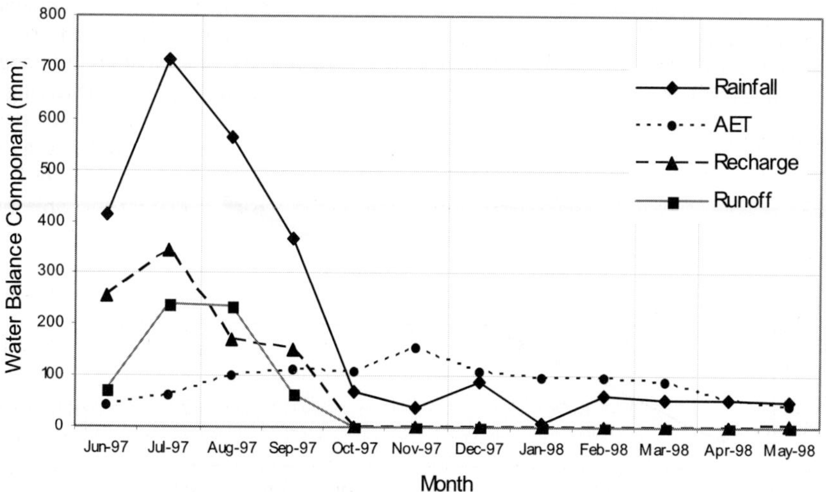

Fig. 3 Recharge rate, runoff, and evapotranspiration during 1997–1998.

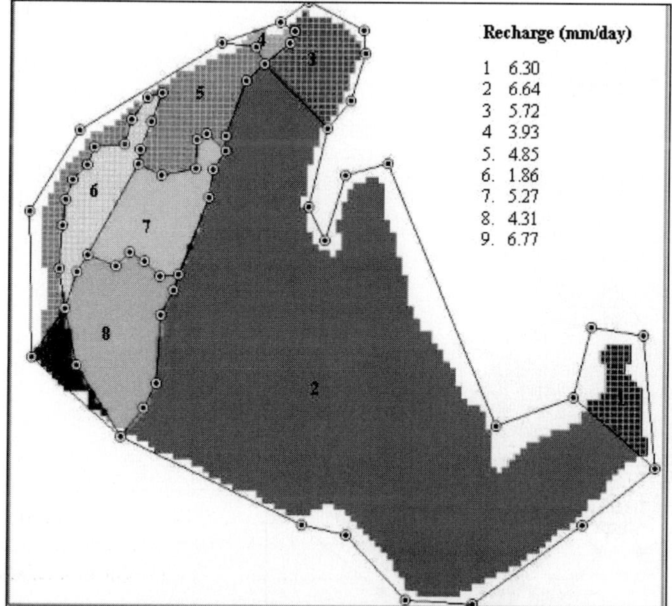

Fig. 4 Spatial variations of recharge rate during monsoon season.

The mass balance method is found to be a more realistic approach to computing recharge, as there are no calibrated empirical models available for the study area. Computation of recharge using the distributed approach is found to be more practical when compared to the lumped value. Computed hydraulic heads during the inverse calibration process have been compared with the field observations at the end of each trial run. Trial runs were conducted until a good correlation between observed and computed heads was obtained, by changing the hydraulic conductivity values spatially. The difference between observed and computed heads varies from 10 cm to a couple of metres. Comparison between computed and measured heads is shown in Fig. 5, from which it can be inferred that there is good correlation between observed and computed heads.

Computed hydraulic conductivity values in the spatial environment using inverse modelling technique are shown in the Fig. 6. From the limited field observations, hydraulic conductivity in this area is found to vary from 10 to 70 m/day (Bartarya, 1995). In the present study the range of calibrated hydraulic conductivities ranges from 6 to 65 m/day, matching well with the field data.

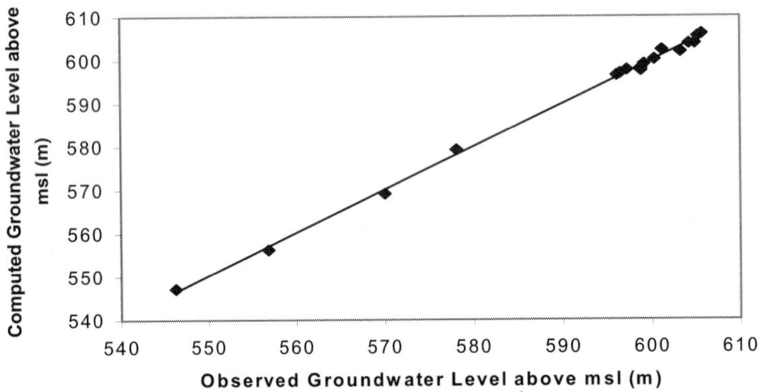

Fig. 5 Comparison between observed and computed groundwater levels.

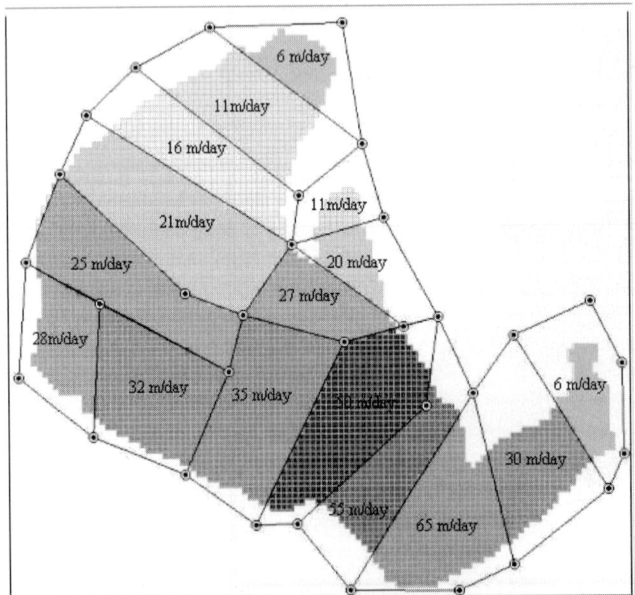

Fig. 6 Computed hydraulic conductivity values using inverse modelling technique.

Groundwater budgeting indicates the accuracy of model calibration. The model was calibrated until the discrepancy in the inflow and outflow flux converges to a possible minimum. The volumetric water budget of the entire model during both the stress periods is shown in the Tables 1 and 2.

Table 1 Groundwater budget in stress period 1.

	In Cumulative volumes (L^3)	Rates for this time step (L^3/T)	Out Cumulative volumes (L^3)	Rates for this time step (L^3/T)
Constant head	0	0	0	0
Wells	7930000.0	65000.0	9733770.00	79785.0
River leakage	0	0	135902128.0	1113951.8
Recharge	137705680.0	1128735.13	0	0
Total	145635680.0	1193735.13	145635904.0	1193736.88
In–out	–224.0 (L^3)		–1.75 (L^3/T)	
Discrepancy (%)	0.0 (L^3)		0.0 (L^3/T)	

Table 2 Groundwater budget in stress period 2.

	In Cumulative volumes (L^3)	Rates for this time step (L^3/T)	Out Cumulative volumes (L^3)	Rates for this time step (L^3/T)
Constant head	0	0	0	0
Wells	23725000.0	65000.0	29121524.0	79785.00
River leakage	0	0	220940608.00	349952.56
Recharge	226338195.40	364742.86	0	0
Total	250063200.0	429742.87	250062128.00	
In–out	1072 (L^3)		5.31 (L^3/T)	
Discrepancy (%)	0.0 (L^3)		0.0 (L^3/T)	

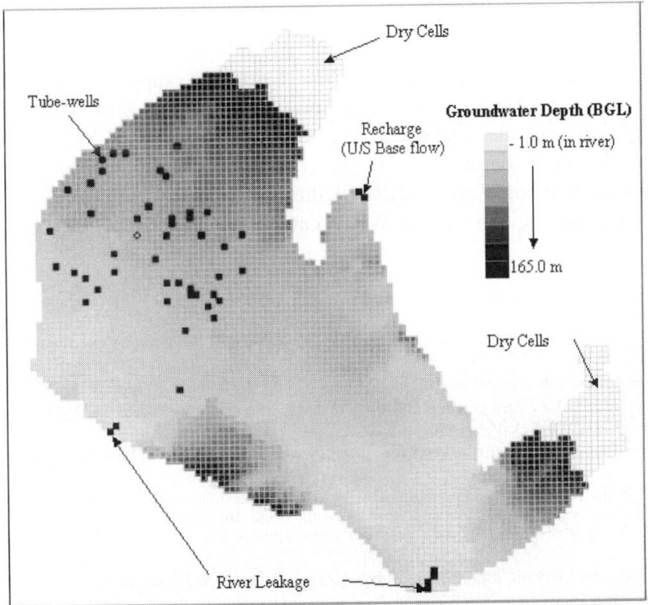

Fig. 7 Groundwater depth variations and other hydraulic conditions of the model.

The results of the calibration should be evaluated both qualitatively and quantitatively. Model verification will help establish greater confidence in the calibration. The calibrated model has been validated for the following measured field parameters: baseflows of Lachiwala and Clement Town have been measured during November 1997 to May 1998 (dry period), average flows at Lachiwala and near Clement Town were about 4.5 and 5.3 m^3/s, respectively. Total out flow discharge at these two locations simulated by the river package of the model is nearly 9.98 m^3/s and matches the field data.

Groundwater depth (b.g.l.) surface has been prepared using the static water levels computed at each pixel location. To give more idea to the user, a groundwater depth map (below ground level) has been prepared by subtracting groundwater elevations (m.s.l.) from the digital elevation model. Groundwater depth variations and other hydraulic conditions of the model are shown in the Fig. 7. These computed groundwater levels match well when compared with the observed water level data in the middle and southern parts of the aquifer. On the northern side, where the water table is very deep, it could not be compared due to the absence of tube wells. Groundwater depth varies from 10 to 165 m from the south to northern parts of the study area. Some places where groundwater table cuts the aquifer bottom are represented as dry cells in the Fig. 7.

CONCLUSIONS

Due to the coarse-grained soil texture, recharge rates in the study area are reasonably high. The mass balance method of estimating recharge may be one of the best methods in areas where field data are not available. The inverse groundwater modelling approach was found to be more effective and realistic in computing any unknown spatial variable such as hydraulic conductivity. Obtaining spatial hydraulic conductivity data using pumping tests in deep aquifers is expensive and time consuming; in such cases inverse modelling is cost and time effective. Discrepancies in the groundwater budget were almost negligible, which is an indication of the accuracy of the model calibration. The calculated depth to groundwater table values showed a good agreement with observed values. The calibrated model may be used to estimate the optimum pumping rates by restricting the pumping to the productive aquifer with a minimum water table drawdown with minimum influence on the downstream irrigation canals.

Acknowledgements The authors sincerely acknowledge the field data support provided by Gadwal Jal Samsthan, Dehradun, Central Water Commission, Central Soil and Water Conservation Research & Training Institute, Dehradun, and The Dehradun City Water Works Department. The authors sincerely acknowledge the support, cooperation, and guidance provided by the Director, National Remote Sensing Centre (NRSC) during the research work. The first author acknowledges the encouragement given by GD, WROG and HWRD of NRSC.

REFERENCES

Anderson, M. P. & Woessner, W. W. (1991) *Applied Groundwater Modelling: Simulation of Flow and Advective Transport.* Academic Press. Inc., New York, USA.

Bartarya, S. K. (1995) Hydro-geology and water resources of intermontane Doon Valley. *J. Himalayan Geol.* **6**(2), 17–28.

Burke, S., Mulligan, M.& Thornes, J. B. (1998) Key parameters controlling recharge at a catchment scale. In: *Hydrology in a Changing Environment*, vol. II, 229–241. Wiley, Chichester, UK.

Chansheng He (1999) Use of hydrologic budget and chemical data for groundwater assessment. *J. Water Resources Plann. Manage. ASCE* **125**(4), 234–238.

D'Agnese, F. A., Faunt, C. C. & Turner, K. A. (1996) Using remote sensing and GIS techniques to estimate discharge and recharge fluxes for the Death Valley regional groundwater flow system, USA. In: *Hydro GIS 96: Application of Geographic Information Systems in Hydrology and Water Resources Management*, 503–511. IAHS Publ. 235. IAHS Press, Wallingford, UK.

Fetter C. W. (1994) *Applied Hydrogeology*, third edition. Prentice Hall, Upper Saddle River, New Jersey, USA.

Ijsbrand, G. H. & Remcod, D. J. (1998) Modelling groundwater and surface water interaction for decision support systems. In: *Shallow Groundwater Systems*, 143–156. Balkema, Rotterdam, The Netherlands.

Lubczynski, M. W. & Gurwin, J. (2005) Integration of various data sources for transient groundwater modelling with spatio-temporally variable fluxes – Sardon study case, Spain. *J. Hydrol.* **306**, 71–96.

Meijerink, A. M. J. (1974) Photo hydrological reconnaissance surveys. PhD Thesis, International institute for Arial Survey and Earth Sciences (ITC), The Netherlands.

Nachtnebel, H. P., Furst, J. & Holzmann, H. (1993) Application of Geographical Information Systems to groundwater modelling. In: *Hydro GIS 93: Application of Geographic Information Systems in Hydrology and Water Resources*, 653–663. IAHS Publ. 211. IAHS Press, Wallingford, UK.

Olsthoorn, T. N., Kamps, P. T. W. J. & Droesen, W. J. (1993) Groundwater modelling using GIS at the Amsterdam water supply. In: *Hydro GIS 93: Application of Geographic Information Systems in Hydrology and Water Resources*, 665–674. IAHS Publ. 211. IAHS Press, Wallingford, UK.

Tain-Shing Ma, Sophocleous, M. & Yun-Sheng Yu (1999) Geo-statistical applications in groundwater modelling in south-central Kansas. *J. Hydrol. Engng* **4**(1), 57–64.

Todd, D. K. (1980) *Groundwater Hydrology*, second edition, John Wiley & Sons, Inc. New York, USA.

Vazeer M. & Durga Rao, K. H. V. (2000) Groundwater modelling using remote sensing and GIS – a case study of Visakhapatnam, India. In: *Proceedings of ICORG-2000, Hyderabad*, vol. I, 162–167.

Watkins, D. W., McKinney, D. C., Maidment, D. R. & Min-Der Lin (1996) Use of Geographic Information System in groundwater flow modelling. *J. Water Resources Plann. Manage. ASCE* **122**(2), 88–96.

A new stochastic framework based on evolutionary algorithms for evaluating potential yield and corresponding risks of optimal deficit irrigation strategies

NIELS SCHÜTZE & GERD H. SCHMITZ
Institute of Hydrology and Meteorology, Dresden University of Technology, Germany
ns1@rcs.urz.tu-dresden.de

Abstract The scarcity of water compared with the abundance of land constitutes the main drawback within agricultural production. Besides the improvement of irrigation techniques (e.g. use of micro-irrigation) a task of primary importance is solving the problem of intra-seasonal irrigation scheduling under limited seasonal water supply. An efficient scheduling algorithm has to take into account the crops' response to water stress at different stages throughout the growing season. It is very difficult to solve the highly multi-dimensional and nonlinear optimization problem for finding the schedule with maximum crop yield for a given water volume. The objective of our research is to assess the risk in yield reduction in view of different sources of uncertainty (e.g. climate, soil conditions and management) based on a new global optimization technique and physically-based modelling for reliable, predictive simulation. In this contribution we introduce a stochastic framework for decision support for the planning of water supply in irrigation. This consists of: (i) a weather generator for simulating regional impacts of climate change; (ii) a new tailor-made evolutionary optimization algorithm for optimal irrigation scheduling with limited water supply; and (iii) mechanistic models for simulating water transport and crop growth in a sound manner. As a result, we present stochastic crop water production functions (SCWPF) for different crops which can be used as basic tools for assessing the impact of climate variability on the risk for the potential yield or, furthermore for generating maps of uncertainty of yield for specific crops and specific agricultural areas. A case study of a French site is used to illustrate these methodologies, and the impacts of predicted climate variability on wheat and maize are discussed.

Key words deficit irrigation; crop water production function; optimal scheduling; risk assessment

INTRODUCTION

A component of improved water use is a better quantitative understanding of the relationship between irrigation practices and grain yield, i.e. crop water production function (CWPF). With this knowledge, the value of each unit of water applied to a field can be estimated and compared with alternative uses within and beyond the agricultural sector. In arid climates, irrigation methods which reduce watering volumes and, at the same time, maximize water use efficiency were developed to coexist with the increasing reduction of water resources and the increase in irrigation costs (Jones, 2004).

A relatively new technique with respect to increased water productivity is controlled deficit irrigation (English, 1990). For deficit irrigation it is necessary to find an optimal irrigation schedule under which crops can sustain an acceptable degree of water deficit and yield reduction. To date, mostly static optimization techniques are applied for providing optimal schedules which maximize yield. This optimization strategy is based on forecasts or scenarios generated by weather generators and simulations of the water balance and crop production of an irrigation system for a whole growing period in advance.

Static optimization leads in general to a mixed integer optimization problem which is difficult to solve, since the number of decision variables (i.e. the number of irrigations) is *a priori* unknown. For this reason, recent studies tend to simplify the optimization problem either by fixing the irrigation dates (Loganathan & Elango, 2004; Shang & Mao, 2006) or the irrigation intervals (Gorantiwar & Smout, 2003; Brown *et al.*, 2006). In addition, heuristic optimizing algorithms such as the Nelder-Mead simplex method (Shang & Mao, 2006) or simulated annealing (Brown *et al.*, 2006) are used which, unfortunately, may fail in practice when: (a) local optimal solutions exist, or (b) when the number of decision variables becomes too large, or (c) require unreasonable computation power and time using brute-force approaches (Raghuwanshi & Wallender, 1997; Singh & Singh, 1997; Scheierling *et al.*, 1997; Shang *et al.*, 2004; Brumbelow & Georgakakos, 2007).

The objective of this study is to demonstrate how to apply an efficient scheduling algorithm for evaluating potential yield and corresponding risks in deficit irrigation strategies, i.e. generating stochastic crop water production functions (SCWPF). In the case study presented, an irrigation experiment under real climate conditions in a Mediterranean region in France (Montpellier) is considered. A statistical analysis based on 500 generated CWPFs was performed for analysing the impact of climate variability on the inherent risk of yield losses under deficit irrigation.

CONSTRUCTION OF STOCHASTIC CROP WATER PRODUCTION FUNCTIONS

There is potentially great value in understanding the role of climatic variability on the relationship between yield and the scale of water consumption for irrigation. In the following we present a methodology which allows for quantifying the impact of climatic variability on CWPF by a two-dimensional probability distribution, which is referred to as Stochastic CWPF (SCWPF). The resulting SCWPF can be a central decision support tool assessing consequences of climate change, which allows evaluating the efficiency of water resources in agriculture under future climate condition can be made.

Stochastic framework

Generally, two components are necessary to generate reliable simulation-based CWPFs (see Fig. 1, loop 1): an irrigation scheduling optimizer and a simulator of plant growth and water transport. The objective of the most inner iteration loop is to maximize yield for a specific climate scenario and a given amount of water, V0, for irrigation during the growing season. With the second loop which iterates over a range of V0, a complete CWPF can be constructed. The generated CWPF represents the maximum yields which can be achieved with a given amount of water and is referred to as the potential CWPF.

The weather generator can be interpreted as a Monte Carlo sampler for providing site specific weather series with a desired number of scenarios. Thus, in the third loop the necessary amount of CWPFs is generated in order to accurately compute the statistical characteristics of the random sample of CWPFs in a non-parametric way, i.e. the resulting SCWPF is an empirical probability function. To be useful for Monte Carlo simulations, the irrigation scheduling optimizer and the simulation model should be highly efficient and effective in finding the global maximum yield in order to generate a SCWPF within an acceptable computation time.

The stochastic weather generator LARS-WG

LARS-WG is a stochastic weather generator based on the series approach; a detailed description is given in Semenov *et al.* (1998). LARS-WG produces synthetic daily time series of maximum and minimum temperature, precipitation and solar radiation. The weather generator uses observed daily weather data for a given site to determine a set of parameters for probability distributions of weather variables, as well as correlations between them. On this basis, this set of parameters is used to generate synthetic weather time series of arbitrary length by randomly selecting values from the appropriate distributions. The use of semi-empirical distributions gives the flexibility to the generator to construct climate change scenarios on the basis of global climate model (GCM) projections in two steps (Semenov, 2007). First, with historical data available for a specific site, the LARS-WG parameters representing the baseline climate will be computed. In a second step, changes in mean and variability of climate variables derived from GCM predictions are taken into account by adjusting the LARS-WG parameter sets for generating daily site-specific weather times consistent with the GCM predictions.

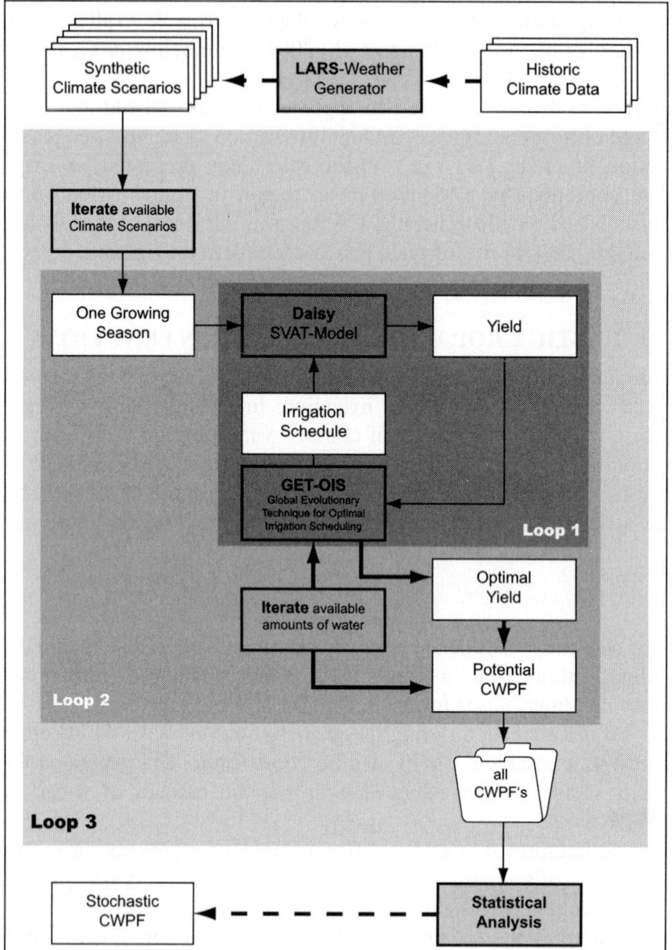

Fig. 1 Construction of stochastic crop water production functions (SCWPF).

The mechanistic irrigation model Daisy

Daisy is a one-dimensional Soil–Vegetation–Atmosphere Transfer (SVAT) model that simulates crop production and crop yield, and water and nitrogen dynamics in agricultural soil based on information on management practices and weather data (daily values), for details see Abrahamsen & Hansen (2000). The water balance model comprises a surface water balance and a soil water balance. The soil water balance includes water flow in the soil matrix as well as in macropores. Furthermore, it includes water uptake by plants and drainage to pipe drains. The heat balance model simulates soil temperature, and freezing and melting in the soil.

The solute balance model simulates transport, sorption and transformation processes. Special emphasis is put on nitrogen dynamics in agro-ecosystems. Mineralization-immobilization, nitrification and denitrification, sorption of ammonium, uptake of nitrate and ammonium, and leaching of nitrate and ammonium are simulated. Degradation, sorption, uptake and transport of agro-chemicals like pesticides are simulated. The crop production model simulates plant growth and development, including the accumulation of dry matter and nitrogen in different plant parts. Furthermore, the development of leaf area index and the distribution of root density are simulated.

and the agricultural management model allows for building of complex management scenarios. The Daisy crop library contains the parameterizations of a number of different crops (maize, wheat, etc.) whose parameters may be calibrated on experimental data sets.

Optimal irrigation scheduling with a new evolutionary technique

Formulation of the scheduling optimization problem The objective of optimal irrigation scheduling is to achieve maximum crop yield *Y* with a given, but limited, water volume V_0, which is to be distributed over the growing season, where the quantity of each irrigation is to be determined. The impact of an irrigation schedule on the crop yield is calculated by Daisy SVAT-model (Abrahamsen & Hansen, 2000). The global optimization problem is:

$$Y^* = \max Y(\mathbf{S}) : \mathbf{S} = \{\mathbf{s}_i\}_{i=1...n} = \{(d_1, v_1), ...(d_i, v_i), ..., (d_n, v_n)\} \quad n, d_i \in \mathbb{N}; \ \mathrm{v_i} \in \mathbb{R} \tag{1}$$

with the optimal solution for maximizing the yield *Y*:

$$\mathbf{S}^* = \arg\max Y(\mathbf{S}) = \arg\max Y(\{(d_i, v_i)\}) \quad i = 1...n \tag{2}$$

where **S** is the schedule for the whole growing season, consisting of $I = 1,...,n$ irrigations each defined by the date d_i and the irrigation depth v_i. The scheduling optimization problem solver (equation (1)) maximizes the yield *Y* of the plant with respect to *n* irrigations, with control parameters optimized for each of the irrigation events. Equation (1) has the specific feature that the number of optimization variables, i.e. the number *n* of irrigation events **s** is not fixed *a priori*.

Solving the optimization problem with GET-OPTIS The GET-OPTIS (Global Evolutionary Technique for OPTimal Irrigation Scheduling) computation starts with a set of solutions, called population, which is, in our case, a random set of schedules. Every member of the set has a fitness value assigned which is directly related to the objective function – its crop yield. In sequential steps, the population of schedules is modified by applying four steps: selection, crossover, mutation, and reconstruction. The details of the algorithm are presented in Schmitz *et al.* (2007). The main features are as follows.

A population X^n is a set of n_{gen} individuals, i.e. irrigation schedules **S**. Each irrigation schedule consists of a group of 2-tuples $\vec{s} = (d_i, v_i)$, where each tuple contains the parameters of an individual irrigation event. The entire group **X** – i.e. all those individuals which are created during the optimization – can be formed by bringing together all the individuals of all generations:

$$\mathbf{X} = \{X^n\}_{n=1...n_{\max}} = \left\{\{\mathbf{S}_j\}^n_{j=1...n_{gen}}\right\}_{n=1...n_{\max}} = \left\{\left\{\{(d_i, v_i)\}^j_{i=1...n_j}\right\}^n_{j=1...n_{gen}}\right\}_{n=1...n_{\max}} \tag{3}$$

The convergence and outcome of GET-OPTIS are determined by the following parameters: the maximum number of evolution loops $n_{\max}$, the stopping criteria, the mutation rate for irrigation dates and volumes, the take-over probability and the crossover probability. A set of best schedules that will be included in the next generation in their unchanged form, employing an elitism procedure to tournament selection. In tournament selection each individual is set once as one participant of the tournament iterating through the whole set of individuals. The other one is randomly chosen. If the set individual wins, then it is retained in the set of best schedules and goes unchanged in the next generation.

The structure of the GET-OPTIS deviates in certain aspects from the standard operators of evolutionary algorithms – selection, crossover and mutation. First, the deviations include a change in the order in which the individual operators are activated. During each generation step the selection is the first operation to be carried out, instead of at the end. Second, an additional reconstruction step rebuilds the created children in order to guarantee feasible solutions which are in compliance with the constraints. For doing this we used available knowledge about irrigation scheduling, e.g. that it is better to irrigate for future crop requirements in advance than to irrigate too late. The implementation of the operators is explained in Schmitz *et al.* (2007).

GET-OPTIS iterates until a certain desired degree of convergence is reached. Evolutionary algorithms usually require many function evaluations for convergence (computational time costly) because they do not use derivative information of the objective function. Other than the common evolutionary algorithms, GET-OPTIS reduces the computational effort by restricting the individuals, which have to be evaluated by simulations, to feasible solutions. In addition, the overall elapsed time of one optimization run can be reduced because the algorithm is set up for extensive parallel processing for the evaluation of the objective function used.

APPLICATION AND EXAMPLE

To illustrate the potential use of the proposed stochastic framework, we analysed the impacts of the climate variability on maize grown and irrigated at a French site (Lavalette, Montpellier).

Construction of the climate scenarios using LARS-WG

Observed daily weather data for 17 years (1991–2007) from the Lavalette climate station were used to set-up the weather generator. The synthetic weather data conform to the distributions estimated from the observed data. For this reason, synthetic data pass a variety of statistical tests, such as the *t*-test and *F*-test for means and variances of weather variables, i.e. minimum and maximum temperature as well as solar radiation. Altogether, 500 realizations of daily weather over a year were generated. A subset of that synthetic weather data which cover the growing period from 1 May to 13 October was then selected for simulation/optimization runs of the Daisy crop model.

Setup of the irrigation model

The set-up of the Daisy model is based on soil data taken from a loamy soil plot (44% silt, 38% sand, 18% clay) at the CEMAGREF experimental site Lavalette in Montpellier, France. The soilprofile of the plot is subdivided into three layers. Soil hydraulic characteristics were determined by inverse methods applied to measurements of capillary tension and moisture content of the soil (Woehling & Mailhol, 2007). Daisy provides a library of crops and assigned crop growth parameter files which are validated and tested in many environments (Abrahamsen & Hansen, 2000). For this study we used silage maize (variety Pioneer). The management description consisted of a sowing date of 1st May with a zero initial water deficit and the irrigation schedule provided by the optimization algorithm GET-OPTIS. During the irrigation events a constant flow rate of 5 mm/h is employed which implies an irrigation efficiency of over 90%.

Monte Carlo simulation optimization for computing SCWPF

The synthetic weather data in combination with the configured Daisy model and the GET-OPTIS algorithm can be used to generate distributions of potential crop yields resulting from climate variability and optimal irrigation scheduling. The optimization runs were distributed among the nodes of a PC-Cluster with 764 nodes and finished within 12 h. Subsequently, we analysed the resulting SCWPF with descriptive statistical methods, i.e. we calculated (0.5, 1, 10, 50, 90, 99, 99.5)-percentiles, median and statistical moments of a sample of 500 achieved yields for each amount of water V_0.

Results and risk analysis on the basis of the generated SCWPF

The CWPFs, which were generated by the simulation optimization runs with GET-OPTIS, are presented in Fig. 2(a). The variation in maize yield reaches up to 3 t/ha and increases dramatically with the available amount of water especially from a water volume of 300 mm on. This upper range of the SCWPF appears to be the most interesting part due to many reasons. First, the impact of the climate variability on potential yield is significant, which indicates that the variability of global radiation is mainly responsible for yield variations under no-rain conditions. This assumption can also be explained by the fact that the distribution of yield is nearly normally distributed at full irrigation (see Fig. 2(d)). Second, the results of the irrigation model Daisy are reasonable for "light" deficit conditions. In contrast, the maize crop simulated by the Daisy model shows an unreasonable behaviour when no or less water is available for irrigation. Then, the crop reduces its photosynthesis, but shows no irreversible damage caused by drought stress and it falls immediately back to full transpiration when stress no longer occurs.

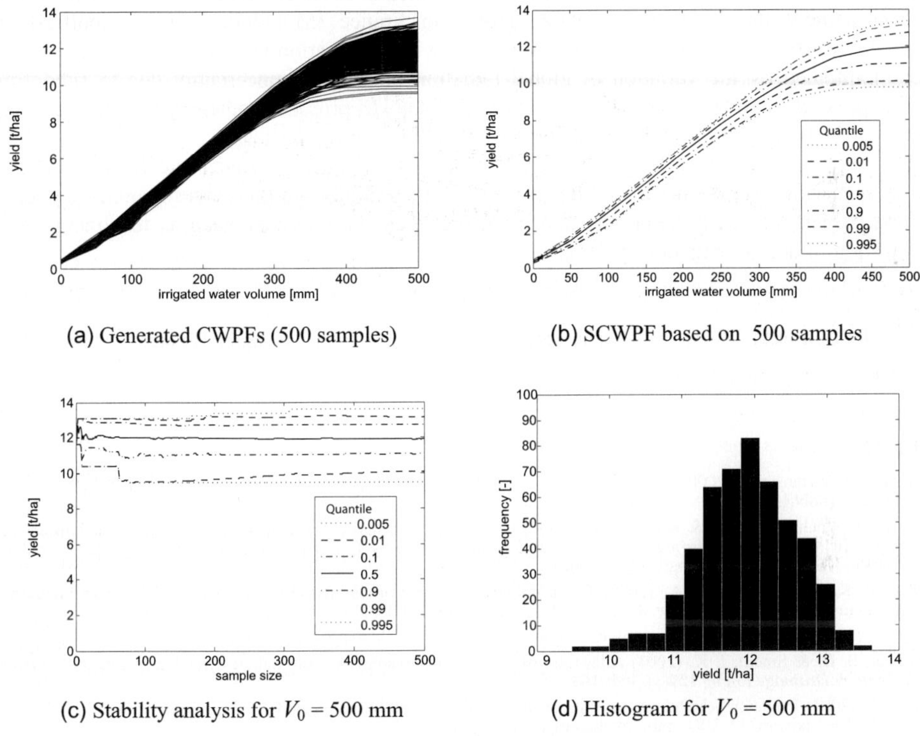

(a) Generated CWPFs (500 samples)

(b) SCWPF based on 500 samples

(c) Stability analysis for $V_0 = 500$ mm

(d) Histogram for $V_0 = 500$ mm

Fig. 2 Results for maize grown at Lavalette site (Montpellier, France).

Third, for optimal irrigation management only the upper part of the SCWPF-function is a subject of interest because the water productivity is decreasing in that range. From Fig. 2(c) it can be seen that the percentiles of yield converge with a sample size of 300, suggesting that samples containing 500 synthetic weather scenarios are adequate for estimating empirical percentiles of the yield distribution with the required precision. This is also supported by the results of Lawless & Semenov (2006), who investigated the impact of climate variability on wheat for different sites in Europe and New Zealand.

Figure 2(a) also reveals that the GET-OPTIS is able to find global solutions for the scheduling optimization problem because all the generated CWPF have a consistent shape, i.e. a monotonic increase in yield, which implies the convergence behaviour shown in Fig. 2(c). This demonstrates that the suggested stochastic framework generates reliable SCWPFs and can be used unconditionally to forecast the impact of climate variability for historical and future climate scenarios on potential yield.

CONCLUSIONS

We have presented a new stochastic framework for constructing daily site-specific stochastic crop water production functions (SCWPF), which are suitable for impact assessment of climate variability on potential yield using process-based models, a weather generator and state-of-the-art global optimization techniques. The tailor-made scheduling optimization algorithm introduced, GET-OPTIS, is the key tool of the stochastic framework because it has some unique features which allow generation of optimal schedules for given amounts of irrigation water in a reliable and computationally efficient way. This is attested by the results of an application example in irrigation practice, namely the SCWPF for maize grown in France (Montpellier) was calculated and statistically analysed. We demonstrated that there is a large variation in crop yield for maize which can be explained by the variation of global radiation because no uncertainty due to suboptimal scheduling is relevant. We could also show that GET-OPTIS provides consistent CWPFs using any arbitrary irrigation model, which makes it most suitable for doing this task.

Future work will focus on applications of the stochastic framework under uncertainty of climate change scenarios and/or soil hydraulic properties. In addition, further investigations are under progress, which include precipitation and more management activities, such as fertilizing, in the optimization of deficit irrigation systems.

Acknowledgements We wish to thank Jean Claude Mailhol from Cemagref (France) for the data he provided.

REFERENCES

Abrahamsen, P & Hansen, S (2000) Daisy: an open soil-crop-atmosphere system model. *Environ. Modelling & Software* **15**(3), 313–330. ISSN 1364-8152.

Brown, P. D., Cochrane, T. A. & Krom, T. D. (2006) Optimal onfarm multi-crop irrigation scheduling with limited water supply. In: *Computers in Agriculture and Natural Resources* (ed. by F. Zazueta, J. Kin, S. Ninomiya & G. Schiefer). Orlando, USA, 4th World Congress Conference.

Brumbelow, K. & Georgakakos, A. (2006) Determining crop-water production functions using yield-irrigation gradient algorithms. *Agric. Water Manage.* **87**(2), 151–161, JAN 24, ISSN 0378-3774. doi:10.1016/j.agwat.2006.06.016.

English, M. (1990) Deficit irrigation, I: Analytical framework. *J. Irrig. & Drainage Engng* **16**(3), 399–412.

Gorantiwar, S. D. & Smout, I. K. (2003) Allocation of scarce water resources using deficit irrigation in rotational systems. *J. Irrig. & Drainage Engng* **129**(3), 155–163.

Jones, H. G. (2004) Irrigation scheduling: advantages and pitfalls of plant-based methods. *J. Exp. Botany* **55**(407), 2427–2436, doi: 10.1093/jxb/erh213. URL http://dx.doi.org/10. 1093/jxb/erh213.

Lawless, C. & Semenov, M. A. (2006) Assessing lead-time for predicting wheat growth using a crop simulation model. *Agric. Forest Met.* **135**:302–3 13, doi:10. 1016/j .agrformet.2006.01 .002.

Loganathan, G. & Elango, K. (2004) Revisiting optimal water-allocation under deficient supply. In: *Assessment & Management of Water Resources – AMWR 2004* (ed. by M. Kumar & M. Sekhar). Department of Civil Engineering, Indian Institute of Science, Bangalore, India,

Raghuwanshi, M. S. & Wallender, W. W. (1997) Economic optimization of furrow irrigation. *J. Irrig. & Drainage Engng* **5**, 377–385.

Scheierling, S. M., Cardon, G. E. & Young, R. A. (1997) Impact of irrigation timing on simulated water crop production functions. *Irrigation Sci. 18*(1), 23–31, doi:10.1 007/s0027 10050041.

Schmitz, G. H., Woehling, T., de Paly, M. & Schütze, N. (2007) Gain-p: A new strategy to increase furrow irrigation efficiency. *Arabian J. Science & Engineering* **32**(1C), 103–114.

Semenov, M. A. (2007) Development of high-resolution UKCIP02-based climate change scenarios in the UK. *Agric. Forest Met.* **144**(1-2), 127–138, ISSN 0168-1923. doi:10.1016/j.agrformet.2007.02.003.

Semenov, M. A., Brooks, R. J., Barrow, E. M. & Richardson, C. W. (1998) Comparison of the WGEN and LARS-WG stochastic weather generators for diverse climates. *Climate Research* **10**(2), 95–107.

Shang, S. & Mao, X. (2006) Application of a simulation based optimization model for winter wheat irrigation scheduling in north china. *Agric. Water Manage.* **85**(3), 314–322.

Shang, S., Li, X., Mao, X. & Lei, Z. D. (2004) Simulation of water dynamics and irrigation scheduling for winter wheat and maize in seasonal frost areas. *Agric. Water Manage.* **68**(2), 117–133.

Singh, R. & Singh, J. (1997) Irrigation planning in wheat (*Triticum aestivum*) under deep water table conditions through simulation modelling. *Agric. Water Manage.* **33**(1), 19–29.

Woehling, Th. & Mailhol, J. C. (2007) Physically based coupled model for simulating 1D surface-2D subsurface flow and plant water uptake in irrigation furrows. *J. Irrig. & Drainage Engng* **33**(6), 548–558.

Modelling of water flow and solute transport in saturated–unsaturated media using a self adapting mesh

ABDELLATIF MASLOUHI[1], HASSAN LEMACHA[1] & MOUMTAZ RAZACK[2]

1 *Laboratoire Interdisciplinaire en Ressources Naturelles et en Environnement, Université Ibn Tofail, Faculté des Sciences de Kenitra, BP133, Morocco*
maslouhi_a@yahoo.com

2 *Dept of Hydrogeology UMR 6532, University of Poitiers, 40 Avenue du Recteur Pineau, F-86022 Poitiers Cedex, France*

Abstract In this work, a mathematical modelling tool was developed to simulate the water and solute transfers in unsaturated–saturated porous medium. We used a mathematical formulation which consists in considering as a single continuum the saturated and the unsaturated zones. The model is based on the finite elements method using the Freefem++ code. This code was adapted to the equations used in this study, namely: Richards equation to study the water flow in the unsaturated zone, the diffusivity equation for the groundwater flow and the transport equation of advection–dispersion type to study the solute transfer. The advantage of the Freefem++ is that it generates self-adapting grids. The modelling tool was validated using experimental data measured on the small-scale physical model. A comparative study with the ADI method was carried out and showed that the use of the finite elements Freefem++ code presents an advantage and provides more accurate results.

Key words modelling; unsaturated–saturated; flow; solute; finite elements; self-adapting grids

1 INTRODUCTION

This paper is focused on the mathematical modelling of the problem related to water flow and solute transport in the unsaturated and saturated zones of unconfined aquifers. Water and solute transfers in a porous unsaturated–saturated environment constitute a difficult problem to tackle because of the nature of the equations governing the transfers between the saturated zone and the unsaturated zone. The difficulty of solving the coupling of the vertical flow in the unsaturated zone and the horizontal flow within the groundwater is linked to the fact that the interface between these two fields (i.e. the water table) is not known *a priori*, and that it takes variable positions with time (Lemacha *et al.*, 2003).

Study of water and solutes transfer in the unsaturated and saturated zones has quite often been carried out by using distinct mathematical models, which are applied separately in each zone (Dupuy *et al.*, 1997a,b). This kind of approach is based on a technique of artificial coupling connecting the two zones, which often generate numerical disturbances at the unsaturated–saturated interface. Diaw *et al.* (2001) performed a study of one-dimensional water flow through the unsaturated–saturated zone. Khanji (1975) developed a steady flow model through the unsaturated–saturated zone constituted by fine sands. In our approach, we developed a mathematical model based on a single flow equation that can be used for both unsaturated and saturated zones which are regarded as a single continuum. The diffusivity equation will be used in a nonlinear form and in a linearized form. Two numerical resolution methods (finite elements method and finite differences method) are used afterwards to solve the mathematical model and their results compared.

The water and solute transfer equations are solved first using the alternating directions implicit method (ADI). This method undertakes a sweep of the discretized domain column by column, then line by line, alternately. The name, method of alternating directions, comes from this procedure. The method offers the advantage of leading to the resolution of linear systems whose matrices are tri-diagonal bands. An implicit scheme with the alternating directions is used to solve the flow equation, and a completely implicit scheme for the resolution of the transport problem.

The water and solute transfer mathematical model is also solved with the use of the finite elements method (FEM). The software used is Freefem++ (Hecht *et al.*, 2003), which is available online and offers the possibility to self-adapt the finite element grid according to the complexity of the problem.

The numerical simulations enabled us to follow the evolution in time of the aquifer water table in unsteady state. The simulations results are validated using experimental results provided in the literature (Khanji, 1975). This work also allowed us to analyse the effect of the self-adapting grid on the accuracy of the results obtained by the finite element method.

2 STUDY DOMAIN

Figure 1 shows the geometry of the study domain. The limits of the flow domain are:

- A horizontal upper surface (ground surface); to part of this surface an initial flow q_0 is applied, causing the infiltration. The infiltration strip can simulate either an irrigation canal, or an artificial recharge reservoir.
- Two tanks located at the same distance from the axis of the infiltration strip in which a water level is imposed, limit the groundwater. Each tank can simulate a drainage ditch.

The problem is thus a plan problem. We consider the system of axes XOZ, axis OX being assimilated with the ground surface, axis OZ being oriented positively towards the bottom and the origin O being in the centre of the ground section of width $2L$. The problem is thus symmetrical compared to OZ. The transfer problem will accordingly be developed considering only one half of the domain.

On the basis of a horizontal unconfined groundwater with a thickness E (m) above an impermeable bottom and a depth e_0, a constant infiltration flow q_0 (m/s) is applied to the ground surface on a strip of $2L_0$ (m) width and infinite length. It is supposed, moreover, that the infiltration strip is centred between the two parallel tanks distant of $2L$, penetrating down to the impermeable layer. A constant piezometric level is maintained in these tanks at a constant depth e_0 (m)

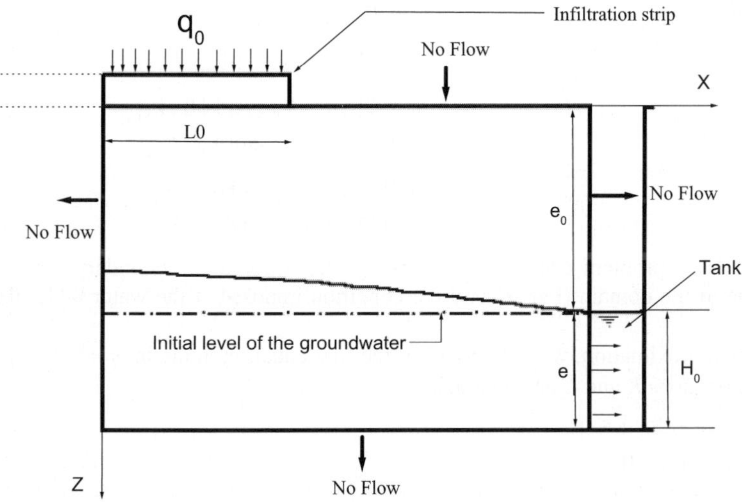

Fig. 1 Schematization of the study domain.

3 MATHEMATICAL MODEL

3.1 Water flow

The flow equation in the unsaturated zone is obtained by coupling the generalized form of Darcy's law which expresses the proportionality law of the flows to the continuity equation (Hillel, 1980). It is assumed that both unsaturated and saturated zones form a medium which can be regarded as only one continuum. The equation used for the two zones has the form of the Richards equation:

$$C\frac{\partial h}{\partial t} = div\left[Kgrad\ (h - z)\right] \tag{1}$$

where C and K are two functions depending on the nature of the medium, h is the water pressure head (m), z is the vertical coordinate (m).

In the unsaturated zone, $C = C(h)$ is the capillary capacity (m^{-1}) and $K = K(h)$ is the unsaturated hydraulic conductivity (m/s). In this case, the flow equation is the classical Richards equation (Richards, 1931):

$$C(h)\frac{\partial h}{\partial t} = div\left[K(h)grad\ (h - z)\right] \tag{2}$$

In the saturated zone, $C = S_S$ is the specific storage coefficient (L^{-1}) and $K = K_s$ is the saturated hydraulic conductivity (m/s). One obtains the diffusivity equation governing the unsteady flow in a saturated medium. This diffusivity equation has a linear form:

$$S\frac{\partial h}{\partial t} = \left(\frac{\partial^2 h}{\partial x^2} + \frac{\partial^2 h}{\partial z^2}\right) \qquad \text{(where } S = \frac{S_S}{K_s}\text{)} \tag{3}$$

When the nonlinear diffusivity equation is assembled with the initial and boundary conditions of the study domain, the following system of equations is obtained:

$$\begin{cases} C(h)\dfrac{\partial h}{\partial t} = \dfrac{\partial}{\partial x}\left(K(h)\dfrac{\partial h}{\partial x}\right) + \dfrac{\partial}{\partial z}\left(K(h)\left(\dfrac{\partial h}{\partial z} - 1\right)\right) & \text{in }]0,T[\ ;\ ]0,X_{max}[\text{ and }]0,Z_0[\\ S\dfrac{\partial h}{\partial t} = \left(\dfrac{\partial^2 h}{\partial x^2} + \dfrac{\partial^2 h}{\partial z^2}\right) & \text{in }]0,T[\ ;]0,X_{max}[\text{ and }]Z_0,Z_{max}[\\ q_0 = K(h(x,0,t))\left(1 - \dfrac{\partial h}{\partial z}\right) & \text{in}]0,T[\ ;\ ]0,X_1\ [\text{ and } z = 0 \\ q_0 = 0 & \text{in}]0,T[\ ;\ ]\ X_1, X_{max}\ [\text{ and } z = 0 \\ q_0 = 0 & \text{in }]0,T[\ ;\ x = X_{max} \text{ and }]\ 0\ ,\ Z_0[\\ h(x,z,t) = z - Z_0 & \text{in}]0,T[\ ;x = X_{max} \text{ and }]\ Z_0,Z_{max}[\\ q_0 = 0 & \text{in }]0,T[\ ;\ x = 0 \text{ and }]0,Z_{max}[\\ q_0 = 0 & \text{in }]0,T[;]0,\ X_{max}[\text{ and } z = Z_{max} \end{cases} \tag{4}$$

Therefore, the formulated problem can be solved taking into account the boundary conditions imposed at the limits of the domain (Fig. 1) and the condition imposed at the water table, that is $h = 0$.

In the numerical approximation, we only consider the first equation in the interval $]0,X_{max}[\ \text{x}\ ]0,Z_{max}[$, where the functions K and C are such that:

$$K = K(h) \text{ and } C = C(h) \text{ for } h < 0 \tag{5}$$

$$K = K_s \text{ and } C = S_S \text{ for } h > 0. \tag{6}$$

3.2 Mass transport

The system of equations shown in equation (7) (opposite) is used.

$$\begin{cases} \theta \dfrac{\partial C}{\partial t} = \dfrac{\partial}{\partial x}\left(\alpha q_x \dfrac{\partial C}{\partial x}\right) - \dfrac{\partial}{\partial x}(qC) + \dfrac{\partial}{\partial z}\left(\alpha q_z \dfrac{\partial C}{\partial z}\right) - \dfrac{\partial}{\partial z}(qC) & \text{in}]0,T[\,;]0,X_{max}[\text{and}]0,Z_0[\\ \theta_s \dfrac{\partial C}{\partial t} = \dfrac{\partial}{\partial x}\left(\alpha q_x \dfrac{\partial C}{\partial x}\right) - \dfrac{\partial}{\partial x}(qC) + \dfrac{\partial}{\partial z}\left(\alpha q_z \dfrac{\partial C}{\partial z}\right) - \dfrac{\partial}{\partial z}(qC) & \text{in}]0,T[;]0,X_{max}[;]Z_0,Z_{max}[\\ C = C_1 & \text{in}]0,T[\,;\,]0,X_1\,[\text{ and } z=0 \\ \Psi_{0z} = 0 & \text{in}]0,T[\,;\,]\,X_1, X_{max}\,[\text{ and } z=0 \\ \Psi_{0x} = 0 & \text{in }]0,T[\,;x = X_{max}\text{ and }]\,0\,,Z_0[\\ \dfrac{\partial C}{\partial x} = 0 & \text{in}]0,T[\,;x=X_{max}\text{ and }]\,Z_0,Z_{max}[\\ \Psi_{0x} = 0 & \text{in }]0,T[\,;\,x=0\text{ and }]0,Z_{max}[\\ \Psi_{0z} = 0 & \text{in }]0,T[;]0,\,X_{max}[\text{ and } z=Z_{max} \end{cases} \quad (7)$$

with C the concentration of the pollutant (mol/L); θ is water content (m^3/m^3); α is dispersivity (m).

4 NUMERICAL MODELLING – FINITE ELEMENT METHOD

In the field of water flow in porous media, the application of the finite element method is rather recent. The advantage of the "concurrent" method of finite differences lies in a greater simplicity of the numerical schemes which leads to lower computing times. However, the finite element methods show greater potential when dealing with the approximations of equations in the vicinity of curved borders on the one hand, and the expressions of numerical schemes of higher degree on the other hand, which represents two advantages over the finite difference methods (Raviart, 1981).

To solve the flow and transport equations by the finite element method, the FreeFem++ (Hecht *et al.*, 2003) software has been used. In order to run this software, a numerical code was developed. This code enables us to define the borders of the domain, to set the number of nodes on each border. The finite element grid is then automatically obtained. The spaces and the degrees of the polynomials are defined and the variational formulations of the equations are posed with the initial and boundary conditions. A time loop and a loop to refine the triangulation at zones sensitive to flow and transport are introduced. Such sensitive zones are found under the infiltration strip and at the level of the capillary fringe.

5 RESULTS AND DISCUSSION

Concerning the numerical model, the infiltration code is adapted to the type of tests which one wants to simulate (infiltration under constant head or with constant flow). All the results obtained with this code (field of pressure head or of water content) can be used to obtain volumes, flows, etc. The pressure head field corresponding to the last step of calculation in the infiltration module is taken again as the initial condition of the recharge module. A treatment unit in this module also makes it possible to obtain the evolution of the water table, the volumes, etc. Thus the numerical model is able to compute, at any time, the pressure head and water content fields and the flows.

To validate the model, experimental data from Khanji (1975) were used. These data relate to fine sand. The parameters of this material and other data used in the model are as follows: X_{max} (length of the study domain) = 300 cm; Z_{max} (depth of the study domain) = 200 cm; q_0 (constant infiltration rate) = 15 cm/h; H_0 (imposed hydraulic head) = 135 cm; K_s (saturated hydraulic conductivity) = 35 cm/h; θ_s (saturated water content) = 0.3 $cm^3\ cm^{-3}$; h_g (scale parameter of water pressure head) =30 cm; ω (effective porosity) = 29 10^{-2}; α (dispersivity) = 10 cm; C_1 (concentration) = 1 mol/L; $S = 8.2\ 10^{-5}$.

Taking into account the heterogeneity of the soil in space, there is at certain vertical positions, a slight difference between the values of the water contents, since it is assumed in our model that the soil is homogeneous and is characterized by the relation θ(h) which is the average of the suction curves measured on each vertical.

5.1 Hydrodynamic properties of the soil in this study

The parameters of the soil are those of the Mnasra zone in Morocco (soil with sandy loam texture with a sandy tendency). The hydrodynamic characteristics of this type of soil are given by Saâdi & Maslouhi (2003) using the relations of Brooks & Corey (1964) and Van Genuchten (1980).

$$K(\theta)=\frac{K_s A}{A+\left[\alpha\left(\frac{\theta_s-\theta}{\theta-\theta_r}\right)\right]^{\delta}} \tag{8}$$

$$\theta(h)=\theta_r+\frac{\alpha(\theta_s-\theta_r)}{\alpha|h|^{\gamma}} \tag{9}$$

The relation between K and h is given by:

$$K(h)=K_s\frac{A}{A+|h|^{\beta}} \quad \left(\text{where } \delta=\frac{\beta}{\gamma}\right) \tag{10}$$

5.2 Water transfer

Figure 2 shows the temporal evolution of the water table profiles. We note that there is a rise of the water table in time. This rising is due primarily to the recharge of the groundwater from the unsaturated zone. We also note that the water table rise still continues after 4 hours. This reflects the flow regime, which is unsteady due to the fact that flow entering the domain by the infiltration strip is greater than that outgoing through the outlet. This will create a head gradient in the saturated zone between the points located beneath the source zone and those close to the fixed head boundary.

Figure 2 indicates a good agreement between the positions of measured and calculated water table profiles during all the recharge periods.

The results obtained by the finite element method (FEM) are closer to the measured values than those obtained by the finite differences method (FDM). This is due primarily to the automatic refinement of the triangular grid at the level of the capillary fringe (Fig. 3) which increases the accuracy and provides a better simulation of the water table.

The comparison between the numerical and experimental results shows that the response of the numerical model developed here is quite satisfactory. Indeed, the model manages to correctly represent the physical processes, if the initial and boundary conditions are expressed on known and

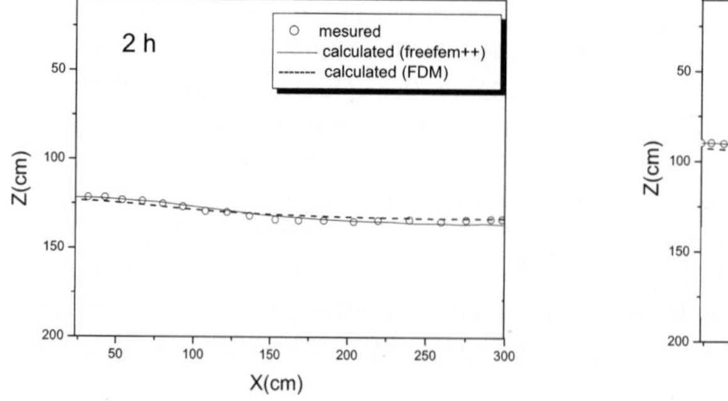

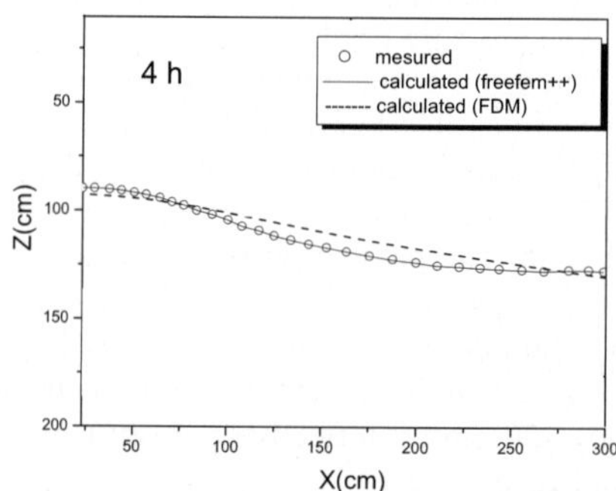

Fig. 2 Comparison between the water-table profiles simulated (finite elements and finite differences) and measured, at times $t = 2$h and 4h.

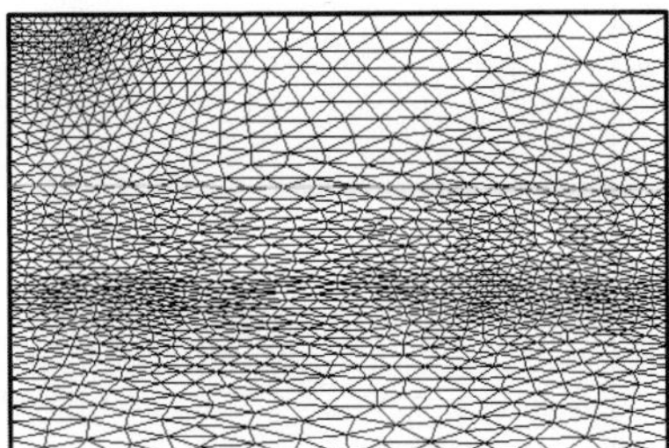

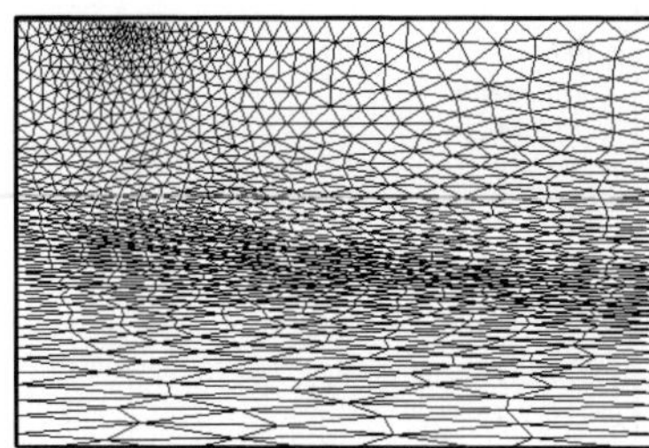

Fig. 3 Self-adapting finite element grid at the level of the wetting front and of the capillary fringe, at the initial state ($t = 0$) and at time $t = 160$ min.

fixed-in-time geometric limits. The mesh adaptivity is performed by computing the space error indicators, refining the mesh where they are larger than the mean value and possibly coarsening the mesh where they are much smaller. Mesh adaptivity based on *a posteriori* error estimation techniques have become an indispensable tool in large-scale scientific computation and much work has been done on finite element discretization of elliptic problems. The main results consist in exhibiting local error indicators which can be computed explicitly as a function of the discrete solution and the data. They are built from the residual of the strong equation and the jumps across the inter-element of the fluxes. Since they are local they provide a good representation of the error distribution, and so they are very efficient tools for mesh adaptivity.

5.3 Solute transport

The study of the solute transport was performed by assuming that water coming from the infiltration strip contains a solute concentration of one mole per litre. It is also assumed that at initial state, the flow domain is characterized by a null concentration. During the recharge phase of the groundwater, a hydraulic gradient is generated between points located beneath the source zone and those close to the fixed head boundary. This involves a water flow towards the groundwater outlet and consequently the movement of the solute.

To study the effect of the self-adapting grid on the quality of the results obtained by the finite element method, Fig. 4 shows the solute concentration contours at various times. The curves illustrating the solute concentration distribution computed without refinement of the grid show some disturbances, whereas those obtained by adaptation of the grid have a very smooth form. This difference is due to the fact that refinement of the grid provides much more accurate results.

6 CONCLUSION

In this paper, we have proposed a mathematical model for the water flow and solute transport in a porous unsaturated–saturated zone. The comparison between the numerical and experimental results shows the positive response of the numerical unstationary model developed in this study to better represent the physical phenomena, when the initial and boundary conditions are expressed on known and fixed-in-time geometric borders. This model allows the simulation of water flow and solute transport in the unsaturated–saturated zone. The validity of the results simulated by this model is checked using experimental laboratory data. Moreover, the comparison between the finite difference scheme and the adaptive finite element approximation used confirms the robustness of the latter method.

The results presented here were obtained at the laboratory scale. It is essential to envisage their extension to the field scale. Passing to the land parcel scale, then to the catchment scale represents the logical extension to this research. Although the mathematical model was designed to allow its application to problems of larger space scales, the main impediment comes from the

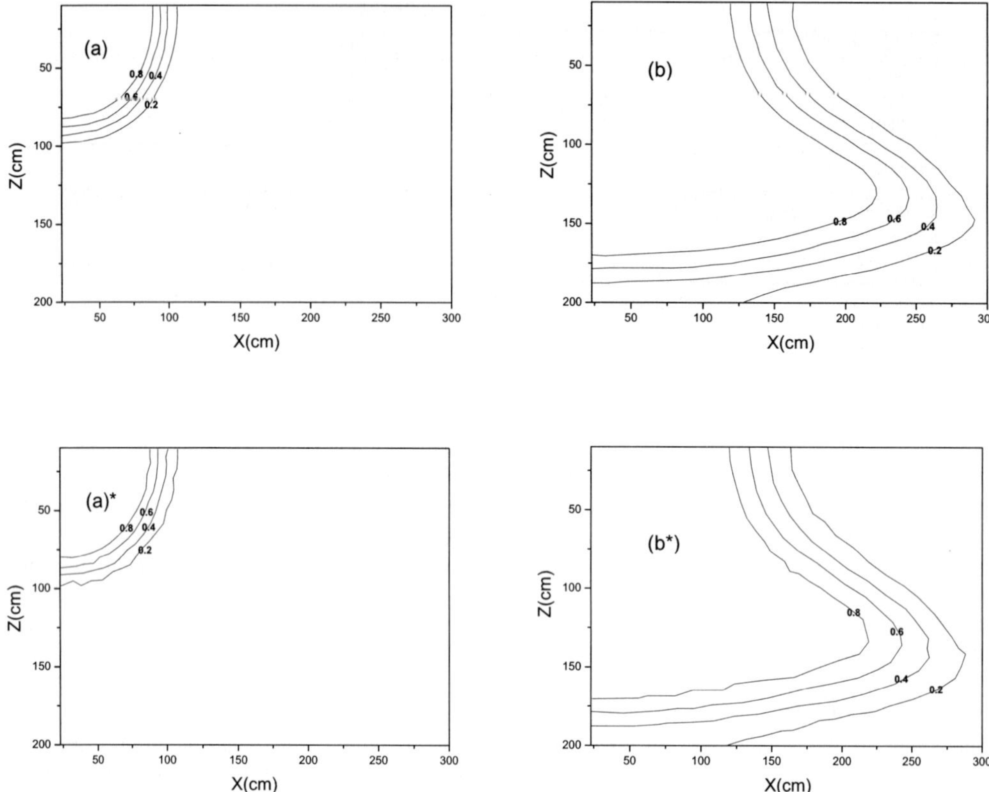

Fig. 4 Distribution in space and time of the solute concentration contours. (a) and (b): with self-adapting grid; (a*) and (b*): without adaptation of the grid.

difficulty in obtaining in the field the soil characteristic relations $K(h)$ and $\theta(h)$ and their strong space variability.

Moreover, moving to the field very often implies the presence of significant heterogeneity, the existence of anisotropy and heat gradients, which indisputably restrict, in certain cases, the applications of results obtained in the laboratory. The presence of a more or less dense vegetation cover also modifies the flow transfer and affects the boundary conditions at the soil surface. These various phenomena will be taken into account in coming modelling work, based on their physical analysis.

Acknowledgements The authors are thankful to the Integrated Actions Morocco-France AI no. 00/221/STU and the project PROTARS III no. D45/06 for providing funding and grants for this research programme.

REFERENCES

Brooks, R. H. & Corey, A T. (1964) Hydraulic properties of porous media. *Hydrol. Paper 3*. Colorado. State Univ., Fort Collins, Colorado, USA.

Diaw, E. B., Lehmann, F. & Ackerer, Ph. (2001) One-dimensional simulation of solute transfer in saturated-unsaturated porous media using the discontinuous finite elements method. *J. Contaminant Hydrol.* **51**, 197–213.

Dupuy, A., Banton, O. & Razack, M. (1997) Contamination nitratée des eaux souterraines d'un bassin versant agricole hétérogène –1. Evaluation des apports à la nappe. *Rev. Sci. Eau* **10**, 23–40.

Dupuy, A., Razack, M. & Banton, O. (1997) Contamination nitratée des eaux souterraines d'un bassin versant agricole hétérogène-2. Evolution des concentrations dans la nappe. *Rev. Sci. Eau* **10**,185–198

Hecht, F., Pironneau, O., Le Hayric, A. & Ohtsuka, K. (2003) FreeFem++. Available at: http://www.freefem.org/.

Hillel, J. (1980) *Fundamentals of Soil Physics*, 220–230. Academic Press, San Diego, USA.

Khanji, D. (1975) Etude de la recharge de nappes à surface libre par infiltration. Thèse Doctorat ès Sciences Physiques, Grenoble, France.

Lemacha, H., Maslouhi, A. & Razack, M. (2003) Modélisation Bidimensionnelle de l'Ecoulement de l'Eau et du Transport de Soluté dans un milieu poreux non saturé-saturé. Sixième Congrès de Mécanique, Tanger, Morocco.

Raviart, P.-A. (1981) *Les méthodes d'éléments finis en mécanique des fluides*. Ed. Eyrolles, Paris, France.

Saâdi, Z. & Maslouhi, A. (2003) Modeling nitrogen dynamics in unsaturated soils for evaluating nitrate contamination of the Mnasra groundwater. *Adv. Environ. Res.* **7**, 803–823.

Van Genuchten, M. Th. (1980) A closed-form equation for predicting the hydraulic conductivity of unsaturated soils. *Soil Sci. Am. J.* **44**, 892–898.

Automatic mapping of a reservoir flood risk map using GIS technology

WANG JUN, LIANG ZHONGMIN & SHI YE

College of College of Hydrology and Water Resource, Hohai University, PO 456, Nanjing 210098, China
wangjun.hhu@gmail.com; wfmingyu@hhu.edu.cn

Abstract A flood risk map is an important non-engineering measure for flood control and management. To draw a flood risk map and design a page layout format for it automatically, a two-dimensional hydraulic model was used to analyse the flood risk of a reservoir when a dam-break is encountered. An automatic method to draw a flood risk map based on GIS technology was used. Meanwhile, taking Shilianghe Reservoir in Jiangsu Province, China, as an example, its flood risk maps were mapped. The results obtained could be used as technical support for flood prevention schemes and organizing the local population to escape from flood disaster.

Key words flood risk map; reservoir; GIS; dam-break flood

INTRODUCTION

Flood risk maps are one kind of thematic map which are used to reflect the risk information of a certain region when floods are encountered (Chinese Office of State Flood Control and Drought Relief Headquarters, 2005). With the development of concepts on water-control of the concision-making department of China, Zhang (2005) proposed that the water-control strategy gradually changed from flood control and prevention, to flood management. As one of the most important means of flood management, flood risk maps and how to generate automatic maps, are gradually being paid more attention. Meanwhile, they play a more and more important role in flood management.

GIS (geography information system) is a rising modern computer technology (ESRI CORPORATION, 2004). It combines the technology of a database, software project, artificial intelligence, network, etc. Li (2002), Dang *et al.* (2003) and Wang *et al.* (2006) agree that due to its powerful ability of spatial analysis, map display, map operation, etc., GIS is widely applied in many fields. It should be emphasized that GIS also pays a crucial role in automatic mapping of flood risk maps. This article shows the process of automatic mapping of flood risk maps.

According to the definition of a flood risk map, it is clear to us that flood information is the basis of a flood risk map, so flood risk analysis is the first step in drawing a flood risk map. Based on the results of flood risk analysis, it manages to automatically generate a flood risk map with interfaces provided by AO (ArcObjects), which is a GIS components kit developed by the ESRI (Environment System Research Institution of America).

FLOOD RISK ANALYSIS

Flood risk analysis of a reservoir can be divided into two sorts, according to the analysis objects: one is flood risk analysis of the region of reservoir, and the other is flood risk analysis of dam-break. There are many methods to analyse flood risk, e.g. hydrological and hydraulic methods. In this article a 2-D hydraulic model was used to analyse the dam-break flood risk because of its powerful function of simulating flood movement. The 2-D hydraulics model of Tang & Xia (2004) simulates flood movement horizontally. It is composed of a successive equation (mass-conservation equation) and momentum equation in the directions x and y; the equations are as follows:

$$\frac{\partial z}{\partial t}+\frac{\partial (uh)}{\partial x}+\frac{\partial (vh)}{\partial y}=0 \qquad (1)$$

$$\frac{\partial u}{\partial t}+u\frac{\partial u}{\partial x}+v\frac{\partial u}{\partial y}+g\frac{\partial z}{\partial x}+g\frac{n^2u\sqrt{u^2+v^2}}{h^{4/3}}=0 \tag{2}$$

$$\frac{\partial v}{\partial t}+u\frac{\partial v}{\partial x}+v\frac{\partial v}{\partial y}+g\frac{\partial z}{\partial y}+g\frac{n^2v\sqrt{u^2+v^2}}{h^{4/3}}=0 \tag{3}$$

where, t denotes time step (unit is s); n is roughness coefficient; u is flow velocity in x direction while v is flow velocity in y direction (unit is m/s).

$$g\frac{n^2u\sqrt{u^2+v^2}}{h^{4/3}}$$

represents current resistance in the x direction while:

$$g\frac{n^2v\sqrt{u^2+v^2}}{h^{4/3}}$$

does so in the y direction; z,h, respectively, represent water level and water depth at the point (x,y). The steps of construction and simulation of the 2-D hydraulic model are shown in Fig. 1.

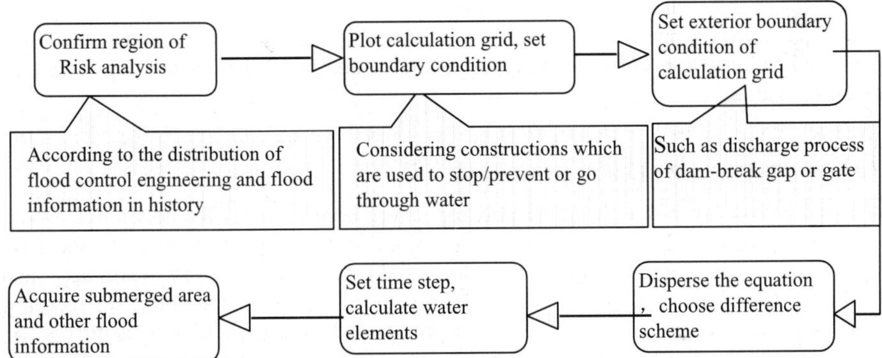

Fig. 1 The common steps to solving the 2-D hydraulic model.

Some of steps shown in the top of Fig. 1 can be automatically run by programming based on GIS technology, and after running the 2-D hydraulic model, hydrological elements in the flood detention area ,such as water depth, submerged area, flow velocity, arrival time of flood, etc., can be obtained.

AUTOMATIC DRAWING OF FLOOD RISK MAP

Different kinds of hydrological elements such as water depth, flow velocity, submerged area, etc. could be obtained after running the 2-D hydraulic model, and on the basis of these simulation results, the flood risk map can be mapped automatically with GIS technology. Today, there are several popular GIS products worldwide, e.g. MapInfo, developed by the MapInfo Company and MapGIS, developed by Wuhai University. The most popular GIS product is ArcGIS, provided by ESRI. As the leading GIS product, the ArcGIS family provides components of the technology to redevelop with AO. The principle of automatic mapping flood risk map is to colour the calculated grid layer according to its attribution value. The specific route of redevelopment was as follows:

(1) Creating a new layer. Utilize Ilayer interface with AO to create a new flood risk layer; this layer is a blank layer which is used to plot into the calculation grid mentioned in the above paragraphs.

(2) Plotting risk layer into grids. Plot the new flood risk layer into the calculation grid, and then a calculation grid layer, of which the attribution table used to store the simulation results of the 2-D hydraulic model, is obtained. In this step, taking the construction which could stop water, prevent water and go through by water into consideration is the key.
(3) Importing risk information into attribution table. Import the simulation results of the 2-D hydraulic model into the attribution table of the flood risk layer with Itable interface. It needs to be pointed out that the simulation results include several kinds of flood risk information such as the largest depth, the largest velocity, the arrival time of flood and different hydrological elements of every moment in the whole process of flood movement.
(4) Recolour the layer according to a designed colour scheme. According to different flood risk map, utilize the IRenderer interface, which is used to describe layer's colour and the IcolorRamp interface, which is used to design the colour scheme to recolour the flood risk layer according to the corresponding field value of hydrological elements in the attribution table.
(5) Overlay basic layers. Add some other basic layer to the flood risk layer, and then a complete flood risk map is finished. The map obtained is the flood risk map of corresponding hydrological element.
(6) Flood risk layer Post-processing. To achieve a better visual effect, it is necessary to modify the flood risk layer by smoothing or merging.
(7) Designing page layout format. A complete flood risk map provided to users should be of a good template and include some information expressed in other formats such as tables, charts, pictures, etc. In the template, a north arrow, legend, scale bar, scale text, map name and some tables with information corresponding to flood risk are inserted into the map layout in the designed location.

POST-PROCESSING OF FLOOD RISK MAP

Since the grid of risk layer is an irregular polygon, every border line of the polygon is a polyline and it causes the edge of the region of the same colour to be sharp. It seems different from the condition of real–submerged. In order to make the visual effect of the flood risk layer approach the real condition, the flood risk layer should be processed through smoothing or merging. A smoothing model has been designed (Fig. 2). In this model, five tools play a main role, so the whole process could be divided into five steps. The specific function of these tools and steps are as follows: (1) the first tool called Times is designed to enlarge the each cell value of the input raster; (2) the second tool called INT is designed to convert each cell value of output raster, which is a result of a previous step, to integer by truncation; (3) the third tool called Raster to TIN is designed to convert the output Raster (2) into TIN; (4) the fourth tool called TIN to Raster is designed to convert output TIN to Raster of smaller cell size; (5) the fifth tool called Reclassify is designed to convert the cell value to an appointed value. Output Raster (4) could convert to SHP, which is a file format of vector, and the SHP is the smooth layer that we want to get.

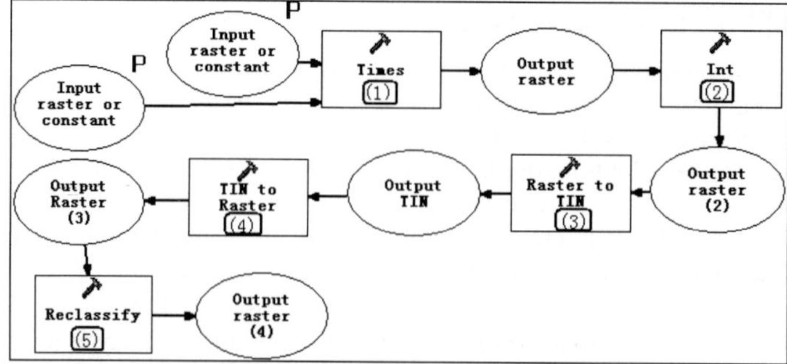

Fig. 2 The model of smoothing flood risk layer.

EXAMPLE

Introduction of study region

ShiLianghe Reservoir is the biggest reservoir in JiangSu province in China. There are several cities distributed in its downstream region. If the dam of ShiLianghe Reservoir breaks, it will threaten the safety of downstream resident and property, so it is clear that drawing a flood risk map of ShiLianghe Reservoir is useful.

Flood risk analysis and flood risk mapping

Based on the history of flooding and construction, the risk region was confirmed, to create a new layer according to the risk region, and then plot the layer into a grid of cell size 500 m × 500 m by Gambit, which is a Professional Software for defining a grid model for fluid simulation. The calculation grid is shown in Fig. 3. A 2-D hydraulic model imports the simulation results into an attribution table of the risk layer and recolours the flood risk layer according to a corresponding colour scheme, to produce an elementary flood risk map. By taking the flood risk map of the max depth distribution map as an example, the elementary flood risk map is shown in Fig. 4. It is obvious that the edge of every colour grade is sharp, so it is necessary to post-process the flood risk layer to achieve a better visual effect.

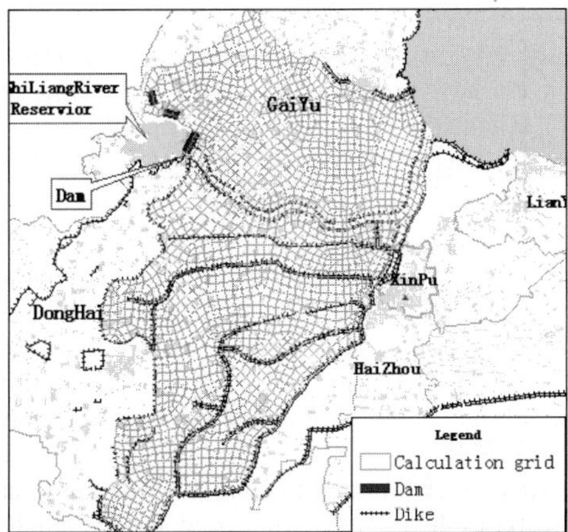

Fig. 3 Schematic diagram calculation grids.

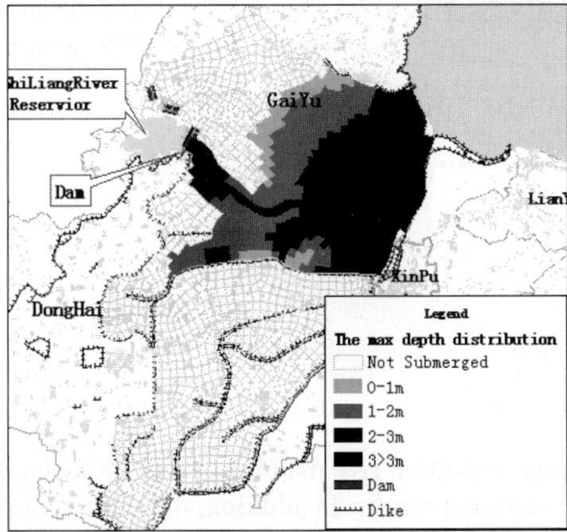

Fig.4 The initial max depth distribution map.

Flood risk map post-processing

According to the technology route shown in Fig. 2, the post-process of the flood risk map obtained from the last step was used to find a smoothing flood risk layer (shown as Fig. 5).

Page layout of flood risk map designing

After post-processing the flood risk map, design the map layout template, insert map elements. As Fig. 6 shows, (I) denotes north arrow, (II) denotes map name, (III) denotes information expressed in table or picture, (IV) denotes legend, and (V) denotes scale bar. Figure 6 is the complete flood risk map, in this example it is the max depth distribution map.

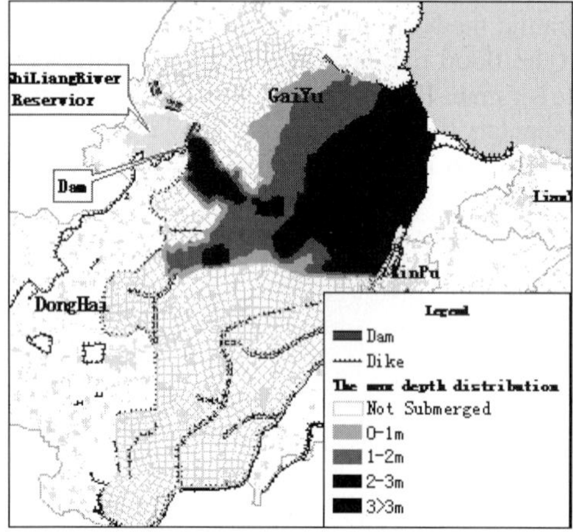

Fig. 5 The max depth distribution of post-processing.

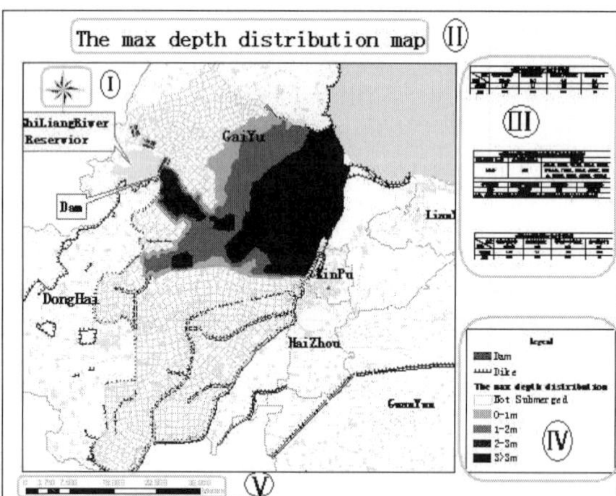

Fig. 6 The complete flood risk map.

CONCLUSIONS

This article presents methods to automatically generate a flood risk map based on GIS technology and the specific principle and steps to achieve the expected results. In addition, the method of post-processing the elementary risk layer was referred to and an example was given and a good visual effect was achieved. It indicates that the methods suggested in this article are reasonable, and the flood risk map obtained can provide a guide to the local flood control and prevention department, leading to the escape of residents during a flood and decreasing the disaster damage caused by the flood.

Although the method proposed in this article is feasible, there are still some aspects which need improvement. For example, importing flood risk information of flood simulation into the attribution table is not automatic. Since the number of results from the 2-D hydraulic model is large, it will take a long time to save and import data into the attribution table. It is therefore difficult to provide a guide for real-time flood management. In future, the authors will do more work to solve this problem.

Acknowledgements This study was supported by the National Basic Research Program of China (also called 973 Program, grant no. 2007CB714104), the Innovation Fund of JiangSu province of China (CX08B_105Z), the National Natural Science Foundation of China (grant no. 50779013), and the Specialized Research Fund for the Doctoral Program of Higher Education (grant no. 20070294018).

REFERENCES

Chinese Office of State Flood Control and Drought Relief Headquarters (2005) Guide of mapping flood risk map..

Dang, Anrong, Jia, Haifeng, *et al.* (2003) *ArcGIS 8 Desktop Application Guide.* Tsinghua Press, Beijing, China.

ESRI Corporation (2004) ArcGIS9 — Get starting with ArcGIS http://www.urbanecology.washington.edu/GIS_Tutorial/Getting_Started_with_ArcGIS.pdf

Li, Na (2002) *GIS-based Flood Risk Management System.* China Institute of Water Resources and Hydropower Research, Beijing, China.

Tang, Yongxiang & Xia, Yufeng (2004) Application 2-D flood analysis of dam-break to mapping flood risk map. *Zhejiang Hydrotechnics* **32–33**.

Wang, Jing, Wang, Jun & Liang, Zhongmin (2006) *Study on GIS-based Flood Risk Map of LianYungang Information Management System.* China WaterPower Press. Beijing, China.

Zhang, Zhidan (2005) Carry out "two-transform" and impel to develop flood risk map. *China Flood & Drought Management J.* **5–6**.

Development of information systems in regulation of water-energy relations between countries of Central Asian transboundary river basins

PARVIZ NORMATOV

Tajik Technical University Republic of Tajikistan, Prospect Acad. Radjabovs, 10, Dushanbe, 734042, Republic of Tajikistan

zar.rakhimov@mail.ru

Abstract This paper describes an information system to support the acceptance of international agreements on using the water-power resource potential in drainage basins, using the example of the basin of the Syrdarya River. The network information system is created to minimize technical control costs due to maximal use of the information resources of departmental monitoring systems integrated to the unique Earth remote sensing (ERP) net. The system includes the organizations: owners of the databases of ground monitoring, environmental managers and administrative parts.

Key word basin; Syrdarya River; Earth remote sensing; digital space information; satellites

INTRODUCTION

Today there is a strong need for information for decision making in the management of water-power resources (WPR) in the Central Asian region, especially the Syrdarya River basin, which is more problematic. In the basin, which covers four states, the problems of WPR use and regulation are rather acute. To support water use regulation in the drainage basin it is necessary to gather and to summarize multilateral information from networks under different supervision, with the attraction of the data of Earth remote sensing (ERP). An effective mechanism to solve this problem is to realize an opportunity for receiving exact and effective information by using modern networked technology.

The main goal of the paper is a system to support the acceptance of international agreements on using the water-power resource potential of drainage basins using the example of the basin of the Syrdarya River. The networked information system was created to minimize technical control costs due to maximal use of the information resources of the departmental monitoring systems integrated to the unique ERP net. The system includes the organization of the owners of the databases, ground monitoring, environment managers, and administrative parts.

The concept of Central Asia (the former name was Middle Asia and Kazakhstan) that is used nowadays includes the republics of CIS: Kazakhstan, Kyrgyzstan, Tajikistan and Turkmenistan and Afghanistan. Hydrographically the region of Central Asia (CA) is distinguished as the Aral Sea basin, which in its turn consists of two basins – those of the Syrdarya and Amudarya rivers.

The main indicators of technical and economic development of Central Asian economic region are given in Table 1.

Table 1 Indicators of macroeconomic development of Central Asian region.

Country	Territory (10^3 km^2)	Population (10^6)	Q (10^3 dollars/man)	D (fuel/man)
Kazakhstan	2636.20	14.95	3.56	3.67
Kyrgyzstan	198.50	4.90	0.68	0.66
Tajikistan	143.10	6.20	0.99	0.84
Turkmenistan	488.00	4.70	1.52	3..30
Uzbekistan	447.36	24.60	2.26	2.70
CA	3913.16	55.35	2.22	2.64

Q: per capita gross inland output by purchasing capacity parity; D: per capita energy consumption, t.

Table 2 Surface water resources of the Aral Sea basin.

Country	*Q* (km^3/year)	*D* (km^3/year)	The Aral Sea basin: (km^3/year)	%
Kazakhstan	–	4.50	4.50	3.9
Kyrgyzstan	1.90	27.4	29.30	25.3
Tajikistan	62.90	1.1	64.00	55.4
Turkmenistan	2.78	–	2.78	2.4
Uzbekistan	4.70	4.14	8.84	7.6
Afghanistan	6.18	–	6.18	5.4
CA	78.46	37.14	115.60	100.0

Q: The Amudarya River basin; *D*: The Syrdarya River basin.

The total water resources of the Aral Sea basin surface waters total 115.6 km^3/year (Table 2). According to approximate evaluations, the underground water resources in the Aral Sea basin total 43.7 km^3/year, of which 15.8 km^3/year (36.2%) is approved as exploitation reserves. Moreover, a large quantity of return waters is formed in the Aral Sea basin – 45.8 km^3/year, a small part of which is repeatedly used for irrigation – 6.0 km^3/year, and a larger part is drained to rivers (23.5 km^3/year) or lost by natural reduction (16.3 km^3/year).

The water-power problems in Central Asia became even more acute after the collapse of the USSR and formation of new independent states, when they obtained intergovernmental status. One such problem is connected with the contradiction between irrigation and hydroelectric engineering. Irrigated agriculture demands maximum use of water during the growing season from April to October. And hydroelectric engineering is concerned with maximum use of river flow in winter, from October to April, the coldest period of a year when rivers contain little water. Thus the irrigation regime requires filling of reservoirs in winter and their use in summer, whereas the needs are *vice versa* for the energy regime: filling of reservoirs is necessary in summer for water to use in winter. It is impossible to combine these two interests with one reservoir.

The common problem of mutual relations between irrigation and hydroelectric engineering in the region is determined by the fact that the countries of the upper basin – Kyrgyzstan and Tajikistan, are interested in the energy regime of river flow use, and the countries of the lower basin – Kazakhstan, Turkmenistan and Uzbekistan, are interested in the irrigation regime. Therefore the top-priority objective today is to overcome all this and form good neighbourly and mutually beneficial relations between the republics of the region.

Management of the water-economic complex, and the making of justified decisions as the basis of agreements, even intergovernmental, can be based on an advanced data control system (river ranges, observation sites) which also presents data processing methods. In Central Asia, among the WPR problems, we note the imperfection of the WPR structural management in each basin [1,2,3]. Today there is no common understanding of the economic rules of the river drainage regulation and pricing setting questions. The data problem necessary for the support of persons accepting decisions (PAD) in the region includes the following main aspects:

Absence of a uniform database The region has no uniform information banks, common guidelines or coordinating centre. As a result, the mode of operation of the waterworks facilities in basins is developed and approved by the Interstate Coordination Water-Economic Commission (ICWEC) with the Basin Water-Economic Association (BWEA) without participation by the hydropower groups. At the same time these works are realized by hydroelectric engineers without water-economic participants. Such a guiding system is ineffective, and besides, each governing body has its own diverse complex networks, which have different tasks, methods, and diverse product indicators. Hence, it is necessary to unite all the available data in a unique informational database. For its creation, we must develop uniform standards.

Incompleteness of available data Today there is no reliable hydrological forecasting, or mode of waterworks facility operation plan, and there is little operational information about its realization. The major problem is hydrological forecasts. After the collapse of the USSR the republics' hydrometeorological services were divided, and many hydrometric stations were liquidated. Despite the fact that in the frame of component "D" of the GEF project new hydrometric stations are being built and the old ones modernized (since 2001, with support of the Swiss government in CAEU, the unification of five republican and one regional centre of hydrological forecasting has begun), none of the practical questions have been solved in information resource changes between governing bodies and the republics' waterworks facilities.

The absence of a uniform coordinated information bank shows in the fact that today there is no operational information exchange between water users, water suppliers and waterworks facilities.

In mountain regions world-wide, basin hydrological information is inadequate because of the complicated water systems and difficulties of data gathering. Many large-scale plans on water management have either failed or have had negative consequences because of the absence of water system information. Development of the necessary hydrometric systems demands big investments that are impossible now. Hence it is necessary to be orientated not only to use of ground-based sites and river gauges, but also to the maximum use of operational ERP data.

Absence of training for PAD information providing Because of the data incompleteness, the total very large number of figures and objects, controlled by departmental networks of monitoring is usually redundant for PAD. It demands "Compression of the information" – receiving of integrated evaluations which PAD can use to place priorities for choosing the various strategies. The important instrument for "curtailing of the information" and providing maintenance of presentation for PAD is a geoinformation system (GIS) which provides informative compact cartographic images.

Maintenance of water use management is a difficult task, demanding supply with information of accepting decision processes. As a first step it is necessary to create parameters – criteria and indicators system [4]. So the topicality of this project is due to:

- the major need to supply information for decision processes connected with river drainage regulation, exhaustion and pollution of water objects in the long-term – and strategic, such as preparing for effective international agreements;
- the absence of a unique database necessary for representational evaluation of WPR in the region and development situation forecasts;
- the possibility to minimize the costs of technical means of control by reduction of the relative density of ground stations by using ERP data and GIS technology.

An effective solution to the decision management problem of the shortage of data and data imperfectness will be an information network database realization which consists of interested organizations. Their materials will contain information from the posts and river ranges, ERP data and also algorithms for processing and modelling on the basis of GIS technology.

In the first stage it is necessary to develop the information database with the example of the River Syrdarya, particularly in the mountain drainage formation zone where there is little need for irrigation water. In the flat zone there are the main irrigated areas and intense river flow consumption, and water abstraction prevails over lateral inflow.

The information is kept in the branches of departments, in administrations, and in private companies. Many of them develop special information systems for decisions on concrete tasks, but they do not provide complex analyses of the territorial process, which is very important for solving the water use questions. Developing lots of different branch databases in the region prevents effective information exchange for complex analyses, and is connected with information incompatibility of GIS because of the difference in the data structure, control system and description language. There is the problem of duplication of information and discrepancies in the data. There is no doubt that it is necessary to create a unique information databank for the region. So, the best way is to support a specialized database on the unique geoinformation platform.

For its development, a conceptually methodical basis for creation of the information database "water-power resources and water use" and the concept of the network informational database

(NIDB) of water-power resources must be created. NIDB will include a supervised network database from the organizations' sites, providing monitoring on departmental programmes and control objects, and ERP materials. Database unification in a uniform network will provide an organization with access and enable it to receive cartographical information. The concept development will determine the nomenclature of observance systems, their sites and river gauges, types of parameters and monitoring frequency. The NIDB functions are: data accumulation on all observance systems, ERP information, exchange of information resources between organizations, information accumulation for data transformation into constantational maps (on GIS basis), and for modelling.

The terrain aspect data of the network information database will include two subsystems – the drainage basin and WPR dispersion zone. In each subsystem it will be necessary to define principal blocks, optimize the quantities in each of them and also the nomenclature and total number of figures, their format, dimension, necessary actualization frequency, etc.

Upper basin subsystem The main subsystem's task is the formation of the quantity characteristics of water-resources. This part of the informational database will include one of the most important system elements of hydroresources possibility control: "Meteostations", "Rivers", "Glaciers", "Hydrology", "Hydrochemistry", etc. On the base of database analysis we will be able to define the hydroresources possibility of the basin for development of schemes of WPR regulation.

Subsystem of dispersion zone (zone of intense consumption of hydropower resources). Its main task is the formation of quantitative characteristics of Syrdarya water resources zone consumption including the offers and costs volume. This zone represents the valley part of the basin, formed from pro-luvial and alluvial flats and includes most of the pumps, irrigated squares and oases. This part will consist from the system elements: "Rivers", "Hydrology", "Hydrochemistry", "Water works facility", etc. We can receive the following information: dynamics of diversion capacity on using water resources, for irrigation, for industrial needs, etc.

Due to the absence of hydrometric stations on the Syrdarya, the ERP data will be used for inflows for water storage evaluation of the basin.

Realization of this task will enable creation of a conceptual-methodical basis for the WPR information database necessary for operational use by PAD. The next step is creation of a cartographic database in GIS.

Concept and working program development of informational NIDB filling – Realization of data gathering about the water-power resources and water supply In accordance with the nomenclature of the observation systems, network observation parameter lists will be optimized, and all data gathered. The IDBS site will be developed and created, its information filling and resource exchange will be organized. The separate work stations and local networks servers will serve like terminals. The data formats, database guiding principles will be defined.

The information filling will be from two directions: data using ground monitoring services and using ERP.

Using of data of ground services Data collection of organizations – networks supervision and ownership will be held according to the main directions of a water policy.

(a) Water resources (figures, defining maintenance and restoration of natural replenishment of water: the total basin area, change of a share of the area of exploited basin, parity of allowable and factually consumed volume of water, quantity of fall outs, the balance of annual replenishment and total value of water consumption).
(b) Off take and water-delivery (industrial-drinking needs data: conduit condition, annual water supply point of superficial and groundwaters, density of hydrological networks, household consumption per head, etc.)
(c) Socio-economic functions of water use (figures of regulation of economic activities: a share of water-power sector in a national product, the relation of volumes of use of the hydroelectric power to the size of its withdrawal for the limits of water basin, the specific charge of water

on a production unit, the investments put in a reservoir, protection of waters and recreation etc.).

(d) Sanitary condition of waters (figures of harmful influences on water objects: a general estimation of pollution, presence of storehouses of waste products, receipt of pollution from the agricultural areas – fertilizers, nitrogen and phosphorus, etc., a condition of the polluted waters – the maintenance of contaminants, etc.).

For realization of estimations of the condition of water-power resources, first, preliminary processing and the analysis of ERP materials on water stores will be carried out.

During the ERP analysis in the upper basin we will detect: hydrological objects, snow surfaces, rocks and vegetation. As red and infrared channels are optimal for spectral brightness division, so the hydrography objects will show in the rather low values of the coefficient of spectral brightness (CSB), snow surface – on the high CSB. Rocks and vegetation take an intermediate position. The green channel can be used for allocation of snow surfaces. The "main areas" with typical characteristics of the upper basin (in the plan of an *a priori* multivariate model) will be selected on the well investigated territories (having better description according to the data of ground services). For providing the maximal distinction of the main classes, providing WPR potential in the zone of water, snow (and glacier) surfaces, rocks and vegetation, the different types of classifications will be studied and both spectral and textured ERP data characteristics used.

For improvement of object perception, reception of their digital characteristics for further complex analysis, reduction of equipment and atmospheric interference, radiometric and geometric corrections, filtering will be carried out, etc. It is expected to carry out the following:

(a) To make the banks analysis of the digital space information received from satellites Terra (MODIS, ASTER) and NOAA (AVHRR) on researched territory, to execute materials selection (materials MODIS and AVHRR) and stages Terra (ASTER) (Landsat ETM +) and Topex-Poseidon.
(b) To process ERP digital materials for mapping geophysical characteristics of a terrestrial surface.
(c) To normalize ERP data, including radiometric and geometric corrections;
(d) To provide ERP information in a cartographical projection at scales 1:500 000 (for all pool) and 1:200 000 on selective territories.
(e) To make ERP data processing about a water stores condition of "main areas".
(f) To make water storage estimations in a basin according to ground services of supervision and to ERP data.
(g) To make possible demonstration of water storage estimates.

A major NIDB component will be the geographical information system (GIS). Its creation will raise work efficiency, reduce losses of information and facilitate coordinated binding of different-level materials on the basis of the standard software (for example, GIS ArcView).

On GIS basis, data transformation into constantation maps will be carried out. GIS includes WPR databases with use of the available archival information – river range and sites of ground services and ERP data. GIS will have a landscape basis and thematic layers, and will answer questions about object place, changes in parameters, etc. Therefore, at this stage it is possible to speak about creation of a preliminary multivariate geoinformation model of WPR potential in a drainage basin, constructed on *a priori* data and on the basis of which the expert estimation of results of processing of the image will be carried out.

Network GIS on a site will provide operational access to the information for its analysis, including "on-line" for reception of estimates of WPR condition and development situation forecasts. For this purpose the GIS concept and architecture will be developed, its base environment will be chosen, the character and volume of the cartographic information, and the list of questions for the analysis of WPR region, which can be interesting for governing bodies, will be determined. Separate fragments of maps (territory of controllable objects, impact zones) can take root in GIS from topographic maps and photomaps of large scale.

ERP materials of three levels of generalization will be used: survey for the analysis of the general laws, the work for actual research; and details necessary for specifications. Optimum scales for each level will be determined.

Further formation-featured spaces will be carried out in which the analysis on fourth elements of a spatial mosaic of all the drainage basin will be made. WPR images received on ERP data will be described in terms of the chosen attributes. The analysis of their mutual relation with parameters of ground services, and also distribution of the revealed attributes on all basins will be carried out. On the basis of an expert estimation of intermediate results, the algorithm of computing operations will be chosen giving an opportunity to raise as much as possible the image self-description that will be applied to all zones. During processing, the changes of feature spaces revealed with the help of "main areas" for final division into districts of territory on a degree of variability of attributes, will be fixed. By the results of analysis recognition and classification of researched objects and WPR phenomena, their generalization and ranking of allocated structures will be carried out.

REFERENCES

1 Ahmadov, A., Shermatov, N. & Nosirov, N. (2005) The main aspects of protection water the objects of mountainous territories. In: Material of republican symposium *Economics and Science of Mountainous – Bodakhshan autonomous Oblast: Past Present and Future*, Khorogh, 190–191.

2 Petrov, G. N. (2004) Water resources of the Tajikistan and used their in hydropower in connection with the problem of climate change. In: *Water Resources of Central Asia* (ed. by I. Sh. Normatov), 176–181. Omu Publ., Dushanbe, Tajikistan.

3 Petrov, G. N., Shermatov, N. & Normatov, I.Sh. (2003) Hydro – power resources of Tajikistan. In: Report of republic applied science conference *The Modern State of Water Resources of Tajikistan – Problems and perspectives of Rational Use*. Dushanbe, 34–36.

4 Shaparev, N. Y. (2004) System of parameters of steady water use. *Vestnik KrasGU – Natural Science* **7**, 135–138.

5 *The Informational System on Biotechnological Potential of the Kirov Region* (ed. by O. M. Leonov, V. M. Syutkin & A. F. Trufanov). Moscow – Kirov: *EXPRESS*, 141.

An e-Science platform for collaborative generation of knowledge and technology in hydrology, hydrogeology and water resources

CARLOS GALVAO[1], RODOLFO NOBREGA[1], FRANCISCO BRASILEIRO[2] & ELIANE ARAUJO[2]

1 *Dept of Civil Engineering, Federal University of Campina Grande, Caixa Postal 505, Campina Grande, PB, 58.100-971, Brazil*
galvao@dec.ufcg.edu.br

2 *Dept of Computer Science, Federal University of Campina Grande, Campus Universitario, Campina Grande, PB, 58.429-900, Brazil*

Abstract *SegHidro* is an enhanced grid computing platform that provides not only capabilities for the integration of computing resources, but also an environment for the sharing and integration of data, models and, ultimately, knowledge. The *SegHidro* platform provides support for: (i) the aggregation of computing and storage resources through the *OurGrid* peer-to-peer grid computing middleware; (ii) the sharing of data through the *OpenDAP* standard; (iii) the discovery of data and resources through *NodeWiz* distributed information service; (iv) the coupling of models through a workflow composition tool; and (v) a set of tools for improving the communication between researchers, the dissemination of scientific results and the sharing of best practices for the use, operation and evolution of the platform. The infrastructure is operational and several applications have already been developed by researchers and governmental water agencies in Brazil, including integration of atmospheric, hydrological, hydrogeological, hydraulic, agricultural and management models.

Key words grid computing; hydrological modelling; groundwater; climate forecasting; information systems; Brazil

INTRODUCTION

A great challenge for scientists and practitioners in the fields of hydrology, hydrogeology and water resources is integration and effective collaboration in the process of generating knowledge and technology. Collaboration between all the participants in the process is mandatory to allow true integration of data, models and knowledge. Moreover, a collaborative approach makes possible the aggregation of the computational resources required to address the highly demanding processing and data storage workloads involved. Although collaboration and integration depends on personal and institutional decision and will, Information and Communication Technologies (ICTs) can provide means for facilitating them.

The term e-Science was coined by John Taylor as follows: "e-science is about global collaboration in key areas of science and the next generation infrastructure that will enable it" (Ridel *at al.*, 2008). The infrastructure that allows access to globally distributed processing and storage resources, enabling large-scale scientific collaboration, is the *computational grid*, or, what is emerging in a more commercial approach, *cloud computing*. An example of e-Science environments and applications is the LEAD project (Droegemeier *et al.*, 2004), which is a "software that enhances the experimental process by automating many of the time consuming and complicated tasks associated with meteorological science", and also a platform developed by the American NSF Science for the Technology Center for Coastal Margin Observation and Prediction (CMOP) (Howe *et al.*, 2008), which integrates a provenance-aware workflow system, 3-D visualization capabilities, and a remote query engine for large-scale ocean circulation models in order to advance science in that ocean observatory.

This paper presents the *SegHidro*, an enhanced grid-computing platform developed for supporting collaboration in e-Science, by providing not only the capabilities for the integration of computing resources, but also an environment for the sharing and integration of data, models and, ultimately, knowledge.

SEGHIDRO PLATFORM

The *SegHidro* approach and related tools provide technology to deal with major challenges in Hydroinformatics, such as model coupling (including pre- and post-processing software), treatment of uncertainties, integration of disciplines and support for decision making.

The *SegHidro* platform provides support for collaboration in knowledge and technology generation and use through: (i) the aggregation of computing and storage resources through the *OurGrid* peer-to-peer grid computing middleware; (ii) the sharing of data through the *OPeNDAP* standard; (iii) the discovery of data and resources through the *NodeWiz* distributed information service; (iv) the coupling of models through a workflow composition tool; and (v) a set of tools for improving the communication between researchers, the dissemination of scientific results and the sharing of best practices for the use, operation and evolution of the platform.

Sharing of computing resources

In the fields of meteorology, hydrology, hydrogeology and water resources, many models – or cascades of models – require high performance computing resources for their execution. These requirements often prevent researchers or operational services from executing them. For some applications, supercomputers or large computer clusters are necessary. However, for several applications, the computing tasks are of the so-called "bag-of-tasks" type. They can be executed independently in the computational grid. Examples are Monte Carlo and other sampling procedures, search procedures in optimization algorithms and ensembles of simulations of atmospheric models, as well as the case of simulations for multiple sites. The grid can also be used to store and provide remote and distributed access to the massive observational data and simulation results.

OurGrid (Cirne Filho *et al.*, 2006) is a peer-to-peer, open, free-to-join grid-computing middleware, which is used by the *SegHidro*. Through *OurGrid*, participants make available their idle resources (computer processing, including clusters, and/or storage) to the community and can access their idle resources when needed.

Sharing and discovery of data and models

SegHidro uses OPeNDAP technology – Open-source Project for a Network Data Access Protocol (Cornillon *et al.*, 2003) to publish and to provide access to distributed data produced and stored in the grid. *NodeWiz* (Basu *et al.*, 2005) is a peer-to-peer grid information service for discovering data distributed geographically and form a catalogue. Not only data, but also models and other software, are registered in the catalogue and can be accessed and used by *SegHidro* users.

Coupling models and data

SegHidro's capability of coupling models, software and data promotes: (i) efficiency in the development of complex systems simulation, by (re)using already existing models and software, (ii) building of interdisciplinary modelling systems, by enabling coupling of models from different disciplines, and (iii) use of pre- and post-processing software, without the need of developing new and time-consuming user interfaces.

SegHidro Core Architecture (SHiCA) (Voorsluys *et al.*, 2007) is responsible for managing the submission of tasks to the grid following a prescribed workflow of models or any computer program. SHiCA enables the coupled models to consume their input data, which can be any input file, including the results of previous models in the workflow.

Sharing of expertise

Primary targets of *SegHidro* are groups working collaboratively, including researchers and practitioners and their institutions. Science and technology demands in integrated water resources management have to be fulfilled by scientists and developers from several areas of knowledge, which form interdisciplinary teams. Coupling models developed by different groups and sharing data and simulations produced by them requires the sharing of knowledge and expertise. *SegHidro* provides tools for enabling communication among its users, such as forums, mail lists, portals and Wiki-based environments.

APPLICATIONS

The *SegHidro* infrastructure is operational and several applications have already been developed by researchers and governmental water agencies in Brazil, including: (i) weather and seasonal climate forecasting using regional atmospheric models for downscaling; (ii) reservoir planning and operation; (iii) rainwater harvesting risk analysis; (iv) integrated surface and groundwater management; (v) operation of water distribution networks; (vi) agricultural planning; (vii) flood forecasting; (viii) soil conservation planning; (ix) regional impact of climate change; and (x) regional collaborative studies on Prediction in Ungauged Basins (PUB).

Figure 1 shows the workflow of models for a class of applications in water resources that use as their main input weather and seasonal forecasts produced by regional atmospheric models (RMs) nested in global ones. Global models (GMs) run on supercomputers and their output can be used for nesting RMs and, thus, to produce high-resolution forecasts for a certain region. RMs can run in computer clusters, which are far more numerous than supercomputers. Computer clusters are accessible by grid computing, reducing greatly the computing time for processing ensemble forecasts. Each of the members of an ensemble triggers the execution of hydrological, hydrogeological and water resources models, in series or in parallel, which may demand an amount of processing time not widely available in research and operational centres, particularly in developing countries.

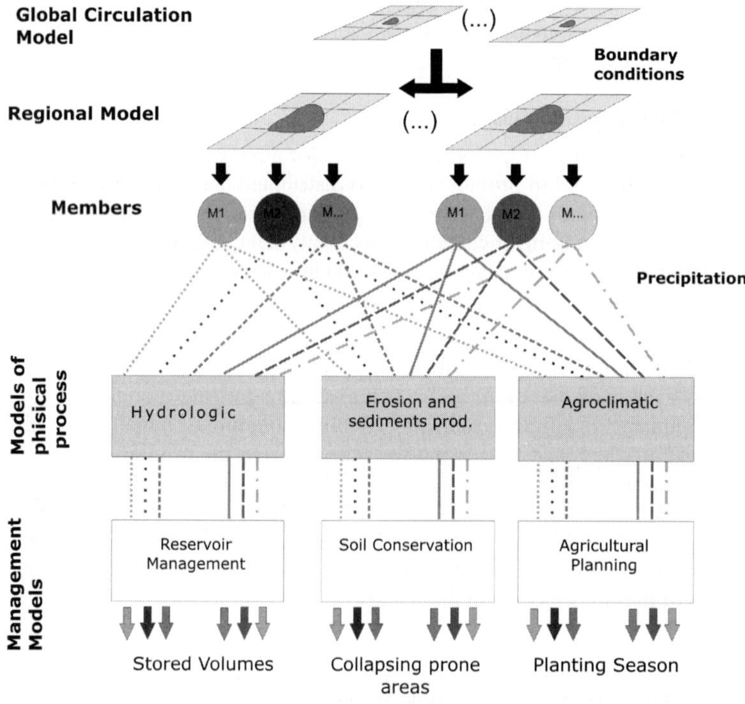

Fig. 1 Workflow of the execution of a cascade of models using *SegHidro*.

On the other hand, different RMs are executed by meteorological centres in a region. Users and also developers of these models can benefit from sharing their results, since they are run for the whole region, and sometimes a continent, such as South America. Sharing simulations means sharing knowledge and also saving computer processing and storage.

One of these applications is the management of reservoirs in the Northeastern region of Brazil. These reservoirs are abundant over the region and are the main source of water for municipal supply, agricultural and industrial production. Seasonal forecasts of rainfall are reliable and used operationally for supporting water management in the region (Galvão *et al.*, 2005). *SegHidro* supports this activity implementing the following workflow: (a) GMs forecast data files are downloaded from the meteorological production centres; (b) RMs are run in the grid using GM's data as input; (c) hydrological models simulate rainfall–discharge process for the reservoir's basins using RM's rainfall forecasts; and (d) reservoirs' behaviour are simulated through management models considering the discharge forecasts. The Water Agency of the State of Paraíba (AESA) applies this system to 115 reservoirs in real time. GM is run by the Centre for Weather Forecast and Climate Studies of the Brazilian National Institute for Space Research (CPTEC/INPE), in São Paulo, and RMs are run by the Ceará Foundation for Meteorology and Water Resources (FUNCEME), in Fortaleza, which makes available their models' outputs in the grid. Hydrological and reservoir simulation models are then fed with the seasonal rainfall forecasts, producing scenarios for reservoir storage levels and management strategies. Similar workflows and models are applied for short-term management using weather forecasts, instead of seasonal forecasts.

AESA also uses *SegHidro* to assess the seasonal risks of the collapse of thousands of rainwater harvesting and storage systems, used in small rural settlements dispersed geographically over the State. The behaviour of the small cisterns are simulated, considering their catchment areas (usually the roofs of the houses), family consumption and rainfall forecasts, producing estimates of storage for one year in advance, which are used by the municipalities and civil defence.

Another application in the State of Paraíba supports small shallow aquifer management, also based on seasonal rainfall forecasts. Its workflow is similar to the reservoirs', but includes, besides the hydrological models, models for estimating aquifer recharge and groundwater flow. This application is not yet operationally used by AESA.

In order to support research on RMs and the use of their weather and seasonal forecasts for reservoir management and flood forecasting, *SegHidro* provides means for simulating forecasts for long time series, to compare them to observations and to assess the performance of hydrological and reservoir models for a variety of basins. Currently, a project with this aim – the Repente Project – is under execution, involving several universities, research centres and operational services in Northeastern Brazil, having *SegHidro* as the e-Science middleware and also supporting its evolution.

Another ongoing collaborative research project, which uses *SegHidro* as an e-Science platform, involves groups from four universities aiming at enhanced techniques for soil and water conservation in the semi-arid region of Brazil. The Agua-Solo project includes field experiments on the techniques under evaluation as well as computer simulation experiments using numerical models of climate, hydrology and sediment yield processes.

One important aspect in all of these applications is the participation of interdisciplinary teams. An e-Science platform, such as *SegHidro*, provides a great opportunity and contribution for knowledge systematization and documentation by every group, so that their models and data can be appropriately handled by the system and effectively adopted by the partners and end-users.

CONCLUSIONS

Grid and cloud computing resources could be widely available for hydroinformatics applications in hydrology, hydrogeology and water resources with the use of tools such as those integrated by the *SegHidro* platform. Regional, national and global-scale research and development initiatives, involving many groups and institutions, can rely on these emerging e-Science technologies for augmenting their collaboration and thus producing more relevant results.

Acknowledgements The research reported in this paper was supported by the Brazilian Ministry of Science and Technology – through agencies CNPq and FINEP, Projects SegHidro, Repente and Agua-Solo, and the Fund for Water Resources (CT-HIDRO) – and by EELA-2 Project (E-Science grid facility for Europe and Latin America). The support from the Ceará Foundation for Meteorology and Water Resources (FUNCEME), from the Centre for Weather Forecast and Climate Studies of the Brazilian National Institute for Space Research (CPTEC/INPE) and from the Water Agency of the State of Paraíba (AESA) is also acknowledged.

REFERENCES

Basu, S., Banerjee, S., Sharma, P. & Lee, S.J. (2005) NodeWiz: peer-to-peer resource discovery for grids. In: *Proc 5th Intl. Workshop on Global and Peer-to-Peer Computing*, 213–220.

Cirne Filho, W., Brasileiro, F., Andrade, N. *et al.* (2006) Labs of the World, Unite!!! *J. Grid Computing*, **4**(3), 225–246.

Cornillon, P., Gallagher, J. & Sgouros, T. (2003) OPeNDAP: Accessing data in a distributed, heterogeneous environment. *Data Science J.* **2**, 164–174.

Droegemeier, K., Chandrasekar, V., Clark, R. *et al.* (2004) Linked environments for atmospheric discovery (LEAD): A cyberinfrastructure for mesoscale meteorology research and education. In: *Am. Met. Soc. 20th Conf. Interactive Info. Processing Systems for Meteorology, Oceanography, and Hydrology*. Seattle, Washington, USA.

Galvão, C. O., Nobre, P., Braga, A. C. F. M. *et al.* (2005) Climatic predictability, hydrology and water resources over Nordeste Brazil. In: *Regional Hydrological Impacts of Climatic Change – Impact Assessment and Decision Making* (ed. by T. Wagener *et al.*), 211–220. IAHS Publ. 295. IAHS Press, Wallingford, UK.

Howe, B. L., Bellinger, P., Anderson, R. *et al.* (2008) End-to-end eScience: integrating workflow, query, visualization, and provenance at an Ocean Observatory. In: *Proc. IEEE Fourth Intl. Conf. on e-Science*, (e-Science '08, Indianapolis, Indiana, USA). IEEE Press, New York, USA.

Riedel, M., Streit, A., Wolf, F., Lippert, Th. & Kranzlmüller, D. (2008) Classification of different approaches for e-Science applications in next generation computing infrastructures. In: *Proc. IEEE Fourth Intl. Conf. on e-Science*, (e-Science '08, Indianapolis, Indiana, USA). IEEE Press, New York, USA.

Voorsluys, W., Araújo, E. C., Cirne Filho, W. *et al.* (2007) Fostering collaboration to better manage water resources. *Concurrency and Computation* **19**, 1609–1620.

Study and practice of small hydropower project development with riverine ecosystem protection

JIEYOU LI, LIJUAN GUO, WANG CHEN, YANG ZHANG & HAO JIANG

State Key Laboratory of Hydrology-Water Resources and Hydraulic Engineering, Hohai University, Nanjing 210098, China

leejy2057@hotmail.com, leejy2057@sina.com

Abstract Rapid economic development and societal growth need an adequate supply of energy resources. Hydropower, as a kind of clean energy, is an important component of energy development and construction. The development of small hydropower projects is enormously beneficial economically, but it has many adverse effects on riverine ecosystems. Based on an analysis of the current situation of a small hydropower project development in Luoding city, in Guangdong Province, south China, as an example, the socio-economic advantages and adverse influence on the riverine ecosystem are discussed, giving particular attention to the following aspects: repair of the riverine ecosystem of a completed small hydropower project; successful development of new small hydropower projects; and improving estimation of the environmental impacts of small hydropower project development.

Key words small hydropower project development; riverine ecosystem; environmental impacts; adverse effects; protection practice

THE CHARACTERISTICS OF SMALL HYDROPOWER SOURCES

Luoding is a city located in the southwest region of Guangdong province, China, and has the characteristic South Asian tropical monsoon climate. The average annual precipitation amount is 1426 mm, the average annual run-off amount is 17.95×10^8 m^3 and runoff depth is 765 mm. The water resources can be divided into local water resources and transit water resources. The former is about 10.33×10^8 m^3, being 57.4% of the whole city average annual runoff amount, and the latter is 7.62×10^8 m^3, which is 42.6% of the whole city average annual runoff amount.

There are many rivers in Luoding. There are 11 strip rivers whose length is more than 100 km, seven secondary-branch rivers and four tertiary-branch rivers, with a combined catchment area greater than 100 km^2. Luoding River, the main stream, has a natural fall of 71 m, of which 64 m can be used for power generation. The maximum natural fall of Silun River is 558 m, and the amount that can be used is 460 m. The theoretical potential water power in Luoding is about 8.9×10^4 kW, of which 6.96×10^4 kW can be developed and used. However, we have developed and used only a proportion of it, 76%, about 5.29×10^4 kW. Based on the statistical information of 2004, the theoretical potential water power is about 0.0829 kW/person and 38.23 kW/km^2.

The major characteristics of the small hydropower resources in Luoding are as follows: those which have already been constructed are located mostly in the middle and upper reaches; the river channel is deep; the natural fall is large; flow is torrential; vegetation cover is good; concentration of suspended load is small; most rivers do not dry up all year; and transportation is relatively convenient.

EFFECTS OF SMALL HYDROPOWER PROJECTS ON SOCIO-ECONOMIC DEVELOPMENT

At present, 179 hydroelectric stations have been constructed across the whole city, and the total installed capacity is 5.867×10^4 kW. There are 24 hydroelectric stations which produce more than 500 kW. In 2002, the power production of the small hydropower projects was 2.53×10^8 kW.h and the amount of electricity that was transferred to the power network was 2.0×10^8 kW.h. According to the present average electricity cost, the output value of power production was CN\$9600 $\times 10^4$. Thus, power is an important source of income for the revenue of Luoding.

Small hydropower projects have played an important part in sustainable economic development, as well as the in GDP increase of Luoding. The reasons are as follows: (a) industrial and agricultural modernization has expanded, and social and economic prosperity accelerated; (b) the small hydropower resource can be used to meet the electric power requirement of city, villages and towns, to ease the balance of supply and demand, and to decrease the loss caused by the electric power of remote transport; (c) city, villages and towns can use the water to support the electricity, use the electricity to promote the water, and use the electricity to support the electricity, use diversification to develop the hydro-economics; (d) the use of electricity instead of firewood or coal can reduce the pollution of the water and the atmosphere, and effectively prevents the destruction of natural plant cover and protects the ecological environment.

IMPACT OF SMALL HYDROPOWER PROJECTS ON THE RIVERINE ECOSYSTEM

Characteristics of the riverine ecosystem

The ecosystem refers to: a natural macrocosm has certain structure and function, which constitutes the biological community and its inorganic environment. In the natural macrocosm, the biology and the environment affect and restrict mutually, evolve unceasingly and form the biological chain; the existence of the biological chain impels the organic matter, inorganic substance and simple inorganic substance of riverine ecosystem circle repeatedly, and reaches the ecological balance in a regular period. The riverine ecosystem is one of the ecosystems that are most close relation with the mankind. Ecosystem of water refers to: A natural macrocosm has certain structure and function, which is constituted of the biological community of river, lake and ocean and inorganic environment that is based on water. Hydrophyte, aquatic animal, micro organism and water environment affect, depend mutually and form the biological chain of the river, lake and ocean. If a certain tache (biological community) of biological chain make their habitats destroyed because of social attribute (human activity), the biological community will grow malignantly or wither away constantly, Which will cause the vicious circle, destroy the ecological balance, lead to the environmental deterioration and influence people's production, life and subsistence finally.

Impacts of reservoir construction on the riverine ecosystem

Building a dam in the river will cause discontinuous diversification along the direction of flow. It will change not only the geometric characteristics (such as river length, chain length, degree of sinuosity, river slope and river transversal section shape) and hydraulic characteristics (such as velocity of flow, water depth, water temperature and condition of fluid flow boundary), but also the structure of the food chain, and the bearing capacity of resisting outside interference and ecological function. These effects would mainly manifest as follows:

(a) The original forest, grassland or farmland would be flooded and land animals would be compelled to migrate.
(b) The former law of the material cycle should change. It is considered that the reservoir plays an important role in intercepting the nutrients in rivers, accelerating algae growth at the water surface and producing algal bloom phenomena.
(c) Because the water depth of the reservoir is higher than that of the river, and photosynthesis is weakened, the biomass of the ecological system and self-resumption ability of the reservoir are relatively weak compared to the river system.
(d) The variation of the sediment deposition in the reservoir and dewatering of the reservoir eroding channel cause ecosystem changes.
(e) Depending on reservoir manual adjustment flow, the hydro-periodical law of plentiful discharge of the original river or low water discharge in a year is changed. This means that the basic situation of the riverine ecosystem along with hydro-periodical law variation and formation impulse-river passageway has been changed (Li, 2000).

The numerous small reservoirs have many negative influences on the riverine ecosystem of Luoding city. Building reservoirs takes over and submerges a large area, which includes rural buildings. The vegetation is destroyed, the terrain is changed, and the soil is eroded. In the upper reaches of Luoding River, soil erosion in constructing the reservoir, is relatively serious.

Impacts of channel construction on the riverine ecosystem

The development of small hydropower projects requires construction of channels to import water. However, channel construction has serious impacts on the riverine ecosystem, which are mainly as follows: (a) the meandering original river is transformed into a straight or folded-line artificial river or river networks; (b) the complicated channel shape of the original river is replaced by a straight or regular geometrical shape, such as trapezoid, rectangle or arc; and (c) the side slope and river bed are gouged out and usually lined with brickwork or concrete slabs. Thus it can be seen that constructing or renovating a channel will change the basic meandering shape of the natural river. The diversified pattern of rivers, from torrent to slow flowing, disappears; and the geometric river transect destroys the varying shapes of the river too. As the heterogeneity of the river ecological environment is removed, it reduces the structure and function of riverine ecosystem. In particular, the diversity of the biome is reduced, which causes the function of the riverine ecosystem to degenerate.

In Luoding city, all sizes of interlaced channels have been built, and the most typical example is Luoding River. Renovating the channel has both straightened the river and shortened the river length. Other rivers and regions also have many similar instances of channel straightening, and enlarging of the transect of the river bed. The variety of channels that have been constructed change the characteristics of the water system and have an adverse effect on the riverine ecosystem.

COUNTERMEASURES OF SMALL HYDROPOWER PROJECT DEVELOPMENT FOR RIVERINE ECOSYSTEM PROTECTION AND SUSTAINABLE DEVELOPMENT

Whilst development of small hydropower projects plays an important role in the social and economic development in Luoding city, it constitutes a threat to the ecosystem. As hydraulic and water-power engineering are a necessity for human progress, both at present and in the future, the question is: How to take into account the advantages and disadvantages on both sides? Therefore, how to conduct harmonious development is a subject worth exploring.

Restoration of the riverine ecosystem

Practice tells us that the primary factor in damaging the water environment is human activity: the result is that sewage and pollutants enter water bodies; certain animals disappear, or their population decreases. Human behaviour destroys the biological chain of the water body and once that has happened, the water body loses its self-purification capacity.

Some people believe that to renovate the riverine ecosystem one must let things take their natural course, and let the living things of the river, the reservoir and the lake freely grow, emerge and perish. The authors believe that aquatic ecosystem restoration is work that has complex theory, multitudinous influencing factors and difficult operation. It not only exerts the function of self-renovation, but also conforms to the science and the actual effect. Through people's endeavour, we will create the multiple environment of waterside and water body, actively connect the biological chain of the river, the reservoir and the lake and thus strengthen the water environment and enhance the self-purification capacity of the water body. In terms of the actual socio-economic development in Luoding city, we should set the Jinyin Reservoir as an example, set Luoding River as a channel model, developing from point to area and complete the ecological restoration of the river. On the technical level, the concrete measures are as follows:

(a) the upper and middle reaches of the water sources are self-controlled and the water sources protected region, and around the reservoir and both banks of river channel plantation should be taken back to forest. This can be done by planting some relatively big trees, including fruit trees, and making them form the forest belt gradually. Lower growing shrubs and grass, which can strengthen the ecological function should be planted. On the one hand, this may reduce the runoff coefficient and delay or postpone the river channel from washing out; on the other hand, it can play an important role in filtering surface runoff, carrying organic matter in the runoff in the form of suspended matter.

(b) In the downstream part of the river, the river channel and the river mouth should be renovated, stabilising the side slope, preventing soil erosion and satisfying flood prevention measures. In the meantime, on the side slopes of the river, we should plant shrubs and grasses (lawn). Because the river slope is a natural transition belt from the water area to the land, and the mixture of shrubs and vegetation and soil may improve the peripheral temperature and humidity, this will create a suitable environment for amphibious animals.

(c) In the reservoir and river channel, various hydrophytes should be planted on the banks. The waterside of the reservoir and river bank is an extremely important part of the riverine ecosystem. The multiple function of waterside hydrophytes can have a major impact: the upward growing root systems can take a deep hold in the submarine silt and absorb large amounts of inorganic salts, while the stem and leaf system emerge above the water surface giving the beneficial effects of photosynthesis. Hydrophytes can change the flow and temperature of the waterside, provide a food source for diverse life forms, as well as provide a habitat for the growth and multiplication wildlife.

In the process of such river restoration projects, all levels of government, and especially the directly benefitting department of small hydropower project development, should invest funds in the initial period, adopt the project measures, the management measures and the biological measures to restore and construct the natural riverine modality and the biological habitat, to allow the maximum effects of the ecosystem's self-repair function to exert itself (Dong, 2003).

Constructing new small hydropower projects

Along with the unceasing scientific and technological development, there are many new and advanced technologies, such as geographic information systems, satellite navigation technology, remote sensing and intelligent automatic control technology, which are widely applied in water conservation. In the process of designing and constructing the hydropower project, we should apply more modern science and technology. The main recommendations are as follows:

(a) In the programming stage, we should make use of satellite remote sensing technology to gain information about the environment and the hydrogeological conditions of the drainage basin and investigate the chief characteristics of the river ecosystem in the new hydropower project region. These should include: the numbers and variety of flora and fauna, the oxygen content, and condition of energy and material exchange between the river and terrestrial environment; the nature of the riverine base, the current of the upper reaches and downstream; and other differences in physical conditions.

(b) A full understanding of the foundations and the natural conditions of the new hydro-power project by modern physical prospecting technology, including geology, landform, rock layer structure, vegetation cover and soil, transportation and so on.

(c) Analysis of the changes of environmental factors around the engineering construction by mathematical physical models, including river width, water depth, velocity of flow, solid deposition, dilution and diffusion and so on.

(d) Institute an optimized scheme for the new project structure and its position by computer expert system. We can overcome the disadvantages that result from the straight vertical shape replacing the river slope. Provisionally, space should be set aside on the river bank to plant trees and vegetation, in contrast to the river channel where flow is quite fast, the slope steep, and shape of the flood wave is big. We should adopt ventilated, pervious and spongy materials to protect the slope. As the same time, we should plant plants and shrubs as far as possible.

Appraisal of small hydropower exploitation of the environment

In the process of exploiting the small hydropower development, to bring valuable changes to the ecological environment, we should make a full appraisal to reduce any drawbacks as far as possible, and take maximum advantage of the economic effect of small hydropower fulfilment to society, the harmonious development of the ecological environment and the overall benefit to Luoding city. Fulfilment of the small hydropower development will result in major changes to the riverine ecosystem, the water resources and environment, which will bring financial benefits. Therefore, to prevent such damaging changes, we need to adopt a variety of project measures, management measures and biological measures. The investment in these measures should be seriously estimated in terms of the overall cost of the economic development in Luoding city, comparing the costs and expenditure. Only then can we determine the investment goal, scale and region, and optimize the investment effect (Liu, 1998).

The development of a small hydropower project has a great impact on the ecology and the environment, and the impact is long-term and potentially harmful. In particular, the diversifications of the structure, function and benefit in the riverine ecosystem, the reservoir system, and the aquatic ecosystem is long-term and slow. Only through long-term monitoring and observation, analysis and research, can we gradually find its inter-relationships and evolution rule. For example, the river mouth is a semi-enclosed water body which is interlinked to the sea, falling under the mixing effect of oceanic tide movement and freshwater inflow. Because dams have been built in the rivers, the outflow is obviously reduced, which could not only cause the landward migration of the interface of salt and freshwater, but also reduce the scouring action of the river mouth, lengthen retention time of the pollutant in the river mouth, and increase the river mouth pollutant concentration. Moreover, the reduction of freshwater flow in the river mouth will reduce the supplement of evaporation losses, which will cause the river mouth to become increasingly saline. These factors will cause the original biotic community in the river mouth region to be destroyed, but its transformation will be long-term and slow. Thus, the influence of the small hydropower development on the ecological environment is of widespread interest. Therefore, let us take restoration of the ecological environment seriously.

CONCLUSIONS

The small hydropower project development brings huge benefit to socio-economic development in Luoding city. At the same time it brings negative influences to the riverine ecosystem. People should follow the harmonious development goal between the small hydropower exploitation and the riverine ecosystem. Regarding the small hydropower projects that have been built and newly constructed, we will take all kinds of project measures, management measures and biological measures, to reduce the disadvantageous effects on the riverine ecosystem as far as possible. We must adhere to safe and clean development, and attach importance to sustainable development.

Acknowledgements This study was financially supported by the Water Conservancy of Guangdong Province, China. The authors do deeply appreciate the Water Conservancy of Luoding City for providing the historical data.

REFERENCES

Dong, Z. R. (2003) The menace about hydraulic engineering to ecosystem. *Water Conservancy and Hydroelectricity Technology* **7**, 37–40.

Li, Z. X. (2000) Hydraulic engineering and ecological environment. *Hydroelectric Power of Shanxi*. **16**(3), 31–32.

Liu, C. M. (1998) *The 21st Century Water Question Tactic*, 121–156. Science Publishing House, Beijing, China.

GIS-based tool to determine streamside forest shelterbelt width

MIKHAIL KORETS & ALEXANDER ONUCHIN

V. N. Sukachev Institute of Forest, Siberian Branch, Russian Academy of Sciences, 50/28, Akademgorodok, 660036, Krasnoyarsk, Russia

mik@ksc.krasn.ru

Abstract Forest areas can intercept surface runoff from upslope bare areas and transfer it to interflow. Therefore, planting protective forests along the banks of rivers, reservoirs, and lakes preserves natural water sources from pollution. Depending on the particular landscape conditions, the streamside forest shelterbelt (SFS) width is often either wider or narrower than the ecologically substantiated width. As a result, either water quality worsens or the ecologically unjustified prohibition of forest use leads to economic losses. The assessment of SFS width using GIS technologies allows considerable simplification of evaluation procedures and their application in practice. DEM processing is integrated into most modern GIS software packages. For example, the popular ESRI ArcGIS package with its Spatial Analyst module provides extra options for calculating a series of relief-based hydrological features, which include calculation procedures for surface flow direction, length of flow-producing slopes and surface flow accumulations. Two algorithms for GIS-based SFS construction were tested for several rivers of the Yenisei basin and Krasnoyarsk Reservoir, Siberia. The first algorithm is technically simple and based on empirical equations of runoff slope length, slope steepness and soil infiltration. The second one includes a three-dimensional flow accumulation procedure and thus it is more sensitive to real surface structure. Both algorithms are ready to be used in practice. The results obtained indicate that, on average, the SFS width along banks of large rivers might be reduced, while in some cases it should be widened along the banks of small streams.

Key words GIS; DEM; streamside forest shelterbelt; surface runoff; Central Siberia

INTRODUCTION

Sustainable forest use principles can only be implemented using modern geographic information system (GIS) technologies that enable all-round optimization of forest resource use based on available ecologically and economically sound forest management models, digital elevation models (DEM), and special software tools. Streamside forest shelterbelt (SFS) width determination and human-caused ecological damage appraisal are among the major challenges involved in sustainable forest management.

There is, in fact, no surface runoff where soil water permeability considerably exceeds precipitation amounts. SFS are capable of intercepting surface runoff coming from the upper non-forested sites and turning it into interflow. This forest function is widely used to protect natural stream channels and still-water bodies from pollution. To do this, SFS need to be delineated and established along river banks, around lakes and water reservoirs.

Ecological parameters, such as erosion rate, forest soil-saving function recovery time, water turbidity in stream channels and still-water bodies, can serve as SFS delineation criteria. Many of these parameters can be estimated using existing empirical models (Onuchin & Burenina, 2000) that consider local forest and hydrological conditions.

A GIS- and DEM-based method to assess spatial and temporal soil erosion patterns was developed for the southeastern Baikal region, Russia. This method can also be used to identify forest sites with high soil erosion danger and estimate ecological impacts of environmental disturbances such as clear cuts and fires (Onuchin *et al.*, 2003).

Geographical condition-specific empirical dependences of SFS width on topography have been known for a fairly long time (Protopopov, 1977; Molchanov, 1977). However, they have not been widely used due to the difficulties of quantification of these parameters and surface runoff development. According to the current official SFS width regulations, the width varies with stream channel length or still-water body size. These regulations do not consider surface runoff development. Therefore, SFS are often wider or narrower than is ecologically required. As a result, water quality can decrease in some cases and ecologically unjustified forest use bans can cause substantial economic losses.

METHODS

Applying GIS technologies simplifies SFS width determination procedures and makes it much more useful for practical decision-making. In this case, slope length (L) and steepness (i) included in empirical models are determined by computer programs using DEM, whereas infiltration is put into the model as a separate layer derived from available soil maps. Most contemporary GIS have built-in functions to work with digital elevation models. The well-known RSRI ArcGIS package with the Spatial Analyst module, for example, has, besides standard procedures for developing the major DEM layers, additional procedures to calculate a number of hydrological characteristics controlled by relief. One of them is a procedure for calculation of slope length, as well as surface runoff direction and accumulation (Jenson & Domingue, 1988). We used these procedures to test two algorithms of computer-aided SFS delineation.

Elements of a digital vector topographic base map, such as linear layers of rivers and elevation isolines, were used as input data. The ESRI ArcGIS 9.2 package with Spatial Analyst extension module allowed us to obtain vector (triangulation – TIN) and raster (interpolation – GRID) relief models that included layers of elevation above sea level (m) and surface steepness (slope). The final DEM pixel size was 10 m × 10 m.

Model 1

The DEM-based slope length (L) and slope (i) were calculated with the help of the ESRI ArcGIS Spatial Analyst function. A soil infiltration (In) map was used as a raster layer with a pixel size similar to that of the DEM. The following empirical equation was further used to determine SFS width:

$$B = \exp(-2.659 + 0.779\ln(L) + 0.884\ln(i) - 0.881\ln(In)) \quad (1)$$

where B is SFS width, m; L is slope length, m; i is slope steepness; and In is soil infiltration, mm/min.

SFS width, B, was calculated using the ESRI Spatial Analyst Map Calculator procedure for each raster layer pixel as a function of L, i, and In layers. To avoid natural logarithm calculation errors at zero input values ($L = 0$ for watersheds and $i = 0$ for level surfaces), these were added by close-to-zero values that did not have any marked influence on the result. A potential SFS width was thus calculated for each pixel, provided that the pixel was assumed to be the end of a slope.

The next step was to build chains of overlapping circle-shaped 50-m-radius polygons lining rivers, with the polygon centres being 50 metres apart from each other and the river banks (Fig. 1). An average SFS width value (Bav) was then calculated for each circle from the pixels that occurred within a given raster using a special built-in GIS function. The average values obtained were downloaded to the circle-shaped polygon layer attribute base. The SFS width value calculated for the circles defined the size of circle's centre buffer polygons, which were then sequentially merged to produce a continuous SFS of varying width (dashed line in Fig. 1).

Model 2

This algorithm of Model 1 could have been put into practice in this form, however, the GIS hydrological component allowed us to improve the methodology. The main drawback of the first model (equation (1)) is that, while it describes separate slope profiles, it does not consider surface runoff accumulation. Natural relief variability, however, makes surface runoff accumulate in some places, i.e. small surface runoff streams merge to form temporary or permanent channels. Therefore, it was decided to use a procedure for calculating surface runoff accumulation (Flow Accumulation procedure of ESRI Spatial Analyst) that develops a DEM-based raster layer for each pixel of which the total number (A) of the "feeding" pixels located above the pixel of interest is determined. The resulting raster layer can reflect the real process of relief-controlled surface runoff development to a certain extent.

A relationship between surface runoff accumulation (A, pixels) and SFS width (B, m) was estimated based on raster layer A and B pixel values using a linear regression separately for sites

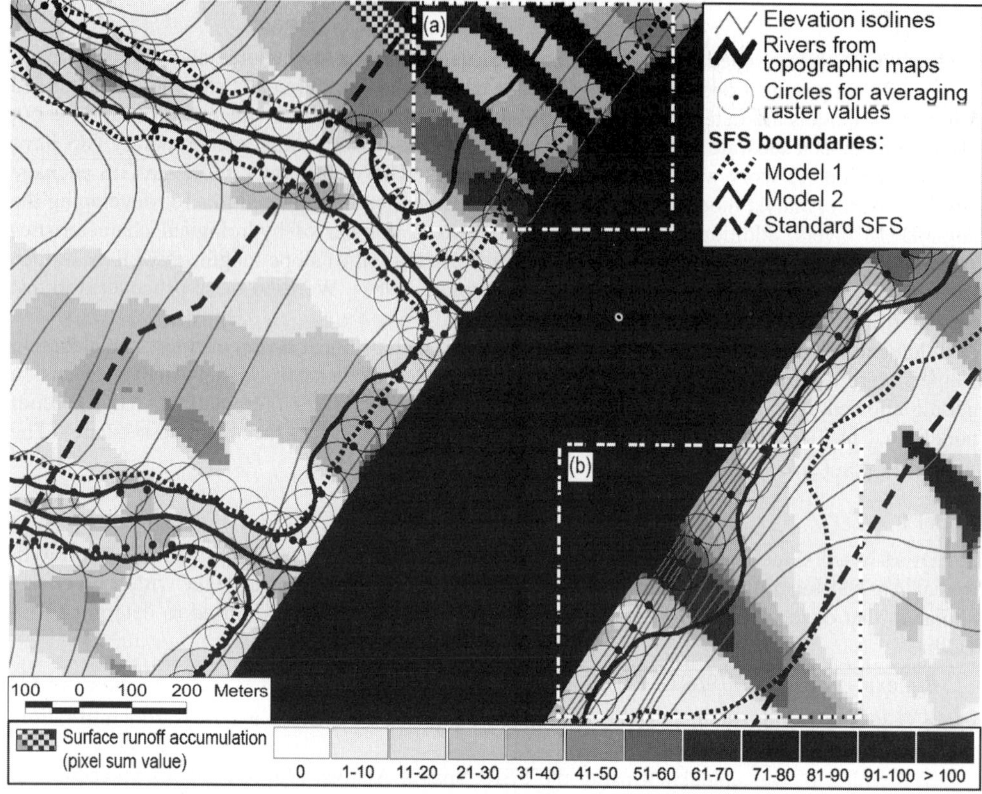

Fig. 1 Streamside forest shelterbelts delineated by different methodologies.

with poorly pronounced relief, and the entire test site. The correlation coefficients between *A* and *B* were found to be 0.8 and 0.3 in the former and latter cases, respectively. This suggested that the models providing values *A* and *B* worked consistently for homogeneous slopes with fairly uniform surface runoff distribution. The conversion coefficient of the correlation between surface runoff accumulation (*A*) and SFS width (*B*) was found to be 1.9 (i.e. $B = 1.9A$).

The algorithm for calculating SFS width using surface runoff accumulation (Model 2) appeared to be technologically similar to that based on *L* and *i* values (Model 1). In both cases, circle-shaped polygon chains are first built and average surface runoff accumulation values (A_{av}) obtained for each polygon. Further, these values are multiplied by conversion coefficients to determine an appropriate average SFS width (B_{av}). The following step is the calculation of SFS solid polygons on the basis of cascade merged polygons forming a SFS (solid line in Fig. 1).

The SFS width obtained appeared to be highly variable depending on the accumulated surface runoff amount, and reached high values along stream channels, even in some not reflected in the topographic base map. For this reason, the initial raster layer was smoothed before A_{av} calculation procedure, using a median filter in an effort to cut off surface runoff accumulation peaks.

RESULTS

These two algorithms of SFS construction are ready to be used in practice. Both models were tested for several rivers of the Yenisei basin and Krasnoyarsk Reservoir (Central Siberia).

The proposed approach to SFS delineation considers surface runoff genesis and topography of riparian zones where SFS are to be established, as compared to the current conventional method resting on SFS width dependence on stream channel length and still-water body size.

Figure 1 shows a 500-m wide SFS (heavy dash line) based on standard regulations. As is clear from Fig. 1, Model 2 widened the SFS where high surface runoff accumulation occurred (Fig. 1(a)) and, conversely, narrowed it for sites of uniformly distributed surface runoff (Fig. 1(b)), while Model 1, which does not consider surface runoff, overestimated SFS width for long and steep slopes (Fig. 1(b)).

The average values of SFS width for the rivers of the Yenisei basin were 80 and 130 m for Model 1 and Model 2, respectively, while the range of SFS width was 50 – 600 m for both models. The results obtained also indicate that, on average, the SFS width along banks of large rivers might be reduced, while in some cases along the banks of small streams it should be widened.

Acknowledgements This study was supported by the Russian Foundation for Basic Research (RFBR projects 07-04-00515-a, 08-05-92502)

REFERENCES

Jenson, S. K. & Domingue, J. O. (1988) Extracting topographic structure from digital elevation data for geographic information system analysis. *Photogram. Engng & Remote Sensing* **54**(11), 1593–1600.

Molchanov, A. A. & Naryshkin, M. A. (1977) An experimental method to assess water-saving forest function efficiency in ETS natural environment. In: *Streamside Forest Shelterbelt Establishment Principles*, 5–27. Nauka Publishing, Moscow, Russia.

Onuchin, A. A. & Burenina, T. A. (2000) Human impacts on forest soil-saving and water-sheltering functions in Siberian mountains. *Lesovedenie* (*J. Forest Science*) **1**, 3–11.

Onuchin, A. A. & Sokolov, V. A. (2002) *Water Protection Zones; Organization of Intensively Protected Areas*, 230–240. Russian Academy Publishing, Novosibirsk, Russia.

Onuchin, A. A., Kosmynin, A. V., Gaparov, K. K. & Korets, M. A. (2003) Modeling and GIS as tools to meet forest hydrology information needs. *Siberian Ecological J.* **6**, 749–754.

Protopopov, V. V. & Kuklin, V. V. (1977) Principles of determination of streamside forest shelterbelt width along central Siberian rivers. In: *Streamside Forest Shelterbelt Establishment Principles*, 92–104. Nauka Publishing, Moscow, Russia.

Monitoring large-scale rainfall variations across India

H. N. SINGH, NITYANAND SINGH & N. A. SONTAKKE
Indian Institute of Tropical Meteorology, Dr. Homi Bhabha Road, Pashan, Pune 411008, India
narendra@tropmet.res.in

Abstract To understand spatial variability of rainfall fields over India, sequential thematic maps have been prepared in a GIS environment (GeoMedia Professional 5.1) using data from 316 raingauge stations for the period 1871–2006. The thematic maps show interannual expansion/contraction of moisture regions (arid, semi-arid, dry subhumid, moist subhumid, humid and perhumid), and very dry, dry, wet and very wet zones of the four seasons and the twelve calendar months. The annual moisture regions and seasonal as well as monthly wet and dry zones exhibit large interannual variation but in an organized manner. The arid area shows a dry period during 1999–2004, semi-arid during 1964–2004, moist subhumid during 1999–2004, humid during 1998–2004 and perhumid 1995–2006, while the dry subhumid shows a wet period during 1984–2006. Overall during recent years (1995–2006) the climate of the country has been drier. To describe spatio-temporal features of the seasonal and monthly rainfall an integrated Rainfall Spatial Distribution Index (RSDI) has been developed by lumping the areas under very dry, dry, wet and very wet conditions. In winter rainfall, a dry period is seen during 1987–2006 and in the summer monsoon rainfall during 1972–2006, while a wet period is seen in summer rainfall during 1977–2006 and in post monsoon rainfall during 1994–1999. In recent years, declining trends are seen in eight months (January, February, March, July, August, September, November and December) and increasing trends in the other four months.

Key words rainfall spatial variability; rainfall monitoring; moisture regions; dry/wet zones; rainfall spatial distribution index

INTRODUCTION

Most natural disasters are of meteorological or hydrometeorological origin. Disasters such as earthquakes and tsunamis are occasional, but rainfall-related hazards, flood or drought, are very frequent and thus call for proper monitoring of large-scale rainfall variation, and better management and planning to combat the problem(s). Variation in rainfall is very large spatially and temporally compared to any other atmospheric parameter, such as pressure or temperature. The large temporal variability in the Indian rainfall has produced frequent flood/drought and periods of above/below normal rainfall for several centuries past. Areas of high rainfall and runoff are flooded in the summer monsoon season whereas others face water stress and drought conditions. The influence of global warming on rainfall variations over India is reported in numerous observational (Singh *et al.*, 1991, 1992) and modelling studies (Lal, 1994). Though considerable rainwater enters the Indian territory from the Himalayan area, the major hydrological activities and water resources of the country are due to *in situ* rainfall occurrences, mostly summer monsoon rainfall during June–September. During a normal hydrological wet season the country receives 2911.064 bcm (billion cubic metres) rainwater of which 1856.947 bcm evaporates and 1054.117 bcm enters into surface hydrological processes (surface storage, streamflow/riverflow and deep aquifer storage) (Ranade *et al.*, 2008). Rainfall events display large spatial variability due to variation in intensity, frequency and geographical location of the rain-producing weather systems, such as circular systems (low, depression, tropical cyclone and mid-tropospheric cyclone), linear systems (trough, meander, squall line and convergence zone) and wave disturbances (easterly and westerly waves). Rainfall in other seasons is equally important to monitor, though restricted to some parts of the country, such as winter rainfall over the extreme north, summer thunder showers over the southwest peninsula and northeast India, and post-monsoon rainfall over the south peninsula.

Indian communities have for long faced rainfall-related disasters and have a history of traditional mechanisms of coping with such calamities by the way of irrigation, bunds, canals and several other techniques during the excess rainfall. Prabhakar & Shaw (2008) have considered the available evidence for climate change over the Indian subcontinent and assessed the existing preparedness and mitigation mechanisms for drought risk reduction in the country. Revi (2008) has considered the likely changes that climate change will bring in temperature, precipitation and extreme rainfall, drought, river and inland flooding, storms/storm surges/coastal flooding, sea-level rise, environmental health risks and the necessary adaptation and mitigation agenda for cities in India where the urban population is rapidly growing. Real-time monitoring of the large-scale rainfall variations can provide vital information in this endeavour. Collection, compilation, analysis of long-period rainfall data and display in the form of maps and graphs on smaller spatial scales showing rainfall variability are required to decide adaptation and mitigation plans and policies. The objectives of the present study are:

- to document climatological and fluctuation features of different moisture regions and dry/wet areas during different seasons and months over the country;
- to report recent tendencies of rainfall spatio-temporal fluctuations; and
- to understand possible causes of recent trends in large-scale rainfall fluctuations.

FEATURES OF LARGE-SCALE RAINFALL VARIATIONS

The spatio-temporal variability of annual, seasonal and monthly rainfall over India is studied using highly quality-controlled data from 316 widely distributed locations (Fig. 1) for the period 1871–2006. Rainfall data up to 1900 were obtained from Eliot (1902) and for the period 1901–2006 from the records of the India Meteorological Department (IMD), Pune. A digitized map of India on 1:6 million scale with the projection system of "Albers Equal Area Projection with Two Standard Parallels, 15°N and 30°N" was utilized in a GIS environment (GeoMedia Professional 5.1) for the spatial analysis of the rainfall data. To understand spatial variation of annual rainfall, expansion/contraction of different moisture regions, viz. arid (rainfall ≤ 560 mm), semi-arid (561–1040 mm), dry subhumid (1041–1420 mm), moist subhumid (1421–1630 mm), humid (1631–2450 mm) and perhumid (≥ 2450 mm), are examined. This climatic classification of the different moisture regions is based on the average annual precipitation as the climatic conditions of any place in the country can be assessed with the average annual precipitation over it (Singh, 1984). Thematic maps of the country showing the normal (1901–2000) position of the moisture regions (Fig. 1) as well as their yearly positions (figures not shown) have been prepared using the GeoMedia GIS. The different moisture regions of the country show large variations from one year to another. On the normal annual isohyetal map the arid region occupies over 15.3% of the area of the country, semi-arid 33.8%, dry subhumid 24.3%, moist subhumid 8.9%, humid 12.4%, and perhumid 5.3%. The standardized value (actual minus mean divided by standard deviation) of the area under different moisture conditions along with its 11-year moving average are plotted in Fig. 2. The sequence of dry and wet periods in the fluctuations of the moisture regions is also given in Table 1. In the low-frequency mode fluctuations (11-year running means), notable features of the moisture regions are: spreading tendency in the area under semi-arid, dry subhumid and moist subhumid conditions from the early 1940s, and a contracting tendency in arid, humid and perhumid regions. Overall the climatic condition of the country shows a drier tendency in recent years/decades.

Table 1 Dry and wet periods in the fluctuations of the six moisture regions over India.

Moisture regions	Dry periods	Wet periods
Arid	1896–1925, 1934–42, 1965–74, 1984–87, 1999–2004	1871–95, 1926–33, 1943–64, 1975–83, 1988–98
Semi-arid	1900–12, 1950–57, 1964–2004	1871–99, 1913–49, 1958–63
Dry subhumid	1914–17, 1948–83	1871–1913, 1918–47, 1984–2006
Moist subhumid	1871–83, 1952–89, 1999–2004	1884–1951, 1990–98
Humid	1899–1913, 1949–53, 1964–89, 1998–2004	1871–98, 1914–48, 1954–63, 1990–97
Perhumid	1898–1910, 1935–42, 1960–87, 1995–2006	1871–97, 1911–34, 1943–59, 1988–94

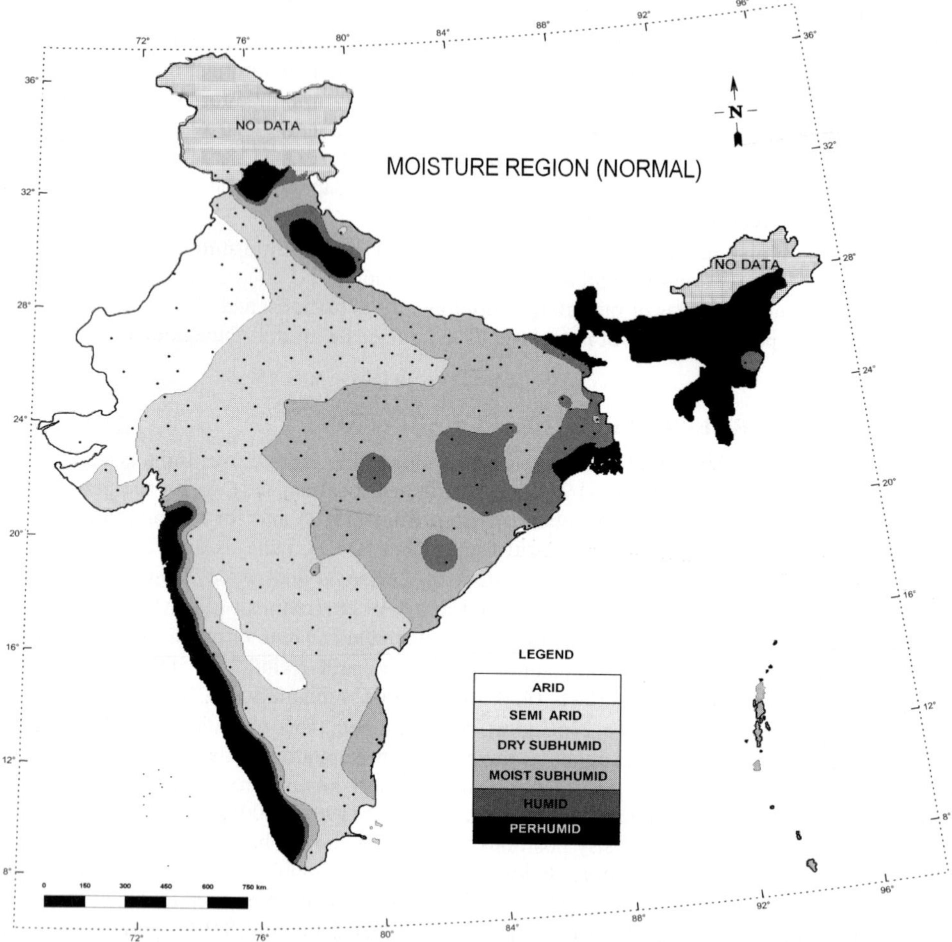

Fig. 1 Map of the country showing climatic moisture regions. Dots indicate the location of 316 rain-gauge stations used.

RAINFALL SPATIAL DISTRIBUTION INDEX (RSDI)

Spatial variation of the seasonal and monthly rainfall has been studied by examining interannual variation of an optimum number of four rainfall zones under different rainfall conditions, viz. very dry (VD), dry (D), wet (W) and very wet (VW), identified on the mean rainfall (1901–2000) chart of the particular period (season/month) of equal size (25%) area of the country. The criterion has been applied to yearly rainfall spatial distribution, and the area of the country under different rainfall conditions for each season and month has been computed. Based on these threshold values for the respective seasons and months, the thematic maps have been prepared showing very dry, dry, wet and very wet areas in each year from 1871 to 2006 for each season and month. In order to understand the rainfall pattern both in terms of amount and areal distribution across the country, a Rainfall Spatial Distribution Index (RSDI) was developed (Singh *et al.*, 2005) by lumping VD, D, W and VW areas. For a particular year, the RSDI is calculated as:

$$RSDI = \frac{1 \cdot a_{VD} + 2 \cdot a_D + 3 \cdot a_W + 4 \cdot a_{VW}}{A}$$

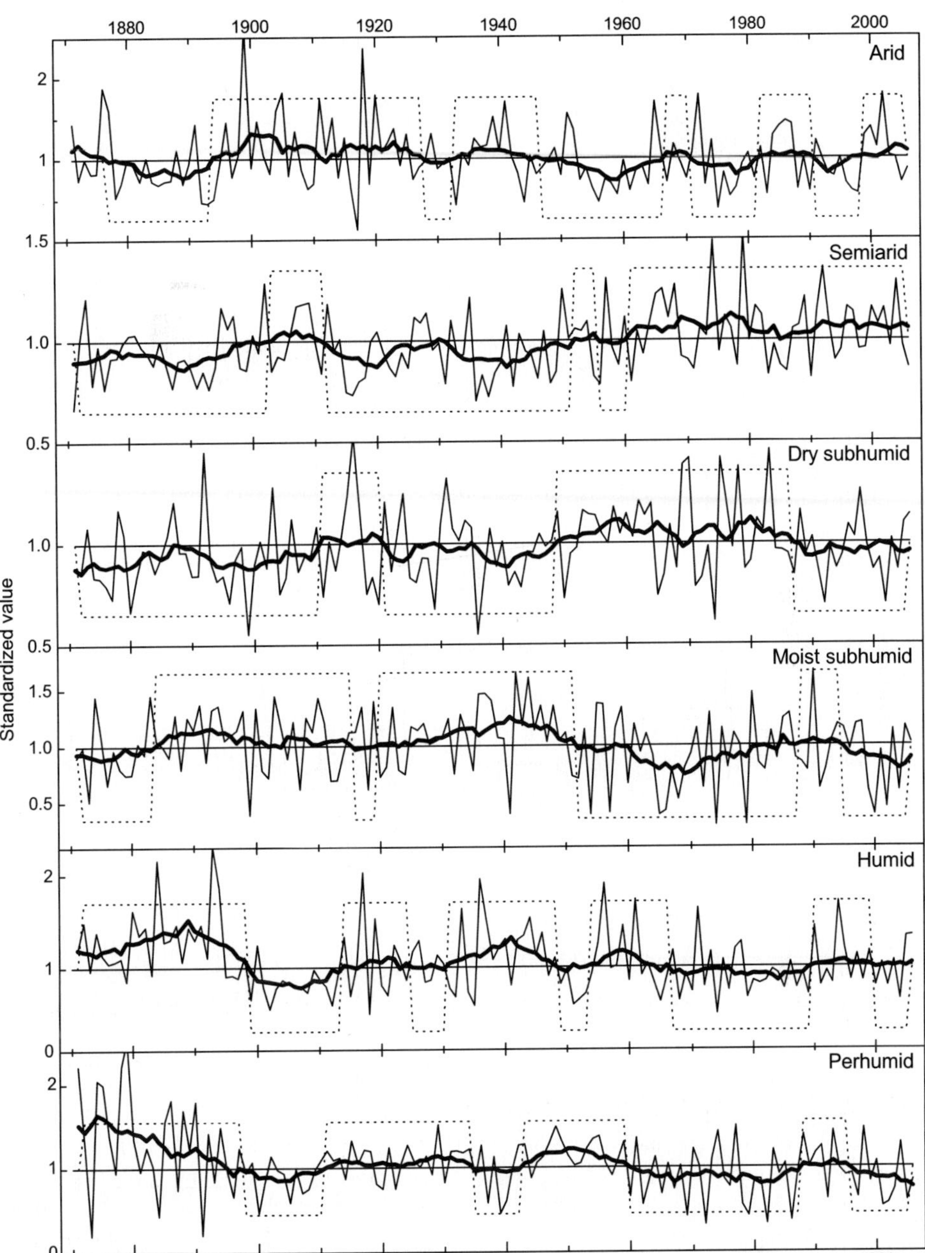

Fig. 2 Standardized value (thin curve) and 11-term running means (thick curves) of the area under six moisture conditions over the country during 1871–2006. The blocks demarcated by dotted line are dry/wet periods in the fluctuation of the respective parameter.

where a_{VD} denotes area of the country under very dry condition, a_D under dry, a_W under wet and a_{VW} under very wet condition, and A is the total geographical area of the country.

A high RSDI value is indicative of enhanced rainfall over large areas, including dry areas, and *vice versa*. The periodic pattern and recent trend of the monthly RSDI fluctuation is shown in Fig. 3. For winter rainfall fluctuations the distinctly dry periods are 1871–1900, 1949–1977 and

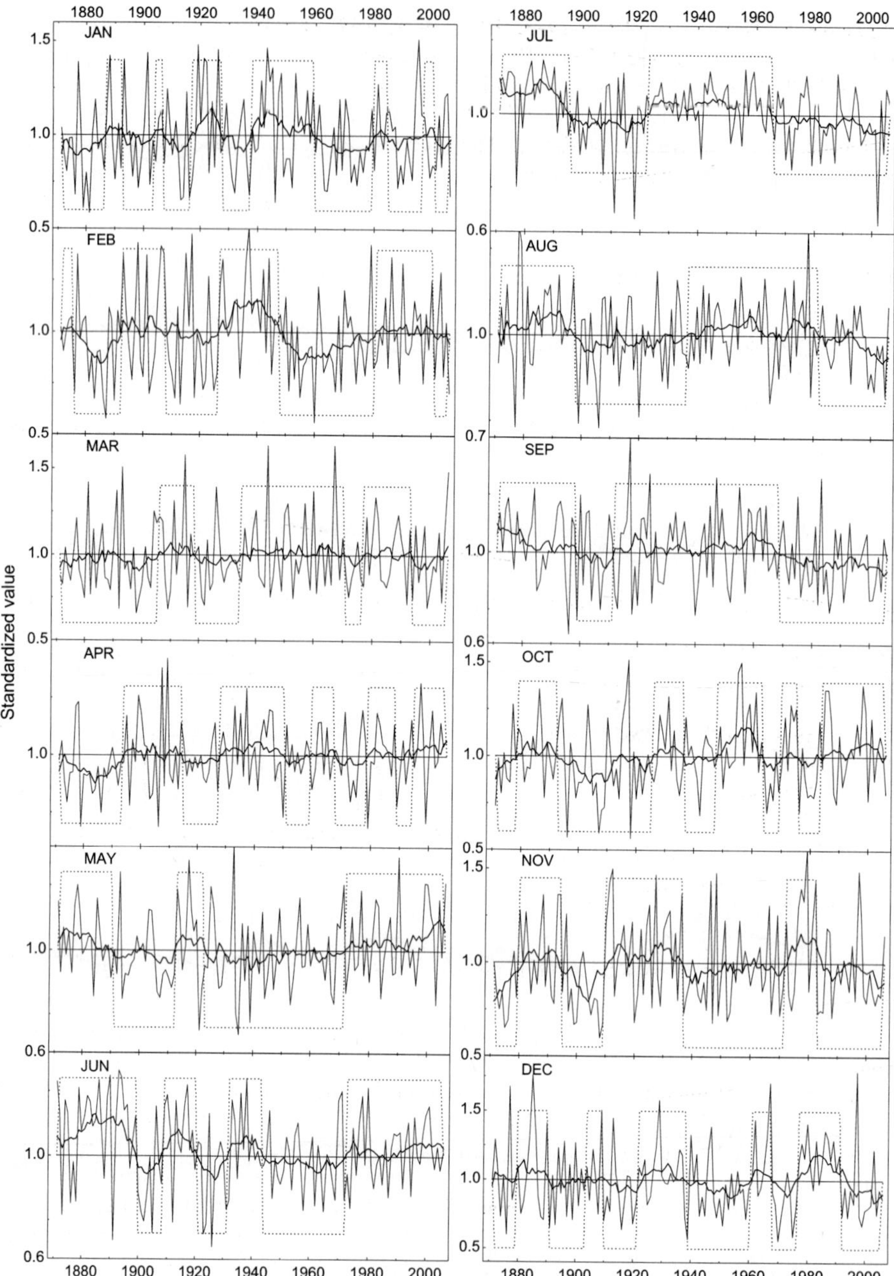

Fig. 3 Standardized value (thin curve) and 11-term running means (thick curves) of the monthly RSDI over the country during 1871–2006. The blocks demarcated by dotted line are dry/wet periods in the RSDI fluctuations.

1987–2006; and wet periods: 1901–1948 and 1978–1986. The dry periods in the summer season occurred 1882–1912, 1921–1935, 1947–1961, 1972–1976 and wet periods: 1877–1881, 1913–1920, 1936–1946, 1962–1971 and 1977–2006. The summer monsoon dry periods are 1899–1925 and 1972–2006 and wet periods 1871–1998 and 1926–1971. And the post monsoon dry periods are 1871–1876, 1895–1926, 1938–1943, 1964–1973, 1988–1993 and 2000–2006 and the wet periods: 1877–1994, 1927–1937, 1944–1963, 1974–1987 and 1994–1999. Over the period of 136 years (1871–2006) seasonal rainfall fluctuations showed four to eleven periods. In recent years/ decades, winter and summer monsoon rainfalls show a decreasing tendency while summer and post monsoon show increasing tendency. The recent tendency of monthly rainfall is: January, February, March, July, August, September, November and December decrease, and others increase. It may be noted that year of start of recent tendency is different for different months (Table 2).

Table 2 Dry and wet periods in the seasonal and monthly rainfall spatial fluctuations.

Period	Dry periods	Wet periods
WIN	1871–1900, 1949–77, 1987–2006	1901–48, 1978–86
SUM	1882–1912, 1921–35, 1947–61, 1972–76	1877–81,1913–20,1936–46,1962–71,1977–2006
MON	1899–1925, 1972–2006	1871–98, 1926–71
POST MON	1871–76, 1895–1926, 1938–43, 1964–73, 1988–93, 2000–06	1877–94, 1927–37, 1944–63, 1974–87, 1994–99
JAN	1871–86, 1896–1918, 1930–40, 1962–79, 1988–2003	1887–95, 1919–29, 1941–61, 1980–87
FEB	1871–92, 1909–26, 1949–78,2001–06	1893–1908, 1927–48, 1979–2000
MAR	1871–1903, 1921–35, 1971–78, 1983–87, 1996–2004	1904–20, 1936–70, 1979–82, 1988–95
APR	1871–94, 1916–24, 1947–58, 1966–75, 1987–95	1895–1915, 1925–46, 1959–65, 1976–88, 1996–2006
MAY	1890–1912, 1921–68	1873–89, 1913–20, 1969–2006
JUN	1900–08, 1923–32, 1943–74	1871–99, 1909–22, 1933–42, 1975–2002
JUL	1895–1918, 1970–2004	1871–94, 1919–69
AUG	1898–1937, 1979–2005	1878–97, 1938–78
SEP	1899–1909, 1927–31, 1965–2004	1871–98, 1910–26, 1932–64
OCT	1871–76, 1895–1927, 1940–54, 1964–70, 1976–82	1977–94, 1928–39, 1955–63, 1971–75, 1983–2005
NOV	1871–79, 1895–1909, 1937–71, 1983–2005	1880–94, 1910–36, 1972–82
DEC	1871–79, 1891–1904, 1910–22, 1939–60, 1968–76, 1992–2006	1880–90, 1905–09, 1923–38, 1961–67, 1977–91

POSSIBLE REASON FOR RECENT CHANGES IN MONSOON RAINFALL OVER THE COUNTRY

The gradient in the geopotential height (GPH) of the upper tropospheric isobaric levels from the Tibetan Anticyclone (TA) region to the Southern Tropical Indian Ocean High (STIOH) region is useful as an index of intensity of the Indian monsoon circulation (Singh *et al.*, 2006; Sontakke *et al.*, 2008). Besides STIOH, the North Pacific High (NPH), the Azores High (AH) and the North Sub Polar Low (NSPL) have been identified and gradients between TA and NPH, TA and AH and TA and NSPL have also been found useful as indexes of intensity of the Indian monsoon circulation. The combined area of the Iranian High (32.5°N–40.0°N; 50.0E–72.5°E) and the West Tibet (27.5°N–37.5°N; 72.5E–85.0°E) is considered as the TA region, the area between parallels 40.0°E and 110°E and meridians 5°S and 20°S as the STIOH region, the area between parallels 135°E and 120°W and meridians 20.0°N and 45.0°N as the NPH region, the area between parallels 60°W and 10°W and meridians 20.0°N and 50.0°N as the AH region, and the area between parallels 40°E and 110°E and meridians 55.0°N and 75.0°N as the NSPL region.

The interannual variation (1949–2006) of the geopotential height gradient (GPHG) at 300 hPa (NCEP/NCAR reanalysis data, Kalnay *et al.*, 1996) from TA to STIOH region for the monsoon season and each of the monsoon months has been examined. Interesting to notice is that the GPHG and the rainfall over northern India (north of 18°N) are in excellent agreement during extreme years and in the latest period from 2001, which suggests the possibility of exploring cause and effect relationships between them.

The two parameters GPH and rainfall for June show a rising tendency from the late 1950s, but for other three monsoon months and the season they show declining tendency. The June GPH of 300 hPa is rising over the TA region at the rate of 5.96 m/decade (significant at 1% level) and over the STIOH region at the rate of 4.93 m/decade (significant at 0.1% level). Therefore the GPHG shows a mild increasing tendency which is producing a slight increase in June rainfall over the country. During July, August, September (200 hPa level) and the whole season the GPH over the STIOH region is rising at the rate of 5.03 m, 5.47 m, 4.59 m and 5.0 m per decade, respectively, all significant at 0.1% level, but over the TA region the rising rate is 3.84 m/decade (significant at 10% level), 5.03 m/decade (significant at 1% level), 1.27 m/decade (not significant) and 4.03 m/decade (significant at 5% level), respectively. The rising tendency in the 200 hPa GPH is at a faster rate over the STIOH region compared to the TA region, therefore, the GPHG from the TA region to STIOH region shows a slight decreasing tendency (statistically not significant). The lower GPH in the upper tropospheric level contributes partially to reduction in rainfall over the country.

The relationship between GPHG at NPH, AH and NSPL with TA and rainfall over the country have been examined for June, July, August, September and the whole summer monsoon season (figures not shown). In brief, the GPHG from the TA region to the STIOH region showed highest correlation with the corresponding rainfall over the Indian region, that from TA to NPH region showed correlation almost of the same order, but from TA to AH region and from TA to NSPL region showed weaker correlation of the same order.

MECHANISM

During heightened summer heating of the Afro-Eurasian dry province (Arabia to Mongolia) and the land area over and around the Tibet-Himalayan highlands, the Tibetan Anticyclone intensifies. A major portion of the outflows from the Tibetan Anticyclone sinks over the North Pacific and intensifies the high over there. The intense North Pacific High produces strong easterlies over the central and western North Pacific. Part of these easterlies turns towards China, Korea and Japan, and part enters into Bay of Bengal after crossing the Vietnam-Laos-Thailand-Cambodia-Mynamar sector (Southeast Asia) where it merges with the main southwest monsoon current and the combined wind blows over the Indo-Gangetic Plains of northern India as an easterly wind parallel to the southern flank of the Himalayas. Part of this easterly wind converges over central India and the rising air flows towards the southern hemisphere via the upper troposphere. Also, the large outflows from the Tibet-Pacific Anticyclone in the upper troposphere, after crossing the equator above the western Indian Ocean, sink over southern Indian Ocean High. Part of this upper tropospheric cross-equatorial flow flows from the northern to southern tropical/subtropical Indian Ocean and while circulating anticyclonically in the lower troposphere it crosses the equator along the Somali coast and becomes southwest monsoon flow, and part merges with the southern temperate westerlies.

The upper tropospheric GPHG from the Tibetan anticyclone region to the NSPL region provides a major intensification of mid-latitude upper tropospheric westerlies. While steep gradients suggest strong poleward-shifted westerlies and a well developed and intense Tibetan Anticyclone, the reverse is true during shallow gradient. In the upper troposphere, the Azores High appears as a westward extension of the Tibetan Anticyclone.

Examination of geopotential height and temperature fluctuation of standard isobaric levels over the five centres of circulation revealed the following notable features:

(i) during June, over TA region the GPH at each standard isobaric level from 925 to 200 hPa is rising (non-linearly) at a faster rate compared to temperature, while over other centres of circulation positive trends in GPH and temperature fluctuations closely parallel each other;

(ii) during the other three monsoon months, over TA region the temperature showed a mild rising tendency in the lower troposphere and mild decreasing trend in the upper troposphere, but the GPH of the different levels showed rising tendency; and

(iii) over the surrounding four centres of circulation, rising trends in the GPH and temperature closely paralleled each other from the surface to the tropopause.

From time series plots of low-frequency mode of fluctuations (9-point Gaussian low-pass filtered values) of the GPH and temperature (not shown) it is clear that within deep highs over oceanic regions, e.g. the STIOH, the NPH and the AH, both GPH and temperature showed rising tendencies at different levels from 925 to 200 hPa. In the TA region, however, the temperature showed a rising trend in the lower troposphere and falling trend in the upper troposphere. Thus the rising trend in the GPH of different levels in the TA region is essentially due to rising trend in the temperature in the lower tropospheric layers. Investigation reveals that the rising tendency in post monsoon rainfall over South Peninsular India is due to rising trends in the sea surface temperatures (SST) of the Arabian Sea and Bay of Bengal (Sontakke *et al.*, 2008). Possible causes of change in the winter and summer rainfall over the country are being investigated.

SUMMARY

1. The arid and semi-arid areas are expanding whereas the moist subhumid, humid and perhumid areas are shrinking. The semi-arid area, which occupies the large and central portion of the country, is expanding significantly. The overall effect is that the country is experiencing dry climate in the recent period with a westward shifting of rainfall.
2. Analysis of the periodic pattern suggests that in the recent years/decades winter and summer monsoon rainfall show a decreasing tendency while summer and post-monsoon shows an increasing tendency. January, February, March, July, August, September, November and December show a decreasing trend in recent years/decades while other months show increasing tendency.
3. Gradient in the GPH of upper tropospheric isobaric levels from the Tibetan Anticyclone (TA) region to southern tropical Indian Ocean High (STIOH) region, to North pacific High (NPH) region, to Azores High (AH), and to North Sub Polar Low (NSPL) region, can be used as index of intensity of the Asian Summer Monsoon Circulation (ASMC).
4. The rising trend in post-monsoon rainfall over South Peninsular India is due to the rising trend in the SST of the oceans around the area.

A declining tendency in monsoon rainfall can be seen from 1965 onwards and a sharp declining trend from 1995. This is essentially due to weakening of the Tibetan Anticyclone associated with cooling of the upper troposphere (600–150 hPa) over the Himalayan region and surrounding atmosphere. The recent sharp increasing trend in the surface air temperature over India is due to a declining trend in the rainfall of winter and summer monsoon seasons (D. R. Kothawale, pers. comm.).

Acknowledgements The authors are extremely grateful to Professor B. N. Goswami, Director, Indian Institute of Tropical Meteorology, Pune, for the necessary facilities to pursue this study. The rainfall data used in this study were provided by the India Meteorological Department, Pune, which is thankfully acknowledged. The research in this paper was funded by the Department of Science and Technology, Government of India (DST Project no. ES/48/ICRP/2000).

REFERENCES

Eliot, J. (1902) Occasional discussions and compilations of Meteorological data: India and the Neighboring Countries. *Indian Meteorological Memoirs* vol. XIV. Published by Order of His Excellency the Viceroy and Governor General of India in Council, Calcutta, India.

Kalnay, E., Kanamitsu, M., Kistler, R., Collins, W., Deaven, D., Gandin, L., Iredell, M., Saha, S., White, G., Woollen, J., Zhu, Y., Chelliah, M., Eblsuzaki, W., Higgins, W., Janowiak, J., Mo, K.C., Ropelewski, C., Wang, J., Leetmaa, A., Reynolds, R., Jenne, R. & Joseph, D. (1996) The NCEP/NCAR 40-Year Reanalysis Project. *Bull. Am. Met. Soc.* **77**, 437–471.

Lal, M. (1994) Water resources of the south-east Asian region in a warmer atmosphere. *Adv. Atmos. Sci.* **11**, 239–246.

NATMO (1986) National Atlas of India Physiographic Regions of India, third edn, Plate 41. Prepared under the Direction of G. K. Dutt, Director NATMO (the National Atlas & Thematic Mapping Organization), Kolkata, India.

Prabhakar, S. V. R. K. & Shaw R. (2008) Climate change adaptation implications for drought risk mitigation: A perspective for India. *Climatic Change* **88**, 113–130.

Ranade A. A., Singh, N., Singh, H. N. & Sontakke, N. A. (2008) On variability hydrological wet season, seasonal rainfall and rainwater potential of the river basins of India (1813–2006). *J. Hydrological Research and Development,* **23**, 79–108.

Revi, A. (2008) Climate change risk: An adaptation and mitigation agenda for Indian cities. *Environment and Urbanization,* **20**(1), 207–229.

Singh, N. (1984) Fluctuations of different moisture regimes in India. *Arch. Met. Geophys. Biocl.* **B35**, 239–256.

Singh, N. & Mulye, S. S. (1991) On the relations of the rainfall variability and distribution with the mean rainfall over India. *Theor. Appl. Climatol.* **44**(3-4), 209–221.

Singh, N., Pant, G. B. & Mulye, S. S. (1991) Distribution and long term features of the spatial variations of the moisture regions over India. *Int. J. Climatol.* **11**, 413–427.

Singh, N., Mulye, S. S. & Pant, G. B. (1992) Some features of the arid area variations over India: 1871–1984. *Pure Appl. Geophys.* **138**, 135–150.

Singh, N., Sontakke, N. A., Singh, H. N. & Pandey, A. K. (2005) Recent trends in spatiotemporal variation of rainfall over India – an investigation into basin-scale rainfall fluctuations. In: *Regional Hydrological Impacts of Climate Change-Hydroclimatic Variability* (ed. by S. Franks, T. Wagener, E. Boegh, H. V. Gupta, L. Bastidas, C. Nobre & C. de O Galvao), 273–282. IAHS Publ. 296. IAHS Press, Wallingford, UK.

Singh, N., Sontakke, N. A. & Singh, H. N. (2006) Global warming, subtropical anticyclones and the Indian summer monsoon. *Proc. National Symposium Tropmet-2006 on Role of Meteorology in National Development*, vol. I, d29–d31. IITM, Pune, India.

Sontakke, N. A., Singh, N. & Singh, H. N. (2008) Instrumental period rainfall series of the Indian region (1813–2005): Revised reconstruction, update and analysis. *The Holocene* **18**, 1055–1066.

Key word index

Climate and the Hydrological Cycle

Edited by
Marc Bierkens *Utrecht University, The Netherlands*
Han Dolman *Vrije Universiteit Amsterdam, The Netherlands*
Peter Troch *University of Arizona, USA*

An in-depth overview of the role of the hydrological cycle within the climate system, including climate change impacts on hydrological reserves and fluxes, and the controls of terrestrial hydrology on regional and global climatology. This book, composed of self-contained chapters by specialists in hydrology and climate science, is intended to serve as a text for graduate and postgraduate courses in climate hydrology and hydroclimatology. It will also be of interest to scientists and engineers/practioners interested in the water cycle, weather prediction and climate change.

IAHS Special Publication 8

(*October 2008*) ISBN 978-1-901502-54-1 (Paperback); 344 + xvi pages
Price £50.00

Sponsored by:

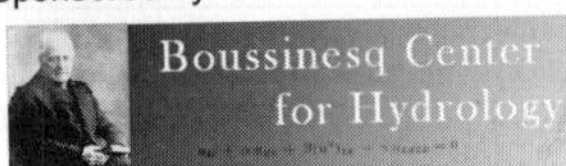

Please send book orders and enquiries to:

Mrs Jill Gash
IAHS Press, Centre for Ecology and Hydrology
Wallingford, Oxfordshire OX10 8BB, UK

jilly@iahs.demon.co.uk
tel.: + 44 1491 692442
fax: + 44 1491 692448/692424

Book prices include postage worldwide. IAHS Members receive discounts on IAHS publications.

See www.IAHS.info for information about IAHS (the International Association of Hydrological Sciences), membership, publications, meetings and other activities.